Lecture Notes in Networks and Systems

Volume 502

The series "Lecture Notes in Networks and Systems" publishes the latest developments in Networks and Systems—quickly, informally and with high quality. Original research reported in proceedings and post-proceedings represents the core of LNNS.

Volumes published in LNNS embrace all aspects and subfields of, as well as new challenges in, Networks and Systems.

The series contains proceedings and edited volumes in systems and networks, spanning the areas of Cyber-Physical Systems, Autonomous Systems, Sensor Networks, Control Systems, Energy Systems, Automotive Systems, Biological Systems, Vehicular Networking and Connected Vehicles, Aerospace Systems, Automation, Manufacturing, Smart Grids, Nonlinear Systems, Power Systems, Robotics, Social Systems, Economic Systems and other. Of particular value to both the contributors and the readership are the short publication timeframe and the world-wide distribution and exposure which enable both a wide and rapid dissemination of research output.

The series covers the theory, applications, and perspectives on the state of the art and future developments relevant to systems and networks, decision making, control, complex processes and related areas, as embedded in the fields of interdisciplinary and applied sciences, engineering, computer science, physics, economics, social, and life sciences, as well as the paradigms and methodologies behind them.

Indexed by SCOPUS, INSPEC, WTI Frankfurt eG, zbMATH, SCImago.

All books published in the series are submitted for consideration in Web of Science.

For proposals from Asia please contact Aninda Bose (aninda.bose@springer.com).

More information about this series at https://link.springer.com/bookseries/15179

Radek Silhavy
Editor

Artificial Intelligence Trends in Systems

Proceedings of 11th Computer Science On-line Conference 2022, Vol. 2

Springer

Editor
Radek Silhavy
Faculty of Applied Informatics
Tomas Bata University in Zlin
Zlin, Czech Republic

ISSN 2367-3370 ISSN 2367-3389 (electronic)
Lecture Notes in Networks and Systems
ISBN 978-3-031-09075-2 ISBN 978-3-031-09076-9 (eBook)
https://doi.org/10.1007/978-3-031-09076-9

This Springer imprint is published by the registered company Springer Nature Switzerland AG
The registered company address is: Gewerbestrasse 11, 6330 Cham, Switzerland

Preface

This volume constitutes the refereed proceedings of the Artificial Intelligence Trends in Systems section of the 11th Computer Science On-line Conference 2022 (CSOC 2022), held on-line in April 2022.

Artificial Intelligence Trends in the scope of systems, systems engineering and software engineering domain are presented in this proceeding. This proceeding is a Vol. 2 of the Computer Science On-line Conference 2022.

CSOC 2022 conference intends to provide an international forum to discuss the latest high-quality research results in all areas related to computer science.

CSOC 2022 is held on-line, which brings equal opportunity to participate for all researchers around the world. CSOC 2022 updates the traditional concept of scientific conferences.

I believe that you find the following proceedings exciting and valuable for your research work.

April 2022 Radek Silhavy

Organization

Program Committee

Program Committee Chairs

Petr Silhavy	Tomas Bata University in Zlin, Faculty of Applied Informatics
Radek Silhavy	Tomas Bata University in Zlin, Faculty of Applied Informatics
Zdenka Prokopova	Tomas Bata University in Zlin, Faculty of Applied Informatics
Roman Senkerik	Tomas Bata UProfessor, niversity in Zlin, Faculty of Applied Informatics
Roman Prokop	Tomas Bata University in Zlin, Faculty of Applied Informatics
Viacheslav Zelentsov	Doctor of Engineering Sciences, Chief Researcher of St. Petersburg Institute for Informatics and Automation of Russian Academy of Sciences (SPIIRAS)
Roman Tsarev	Department of Information Technology, International Academy of Science and Technologies, Moscow, Russia
Stefano Cirillo	Department of Computer Science, University of Salerno, Fisciano (SA), Italy

Program Committee Members

Boguslaw Cyganek	Department of Computer Science, University of Science and Technology, Krakow, Poland
Krzysztof Okarma	Faculty of Electrical Engineering, West Pomeranian University of Technology, Szczecin, Poland
Monika Bakosova	Institute of Information Engineering, Automation and Mathematics, Slovak University of Technology, Bratislava, Slovak Republic
Pavel Vaclavek	Faculty of Electrical Engineering and Communication, Brno University of Technology, Brno, Czech Republic
Miroslaw Ochodek	Faculty of Computing, Poznan University of Technology, Poznan, Poland
Olga Brovkina	Global Change Research Centre Academy of Science of the Czech Republic, Brno, Czech Republic & Mendel University of Brno, Czech Republic
Elarbi Badidi	College of Information Technology, United Arab Emirates University, Al Ain, United Arab Emirates
Luis Alberto Morales Rosales	Head of the Master Program in Computer Science, Superior Technological Institute of Misantla, Mexico
Mariana Lobato Baes	Research Professor, Superior Technological of Libres, Mexico
Abdessattar Chaâri	Laboratory of Sciences and Techniques of Automatic control & Computer engineering, University of Sfax, Tunisian Republic
Gopal Sakarkar	Shri. Ramdeobaba College of Engineering and Management, Republic of India
V. V. Krishna Maddinala	GD Rungta College of Engineering & Technology, Republic of India
Anand N. Khobragade (Scientist)	Maharashtra Remote Sensing Applications Centre, Republic of India
Abdallah Handoura	Computer and Communication Laboratory, Telecom Bretagne—France

Technical Program Committee Members

Ivo Bukovsky	Czech Republic
Maciej Majewski	Poland
Miroslaw Ochodek	Poland
Bronislav Chramcov	Czech Republic
Eric Afful Dazie	Ghana
Michal Bliznak	Czech Republic
Donald Davendra	Czech Republic
Radim Farana	Czech Republic
Martin Kotyrba	Czech Republic
Erik Kral	Czech Republic
David Malanik	Czech Republic
Michal Pluhacek	Czech Republic
Zdenka Prokopova	Czech Republic
Martin Sysel	Czech Republic
Roman Senkerik	Czech Republic
Petr Silhavy	Czech Republic
Radek Silhavy	Czech Republic
Jiri Vojtesek	Czech Republic
Eva Volna	Czech Republic
Janez Brest	Slovenia
Ales Zamuda	Slovenia
Roman Prokop	Czech Republic
Boguslaw Cyganek	Poland
Krzysztof Okarma	Poland
Monika Bakosova	Slovak Republic
Pavel Vaclavek	Czech Republic
Olga Brovkina	Czech Republic
Elarbi Badidi	United Arab Emirates

Organizing Committee Chair

Radek Silhavy — Tomas Bata University in Zlin, Faculty of Applied Informatics,
email: radek@silhavy.cz

Conference Organizer (Production)

Silhavy s.r.o.
Website: https://www.openpublish.eu
Email: csoc@openpublish.eu

Conference website, Call for Papers

https://www.openpublish.eu

Contents

AIC Approach for Business Students' Entrepreneurial Intent

Dam Tri Cuong(✉)

Industrial University of Ho Chi Minh City, Ho Chi Minh City, Vietnam
damtricuong@iuh.edu.vn

Abstract. Entrepreneurship has contributed to the social and business development of the country, stimulating knowledge growth, technological evolution, competitiveness, and innovation. Besides, entrepreneurial education has been debated as a useful way to support and encourage passion entrepreneurship among undergraduate students. Familiarity with this knowledge may positively inspire attitudes to entrepreneurship among undergraduates. So, this study's purpose is to an empirical application by utilizing the Akaike information criterion (AIC) approach for business students' entrepreneurial intent. AIC is one of the most commonly applied model choice methods in statistical software. Study data were collected through the survey online from 242 business students in Ho Chi Minh City, Vietnam by Google form. The sample was gathered by a convenient sampling method. The research findings indicated that the best model for business students' entrepreneurial intent including four factors (attitude toward entrepreneurship, subjective norm, perceived behavior control, entrepreneurial education) that influenced positively business students' entrepreneurial intent; in which the attitude toward entrepreneurship was the most vital factor to business students' entrepreneurial intent.

Keywords: AIC approach · Entrepreneurial intent · Business students · Theory of planned behavior

1 Introduction

Entrepreneurship has contributed to the social and business development of the country, stimulating knowledge growth, technological evolution, competitiveness, and innovation [1]. Likewise, entrepreneurship is recognized as the economic strategy development to enhance the nation's economy and sustain its competitiveness in challenging the expanding globalization tendency [2]. Hassan et al. [3] disclosed that the increase in the expansion of multiple firms is not only essential but also crucial as they create jobs, lead innovation, and improve efficiency in different economic areas. Besides, entrepreneurial education has been debated as a useful way to support and encourage passion entrepreneurship among undergraduate students. Furthermore, familiarity with this knowledge may positively inspire attitudes to entrepreneurship among undergraduates [4, 5]. Hou et al. [6] also demonstrated that undergraduate students are frequently recognized potential entrepreneurs, the entrepreneurial intention has been the antecedent to forecast the entrepreneurial behavior of undergraduate students.

R. Silhavy (Ed.): CSOC 2022, LNNS 502, pp. 1–9, 2022.
https://doi.org/10.1007/978-3-031-09076-9_1

Moreover, in Vietnam, the government also acknowledged the significance of entrepreneurship and gave Decision 1665 the project 'Supporting students to entrepreneurial until 2025'. The general purpose of encouraging the students' entrepreneurial spirit while learning at the universities. Creating convenient conditions to assist undergraduates in building and realizing their thoughts and start-up projects, adding to making jobs for undergraduates and graduates [7].

Prior research has revealed that entrepreneurial intention has been the best antecedent of entrepreneurial behavior. Consequently, scholars have been more attention increasingly to entrepreneurial intention and its affecting determinants [8]. Previous researches have been debated some factors impacting students' entrepreneurial intention such as entrepreneurial attitude [8, 9], subjective norms [8–10], perceived behavior control [9, 10], entrepreneurial education [11–13], previous work experience [14, 15]. Yet, there are very few studies utilizing the Akaike information criterion (AIC) approach for business students' entrepreneurial intent in Vietnam. Thus, this research is trying to fill this gap.

2 Literature Review

2.1 Theory of Planned Behavior

Theory of planned behavior (TPB) is recognized as the fundamental theory that describing the students' entrepreneurial intent [9, 16–19]. TPB was introduced by Ajzen [20], according to TPB, entrepreneurial intent is meant the attempt that the individual will start to perform entrepreneurial behavior. TPB is applied to investigate determinants influencing intent and to forecast intent, which explains TPB as a valuable instrument in the study. The theory proposes an important and relevant framework to describe individual planned behavior and hence enables scholars to predict entrepreneurial intent including the consideration simultaneously of both individual and social factors [9, 21]. TPB includes three factors that influence entrepreneurial intent as attitude toward entrepreneurship, subjective norm, and perceived behavioral control.

Attitude toward entrepreneurship is defined as the individual's positive or negative valuation of entrepreneurship behavior [8]. Attitude also exhibits the individual's beliefs that the individual may have to a particular action or behavior such as entrepreneurship. Such beliefs could be positive or negative, and as a result, influence how individuals approach that specific behavior, if the individual act or does not act [19].

Subjective norm refers to an individual's viewpoints on how people surrounding them or influential other people, for instance, relatives, friends, or colleagues, think of a particular behavior such as starting business enterprises [9]. Taha et al. [21] indicate that undergraduates are prompted to be entrepreneurs since others (relatives, spouse, colleagues, and people whose ideas were considered) that correlated to them in the choices to be entrepreneurs.

Perceived behavioral control is considered as the thoughts of the easiness and difficulties to be an entrepreneur [21]. The perceived behavioral control depends on beliefs concerning the availability or lack of necessary resources and chances to achieve specific activities. Generally, there is a view that the higher the perceived behavioral control the more powerful the individuals' entrepreneurship intent [10].

Furthermore, previous scholars also affirmed that these preditors (attitude toward entrepreneurship, subjective norms, and perceived behavioral control) were positively related to students' entrepreneurial intent [10, 16, 18, 19].

On the other view, former empirical studies also discussed that other predictors also influenced students' entrepreneurial intent (e.g., entrepreneurial education, previous work experience [11–15]).

2.2 Entrepreneurial Education

Entrepreneurship education is portrayed as entrepreneurship education of the subjects that teach the theory and entrepreneurship practice at school with the purpose aids students to have the capacity in knowing and running entrepreneurship, creating changes of attitudes, norms, and behavior [22]. Entrepreneurial education could be recognized as the skills that could be taught to individuals that will allow them to build new and innovative plans [23]. Entrepreneurship education is also characterized as the entrepreneurial climate for undergraduates which involves the curriculum designing that would match the need of undergraduates to be entrepreneurs [2]. Likewise, Huang et al. [24] demonstrated that entrepreneurship education aims to give knowledge, experiences, and skills of entrepreneurship to university students through courses, discussions of entrepreneurship, and other programs, to promote new opinions, and commercial environments. Moreover, together with the scientific and technological growth, entrepreneurship education will prompt potential entrepreneurs to have the quality of innovation, match the trends of the period, and promote economic growth [25]. Israr & Saleem [26] stated that entrepreneurial education develops the knowledge and skills of individuals and increases entrepreneurship intent as well. Earlier scholars designated that entrepreneurial education was the antecedent of undergraduates' entrepreneurial intent [27, 28].

2.3 Previous Work Experience

Yuan et al. [29] stated that previous work experience was any experience that the individual gained while engaging in a certain occupation or field. Prior work experience had from employment, potential businesspeople enhance and gain their knowledge, experiences, and commercial perception. Likewise, the past and existing experience give incubator opportunities for potential entrepreneurs. Moreover, it's creating a fundamental and critical influence on potential entrepreneurs' perception to identify and exploit the opportunities [30]. If individuals are insufficient work experience, individuals may face difficulties in performing the first stages to turning an entrepreneur. Therefore, individuals could be influenced by former work experience to be potential entrepreneurs [31]. Fatoki [32] identified that business students with prior work experience have a higher level of entrepreneurship intention (versus undergraduates taking no work experience). Yet, this difference is not statistically significant. Moreover, other scientists (e.g. Yuan et al. [29]) affirmed that there was a connection between past work experience and entrepreneurial intention. Likewise, previous studies revealed that prior work experience related to students' entrepreneurial intent [15, 33].

Based on the above analysis, in this research, we have considered empirical research to identify the ultimate model for determinants of business students' entrepreneurship intent in Vietnam by applying the AIC approach as follows:

$$\mathrm{SEI} = \beta 0 + \beta 1 ATE + \beta 2 SN + \beta 3 PBC + \beta 4 EE + \beta 5 PWE + \varepsilon \qquad (1)$$

where:

SEI: Business students' entrepreneurship intent, ATE: Attitude toward entrepreneurship, SN: Subjective norm, PBC: Perceived behavioral control, EE: Entrepreneurial education, PWE: Previous work experience, $\mathcal{E}$: Error

3 Methodology

3.1 Analytical Method

Akaike information criterion (AIC) is introduced by Akaike [34]. AIC is the approach of an information theory for model (variable) choice, via Kullback-Leibler divergence. In statistical software, AIC is one of the most commonly used model selection methods [35]. Furthermore, the model selection index via AIC has become one of the popular tools to utilize for the regression model. The reason for this is the scientist can apply the model choice criteria exactly for choosing among alternative models [36]. The ultimate model is the model with the smallest AIC index [37].

In this research, the AIC index by RStudio software is applied to compute the optimal model. AIC for the candidate model is calculated as follows [35]:

$$\mathrm{AIC} = -2\ln(\mathrm{L}) + 2\mathrm{p} \qquad (2)$$

where: L regards the likelihood under the candidate model, p is the number of parameters in the model.

3.2 Data and Sample

Study data were collected through the survey online from 242 business students in Ho Chi Minh City, Vietnam by Google form. A convenient methodology was applied for selecting the population sample. A five-point Likert scale was utilized to measure the factors (from 1 = completely oppose to 5 = completely assent). The questionnaire was adopted to estimate the variables of the research: Attitude toward entrepreneurship (four items) from [19], subjective norms (three items) and perceived behavioral control (four items) from [21], entrepreneurial education (four indicators) from [38], previous work experience (four indicators) and business students' entrepreneurship intent (three indicators) from [39].

4 Results and Discussion

4.1 Results

Descriptive Statistics

The demographic traits of the sample as exhibited in Table 1.

Table 1. Demographic traits of the sample

Features	Classifications	Frequency	Percent
Gender	Female	135	55.8
	Male	107	44.2
	Total	242	100.0
School-year	First-year	30	12.4
	Second-year	47	19.4
	Third-year	99	40.9
	Final-year	66	27.3
	Total	242	100.0

As shown in Table 1, about gender: the sample has 135 students' female account for 55.8% and 107 students' male account for 44.2%. About school-year: 30 first-year students account for 12.4%, 47 s-year students account for 19.4%, 99 third-year students account for 40.9%, and 66 final-year students account for 27.3%.

Akaike Information Criterion (AIC)

The AIC finding can be compiled in Table 2.

Table 2 demonstrates the AIC results by per step to searching the final model. The initial step is the model with five predictor variables and AIC = 430.43. The second step, also the final step, is the model comprising four independent variables and AIC = 428.48.

Table 2. AIC

Model	AIC
SEI ~ ATE + SN + PBC + EE + PWE	430.43
SEI ~ ATE + SN + PBC + EE	428.48

The results with two steps to explore the optimal model, RStudio stops at the ultimate model includes four independent variables. That means this analysis disclosed the optimal model is the model with four independent variables as it has the smallest AIC index.

Model' Predictor Coefficients Testing

Table 3 exhibits the result of predictor coefficients in the ultimate model.

Table 3 uncovers the coefficients of independent variables in the final model have the p-value < 0.05 and is statistically meaningful. Hence, the assumption of predictor variables in the final model is verified. Therefore, the optimal model is as below:

$$Y = -1.549 + 0.695ATE + 0.162SN + 0.227PBC + 0.340EE + \varepsilon \quad (3)$$

Besides, the R-squared in the optimal model is 0.6412, which means, 4 independent variables (attitude toward entrepreneurship, subjective norm, perceived behavioral control, and entrepreneurial education) describe 64.12% of the variance of the business students' entrepreneurial intent.

Table 3. Predictor coefficients in the final model

	Estimate	Std. Error	t-value	p-value
(Intercept)	−1.54922	0.27831	−5.566	0.0000
ATE	0.69532	0.07659	9.079	0.0000
SN	0.16202	0.07035	2.303	0.0221
PBC	0.22754	0.05746	3.960	0.0000
EE	0.34015	0.07026	4.841	0.0000

Residual standard error: 0.5781 on 237 degrees of freedom				
Multiple R-squared: 0.6412	Adjusted R-squared: 0.6351			
F-statistic: 105.9 on 4 and 237 DF, p-value < 0.00000				

Variance Inflation Factor (VIF) Testing
VIF testing in the optimal model is exhibited in Table 4.

Table 4. The VIF

	ATE	SN	PBC	EE
VIF	1.746896	1.299778	1.311482	1.731671

As presented in Table 4, the VIF criterion of the predictor variables is under 5, so there is no multicollinearity between the independent variables [40].

4.2 Discussion

The finding exhibited that the ultimate model by implementing the AIC approach including 4 antecedent variables (attitude toward entrepreneurship, subjective norm, perceived behavioral control, and entrepreneurial education) and the dependent variable (business students' entrepreneurial intent). This final model confirmed that 4 independents variables (attitude toward entrepreneurship, subjective norms, perceived behavioral control, and entrepreneurial education) positively influenced and were statistically significant on business students' entrepreneurial intent.

Besides, as illustrated in Table 3, attitude toward entrepreneurship (ATE) was the most critical impact on business students' entrepreneurial intent ($\beta = 0.695$). That meant if the attitude toward entrepreneurship raised by 1 point, business students' entrepreneurial intent increased by 0.695. This finding has been consistent with these researches [7, 8, 18], also the strongest antecedent affected on students' entrepreneurial intent. This could be described when we examine the similarity between the notion of entrepreneurial intent and attitude toward entrepreneurship. In the TPB theory, some scholars confirm that entrepreneurial intent is the person's attitude to the participant in entrepreneurial activities [8]. Thus, policymakers and university managers expect to increase business students' entrepreneurial intent should be improved their attitude toward entrepreneurship. Likewise, entrepreneurial education was the second essential impact on business students' entrepreneurial intent ($\beta = 0.340$). That meant if entrepreneurial education raised by 1 point, business students' entrepreneurial intent increased by 0.340. This outcome has been confirmed by scholars (e.g. Kabir et al. [11], Mukhtar et al. [41]). Next, perceived behavioral control was the third important influence on business students' entrepreneurial intent ($\beta = 0.227$). That meant if perceived behavioral control increased by 1 point, business students' entrepreneurial intent increased by 0.227. This outcome has been proved by the researchers (e.g. Zanabazar & Jigjiddorj [16], Su et al. [9]). Finally, the subjective norm was the fourth vital influence on business students' entrepreneurial intent ($\beta = 0.162$). That meant if subjective norm increased by 1 point, business students' entrepreneurial intent increased by 0.162. This outcome has been affirmed by the researchers (e.g. Kabir et al. [11], Zanabazar & Jigjiddorj [16]).

5 Conclusions and Limitations

This study verified the ultimate model for business students' entrepreneurial intent by the AIC approach. The research outcomes in the best model exhibited that four determinants (attitude toward entrepreneurship, subjective norm, perceived behavioral control, and entrepreneurial education) influenced positively business students' entrepreneurial intent. Consequently, policymakers and university managers should note these factors if they would like to promote business students' entrepreneurial intent.

This research has just been tested with business students in Ho Chi Minh City, Vietnam, thus the research was restricted for the generalization from the research results. Future researches should be utilized this model for other cities in Vietnam.

References

1. Cardella, G.M., Hernández-Sánchez, B.R., García, J.C.S.: Entrepreneurship and family role: a systematic review of a growing research. Front. Psychol. **10**, 1–17 (2020). https://doi.org/10.3389/fpsyg.2019.02939
2. Shamsudin, S.F.F., Mamun, A., Nawi, N.B., Nasir, N.A.B., Zakaria, M.N.: Factors affecting entrepreneurial intention among the malaysian university students. The J. Developing Areas **51**, 423–431 (2017). https://doi.org/10.1353/jda.2017.0111
3. Hassan, A., Saleem, I., Anwar, I., Hussain, S.A.: Entrepreneurial intention of Indian university students: the role of opportunity recognition and entrepreneurship education. Educ. Train. **62**, 843–861 (2020). https://doi.org/10.1108/ET-02-2020-0033
4. Basu, A., Virick, M.: Assessing entrepreneurial intentions amongst students: A comparative study. In: 12th Annual Meeting of the National Collegiate, pp. 79–86. (2008)

5. Tong, X., Tong, D.Y., Loy, L.: Factors influencing entrepreneurial intention among university students. Int. J. Soc. Sci. Humanit. Stud. **3**, 487–496 (2011)
6. Hou, H.F., Su, Y., Lu, M., Qi, M.: Model of the entrepreneurial intention of university students in the Pearl River Delta of China. Frontiers Psychol. **10**, 916 (2019). https://doi.org/10.3389/fpsyg.2019.00916
7. Prime Minister of Vietnam: Decision 1665 approving the project "Supporting students to entrepreneurial until 2025" (Vietnamese) (2017)
8. Genshu, L., Song, Y., Pan, B.: How university entrepreneurship support affects college students' entrepreneurial intentions: an empirical analysis from China. Sustain. **13**(6), 3224 (2021). https://doi.org/10.3390/su13063224
9. Su, Y., et al.: Factors influencing entrepreneurial intention of university students in china: Integrating the perceived university support and theory of planned behavior. Sustain. (Switzerland) **13**, 1–17 (2021). https://doi.org/10.3390/su13084519
10. Venkatapathy, R., Pretheeba, P.: Gender, family business background and entrepreneurial intentions in an emerging economy. Int. J. Bus. Emerg. Markets **6**(3), 217 (2014). https://doi.org/10.1504/IJBEM.2014.063890
11. Kabir, S., Ahasanul, H., Sarwar, A.: Factors affecting the intention to become an entrepreneur: a study from bangladeshi business graduates perspective. Int. J. Eng. Inform. Syst. **1**, 10–19 (2017)
12. Farashah, A.D.: The process of impact of entrepreneurship education and training on entrepreneurship perception and intention: study of educational system of Iran. Educ. Train. **55**, 868–885 (2013). https://doi.org/10.1108/ET-04-2013-0053
13. Solomon, G., Dickson, P.H., Solomon, G.T., Weaver, K.M.: Entrepreneurial selection and success: does education matter? J. Small Bus. Enterp. Dev. **15**, 239–258 (2008). https://doi.org/10.1108/14626000810871655
14. Yuan, L.W., Qalati, S.A., Iqbal, S., Hussain, R.Y., Ali, S.: Impact of prior work experience on entrepreneurial intention and theory of planned behaviour in the context of Pakistan. J. Entrepreneurship Organ. Manag. **8**, 1–9 (2019). https://doi.org/10.4172/2169-026X.1000268
15. Masoomi, E., Zamani, N., Bazrafkan, K., Akbari, M.: An investigation of the factors influencing entrepreneurial intention of senior agricultural students at Shiraz University. Int. J. Agric. Manag. Dev. (Ijamad) **6**, 431–437 (2016)
16. Zanabazar, A., Jigjiddorj, S.: The factors effecting entrepreneurial intention of university students: case of Mongolia. SHS Web of Conf. **73**, 01034 (2020). https://doi.org/10.1051/shsconf/20207301034
17. Al-Jubari, I., Hassan, A., Liñán, F.: Entrepreneurial intention among University students in Malaysia: integrating self-determination theory and the theory of planned behavior. Int. Entrepreneurship Manag. J. **15**(4), 1323–1342 (2018). https://doi.org/10.1007/s11365-018-0529-0
18. Gird, A., Bagraim, J.J.: The theory of planned behaviour as predictor of entrepreneurial intent amongst final-year university students. S. Afr. J. Psychol. **38**, 711–724 (2008). https://doi.org/10.1177/008124630803800410
19. Al-Jubari, I.: College students' entrepreneurial intention: testing an integrated model of SDT and TPB. SAGE Open **9**, 1–15 (2019). https://doi.org/10.1177/2158244019853467
20. Ajzen, I.: The theory of planned behavior. Organ. Behav. Hum. Decis. Process. **50**, 179–211 (1991)
21. Linan, F., Chen, Y.-W.: Development and cross-cultural application of a specific instrument to measure entrepreneurial intentions. Entrepreneurship: Theory Pract. **33**, 593–617 (2009)
22. Ginanjar, A.: Entrepreneurship education and entrepreneurial intention on entrepreneurship behavior: a case study. Adv. Econ. Bus. Manag. Res. **15**, 683–686 (2016). https://doi.org/10.2991/gcbme-16.2016.129

23. Jones, C., English, J.: A contemporary approach to entrepreneurship education. Education + Training **46**(8/9), 416–423 (2004). https://doi.org/10.1108/00400910410569533
24. Huang, Y., An, L., Wang, J., Chen, Y., Wang, S., Wang, P.: The role of entrepreneurship policy in college students' entrepreneurial intention: the intermediary role of entrepreneurial practice and entrepreneurial spirit. Frontiers Psychol. **12**, 585698 (2021). https://doi.org/10.3389/fpsyg.2021.585698
25. Henry, C., Hill, F., Leitch, C.: Developing a coherent enterprise support policy: a new challenge for governments. Eviron. Plann. C. Gov. Policy **21**, 3–19 (2003). https://doi.org/10.1068/c0220
26. Israr, M., Saleem, M.: Entrepreneurial intentions among university students in Italy. J. Glob. Entrep. Res. **8**(1), 1–14 (2018). https://doi.org/10.1186/s40497-018-0107-5
27. Issa, H.-E.: The Influence of Education and Environmental Support on Entrepreneurial Intentions. Int. J. Econ. Bus. Adm. **VIII**, 946–965 (2020). https://doi.org/10.4018/978-1-7998-1981-3.ch006
28. Ajike, E., Kelechi, N., Hamed, A., Onyia, V., Kwarbai, J.: Entrepreneurship education and entrepreneurial intentions: the role of theory of planned behaviour. Int. J. Adv. Res. Soc. Eng. Dev. Strat. **3**, 118–135 (2015)
29. Yuan, L., Qalati, S., Iqbal, S., Hind, H., Ali, S.: Impact of prior work experience on entrepreneurial intention and theory of planned behavior in context of Pakistan. Int. J. Adv. Res. **7**(4), 874–887 (2019). https://doi.org/10.21474/IJAR01/8900
30. Cooper, S.Y., Park, J.S.: The impact of "incubator" organizations on opportunity recognition and technology innovation in new, entrepreneurial high-technology ventures. Int. Small Bus. J. **26**, 27–56 (2008). https://doi.org/10.1177/0266242607084658
31. Kim, P.H., Aldrich, H.E., Keister, L.A.: Access (not) denied: the impact of financial, human, and cultural capital on entrepreneurial entry in the United States. Small Bus. Econ. **27**, 5–22 (2006). https://doi.org/10.1007/s11187-006-0007-x
32. Fatoki, O.: The entrepreneurial intention of undergraduate students in South Africa: The influences of entrepreneurship education and previous work experience. Mediterr. J. Soc. Sci. **5**, 294–299 (2014). https://doi.org/10.5901/mjss.2014.v5n7p294
33. Sahinidis, A.G., Xanthopoulou, P.I., Tsaknis, P.A., Vassiliou, E.E.: Age and prior working experience effect on entrepreneurial intention. Corp. Bus. Strategy Rev. **2**, 18–26 (2021)
34. Akaike, H.: A new look at the statistical model identification. IEEE Trans. Autom. Control **19**, 716–723 (1974). https://doi.org/10.1109/TAC.1974.1100705
35. Chaurasia, A., Harel, O.: Model selection rates of information based criteria. Electron. J. Stat. **7**, 2762–2793 (2013). https://doi.org/10.1214/13-EJS861
36. Karlsson, P.S., Behrenz, L., Shukur, G.: Performances of model selection criteria when variables are Ill conditioned. Comput. Econ. **54**(1), 77–98 (2017). https://doi.org/10.1007/s10614-017-9682-8
37. Bulteel, K., Wilderjans, T.F., Tuerlinckx, F., Ceulemans, E.: CHull as an alternative to AIC and BIC in the context of mixtures of factor analyzers. Behav. Res. Methods **45**(3), 782–791 (2013). https://doi.org/10.3758/s13428-012-0293-y
38. Walter, S.G., Block, J.H.: Outcomes of entrepreneurship education: an institutional perspective. J. Bus. Ventur. **31**, 216–233 (2016). https://doi.org/10.1016/j.jbusvent.2015.10.003
39. Miralles, F., Giones, F., Riverola, C.: Evaluating the impact of prior experience in entrepreneurial intention. Int. Entrepreneurship Manag. J. **12**(3), 791–813 (2015). https://doi.org/10.1007/s11365-015-0365-4
40. Hair, J.F., Hult, G.T.M., Ringle, C.M., Sarstedt, M.: A Primer on Partial Least Squares Structural Equation Modeling (PLS-SEM). SAGE Publications Inc, Los Angeles, CA (2017)
41. Mukhtar, S., Wardana, L.W., Wibowo, A., Narmaditya, B.S.: Does entrepreneurship education and culture promote students' entrepreneurial intention? the mediating role of entrepreneurial mindset. Cogent Education. **8**, 1–18 (2021). https://doi.org/10.1080/2331186X.2021.1918849

Development of Adaptive System for Detection of People with Increased Temperature Under Different Climatic Conditions

Shchelkunov Alexandr(✉) and Sergeev Nikolay

Southern Federal University, Rostov-on-Don 344006, Russia
{shelkunov,nesegeev}@sfedu.ru

Abstract. The paper presents an overview of existing methods for controlling the temperature of people in the stream using infrared cameras. The advantages and disadvantages of existing methods are presented. The developed adaptive method for detecting people with elevated temperature is described, this method eliminating the same shortcomings of existing methods. Presented are estimates of the accuracy of detecting people with elevated temperature using this method, as well as estimates of the probability of false alarms. A procedure has been developed for the formation of training data for the algorithm for detecting a human face on a thermal imaging frame, and semantic segmentation detected area into 18 classes, which is necessary to select a temperature control area on the surface of a human face. The procedure for selecting the area of temperature analysis based on international medical standards is described. The block diagram of the software product based on the developed method is presented, algorithms are presented that allow to increase the assessment of the accuracy of decision-making, such as nonlinear contrasting of the infrared image, the choice of the area for analyzing the temperature on the face surface based on the results of the algorithms for detecting and semantic segmentation of the human face thermogram. The architectures of the used neural networks are presented. Estimates of the probability of a false alarm and correct detection of a neural network algorithm for detecting a human face in a thermal imaging image are presented.

Keywords: Face detection · Temperature estimation · Face semantic segmentation · Neural network · Data analysis · Nonlinear image contrast

1 Introduction

The development of effective means of objectively controlling the temperature in the flow in crowded places is necessary to prevent the further development of the SARS COV-19 pandemic and other acute respiratory infections. One of the objective signs of SARS COV-19 is an increased body temperature of the patient. The use of common measuring instruments for remote control of temperature (pyrometers, thermal imagers) does not always make it possible to detect people with abnormal temperatures, especially in dynamically changing climatic conditions. The radiation temperature of the

R. Silhavy (Ed.): CSOC 2022, LNNS 502, pp. 10–21, 2022.
https://doi.org/10.1007/978-3-031-09076-9_2

surface of human skin, determined with the help of pyrometers and thermal imagers, is highly dependent on climatic conditions, as well as the accuracy of measuring instruments, often designed to operate under normalized climatic conditions. Considering the above, the development of a hardware-software complex for epidemiological control is an urgent task. The aim of the work is to research and develop a system for identifying people with elevated temperatures in various environmental conditions, as well as to determine ways for the further development of such systems.

2 Methods for Controlling the Temperature of People in the Stream

There are three main methods for controlling the temperature of people in the stream:

1) Estimation of temperature using a blackbody (an object with a known reflectance close to 1 in the IR range and a known stable temperature) [1].
2) Analysis of the thermogram of the face, classification of thermograms according to the distribution of brightness into normal and with increased temperature [2].
3) Statistical analysis, measured brightness, detection of anomalies in the flow.

In the first method, knowing the value of the blackbody temperature, the brightness of the thermogram on the surface of the blackbody and the brightness of the thermogram of a person's face, the human temperature is calculated. The use of this approach makes it possible to accurately estimate the temperature of the skin of a person ± 0.1 °C, however, accurate measurement of the temperature of the skin of a person does not allow identifying people with an elevated temperature in the flow due to large fluctuations in the temperature of the skin when the external conditions change.

The second method uses an analysis of the distribution of brightness on a thermogram to identify a person with an elevated temperature [2]. The disadvantage of this method is that the classification accuracy decreases when this method is operated in unstable temperature conditions, since the classification algorithm was trained on data obtained in room environmental conditions. The condition of the evaluated person also affects the classification accuracy, as the brightness distribution on the thermogram of a person's face changes significantly after physical exertion.

To reduce the influence of the disadvantages of the above methods in the problem of recognizing a person with an elevated temperature in conditions of environmental variability, an adaptive method was developed based on a statistical analysis of the brightness of thermograms of human faces in time. Based on the fact that the temperature distribution of people is close to the normal distribution, with an estimate of the expected value (EV) of 36.6 °C and an estimate of the standard deviation (RMS) of 0.4 °C [4], as well as from the assumption that the temperature of the skin is linearly related to the brightness of the thermogram of the face to estimate the temperature, we can use

$$T = 36.6 + \frac{(M - b) \cdot \sigma}{0.4} \quad (1)$$

where M – EV of the brightness of the thermograms of people, b – estimated brightness of the thermogram of a person, σ – RMS of the brightness of the thermograms of people.

The EV of the brightness of thermograms of people strongly depends on environmental conditions, which can change over time. For this reason, it is necessary to analyze the change in the M and σ brightness over time. For this, the measurements for the last 30 min are analyzed and the values M and σ of the brightness of people are calculated. To reduce fluctuations in estimates of M and σ, after obtaining an estimate of the brightness of a person, the average brightness and RMS are averaged according to the formulas:

$$\sigma_t = \frac{\sigma_{t-1}\beta N}{N_{\max}} + \sigma(1 - \frac{\beta N}{N_{\max}}) \tag{2}$$

$$M_t = \frac{M_{t-1}\beta N}{N_{\max}} + M(1 - \frac{\beta N}{N_{\max}}) \tag{3}$$

where σ_t – standard deviation at the output of the low-pass filter, σ_{t-1} is the standard deviation at the output of the low-pass filter at the previous time, σ – standard deviation of the brightness of the thermograms for the last 30 min, N – number of evaluated people in the last 30 min, Nmax – maximum number of people that can analyze the system in 30 min, M_t – average brightness at the output of the low-pass filter, M_{t-1} – average brightness at the output of the low-pass filter at time t−1, M is the average brightness over the last 30 min. The use of such a filter is because the higher the flux density of people, the more accurately the value of EV and RMS of brightness is determined and, therefore, the degree of confidence in these estimates increases. At low flux density values, the filter becomes more inertial.

3 System Development

To improve the accuracy of detecting people with elevated temperatures and to reduce the likelihood of false alarms, the temperature of people should be measured in the most thermally stable areas of a person's face. In [5] it is indicated that the most stable area for measuring the temperature of the skin of the human face is the area of the lacrimal sac. To assess the brightness of the thermogram in this area, two neural networks were used. The first detects the area of the person's face on the thermogram, the second classifies the pixels into 18 classes in the detected area, presented in Table 1. After that, the obtained areas are analyzed, and the temperature control point is selected. In (Fig. 1) shows the results of processing human thermograms to select the area of analysis, the selected area of analysis is highlighted in red.

Table 1. Classes of the semantic segmentation algorithm for a thermogram of a human face

Class ID	Class name	Class ID	Class name
1	Skin	10	Mouth
2	Left eyebrow	11	Upper lip
3	Right eyebrow	12	Under lip
4	Left ear	13	Cloth
5	Right ear	14	Hair
6	Left eye	15	Headdress
7	Right eye	16	Neck
8	Glasses	17	Background
9	Nose	18	Medical mask

Fig. 1. Temperature control area selection

To detect a human face used neural network detection algorithm with the YoloV4 architecture [7]. To train this network, 15,000 images of thermograms of people in which there are faces of people and 40,000 images of thermograms without faces were used. Faceless thermograms were used as negative samples to reduce the likelihood of false alarms. Open datasets of thermal imaging images [8], which were filtered for the presence of human objects in the images. A neural network classification algorithm based on the BiseNet architecture [9]. For training, pre-processed television images with marked masks and marked images were used in a total of 40,000 images [10].

To improve the assessments of the algorithms for detecting and segmenting the face, as well as displaying it to the operator, the nonlinear contrasting algorithm is used [11]. The use of this algorithm is necessary because on the thermogram the temperatures characteristic of the skin, people correspond to a rather narrow range of brightness. The temperature ranges from 20 °C to 40 °C accounts for about 4000 gradations of brightness on the original image of the infrared camera, with a color depth of 16 bits per pixel, i.e., the entire temperature range of interest accounts for 7% of the dynamic range of camera brightness. In (Fig. 2.a) and (Fig. 2.b) show images of thermograms before the contrasting algorithm and after, respectively.

a. b.

Fig. 2. The image received from the IR camera. a - before contrasting, b - at the output of the nonlinear contrasting algorithm.

After processing the image with the contrasting algorithm, the image must be normalized; for normalization, the formulas are used

$$Min_t = Min_{t-1}\beta + min \cdot (1 - \beta), \tag{2.1}$$

$$Max_t = Max_{t-1}\beta + max \cdot (1 - \beta), \tag{2.2}$$

$$\mathrm{Im}_i = \begin{cases} \frac{\mathrm{Im}_i - Min_{t-1}}{Max_{t-1} - Min_{t-1}}, & Min_{t-1} < \mathrm{Im}_i < Max_{t-1} \\ 1, & \mathrm{Im}_i > Max_{t-1} \\ 0, & \mathrm{Im}_i < Min_{t-1} \end{cases}, \tag{2.3}$$

where Min_t – minimum brightness at the output of the low-pass filter at time t, Max_t – maximum brightness at the output of the low-pass filter at time t, β – transfer function coefficient of the low-pass filter, max – maximum brightness in the human face in the current image, min – minimum brightness in the person's face in the current image, Im_i – pixel brightness at the output of the preprocessing algorithm. (Fig. 3) shows the structure of the epidemiological control system software. Bold lines show image data stream and thin lines show data stream.

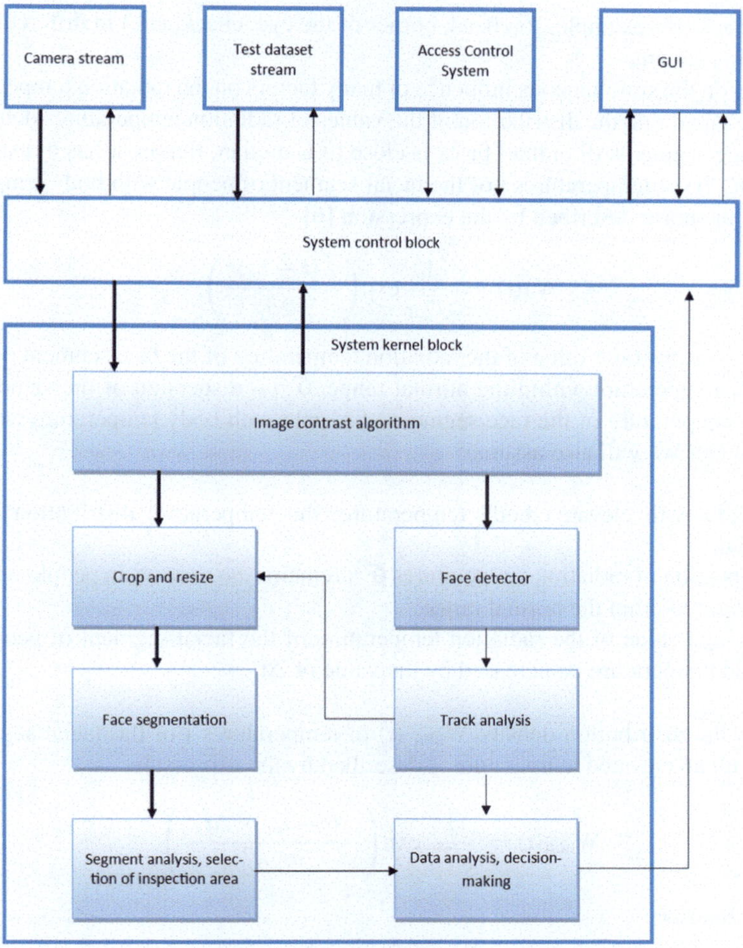

Fig. 3. Epidemiological control system software structure

4 Probabilistic Characteristics of Detecting People with an Abnormally High Temperature Using Thermal Imaging Devices

The brightness of the pixels of the image formed by a thermal imaging device, in the field of view of which there are human faces, is proportional to the radiation temperature of the corresponding areas of the skin. The radiation temperature of the skin of people depends on many factors, such as the external temperature of the environment, air speed, skin type, the intensity of sweating, the presence of cosmetics, the spread of the body temperature of people within the normal range, the angle of rotation of the head during the control period, etc. As a result, the radiation temperature of the same parts of the face of different people has a certain spread, the value of which is determined both by the reasons mentioned above, and by the accuracy of localization of the same segments of a person's face (for example, forehead, corner of the eye, cheek, etc.) in different people in a moving stream.

Based on the simultaneous influence of many factors on the radiation temperature, it can be assumed that the distribution of the values of radiation temperatures determined in the same segments of human faces is close to Gaussian, that is, it has a distribution density W_n (t) of temperatures t of the facial segment of people with body temperature in within the norm described by the expression [6].

$$W_n(t) = \frac{1}{\sqrt{2\pi}} \exp\left(-\frac{(t - t_m)^2}{2\sigma_t^2}\right), \tag{3.1}$$

where t_m – the average value of the radiation temperature of the face segment of people with body temperature within the normal range, $\sigma^2{}_t$ – dispersion of the values of the radiation temperature of the face segment of people with body temperature within the normal range. We will also assume:

- in people with elevated body temperature, the temperature distribution remains Gaussian.
- the dispersion of radiation temperatures $\sigma^2{}_t$ remains the same as in people with body temperature within the normal range.
- the average value of the radiation temperature of the facial segment of people with elevated temperature is increased by the value of Δt.

Thus, the distribution density W_{ubn} (t) of temperatures t of the facial segment of people with an elevated temperature is described by the expression.

$$W_{ubn}(t) = \frac{1}{\sqrt{2\pi}} \exp\left(-\frac{(t - t_m - \Delta t)^2}{2\sigma_t^2}\right), \tag{3.2}$$

Decision rule

The decision to detect people with an elevated temperature is made by determining the radiation temperature t^* of the selected facial segment and comparing the obtained value with the threshold value T. The decision to detect a person with an elevated temperature is made if:

$$t^* > T. \tag{3.3}$$

Under the above assumptions, the probabilities of false alarms P_{fp} and correct detection D are determined by the expressions.

$$P_{fp}(T, t_m, \sigma_t) = \int_T^{\infty} Wn(t)dt = 1 - F\left(\frac{T - t_m}{\sigma_t}\right), \tag{3.4}$$

$$D(T, t_m, \sigma_t, \Delta t) = \int_T^{\infty} Wubn(t)dt = 1 - F\left(\frac{T - t_m - \Delta t}{\sigma_t}\right), \tag{3.5}$$

where F (X) – integral Gaussian probability distribution function,

$$F(X) = \int_{-\infty}^{X} \frac{1}{\sqrt{2\pi}} \exp\left(-\frac{y^2}{2}\right) dy. \tag{3.6}$$

The probability of false alarms is determined by the variance $\sigma^2{}_t$ - values of the radiation temperature of the face segment of people, as well as the difference T-tm of the temperature threshold for decision-making T and the average value tm of the temperature distribution of people with body temperature within normal limits. The decision-making threshold T is selected for reasons of ensuring the desired level of false alarms no more than a given one, for example, $P_{fp} = 0.03$. For the selected value T of the decision threshold, the probability of correct detection D(T, tm, σt, Δt) additionally depends on the magnitude of the temperature anomaly Δt. All other things being equal, the probability of a correct detection of D is the higher, the greater the value of the temperature anomaly Δt. It is also obvious that the probabilistic characteristics of detecting people with an elevated temperature will be the better, the more uniform the flow of people in terms of the radiation temperature of the face segment, i.e. the smaller is the variance $\sigma^2{}_t$. Experimentally, under the conditions of winter temperatures of the external environment – 5°… 0 °C at the aircraft, the value of RMS t = 0.34° was obtained. In other climatic conditions, one can expect a change in σt in the range of 0.25°… 0.4 °C. Below in Figs. 4 and 5, the graphs of the D (P_{fp}) dependences are shown calculated for the RMS values σt = 0.34° and σt = 0.25° at various values of Δt = 0.8°, Δt = 1.0° and Δt = 1, 2° of detectable temperature anomaly. From a comparison of the graphs (see Fig. 4) and (see Fig. 5), the characteristics of detecting temperature anomalies significantly improve with a decrease in the RMS σt of the temperature distribution in the flow of people. For example, at σt = 0.34° and the probability of false alarms is not more than Plt = 0.03, the probability of correct detection exceeding D = 0.9 is achieved at Δt > 1.1°, while at σt = 0.25° the same values of P_{fp} and D are achieved at Δt > 0.8 °C.

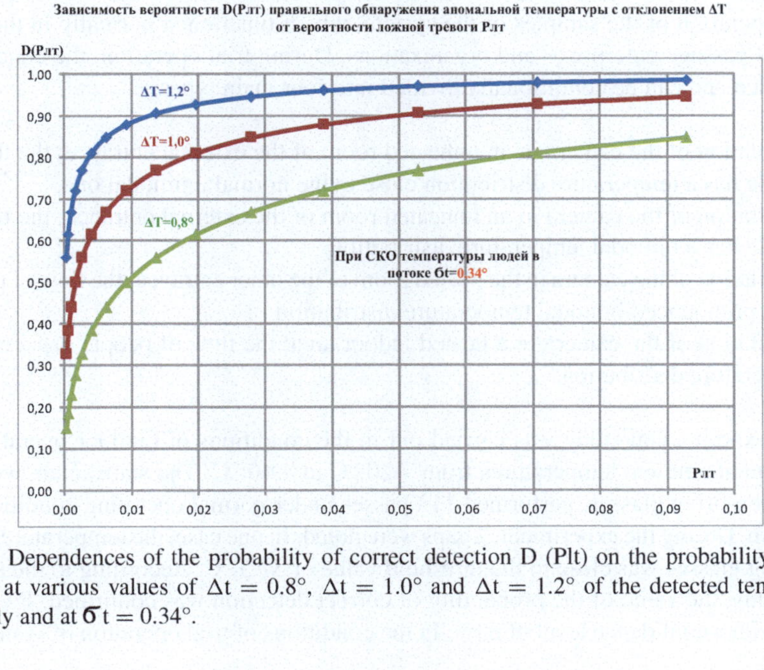

Fig. 4. Dependences of the probability of correct detection D (Plt) on the probability of false alarms at various values of Δt = 0.8°, Δt = 1.0° and Δt = 1.2° of the detected temperature anomaly and at σt = 0.34°.

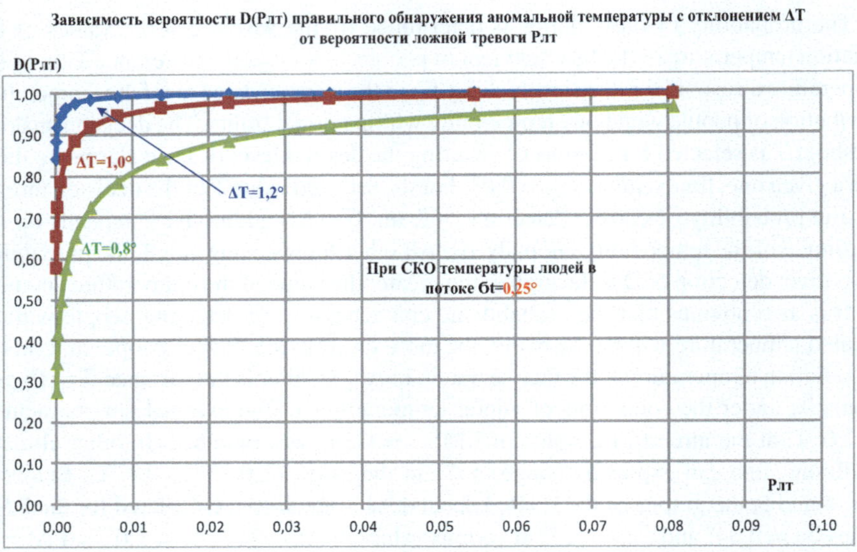

Fig. 5. Dependences of the probability of correct detection D (Plt) on the probability of false alarms at various values of Δt = 0.8°, Δt = 1.0° and Δt = 1.2° of the detected temperature anomaly and at σt = 0.25°.

5 Results of Trial Operation of the System

Experimental checks of the characteristics of the system were carried out in the conditions of trial operation of the samples of the system, which functioned normally in the conditions of passage enterprises and organizations. During trial operation, the conditions were varied and can be conditionally divided into four main types:

1. Installation of the camera in an unheated room of the external entrance; the flow of people has a temperature distribution close to the normal unimodal one.
2. Installation of the camera in an unheated room of the external entrance; the flow of people has a bimodal temperature distribution.
3. Installation of the camera in the heated room of the inner entrance; the flow of people has a pronounced bimodal temperature distribution.
4. Installation of the camera in a heated indoor area; the flow of people has a normal temperature distribution.

The experimental study was carried out in the conditions of outdoor operation of the system at ambient temperatures from −20 °C to +40 °C. The statistician, wearing heated measuring glasses, performed 113 passes under normal operating conditions of the system. During the experiment, 2 gaps were noted. In one case, the temperature of the measuring glasses was close to the threshold value of 37.5 °C. According to the results of the study, the value of the probability of correct detection was confirmed: P = 0.98 ± 0.02 with a confidence level of 85%. In the conditions of trial operation of systems at enterprises, real cases of detecting people with elevated temperatures of 37.1° and 38° were recorded, confirmed by temperature measurements.

5.1 Estimation of the Frequency of False Detections of a Person with Elevated Temperature in a Stream of People with a Unimodal Temperature Distribution

False detection observations were summarized based on data from all enterprises where the system is installed. The number of people passing through the checkpoints during the day varies from 600 to 2500 people. According to the observation results, it was noted that the relative frequency of false alarms on average does not exceed 3% under conditions of normal operation.

In the case of checking the system in the second and third of the above conditions (bimodal distribution of temperatures of people) in the absence of adjustments to the decision-making threshold, an increase in the level of false alarms was noted several times under conditions of a significant temperature difference between the two environments in which people stayed for a long time. This circumstance led to the need to refine the decision-making mechanism. This refinement made it possible, in conditions of a bimodal temperature distribution of a mixed flow of people, to reduce the frequency of false alarms to acceptable values (3%).

Due to the complexity of considering the mechanisms of the influence of inhomogeneous conditions of stay of people during the temperature control of people in the combined flow, research work in terms of optimizing decision rules for detecting temperature anomalies must be continued.

5.2 Estimation of the Relative Frequency of Correct Face Detection

To confirm the characteristics of the face detector, a test dataset was assembled, consisting of 10,000 images, each of which contains the face of one person. The dataset was collected from samples of the summer, autumn and winter periods. The images show people wearing hats, masks, glasses, etc. This sample was passed through an automatic ROI detection algorithm. The results of the study of the assessment of the relative frequency of correct detections from the threshold value of the confidence of the face detector (the threshold value lies in the range [0, 1]) are presented in (Fig. 6).

The estimation of the false alarm rate was carried out on a test dataset consisting of 58815 thermal images containing objects of non-natural origin. With a given face detector confidence threshold equal to 0.6, the false alarm rate was $3{,}1 \times 10^{-4}$. (Fig. 7) shows the dependence of the estimate of the false alarm probability of the face detector on the threshold.

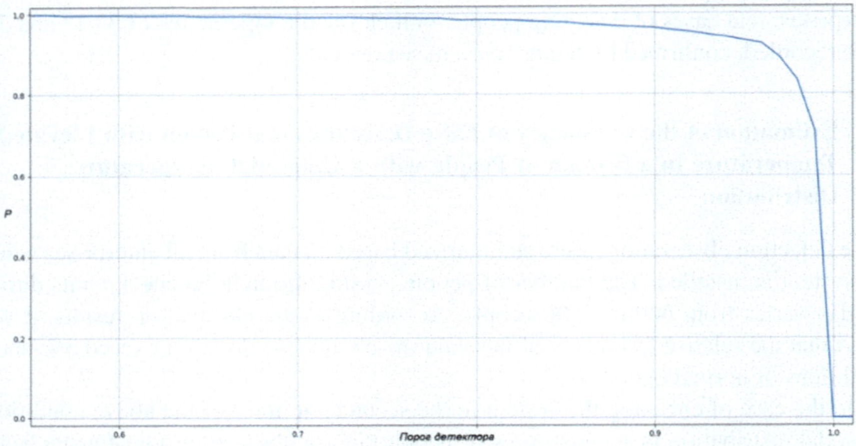

Fig. 6. Estimation of the probability of correct detection from the confidence threshold of the face detector.

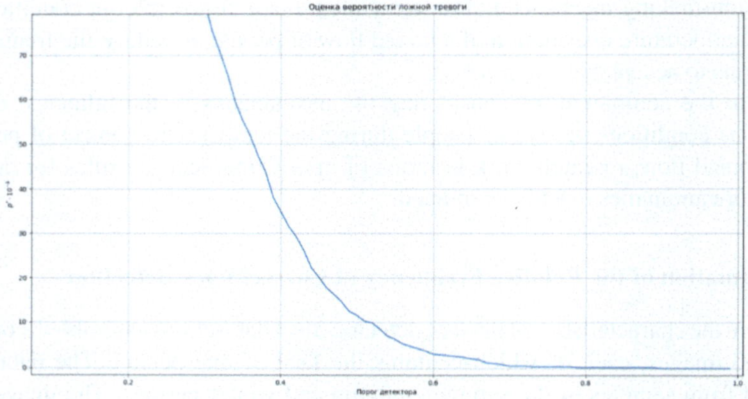

Fig. 7. Dependence of the relative frequency of false alarms on the threshold value of the face detector confidence

6 Conclusion

Accurate measurement of the temperature of the skin of people does not allow the identification of people with an elevated temperature in the flow due to large fluctuations in the temperature of the skin when external conditions change. For this reason, it is necessary to develop methods for detecting abnormal temperatures in the current operating conditions of the system. To reduce the influence of external conditions on the accuracy of temperature measurement, it must be assessed in the most stable area of a person's face - canthus. To highlight this area, the developed system uses two neural network algorithms for face detection and segmentation based on a thermal imaging image. The article presents the results of trial operation of the system, assessing the accuracy of detecting people with elevated temperatures, assessing the probability of missing, false alarms. In the future, the authors plan to optimize the architecture of the networks used in order to reduce the computing resources required for the operation of the system. Optimization will be performed both by choosing other network architectures, and by using quantization methods and other methods that reduce the processing time of one frame. This is necessary to increase the estimates of the accuracy of decision making. During trial operation, the average time for assessing the temperature of one person is 4.6 s (this time a person is in the field of view of the camera), in order to increase the accuracy of decision making, it is necessary to analyze as many frames as possible, which leads to the conclusion that it is necessary to reduce the processing time of one frame.

References

1. Zhang, Y., Li, Q., Zhou, H.: Theoretical foundation and basic properties of thermal radiation. In: Theory and Calculation of Heat Transfer in Furnaces, pp. 1–43. Elsevier (2016). https://doi.org/10.1016/B978-0-12-800966-6.00001-6
2. Masaki, A., Nagumo, K., Lamsal, B., Oiwa, K., Nozawa, A.: Anomaly detection in facial skin temperature using variational autoencoder. Artif. Life Robot. **261**(1), 122–128 (2020)
3. Nagumo, K., Kobayashi, T., Oiwa, K., Nozawa, A.: Face alignment in thermal infrared images using cascaded shape regression. Int. J. Environ. Res. Public Health **18**(4), 1776 (2021). https://doi.org/10.3390/ijerph18041776
4. Obermeyer, Z., Samra, J.K., Mullainathan, S.: Individual differences in normal body temperature: longitudinal big data analysis of patient records. BMJ **359**, j5468 (2017). https://doi.org/10.1136/bmj.j5468
5. ISO/TR 13154 Medical electrical equipment—Deployment, implementation and operational guidelines for identifying febrile humans using a screening thermograph (2017)
6. Ventsel, E.S., Ovcharov, L.A.: Teoriya veroyatnosti i ee inzhenernye prilozheniya. Ucheb. Posobie dlya vuzov — 2-e izd. - M.: Vyssh. shk. 480 s (2000)
7. Wang, C.-Y., Bochkovskiy, A., Mark Liao, H.-Y.: Scaled-YOLOv4: scaling cross stage partial network. In: Proceedings of the IEEE/CVF Conference on Computer Vision and Pattern Recognition (CVPR), pp. 13029–13038 (2021)
8. Krišto, M., Ivašić-Kos, M.:Thermal imaging dataset for person detection. In: 2019 42nd International Convention on Information and Communication Technology, Electronics and Microelectronics (MIPRO), pp. 1126–1131 (2019)
9. Fang, W., Xiaoyan, L., Qixiong, W., Lu, L.: Aerial-BiSeNet: a real-time semantic segmentation network for high resolution aerial imagery. Chin. J, Aeronaut. **34**(9), 47–59 (2021)
10. Shchelkunov, A.E.: Vyyavlenie lyudei s povyshennoi temperaturoi v potoke. In: Sbornik statei VII Vserossiiskaya nauchno-tekhnicheskaya konferentsiya "FPAKTIB", pp. 376–380 (2021)
11. Shchelkunov, AE, Kovalev, V.V., Morev, K.I., Sidko, I.V.: The metrics for tracking algorithms evaluation. Izvestiya SFedU. Eng. Sci. **1**, 233–245 (2020)
12. Everingham, M., Van Gool, L., Williams, C.K.I., Winn, J., Zisserman, A.: The pascal visual object classes (VOC) challenge. Int. J. Comput. Vis. **88**(2), 303–338 (2009)

Proximity-Based MAENS for Capacitated Multiple Traveling Salesmen Problem

Ruo-Yu Zhu[1,2], Chuan Wang[3(✉)], Qiang Yang[4(✉)], Xiao-Fang Liu[5], Dong Liu[1,2], Lin Sun[1,2], Hua Wang[6], and Jun Zhang[6,7]

[1] College of Computer and Information Engineering, Henan Normal University, Xinxiang, China

[2] Key Laboratory of Artificial Intelligence and Personalized Learning in Education of Henan Province, Xinxiang, Henan, China

[3] College of Software, Henan Normal University, Xinxiang, China
wangch@htu.edu.cn

[4] School of Artificial Intelligence, Nanjing University of Information Science and Technology, Nanjing, China
mmmyq@126.com

[5] College of Artificial Intelligence, Nankai University, Tianjin, China

[6] Institute of Sustainable Industries and Livable Cities, Victoria University, Footscray, Australia

[7] Chaoyang University of Technology, Taichung, Taiwan

Abstract. Capacitated Multiple Traveling Salesmen Problem (CMTSP) is an extension of the traditional Multiple Traveling Salesmen Problem (MTSP). In this problem, each city has a good with a certain weight to serve, and each salesman has a maximum weight capacity to serve the goods. To solve this problem effectively, this paper proposes a proximity-based crossover strategy for an existing and widely utilized method, namely the Memetic Algorithm with Extended Neighborhood Search (MAENS), leading to a new version of MAENS, which we name as PMAENS. In particular, this paper devises a similarity measure based on the common edges to calculate the similarity between two solutions. Then, during the selection of parents for crossover in MAENS, the similarity between any two solutions is taken into consideration so that two similar solutions preserve a high probability to crossover. In this way, the algorithm could achieve a promising balance between exploration and exploitation and thus is expected to obtain promising performance. Extensive experiments on the generated CMTSP instances demonstrate that the proposed PMAENS achieves competitive or even much better performance than MAENS with respect to both the convergence speed and the solution quality.

Keywords: Capacitated Multiple Traveling Salesmen Problem · Memetic algorithm · Traveling salesman problem · Local search · Proximity based crossover

1 Introduction

Capacitated Multiple Traveling Salesmen Problem (CMTSP) is a typical extension of the traditional multiple traveling salesmen problem (MTSP). In such a kind of problems,

R. Silhavy (Ed.): CSOC 2022, LNNS 502, pp. 22–33, 2022.
https://doi.org/10.1007/978-3-031-09076-9_3

each city has a good with a certain weight to serve, and each salesman has a maximum capacity. Then, multiple salesmen with different capacities cooperatively serve all goods in the cities by visiting each city once and only once. Such problems aim to seek an optimal route for each salesman to serve all goods in the cities in a given road network so that the capacity constraints of all salesmen are satisfied, and at the same time, the total cost of serving all goods is minimized. Actually, this kind of optimization problems widely exist in many applications in the real world, such as waste collection [1, 2] and logistics scheduling [3].

In the literature, evolutionary algorithms, like particle swarm optimization algorithms [4–6], differential evolution algorithms [7, 8], estimation of distribution algorithms [9], genetic algorithm (GA) [10], and ant colony optimization (ACO) [11–13], have shown promising performance in solving various optimization problems, including traditional traveling salesman problems (TSP) where only one salesman is involved [14]. To take advantage of this merit of EAs, researchers have made attempts to adapt EAs to solve MTSP [15, 16].

First, with more than one salesman involved, the representation of solutions in MTSP is very different from those in traditional TSP. Therefore, to effectively deal with MTSP, researchers have been devoted to designing novel representation schemes for EAs. As far as we know, in the literature, two popular representation schemes exist, namely the path representation approach [17] and the double-vector representation method [18]. The former mainly utilizes the routes of all salesmen to represent an individual, while the latter encodes an individual by two vectors, with one denoting the indexes of the cities and the other representing the associated salesmen that the corresponding cities in the first vector are assigned.

Then, based on the above two representation schemes, different novel operations have been developed for different EAs to solve MTSP effectively. For instance, based on the path representation approach, a new neighbor solution generation was devised for the artificial bee colony (ABC) algorithm in [19] to solve MTSP. Based on the double-vector representation method, novel crossover and mutation operators were designed for genetic algorithms (GAs) to solve MTSP effectively [20]. Likewise, based on the path representation approach, an ant-team-based ant colony optimization (ACO) algorithm [21] and a novel pheromone update method were devised to solve MTSP.

Although some EAs have been employed to cope with MTSP, compared with traditional TSP with a single salesman, MTSP is less extensively researched. Let alone the CMTSP. Compared with traditional MTSP, CMTSP is more difficult to optimize due to capacity constraints. Though CMTSP is an extension of MTSP, directly utilizing existing approaches for MTSP to solve CMTSP is far from effectiveness and efficiency, leading to that few studies exist on CMTSP.

Taking deep observation on CMTSP, we find that it is a little similar to another route optimization problem, namely the capacitated arc routing problem (CARP), which dispatches a fleet of vehicles with the same capacity to serve goods on the arcs [22]. Based on this observation, we adapt a popular and widely used algorithm for CARP to solve CMTSP. This method is the memetic algorithm with extended neighborhood search (MAENS) [23], which has been demonstrated to solve CARP effectively and is widely researched in the literature [24, 25].

Specifically, this paper first devises a proximity measure based on the common edges to calculate the similarity between two solutions. Then, based on this proximity measure, a novel crossover strategy is developed. In particular, in this crossover strategy, individuals preserve higher probabilities to crossover with those closer to themselves. In this way, the effectiveness of the crossover can be improved because more common promising city sequences between two close individuals can be preserved to generate offspring. Particularly, when the solution space is large, such as a large number of cities or a large number of salesmen are involved, such a crossover strategy is very beneficial for accelerating the convergence. Embedding the proposed proximity-based crossover into MAENS, we develop a new version, which we name as proximity-based MAENS (PMAENS), to solve CMTSP.

To substantiate the effectiveness and efficiency of the proposed PMAENS, we first generate some instances of CMTSP and then conduct comparative experiments by comparing PMAENS with the original MAENS.

The rest of this paper is organized as follows. Section 2 introduces the background of CMTSP briefly and formulates CMTSP mathematically. In Sect. 3, the related MAENS and the proposed method are elucidated. Subsequently, extensive experiments are conducted to verify the effectiveness and efficiency of the proposed PMAENS in Sect. 4. Finally, conclusions are drawn in Sect. 5.

2 Capacitated Multiple Traveling Salesmen Problem

In reality, many routing optimization problems can be transformed into MTSP [26], such as goods transportation and delivery [27], logistics scheduling [28], and crop monitoring [29]. However, such transformation is far from practice because many real-world applications involve some critical constraints, like capacity constraints. To more accurately model optimization problems in real-world applications, this paper introduces the capacitated MTSP (CMTSP). Specifically, similar to MTSP, in the CMTSP, multiple salesmen start from the depot and then collaboratively visit all cities, with each city visited once by only one salesman. At last, all salesmen return to the depot. However, unlike MTSP, in the CMTSP, each city has a good with a specific weight that needs to be served by only one salesman, and each salesman has a maximum capacity. Besides, different salesmen could preserve different capacities. For each salesman, the total weight of the goods on its route must be less than its capacity. Finally, CMTSP aims to seek an optimal route for each salesman to serve all goods in the cities in a given road network on the condition that the capacity constraints of all salesmen are satisfied and at the same time, the total cost of serving all goods by all salesmen is minimized.

Mathematically, a typical CMTSP can be formulated as follows. First, suppose the road map is an undirected graph $\boldsymbol{G} = (\boldsymbol{V}, \boldsymbol{E})$, where $\boldsymbol{V} = \{1, 2, \ldots, N\}$ is the set of cities, N is the number of cities; $\boldsymbol{E} = \{e_{ij} \mid i, j \in \boldsymbol{V}, i \neq j\}$ is the edge set. In this map, each node (city) c_i has a good with weight w_i, and each edge e_{ij} is associated with a traveling cost TC_{ij}, which indicates the cost between the city i and the city j. Suppose M salesmen are involved in CMTSP to serve all goods in the cities. Each salesman has a maximum capacity denoted as Q_i, and a cost denoted as SC_i.

Then, based on the path representation approach, we denote each solution to CMTSP as M routes with each associated with a salesman as shown in Fig. 1. Thereby, the aim of

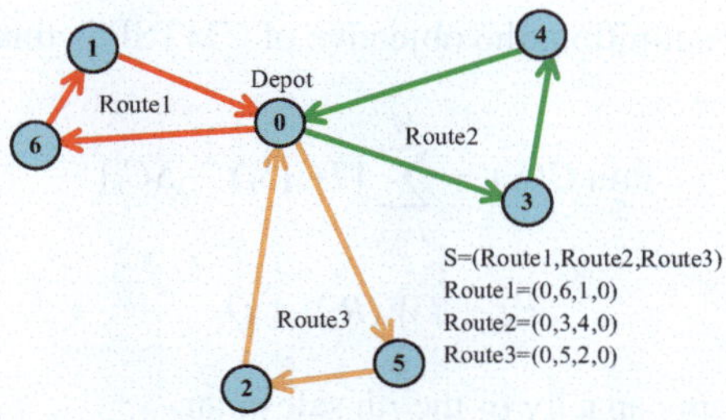

Fig. 1. An illustration of the solution representation.

CMTSP is to seek for an optimal path for each salesman to serve all goods in the cities so that the total cost (the cost of all edges in the routes plus the cost of all salesmen) is minimized with the following constraints satisfied:

1. Each salesman must start from and at last return to the depot.
2. Each city can only be traveled by one salesman and must be traveled exactly once.
3. The total weight of goods served by each salesman on its route should be less than the maximum capacity of the salesman.

Given a feasible solution to CMTSP is $\boldsymbol{S} = (R_1, R_2..., R_M)$, where $\boldsymbol{R_i}$ is the route of the ith salesman. As shown in Fig. 1, each route in the solution starts and ends at the depot, indexed as 0. In addition, each route is associated with a triplet $\boldsymbol{SM_i} = (S_i, Q_i, SC_i)$, where S_i denotes the index of the salesman, Q_i is the maximum capacity of the salesman, and SC_i is the cost of the salesman. Based on the above representation, the total cost of a feasible solution $\boldsymbol{S}$ can be formulated as

$$C(S) = \sum_{i=1}^{M} [TC(S_i) + SC_i] \tag{1}$$

where $TC(S_i)$ is the total traveling cost of the ith salesman, which is calculated by

$$TC(S_i) = \sum_{j=0}^{length(R_i)-1} TC_{j,j+1} + TC_{length(R_i)-1,0} \tag{2}$$

where $length(\boldsymbol{R}_i)$ is the number of the cities in $\boldsymbol{R}_i$, $TC_{j,j+1}$ denotes the traveling cost from the jth city to the $j+1$th city in the route R_i. It should be mentioned that the last item is added because the salesman must return to the depot at last.

In particular, the total weight of the goods served in the route $\boldsymbol{R}_i$ of the ith salesman can be computed as follows:

$$TW(R_i) = \sum_{k=1}^{length(R_i)-2} w_k \tag{3}$$

where $TW(R_i)$ is the total weight of the goods served in the route $\boldsymbol{R}_i$. It should be noticed that the depot contains no good to serve, and thus the start and end cities of the route R_i should be excluded because they are the depot.

Based on the above definition, the objective of CMTSP is finally obtained:

$$\min_{S} C(S) = \sum_{i=1}^{M} [TC(S_i) + SC_i] \tag{4}$$

$$s.t.: TW(R_i) \leq Q_i \tag{5}$$

where Q_i is the maximum capacity of the ith salesman.

From the above objective, it is found that CMTSP aims to find the optimal route of each salesman to minimize the total traveling cost of all salesmen on the condition that the total weight of served goods by each salesman does not exceed the capacity of the associated salesman. In particular, such optimization not only involves the assignment of cities to each salesman, but also involves optimizing the traveling sequence concerning the assigned cities. Due to the capacity constraint, the assignment of cities becomes much harder than traditional MTSP. Therefore, it is very challenging to solve CMTSP with high effectiveness and efficiency.

3 The Proposed Method

Comparing CMTSP with the widespread CARP [22], we find that these two problems share some common features, like the capacity constraint and the objective of minimizing the total traveling cost. The difference between them mainly lies in two aspects: 1) In CMTSP, the number of salesmen is known in advance, while that in CARP is unknown and needs optimizing; and 2) In CMTSP, the goods lie in the cities, namely vertices in the constructed graph $\boldsymbol{G}$, while in CARP, the goods lie along the arcs, namely edges in the constructed graph $\boldsymbol{G}$. Compared with CARP, CMTSP is more common in real-world applications, especially in logistics areas.

Based on the above similarity between CMTSP and CARP, this paper aims to adapt a popular and widely used algorithm for CARP to solve CMTSP. This method is the memetic algorithm with extended neighborhood search (MAENS) [23]. In particular, this algorithm has been verified to solve CARP effectively and is widely researched in the literature [24, 25]. To improve its effectiveness in solving CMTSP, this paper proposes a proximity measure for MAENS to select individuals to crossover, and thus an improved version of MAENS, which we name as PMAENS, is developed for CMTSP.

3.1 MAENS

Before the elucidation of the proposed PMAENS, this subsection first introduces MAENS [23]. In fact, MAENS is a hybrid of genetic algorithm (GA) and a kind of local search operator [23]. The crucial innovation in MAENS is the proposed local search operator, namely the Merge-Split (MS) operator. It plays a similar role with the mutation strategy in GA and is mainly conducted on the offspring after the crossover operator to generate new individuals.

As a whole, the overall framework of MAENS is outlined in Algorithm 1, where instead of selecting individuals based on proximity in Lines 13~15, MAENS selects individuals randomly to crossover. Specifically, in Lines 1~9, the algorithm randomly initializes a population (***pop***) of different *psize* individuals. Then, in Lines 12~27, the algorithm generates *spsize* offspring based on the crossover operator and the local search operator. After generating a new offspring, the algorithm compares it with those stored in a temporary population ($\boldsymbol{pop_t}$), which contains the individuals in the parent population ***pop*** and the already generated offspring. The generated offspring can only be inserted into $\boldsymbol{pop_t}$ when it is not a duplicate of any solution in $\boldsymbol{pop_t}$. In this way, the stored individuals in $\boldsymbol{pop_t}$ are different from each other. After the generation of the offspring, all individuals in $\boldsymbol{pop_t}$ are sorted based on their fitness (Line 28) and then the top best *psize* individuals are selected to form a new population ***pop*** for the next generation (Line 29). The above iteration goes on until the stopping criterion is met. In this paper, we utilize the maximum number of generations as the stopping criterion.

3.2 Proximity-Based MAENS

Taking deep comparison on the crossover operator for continuous optimization and discrete optimization, we find that the effectiveness of the crossover operator is very different on these two kinds of optimization problems. In particular, we empirically find that it is more likely for the generated offspring to inherit the promising evolutionary information of the two parents used to crossover in continuous optimization than in discrete optimization. Taking TSP problems as an example, we find that if we randomly select parents to crossover, it is hardly possible to obtain better solutions after crossover. This is because the visiting sequences of the randomly selected two individuals are very likely different. After crossover and conflict correction, many promising visiting sequences are destroyed, leading to that the generated offspring can hardly inherit the promising visiting sequences in the two parents. Taking inspiration from the above findings, we propose a proximity-based crossover for MAENS to improve its effectiveness in solving CMTSP.

In TSP variants, including CMTSP, the visiting sequence of cities is actually a permutation of all cities. Since the visiting order of cities is crucial, it is natural to consider the proximity between two permutations based on the visiting order of cities. To this end, this paper devises a proximity measure based on the number of common edges for two permutations. Specifically, suppose π_i and π_j are two permutations associated with two feasible solutions, while $\Phi(\pi_i)$ and $\Phi(\pi_j)$ denote their edge sets respectively. Then, the similarity between π_i and π_j is calculated as follows:

$$P(\pi_i, \pi_j) = \frac{|\Phi(\pi_i) \cap \Phi(\pi_j)|}{N - 1} \tag{6}$$

where $|\cdot|$ denotes the number of common edges in the two permutations. N–1 is the total number of edges in one permutation.

From Eq. (6), we find that the proximity is actually the ratio of the common edges in the two solutions out of all edges (in one solution). Particularly, the larger the proximity value is, the more common edges exist in the associated two solutions. Therefore, it is likely that the two solutions are closer to each other. It should be mentioned that such a proximity index can only roughly measure the similarity between two solutions.

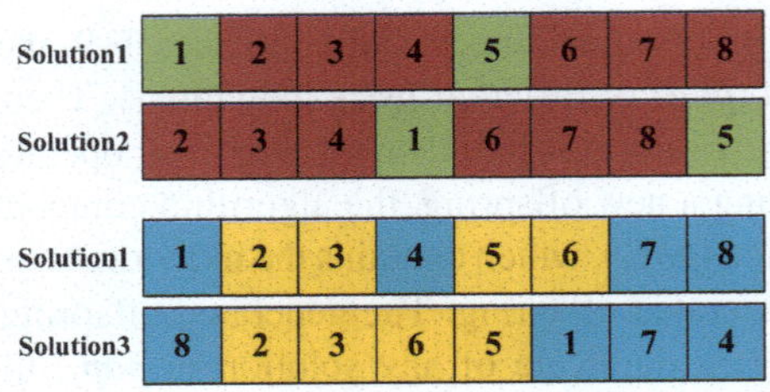

Fig. 2. An example of the proximity measure.

Though such a roughness may result in incorrect similarity, it brings some diversity for evolutionary algorithms in turn, which is beneficial for the population to escape from local areas.

Algorithm 1: The pseudocode of PMAENS

Input: population size *psize*, number of offspring *spsize*, maximum number of failure trials *ltrial*, and local search probability P_{ls}

Output: The best found solution S_{bf}

```
Set the current population pop = Ø;
while | pop | < psize do;
  Set the trial counter ntrial = 0;
  repeat
    Generate an initial solution S_init;
    ntrial ← ntrial + 1;
  until S_init is not a clone of any solution S ∈ pop or ntrial == ltrial;
  pop ← pop ∪ S_init;
end
while stopping criterion is not met do
  pop_t = pop;
  for i = 1 → spsize do
    Randomly select a solution S_1 from pop;
    Calculate the proximity between S_1 and the other solutions in pop based on Eq. (6);
    Randomly select another different solution S_2 from pop based on the roulette wheel
    selection according to the calculated proximity;
    Apply the crossover operator on S_1 and S_2 to generate S_x;
    if rand(0,1) < P_ls then
      Apply local search on S_x to generate S_ls;
      if S_ls is not a clone of any S ∈ pop_t then
        pop_t = pop_t ∪ S_ls
      else if S_x is not a clone of any S ∈ pop_t then
        pop_t = pop_t ∪ S_x
      end
    else if S_x is not a clone of any S ∈ pop_t then
      pop_t = pop_t ∪ S_x
    end
  end
  Sort the solutions in pop_t based on their fitness;
  Select the top best psize solutions in pop_t to form a new pop;
end
return the best solution S_bf in pop;
```

Taking the three feasible solutions shown in Fig. 2 as an example, we can see that the proximity between Solution 1 and Solution 2 is 4/7 based on Eq. (6), because there are four common edges in these two solutions, namely e_{23}, e_{34}, e_{67}, and e_{78}. As for Solution 1 and Solution 3, we can find that there are only two common edges, namely e_{23} and e_{56} (e_{65}). Hence, the proximity between Solution 1 and Solution 3 is 2/7. As a result, we can see that the proximity between Solution 1 and Solution 2 is larger than Solution 1 and Solution 3. Therefore, Solution 2 preserves a higher probability to be selected than Solution 3, to crossover with Solution 1 during the evolution.

As for CMTSP introduced in this paper, we adopt the path representation scheme to encode individuals in MAENS. Though in the encoding scheme, the route of each salesman is separated from others, the edge set of all salesmen can be obtained and then is used to calculate the similarity between two solutions.

This paper utilizes the devised proximity measure to select parents for the crossover operator, leading to the improved MAENS, which we name PMAENS. Specifically, the overall framework of PMAENS is very similar to MAENS, as shown in Algorithm 1. The only difference lies in Lines 13–16. During the offspring generation in MAENS, we first randomly select a parent from the current population ***pop***, and then the proximity between the selected individual and the others in ***pop*** is calculated based on Eq. (6). Subsequently, we randomly select another different individual for crossover by the roulette wheel selection method based on the selection probability computed on the basis of the calculated proximity. After selecting the two parents, the crossover operator is conducted to generate a new solution.

With the above manner, the larger the value of the proximity between two individuals, the higher the probability that these two individuals are conducted the crossover operator. In this way, the generated new solution is more likely to inherit promising city sequences in these two parents. Hence, the chance of generating more promising solutions becomes higher, which is beneficial for MAENS to find promising areas fast.

4 Experimental Studies

4.1 Experimental Setting

To verify the effectiveness of the proposed PMAENS, based on the fact that CMTSP shares some similar features with CARP, we first generate 12 instances with different numbers of salesmen based on the CARP sets. Specifically, we select the CARP set with 77 cities and 140 cities as the benchmark sets and then allocate different numbers of salesmen with different capacities to each set. The brief information of the generated instances is listed in Table 1.

To compare with the proposed PMAENS, we select the original MAENS as the compared method. To make fair and comprehensive comparisons, on the one hand, we directly utilize the same parameter settings in MAENS; on the other hand, we utilize the best and the mean values over 10 independent runs to evaluate the performance of both algorithms. In addition, it deserves attention that both algorithms are implemented with C++ language and run on a PC with 4 Intel(R) Core (TM) i5-10210U 1.60 GHz CPUs, 16 Gb memory, and 64-bit Ubuntu 12.04 LTS system.

Table 1. Comparison results between PMAENS and MAENS on the generated 12 instances.

Name	N	M	MAENS		PMAENS	
			best	mean	best	mean
Instance 1	77	5	**3548**	3548.0	**3548**	3548.0
Instance 2	77	7	**5018**	5018.4	**5018**	5018.0
Instance 3	77	9	**6476**	6490.2	**6476**	6483.1
Instance 4	77	10	**5595**	5611.3	**5595**	5600.4
Instance 5	77	12	7788	7810.2	**7778**	7795.1
Instance 6	77	14	9024	9052.3	**9013**	9055.4
Instance 7	77	20	**11626**	11714.9	11628	11743.4
Instance 8	140	10	6394	6435.3	**6388**	6433.7
Instance 9	140	14	8518	8526.5	**8502**	8521.2
Instance 10	140	19	12446	12516.7	**12439**	12480.3
Instance 11	140	27	**16493**	16564.3	16495	16539.9
Instance 12	140	36	20933	21008.6	**20878**	20970.0

4.2 Comparison Between PMAENS and MAENS

Table 1 presents the total traveling cost of all salesmen associated with the final solutions obtained by both algorithms on 12 instances. In the table, *N* represents the number of cities, and *M* denotes the number of salesmen. The best value represents the minimum total cost value in the 10 independent runs, and the mean value represents the average cost over the 10 runs. In addition, in this table, the best results between the two algorithms with respect to the best value are highlighted in bold, while the best results in terms of the mean value are shaded.

From Table 1, it is found that: 1) Concerning the best value in the 10 independent runs, the proposed PMAENS achieves the best results on 10 instances, while the original MAENS achieves the best results on only 6 instances. 2) From the mean value over the 10 independent runs, PMAENS achieves the best results on 10 instances, while MAENS only obtains the best results on 3 instances. 3) With the number of salesmen increasing, no matter from the perspective of the best value or the perspective of the mean value, the superiority of PMAESN to MAENS becomes more and more notable.

To further demonstrate the effectiveness of the proposed PMAENS, we make the convergence comparison between PMAENS and MAENS on the above 12 instances. Figure 3 displays the convergence curves of the two algorithms with the maximum number of generations set as 100.

From this figure, it is found that on most instances, the proposed PMAENS not only converges faster than MAENS, but also obtains much better solutions than MAENS.

Based on the above experimental results, it is demonstrated that the proposed PMAENS is much superior to the original MAENS. Such superiority of PMAENS mainly benefits from the proposed proximity strategy for crossover. In particular, in PMAENS, individuals with higher proximity are more likely to be selected to conduct the crossover operator. In this way, the promising visiting sequences in two closer parents are more likely to be inherited by the generated offspring. This results in that the chance of generating new better solutions becomes higher. Therefore, with this mechanism, the proposed PMAENS is expected to achieve faster convergence and better solutions than MAENS.

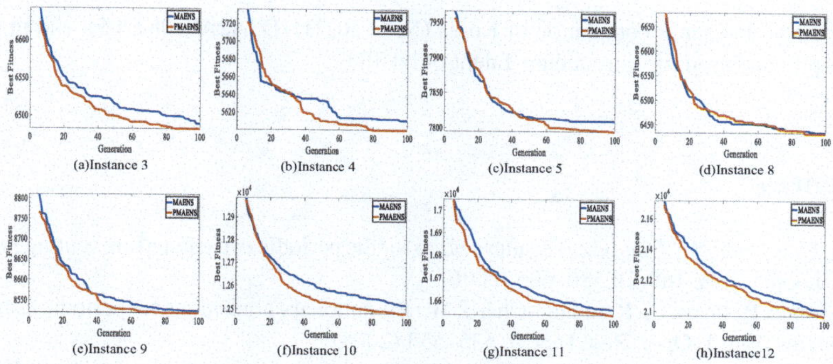

Fig. 3. Convergence comparison between PMAENS and MAENS.

5 Conclusion

This paper first introduces the capacitated multiple traveling salesmen problems (CMTSP), which are extensions of traditional multiple traveling salesmen problems (MTSP). In such a kind of problems, each city has a good with a certain weight to serve, and each salesman has a maximum capacity. Then, multiple salesmen with different capacities cooperatively serve all goods in the cities. This problem aims to seek an optimal route for each salesman so that the traveling cost of all salesmen is minimized on the condition that the total weight of goods on the route of each salesman does not exceed the capacity of the salesman. Compared with MTSP, CMTSP is more difficult to optimize because of the additional capacity constraints.

To effectively solve this problem, we adapt an existing method named the memetic algorithm with extended neighborhood search (MAENS) to solve CMTSP. In particular, to further improve its effectiveness, we propose a proximity measure to calculate the similarity between any two solutions based on the common edges. With this proximity measure, we use the roulette wheel selection method to select two individuals to conduct the crossover operator, leading to that closer individuals have a higher probability to be selected. In this way, the promising visiting sequences in two parents could be more likely to be inherited by the generated offspring.

Extensive experiments on the 12 generated CMTSP instances have demonstrated that the proposed PMAENS achieves faster convergence and better solutions than the original MAENS on most instances. This implicitly demonstrates the effectiveness of the proposed proximity measure for the selection of parents to conduct the crossover operator.

In this paper, we mainly tackle small-scale CMTSP. In the future, we will utilize large-scale evolutionary algorithms [30–32] to tackle large-scale CMTSP, since large-scale CMTSP is becoming more and more common in the era of big data and internet of things.

Acknowledgment. This work was supported in part by the National Natural Science Foundation of China under Grant 62006124, 62072160, and 61873097, in part by the Natural Science Foundation of Jiangsu Province under Project BK20200811, in part by the Natural Science Foundation of the Jiangsu Higher Education Institutions of China under Grant 20KJB520006, in part by the National Research Foundation of Korea (NRF-2021H1D3A2A01082705), and in part by the Startup Foundation for Introducing Talent of NUIST.

References

1. Chu, F., Labadi, N., Prins, C.: A scatter search for the periodic capacitated arc routing problem. Eur. J. Oper. Res. **169**(2), 586–605 (2006)
2. Lacomme, P., Prins, C., Ramdane-Chérif, W.: Evolutionary algorithms for periodic arc routing problems. Eur. J. Oper. Res. **165**(2), 535–553 (2005)
3. Bektas, T.: The multiple traveling salesman problem: an overview of formulations and solution procedures. Omega **34**(3), 209–219 (2006)
4. Yang, Q., Chen, W.N., Gu, T., Jin, H., Zhang, J.: An adaptive stochastic dominant learning swarm optimizer for high-dimensional optimization. IEEE Trans. Cybern. **52**, 1–17 (2020)
5. Wei, F.F., et al.: A classifier-assisted level-based learning swarm optimizer for expensive optimization. IEEE Trans. Evol. Comput. **25**(2), 219–233 (2021)
6. Yang, Q., et al.: A distributed swarm optimizer with adaptive communication for large-scale optimization. IEEE Trans. Cybern. **50**(7), 3393–3408 (2020)
7. Yang, Q., Xie, H., Chen, W., Zhang, J.:Multiple Parents Guided Differential Evolution for Large Scale Optimization. In: IEEE Congress on Evolutionary Computation, pp. 3549–3556 (2016)
8. Yu, W.J., Ji, J.Y., Gong, Y.J., Yang, Q., Zhang, J.: A tri-objective differential evolution approach for multimodal optimization. Inf. Sci. **423**, 1–23 (2018)
9. Yang, Q., Chen, W., Li, Y., Chen, C.L.P., Xu, X., Zhang, J.: Multimodal estimation of distribution algorithms. IEEE Trans. Cybern. **47**(3), 636–650 (2017)
10. Potvin, J.-Y.: Genetic algorithms for the traveling salesman problem. Ann. Oper. Res. **63**(3), 337–370 (1996). https://doi.org/10.1007/BF02125403
11. Dorigo, M., Gambardella, L.M.: Ant colony system: a cooperative learning approach to the traveling salesman problem. IEEE Trans. Evol. Comput. **1**(1), 53–66 (1997)
12. Chen, W.N., Tan, D.Z., Yang, Q., Gu, T., Zhang, J.: Ant colony optimization for the control of pollutant spreading on social networks. IEEE Trans. Cybern. **50**(9), 4053–4065 (2020)
13. Yang, Q., et al.: Adaptive multimodal continuous ant colony optimization. IEEE Trans. Evol. Comput. **21**(2), 191–205 (2017)
14. Dutta, M.:TSP solution using dimensional ant colony optimization. In: International Conference on Advanced Computing & Communication Technologies, pp. 506–512. IEEE, Haryana (2015)
15. Yuan, S., Skinner, B., Huang, S., Liu, D.: A new crossover approach for solving the multiple travelling salesmen problem using genetic algorithms. Eur. J. Oper. Res. **228**(1), 72–82 (2013)
16. Dubey, I., Gupta, M.:Uniform mutation and SPV rule based optimized PSO algorithm for TSP problem. In: International Conference on Electronics & Communication Systems, pp. 168–172. IEEE, Coimbatore (2017)
17. Tang, J.L.L., Rong, A., Yang, Z.: A multiple traveling salesman problem model for hot rolling scheduling in shanghai baoshan iron and steel complex. Eur. J. Oper. Res. **124**(2), 267–282 (2000)
18. Chen, S.H., Chen, M.C.:Operators of the two-part encoding genetic algorithm in solving the multiple traveling salesmen problem. In: International Conference on Technologies and Applications of Artificial Intelligence, pp. 331–336. IEEE, Chung Li (2011)
19. Pandiri, V., Singh, A.: A swarm intelligence approach for the colored traveling salesman problem. Appl. Intell. **48**(11), 4412–4428 (2018). https://doi.org/10.1007/s10489-018-1216-0

20. Singh, D.R., Singh, M.K., Singh, T.:Multiple traveling salesman problem using novel crossover and group theory. In: International Conference on Computing, Communication and Automation, pp. 368–372, IEEE, Greater Noida (2017)
21. Lu, L.C., Yue, T.W.: Mission-oriented ant-team ACO for min-max MTSP. Appl. Soft Comput. **76**, 436–444 (2018)
22. Golden, B.L., Wong, R.T.: Capacitated arc routing problems. Networks **11**(3), 305–315 (2010)
23. Tang, K., Mei, Y., Yao, X.: Memetic algorithm with extended neighborhood search for capacitated arc routing problems. IEEE Trans. Evol. Comput. **13**(5), 1151–1166 (2009)
24. Mei, Y., Li, X., Yao, X.: Cooperative coevolution with route distance grouping for large-scale capacitated arc routing problems. IEEE Trans. Evol. Comput. **18**(3), 435–449 (2014)
25. Mei, Y., Tang, K., Yao, X.: Decomposition-based memetic algorithm for multiobjective capacitated arc routing problem. IEEE Trans. Evol. Comput. **15**(2), 151–165 (2011)
26. Cheikhrouhou, O., Khoufi, I.: A comprehensive survey on the multiple traveling salesman problem: applications, approaches and taxonomy. Comput. Sci. Rev. **40**(4), 100369 (2021)
27. Kitjacharoenchai, P., Ventresca, M., Moshref-Javadi, M., Lee, S., Tanchoco, J.M., Brunese, P.A.: Multiple traveling salesman problem with drones: mathematical model and heuristic approach. Comput. Ind. Eng. **129**, 14–30 (2019)
28. Qian, W.W., Zhao, X., Ji, K.: Region division in logistics distribution with a two-stage optimization algorithm. IEEE Access **8**, 212876–212887 (2020)
29. Conesa-Muñoz, J., Bengochea-Guevara, J.M., Andujar, D., Ribeiro, A.: Route planning for agricultural tasks: a general approach for fleets of autonomous vehicles in site-specific herbicide applications. Comput. Electron. Agric. **127**, 204–220 (2016)
30. Yang, Q., Chen, W., Deng, J.D., Li, Y., Gu, T., Zhang, J.: A level-based learning swarm optimizer for large-scale optimization. IEEE Trans. Evol. Comput. **22**(4), 578–594 (2018)
31. Yang, Q., et al.: Segment-based predominant learning swarm optimizer for large-scale optimization. IEEE Trans. Cybern. **47**(9), 2896–2910 (2017)
32. Yang, Q., Chen, W., Zhang, J.: Evolution consistency based decomposition for cooperative coevolution. IEEE Access **6**, 51084–51097 (2018)

Historical Alphabet Transliteration Software Using Computer Vision Classification Approach

Adéla Hamplová(✉), David Franc, and Jan Tyrychtr

Czech University of Life Sciences in Prague, Kamýcká 129,
165 00 Praha-Suchdol, Czech Republic
{hamplova,francd,tyrychtr}@pef.czu.cz

Abstract. The article presents the problem of developing mobile software for classification and automatic transliteration of historical alphabets to Latin alphabet using OCR Computer Vision algorithms and is presented on Palmyrene Alphabet. Our suggested solution of semi-automatic transliteration of historical alphabets speeds up and simplifies the process of ancient text analysis and makes reading historical alphabets available to the public. We created a mobile application template for field use and proved the functionality on our own photographic and digitized hand-written datasets of Palmyrene letters, using a MobileNet Artificial Neural Network for character recognition. Such an application helps archaeologists with a faster character transliteration of newly discovered, archived, but untranslated tablets, columns etc., and for checking hand-transliterated texts.

Keywords: Artificial intelligence · Classification · Historical alphabet · Software development

1 Introduction

The technology of Optical Character Recognition, commonly known as OCR, has developed rapidly in the last years, and its applications can be found in various areas of human activities. Applications vary from getting text from scanned documents to automatic registration plate-number recognition. However, not all alphabets have this technology available, and transliteration has to be done manually.

Archaeology, like any other human activity and research area, can profit from such technology. There is a large number of historical documents in different alphabets that archaeologists need to analyze, rewrite by hand to paper and then translate, one of which is Palmyrene.

2 Theoretical Overview

2.1 Image Classification

Artificial Intelligence is an area, which first appeared in the 1950s. It includes a wide area of computer science tasks that don't use rules and data to get answers, but uses data and answers to create rules [1].

R. Silhavy (Ed.): CSOC 2022, LNNS 502, pp. 34–45, 2022.
https://doi.org/10.1007/978-3-031-09076-9_4

The input data points are taken as a training set, and the learning algorithm (also called fitting algorithm) is trying to map these inputs to the expected outputs, by alternating neuron weights and keeping the same activation functions in the network during the process. This process is called learning, and the quality of learning can be measured by computing error – the loss. Loss can be simplified as the difference between predicted and real outputs. In Artificial Intelligence terminology, the trained network is called a model. It consists of network architecture and trained weights [1].

Image classification is using deep learning architectures with Convolutional Neural Networks. CN Networks accept input tensors in the shape of (image_height, image_width, image_depth). Deeper in the network the dimensions are lower and lower, since the Convolution operation is reducing the image to local connections by using feature maps. The last but one layer of CNN needs to be flattened by using the Flatten layer. The output is provided by a Dense layer, which returns an integer – index of class, where the image is predicted to [1].

The classification success rate can be measured by standard metrics. They are called precision, recall, F-measure and accuracy and are explained in Methodology [2]. These metrics are calculated usually on the training set, which is a subset of images, which have not been used during the training.

2.2 OCR

As early as the 19th century, the first attempts to read letters by machine were recorded, until a technology reading exactly one character was developed [3]. In the 19th century, OCR was already widely used and was getting known to the general public [4]. In 2020, OCR was even used for biometric identification with cross-entropy fitness function to get the smallest squares [5]. Among the latest advances in this area is a specially developed Artificial Neural Network called AutOCR. In comparison to other applications, there is an automated process of training parameter tuning, and the latest research of OCR applies LSTM (Long Short-Term Memory) layers to neural network architectures for processing words – sequences of characters [6].

Although the technology for reading characters already exists, there is still space for experiments with the optimization of a suitable neural network architecture, special for historical alphabets recognition from either hand-written sources or directly from photographic records.

2.3 Historical Alphabet Digitisation Including Palmyrene

There are multiple researches specialized in historical alphabet character recognition, such as Persian [7] or Bangladeshi [8], as well as hand-written transliterations of cuneiform [9]. However, there is still currently no Palmyrene transliteration available.

At present, there are a large number of published texts in Palmyrene Aramaic, written in a special alphabet, that is similar to but slightly different from Aramaic at the same time. Palmyrene was used in years 100–400 A.D. in the Syrian city of Palmyra (otherwise called Tadmur) and in its surroundings. This language was derived from classical Aramaic but uses its own dialect and alphabet. Palmyrene Aramaic was spoken in western parts of Syria in the beginning of the first century, on the other hand in eastern parts classical Aramaic was spoken.

The study of texts in Palmyrene Aramaic contributes not only to Palmyrene studies, but also Semitic and Biblical studies, the study of ancient history and art [10]. Ones of the most extensive anthologies were published in years 1996 and 2001 [10, 11].

In 2010, a standard coding for Palmyrene letters was proposed and adopted [12], and it is currently a part of an extended Unicode character table. The Palmyrene alphabet contains 32 characters, 10860-1087F Unicode [13]. The script is written horizontally, right to left, letters are not divided by a blank space, apart from "y" there is no wovel, only consonants. Writing numbers is similar to the Roman way, but there are only characters for number 1, 5, 10 (also used for hundreds) and 20. Numbers 2, 3 and 4 are depicted as a multiple of 1.

Currently, there is a research gap in the field of automatic character reading of historical alphabets and transliterating them to Latin alphabet, part of it means there is also no Palmyrene alphabet transliterator available yet.

2.4 Android Software Development

Android is an operating system based on Linux, mainly focused on mobile devices. Its apps are developed in Java mainly using Android Studio as IDE. There are also other multi-platform development tools, which can be used for Android as well.

A set of basic libraries that are included in each version of Android API is upgraded with every Android system update and some methods in these libraries are added and improved.

Image classification is not included in standard Android API libraries. Premade classificators using CNN are available in Tensorflow Lite. The most common models for image classification use TFLite MobileNet. Current research recommends using efficient_lite0 network [14]. The architecture of this model is visible in Fig. 1. The dropout layer randomly sets some neuron weights to 0 to prevent overfitting.

```
_________________________________________________________________
Layer (type)                 Output Shape              Param #
=================================================================
hub_keras_layer_v1v2 (HubKer (None, 1280)              3413024
_________________________________________________________________
dropout (Dropout)            (None, 1280)              0
_________________________________________________________________
dense (Dense)                (None, 28)                35868
=================================================================
Total params: 3,448,892
Trainable params: 35,868
Non-trainable params: 3,413,024
```

Fig. 1. MobileNet efficientnet_lite0 model summary

3 Objective

OCR technology is still being developed and improved and currently there is no tool that would help researchers or the general public transliterating Palmyrene characters to Latin.

The main objective of this research is therefore to create a mobile application using artificial intelligence methods (OCR) for historical character classification and transliteration and verify the research on the use case of Palmyrene alphabet. For performance tests, we also need to create two dataset – hand-written and photographic and make a tool for it. The required test accuracy is above 70% in order to consider the classificator satisfactory.

The target group of users are archaeologists, who would use the software for faster Palmyrene Aramaic texts transliteration, the secondary focus group are other researchers and people outside the scientific community that could use the transliteration for educational purposes. The Android application can also be modified if another historical alphabet model is trained. Therefore, it can be used for further semi-automatic transliteration of any alphabet.

4 Methodology

4.1 Acquiring Data

For hand-written dataset creation, we created a software tool called "Write Character" for computer hand-writing by modifying AxelThevenot's Python-Interface-to-Create-Handwrittendataset" tool available at Github [17]. It creates a dataset, which contains the same number of characters in each class.

In order to get a photographic dataset, a sufficient number of images containing the required script, such as the one in Fig. 2, needs to be obtained and pre-processed. If there are not enough characters available from the photos, the remaining count is filled by an imitation of sandstone tablets – writing letters to sand.

For the acquisition of such photographs, our team established a cooperation with two museums – Musée du Louvre in Paris [15] and Virtual Museum Of Palmyra [16]. Both museums provided images of tablets with Palmyrene inscriptions.

The image pre-processing consists of cutting input images to squares, each square containing one letter and sorting them to folders. If images have low contrast, a gradient edge highlighting method should be used.

4.2 Picking Architecture and Training OCR Model

One of the current recommended neural networks used in Android applications is TFLite MobileNet efficient_lite0 [14]. The final activation function needs to be logistic, as this is a multiple category classification. Using optimisation methods like dropout is advisable.

4.3 Model Evaluation

For model evaluation, following measures are used, two counted from the classification results, one is derived and one is dependent on network output [2]:

Precision

$$p = \frac{TP}{TP + FP} \tag{1}$$

Fig. 2. Example of input image containing Palmyrene script, Antiquités orientales, AO2204, 08-516087, Paris, musée du Louvre, photo (C) RMN-Grand Palais (musée du Louvre) - © Franck Raux, Dalle portant une inscription funéraire palmyrénienne

Recall (sensitivity)

$$s = \frac{TP}{TP + FN} \tag{2}$$

Quality of algorithm (F-measure)

$$F = \frac{2 \cdot (s \cdot p)}{s + p} \tag{3}$$

where

TP = true positives, the number of correctly classified objects as positive.
FP = false positives, the number of incorrectly classified objects as positive.
FN = false negatives, the number of incorrectly classified objects as negative.
If $s = p$, then F-measure $= s = p$.

The values of TP, FP, FN as well as TN (true negatives, the number of correctly classified objects as negative) can be written in a 2-dimensional matrix called "confusion matrix" [2]. If there is just one class, i.e. the image contains the object of interest or not, there is a confusion matrix with 2 columns and rows as visible in Table 1.

Table 1. 2-dimensional confusion matrix for 1 class classification

Actual/predicted	Positive	Negative
Positive	*TP*	*FN*
Negative	*FP*	*TN*

For multiclass classification, the confusion matrix has as many columns and rows as there are classes, as visible in Table 2, with actual values in the column and predicted values in the row as well. The *TP*, *FN*, *FP* and *TN* values are calculated from the matrix as a sum of corresponding values.

Table 2. 2-dimensional confusion matrix for 4 class classification

Actual/predicted	Class 1	Class 2	Class 3	Class 4
Class 1	*TP* for class 1	*FN* for class 1, *FP* for class 2	*FN* for class 1, *FP* for class 3	*FN* for class 1, *FP* for class 4
Class 2	*FN* for class 2, *FP* for class 1	*TP* for class 2	*FN* for class 2, *FP* for class 3	*FN* for class 2, *FP* for class 4
Class 3	*FN* for class 3, *FP* for class 1	*FN* for class 3, *FP* for class 2	*TP* for class 3	*FN* for class 3, *FP* for class 4
Class 4	*FN* for class 4, *FP* for class 1	*FN* for class 4, *FP* for class 2	*FN* for class 4, *FP* for class 3	*TP* for class 4

Confidence Score

The confidence score indicates the ratio of correct predictions in percentages. It is calculated separately on training, validation and testing dataset using "accuracy" as scoring method during training and predicting. The testing accuracy shows how the model generally performs on images outside of the training set. It is calculated during the training and evaluation in the code.

4.4 Application Development

For the mobile application, we use a wide set of Android libraries, i.e. a native library android.graphics, which contains tools that handle standard graphic operations such as picture resizing. It is needed to deliver images from drawing or camera inputs in specific format to image classificator, which accepts explicitly $224 \times 224 \times 3$ inputs to the first convolutional layer in accordance with MobileNet architecture. These parameters, along with normalisation methods, authors information and license, are determined in model metadata.

For image classification, the trained models in.tflite format are loaded. The input image is transformed, analysed and the output category with confidence score will be returned and displayed in the form of a table with three results with the greatest confidence.

5 Results

5.1 Dataset Creation Results

The "Write Character" tool has been used for the creation of a dataset consisting of 56207 handwritten images belonging to 28 classes. A separate class for number 5 "ɣ"

Table 3. Palmyrene system font "Palmmene" with transcription to Latin

Palmyrene	Latin	Transcription to Latin
𐡠	Aleph	'/A
𐡡	Beth	B
𐡢	Gimmel	G
𐡣	Daleth	D
𐡤	He	H
𐡥	Waw	V
𐡦	Zayin	Z
𐡧	Heth	KH
𐡨	Teth	T
𐡩	Yodh	Y
𐡪	Kaph	K/C
𐡫	Lamedh	L
𐡬	Mem	M
𐡭	Final Nun	M
𐡮	Nun	N
𐡯	Samekh	S
𐡰	Ayin	`/E
𐡱	Pe	P
𐡲	Sadhe	TS
𐡳	Qoph	Q
𐡴	Resh	R
𐡵	Shin	SH
𐡶	Taw	TH
𐡷	left	←
𐡸	right	→
𐡹	1	1
𐡺	2	2
𐡻	3	3
𐡼	4	4
𐡽	5	5
𐡾	10, 100	10, 100
𐡿	20	20

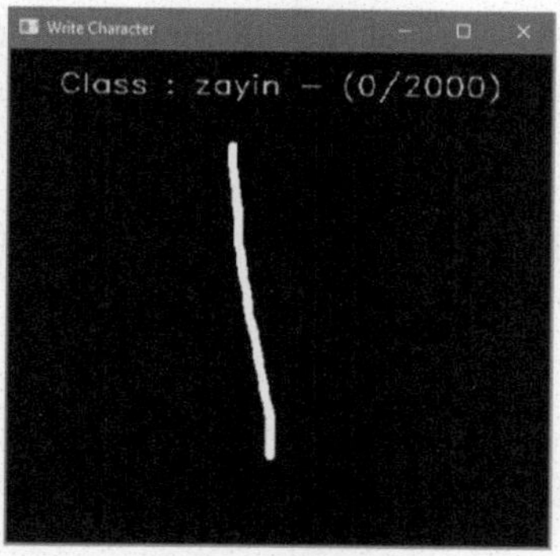

Fig. 3. User interface of the dataset handwriting tool

has been left out, as the character resembles very highly to the character ayin " 𐤏", the used system font was "palmmne" (Table 3).

We created the photographic dataset by cutting photos provided by museums and by creating sandstone tablet imitations; each class contains roughly 440 photos. The total number of photos was 12351 images (Fig. 4).

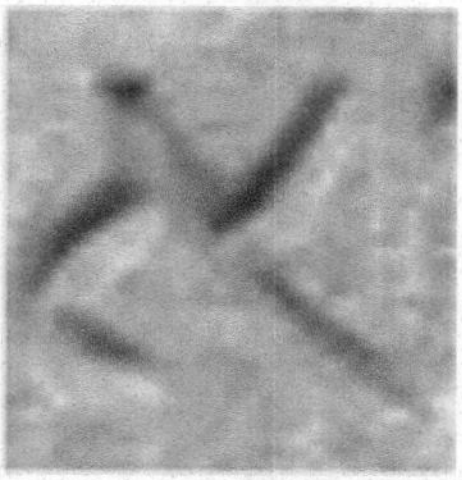

Fig. 4. Example of data from photographic dataset: letter Aleph

5.2 Model Training

The training of both models was conducted for 5 epochs using the graphic card Nvidia GeForce GTX 1650. The train/test split was 90% for training and 10% for testing. We used the library tflite-model-maker for the efficient_lite0 model creation, without any further alteration of the architecture.

Photographic Model
The training accuracy improved from 0.4328 in the first epoch to 0.7523 in the last epoch. The testing accuracy was 0.7712.

Hand-Written Model
The training accuracy was very high even after the first epoch, it reached 0.902, and in the last epoch it rached 0.993. The testing accuracy was 0.977.

Model Evaluation
The testing set for evaluating both hand-written and photographic models consisted of 112 images (4 of each class). We conducted the test directly in the developed Android application, we drew the images in the "draw letter" module and took photos of new sandstone imitation data, manually checked it, and summarised the prediction results in confusion matrices and shortened them to basic confusion matrices, as if it was a true/false classification, and divided the false predictions between *FN* and *FP*. The confusion matrices are visible in Table 4.

Table 4. Shortened confusion matrices of hand-written and photographic models

Actual/predicted	Positive	Negative
Positive	$TP_{HW} = 81$	$FN_{HW} = 20$
	$TP_{P} = 68$	$FN_{HW} = 27$
Negative	$FP_{HW} = 20$	$TN_{HW} = 0$
	$FP_{HW} = 27$	$TN_{HW} = 0$

The calculated precision, recall and F-values are 80,2% for the hand-written set and 71,96% for the photographic set. The most common errors were false positives in the class "pe".

5.3 Android Software Template and Palmyrene Transliterator

Our team created a software package – an Android application template, implemented on the use case of Palmyrene letters, using image processing libraries as well as org.tensorflow.lite. The module for image analysis uses TFLite MobileNet architecture. There are 2 main usages – drawing a letter and analysing it with the first MobileNet model for hand-written letter classification. The second one starts with taking a picture, and the camera input is fed into the other MobileNet model and returns the same output – three classes with the highest confidence score (Fig. 5).

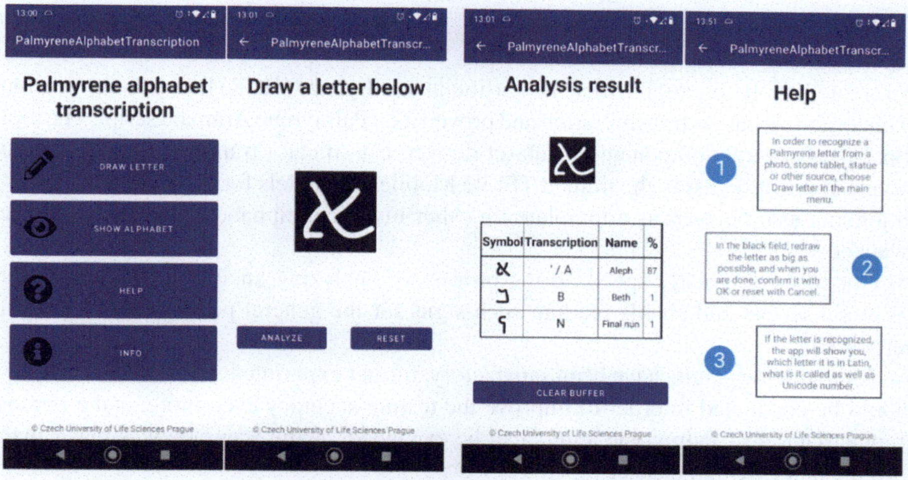

Fig. 5. Mobile application user interface

6 Discussion

In relation to a resemblance of some Palmyrene letter signs, the results of the image classification were satisfactory, as the quality of the algorithm reached 80,2% for hand-written dataset and 71,96% for photographic dataset. The difference is caused by the number of images in the training dataset as well as the difficulty of the input. While hand-written characters are only black and white, the photos vary from black to white in different shades of grey and there is much more noise in the pictures.

In GigaMesh, the analysis of 3D scans of cuneiform tablets has been conducted, and the character extraction success rate has not been published in the article [18].

Also the recognition of handwritten cuneiform vowels published by Yamauchi [9] has not shown any tables with classificators results.

The character recognition of Ghosh et al.'s model of Bangladeshi signs reached 96.46% testing accuracy on MobileNet [8], and Sazal et al. reached 90.27% accuracy on a 60-class model of hand-written Bangladeshi script [19].

From the artificial intelligence point of view, different authors have used object recognition to achieve other tasks. Cho Junghwan et al. has reached about 97% accuracy on GoogLeNet network architecture (Inception v1) using a dataset of about 4000 very high-quality images of CT body scans, and the decrease rate for lower number of images has been described there accordingly [20]. A smaller success rate from this research has been reached in Great Tits recognition from Smart Nest Boxes using YOLOv3 architecture, where the F-measure reached 83% due to difficult object detection [21].

This fact indicates that the results of Palmyrene letter classifications are satisfactory (reached above 70% of F-measure as specified in the initial criteria) and are comparable with other authors' works, but still have room for improvement, as the Palmyrene alphabet has some letters, which are very similar to each other.

7 Conclusion

We created a software tool, which uses artificial intelligence (OCR) for semi-automation of historical alphabets transliteration and proved it on Palmyrene Aramaic script. The tool consists of several components – dataset drawer, classificator training and an Android mobile application using the trained TFLite MobileNet models for field use. This application can also be used as a template for other historical alphabets, if the models are swapped.

Our tool can greatly speed up the process of analyzing ancient texts in different historical scripts and enable reading such signs for the general public for educational purposes.

Although the results have been satisfactory, further experiments with network layers should be conducted in order to improve the testing accuracy even more, and a photographic dataset of Palmyrene letters can be expanded, as the accuracy was lower than with the hand-written dataset.

Acknowledgment. This article was created with the support of the Internal Grant Agency (IGA) Faculty of Management and Economy, Czech University of Life Sciences in Prague, 2021A0004 – "Reading the characters of Palmyrene alphabet using artificial intelligence tools".

References

1. Chollet, F., Pecinovský, R.: Deep learning v jazyku Python. pp. 22, 24, 118. Grada Publishing (2019)
2. Géron, A.: Hands-on machine learning with Scikit-Learn and TensorFlow, 2nd ed., pp. 88–92. O'Reilly (2019)
3. Schantz, H.F.: The History of OCR, Optical Character Recognition. Recognition Technologies Users Association, Manchester Center, Vt. (1982)
4. Mori, S., Nishida, H., Yamada, H.: Optical Character Recognition. John Wiley and Sons, New York (1999)
5. Akhtar, Z., Lee, J.W., Attique Khan, M., Sharif, M., Ali Khan, S., Riaz, N.: Optical character recognition (OCR) using partial least square (PLS) based feature reduction: an application to artificial intelligence for biometric identification. J. Enterprise Inform. Manage. (2020). https://doi.org/10.1108/JEIM-02-2020-0076
6. Zhao, Z., Jiang, M., Guo, S., Wang, Z., Chao, F., Tan, K.C.: Improving deep learning based optical character recognition via neural architecture search. In: IEEE Congress on Evolutionary Computation (CEC), Glasgow, United Kingdom, pp. 1–7 (2020). https://doi.org/10.1109/CEC48606.2020.9185798
7. Hajihashemi, V., Arab Ameri, M.M., Alavi Gharahbagh, A., Bastanfard, A.: A pattern recognition based Holographic Graph Neuron for Persian alphabet recognition. In: International Conference on Machine Vision and Image Processing (MVIP), pp. 1–6 (2020). https://doi.org/10.1109/MVIP49855.2020.9116913
8. Ghosh, T., et al.: Bangla handwritten character recognition using MobileNet V1 architecture. Bull. Electr. Eng. Inform. **9**(6), 2547–2554 (2020)
9. Yamauchi, K., Yamamoto, H., Mori, W.: Building a handwritten cuneiform character imageset. In: Proceedings of the Eleventh International Conference on Language Resources and Evaluation. [online] LREC 2018 (2018). http://aclweb.org/anthology/L18-1115

10. Hillers, D., Cussini, E.: Palmyrene Aramaic Texts. Johns Hopkins Univ. Press, Baltimore (1996)
11. Taylor, D.G.K.: An annotated index of dated Palmyrene Aramaic texts, J. Semitic Stud. **XLVI**(2), 203–219 (2001). https://doi.org/10.1093/jss/XLVI.2.203
12. Everson, M.: Proposal for Encoding the Palmyrene Script in the SMP of the UCS. Department of Linguistics, UC Berkeley. https://escholarship.org/uc/item/27b327h7 (2010)
13. Palmyrene, Range: 10860–1087F. https://www.unicode.org/charts/PDF/U10860.pdf (2010)
14. Gupta, D.: Mobile application for bird species identification using transfer learning. In: 2021 IEEE International Conference on Artificial Intelligence in Engineering and Technology (IICAIET), pp. 1–6. IEEE (2021)
15. Le Louvre: Louvre Museum Official Website. https://www.louvre.fr/en. Last accessed 1 Dec. 2021
16. Palmyra Archaeological Museum: The Archaeological Museum of Palmyra. https://virtual-museum-syria.org/palmyra/ (2021). Last accessed 1 Dec. 2021
17. Thevenot, A.: GitHub – AxelThevenot/Python-Interface-to-Create-Handwritten-dataset: Python Interface to Create Handwritten Dataset. GitHub (2019)
18. Mara, H., Krömker, S., Jakob, S., Breuckmann, B.: GigaMesh and Gilgamesh – 3D Multiscale Integral Invariant Cuneiform Character Extraction. In: The 11th International Symposium on Virtual Reality, Archaeology and Cultural Heritage. VAST 2010 (2010)
19. Sazal, M.M.R., Biswas, S.K., Amin, M.F., Murase, K.: Bangla handwritten character recognition using deep belief network. In: Int. Conf. Electr. Inf. Commun. Technol. EICT 2013 pp. 1–5 (2013)
20. Junghwan, Ch., et al.: How much data is needed to train a medical image deep learning system to achieve necessary high accuracy? 1511.06348 (2015)
21. Hamplová, A., Pavlíček, J.: Object recognition tool for "smart nest boxes." In: Proceedings of IAC in Budapest 2020, IAC-ETITAI. Czech Institute of Academic Education, Prague, Czech Republic, pp.105–109 (2020)

The Impact of Foreign Direct Investment on the CIS: Econometrics Modeling

Irina Yarygina, Vladimir Gisin(✉), Tural Mamedov, and Boris Putko

Financial University Under the Government of the Russian Federation, Leningradsky pr. 49, Moscow 125993, Russia
vgisin@fa.ru

Abstract. The article justifies the role of foreign direct investment in the development of CIS economies and deepening their economic cooperation. To assess the contemporary environment, econometric calculations are used, that identify the causal relationship between foreign direct investment, foreign trade and GDP of developing countries. The application of the Granger model shows that there is a long-run relationship between FDI variables and foreign trade for the CIS countries. The Vector Autoregression model indicates that there is a cause-and-effect relationship between the variables across the CIS partner countries: foreign direct investment flows from Russia to CIS countries contribute to the growth of foreign trade. The article proves the positive role of foreign direct investments in the development of CIS economies and deepening their economic relations. The results of the study justify the role of direct investments in the CIS countries economy. The proposals for creating a favorable investment climate in the CIS are presented and boost the practical value of the paper.

Keywords: Foreign direct investment · CIS econometric modeling · Vector autoregression · Economic cooperation

1 Introduction

Commonwealth of Independent States (CIS) was created in December 1991. In the adopted Declaration, the participants of the Commonwealth (Belarus, Kazakhstan, Kyrgyzstan, Russia) declared their interaction on the basis of sovereign equality. In September 1993 the Heads of the CIS States signed an Agreement on a creation of Economic Union to form common economic space grounded on free movement of goods, services, labor force, capital; to elaborate coordinated monetary, tax, price, customs, external economic policy; to bring together methods of regulating economic activity and create favorable conditions for the development of direct production relations. At present, the CIS unites Azerbaijan, Armenia, Belarus, Kazakhstan, Kyrgyzstan, Moldova, Russia, Tajikistan, Turkmenistan and Uzbekistan.

Please note that the first paragraph of a section or subsection is not indented. The first paragraphs that follows a table, figure, equation etc. does not have an indent, either.

There is no denying the fact, that effective finance and relevant investment strategy depends on accurate assessment of common problems and a search for mutual solutions

R. Silhavy (Ed.): CSOC 2022, LNNS 502, pp. 46–58, 2022.
https://doi.org/10.1007/978-3-031-09076-9_5

to improve quality of life of people, suffering from Covid-19 pandemic situations. From experience, fast liberation of financial-economic relationship in the period of after global turbulence leads to decline in strategic areas of national economy. At the same time, a slowdown of government regulation of key industrial areas leads to increase of bank speculations and increase the risk level of economic and financial environment. To find the way in turbulence is the main task for the governments, which has to use different vehicles, intergovernmental monetary and fiscal measures including.

Actually, economic development plays the key strategic role in banking and finance of developing countries. The analysis has proved that banks are able to support different sectors of economy and the growth of partnership by backing economic interest of countries and encouraging trade development that increases cooperation. Financial support for the least protected market participants – SMEs and entrepreneurs – driver and buffer of any economy – creates favorable conditions for effective operations, employment, production of goods and value chains, that are under market demand.

At the same time, rating of the market demands of developing countries shows that long-term financing, project financing, syndicated loans and transparent information accelerate economic developments.Effective cooperation between entities are built on foundation of common interests and public support. Expertise from Central and Eastern Europe has proved that public support minimize shocks and promote progressive actions in restoration of economic relations. Another important lesson is the following: population is not concerned with prospect of any kind of reforms, but extremely interested in how they are going to be implemented in reality. Such kind of an experience is important in the post covid era and should be taken into consideration.

According to the agreements the CIS member – countries are going along the road of developing foreign direct investment (FDI) and harmonization of rules and regulations, that help to minimize transactional costs and contribute to risk management. Practice has given positive examples of FGI as a key factor in the development of national and international economy. The inflow of foreign investments contributes to the sustainable economic growth and helps to introduce innovative technologies into real sector of the economy.

In the context of current Western sanctions against Russia and the COVID-19 pandemic, it seems crucial to develop investment cooperation in the post-Soviet space. Deepening financial relations within the CIS (Russia, Belarus, Kazakhstan, Kyrgyzstan, Armenia, Azerbaijan, Tajikistan, Uzbekistan, Ukraine and Turkmenistan) will allow the development of new formats for regional economic integration of the member states under study. The expansion of Russia's multilateral cooperation with partner countries is one of the key factors in attracting FDI to the national economy.

The issue of FDI attraction in the CIS is of prominence, as there is a low level of investment activity in the countries of the union. For example, in 2020 FDI flows to the CIS decreased by 38% compared to the previous year [1]. The principal drivers of the decline in investment in new projects were pandemic and quarantine measures, falling commodity prices and unfavorable forecasts of economic growth. It is to be noted that interaction between individual CIS countries in the sphere of direct investment is practically unsupported. Kyrgyzstan, Tajikistan, Turkmenistan and Moldova have little cooperation with other CIS countries. FDI flows are currently declining in all regions, which is in line with the trends of slowing investment activity in the global economy.

In this context, it seems vital to intensify the role of long-term investments in financing major business projects in the region. This would enable the financing of joint projects, the development of joint private equity funds and the utilization of foreign management experience.

We believe that the encouragement of mutual direct investment between CIS member states is a priority trend of economic integration in the current framework. This is since the interstate association – has a huge investment potential. Creation of new forms of cooperation, gaining access to financial resources, implementation of innovative projects and attraction of long-term investments in the real sector of economy are of particular importance in the sphere of FDI.

Economic scholarship has conducted a great deal of research on cross-border direct investment flows and their impact on the recipient economies. As of today, there is no single theory that comprehensively reveals the factors for attracting long-term investment into the national economy [2].

For example, foreign studies demonstrate that FDI is an important factor that positively affects economic growth in industrialized countries [3].

In 1970, Japanese researcher K. Kojima, based on empirical calculations in developing countries, revealed that international trade and FDI are drivers of economic growth in developing countries [4].

2 Study

In order to demonstrate the positive impact of long-term investment on the national economies of the countries studied, it is crucial to identify the cause-and-effect relationship between FDI inflows and economic growth in the CIS countries. The growth of any economy is primarily reflected in the volume of GDP.

The time series used are the systematic annual FDI and GDP figures taken from UNCTAD's official website between 2000 and 2020. Data are presented in millions of US dollars.

To deal with such variables, the extended Dicky-Fuller test and the Johansen method are used to identify the co-integration relationship between time series. This methodology includes the construction of an econometric assessment model to study the impact of foreign direct investment on the economic growth of CIS member states. The calculations have been carried out in GRETL software.

In the first stage of the estimation process, the study examines the stationarity properties of data series. In fact, most economic variables show a trend and therefore most cases are not stationary. Thus, the keyway to test for non-stationarity is to check for the presence of a unit root.

The study uses the extended Dickey-Fuller criterion (ADF) for unit roots. The ADF test includes additional conditions dependent variables in order to eliminate autocorrelation.

The lag length at this exposure term is determined by means of the Akaike information criterion (AIC) and the Schwarz criterion (SC) in GRETL software.

The ADF test has the following regression equation

$$\Delta Y_{t-1} = \alpha_0 + \gamma Y_{t-1} + \alpha_2 t + \sum\nolimits_{i=1}^{p} \beta_i \Delta Y_{t-1} + E_t, \quad (1)$$

where p, α, t are the variables that govern autocorrelation errors.

The ADF regression test for unit root Y_t is presented in logarithm format for the FDI and GDP variables at time t. The variable Y_{t-1} expresses the first difference with lags.

The null and alternative hypothesis of the existence of a unit root in the variable Y_t:

$$H0: \sigma 2 = 0;$$

$$H1: \sigma 2 < 0.$$

The results of the augmented Dickey-Fuller test (ADF) suggest that for the countries under study the null hypothesis of non-stationarity is completely rejected at the 5% significance level (for a series of countries at the 10% level), while the first difference time series eliminates non-stationarity (with constant; with constant and trend). The time series contains a trend. This confirms that the FDI and GDP time series are integrated oversampled and have a unit root, as reflected in Tables 1 and 2.

Table 1. Augmented Dickey-Fuller test (ADF) on FDI.

Sample = 18	Level y_t, 5% and 10%		First difference, Δy_t	
Country	With constant	With constant and trend	With constant	With constant and trend
Russia	−2,34233 (0,1711)	−2,22092 (0,4497)	−3,78296 (0,01301)	−3,67009 (0,0556)
Azerbaijan	−1,83832 (0,3508)	−1,98754 (0,5664)	−3,75689 (0,01368)	−3,59882 (0,06256)
Belarus	−1,5224 (0,5224)	−2,79079 (0,2185)	−7,14546 (2,40)	−7,19487 (0,00015)
Kazakhstan	−1,58147 (0,4699)	−0,905016 (0,9314)	−3,6984 (0,0153)	−3,99844 (0,03189)
Kyrgyzstan	−2,79627 (0,07963)	−5,32076 (4,054)	−4,61119 (0,0001)	−4,1174 (0,00587)
Ukraine	−2,03764 (0,2696)	−1,79645 (0,6612)	−4,6087 (0,002697)	−4,78402 (0,00819)
Moldavia	−2,35459 (0,1679)	−2,24634 (0,4374)	−4,44982 (0,000237)	−4,53533 (0,00127)
Armenia	−1,50038 (0,5092)	−1,09634 (0,8993)	−3,73209 (0,01434)	−4,11696 (0,026)

(continued)

Table 1. (*continued*)

Sample = 18	Level y_t, 5% and 10%		First difference, Δy_t	
Country	With constant	With constant and trend	With constant	With constant and trend
Uzbekistan	−1,61053 (0,4559)	−1,28684 (0,8556)	−3,88658 (0,01067)	−4,12647 (0,00569)
Tajikistan	−3,06134 (0,04916)	−3,03789 (0,1512)	−5,00742 (0,001272)	−4,92386 (0,00646)

Values in parentheses are *p*-values.
Source: Compiled by the authors.

Table 2. Augmented Dickey-Fuller test (ADF) on GDP

Sample = 18	Level yt, 5% and 10%		First difference, Δyt	
Country	With constant	With constant and trend	With constant	With constant and trend
Russia	−1,98152 (0,2912)	−2,66718 (0,2595)	−5,0105 (0,001265)	−4,93276 (0,006359)
Azerbaijan	−1,58604 (0,4898)	−1,72448 (0,7406)	−2,40554 (0,1556)	−2,60032 (0,2843)
Belarus	−1,59317 (0,4643)	−0,970068 (0,9217)	−3,34285 (0,03001)	−3,61967 (0,06047)
Kazakhstan	−1,4171 (0,5755)	−2,39238 (0,3835)	−2,63979 (0,1059)	−2,71665 (0,2433)
Kyrgyzstan	−0,529608 (0,8623)	−2,74638 (0,2177)	−3,8205 (0,00271)	−3,74832 (0,01928)
Ukraine	−1,77065 (0,381)	−1,25212 (0,8646)	−3,22158 (0,03758)	−3,49292 (0,04007)
Moldavia	−0,829095 (0,7847)	−2,29222 (0,4376)	−3,73134 (0,003713)	−3,89723 (0,01215)
Armenia	−1,34453 (0,5837)	−1,37256 (0,8312)	−3,59029 (0,01881)	−3,67268 (0,05536)
Uzbekistan	−3,177 (0,02136)	1,41386 (1)	0,804523 (0,9942)	1,9799 (1)
Tajikistan	−1,25364 (0,6532)	−2,2594 (0,4558)	−2,57974 (0,1172)	−2,66641 (0,2603)

Source: Compiled by the authors.

Therefore, the reliability of the results allows us to determine whether there is cointegration between the variables under study.

For the case of rejecting the null hypothesis of the presence of a unit root, the t statistic must be less than the threshold value. The threshold value for the ADF has an eigenvalue distribution. Table 3 shows the values for different sample sizes, by time series with and without trend [5].

Table 3. Critical values for Dickie Fuller tests

Sample size	With constant, no trend			With a constant and a trend		
	1%	5%	10%	1%	5%	10%
$T = 25$	−3,75	−3,00	−2,63	−4,38	−3,60	−3,24
$T = 50$	−3,58	−2,93	−2,60	−4,15	−3,50	−3,18
$T = 100$	−3,51	−2,89	−2,58	−4,04	−3,45	−3,15
$T = \infty$	−3,43	−2,86	−2,57	−3,96	−3,41	−3,13

The Johansen test is defined based on a vector error correction model. It should be noted that, for two variables, this system of two equations, as shown in the following formulae:

$$\Delta \mathrm{FDI}_t = P + \sum\nolimits_{i=1}^{p-1} \alpha_i \Delta \mathrm{FDI}_{t-1} + \sum_{j=1}^{p-1} \beta_j \Delta \mathrm{GDP}_{t-j} + \gamma(\mathrm{FDI}{-}{-}\mathrm{GDP})_{t-1} + e_t, \quad (2)$$

$$\Delta \mathrm{GDP}_t = P^* + \sum\nolimits_{i-1}^{p-1} \alpha_i \Delta \mathrm{FDI}_{t-1} + \sum\nolimits_{j=1}^{p-1} \beta_j \Delta \mathrm{GDP}_{t-1} + \varsigma(\mathrm{FDI} - \mathrm{GDP})_{t-1} + e_t^*, \quad (3)$$

where P and P^* are constants;

Δ is the difference operator (e.g.: $\mathrm{FDI}_t - \mathrm{FDI}_{t-1} = \Delta\mathrm{FDI}_t$);

e_t and $e_t{}^*$ are errors (residual of vector regression);

γ and ς are error correction coefficients measuring the rate of return systems in <<FDI-GDP>> to equilibrium.

If hypothesis $r = 0$ is not rejected at the first step, then the process is non-stationary and there is no cointegration.

The aim of the econometric model is to empirically test the dependence of GDP on FDI inflows. The authors assume that direct investment has a positive effect on the growth of CIS economies. If the null hypothesis is confirmed, there is no cointegration correlation. If the maximum number (γmax) and/or trace of the matrix (γTrace) exceeds the CC (Pearson criterion critical value), then there is a cause-and-effect relationship between the time series under study.

The trace of the matrix (γTrace) has the following regression equation

$$\gamma_{\mathrm{Trace}} = n \sum\nolimits_{i=r+1}^{p} \ln(1 - \gamma_i), \quad (4)$$

where γ_i is the largest eigenvalue of the defined matrix.

The value γmax is the logarithm of the likelihood function. Looking for cointegration r, Johansen's test tested the null hypothesis and the alternative hypothesis (H0): $r = 0$ and $r \leq 1$.

Therefore, based on the test performed, two statistics were obtained: the maximum number and the trace of matrix P. The results of the dependencies are presented in Table 4.

The estimation indicated that a long-run relationship was found between the resulting economic integrated variables (FDI inflows and current-price GDP) in 5 out of 10 CIS countries are cointegrated.

The statistic (γTrace) in Russia was 16.186, which exceeded the critical value, thus refuting H0: $r = 0$ and not refuting H0: $r \leq 1$. Consequently, the FDI and GDP variables are cointegrated time series.

Table 4. Johansen cointegration test results.

Country	H_0	H_1	Maximal number (γmax)	5% – critical value (K3)	Matrix trace (γTrace)	5% – critical value (K3)
Russia	$r = 0$ $r \leq 1$	$r = 1$ $r = 2$	11,6240 4,5620	14,26 3,84	16,1860 4,5620	15,49 3,84
Azerbaijan	$r = 0$ $r \leq 1$	$r = 1$ $r = 2$	13,807 2,315	14,26 3,84	16,1230 2,3152	15,49 3,84
Belarus	$r = 0$ $r \leq 1$	$r = 1$ $r = 2$	17,571 2,6244	14,26 3,84	20,1960 2,6244	15,49 3,84
Kazakhstan	$r = 0$ $r \leq 1$	$r = 1$ $r = 2$	9,4783 3,1048	14,26 3,84	2,5830 3,1048	15,49 3,84
Kyrgyzstan	$r = 0$ $r \leq 1$	$r = 1$ $r = 2$	16,7050 0,4473	14,26 3,84	17,1520 0,44727	15,49 3,84
Ukraine	$r = 0$ $r \leq 1$	$r = 1$ $r = 2$	12,989 4,385	14,26 3,84	17,3740 4,3851	15,49 3,84
Moldavia	$r = 0$ $r \leq 1$	$r = 1$ $r = 2$	6,1839 0,4667	14,26 3,84	6,65060 0,46670	15,49 3,84
Armenia	$r = 0$ $r \leq 1$	$r = 1$ $r = 2$	2,73870 0,43970	14,26 3,84	3,17840 0,43970	15,49 3,84
Uzbekistan	$r = 0$ $r \leq 1$	$r = 1$ $r = 2$	10,731 2,170	14,26 3,84	12,9010 2,1700	15,49 3,84
Tajikistan	$r = 0$ $r \leq 1$	$r = 1$ $r = 2$	8,8201 1,1770	14,26 3,84	9,9971 1,1770	15,49 3,84

a. Number of observations – 19.
b. The number of equations is 2.
c. Gap order – 1.
Source: Compiled by the authors.

In addition to the Johansen test, the Engle-Granger test can be applied. It is two-step procedure: firstly the simple ls-procedure is used to estimate the regression $Y2 = \alpha + \beta Y1 + \varepsilon$ where $Y1$, $Y2$ are the time-series for cointegration, and then regression on residuals is checking for unit root existence, wherein the test statistics follow the nonstandard Phillips-Ouliaris (PO) distributions. Note that the Johansen test is more common because it has more capabilities than the Engle-Granger test. We also consider its results to be preferable.

The positive impact of FDI on cooperation in the CIS countries can be explained by the fact that foreign direct investment is a source of growth in employment and the population's ability to pay, allows the financing of large investment projects within the union and replenishes national budgets. These factors contribute to increase in GDP and GDP per capita of FDI recipient countries.

As the world practice shows, one of the important directions of development of economic cooperation between the CIS countries is foreign trade.

To determine the impact of the FDI mechanism on the development of foreign trade in the CIS countries, the calculations in the GRETL software were carried out by using the Granger causality test [6]. The annual indicators on the volume of attracted FDI and foreign trade of the studied countries for the period from 2000 to 2020 are used as a time series. The data is presented in millions of US dollars.

The Granger test has the following regression equation:

$$F_t = \alpha_0 + \sum \alpha_j \mathrm{FDI}_{t--\mathrm{j}} + \sum \beta_\mathrm{j} \mathrm{TR}_{t--j} + e_t, \qquad (5)$$

where TR and FDI are variables (time series); α_0, α_j, β_j are the coefficients calculated from regression, and e_t are the errors (the residuals of the vector regression), $j = 1, 2, 3\ldots n$.

According to the Granger test, if changes in the variable TR (foreign trade) cause changes in FDI (FDI), therefore changes in TR precede changes in FDI. Annual FDI and foreign trade turnover figures are used as a time series. The data for the calculation is taken from the official websites of the Bank of Russia.

If P-value < 0.05, the null hypothesis is rejected. It is important that one of the hypotheses is rejected in order to identify a causal relationship. This fact will indicate that one variable is significant in predicting the other.

Rejecting two hypotheses means that the variables are interrelated and possibly influenced by a third variable. The main results of the dependencies are presented in Table 5.

FDI inflows and the volume of foreign trade are mutually related in all countries of the union. Using a vector autoregression model, it can be concluded that there is a causal relationship between these variables across the CIS countries. FDI flows from Russia to some CIS countries increase their foreign trade and, consequently, the economic cooperation. The findings of the Granger test support the hypothesis of the research study.

Therefore, the study confirms the feasibility of attracting FDI to the national economies of the countries.

Table 5. Results of the Granger Causation test.

Zero hypothesis	*t*-statistics, 5%	P-value	Inference
Kazakhstan			
a. FDI does not affect TR	7,3300	2.51e−0.5	Rejected
b. TR does not affect FDI	0,3975	0,6994	Accepted
Belarus			
a. FDI does not affect TR	11,0200	6.46e−0.7	Rejected
b. TR does not affect FDI	−0,6331	0,5409	Accepted
Armenia			
a. FDI does not affect TR	6,182	0,0001	Rejected
b. TR does not affect FDI	2,059	0,0665	Rejected
Kyrgyzstan			
a. FDI does not affect TR	13,850	$7{,}48e^{-08}$	Rejected
b. TR does not affect FDI	1,065	0,3118	Accepted
Moldavia			
a. FDI does not affect TR	4,9400	$3{,}62e^{-08}$	Rejected
b. TR does not affect FDI	0,2767	0,7876	Accepted
Tajikistan			
a. FDI does not affect TR	26,8200	$1{,}20e^{-010}$	Rejected
b. TR does not affect FDI	−0,1414	0,8904	Accepted
Uzbekistan			
a. FDI does not affect TR	18,780	$3{,}96e^{-09}$	Rejected
b. TR does not affect FDI	1,195	0,2595	Accepted
Azerbaijan			
a. FDI does not affect TR	4,6570	0,0012	Rejected
b. TR does not affect FDI	0,6561	0,5265	Is accepted
Ukraine			
a. FDI does not affect TR	6,30800	$8{,}82e^{-05}$	Rejected
b. TR does not affect FDI	0,07628	0,9407	Is accepted
Turkmenistan			
a. FDI does not affect TR	7,57200	$1{,}90e^{-05}$	Rejected
b. TR does not affect FDI	0,02910	0,9774	Is accepted

Source: Compiled by the authors.

3 Measures to Attract Foreign Investments

The formation of a mutual economic strategy depends on effective consideration of the experience of CEE countries, which overcome the consequences of market reforms in the economy. In this regard, presence of objectivity in progressive actions towards available financial resources is very important. Governments support of entrepreneurs who are dealing in the area of international trade, unification of investment activity and policies will create favorable conditions for CIS economic cooperation. Growth of economic ties of partner-countries based on common values and development goals suggests a long-term strategic alliance of private and public institutions. Sound financial management, modernization of technology will also contribute to the development of the financial markets. Support of the balance of interests of public and private structures, growing capacity of joint investment financing, strengthening the legal framework and improving the legal support of business – processes contributes to the development of economic cooperation of economic entities. In addition, financial activity, supporting CIS economic cooperation should be transparent and accessible to the customer, regardless of the place of accreditation if their business. As an example, one of the important directions of banks' activity – financial intermediaries - is a support programs of trade relations («Trade – Finance Promotion»), which assists customers in selecting contractors for International Cooperation. Currently, small and medium-sized enterprises of CIS countries have to apply adverse cooperation schemes, including an advance payment for imports. In other cases, the completion of foreign trade contracts is prevented by excessive pricing conditions for underwriting and insurance payments. In this regard, using interstate status, a number of development banks can help to achieve favorable agreements with banks – exporters to improve the conditions for international investments and trade.

One of the most important trends of investments support is a provision of payment guarantees to exporters, ensuring pre-export financing and cooperation with insurance companies. It seems appropriate for financial institutions in a post Covid era to move towards the development of investment activities of economic entities and to provide customers and their counterparties with a variety of services on the agreed "flexible" terms of cooperation. It can be provision of documentary operations, syndicates for investments finance, advising stakeholders, assist clients in using the fiscal opportunities.

While investing, financial institutions, mostly banks, are expedient also to accept drafts drawn by exporter/importer, finance exporters against documentary letters of credit issued by reputable banks, to provide the agreed overdrafts, to ensure the implementation of credit lines for exporters/importers. Financing of drafts drawn by exporters or documents against acceptance as well as financing of goods supply in the initial stage of the implementation of contracts (for up to 180 days) and financing of importers/exporters, financing of the fulfilment of contractual obligations are also advisable. Actually, banks can accelerate provision of specialized services to business entities and convey clearing, maintain multicurrency accounts with the possibility of strengthening account liquidity, manage clients' short-term liquidity in the national currencies of developing countries, if they are not members of currency unions, manage accounts in national currencies through telecommunication systems and maintain the "escrow" accounts. Considerable attention of investors is to be payed to their corporate partners and their need of consultations on financial conditions of investment contracts, currency regulation and control, including syndicated loans, hedge currency risks management, currency arbitrage, trust operations on behalf of their clients.

An important component of creating a common economic space of the CIS countries is a presence of special intergovernmental insurance program with the involvement of government agencies for export insurance and guarantee of international operations. Specialized agencies are also able to promote multilateral investment cooperation for business partners.

Financial institutions' involvement in encouraging formation of multinational leasing companies with support of public institutions also promotes investments. An important area of international cooperation is discounted (governments – sponsored) investments in innovative and knowledge-intensive projects, as well as socio-economic programs, such as environmental protection, infrastructure development, energy – supply and health support. At the same time public institutions play an important role in engineering support of joint investments and credit lines in order to ensure the economic benefits and enhance the living standards of countries, suffering from the post COVID turbulence of national economies.

In view of optimizing the investment interaction of the CIS countries, it is important to take targeted measures aimed at creating a favorable investment climate. It is necessary to develop targeted programs, form special funds of all forms of ownership, create agencies for attracting FDI. One of the essential motives for enhancing cooperation in the CIS space is the creation of regional collective funds of financial resources. The analysis of the volume of mutual FDI suggests that the funds in the post-Soviet space cannot solve the problem of mass capital inflows. It is important to create additional private equity agencies and with the help of government support intensify the activities of the funds in all CIS member states. The established agencies should be most active in informing foreign investors about the opportunities for profitable FDI within the union.

Establish specialized Small and Medium-sized Business Development Centers and Funds, namely guarantee funds that operate on market principles to offset the risks of financial and credit institutions. The funds could be set up within the CIS using EDB funds.

Draw up a multilateral Agreement on the Promotion and Reciprocal Protection of Investments of the CIS member states and consider eventual accession of the union countries to the 1976 OECD Declaration on International Investment and Multinational Enterprises. Today bilateral investment agreements have been signed and entered into force between Russia and CIS countries. However, the preparation of a multilateral document would allow the countries of the regional association to form common rules for the regulation of FDI. Actually, Kazakhstan is currently the only CIS country to be a party to the Declaration. Accession to the Agreement on Foreign Investment will create new standards for FDI regulation based on the use of foreign experience.

Another favorable instrument is the establishment of free trade zones (FTAs) between the CIS member states and their partners that have not joined the CIS FTA of 2011, while at the same time rehabilitating the real sector of economy.

Creation of cross-border industrial and innovation clusters in the CIS space is also reasonable. It is necessary to identify and designate the backbone industries of partner countries within the CIS, to create joint competitive enterprises that are in line with the global level. In order to promote the investment process, a Common CIS Cluster Policy Council should be established, which will make it possible to expand cooperation among CIS partner countries in the sector of FDI. It will help to attract mutual direct investment in high-tech industries through the implementation of multilateral projects in the region. It is also recommended to create a common space of interconnected economy, which will open new competitive advantages for all partner countries in the modern environment. The key functions of the Council can be developing a unified management system for evaluating clusters within the union, improving management efficiency, intensifying the inflow of mutual FDI between the CIS partners, forming generally accepted standards, criteria and ratings in this area as well as enhancing the role of CIS partner countries in the implementation of a single coordination mechanism for cluster members.

4 Conclusion

The results of the study provide a case for investments study within CIS economics in contemporary environment of post Covid era. Recommendations can be taken into consideration while constructing economic and political approach to an international investments infrastructure. Meeting the economic needs of population and mutual interests are certainly a step in the right direction of cooperation. However, the setting of needs depends on public strategy and relevant activity of intermediaries. It is strongly recommended to take into consideration the international experience and to introduce the required instruments to be successful within investments procedures.

Following the adoption of various declarations by partner – countries, more and more attention is focused on the objective need for infrastructure development. At the same time, the insufficient level of development of infrastructure, mainly transport and logistics, significantly hinders the expansion and deepening of economic cooperation between countries. It is worthwhile drawing the attention to the fact, that official documents of CIS countries indicate insufficient funding for such areas as infrastructure development and relevant financial services. In addition to the above obstacles that hinder direct investments in a multilateral format, problems in the field of customs cooperation and industrial development also need to be addressed. Despite the dominance of bilateral economic relations within contemporary framework, many of these problems should be solved jointly, in a combination of three, four partner – countries, depending on the degree of complexity of the problem and its significance for each country.

The development of electronic commerce and safe financial services are to be highlighted as one of the most promising solutions in the post Covid era. The expediency of this has been emphasized repeatedly at high level CIS meetings. Actually, the government-to-business (G2B) business model has become most required. In the nearest future, the introduction of other business models, such as B2B, B2C, C2C is also important, in order to ensure full investment cooperation and realization of the interests of all the main participants: government, business, financial institutions, consumers in each category of the population.

At the same time, it is worthwhile mentioning, that on the bases of time series analysis we have come to conclusion, that there is a causal relationship between the total inflow of direct investment, foreign trade and GDP of the CIS countries. It is revealed that FDI are the key factors of the economic development of the CIS economies. There is proved the impact of FDI in the deepening the economic cooperation of the member states of the CIS. The findings of the Granger causality test support that capital flows to CIS countries from Russia contribute to mutual growth of foreign trade relations.

Russia is currently the main investor in the region. However, there is a weak investment cooperation among the rest of CIS countries. In this regard, it is necessary to intensify the inflow of FDI to the CIS countries, especially in the context of COVID-19.

Presented results of the study can benefit regulators and banks by providing new instruments for economic developments in post Covid period. Further novel concept introduced in this study relates to the usage of variety techniques in short-term and long-term relationship, backed by real requirements of legal entities.

The concept, presented in the study can be generalized in assessing trends and spheres of cooperation.

References

1. World Investment Report 2021: Investing in Sustainable Recovery. https://unctad.org/system/files/official-document/wir2021_en.pdf. Last accessed 20 Nov 2021
2. Dinkar, N., Choudhury, R.N.: Selective review of foreign direct investment theories. Asia-Pacific Research and Training Network On Trade. Working Paper 143, pp. 2–27 (2014)
3. Boudreau, K.J., Lacetera, K., Lakhani, K.N.: Incentives and problem uncertainty in innovation contests: an empirical analysis. Manage. Sci. **57**(5), 843–863 (2016)
4. Kojima, K., Ozava, T.: Micro- and macroeconomic models of direct foreign investment: toward a synthesis. Hitotsubashi J. Econ. **25**, 1–20 (1984)
5. Johansen, S., Juseliu, K.: Maximum likelihood estimation and inference on cointegration with applications to the demand for money. Oxford Bull. Econ. Stat. **52**(2), 169–210 (1990)
6. Hemstra, C., Jonathan, D.: Testing for linear and nonlinear Granger causality in the stock price-volume relation. J. Financ. **49**(5), 1639–1664 (1994)

Analysis of Artificial Neural Network Settings in Determining «Significance» Stage-to-Stage Transitions on the Grid Graphs of Business Processes in Banks

Andrey A. Bugaenko(✉)

Department of Finance, Sberbank PJSC, Process Mining Team, 19 Street Vavilova, Moscow 105275, Russia
germes86@mail.ru

Abstract. The article analyzes which settings the artificial neural network line had as it determined the 'significance' level of stage-to-stage transitions on the grid graphs of business processes in banks. The analysis defined settings that give almost the same outcome as AutoML (greed search) logging for all network settings mixes, but having the GPU (CPU) load several orders of magnitude lower.

1 Problem Statement

The current task of analytical offices of banks is to find ways to optimize business processes. One of the most successful techniques for identifying low-efficiency areas is to apply reinforcement learning [1, 2] to grid graphs obtained in an process mining, and to clustering business stage-to-stage transitions into 'problematic' and 'effective' adjusted for the 'significance' and 'redundancy' criteria. [3, 4] The technique is most successful in large banks, where analysts are unlikely to analyze grid graphs using BI tools manually due to the huge number of process stages and hundreds and thousands of stage-to-stage transitions. This technique does have a drawback, though. When determining the 'significance' attribute, a model is built based on the number of process stage-to-stage transitions depending on the process execution time, where each feature is a one-of-a-kind form of transition between two distinct process stages. The problem is that there is no generic way to build a model which will be the most powerful for any business process [3, 4]. Based on this, using Auto Machine Learning (AutoML, greed search), the machine learning method must be selected individually for each business process. However, trying all the 'hyperparameters' mixes, the number of which can be tens of thousands, to find a proper model is much more costly from the GPU (CPU) load point of view.

R. Silhavy (Ed.): CSOC 2022, LNNS 502, pp. 59–69, 2022.
https://doi.org/10.1007/978-3-031-09076-9_6

Alternative approaches to anomalies analysis are discussed in the literature [5–9], but their usage is poorly relevant to our task due highly 'sparse' dataset structure. Most factors are zero, and the number of factors can reach hundreds or even thousands, for each case of the process, i.e.

$$
\begin{aligned}
t_1 &= f(n_{11}, n_{12}, \ldots n_{1w}) = f(n_{11}, n_{12}, \ldots n_{1ij}) \\
&\ldots \\
t_k &= f(n_{k1}, n_{k2}, \ldots n_{kw}) = f(n_{k1}, n_{k2}, \ldots n_{kij}) \\
&\ldots \\
t_m &= f(n_{m1}, n_{m2}, \ldots n_{mw}) = f(n_{m1}, n_{m2}, \ldots n_{mij})
\end{aligned}
\tag{1}
$$

where t_k is the implementation time of the kth process instance, h is the total number of stages per process, w is the total number of stage-to-stage transitions per process, m is the total number of process instances per data set, i, j $\in$ [1, h], k $\in$ [1, m], $n_{kij} = n_{kw}$ is the number of ith to jth stage transitions in a kth process instance, $\forall n_{11} .. n_{mw} \in N$. Alongside this, approximately 70%–80% of all $n_{kw} = 0$.

2 Analysis of Artificial Neural Network Settings

The rational solution to this issue is to try to find the optimal structure of the hyperparameters settings that are most often the best for the machine learning methods and are the most costly in terms of GPU (CPU) load. It would remove the need to use greed search for certain hyperparameters, significantly raising the computational load. One of frequent the approach to create model, which takes too much GPU (CPU) power and selected like the best method, is artificial neural network. Some articles [10–13] address the features of network settings; but these do not apply due to the nature of the task and the features of the dataset structure.

To analyze neural network settings, datasets were taken for the business processes of the banking operations, namely: accepting payments from natural persons (NP); depositing/withdrawing cash to/from an account; issuing, exchanging and closing bank cards; selling coins and gold bars; NP's bank transfers; paying off regular loan payments; opening/closing/adding deposits; exchanging NP's currency; renting safety deposit boxes; security and certificate sale; cash change and damaged banknote exchange; banknote counterfeiting tests; indemnity and payouts to non-community bank depositors.

Basic settings of neural network are:

- hidden layer sizes – the number of neurons in each hidden layer (unique to each layer);
- number of layers;
- activation – activation function;
- solver – weight optimization algorithm;
- learning rate – learning algorithm for updating weights;
- epoch – number of epoch.

The list of settings was selected based on the set that made the greed search dictionary, when using AutoML.

Activation features, which were tested:

- identity – linear function: f(x) = x;
- logistic – logistic sigmoid: f(x) = 1/(1 + exp(−x));
- tanh – hyperbolic function: f(x) = tanh(x);
- relu – modified linear function f(x) = max(0, x).
- Weight optimization algorithms, which were tested:
- lbfgs – quasi Newton method;
- sgd – stochastic gradient descent;
- adam – stochastic gradient optimizer proposed by Kingma, Diederik, and Jimmy Ba.

Learning algorithm for updating weights, which were tested:

- Constant – constant update rate 0.001,
- Invscaling – changing update rate from 0.001 with a change factor of 0.5,
- Adaptive – adaptive algorithm: the update start starts from 0.001 and every time adding 2 epoch in a row cannot improve accuracy, the factor is multiplied by 0.2.

The accuracy metric was MAPE (the mean absolute percentage error), since all values in the data set are positive.

Initially, 10% of the data was randomly extracted from the set (for each business process) for final testing of the result.

Obviously, the accuracy of neural network will vary from epoch and number layers of neural network (this thesis will be checked later), so the analysis was initially performed for 5 epoch (you may take every other number, as it does not affect the start of the analysis) and for 1 layer. Again, «constant» was taken as learning algorithm for updating weights. The average accuracy for all business processes for the test set was plotted on the graph (Fig. 1, 2, 3, 4, 5, 6, 7, 8, 9, 10, 11, 12 and Fig. 13).

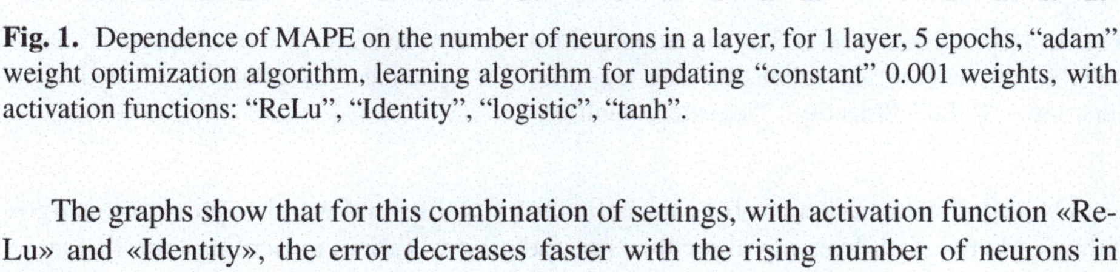

Fig. 1. Dependence of MAPE on the number of neurons in a layer, for 1 layer, 5 epochs, "adam" weight optimization algorithm, learning algorithm for updating "constant" 0.001 weights, with activation functions: "ReLu", "Identity", "logistic", "tanh".

The graphs show that for this combination of settings, with activation function «ReLu» and «Identity», the error decreases faster with the rising number of neurons in layer.

The graphs show that the average accuracy for weight optimization algorithm «lbfgs» is higher than for «adam». Moreover, with «lbfgs», the error MAPE decreases with the growth of number of neurons in layer to some N_{max}, and then changes little, if at all.

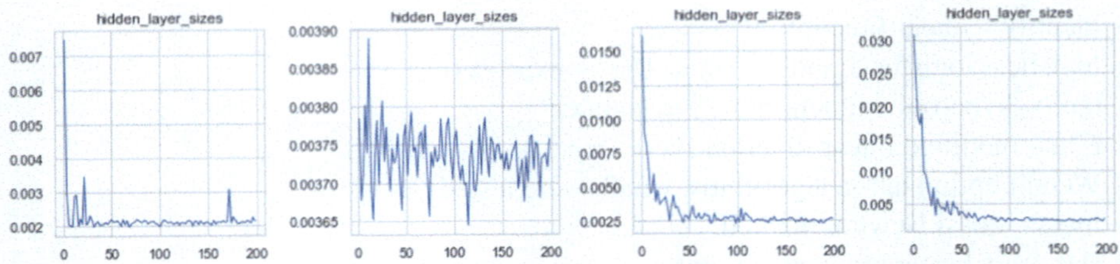

Fig. 2. Dependence of MAPE on the number of neurons in a layer, for 1 layer, 5 epochs, "lbfgs" weight optimization algorithm, learning algorithm for updating "constant" 0.001 weights, with activation functions: "ReLu", "Identity", "logistic", "tanh".

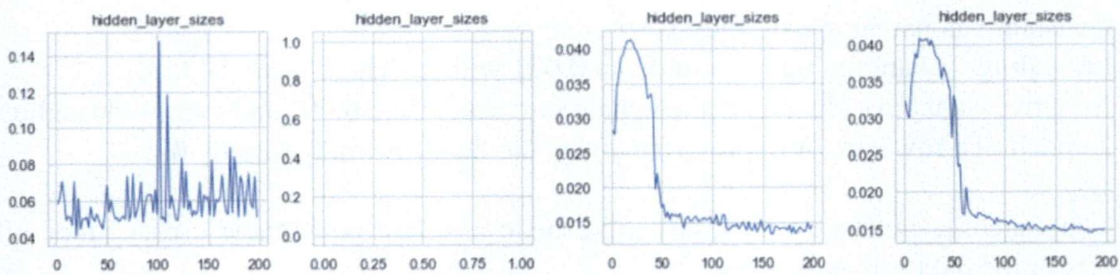

Fig. 3. Dependence of MAPE on the number of neurons in a layer, for 1 layer, 5 epochs, "sgd" weight optimization algorithm, learning algorithm for updating "constant" 0.001 weights, with activation functions: "ReLu", "Identity", "logistic", "tanh".

When using weight optimization algorithm «sgd», with the same value of number of neurons in layer, the accuracy was lower than with «lbfgs». Another important point is that with «sgd», activation function «identity» never gave a single adequate solution. In addition, the accuracy with «logistic» and «tanh» was higher than that of «relu».

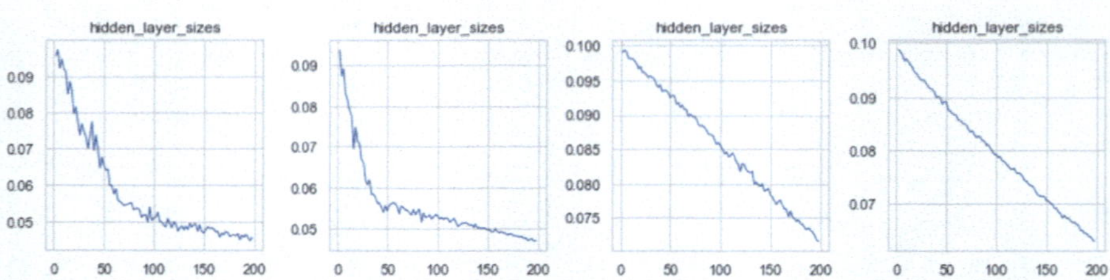

Fig. 4. Dependence of MAPE on the number of neurons in a layer, for 1 layer, 5 epochs, "adam" weight optimization algorithm, learning algorithm for updating "invscaling", with activation functions: "ReLu", "Identity", "logistic", "tanh".

Overall, the dependencies for each activation function and weight optimization algorithm «adam», with learning algorithm for updating weights «invscaling» are the same as with «constant». The error decreases with the rising number of neurons in layer.

Again, for «lbfgs», replacing learning algorithm for updating weights with «invscaling» did not significantly change the MAPE - number of neurons in layer dependence, but the error MAPE increased in absolute terms and the number of neurons in layer required to attain Nmax was now greater.

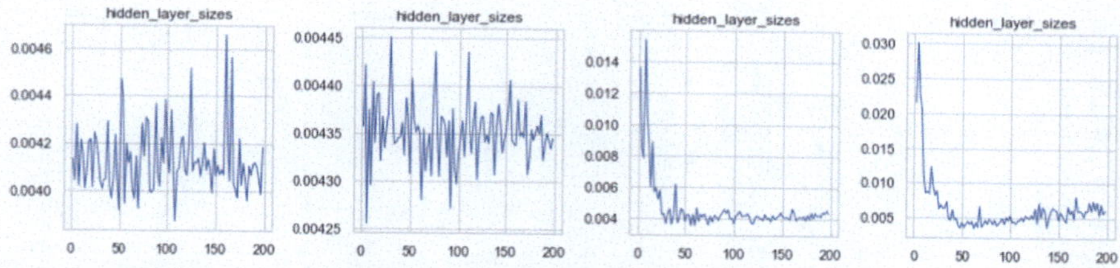

Fig. 5. Dependence of MAPE on the number of neurons in a layer, for 1 layer, 5 epochs, "lbfgs" weight optimization algorithm, learning algorithm for updating "invscaling", with activation functions: "ReLu", "Identity", "logistic", "tanh".

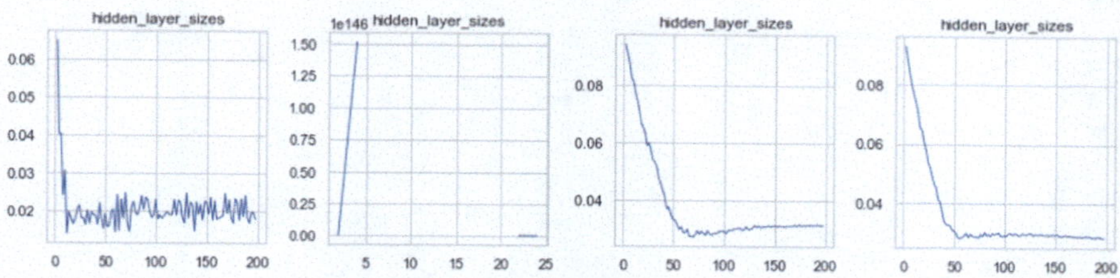

Fig. 6. Dependence of MAPE on the number of neurons in a layer, for 1 layer, 5 epochs, "sgd" weight optimization algorithm, learning algorithm for updating "invscaling", with activation functions: "ReLu", "Identity", "logistic", "tanh".

Switching to «invscaling» did not significantly change the number of neurons in layer – MAPE dependence for «sgd» either. With activation function «relu», the average error has fallen.

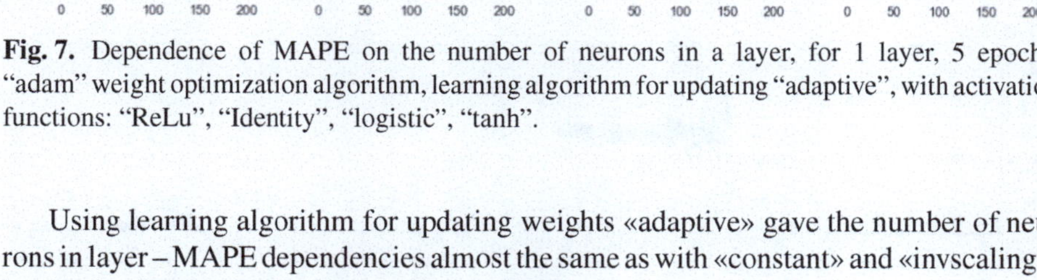

Fig. 7. Dependence of MAPE on the number of neurons in a layer, for 1 layer, 5 epochs, "adam" weight optimization algorithm, learning algorithm for updating "adaptive", with activation functions: "ReLu", "Identity", "logistic", "tanh".

Using learning algorithm for updating weights «adaptive» gave the number of neurons in layer – MAPE dependencies almost the same as with «constant» and «invscaling», but in absolute terms the average error is higher.

Conclusion 1: the functional dependence of the model accuracy from number of neurons in layer is weakly related to the learning algorithm for updating weights type, but the highest accuracy for all the settings mixes tested was obtained with the «constant» algorithm.

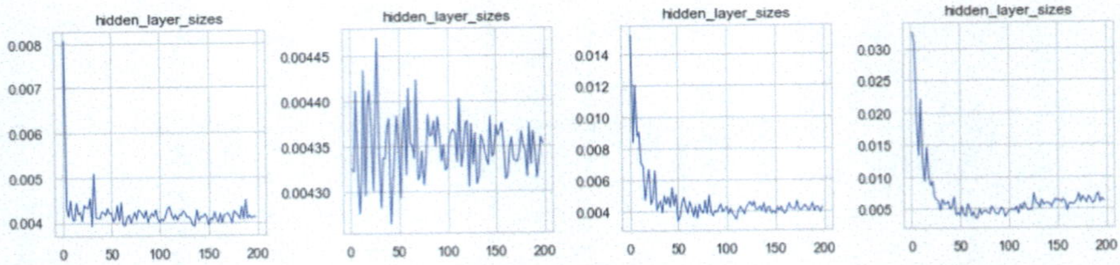

Fig. 8. Dependence of MAPE on the number of neurons in a layer, for 1 layer, 5 epochs, "lbfgs" weight optimization algorithm, learning algorithm for updating "adaptive", with activation functions: "ReLu", "Identity", "logistic", "tanh".

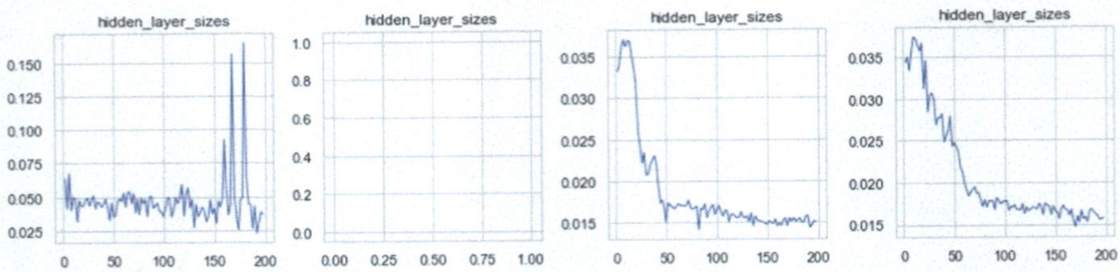

Fig. 9. Dependence of MAPE on the number of neurons in a layer, for 1 layer, 5 epochs, "sgd" weight optimization algorithm, learning algorithm for updating "adaptive", with activation functions: "ReLu", "Identity", "logistic", "tanh".

Conclusion 2: in all cases with weight optimization algorithm «lbfgs», the accuracy was higher.

Conclusion 3: for different data sets and different business processes, different activation function (ReLu, identity, Logistic, tanh) were the best, which is why it is difficult to define one which would be standardized for any business process.

In all the examples considered, number of neurons in layer ranged from 1 to 200 with an 1 increment. Besides, in all cases, the error decreased as number of neurons in layer grew. Nonetheless, there is a possibility this dependence may have a extremum(minimum), so after a certain point, the error may rise again. To test this hypothesis, the 0 to 2000 range of number of neurons in layer was analyzed.

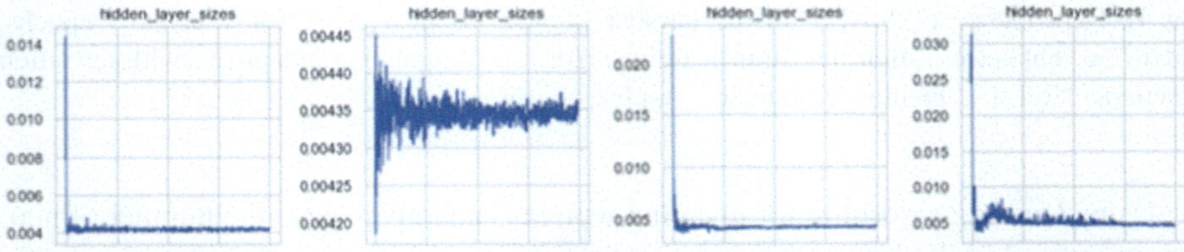

Fig. 10. Dependence of MAPE on the number of neurons in a layer, for 1 layer, 5 epochs, "lbfgs" weight optimization algorithm, "constant" learning algorithm for updating weights, with activation functions: "ReLu", "Identity", "logistic", "tanh".

The graph shows that when number of neurons in layer rises to 2000 on average, the dependence will be the same, i.e. number of neurons in layer rises and the error decreases, but after a certain value of number of neurons in layer, this decrease is within the random error range, it is minor in essence and does not change greatly.

Conclusion 4: if to maximize number of neurons in layer to a certain peak value, the model error can be made lower, but after this peak value, the model accuracy neither falls nor rises, fluctuating within a random error.

In all the settings mixes considered herein, number of layer has only a constant value of 1. However, the accuracy may depend on number of layers. The dependence of accuracy from number of layers was the point of analysis.

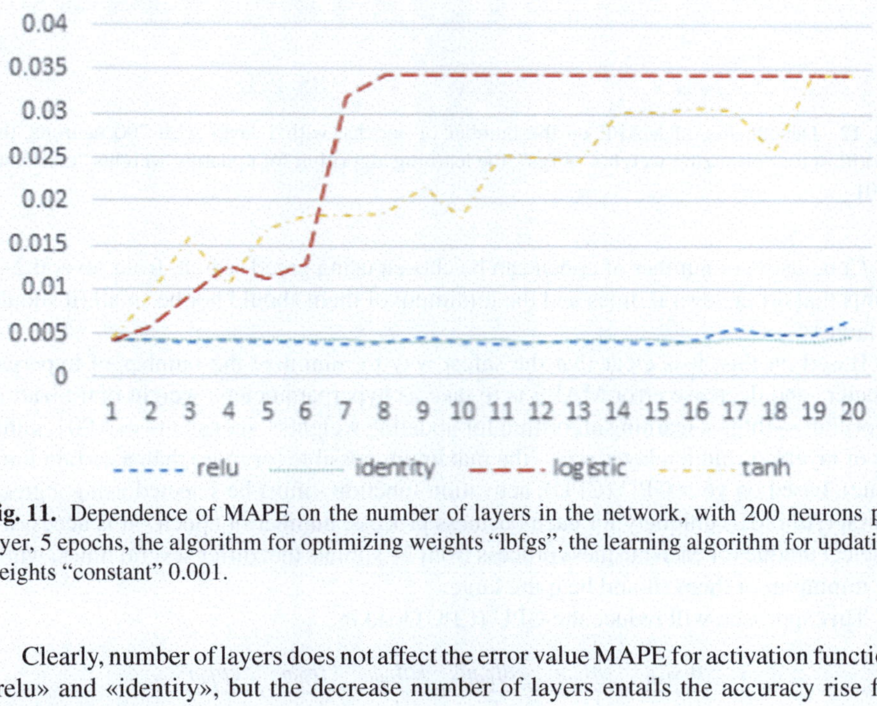

Fig. 11. Dependence of MAPE on the number of layers in the network, with 200 neurons per layer, 5 epochs, the algorithm for optimizing weights "lbfgs", the learning algorithm for updating weights "constant" 0.001.

Clearly, number of layers does not affect the error value MAPE for activation function «relu» and «identity», but the decrease number of layers entails the accuracy rise for «logistic» and «tanh».

Conclusion 5: when building a dependence model of the business process execution time vs the number of stage-to-stage transitions, with neural network applied, the constant number of layers of 1 will be sufficient.

Another parameter that has been taken constant in all previous cases and that can affect the accuracy is number of epoch. Before this, the settings of neural network with 5 epoch were taken for all the cases, but dependence of accuracy by the number of epoch is worth analyzing.

With the rising number of epoch, the error first decreases dramatically, but it stands at about the same point for a certain time and can increase marginally after attaining a certain peak value.

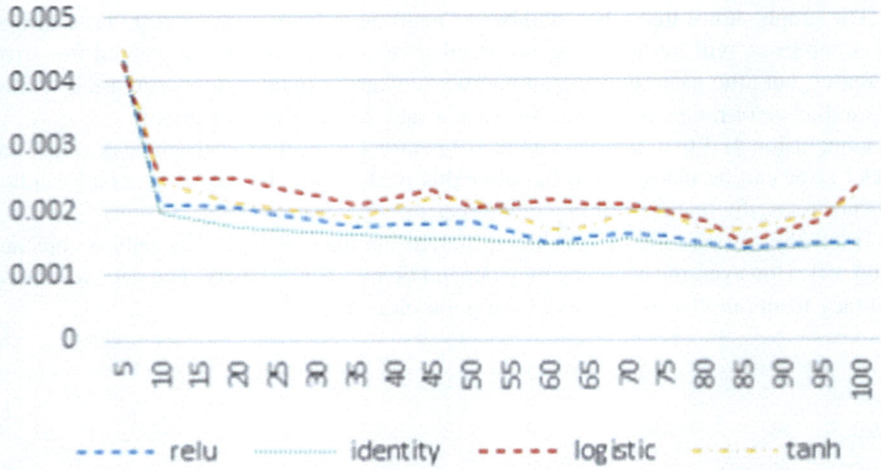

Fig. 12. Dependence of MAPE on the number of epochs, with 1 layer with 200 neurons, the algorithm for optimizing weights "lbfgs", the learning algorithm for updating weights "constant" 0.001.

Conclusion 6: number of epoch can be chosen using greed search, from several 2–3 points that differ several times and the minimum of them should not be small (it should be large).

Based on this, it is clear that the safest way to minimize the number of hyperparameters and decrease error MAPE is to take as hyperparameters: weight optimization algorithm - «lbfgs», learning algorithm for updating weights - «constant» = 0.001, number of layers -1, hidden layer size – the maximum possible (or more than a certain limit value), based on your GPU (CPU), activation function - must be selected using a greed search (AutoML) uniquely for each business process, number of epoch - it is necessary to select unique for each business process from 2–3 points that differ several times, while the minimum of them should be quite large.

This approach will reduce the GPU (CPU) load in:

$$\frac{hls_{full}}{hls_{opt}} \times \frac{hl_{full}}{hl_{opt}} \times \frac{af_{full}}{af_{opt}} \times \frac{sol_{full}}{sol_{opt}} \times \frac{lr_{full}}{lr_{opt}} \times \frac{ep_{full}}{ep_{opt}}$$
$$= \frac{hls_{\max}}{1\frac{hl_{full}}{1}\frac{af_{full}}{af_{full}}\frac{4}{3}\frac{3}{1}\frac{ep_{full}}{3}} \tag{2}$$
$$= 4 \times hlsfull_{full_{max}}$$

where $hls_{full} = hls_{max}$ is the number of various options of the layer neurons tested, $hls_{opt} = 1$ is the optimal number of options of the layer neurons tested, nl_{full} is the number of various options of the number of layers tested, $nl_{opt} = 1$ is the optimal number of options of layers tested, af_{full} is the total number of activation functions tested, af_{opt} is the optimal number of activation functions tested, sol_{full} is a total number of various options of scales optimization algorithms tested (we chose 4 for analysis), $sol_{opt} = 1$ is the optimal number of various options of scales optimization algorithms tested, lr_{full}

is the total number of various options of scales update learning algorithms tested (we chose 3 for analysis), $lr_{opt} = 1$ is the optimal number of various options of scales update learning algorithms tested, ep_{full} is the total number of options of the number of epochs tested, ep_{opt} is the optimal number of options of the number of epochs tested. For this task we took 200 options of the number of neurons tested (1 to 200 with 1 increment), 20 option of the number of layers (1 to 20 with 1 increment), and 20 option of the number of epochs (5 to 100 with 5 increment). Assuming that each option of the hyperparameters combination acted with the same GPU (CPU) load (at the same speed), we get a $4 \times 200 \times 20 \times 20 = 320000$ times decrease in the GPU (CPU) load.

Then the remainder of the set (10%) was used, which was put aside purposefully at the beginning. This remainder was arbitrarily broken into 5 portions for business process instances. 5 models were built per portion, 4 portions for learning and 1 for testing. Each of the 5 model types was built in two options: one using AutoML (greed search for hyperparameters), the other with the hyperparameters defined as the most optimal. In this study (Hidden layer sizes = 200, number of layers = 1, activation – individually selected for each model (identity, logistic, tanh, relu), solver = «lbfgs», learning rate = «constant», number of epoch is chosen for each model for logic considerations ω_0, $\omega_1 = 2 \times \omega_0$, $\omega_2 = 3 \times \omega_0$ in our case [20, 40, 60]). Next, the accuracy values obtained for the models were averaged.

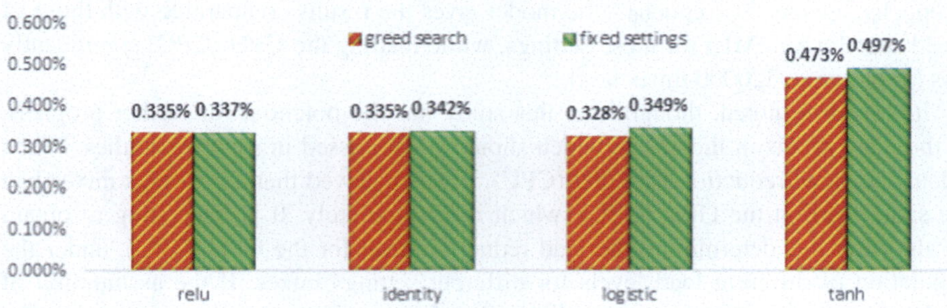

Fig. 13. Graph comparing MAPE errors for various activation functions when using the selected settings (fixed setting) and network settings using greed search.

The graph shows that the error is lower when using greed search. With activation function «relu», MAPE is lower by 0.002 pp; with «identity» – by 0.007 pp; with «logistic» – by 0.02 pp; with «tanh» – by 0.024 pp, which is negligible in all the cases reviewed. But the speed at which this part of the script is work has been 320000 times faster.

3 Conclusion

The article has analyzed the settings of the artificial neural network determining the 'significance' level of process stage-to-stage transitions. The outcomes thus derived are indicative of various regularities which can be the base for a number of setting principles for neural network to be distiguished to fit even our task. First, functional dependence of the model accuracy by the number of neurons in layer is weakly related to the learning algorithm for updating weights type, but the highest accuracy for all the settings mixes tested was obtained with the «constant» algorithm. In addition, in all cases using weight optimization algorithm "lbfgs" the accuracy was higher. Similarly, for different data sets and different business processes, various activation function («relu», «identity», «logistic», «tanh») were the best, which is why it is infeasible to define one which would be standardized for any business process. Another point is that by maximizing number of neurons in layer to a certain peak value, the model error can be reduced, but after this extremum(minimum), the model accuracy neither falls nor rises, fluctuating within a random error. Next, when building a dependence model of the business process execution time vs the number of stage-to-stage transitions, with neural network applied, the constant number of layers of 1 will be sufficient. Finally, number of epoch can be chosen by means of greed search from 2–3 points chosen according to the principle: some sufficiently large initial value of $epoch_0$ is set, and more 2 points are chosen as $epoch_1 = 2 \times epoch_0$, $epoch_2 = 3 \times epoch_0$.

Based on the results, we can infer that when using a combination of settings for neural network: hidden layer sizes – 200 and more, number of layers – 1, function activation – selection unique for each case, weight optimization algorithm - «lbfgs», learning algorithm for updating weights - «constant», number of epoch – selected unique for each model from three options: a sufficiently large initial value $epoch_0$, $epoch_{1}= 2 \times epoch_0$, $epoch_{2}=3 \times epoch_0$ - the model gives the results comparable with those of greed search (AutoML) for these settings, while loading the GPU (CPU) significantly less (in our case, 320000 times less).

It should be noted, though, that this study has the potential for further progress, as there are points in the article which should be discussed in separate studies. When calculating load reduction for GPU (CPU), it was believed that all settings mixes had the same load on the GPU (CPU), which is very unlikely. It is interesting to obtain an algorithm for determining the load reduction level for the GPU (CPU), under the stipulation of different load levels for different settings mixes. If the assumption of load constancy is omitted, it is very likely that the resulting reduction figure due to the replacement of greed search with stationary settings would be a little lower, or not. In addition, it is interesting to analyze other various machine learning methods turning out to be the best from time to time when building a model of the dependence of the business process execution time on the number of unique stage-to-stage transitions in the sense of shortening the number of hyperparameters settings mixes, and lower GPU (CPU) load.

References

1. Bugaenko, A.A.: Replacing the reinforcement learning (RL) to the auto reinforcement learning (AutoRL) algorithms to find the optimal structure of business processes in the bank. In: Silhavy, R., Silhavy, P., Prokopova, Z. (eds.) CoMeSySo 2021. LNNS, vol. 232, pp. 15–22. Springer, Cham (2021). https://doi.org/10.1007/978-3-030-90318-3_2
2. Bugaenko, A.A.: Application of reinforcement learning to optimize business processes in the bank. Turk. J. Comput. Math. Educ. (TURCOMAT) **12**(6), 1638–1644 (2021)
3. Panfilov, M., Goncharenko, I., Bugaenko, A.: Application of DS methods for solving applied task in finance. In: AI Journey Conference. Kaliningrad (2019)

4. Bugaenko, A.A.: Application of machine learning for post process mining analysis and problem detection in bank. In: Silhavy, R. (ed.) CSOC 2021. LNNS, vol. 229, pp. 417–427. Springer, Cham (2021). https://doi.org/10.1007/978-3-030-77445-5_39
5. Sarno, R., Sinaga, F., Sungkono, K.R.: Anomaly detection in business processes using process mining and fuzzy association rule learning. J. Big Data **7**(1), 1–19 (2020). https://doi.org/10.1186/s40537-019-0277-1
6. Tang, P., Qiu, W., Huang, Z., Lian, H., Li, Z.: Anomaly detection in electronic invoice systems based on machine learning. Inf. Sci. **535**, 172–186 (2020)
7. Angiulli, F., Basta, S., Lodi, S., Sartori, C.: Reducing distance computations for distance-based outliers. Expert Syst. Appl. **147**, 113215 (2020)
8. Liu, C., Yang, J., Wu, J.: Web intrusion detection system combined with feature analysis and SVM optimization. EURASIP J. Wirel. Commun. Netw. **2020**(1), 33 (2020)
9. Gdanskii, N.I., Karpov, A.V., Bugaenko, A.A.: Algorithm for dynamic analysis of optimal sensor-interrogation frequency for determination of work performed by an electric motor. Chem. Petroleum Eng. **48**(5–6), 301–305 (2012)
10. Pourbafrani, M., Zelst, S.J., Aalst, W.M.P.: Supporting decisions in production line processes by combining process mining and system dynamics. In: Ahram, T., Karwowski, W., Vergnano, A., Leali, F., Taiar, R. (eds.) IHSI 2020. AISC, vol. 1131, pp. 461–467. Springer, Cham (2020). https://doi.org/10.1007/978-3-030-39512-4_72
11. Dudarenko, D.M., Smirnov, P.A.: Setting artificial neural network hyperparameters for mobile platform navigation. Proc. Southwest State Univ. **23**(6), 115–132 (2019). (In Russ.)
12. Tso, W.W., Burnak, B., Pistikopoulos, E.N.: HY-POP: hyperparameter optimization of machine learning models through parametric programming. Comput. Chem. Eng. **139**, 106902 (2020)
13. O'Mahony, N., Campbell, S., Krpalkova, L., Riordan, D., Walsh, J., Murphy, A., Ryan, C.: Deep learning for visual navigation of unmanned ground vehicles. In: 2018 29th Irish Signals and Systems Conference (ISSC), no. 1–6 (2018)
14. Process Mining: Discovery, Conformance and Enhancement of Business Processes by van der Aalst, W.M.P. Springer, Verlag (2011). ISBN: 978-3-642-19344-6
15. Leemans, S.J.J., Fahland, D.: Information-preserving abstractions of event data in process mining. Knowl. Inf. Syst. **62**(3), 1143–1197 (2019)
16. Mitsyuk, A.A., Shugurov, I.S., Kalenkova, A.A., van der Aalst, W.M.P.: Generating event logs for high-level process models. Simul. Model. Pract. Theory **74**, 1–16 (2017)

Organization of Three-Dimensional Gesture Control Based on Machine Vision and Learning Technologies

Artem Obukhov(✉), Kirill Patutin, Ekaterina Surkova, and Alexandra Nazarova

Tambov State Technical University, 392000 Tambov, Russian Federation
obuhov.art@gmail.com

Abstract. The paper presents an analysis of the problem of organizing gesture control based on machine vision and learning technologies. An approach based on a combination of the MediaPipe machine vision library for recognizing hands and fingers in a frame is proposed to develop a neural network model for gesture classification. This approach, through the use of data on the distance to the user's hands, makes it possible to increase the accuracy of gesture recognition and reduce the percentage of erroneous recognitions. The structure of the three-dimensional gesture control system is formulated. Machine vision techniques and a trained neural network for gesture classification are integrated into the proposed system structure. The developed software successfully recognized the declared gestures, and the accuracy of the trained neural network was more than 99.7%. The novelty of the research lies in the use of data on three-dimensional coordinates of the user's key points, a combination of proven tools (MediaPipe) with the author's neural network model to solve the problem of organizing gesture control. Further research is related to the implementation of dynamic gestures.

Keywords: Neural networks · Machine learning · Computer vision · Stereo vision · Gesture control · Object recognition

1 Introduction

With the development of information technology and hardware solutions, ideas about how to interact with a computer and software are changing. The "classic" keyboard and computer mouse have been replaced by touchpads and touch screens. There are new types of controllers and content transmission through virtual and augmented reality [1]. With the advent of augmented reality devices such as Google Glass, there is a need for interfaces that can harmoniously fit into this concept and provide convenient control for the user. Examples of such interfaces are such devices as Intel Creative Camera [2], Microsoft Kinect [3, 4], Leap Motion [5] and others. The main control elements in these devices are the user's hands.

An important step in improving the system is to improve the reliability of its operation. It is necessary to refine the recognition strategy in such a way as to avoid performing

R. Silhavy (Ed.): CSOC 2022, LNNS 502, pp. 70–81, 2022.
https://doi.org/10.1007/978-3-031-09076-9_7

false gestures and background gestures (gestures that do not belong to the user who is directly working with the system), to separate the work area from the background. An analysis of existing approaches revealed that positioning the hand only along two axes allows the appearance of erroneous gesture recognition from unauthorized users, and limits the functionality and accuracy of the control system.

Therefore, within the framework of the study, one of the most important components of the system being developed is to obtain data on the distance to the user's hands and fingers.

The implementation of this system can be used in the following areas:

- use in graphic editors and 3D modeling environments to improve the efficiency of artists, sculptors, architects, 3D-modelers;
- integration into VR systems to replace controllers and capture human finger and hand movements. Such interfaces are used to organize control in VR applications, as well as in various robotic systems, for example, remotely controlled medical manipulators.
- implementation of a digital environment for people with disabilities and people with disabilities. Gesture control systems can be useful for people with speech or hearing impairments [6], musculoskeletal system [7].

The listed areas substantiate the relevance of this area of research in various spheres of human activity.

1.1 Overview of OpenCV Hand Recognition Algorithms

One of the popular tools for recognizing hands in a frame is the use of OpenCV algorithms. These methods use two-dimensional features of the image to model the visual appearance of the hand and compare these parameters with the same features extracted from the input image. Their feature is to analyze only the appearance (shape, position, etc.) of the target object. For recognition, there is no need to store information about the physical properties of the object in question. This category includes methods that use such characteristics as:

- Skin color [8–10] - for this approach, the choice of color space often plays a decisive role. To achieve resistance to changes in lighting, it is usually preferable to use those spaces in which the chromaticity and luminance components of the colors are effectively distinguished. For this reason, methods of detecting hands by skin color most often use the HSV [11–15], YCrCb [16, 17] spaces. Despite the high processing speed and simplicity in implementation, algorithms and methods based on skin color have major drawbacks: they cannot distinguish objects with a similar color (for example, a face, hands, and other parts of the body); their efficiency is highly dependent on lighting and sharply decreases when it changes.
- Local features [18] - shape is that characteristic feature of the hand, which does not depend on skin color or lighting. The main features can be the contours, the edges of the constituent parts of the hand, or the points of the fingertips. The obvious and most useful piece of information drawn from the shape of the hand is its outline. The hand contour extraction method is applied on a uniform background in real time. However,

extracting the contour of a hand from an image with a complex background is not an easy task. edge detector has difficulty distinguishing the edges of an object of interest from the edges of other objects in the background. Therefore, this approach to hand detection usually requires complex post-processing to improve operational reliability and is often used in conjunction with color information or background subtraction.
- The Jones-Viola method [19–23] was developed for the problem of detecting human faces in images, then it was successfully applied to detect a hand and other types of objects. The main advantage of this method is the high speed of its execution, which allows it to work in real time. In the Jones-Viola detector, it is proposed to use a new type of features based on Haar wavelets. For high-speed computation of features, the concept of "integral image" has been introduced, with the help of which it is also possible to abandon the need to create a pyramid of images, which leads to a significant reduction in the number of calculations.

Thus, the approaches considered are highly dependent on external conditions (illumination and background) and are resource-demanding.

1.2 Overview of Neural Network Algorithms for Determining Key Points of Hands

In view of the significant drawbacks of the approaches discussed in the previous section, methods based on machine learning technologies are very promising. These approaches are focused on determining key points of fingers and hands using neural networks.

Currently, the best results among alternative solutions have been shown by the neural network MediaPipe Hands from Google [24, 25], which uses the ML pipeline (Machine Learning pipeline), which consists of several interconnected models of neural networks:

- palm detection model that works with the full image and returns the oriented bounding box of the hand;
- hand landmark model that works with the cropped area of the image detected by the palm detector and returns the 3D key points of the hand.

Providing a properly cropped image for the hand landmark model dramatically reduces the need for augmentation of data (e.g., obtained by rotating, moving, and scaling the original frames) and instead allows the neural network to more accurately determine the coordinates of the hand.

1.3 Overview of Stereo Vision Algorithms for Determining the Distance to Objects in the Frame

Stereo vision systems (SV) are often used to obtain spatial information about the position of the user's hands. These systems use two or more cameras to capture images of an object from different angles. In the resulting images, the object is in different positions. The relative displacement value of the object in these images can be used to calculate distance and estimate depth. The use of the third dimension in solving the problem of gesture recognition allows both expanding its functionality (by implementing three-dimensional gestures) and increasing its accuracy (by ignoring extraneous gestures in the background).

For this task, both special stereo cameras, such as Stereolabs ZED, Intel RealSense, Microsoft Kinect, and homemade systems consisting of several simple cameras can be used [26].

Stereo vision systems provide more accurate information about an object and have a wider viewing angle. The depth value of the target object makes it possible to separate the user's gesture from the background [27]. Using the depth information, you can accurately determine the position and direction of the hand and fingers in space. This makes it possible to use more complex gestures [28].

The images received from cameras in their original form cannot be used to calculate the distance to an object in the frame. At the first stage, it is necessary to calculate the disparity map - the values of the difference in the positions of the common points of the scene on the left and right images. The search for matching pairs of points is performed on one line, which must be parallel to one of the sides of the image. Since camera lenses have radial and tangential distortion, the original frames are distorted, which complicates the search task. Therefore, it is necessary to align the images. This leveling process is called rectification. In order to rectify images, you need to know the system parameters and distortion coefficients. To obtain all the required values, the stereo camera is calibrated.

The process of calibrating a stereo camera is to obtain external, related to the position of the system in space, and internal (focal length, camera centers) camera parameters. First, a reference image is taken from different sides to automatically determine its control points. Typically, a checkerboard pattern or circular grid is used as a reference for calibration. After finding the reference points on the images from two cameras, the internal and external parameters of the cameras, the lens distortion coefficients are calculated [29].

Image matching algorithms are used to calculate the disparity map. There are many similar algorithms nowadays. As a rule, matching algorithms are divided into local and global. Local algorithms only calculate inconsistencies in areas where there are similar areas. Global algorithms calculate inconsistencies for each point in the image. Local methods are inferior in accuracy to global methods, but at the same time they have higher performance. In [30], the Hirschmüller Semi-Glob Block Matching algorithm is used, which combines the principles of global and local algorithms. This algorithm is implemented in the OpenCV library and allows you to calculate inconsistencies with high performance, providing a sufficiently high accuracy.

After the disparity maps are defined, it is possible to calculate the distance from the target object to the camera. Thus, by combining stereo vision technologies with finger and palm positioning mechanisms, we can expand the possibilities for the accuracy of recognition of user gestures.

2 Methods

Based on the analysis, the following research task was posed: to form the necessary mathematical and structural models for the gesture control system, which, based on information about the three-dimensional position of the user's hand (or hands) in the frame, make it possible to recognize a gesture and form a control command. To do this, at the first stage, it is necessary to carry out a system analysis of the subject area, after which to form the structure of the gesture control system, and then implement, on the basis of the obtained theoretical results, software that solves the problem of gesture control.

2.1 Development of a Three-Dimensional Gesture Control Method Based on Machine Vision and Neural Networks

To implement gesture control and solve the problem of recognizing a hand in a frame, the following model was formulated:

$$M : F \to C, \tag{1}$$

where M is a gesture control method;

$F = f_1, f_2, \ldots, f_n$ - a frame (or some n - a sequence), each of which has the dimensions of $S_x \times S_y$ pixels;
$C = c_1, c_2, \ldots, c_m$ - many control commands.

The gesture control method can be implemented on the basis of various approaches:

m_1 - 2D frame processing methods;
m_2 - MediaPipe method;
m_3 - stereovision processing method;
m_4 - combined method (our).

The choice of the method determines the accuracy A of the gesture control system. Each f_k frame has a set of characteristics:

$$f_k = (P_k, S_x, S_y), \tag{2}$$

where P_k is the $S_x \times S_y$ value of the frame pixels.

In every frame f_k, there may be some $g_j \in G$ gesture, where G is the set of all possible gestures.

Then, to solve the problem of gesture control, it is necessary to choose such a M^* method that achieves the maximum accuracy of gesture recognition G while ensuring the required level of performance of the $FPS_{\min}$ system:

$$\begin{aligned} &A(M^*) \to \max, \\ &FPS(M^*) \geq FPS_{\min} \end{aligned} \tag{3}$$

where $FPS(M^*)$ is an estimate of the system performance using the method, frames per second.

Let's consider the implementation of the method of three-dimensional gesture control based on machine vision and m_4 neural networks. This method will include the following data transformation steps:

1. Receiving a sequence of frames from two cameras.
2. Hand and finger recognition using Mediapipe NN_{MP}:

$$NN_{MP}(F) = H, \tag{4}$$

where $H_i = \{h_{ij}\}$ is a recognized hand, consisting of 21 key points ($j = 1 \ldots 21$).

3. Determination of the coordinates of each point of the hand

$$h_{ij} = (x_{ij}, y_{ij}, z_{ij}), \tag{5}$$

$$x_{ij} = xm_{ij} \times S_x, \tag{6}$$

$$y_{ij} = ym_{ij} \times S_y, \tag{7}$$

where xm_{ij}, ym_{ij} is a normalized values of the coordinates of key points (vary from 0 to 1).

To calculate the distance, the ratio of the product of the baseline of a stereo camera (the distance between the centers of two cameras) and the focal distance to the discrepancy value obtained for it is often used. In some cases, for more accurate information about the position of an object in space, a three-dimensional reconstruction is used.

$$z_{ij} = \frac{B * f}{x_{ij}^l - x_{ij}^r} \tag{8}$$

where B is the baseline of the cameras,

f - focal length,
x_{ij}^l - coordinate of the object in the left image,
x_{ij}^r - the coordinate of the object in the right image.

4. Classification of the $g_j \in G$ gesture based on the processing of the relative coordinates of the hand points by the neural network:

$$NN(H) = g_j, \tag{9}$$

5. Selecting a command from the database for a recognized gesture, depending on the environment ($s_t \in S$, S - the set of all applications of the user's operating system):

$$(g_j, s_t) \rightarrow c_j, \tag{10}$$

6. Performing an action in the $s_t \in S$ application based on the $c_j \in C$ command.

The presented method will be implemented in the software of the gesture control system to solve the problem.

2.2 Software and Hardware Structure

To implement the presented approach, a software structure has been developed, which consists of the following components:

1 Used to work with database tables (images, gestures, commands, available applications).
2 Subsystem of input-output. Used to enter gesture control settings and display results.
3 Subsystem of analysis. Used to recognize hands, palms and to identify gestures. After recognition, the gesture is sent to the control subsystem.
4 Application management subsystem. This subsystem includes tools and technologies for integrating the system into the operating system or with other equipment (VR-helmet, medical simulators), as well as tools for managing third-party applications (game applications, simulation applications). This subsystem receives recognized gestures from the analysis subsystem and generates a control command.
5 System-wide software. Includes the software used for the development of the control system: Python programming language, Pycharm development environment, SQL programming language, SQLite database management system, Python library package (Anaconda).
6 General system technical support. Includes a computer with the following specifications: 2.8 GHz processor, 8 GB RAM. Camera specifications: video stream resolution - 640 × 480, frames per second - 30 frames.

3 Results

To implement the mathematical model of the gesture control system developed in Sect. 2 and to solve the problem, it is necessary to carry out appropriate experimental studies. At the first stage, the problem of positioning the hand and fingers in three-dimensional space will be solved, at the second - the problem of gesture recognition.

3.1 Determining the Position of the Hand in Three-Dimensional Space

In the Python programming language, stages 1–3 of the proposed method were implemented using the MediaPipe (MP) library and pre-trained neural networks. First, using MediaPipe Hands, the key points of the user's hands are determined in both images, and coordinates in pixels are extracted from each point. Then, having obtained a discrepancy for each point of the arm, the distances for each point of the arm and the average distance to the entire arm are calculated (Fig. 1).

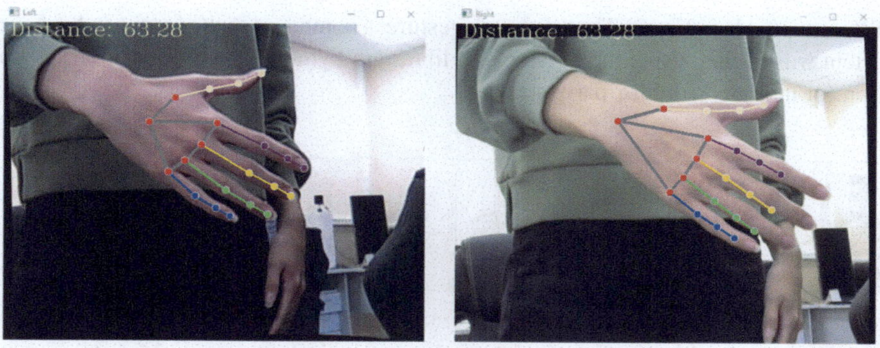

Fig. 1. Hand keypoints recognition using MediaPipe.

3.2 Implementation of a Neural Network-Based Gesture Recognition System

The sets of points for each hand obtained during the operation of the algorithm (21 points with three-dimensional coordinates) are fed to the input of the neural network. The neural network was trained on a dataset collected during the experiments, consisting of 121,000 frames of various gestures. The data was categorized into 6 categories, 5 of which correspond to specific gestures and 6 to arbitrary hand position. Examples of gestures are shown in Fig. 2.

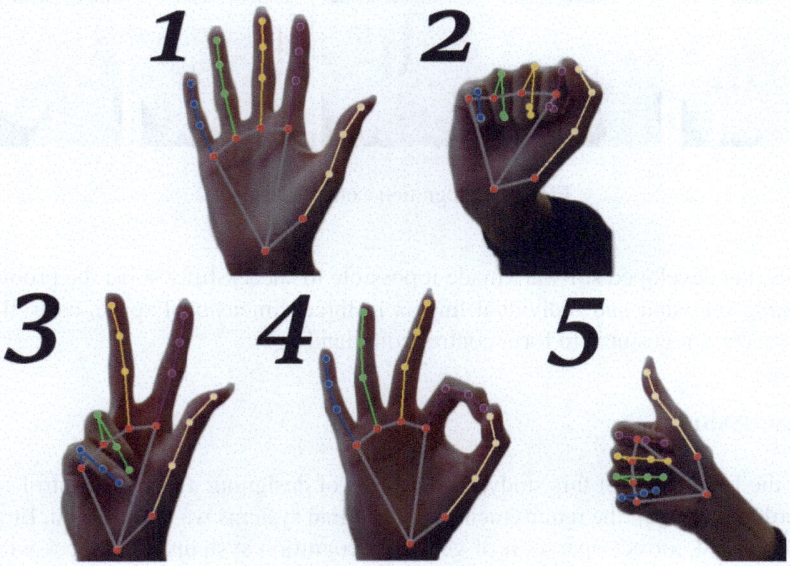

Fig. 2. Recognizable gestures.

As a result of training for 30 epochs, the final accuracy on the test set was obtained, equal to 99.7%. During the control check, the following results were obtained: the neural network successfully recognized all the gestures (a fragment of this test is presented for

5 different people in Fig. 3) Then, for each gesture, a command in the form of a keyboard shortcut was formed using the PyAutoGUI library in Python, which performs a specific action for the selected third-party software.

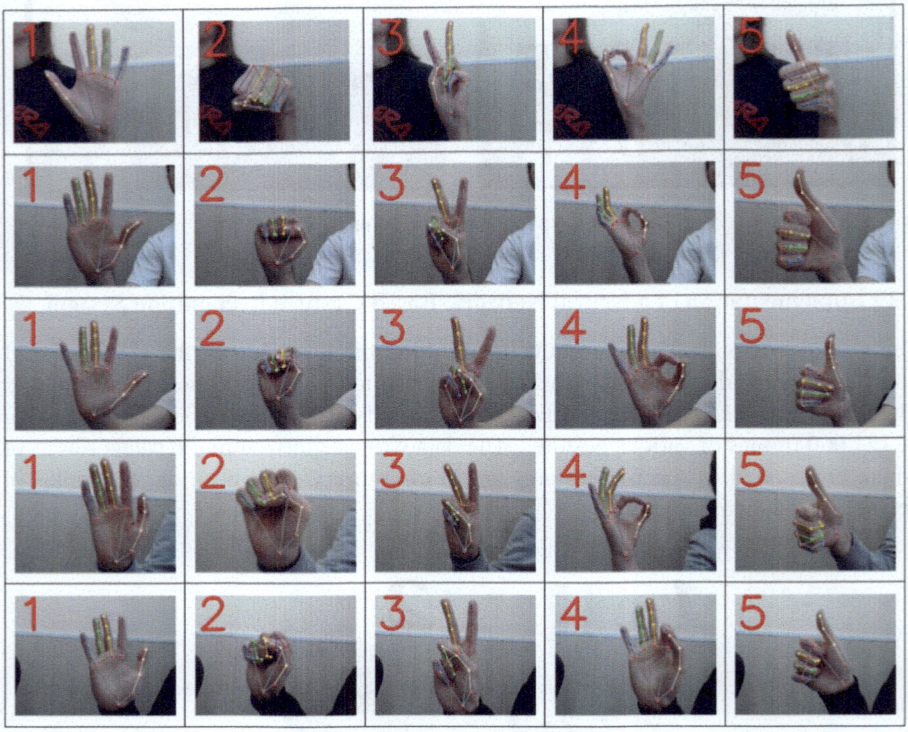

Fig. 3. Recognition examples table

Thus, the developed software made it possible to successfully solve the problem of positioning the hand and individual fingers in three-dimensional space, as well as to recognize certain gestures to form control commands.

4 Discussion

Within the framework of this study, the problem of designing a gesture control system in accordance with all the requirements for real-time systems was considered. Ensuring the smooth and correct operation of gesture recognition systems on devices with low computing capabilities is a non-trivial problem.

Existing sign control systems in the process of their implementation and development faced the following problems:

- the problem of hand recognition in the frame due to quality, camera resolution, illumination, background heterogeneity;

- low efficiency of machine vision algorithms and low performance of hand recognition systems;
- modern solutions of virtual reality (for example, Oculus Quest 2) allow recognition of hands and fingers in the frame, but the tracking area is limited by the direction of the user's gaze;
- the presence of false gestures and background gestures-noise arising from the appearance of unauthorized users in the frame.

Therefore, to solve the above problems, it is necessary to develop a new method of organizing gesture control that provides sufficient performance for working in a real-time system, positioning the user's fingers in three-dimensional space, and implementing software for executing commands based on recognized gestures. The use of a MediaPipe tool approach eliminated these problems.

At this stage, the system includes only the simplest - static gestures (which do not require hand movements). In the future, gestures of a dynamic format (gestures with hand movements) should be added to improve the naturalness of interaction with the systems of the virtual world. Integration with virtual reality platforms, especially with Oculus, where the mechanism of the simplest gestures is already implemented, with the help of which navigation is carried out, also seems to be the main development prospect. Since the existing virtual reality controllers have limited functionality and freedom of interaction, require constant charging, restrict the natural movement of hands and interaction with objects, the use of the proposed system and the extension of the gesture database will completely replace the controllers and thereby simplify the packaging of virtual reality kits.

5 Conclusion

Within the framework of this research, the problem of gesture control with the use of machine vision is considered.

At the beginning of the work, the analysis of existing approaches to positioning and recognition of human hands using machine vision technologies and neural networks was carried out. Further, a method for organizing gesture control is formalized, which is distinguished by taking into account the three-dimensional coordinates of the user's hands, as well as a combination of machine vision technologies and neural networks. The structure of the control system has been formulated. The obtained theoretical results are implemented in the software of the control system, it was tested, which confirmed the effectiveness of the proposed solution.

In further studies, it is necessary to expand the list of gestures, realize their recognition in dynamics and integrate the control system into virtual reality complexes.

This work was supported by the laboratory of medical VR simulation systems for training, diagnostics and rehabilitation.

References

1. Bachmann, D., Weichert, F., Rinkenauer, G.: Review of three-dimensional human-computer interaction with focus on the leap motion controller. Sensors **18**(7), 2194 (2018)

2. Siena, F., Byrom, B., Watts, P., Breedon, P.: Utilising the intel realsense camera for measuring health outcomes in clinical research. J. Med. Syst. **42**(3), 1–10 (2018)
3. Napoli, A., Glass, S., Ward, C., Tucker, C., Obeid, I.: Performance analysis of a generalized motion capture system using microsoft kinect 2.0. Biomed. Sig. Process. Control **38**, 265–280 (2017)
4. Capece, N., Erra, U., Romaniello, G.: A low-cost full body tracking system in virtual reality based on microsoft kinect. In: De Paolis, L.T., Bourdot, P. (eds.) Augmented Reality, Virtual Reality, and Computer Graphics: 5th International Conference, AVR 2018, Otranto, Italy, June 24–27, 2018, Proceedings, Part II, pp. 623–635. Springer International Publishing, Cham (2018). https://doi.org/10.1007/978-3-319-95282-6_44
5. Lu, W., Tong, Z., Chu, J.: Dynamic hand gesture recognition with leap motion controller. IEEE Signal Process. Lett. **23**(9), 1188–1192 (2016)
6. Starodubtsev, I., Averbukh, V., Averbukh, N., Tobolin, D.: Professional natural interfaces for medicine applications. In: Stephanidis, C. (ed.) HCI 2014. CCIS, vol. 435, pp. 435–439. Springer, Cham (2014). https://doi.org/10.1007/978-3-319-07854-0_76
7. Ding, J., Lin, R., Lin, Z.: Service robot system with integration of wearable Myo armband for specialized hand gesture human–computer interfaces for people with disabilities with mobility problems. Comput. Electr. Eng. **69**, 815–827 (2018)
8. El Sibai, R., Abou Jaoude, C., Demerjian, J.: A new robust approach for real-time hand detection and gesture recognition. In: 2017 International Conference on Computer and Applications (ICCA), pp. 18–25 (2017)
9. Edwin, G., Supriana, I.: Hand detection for virtual touchpad. In: Proceedings of the 2011 International Conference on Electrical Engineering and Informatics, pp. 1–5 (2011)
10. Kakkoth, S., Gharge, S.: Real time hand gesture recognition & its applications in assistive technologies for disabled. In: 2018 Fourth International Conference on Computing Communication Control and Automation (ICCUBEA), pp. 1–6 (2018)
11. Paterson, J., Aldabbagh, A.: Gesture-controlled robotic arm utilizing OpenCV. In: 2021 3rd International Congress on Human-Computer Interaction, Optimization and Robotic Applications (HORA), pp. 1–6 (2021)
12. Gourob, J., Raxit, S., Hasan, A.: Robotic hand: controlled with vision based hand gesture recognition system. In: 2021 International Conference on Automation, Control and Mechatronics for Industry 4.0 (ACMI), pp. 1–4 (2021)
13. Harini, V., Prahelika, V., Sneka, I., Adlene Ebenezer, P.: Hand gesture recognition using OpenCv and Python. In: Smys, S., Iliyasu, A.M., Bestak, R., Shi, F. (eds.) ICCVBIC 2018, pp. 1711–1719. Springer, Cham (2020). https://doi.org/10.1007/978-3-030-41862-5_174
14. Kumar, R.: An improved hand gesture recognition algorithm based on image contours to identify the american sign language. In IOP Conf. Ser.: Mater. Sci. Eng. **1116**(1), 012115 (2021)
15. Fayyaz, S., Bukhsh, R., Khan, M., Gondal, H., Tahir, S.: Adjustment of bed for a patient through gesture recognition: an image processing approach. In: 2018 IEEE 21st International Multi-Topic Conference (INMIC), pp. 1–8 (2018)
16. Phung, S., Bouzerdoum, A., Chai, D.: A novel skin color model in ycbcr color space and its application to human face detection. In: Proceedings International Conference on Image Processing, vol. 1, pp. I-I (2002)
17. Sharma, S., Jain, S.: A static hand gesture and face recognition system for blind people. In: 2019 6th International Conference on Signal Processing and Integrated Networks (SPIN), pp. 534–539 (2019)
18. Oka, K., Sato, Y., Koike, H.: Real-time fingertip tracking and gesture recognition. IEEE Comput. Graphics Appl. **22**(6), 64–71 (2002)
19. Ismail, A., Abd Aziz, F., Kasim, N., Daud, K.: Hand gesture recognition on python and opencv. In IOP Conf. Ser.: Mater. Sci. Eng. **1045**(1), 012043 (2021)

20. Islam, M., Islam, M., Islam, M.: An efficient human computer interaction through hand gesture using deep convolutional neural network. SN Comput. Sci. **1**(4), 1–9 (2020)
21. Chetan, H., Praveen, S., Shreyas, S., Singh, S., Urvi, R.: Relative study between technology to perceive hand gestures. In: Fong, S., Dey, N., Joshi, A. (eds.) ICT Analysis and Applications. LNNS, vol. 154, pp. 621–628. Springer, Singapore (2021). https://doi.org/10.1007/978-981-15-8354-4_62
22. Taban, D., Al-Zuky, A., Kafi, S., Al-Saleh, A., Mohamad, H.: Smart electronic switching (ON/OFF) system based on real-time detection of hand location in the video frames. In J. Phys.: Conf. Ser. **1963**(1), 012002 (2021)
23. Muralidhar, P., Saha, A., Sateesh, P.: Customizable dynamic hand gesture recognition system for motor impaired people using siamese neural network. In: 2019 International Conference of Artificial Intelligence and Information Technology (ICAIIT), pp. 354–358 (2019)
24. Zhang, F., et al.: Mediapipe hands: on-device real-time hand tracking, arXiv preprint arXiv: 2006.10214 (2020)
25. Bagby, B., Gray, D., Hughes, R., Langford, Z., Stonner, R.: Simplifying sign language detection for smart home devices using google mediapipe (2021). https://bradenbagby.com/Portfolio/Resources/PDFs/ResearchPaper.pdf. Accessed 20 Jan 2022
26. Marin, G., Agresti, G., Minto, L., Zanuttigh, P.: A multi-camera dataset for depth estimation in an indoor scenario. Data Brief **27**, 104619 (2019)
27. Jiang, D., et al.: Gesture recognition based on binocular vision. Clust. Comput. **22**(6), 13261–13271 (2018)
28. Jadooki, S., Mohamad, D., Saba, T., Almazyad, A.S., Rehman, A.: Fused features mining for depth-based hand gesture recognition to classify blind human communication. Neural Comput. Appl. **28**(11), 3285–3294 (2016)
29. Zhang, Z.: A flexible new technique for camera calibration. IEEE Trans. Pattern Anal. Mach. Intell. **22**(11), 1330–1334 (2000)
30. Hirschmuller, H.: Stereo processing by semiglobal matching and mutual information. IEEE Trans. Pattern Anal. Mach. Intell. **30**(2), 328–341 (2007)

A Method of Automatic Social Tension Detection on the Internet Based on Cognitive Analysis and Individual's Emotional State

Korovin Iakov[1], Anna Klimenko[1](✉), and Irina Safronenkova[2]

[1] Scientific Research Institute of Multiprocessor Computer Systems, Southern Federal University, 2, Chekhov Street, 347928 Taganrog, Russian Federation
anna_klimenko@mail.ru

[2] Federal Research Center the Southern Scientific Center, Russian Academy of Sciences, 41, Chekhov Street, 344006 Rostov-on-Don, Russian Federation

Abstract. The purpose of this study is to develop a method for automatic detection of hotbeds of social tension on the Internet. The main problem with such detection is that currently there is no general methodology for detecting signs of social tension in the digital space, while more and more communication processes are moving to the network. Within the framework of this study, based on the analysis of sources, attributes of social tension were formed that are characteristic of expression in the digital space. Cognitive models and an ontology of attributes were developed, on the basis of which attributes of social tension, characteristic of the initial stages of the development of the situation, were formed. The formed complex of attributes is based on the methods of automatic detection of psychoemotional states of the individual. The developed method of automatic social tension detection on the Internet makes it possible to detect hotbeds of information dissemination that are suspicious from the point of view of intensifying social tension with the aim of further information countermeasures.

Keywords: Edge computing · Reliability · Distributed systems · Reconfiguration · Information systems management · Decentralized control · Distributed leader

1 Introduction

Detecting in the network sources provoking an increase in social tension is an urgent task of our time. This is due to the fact that communication processes are intensively moving to the Internet, which, due to the implementation of the properties of physical non-representation and anonymity [1], has become an exclusively breeding ground for the spread of rumors, all sorts of fake information, appeals and materials of extremist orientation. In addition, the Internet provides the ability to create multiple false accounts and register on many publishing platforms. This makes it extremely difficult to develop adequate strategies and methods of information countermeasures, including the detection of sources that provoke an increase in social tension in the early stages.

R. Silhavy (Ed.): CSOC 2022, LNNS 502, pp. 82–93, 2022.
https://doi.org/10.1007/978-3-031-09076-9_8

The Internet has another important feature: large amounts of circulating data. This leads to low efficiency of data analysis in the context of time constraints, as well as to the uncertainty of the timing of the required data discovery.

The urgency of the problem of detecting hotbeds of social tension on the Internet is exacerbated by the fact that to date, no general methodology has been formed for identifying social tension on the Internet. The transfer of communication processes to the network is a relatively new phenomenon, and therefore, despite the existence of separately developed methods [2–5], work has not been carried out to draw analogies between the offline space and the network space, between the manifestations of social tension offline (which was previously recorded by analyzing the collected questionnaires) and those on the network. The last could serve as a justification for a number of attributed that should be paid attention primarily in the context of the development of a situation of tension.

The aim of this study is to form a method of automatic social tension detection on the Internet.

2 Social Tension

Social tension is a psychological, social phenomenon and has a fairly wide range of interpretations.

Social tension is defined as the state of a social system, the content of which is the process of the emergence and development of inconsistency in relations, interests, actions of people, social groups, institutions and society as a whole [6].

There is also another point of view, within which social tension is displayed in the minds of individuals as objective characteristics of reality. Social tension is fixed in negative, stable and dynamic emotions, feelings, sensations and representations, acting as a function of reflection [7].

Thus, in Russian sociology, social tension is seen as one of the most significant characteristics of intragroup and intergroup relations.

Within the framework of foreign approaches to the study of social tension, the studies [8–14] should be highlighted. In the listed studies, the reasons for the emergence of social tension are based on ethnic heterogeneity, as well as the economic segregation of population groups.

The study [15] contains a number of theoretical provisions that can be used as detection of early signs of social tension in society. Attention is paid to intra-group and inter-group interaction, and the following theoretical conclusions are obtained:

- Intergroup social tension arises more often and more easily than intragroup one.
- A decrease in the size of a large group of individuals, along with an increase in the size of a small group, leads to a decrease in social tension in the larger group and to an increase in social tension in the smaller group.
- Segregation of groups leads to a decrease in social tension within groups.
- Segregation increases intergroup social tensions.
- An increase in social mobility leads to an increase in social tension within classes and to a decrease in social tension between classes.

At present, in addition to observing objective reality, population polls [16, 17] with subsequent modeling are the standard method for determining the degree of development of social tension.

However, this approach, firstly, takes a lot of time, from collecting data from experts to modeling, while the formation and development of social tension is a dynamic process. Secondly, the experts' answers are often not frank, and the sample of respondents may be unrepresentative and, thirdly, such methods of assessment do not take into account the enhanced possibilities of anonymous communication, which is provided by the World Wide Web.

Therefore, the task of automatic detection of the development of social tension and its sources in the information space becomes urgent, based on classical methods for detecting levels of social tension, but taking into account the specifics of the digital environment.

3 The Formation of Social Tension Attributes for the Digital Space on the Internet

We will consider the listed attributes of social tension in relation to their possible expression in social networks.

1. Aggression. The individual's state of aggression is reflected in the network as follows:

   - in the form of posts containing keywords-markers of aggression,
   - implemented through cyberbullying,
   - in the form of images and videos containing an image of an individual with non-verbal attributes of aggression.

2. Depression. Depressive moods are displayed in a similar way:

   - posts, graphics, videos that contain markers of depressive states. Depressive states are also typical for victims of bullying.

3. Anger. It is identified by analyzing posts for keywords, as well as by fixing non-verbal signs of anger.
4. Fatigue. It is revealed similarly to the previous feature: based on the identification of non-verbal markers of fatigue in the texts and images of the individual. The attributes listed above relate to the emotional sphere of individuals and should be used in conjunction with indirect attributes related to the environment of the individual in the network space.
5. The presence of closely related populations with little external interaction. The presence of such groups in the network space first of all means the presence of a group of individuals united on the basis of some idea or occupation. Therefore, when detecting social tension in the network, it is advisable to fix groups, especially strongly connected (which can be determined by the frequency of reposts, by the speed of dissemination of opinions and by the aggressive mood towards individuals who are not members of the group).

6. A sharp increase in any social group and a decrease in the size of another. This feature, similar to the previous one, is more related to offline-formed social groups, but in the same way it can serve as an indirect attribute of the creation of a group on the Internet, which is actively recruiting new members.
7. Dissatisfaction with something. It is expressed in texts, graphics, videos and emotionally comes down to attributes 1–4. In addition, the expression of dissatisfaction with their material condition, government, etc. reflected by creating relevant posts and reposts.
8. The presence of protest moods/formation of the image of the enemy/protest calls. The above is revealed by analyzing text content using dictionaries. Also, as mentioned above, CTA (Calls to Action) values have.
9. The circulation of rumors in social groups, as a rule, does not correspond to reality and contributes to the growth of social tension.

Let us correlate the markers of social tension listed above with the markers that make it possible to identify an individual - a source of increased social tension in the digital environment.

1. Emotional markers (aggression, depression, anger, fatigue) are revealed in texts, images and videos by fixing non-verbal markers and by determining the sentiments of texts.
2. The attributes of the presence closely related groups in relation to one individual is expressed by a limited circle of communication in social networks, regular reposting of information from a limited circle of communities, regular visits to the same sites (profiles, forums, etc.)
3. The attributes of the presence of a sharp growth of any group relative to an individual is determined as follows:

   - the individual takes an active position as a recruiter in the group by regularly sending invitations, possibly with calls for protest actions,
   - an individual is actively reposting information published in a group,
   - the individual joins a growing group and takes on the role of a recruiter (opinion leader) or disseminator of information.

4. Dissatisfaction of any kind is expressed in non-verbal signs (aggression, depression, anger, fatigue) in conjunction with posts of characteristic content, as well as the generation of false rumors, calls for active protests.
5. The presence of protest moods/formation of the image of the enemy/protest calls of the individual is expressed in the publication of posts on the relevant topic, and is also recorded by non-verbal attributes.
6. The generation of rumors in the context of one individual is recorded as follows:

   - the individual himself is the source of rumors, while using fake content and fake accounts,
   - the individual is reposting rumors.

Here it is advisable to divide the individuals who spread social tension into two groups: opinion leaders and followers.

Opinion leaders engage in direct stimulation of tension in some way and constantly increase the number of followers. The followers are more involved in reposting authoritative opinions and ultimately also increase the number of followers.

Thus, systematizing the selected features, they can be classified as follows.

1. By the type of expression: 1.1. Non-verbal: 1.1.1. Anger, 1.1.2. Aggression, 1.1.3. Depression, 1.1.4. Fatigue; 1.2. Verbal: 1.2.1 CTA, 1.2.2. Expression of dissatisfaction, 1.2.3. Recruiting, 1.2.4. The presence of closely related groups; 1.3. Based on Internet technologies: 1.3.1 Recruiting, 1.3.2. Reposting, 1.3.3. Fake accounts, 1.3.4. Publication of rumors.
2. By the type of implementation: 2.1. Text; 2.2. Graphics; 2.3. Video.
3. By the type of individual participation: 3.1. Initiator; 3.2. Follower.

Non-verbal attributes can be fixed by methods, depending on the type of implementation. For example, the marker "anger" in the video is fixed by detecting marker states of the individual in accordance, for example, with the study of Paul Ekman [27], in the text - by determining the emotional coloring of the message. The features allocated to the group "Based on Internet technologies" are specific to the network environment and can be fixed by the methods used to identify reposting, recruiting, fake accounts, etc. However, it should be noted that features from group 1.3, namely features 1.3.1, 1.3.2., 1.3.4., 1.3.5 are elements of published (sent out) content, and therefore can be identified by means of signs of groups 1.1. and 1.2. It should also be noted that recruiting is included in group 1.2. and 1.3. for the reason that it is a separate type of content and implies a very specific content, while at the same time being the implementation of the capabilities of Internet technologies.

4 The Cognitive Model of the Situation Development

Let us represent the processes and groups of individuals influencing the level of their intensity in the form of cognitive map concepts.

1. Opinion leader. We will assume that there is only one such leader and begins to form the group. In reality, there are several leaders organizing preparations for protest actions, but within the framework of modeling, we will assume that the actions they take are identical, and we can consolidate a group of leaders, considering it as one opinion leader.
2. Dissatisfaction with the situation. It is the leading concept of the model, since it has a formative influence on opinion leaders (it is unlikely that a person who is satisfied with everything will begin to produce tension). Dissatisfaction with the situation forms opinion leaders, is the reason for the non-verbal manifestation of non-verbal signs of tension, is an additional factor in the growth of the number of followers (people who are satisfied with everything are rarely provoked).

3. Followers. These are people who, under the pressure of dissatisfaction with the situation, ongoing recruiting and reposting of opinions, begin to repost, rumors, etc.
4. Opponents. This is a group that do not share the opinions of opinion leaders and followers, however, under information pressure in the aggregate of dissatisfaction with the situation, bullying by opinion leaders and their followers, reposting opinions, rumors and CTA, they move to the group of followers (or at least lose the ability to resist followers and opinion leaders).
5. Expression of dissatisfaction among opinion leaders and followers produces: non-verbal signs of social tension (which is expressed in markers of negative and destructive emotions), reposting opinions, rumors, CTA, contributes to increasing dissatisfaction with the situation.
6. Non-verbal attributes of tension produce dissatisfaction with the environment.
7. Reposting - carried out by followers and implements the dissemination of information in networked environments, is also a manifestation of an expression of discontent on the network.
8. Rumors are the result of the expression of dissatisfaction on the network.
9. CTA is also a way of expressing dissatisfaction.
10. Bullying is carried out by opinion leaders and followers in relation to opponents in order to deactivate alternative opinions on the network.
11. Recruiting is carried out by opinion leaders and their followers. At the same time, recruiting affects the number of followers.
12. Fake accounts - used in rumors, recruiting and bullying.

The developed cognitive model is as follows (see Fig. 1).

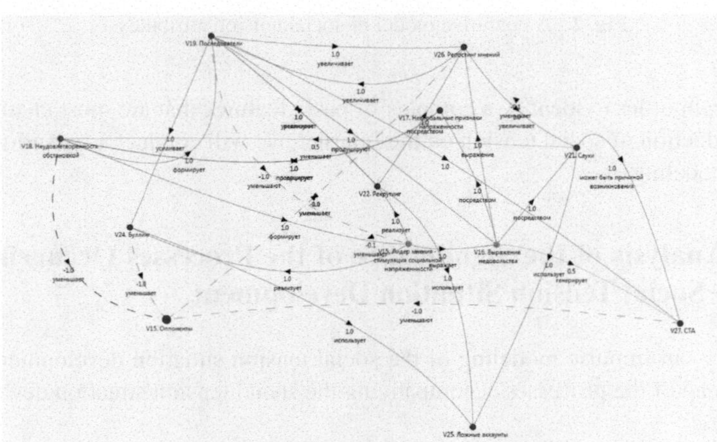

Fig. 1. A cognitive model of processes characterizing social tension.

It should be noted that the given cognitive map of the processes characterizing social tension was developed for the network environment: unlike the currently used questionnaires, first of all, the model takes into account the processes that are implemented exclusively by means of the Internet, namely: reposting, bullying, CTA, false accounts.

5 Cognitive Model of Attributes Accompanying the Social Tension Situation Development

The main difficulty in identifying the key features that make it possible to identify the source of social tension is the close mutual influence of attributes of social tension in the network. The links between the features are shown in Fig. 2.

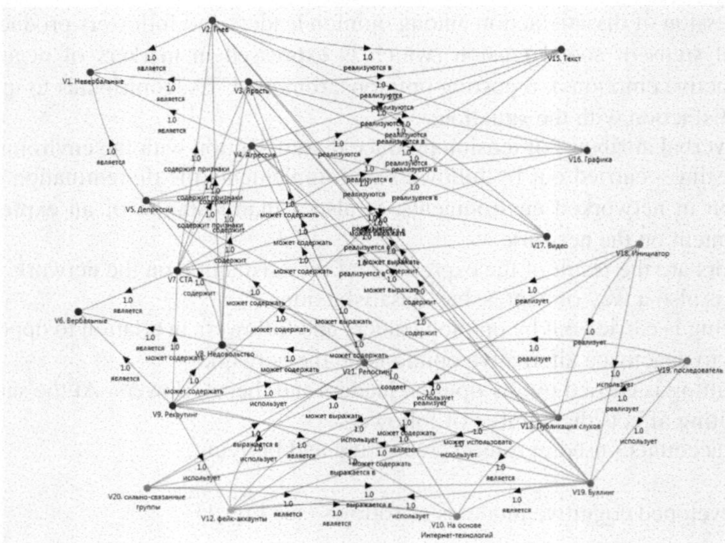

Fig. 2. A cognitive model of social tension attributes.

Further, in order to identify a complex of basic features that are most characteristic for the production of social tension on the Internet, we will conduct a scenario analysis (impulse modeling).

6 The Analysis of the Significance of the Processes Occurring in the Social Tension Situation Development

Let us carry out impulse modeling of the social tension situation development on the cognitive map of the processes accompanying the social tension situation development (see Fig. 3).

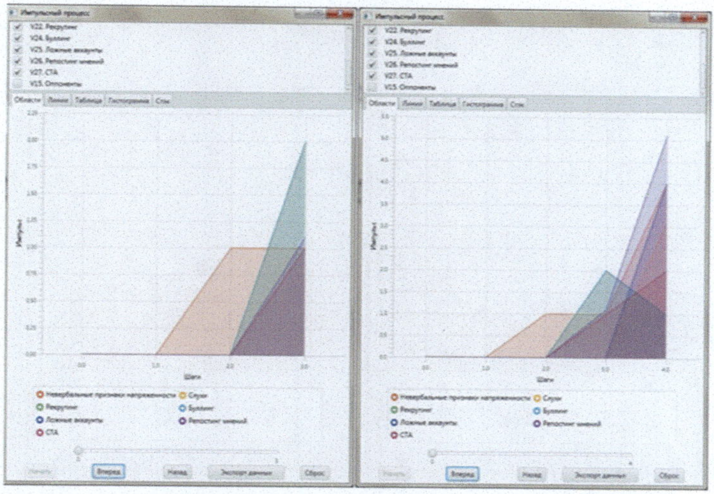

Fig. 3. The initial stage of the situation development in accordance with the process of impulse modeling.

At the beginning of the situation development (Fig. 3), there is a sharp significant increase in the level of recruiting, while non-verbal signs of tension, CTA and opinion reposting remain at the same level. It should be noted that the very first process that develops within the framework of the development of social tension is the manifestation of non-verbal attributes of social tension. Based on the results of the previous modeling of the development of the situation, we identified the following processes developing at the initial stage, namely: recruiting, non-verbal manifestations of tension, CTA, reposting.

Let us consider each of these processes from the point of view of features and, in particular, features that are revealed primarily in the listed processes (see Fig. 4 and Fig. 5).

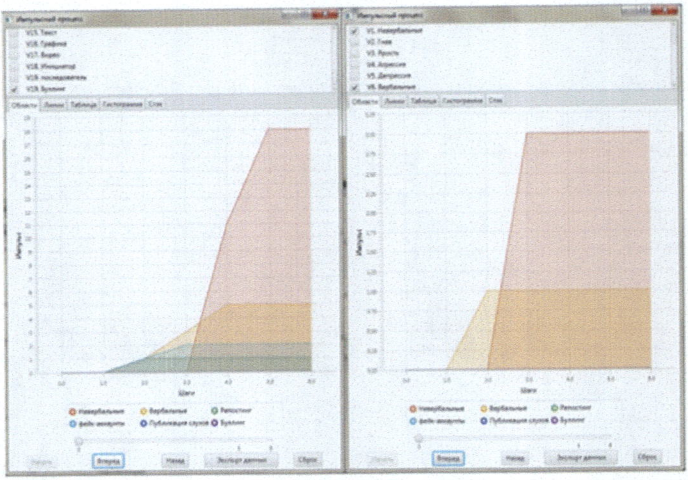

Fig. 4. Involved attributes of ST when giving impulse to the concept "Recruiting" and "CTA".

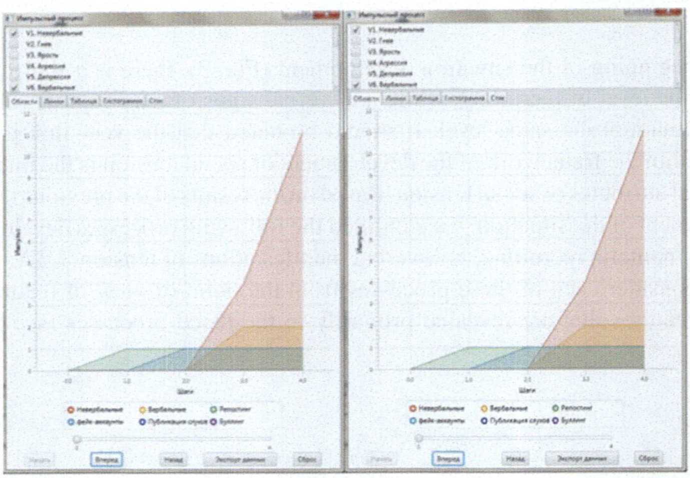

Fig. 5. The attributes used when giving an impulse to the concept of reposting and the consequent influence on the concepts of "recruiting", "CTA", "reposting" also during modeling confirms the importance of non-verbal manifestations of social tension.

Thus, the modeling performed allows us to conclude that among the attributes of social tension that are implemented on the Internet, as a result of the expression of attributes through each other, non-verbal manifestations of such negative emotions as anger, rage, depression accompany all the processes of social expression that we have identified. tension. Further, the automatic detection of sources of social tension in the digital space is reduced to the detection of the corresponding psychoemotional states of the individual.

7 A Method of Automatic Social Tension Detection on the Internet Based on Cognitive Analysis and Individual's Emotional State

The study allows us to conclude that the centers of social tension in the network are primarily characterized by a high intensity of posts, accompanied by negative emotions, expressed both verbally and non-verbally.

Therefore, the method of automatic social tension detection on the Internet will be two-component and rely on a set of methods:

1. Verbal expression of negative emotions includes:

- analysis of the sentiment of texts;
- analysis of texts for the presence of CTA;
- analysis of images for the presence of CTA.

2. Non-verbal expression of negative emotions includes:

- analysis of images for the psychoemotional state of the individual.

  The identification of negative psychoemotional states is based on the methods of P. Eckmann, which is implemented at the level of automatic detection of the individual's emotional states.

  Also, since the method is focused on working in the digital environment of the Internet, it is necessary to highlight its main stages:

- data collection,
- - data processing,
- - providing the result.

The proposed method is shown schematically in Fig. 6.

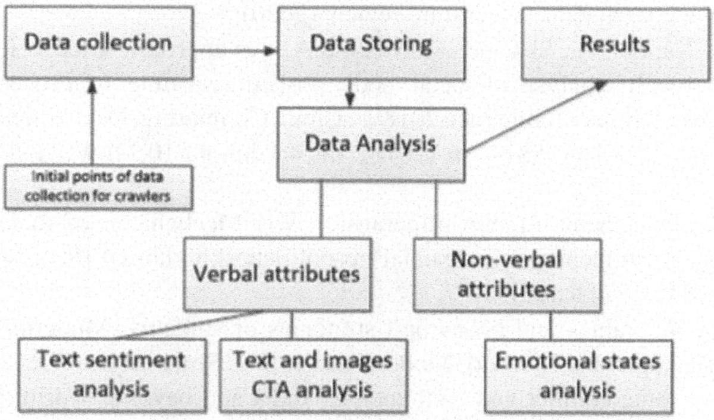

Fig. 6. A method of automatic social tension detection in the Internet.

8 Conclusion

This study is devoted to the issue of automatic detection of sources of stimulation of social tension on the Internet. Since the transfer of communication processes to the network space is a relatively new phenomenon, to date, apart from individual techniques, no complex of attributes of social tension specific to the digital space has been developed, as well as methods of their automatic detection.

Within the framework of this study, based on methods for detecting social tension outside the network space, as well as on the results of cognitive modeling, a set of features has been formed that allow detecting foci that stimulate social tension at the early stages of its development. The fixation in the digital space of hotbeds with a predominance of non-verbal expression of negative emotions accompanying a situation of social tension allows, in fact, to identify the foci themselves that stimulate social tension and, therefore, apply well-known methods of information countermeasures.

Acknowledgments. This study is conducted with the financial support of RFBR grants № 20-04-60485, № 18-29-22093.

References

1. Melnik, E., Korovin, I., Klimenko, A.: A cognitive assistant functional model and architecture for the social media victim behavior prevention. In: Silhavy, R. (ed.) CSOC 2019. AISC, vol. 985, pp. 51–61. Springer, Cham (2019). https://doi.org/10.1007/978-3-030-19810-7_6
2. Thakur, P., Shrivastava, D.R.: A review on text based emotion recognition system. Int. J. Adv. Trends Comput. Sci. Eng. **7**(5), 67–71 (2018). https://doi.org/10.30534/ijatcse/2018/01752018
3. Oueslati, O., Khalil, A.I.S., Ounelli, H.: Sentiment Analysis for Helpful Reviews Prediction. Int. J. **7**(3), 34–40 (2018). https://doi.org/10.30534/ijatcse/2018/02732018
4. Nurul, S.J., Siti, J., Kamaruddin, S.S., Ahmad, F.K., Angeli, A.C.: Social tension detection on social media textual data: a literature review (2020)
5. Sachdeva, A., Kapoor, R., Sharma, A., Mishra, A.: An approach towards identification and prevention of riots by analysis of social media posts in real-time. In: Proceedings of the 8th International Conference Confluence 2018 on Cloud Computing, Data Science and Engineering, Confluence 2018, pp. 581–586 (2018). https://doi.org/10.1109/CONFLUENCE.2018.8442714
6. Danakin, N.S., Dyatchenko, L.Y.A., Speranskij, V.I.: Mechanisms of social regulation and conflict management technologies. Social'no-politicheskij zhurnal [Socio-political journal] (3), 151–162 (1996) (In Russ.)
7. Muhanova, M.N.: Values and behavioral strategies of students (Modernizaciya social'noj struktury rossijskogo obshchestva). Modernization Soc. Struct. Russ. Soc. 271–265 (2008)
8. Varshney, A.: Ethnic conflict and civil society: India and beyond. World Pol. **53**, 362–398 (2001)
9. Horowitz, D.: Direct, displaced and cumulative ethnic aggression. Comp. Polit. **6**, 1–16 (1973)
10. Horowitz, D.: Ethnic Groups in Conflict, Second Edition [1985]. University of California Press, Berkeley (2000)
11. Horowitz, D.: The Deadly Ethnic Riot. University of California Press, Berkeley (2001)

12. Jha, S.: Trade, institutions and religious tolerance: evidence from India (2005). mimeo, Stanford University Justino, Patricia (2008)
13. Lester, A.: Essays in the theory of economic growth. PhD Thesis, MIT, Chapter 1 (2005)
14. Montalvo, J., Reynal-Querol, M.: Ethnic polarization, potential conflict, and civil wars. Am. Econ. Rev. **95**, 796–815 (2005)
15. Rohner, D.: Reputation, group structure and social tensions. J. Dev. Econ. **96**(2), 188–199 (2010). https://doi.org/10.1016/j.jdeveco.2010.10.008
16. Baranova, G.V., Alekhin, E.I.: Formation of social tension in the regions of the Russian Federation. Methods of analysis and forecasting of social tension. Vestnik VGU. Seriya: Lingvistika i mezhkul'turnaya kommunikaciya [Bull. VSU. Series: Linguist. Intercultural Commun. (2–2), 149–154 (2007)
17. Rich, G.: Ekman Paul (2013) https://doi.org/10.1002/9781118339893.wbeccp187

Predicting Diarrhoea Among Children Under Five Years Using Machine Learning Techniques

Elliot Mbunge[1(✉)], Garikayi Chemhaka[2], John Batani[3], Caroline Gurajena[1], Tafadzwa Dzinamarira[4,5], Godfrey Musuka[5], and Innocent Chingombe[5]

[1] Department of Computer Science, Faculty of Science and Engineering, University of Eswatini, Kwaluseni Campus, Kwaluseni, Eswatini
mbungeelliot@gmail.com

[2] Department of Statistics and Demography, Faculty of Social Sciences, University of Eswatini, Kwaluseni Campus, Kwaluseni, Eswatini

[3] Faculty of Engineering and Technology, Botho University, Maseru 100, Lesotho

[4] School of Health Systems and Public Health, University of Pretoria, Pretoria 0002, South Africa

[5] ICAP at Columbia University, Harare, Zimbabwe

Abstract. Globally, diarrhoea remains a significant cause of death among children under five years. Several preventive interventions such as hygiene practice, safe drinking water, rotavirus vaccination and health promotion were implemented to reduce the catastrophic impact of diarrhoea. However, effective tackling of the diarrhoeal disease requires robust preventive interventions and computational techniques to predict diarrhoea among children under five years using risk factors. Therefore, this study applied a decision tree classifier, logistic regression and support vector machines to predict diarrhoea among children under five years using the recent Zimbabwe Demographic Health Survey dataset. The study revealed that logistic regression out-performed other diarrhoea predictive models with the prediction accuracy of 85%, precision of 86%, recall of 100% and the F1-score of 94%. Support vector machines also performed better in predicting diarrhoea with predicting accuracy of 84%, precision of 85%, recall of 100% and F1-score of 92%. The study also revealed that understanding risk factors such as climatic or meteorological, socioeconomic and demographic factors plays a tremendous role in tackling diarrhoea among under-fives. The application of machine learning techniques can assist policymakers in designing effective and adaptive diarrhoea preventive interventions, control programmes and strategies for tackling diarrhoea.

Keywords: Diarrhoea · Children under-five · Machine learning · Prediction · Zimbabwe

1 Introduction

Diarrhoea remains a public health problem and the leading cause of death among children under five years of age in many developing countries [1]. Approximately 1.7 billion cases of childhood diarrhoeal disease occur every year globally, resulting in the death

R. Silhavy (Ed.): CSOC 2022, LNNS 502, pp. 94–109, 2022.
https://doi.org/10.1007/978-3-031-09076-9_9

of approximately 525 000 children [2, 3]. Such high under-five mortality and morbidity primarily result from contaminated food and a lack of improved drinking water sources. For instance, a report from WHO stated that globally, about 780 million people do not have access to safe drinking water and 2.5 billion lack adequate sanitation [2]. Developing countries have a high rate of diarrhoea infections, which contribute significantly to the hospitalization of children under the age of five years with nearly four-fifths of under-five mortality occurring in sub-Saharan Africa (SSA) South Asia [4]. However, a significant fraction of diarrhoeal incidences can either be prevented or treated through interventions such as the provision of safe drinking water, rotavirus vaccination [5], adequate sanitation [6], hygiene, health education [7] and improving access to healthcare. Strengthening health systems initiatives and health policy formulation to provide cost-effective and affordable interventions for children will potentially reduce child mortality by decreasing the risk of diarrhoeal disease.

The third goal of the United Nations' 2030 Agenda for Sustainable Development which was adopted by its member states in 2015 targets to "*ensure healthy lives and promote well-being for all at all ages*" [8]. One of these goals aims to end preventable deaths of newborns and under-five children by 2030 [8]. In response, Zimbabwe, as one of the affected countries in the SSA region, also developed intervention programs and strategies to prevent, control and reduce diarrhoeal mortality and morbidity. Such interventions include allocating limited resources and securing commodity supply chains to avoid stock-outs [9], training and deploying staff to remote communities. However, some of the barriers encountered in this approach are the inability to reach high-risk children and the failure to provide sick children with piecemeal services [10].

Contribution of the Study

Despite remarkable progress in reducing under-five mortality, childhood diarrhoeal disease is still of the leading causes of child mortality and morbidity [11]. Indeed, the current approaches are insufficient and need other new strategies to reduce child mortality. Techniques such as the integration of emerging technologies like machine learning (ML) models and other computational intelligence-based models to predict diarrhoea incidences and child mortality can benefit health systems and health policymakers. Such models require training and testing data to predict diarrhoea incidences which are not always available accurately. In addition, some scholars, including [11], highlighted a significant gap in determining the prevalence and determinants of diarrhoea among under-five children. To address this challenge, some studies examined diarrhoea prevalence and risk factors associated with under-five diarrhoeal diseases using logistic regression, which may present interpretation difficulties [12–14]. However, the risk factors are numerous and interrelated in sophisticated ways that make it cumbersome to predict diarrhoea using traditional statistical methods [10]. The majority of these researches have either of the following significant drawbacks: (i) focus on one kind of infectious diarrhoea [15], (ii) failing to include all risk factors (climatic, socioeconomic and demographic) in predicting diarrhoea among children under-five years which contribute significantly towards child mortality, and (iii) applied traditional statistical methods that could not effectively deal with complex relationships between interrelated childhood diarrhoea risk factors. Understanding key factors that determine diarrhoea prevalence and their impact can strongly highlight the need for family planning, adoption of digital interventions, "*improvement in maternal education and addressing regional disparities in child health which could help inform health policy and intervention strategies to improve child survival*" [16]. Therefore, this study applied machine learning techniques to predict diarrhoea among children under-five years using 2015–2016 Zimbabwe Demographic and Health Survey data. Machine learning techniques can extract meaningful hidden insights, unknown patterns and relationships between risk factors associated with diarrhoea incidences. Machine learning-based diarrhoea prediction models can facilitate early detection, prevention and remedy of diarrhoea to save lives and resources.

2 Methodology

2.1 Study Area and Data Source

The study was conducted in Zimbabwe, a landlocked country that is divided into ten administrative provinces: Manicaland, Mashonaland Central, Mashonaland East, Mashonaland West, Matabeleland North, Matabeleland South, Midlands, Masvingo, Harare, and Bulawayo. The 2015 Zimbabwe Demographic and Health Survey (ZDHS) was implemented by the Zimbabwe National Statistics Agency (ZIMSTAT), with participants over 11,000 households. The 2015–2016 ZDHS is the latest national survey after 2010–2011 ZDHS. ZDHS surveys provide updated health indicators and basic demographic and health indicators. The study design of the 2015–2016 ZDHS was described in the Zimbabwe DHS final report. The 2015–2016 ZDHS dataset used in the study can be accessed on the following link (http://www.measuredhs.com) upon reasonable request.

2.2 Risk Factors Associated with Diarrhoeal Disease Among Children Under-Five Years

Several studies in developing countries examined the association between diarrhoea incidences and climatic or meteorological factors [15, 17], socioeconomic and demographic factors. For instance, Wang et al. [18] applied Artificial neural networks (ANN) to predict diarrhoea using meteorological factors. Abubakar and Olatunji [10] also applied ANN to predict diarrhoea using demographic and health survey data. Research done in Ghana showed that household wealth status, coupled with residence, correlates with the incidence of childhood diarrhoea [19]. Another study done in Ethiopia showed that sanitation facilities and hygienic conditions had been associated with diarrhoeal disease [12, 20, 21]. Santika et al. [22] considered social factors, child care practice, and environmental factors determinants of diarrhoea in Indonesia. A study done in Ethiopia in 2020 showed that risk factors associated with diarrhoea in children under the age of five include birth order, rotavirus vaccination, starting complementary feeding and feeding children using hands [23]. A cross-sectional study conducted in Senegal considered sociodemographic (age, educational level, number of household members, wealth status, drinking water sources, water storage, toilet availability, type of toilet facilities, toilet shared with others households), environmental and hygiene behavioural factors (duration of water storage, treatment water stored) [24]. A study [1] highlighted the influence of such socioeconomic and demographic factors on the outbreak of diarrhoea in Zimbabwe.

2.3 Data Pre-processing

In this study, we utilised socioeconomic and demographic factors such as age, sex of the child, age of the mother, wealth status, household size, place of residence, source of drinking water, type of toilet facility, vitamin A given in the last six months, maternal education and diarrhoea reported recently (target variable). We split up the dataset into training and testing using the ratio of 70% and 30%, respectively, as shown in Fig. 1.

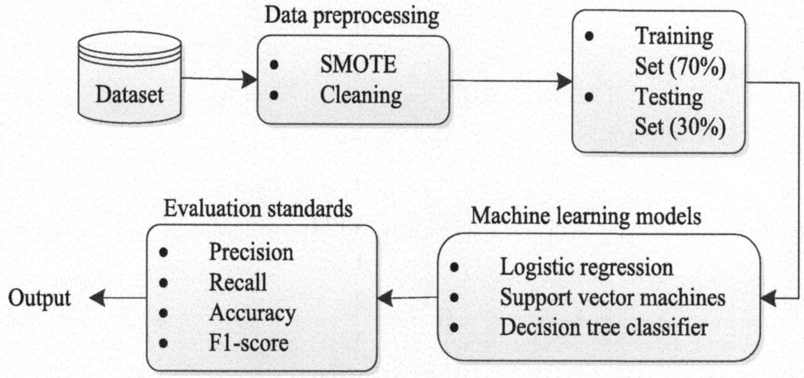

Fig. 1. Machine learning-based diarrhoea predictive models

Independent variables include child characteristics (age of the child- B19, sex of the child- B4, vitamin A supplemented- H34), mother's characteristics (mother's age – V012, maternal education- V106), household characteristics (household wealth- V190, family/household size- V136, source of drinking water – V113, toilet facility – V116), place of residence – V102 (urban and rural residence). These independent variables and dependent variables (presence of diarrhoea-H11) were coded as shown in Table 1. Data coding is paramount to ensure that all risk factors are in the acceptable format for the predictive models.

However, the synthetic minority oversampling technique (SMOTE) was applied to address the class imbalance problem. The heatmap plot for Pearson's correlation coefficient between relevant independent variables is shown in Fig. 2.

Table 1. Risk factors data coding

Risk factor	Coding description
Age of child in months	B19
50–11	1
12–23	2
24–35	3
36–47	4
48–59	5
Sex of child	B4
Male	1
Female	0
Age of Mother	V012
15–24	1
25–34	2
35–49	3
Wealth status	V190
Poor	1
Medium	2
Rich	3
Maternal education	V106
None	1
Primary	2
Secondary or higher	3
Household size	V136
5 or less	1
6 and above	2
Place of residence	V102
Urban	1
Rural	0
Source of drinking water	V113
Not improved	0
Improved/protected	1
Type of toilet facility	V116
No facility/other	0

(*continued*)

Table 1. (*continued*)

Risk factor	Coding description
Pit or flash toilet	1
Vitamin A supplemented	H34
No	0
Yes	1
Presence of diarrhoea	**H11**
No	0
Yes	1

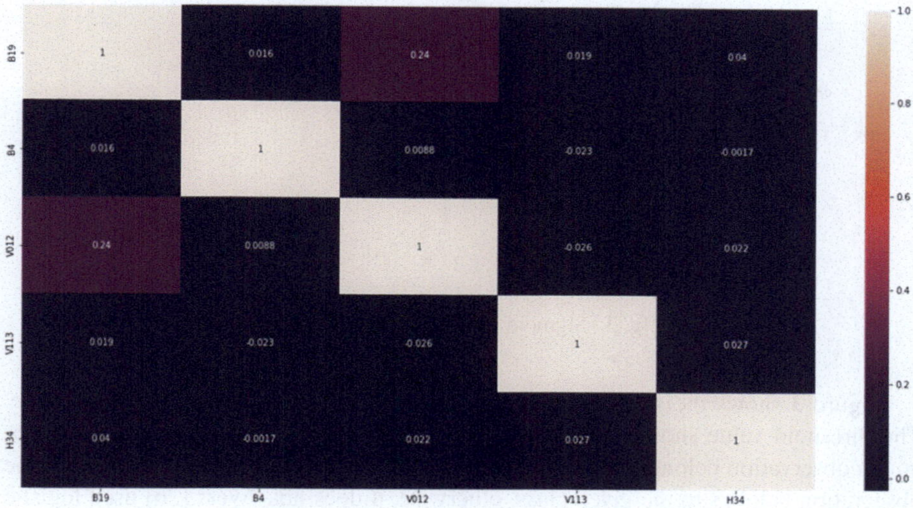

Fig. 2. Heatmap plot for Pearson's correlation

2.4 Machine Learning-Based Diarrhoea Prediction Models

The study applied decision tree classifier, logistic regression, and support vector machines to predict diarrhoea among children under five years old.

A decision tree classifier is a simple but widely used classification algorithm [25]. It solves a classification problem by asking a series of questions until a class is found. It uses a tree data structure with three types of nodes: root, internal and terminal/leaf nodes [25]. Each terminal/leaf node represents a class or label, while the root and internal nodes contain attribute test conditions to separate records with different characteristics. Once a decision tree is constructed, classifying an item becomes more accessible and straightforward. [26] used a decision tree classifier to delineate viral causes of diarrhoea from bacterial or parasitic ones, while [27] used the decision tree algorithm to derive dehydration scoring for children with diarrhoea empirically.

Logistic regression is a supervised learning machine learning algorithm used for classification problems [28]. Given multiple observations, this algorithm is used to assign them to a discrete set of categories or classes [29]. It uses the sigmoid function to return the probability of an observation falling into a specific class; thus, the range of output values of the sigmoid activation function is squashed between 0 and 1. The sigmoid activation function's formula is shown in Eq. 1.

$$f(x) = \frac{1}{1 + e^{-x}} \tag{1}$$

where x is the weighted sum of input for a neuron.

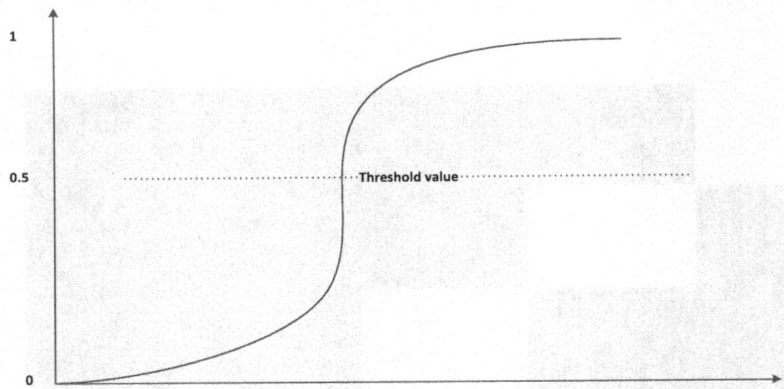

Fig. 3. Sigmoid activation function curve

Figure 3 shows the sigmoid activation function's curve with the threshold value. The threshold value shows a boundary value upon which the decision of whether or not an observation belongs to the class is based. If the output value is 0.5 or above, the observation belongs to the given class, otherwise, it does not. Ayers [26] used logistic regression to assess diarrhoea severity in rural Kenyan under-five children while [30] used logistic regression for determining diarrhoea high-risk factors.

Support Vector Machines (SVM) is a supervised learning machine learning algorithm used for both classification and regression problems, albeit commonly used for classification. Support vectors are coordinates of observations on a Cartesian plane.

Figure 4 shows a diagrammatic representation of the concept of the SVM. Given observations represented on a Cartesian plane, a hyperplane can be drawn to classify the observations or categories. Kurisu et al. [9] used three ML algorithms including SVM to determine the relationship between risk factors and the onset of nosocomial diarrhoea while [30] applied the SVM algorithm to predict the diarrhoea status of children in Bangladesh.

3 Discussion of Results

For univariate analysis, the percentage distribution of sample characteristics was presented. A Chi-square test was used to assess the association between the diarrhoeal

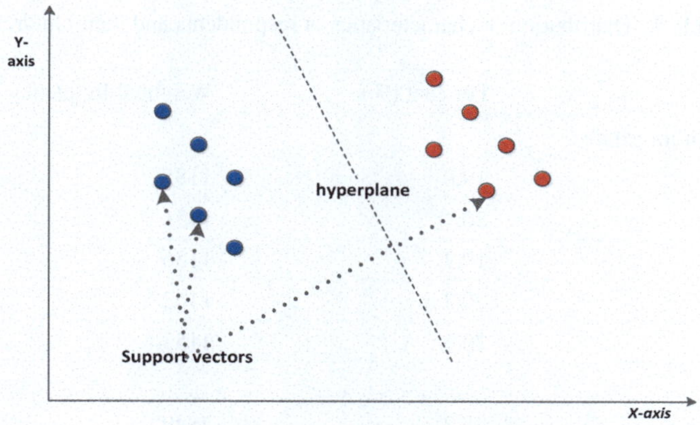

Fig. 4. Support vector machines

presence and each of the selected independent variables. Bivariate and multivariable logistic regression analyses were used to investigate associations between diarrhoeal presence and independent variables. Crude odd ratios (COR) were computed for the individual variables separately. Adjusted odd ratios (AOR) and 95% confidence intervals (CI) were run for the final model with all variables. Statistical significance was considered at $p < 0.05$. Multicollinearity of the independent variables was tested using the variance inflation factor (VIF), in which all variables were less than a value of 10. All analyses excluded missing responses. Stata 15 was used for analysis taking survey sampling weights into account. Table 1 presents the characteristics of the sample, demographic and household information of the child and mother, and geographic characteristics. Except for Vitamin A supplementation, child characteristics of age and sex were similar. Most women aged 25–34 years, had secondary or higher education, were poor and resided in rural. The majority of households were 5 or less in family size and had improved sources of drinking water and an improved toilet facility. Table 2 shows the descriptive statistics and the determinates of diarrhoeal infections among children under the age of five.

Table 3 shows the percentage of women reporting diarrhoea among under-five children and unadjusted odds for diarrhoea. Noteworthy differences in diarrhoeal presence were observed for the child's age, sex of the child, mother's age, source of drinking and vitamin A supplemented. After adjusting for all other independent variables through logistic regression, age of the child, sex of the child, age of mother, source of drinking water and vitamin A given prior 6 months, the survey remained statistically significantly correlated with the presence of diarrhoea among children under the age of 5 years. All other variables also remained not associated with diarrhoeal presence. As expected, adjusting for other variables, the odds of diarrhoeal presence were highest among children aged 12–23 months (AOR = 1.63, CI: 1.3–2.04) and lowest among those aged 48–59 months (AOR = 0.30, CI: 0.22–0.43) when compared to those age 0–11 months. Compared to male children, females had lower rates of getting diarrhoea (AOR = 0.84, CI: 0.71–1.00). Older women (age 35–49) were less likely to have had a child getting diarrhoea when compared to younger women aged 15–24 (AOR = 0.64, CI: 0.49–0.85).

Table 2. Distribution of characteristics of respondents and their children

	Per cent (%)	Weighted frequency
Age of child in months		
50–11	19.8	1185
12–23	20.1	1151
24–35	19.7	1156
36–47	20.2	1162
48–59	20.3	1153
Sex of child		
Male	48.7	2849
Female	51.3	2958
Age of Mother		
15–24	29.7	1800
25–34	50.8	2893
35–49	19.5	1114
Wealth status		
Poor	42.3	2183
Medium	16.8	893
Rich	41	2731
Maternal education		
None	1.1	56
Primary	31.1	1679
Secondary or higher	67.7	4072
Household size		
5 or less	57.8	3449
6 and above	42.2	2358
Place of residence		
Urban	32	2216
Rural	68	3591
Source of drinking water		
Not improved	27.8	1374
Improved/protected	72.2	4433
Type of toilet facility		
No facility/other	30.6	1661

(*continued*)

Table 2. (*continued*)

	Per cent (%)	Weighted frequency
Pit or flash toilet	69.4	4146
Vitamin A supplemented		
No	39.1	2255
Yes	60.9	3552
Presence of diarrhoea		
No	83.3	4876
Yes	16.7	931
Total	100	5807

Compared to the unimproved source of water, improved or protected water was more likely to reduce the presence of diarrhoea among children under five (AOR = 0.80, CI: 0.66–0.97). Finally, contrary to expectation, relative to Vitamin A deficiency, Vitamin A supplementation was more likely to increase by 1.37 times the chance of getting diarrhoea among the children, net of all other variables.

The study revealed that 17% of children under five years had a diarrhoeal episode 2 weeks prior to the survey. The frequency of diarrhoea infections increases from 19.9% of babies under 11 months to a peak of 30.4% among toddlers aged between 12–23 months. This showed that the risk of diarrhoea increases when children start to walk, which might be due to the increased risk of contamination from the environment. The increase might also be attributed to the introduction of complementary foods, facilitating the spread of disease-causing microbes. Diarrhoea is slightly more prevalent among children from households that lacked improved drinking-water sources (18.9%) than children from households with access to improved drinking-water sources (15.9%).

Similarly, the prevalence of diarrhoea is higher among children whose households do not have toilet facilities (17.4%) than households with toilet facilities (16.5%). The prevalence of diarrhoea varies by province; it is highest in Mashonaland West (23%) and lowest in Matabeleland South (9%). Also, the study revealed that diarrhoea is slightly increased (with 23.6%) among children with mothers who do not have formal education compared to children with mothers who do have primary and secondary education.

To evaluate the performance of decision tree classifier, logistic regression and support vector machines to predict diarrhoea among children under five years, the study used performance evaluation standards, namely accuracy, precision, recall and F1-score. These metrics are calculated using the confusion matrix, which consists of True Positive (TP), False Positive (FP), True Negative (TN) and False Negative (FN) [31]. True Positive refers to the actual diarrhoea incidences that were correctly classified as diarrhoea incidences [32]. False positive refers to the non-diarrhoea incidences that were incorrectly classified as diarrhoea incidences. False negative refers to the diarrhoea incidences that were incorrectly classified as non-diarrhoea incidences. True negative means non-diarrhoea incidences that were correctly classified as non-diarrhoea incidences.

Table 3. Regression analysis of diarrhoea among under-five children, ZDHS 2015–16

	Diarrhoea					
	No	Yes	Total	P-value	Crude OR (95% CI)	Adjusted OR (95% CI)
Age of child in months						
0–11	80.2	19.9	1185	0.000	1	1
12–23	69.6	30.4	1151		1.76 (1.42–2.18)***	1.63 (1.3–2.04)***
24–35	82.7	17.3	1156		0.84 (0.65–1.1)	0.8 (0.61–1.06)
36–47	90.4	9.6	1162		0.43 (0.33–0.57)***	0.43 (0.33–0.57)***
48–59	93.2	6.8	1153		0.29 (0.21–0.41)***	0.3 (0.22–0.43)***
Sex of child						
Male	81.8	18.2	2849	0.016	1	1
Female	84.6	15.4	2958		0.82 (0.7–0.96)*	0.84 (0.71–1)*
Age of mother						
15–24	79.4	20.6	1800	0.000	1	1
25–34	83.6	16.4	2893		0.76 (0.63–0.9)**	0.87 (0.72–1.04)
35–49	88.2	11.8	1114		0.52 (0.4–0.67)***	0.64 (0.49–0.85)**
Wealth status						
Poor	82.5	17.5	2183	0.573	1	1
Medium	83.7	16.3	893		0.92 (0.74–1.16)	0.98 (0.76–1.27)
Rich	83.9	16.2	2731		0.91 (0.75–1.11)	0.96 (0.7–1.32)
Maternal education						
None	76.4	23.6	56	0.223	1	1
Primary	82.3	17.7	1679		0.69 (0.35–1.39)	0.68 (0.3–1.52)
Secondary or higher	83.8	16.2	4072		0.63 (0.31–1.25)	0.59 (0.26–1.35)
Household size						
5 or less	83.2	16.8	3449	0.837	1	1
6 and above	83.4	16.6	2358		0.98 (0.83–1.16)	1.03 (0.86–1.24)
Place of residence						
Urban	83.4	16.6	2216	0.860	1	1
Rural	83.2	16.8	3591		1.02 (0.82–1.26)	0.83 (0.59–1.16)

(*continued*)

Table 3. (*continued*)

	Diarrhoea					
	No	Yes	Total	P-value	Crude OR (95% CI)	Adjusted OR (95% CI)
Source of drinking water						
Not improved	81.2	18.9	1374	0.017	1	1
Improved/protected	84.1	15.9	4433		0.82 (0.69–0.96)*	0.8 (0.66–0.97)*
Type of toilet facility						
No facility/other	82.6	17.4	1661	0.420	1	1
Pit or flash toilet	83.6	16.5	4146		0.94 (0.79–1.1)	1.01 (0.84–1.22)
Vitamin A supplemented						
No	86.6	13.5	2255	0.000	1	1
Yes	81.2	18.9	3552		1.5 (1.25–1.78)***	1.37 (1.13–1.66)***

Accuracy is the general percentage of samples that were classified correctly [33]. It is calculated as follows:

$$Accuracy = \frac{TP + TN}{TP + TN + FP + FN} \tag{2}$$

Recall which is also known as true positive rate [34] and it is calculated as follows:

$$Recall = \frac{TP}{TP + FN} \tag{3}$$

Precision is the number of accurate positive predictions as a percentage of the total number of positive predictions is known as precision [35]:

$$Recall = \frac{TP}{TP + FN} \tag{4}$$

F1-score represents the average of recall and precision [36]. It serves as a red flag of incorrectly classified performance [37]. The higher values of the F1 score specify the excellent performance of the model [38].

$$F1 - score = 2 * \frac{Recall * Precision}{Reacll + Precision} \tag{5}$$

The results in Table 4 show that machine learning techniques play a significant role in predicting diarrhoeal disease among children under-five years by using risk factors. Logistic regression out-performed other diarrhoea predictive models with the prediction accuracy of 85%, precision of 86%, recall of 100% and the F1-score of 94%. Support vector machines also performed better in predicting diarrhoea with predicting accuracy of 84%, precision of 85%, recall of 100% and F1-score of 92%.

Table 4. Diarrhoea prediction models evaluation metrics

Machine learning model	Accuracy	Precision	Recall	F1-score
Logistic regression	0.85	0.86	1.00	0.94
Decision trees classifier	0.74	0.86	0.83	0.85
SVM	0.84	0.85	1.00	0.92

4 Conclusion

Diarrhoea among children under five years remains a significant public health problem in many developing countries in the sub-Saharan region, including Zimbabwe. Tackling diarrhoea among under-fives requires a thorough understanding of the risk factors, including climatic or meteorological, socioeconomic and demographic factors. These factors contribute significantly and sometimes have seasonal variations that usually influence trends and patterns of diarrhoeal disease among under-fives. Among other factors maternal education, lack of toilet facilities [9], lack of handwashing facilities, lack of separate feeding materials for children [39], not breastfeeding exclusively and handwashing practice, wealth status, source of drinking water [1] and residential place found significantly associated with the diarrhoeal disease among under-fives. This study applied a decision tree classifier, logistic regression and support vector machines to predict diarrhoea among children under-five years. The study shows that logistic regression out-performed well in predicting diarrhoea among under-fives, with prediction accuracy of 85%, precision of 86%, recall of 100% and the F1-score of 94%. The application of machine learning techniques can improve forecasting of diarrhoea incidences. Understanding the risk factors can assist policymakers in designing diarrhoea preventive, control programmes and strategies for tackling diarrhoea disease among under-fives. Therefore, we recommend devising proactive strategies that incorporate computational models to forecast diarrhoea incidences among under-fives, provide clean drinking water, and improve handwashing facilities and practices. We also recommend integrating feasible digital interventions to improve healthcare services delivery, especially to under-fives. Possible digital interventions are required to inform policies and strategies to reduce diarrhoea incidences, morbidity, and under-five mortality. Therefore, using the same ZDHS 2015–2016 dataset, the future work will focus on determining the feasible digital technologies/interventions for effective monitoring, improving health promotion and enhance communication between healthcare professionals and the vulnerable communities to reduce under-five mortality.

References

1. Musuka, G., et al.: Associations of diarrhea episodes and seeking medical treatment among children under five years: insights from the Zimbabwe Demographic Health Survey (2015–2016). Food Sci. Nutr. **9**, 6335–42 (2021). https://doi.org/10.1002/FSN3.2596
2. UNICEF: Diarrhoea - UNICEF data. UNICEF off page 2017. https://data.unicef.org/topic/child-health/diarrhoeal-disease/ Accessed 3 Dec 2021

3. Mosisa, D., Aboma, M., Girma, T., Shibru, A.: Determinants of diarrheal diseases among under five children in Jimma Geneti District, Oromia region, Ethiopia, 2020: a case-control study. BMC Pediatr. **21**, 532 (2021). https://doi.org/10.1186/s12887-021-03022-2
4. Jiwok, J.C., Adebowale, A.S., Wilson, I., Kancherla, V., Umeokonkwo, C.D.: Patterns of diarrhoeal disease among under-five children in Plateau State, Nigeria, 2013–2017. BMC Public Health **21**, 1–9 (2021). https://doi.org/10.1186/s12889-021-12110-y
5. Tian, Y., et al.: Prevalence and genotypes of group a rotavirus among outpatient children under five years old with diarrhea in Beijing, China, 2011–2016. BMC Infect. Dis. **18**, 1–11 (2018). https://doi.org/10.1186/S12879-018-3411-3/FIGURES/6
6. Ali, M., Abbas, F., Shah, A.A.: Factors associated with prevalence of diarrhea among children under five years of age in Pakistan. Child. Youth Serv. Rev. **132**, 106303 (2022). https://doi.org/10.1016/j.childyouth.2021.106303
7. Hart, C.A., Umar, L.W.: Diarrhoeal disease. Trop. Doct. **30**, 170–2 (2000). https://doi.org/10.1177/004947550003000321
8. United Nations: The 17 Sustainable Development Development Goals (2015)
9. Kurisu, K., Yoshiuchi, K., Ogino, K., Oda, T.: Machine learning analysis to identify the association between risk factors and onset of nosocomial diarrhea: a retrospective cohort study. PeerJ **7**, e7969 (2019). https://doi.org/10.7717/peerj.7969
10. Abubakar, I.R., Olatunji, S.O.: Computational intelligence-based model for diarrhea prediction using demographic and health survey data. Soft. Comput. **24**, 5357–66 (2020). https://doi.org/10.1007/S00500-019-04293-9/TABLES/4
11. Alebel, A., Tesema, C., Temesgen, B., Gebrie, A., Petrucka, P., Kibret, G.D.: Prevalence and determinants of diarrhea among under-five children in Ethiopia: a systematic review and meta-analysis. PLoS ONE **13**, e0199684 (2018). https://doi.org/10.1371/JOURNAL.PONE.0199684
12. Solomon, E.T., Robele, S., Kloos, H., Mengistie, B.: Effect of household water treatment with chlorine on diarrhea among children under the age of five years in rural areas of Dire Dawa, eastern Ethiopia: a cluster randomized controlled trial. Infect. Dis. Poverty **9**, 1–13 (2020). https://doi.org/10.1186/S40249-020-00680-9/TABLES/4
13. Dagnew, A.B., et al.: Prevalence of diarrhea and associated factors among under-five children in Bahir Dar city, Northwest Ethiopia, 2016: a cross-sectional study. BMC Infect. Dis. **19**, 1–7 (2019). https://doi.org/10.1186/S12879-019-4030-3/TABLES/5
14. Anteneh, Z.A., Andargie, K., Tarekegn, M.: Prevalence and determinants of acute diarrhea among children younger than five years old in Jabithennan District, Northwest Ethiopia, 2014. BMC Public Health **17**, 1–8 (2017). https://doi.org/10.1186/S12889-017-4021-5/TABLES/4
15. Yang, X., Xiong, W., Huang, T., He, J.: Meteorological and social conditions contribute to infectious diarrhea in China. Sci. Reports **11**, 1–13 (2021). https://doi.org/10.1038/s41598-021-00932-0
16. Aheto, J.M.K.: Predictive model and determinants of under-five child mortality: evidence from the 2014 Ghana demographic and health survey. BMC Public Health **19**, 1–10 (2019). https://doi.org/10.1186/S12889-019-6390-4/FIGURES/2
17. Fang, X., et al.: Forecasting incidence of infectious diarrhea using random forest in Jiangsu Province, China. BMC Infect. Dis. **20**, 1–8 (2020). https://doi.org/10.1186/S12879-020-4930-2/FIGURES/3
18. Wang, Y., Li, J., Gu, J., Zhou, Z., Wang, Z.: Artificial neural networks for infectious diarrhea prediction using meteorological factors in Shanghai (China). Appl. Soft Comput. **35**, 280–90 (2015). https://doi.org/10.1016/J.ASOC.2015.05.047
19. Kumi-Kyereme, A., Amo-Adjei, J.: Household wealth, residential status and the incidence of diarrhoea among children under-five years in Ghana. J. Epidemiol. Glob. Health **6**, 131–40 (2016). https://doi.org/10.1016/J.JEGH.2015.05.001

20. Adane, M., Mengistie, B., Kloos, H., Medhin, G., Mulat, W.: Sanitation facilities, hygienic conditions, and prevalence of acute diarrhea among under-five children in slums of Addis Ababa, Ethiopia: baseline survey of a longitudinal study. PLoS ONE **12**, e0182783 (2017). https://doi.org/10.1371/JOURNAL.PONE.0182783
21. Adane, M., Mengistie, B., Mulat, W., Medhin, G., Kloos, H.: The most important recommended times of hand washing with soap and water in preventing the occurrence of acute diarrhea among children under five years of age in slums of Addis Ababa, Ethiopia. J. Community Health **43**, 400–5 (2018). https://doi.org/10.1007/S10900-017-0437-1/TABLES/3
22. Santika, N.K.A., Efendi, F., Rachmawati, P.D., Has, E.M.M., Kusnanto, K., Astutik, E.: Determinants of diarrhea among children under two years old in Indonesia. Child. Youth Serv. Rev. **111**, 104838 (2020). https://doi.org/10.1016/j.childyouth.2020.104838
23. Shine, S., Muhamud, S., Adanew, S., Demelash, A., Abate, M.: Prevalence and associated factors of diarrhea among under-five children in Debre Berhan town, Ethiopia 2018: a cross sectional study. BMC Infect. Dis. **20**, 1–6 (2020). https://doi.org/10.1186/S12879-020-4905-3/TABLES/4
24. Thiam, S., et al.: Prevalence of diarrhoea and risk factors among children under five years old in Mbour, Senegal: a cross-sectional study. Infect. Dis. Poverty **6**, 1–12 (2017). https://doi.org/10.1186/S40249-017-0323-1/TABLES/3
25. Tan, P.-N., Steinbach, M., Karpatne, A., Kumar, V.: Introduction to Data Mining. 2nd edn (2019)
26. Ayers, T.L.: Machine Learning Approaches for Assessing Moderate-to-Severe Diarrhea in Children < 5 Years of Age, Rural Western Kenya 2008–2012. Georgia State University (2016)
27. Levine, A.C., et al.: Empirically derived dehydration scoring and decision tree models for children with Diarrhea: assessment and internal validation in a prospective cohort study in Dhaka, Bangladesh. Glob. Heal. Sci. Pract. **3**(3), 405–418 (2015). https://doi.org/10.9745/GHSP-D-15-00097
28. Carnegie Mellon University: Introduction to Machine Learning. Carnegie Mellon University Website (2020)
29. Levy, J.J., O'Malley, A.J.: Don't dismiss logistic regression: the case for sensible extraction of interactions in the era of machine learning. BMC Med. Res. Methodol. **20**, 171 (2020). https://doi.org/10.1186/s12874-020-01046-3
30. Maniruzzaman, Shaykhul, I., Menhazul, A., Amanullah, Sadiq, H.: Prediction of childhood Diarrhea in Bangladesh using machine learning approach. Insights Biomed. Res. **4** (2020). https://doi.org/10.36959/584/456
31. Mbunge, E., Fashoto, S.G., Bimha, H.: Prediction of box-office success: a review of trends and machine learning computational models. Int. J. Bus. Intell. Data Min. **20**, 192–207 (2022). https://doi.org/10.1504/IJBIDM.2022.120825
32. Akinnuwesi, B.A., et al.: Application of intelligence-based computational techniques for classification and early differential diagnosis of COVID-19 disease. Data Sci. Manag. **4**, 10–8 (2021). https://doi.org/10.1016/J.DSM.2021.12.001
33. Mbunge, E., Muchemwa, B.: Deep Learning and Machine Learning Techniques for Analyzing Travelers' Online Reviews: A Review, 1AD, pp. 20–39. https://doi.org/10.4018/978-1-7998-8306-7.CH002
34. Mbunge, E., Fashoto, S., Mafumbate, R., Nxumalo, S.: Diverging Hybrid and Deep Learning Models into Predicting Students' Performance in Smart Learning Environments – A Review, pp. 182–202 (2021). https://doi.org/10.1007/978-3-030-93314-2_12
35. Buraimoh, E., Ajoodha, R., Padayachee, K.: Application of machine learning techniques to the prediction of student success. In: 2021 IEEE Int. IOT, Electron Mechatronics Conf. IEMTRONICS 2021 – Proc. (2021). https://doi.org/10.1109/IEMTRONICS52119.2021.9422545

36. Buraimoh, E., Ajoodha, R., Padayachee, K.: Importance of data re-sampling and dimensionality reduction in predicting students' success. In: 3rd Int. Conf. Electr. Commun. Comput. Eng. ICECCE 2021 (2021). https://doi.org/10.1109/ICECCE52056.2021.9514123
37. Govender, P., et al.: The application of machine learning to predict genetic relatedness using human mtDNA hypervariable region I sequences. PLoS ONE **17**, e0263790 (2022). https://doi.org/10.1371/JOURNAL.PONE.0263790
38. Elujide, I., Fashoto, S.G., Fashoto, B., Mbunge, E., Folorunso, S.O., Olamijuwon, J.O.: Application of deep and machine learning techniques for multi-label classification performance on psychotic disorder diseases. Inform. Med. Unlocked **23**, 100545 (2021). https://doi.org/10.1016/J.IMU.2021.100545
39. Oloruntoba, E.O., Folarin, T.B., Ayede, A.I.: Hygiene and sanitation risk factors of diarrhoeal disease among under-five children in Ibadan, Nigeria. Afr. Health Sci. **14**, 1001 (2014). https://doi.org/10.4314/AHS.V14I4.32

Restriction Set Design for Travel Demand Values in an Urban Road Network

Anastasiya Raevskaya(✉), Alexander Krylatov, Petr Ageev, and Darya Kuznetsova

Saint-Petersburg State University, Universitetskaya emb. 7-9, 199034 Saint-Petersburg, Russia
{a.raevskaya,a.krylatov}@spbu.ru

Abstract. Efficient traffic management in a modern urban road network seems impossible today without the support of artificial intelligence systems that use accurate travel demand data for predicting traffic congestions. However, despite researchers being equipped with different approaches and techniques to cope with travel demand estimation, there is still a gap between up-to-date accuracy requirements and available methods. The present paper is devoted to this urgent problem and investigates evolutionary strategies for the travel demand search task, formulated as an inverse traffic assignment problem. We develop polynomial regression models to estimate overall demand by observed congestions. The overall demand value allows one to restrict the set of feasible travel demand matrices. Eventually, we offer the travel demand estimation problem minimizing the deviation of both congestion and time on the simplex.

Keywords: Bilevel programming · Evolutionary optimization · Regression

1 Introduction

Efficient traffic management in a modern urban road network seems impossible today without the support of artificial intelligence systems that use accurate travel demand data for predicting traffic congestions. Researchers deal with travel demand estimation (or origin-destination (OD) matrix estimation) by virtue of different approaches and techniques. From the mathematical point of view the simplest approach may refer to the gravity methods, which have been well described in context of spatial regional analysis [7]. Another large class of approaches can be associated with *a priori* economic theorizing (see, for instance, [5]). Such basic concepts have been developed and implemented in various ways by scientists from various fields. Moreover, development of technical

A. Raevskaya—The work was supported by a grant from the Russian Science Foundation (No. 20-71-00062 Development of artificial intelligence tools for estimation travel demand values between intersections in urban road networks in order to support operation of intelligent transportation systems).

R. Silhavy (Ed.): CSOC 2022, LNNS 502, pp. 110–120, 2022.
https://doi.org/10.1007/978-3-031-09076-9_10

facilities and devices, such as traffic counters or plate scanning sensors has not completely solved the above-mentioned problem, but has initiated investigations of emerging problem issues. Thus, researchers have developed a technique which could possibly help to overcome the incompleteness of data gathered by plate scanners [13]. In fact, even back then, incomplete data was a routine input condition when solving the travel demand estimation problem.

First bi-level formulations of the OD-matrix estimation problem employing the user equilibrium assignment principle were made in 1980's [6,14]. In the beginning of 1990's, this approach was reinforced by clear demonstration of various alternative bi-level problems with user equilibrium assignment (in lower-level), which reflected the travel demand estimation process [19]. Since those bi-level programs which had appeared, were actually NP-hard, a single-level optimization model was based on the stochastic user equilibrium concept to estimate path flows and hence, the travel demand [2]. In turn, a method for estimating the set of routes and their flows under the static user-equilibrium assignment principle was presented [1]. Researchers offered to cope with complication of the bi-level problem by means of heuristic methodology [12]. Moreover, convex approximation of the bi-level program by single-level optimization was offered [17]. The stochastic user equilibrium principle was also employed researchers in order to avoid bi-level formulation of the travel demand estimation problem [18].

Within the present paper we concentrate on development of approaches for solving bi-level formulation of travel demand estimation problem employing the user equilibrium assignment principle. A deep understanding of the relationship between traffic assignment and travel demand estimation allows one to develop artificial intelligence tools of high performance and efficiency for traffic management support. The traffic assignment problem is formulated in a form of non-linear optimization program in Sect. 2. Bilevel travel demand estimation problem is given in Sect. 3. Up-to-date issues and challenges are discussed. Section 4 is devoted to evolutionary strategies for travel demand estimation. Polynomial regression for overall demand estimation is investigated in Sect. 5. Travel demand estimation under given overall demand value is discussed in Sect. 6. Finally, the last section is devoted to conclusions and further plans.

2 Traffic Assignment Problem

Let us consider an urban road network presented by a directed graph $G = (E, V)$, where V represents a set of intersections, while E represents a set of available roads between the adjacent intersections. Define $W \subseteq V \times V$ as the ordered set of pairs of nodes with non-zero travel demand $F^w > 0$, $w \in W$. W is usually called as the set of origin-destination pairs (OD-pairs), $|W| = m$. Any set of sequentially linked arcs initiating in the origin node of OD-pair w and terminating in the destination node of the OD-pair w we call *route* between the OD-pair w, $w \in W$. The ordered set of all possible routes between nodes of the OD-pair w we denote as R^w, $w \in W$, and the ordered set of all possible routes between all OD-pairs we denote as R, $R = \cup_{w \in W} R^w$. Demand $F^w > 0$ seeks to be assigned between the available routes R^w: $\sum_{r \in R^w} f_r^w = F^w$, where f_r^w is a variable corresponding to a traffic flow through route $r \in R^w$ between nodes of OD-pair $w \in W$. Introduce the vector of demand $F = (F^1, \dots, F^m)^{\mathrm{T}}$ and the vector f associated with the route-flow traffic assignment pattern, such that $f = f_R = (\dots, f_r^w, \dots)^{\mathrm{T}}$ is actually a vector of $\{f_r^w\}_{r \in R^w}^{w \in W}$.

Let us introduce differentiable strictly increasing functions on the set of real non-negative numbers $t_e(\cdot)$, $e \in E$. We suppose that $t_e(\cdot)$, $e \in E$, are non-negative and their first derivatives are strictly positive on the set of real non-negative numbers. By x_e we denote traffic flow on the arc e, while x is an appropriate vector of arc-flows, $x = (\dots, x_e, \dots)^{\mathrm{T}}$, $e \in E$. Defined functions $t_e(x_e)$ are used to describe travel time on arcs e, $e \in E$, and they are commonly called arc *delay*, *cost* or *performance* functions. In this section we assume that the travel time function of the route $r \in R^w$ between OD-pair $w \in W$ is the sum of travel delays on all arcs belonging to this route. Thus, we define travel time through the route $r \in R^w$ between OD-pair $w \in W$ as the following separable function

$$t_r^w(f) = \sum_{e \in E} t_e(x_e)\delta_{e,r}^w \quad \forall r \in R^w, w \in W, \tag{1}$$

where, by definition,

$$\delta_{e,r}^w = \begin{cases} 1, \text{ if arc } e \text{ belongs to the route } r \in R^w, \\ 0, \text{ otherwise.} \end{cases} \quad \forall e \in E, w \in W,$$

while, naturally,

$$x_e = \sum_{w \in W} \sum_{r \in R^w} f_r^w \delta_{e,r}^w \quad \forall e \in E, \tag{2}$$

i.e. traffic flow on the arc is the sum of traffic flows through all routes that include this arc.

The *equilibrium traffic assignment problem* under separable travel time functions has the following form of optimization program [15,16]:

$$\chi(F) = \min_x \sum_{e \in E} \int_0^{x_e} t_e(u)du, \tag{3}$$

subject to

$$\sum_{r \in R^w} f_r^w = F^w, \quad \forall w \in W, \tag{4}$$

$$f_r^w \geq 0 \quad \forall r \in R^w, w \in W, \tag{5}$$

where, by definition,

$$x_e = \sum_{w \in W} \sum_{r \in R^w} f_r^w \delta_{e,r}^w \quad \forall e \in E. \tag{6}$$

An arc-flow assignment pattern x and corresponding route-flow assignment pattern f, satisfying (3)–(6), are proved to reflect *user equilibrium* traffic assignment or such an assignment that

$$t_r^w(f) = \sum_{e \in E} t_e(x_e)\delta_{e,r}^w \begin{cases} = \mathrm{t}^w, \text{ if } f_r^w > 0, \\ \geq \mathrm{t}^w, \text{ if } f_r^w = 0, \end{cases} \quad \forall r \in R^w, w \in W, \tag{7}$$

where t^w is *equilibrium travel time* or travel time on actually used routes between OD-pair w, $w \in W$. Herewith, in case x is searched as a result of solving (3)–(6), then one deals with *arc-flow equilibrium traffic assignment problem* [9]. Otherwise, if f is searched, then *route-flow equilibrium traffic assignment problem* is under consideration [10].

3 Travel Demand Estimation Problem

Travel demand estimation and traffic assignment search are the problems that can be considered as mutually inverse. Indeed, the solution variables of the travel demand estimation problem are input data for traffic assignment search, while the solution variables of the traffic assignment problem are input data for travel demand estimation (Table 1).

Table 1. Input data and solution variables.

	Locations of OD-pairs	Travel demand values	Traffic congestions
Traffic assignment problem	Input data	Input data	Solution variables
Travel demand estimation	Solution variables	Solution variables	Input data

The first consideration of the travel demand estimation problem as the inverse traffic assignment problem was made by [4]. However, he formulated the inverse problem in very general terms where even a matrix of route choice were believed to be given. One of the main purposes within the present paper is to eliminate the excessive use of pre-given and *a priori* information when estimating the travel demand. Indeed, when using an a priori matrix in the objective function of the demand evaluation task, the travel demand estimation problem is reduced to the standard deviation search. In fact, the problem of travel demand estimation is reduced to the problem of iterative updating of the travel demand information for a given set of origin-destination pairs.

We believe the assumption on existence of *a priori* OD-matrix is too strict when considering the travel demand estimation and we formulate this problem in such a way so we could avoid the pre-given or *a priori* information. Actually, we formulate the inverse traffic assignment problem, where the equilibrium assignment pattern $\bar{x}$ is given, while locations of OD-pairs and travel demand values are to be found [8]. Therefore, the *travel demand estimation problem* has the following form of bi-level optimization program:

$$\min_{F} ||\chi(F) - \bar{x}|| \quad (8)$$

subject to

$$F^w \geq 0 \quad \forall w \in V \times V, \tag{9}$$

where mapping $\chi(F)$ is given by the optimization program (3)–(5) with definitional constraint (6).

Let us consider a transportation network presented by a directed graph with 4 nodes and 4 arcs (Fig. 1). We suppose that there is a traffic flow on each arc, which is equal to 10. A solution to the corresponding inverse traffic assignment problem does exist. Moreover, there are several solutions for the travel demand estimation problem: for example, $F^{(1,3)} = 10$ and $F^{(3,1)} = 10$ or $F^{(2,4)} = 10$ and $F^{(4,2)} = 10$ (Fig. 1).

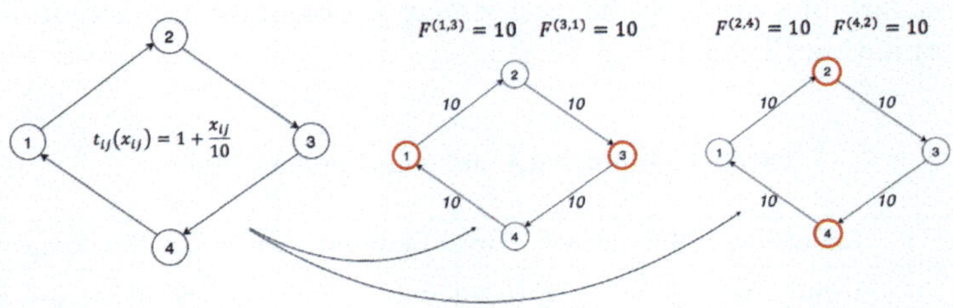

Fig. 1. Transportation network of 4 nodes

Thus, one can see that generally the travel demand estimation problem has multiple solutions. Actually, it has been already revealed that [8]:

- generally, the inverse equilibrium traffic assignment problem has no unique solutions;
- if the location of the OD-pairs is known, then the inverse equilibrium traffic assignment problem has the unique solution.

Hence, once the travel demand locations are specified, then the obstacles concerning OD-pairs search are disappeared. Otherwise, the upper-level goal function demonstrates fuzzy behavior and continuous optimization cannot cope with the problem. Let us test evolutionary strategies to cope with the travel demand estimation problem.

4 Evolutionary Strategies for Travel Demand Estimation

We tend to optimize goal function (8), i.e. $Q(F) = ||\chi(F) - \bar{x}||$, with respect to decision variables F^w, $w \in W$, where mapping $\chi(F)$ is given by the optimization program (3)–(5) with definitional constraint (6). Our computational technique is based on evolution strategies [3]:

1. Change all variables at a time, mostly slightly and at random.
2. If the new set of variables does not diminish the goodness of the device, keep it, otherwise return to the old status.

Mention that evolution strategies operate on populations $\mathfrak{P}$ of individuals $\mathfrak{a}$ such that $\mathfrak{a}_k = (F_k, s_k, Q(F_k))$, where s_k is the endogenous strategy parameter. The way we generate the offspring population is given by the $(\mu/\rho \overset{+}{,} \lambda)$-ES algorithm.

Algorithm 1. $(\mu/\rho \overset{+}{,} \lambda)$-ES

$i \leftarrow 0$

$\mathfrak{P}_p^{(0)} \leftarrow \left\{ \left(F_m^{(0)}, s_m^{(0)}, Q(F_m^{(0)}) \right),\ m = 1, \dots, \mu \right\}$

repeat

  for $j = 1, \dots, \lambda$ **do**

    $\mathfrak{M}_j \leftarrow \text{marriage}\left(\mathfrak{P}_p^{(i)}, \rho \right)$

    $s_j \leftarrow s \text{ recombination}(\mathfrak{M}_j)$

    $F_j \leftarrow F \text{ recombination}(\mathfrak{M}_j)$

    $\tilde{s}_j \leftarrow s \text{ mutation}(s_j)$

    $\tilde{F}_j \leftarrow F \text{ mutation}(F_j, \tilde{s}_j)$

    $\tilde{Q}_j \leftarrow Q(\tilde{F}_j)$

  end for

  $\mathfrak{P}_o^{(i)} \leftarrow \left\{ \left(\tilde{F}_j^{(0)}, \tilde{s}_j^{(0)}, \tilde{Q}_j \right),\ j = 1, \dots, \lambda \right\}$

  $\mathfrak{P}_p^{(i+1)} \leftarrow \text{selection}\left(\mathfrak{P}_o^{(i)}, \mathfrak{P}_p^{(i)} \right)$

  $i \leftarrow i + 1$

until termination criterion

We apply these optimization strategies to solve a travel demand estimation problem for the Sioux-Falls test road network (input data for test road networks are available at https://github.com/bstabler/TransportationNetworks). Figure 2 demonstrates corresponding computational results.

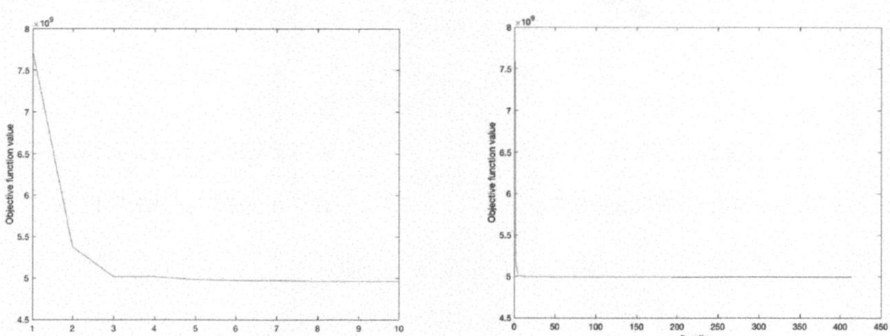

First 10 iterations   Reduction of the mutation impact

Fig. 2. Progress of fitness function (8) value

One can see the reduction of the mutation impact after the very first iterations. In other words, the algorithm converges into the pseudo-local minimum, where mutations degrade to the invariant population. Actually, since problem (8) has the only restriction (9), there are a lot of feasible pseudo-optimal solutions to the bi-level problem. Therefore, we need to design the restriction set as tiny as possible. Let us start with the restriction on the sum of travel demand values:

$$\sum_{w \in W} F^w = D \quad \forall w \in V \times V. \tag{10}$$

Generally, the value D is unknown, as well as travel demand values F^w, $w \in W$. However, fortunately, we can develop a machine learning technique based on polynomial regression, which allows us to estimate overall demand D by given congestion values.

5 Polynomial Regression for Overall Demand Estimation

Such a machine learning technique as polynomial regression can appear to be very useful for evaluation overall travel demand in an urban road network. Indeed, due to optimization program (3)–(6), we have mapping $\chi(F)$: from the set of feasible demand values to the set of equilibrium traffic assignment patterns. Hence, we are able to generate any size datasets to learn and test the regression model, which allows us to reconstruct overall travel demand by observed traffic. Firstly, we generated 400 000 pairs of the traffic assignment pattern and the overall demand value. Then, we developed the linear regression model (Fig. 3).

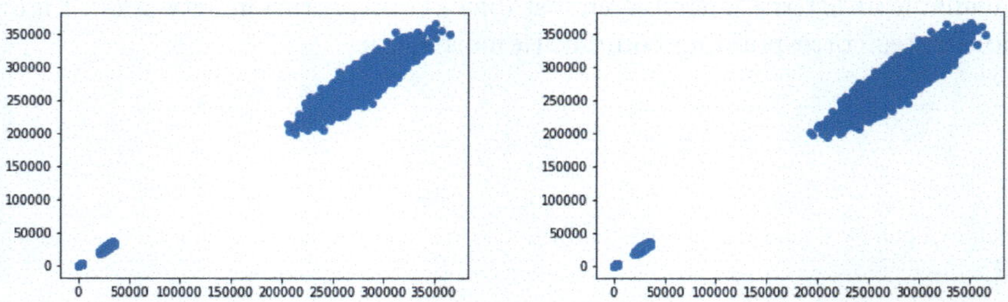

R^2: 0.9978288568823274 R^2: 0.9977746187284728

Fig. 3. Linear regression

As one can see, the model works very well: R^2 value is close to 1, i.e. the regression predictions perfectly fit the data. Moreover, the left picture demonstrates results for the model learned by 70% of data and tested by 30% of data, while the right picture demonstrates results for the model learned by only 20% of data. Therefore, the model does not require extra-large datasets to give a valuable response.

At the next step, we developed the second-degree polynomial regression model (Fig. 4).

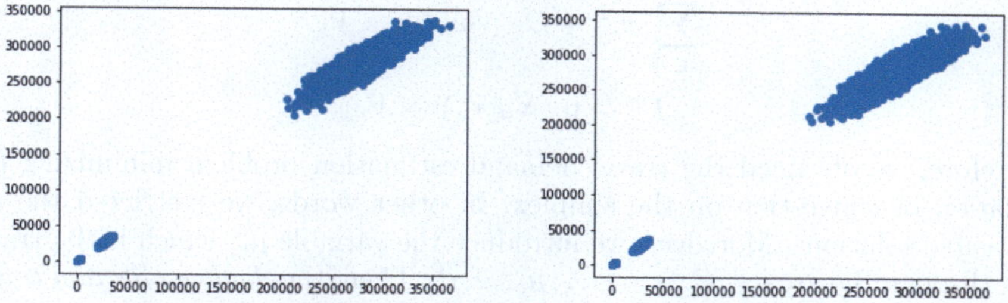

R^2: 0.9982787603645912 R^2: 0.9981649141146739

Fig. 4. Polynomial regression 2°

The model appeared to work even better than the first one. Thus, we developed the third-degree polynomial regression model (Fig. 5).

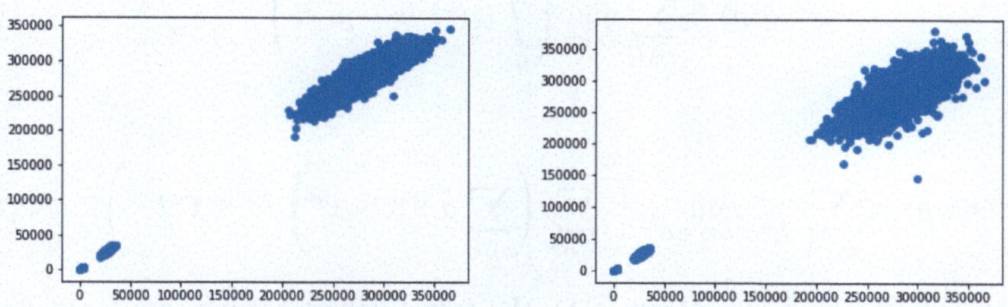

R^2: 0.9980645257730227 R^2: 0.9965214104118396

Fig. 5. Polynomial regression 3°

According to R^2 values, we conclude that the second-degree polynomial regression model demonstrates the best results. Eventually, we can summarize all obtained findings as follows:

- the second-degree polynomial regression model fits the best for the overall demand estimation by observed traffic;
- the polynomial regression model does not require large-size learn datasets to work well.

6 Travel Demand Estimation Under Given Overall Demand Value

Due to polynomial regression model for the overall demand estimation by observed traffic, the *travel demand estimation problem* takes the following form of a bi-level optimization program:

$$\min_{F} ||\chi(F) - \bar{x}||,$$

subject to

$$\sum_{w \in W} F^w = D \quad \forall w \in V \times V,$$

$$F^w \geq 0 \quad \forall w \in V \times V.$$

Therefore, we obtained the travel demand estimation problem minimizing the deviation of congestion on the simplex. In other words, we restricted the set of feasible solutions. Moreover, we introduce the variable μ_e, which is the travel time through the arc $e \in E$, $\mu = (\dots, \mu_e, \dots)^{\mathrm{T}}$. Then, the *dual equilibrium traffic assignment problem* under arc-additive travel time functions has the following form of optimization program [11,15]:

$$(\mathrm{t}^*, \mu^*) = \arg\max_{\mathrm{t},\, \mu} \left[\theta(\mu) + \Theta(\mathrm{t}, \mu)\right], \tag{11}$$

subject to

$$\theta(\mu) = \sum_{e \in E} \min_{x_e \geqslant 0} \left(\int_0^{x_e} t_e(s)\, ds - \mu_e x_e \right), \tag{12}$$

and

$$\Theta(\mathrm{t}, \mu) = \sum_{w \in W} \min_{f_r^w \geqslant 0, \forall r \in R^w} \left(\sum_{r \in R^w} \left(\sum_{e \in E} \mu_e \delta_{e,r}^w - \mathrm{t}^w \right) f_r^w + \mathrm{t}^w F^w \right). \tag{13}$$

Let us mention that in problem (11), the travel demand values are input data, while the travel time through an arc and the equilibrium travel time between an OD-pair are unknown variables. Thus, the inverse problem searches the set of OD-pairs as well as travel demand values by the given congestions $\hat{x}$ and observed travel times $\bar{\mathrm{t}}$. Indeed, we introduce mapping $\tau(F)$ such that

$$\tau(F) = \mathrm{t}^*, \tag{14}$$

where t^* is obtained from (11)–(13). Then, the *travel demand estimation problem* gets the following form of bi-level optimization program:

$$\min_{F} ||\chi(F) - \bar{x}|| + \beta\, ||\tau(F) - \bar{\mathrm{t}}||, \tag{15}$$

subject to

$$\sum_{w \in W} F^w = D \quad \forall w \in V \times V, \tag{16}$$

$$F^w \geq 0 \quad \forall w \in V \times V, \tag{17}$$

where $\beta \geq 0$ is the weight. Eventually, we obtain the travel demand estimation problem minimizing the deviation of both congestion and time on the simplex.

7 Conclusion

Despite researchers being equipped with different approaches and techniques to cope with travel demand estimation, there is still a gap between up-to-date accuracy requirements and available methods. The present paper was devoted to this urgent problem and investigated evolutionary strategies for the travel demand search task, formulated as an inverse traffic assignment problem. We developed polynomial regression models to estimate overall demand by observed congestions. The overall demand value allows one to restrict the set of feasible travel demand matrices. Eventually, we offer the travel demand estimation problem minimizing the deviation of both congestion and time on the simplex.

References

1. Bar-Gera, H.: Primal method for determining the most likely route flows in large road network. Transport. Sci. **40**(3), 269–286 (2006)
2. Bell, M.G., Shield, C.M., Busch, F., Kruse, C.: A stochastic user equilibrium path flow estimator. Transport. Res. Part C **5**(34), 197–210 (1997)
3. Beyer, H.-G., Schwefel, H.-P.: Evolution strategies. Nat. Comput. **1**, 3–52 (2002)
4. Bierlaire, M.: The total demand scale: a new measure of quality for static and dynamic origin-destination trip tables. Transp. Res. Part B **36**, 837–850 (2002)
5. Fisher, F.: A Priori Information and Time Series Analysis. North-Holland Publishing Co. (1962)
6. Fisk, C.: On combining maximum entropy trip matrix estimation with user optimal assignment. Transport. Res. Part B **22**(1), 69–73 (1988)
7. Isard, W.: Methods of Regional Analysis: An Introduction to Regional Science. Wiley (1960)
8. Krylatov, A., Raevskaya, A., Zakharov, V.: Travel demand estimation in urban road networks as inverse traffic assignment problem. Transp. Telecommun. **22**(3), 287–300 (2021)
9. Krylatov, A.: Reduction of a minimization problem for a convex separable function with linear constraints to a fixed point problem. J. Appl. Ind. Math. **12**(1), 98–111 (2018)
10. Krylatov, A.Y., Shirokolobova, A.P.: Projection approach versus gradient descent for network's flows assignment problem. In: Battiti, R., Kvasov, D.E., Sergeyev, Y.D. (eds.) LION 2017. LNCS, vol. 10556, pp. 345–350. Springer, Cham (2017). https://doi.org/10.1007/978-3-319-69404-7_29
11. Krylatov, A., Shirokolobova, A., Zakharov, V.: OD-matrix estimation based on a dual formulation of traffic assignment problem. Informatica (Slovenia) **40**(4), 393–398 (2016)
12. Lundgren, J.T., Peterson, A.: A heuristic for the bilevel origin-destination matrix estimation problem. Transp. Res. Part B **42**, 339–354 (2008)
13. Makowski, G.G., Sinha, K.C.: A statistical procedure to analyze partial license plate numbers. Transpn. Res. **10**, 131–132 (1976)
14. Nguyen, S.: Estimating an OD matrix from network data: a network equilibrium approach. Publication 60, Centre de Recherche sur les Transports, Universitét de Motréal. (1977)
15. Patriksson, M.: The Traffic Assignment Problem: Models and Methods. VSP, Utrecht (1994)

16. Sheffi, Y.: Urban Transportation Networks: Equilibrium Analysis with Mathematical Programming Methods. Prentice-Hall Inc., Englewood Cliffs (1985)
17. Shen, W., Wynter, L.: A new one-level convex optimization approach for estimating origin-destination demand. Transp. Res. Part B **46**, 1535–1555 (2012)
18. Wei, C., Asakura, Y.: A Bayesian approach to traffic estimation in stochastic user equilibrium networks. Transport. Res. Part C **36**, 446–459 (2013)
19. Yang, H., Sasaki, T., Iida, Y., Asakura, Y.: Estimation of origin-destination matrices from link traffic counts on congested networks. Transp. Res. Part B **26**(6), 417–434 (1992)

Deep Learning Techniques for Natural Language Processing

Sergey Rodzin, Victoria Bova, Yury Kravchenko, and Lada Rodzina(✉)

Southern Federal University, Taganrog, RU 344090, Russia
srodzin@sfedu.ru, lada.rodzina@gmail.com

Abstract. The article analyzes problems of deep learning (DL) approaches to natural language processing (NLP). The research covers various NLP tasks implemented with artificial neural networks (ANNs), convolutional neural networks (CNNs), and recurrent neural networks (RNNs). These architectures allow solving a wide range of natural language processing problems that previously did not lend themselves to effective solutions: sentence modeling, semantic role marking, named entity recognition, answering questions, text categorization, machine translation. Along with the advantages of using CNN for solving NLP problems, there are problems associated with many variable network parameters and the choice of its architecture. Authors propose an evolutionary algorithm for optimizing the architectures of convolutional neural networks, which has the advantage of a small population size of networks. It turned out to be important because, due to the complexity of CNN, it takes a long time to train one network. The experimental results showed that the proposed neuroevolutionary approach could quickly find an optimized CNN architecture for a given dataset with acceptable accuracy.

Keywords: Deep learning · Natural language processing · Neural networks · Convolutional neural networks · Recursive neural networks · Representation learning

1 Introduction

1.1 Background

Machine Learning (ML) is a class of artificial intelligence methods, where is no straightforward solution to a problem but learning through solutions to many similar issues. Distinguish between learning by precedents, based on identifying empirical patterns in data, and deductive learning, which involves formalizing expert knowledge and transferring it to a computer in the form of a knowledge base. Deductive learning usually belongs to the domain of expert systems. Therefore, machine learning and use case learning can be considered synonymous.

The general idea of the learning task by use cases is as follows. There are many objects and many possible answers. There is some relationship between responses and objects, but unknown. Only a finite set of precedents is known—"object/response" pairs, called "a training set". Based on these data, it is required to restore the implicit dependence, that is,

R. Silhavy (Ed.): CSOC 2022, LNNS 502, pp. 121–130, 2022.
https://doi.org/10.1007/978-3-031-09076-9_11

to build an algorithm capable of giving a sufficiently accurate classifying answer for any possible input object. This relationship is not necessarily expressed analytically, and here artificial neural networks (ANN) implement the principle of an empirically generated solution. An important feature is the ability of the learning system to generalize, that is, to adequately respond to data that goes beyond the existing training sample. Quality scoring functional is used to measure the accuracy of responses.

It is a generalization of the classical problems of approximation of functions. The objects in them are real numbers or vectors. In the tasks of processing linguistic information, the input data is textual. These features lead to a wide variety of deep learning methods: supervised, unsupervised, reinforced. Deep learning methods are based on teaching representations rather than specialized algorithms for specific tasks. These methods were known back in the 1980s. Still, the results were not impressive until advances in the theory of artificial neural networks and computational power created complex technological neural network architectures with acceptable performance. These architectures made it possible to solve a wide range of problems that previously did not lend themselves to practical solutions, for example, computer vision, machine translation, speech recognition, and natural language processing (NLP). At the same time, the quality of the solution is often comparable and, sometimes, exceeds the effectiveness of a person.

Deep Learning (DL) includes multiple layers of data processing that allow a machine to learn from data using different levels of abstraction for a specific task, without human intervention, or using previously acquired knowledge. Research into the suitability of DL approaches for NLP problems has received much attention from the ML and NLP research communities [1]. Notable results have been obtained for the following NLP problems: sentence modeling [2], semantic role labeling [3], named entity recognition [4], answering questions [5], text categorization [6], and machine translation [7].

Next, we will present an analysis of the main DL approaches and the prospects for solving NLP problems.

2 Methods

2.1 Deep Learning and Artificial Neural Network Approaches

ANNs are not programmed in the usual sense of the word; they are trained. Learning is one of the main advantages of ANN over traditional algorithms. Technically, training consists in finding the coefficients of connections between neurons. While training, ANN can identify complex dependencies between input and output data and perform generalization. It means that in case of successful training, the network will return the correct result based on data absent in the training sample and incomplete and or "noisy", partially distorted data. By the nature of the connections, ANNs are divided into two types of architectures: feedforward networks and recurrent neural networks (RNNs). In the architecture of a feedforward network, all connections are directed strictly from input neurons to output neurons. On the contrary, in RNN, the signal from the output neurons or neurons of the hidden layer is partially transmitted back to the inputs of the neurons of the input layer (feedback). For example, the Hopfield recurrent network "filters" the input data, returning to a steady-state and allowing solving data compression problems and building associative memory.

ANNs used to solve NLP problems consider the syntactic features of the text as part of semantic analysis [8]. In [9], new models for learning neural networks were proposed, which were applied to various natural language text processing tasks, such as marking semantic roles and recognizing named entities. The advantage of these approaches is that we can avoid prior training. The use of ANN made it possible to achieve high performance in marking using small computing resources [3]. In [2], the authors propose a Dynamic Convolutional Neural Network (DCNN) for solving the semantic modeling of sentences, which includes the extraction of features from sentences. DCNN is trained by selectively weighting functions between connected layers of the network. In [10], an approach was proposed for image quality control. This approach combines NLP and image representation for a textual image answering system. However, this system gave incorrect answers to questions regarding the size of objects in the image and questions related to spatial thinking ("which object is the largest?"). In the study [11], a new approach was proposed for comparing requests with possible answers (sentences) depending on their semantic meaning. This approach uses deep learning without the need for linguistic knowledge. Note that modern distribution models based on neural networks have achieved high performance in solving such NLP problems as document classification [12], disambiguation, and entity recognition [9].

Study [5] uses RNN in a system capable of answering paragraph questions. In [13], a neural machine was proposed for generating short textual dialogues. RNN models perform well in classifying different opinions [12, 14] and answering questions [5]. The adaptive RNN (ADARNN) model for categorizing other ideas is used in Twitter promotions [15].

Another approach to solving NLP problems is the ANNABEL model [7]. This model can process verbal and non-verbal natural language without prior knowledge.

2.2 Deep Learning and Convolutional Neural Network Approaches

Convolutional Neural Network (CNN) was initially developed in machine vision research. The first application of CNN is associated with image classification tasks, learning to detect the edges of an image to identify such objects as a human face [16]. The idea behind CNN is to alternate convolutional layers and downsampling layers. The structure of the network is unidirectional and fundamentally multilayer. The network architecture got its name due to the convolution operation, the essence of which is that each fragment of the image is multiplied by the convolution matrix (kernel) element by element, and the result is summed and written to a similar position in the output image. CNN's work is interpreted as a transition from concrete features of the image to more abstract details and even more abstract information, up to highlighting high-level concepts. At the same time, the network self-adjusts and develops the necessary hierarchy of abstract features (sequences of feature maps), filtering unimportant details and highlighting essential.

This interpretation is somewhat symbolic or illustrative. The "features" generated by a complex CNN are poorly understood and difficult to interpret so much that in practice, they do not even try to understand the essence of these features, let alone "correct", but instead change the structure and architecture of the network to improve the recognition results. So, ignoring the system of some significant phenomena can indicate that either there is not enough data for training, or the network structure has drawbacks, and the system cannot develop effective signs for these phenomena.

A typical CNN architecture is shown in Fig. 1.

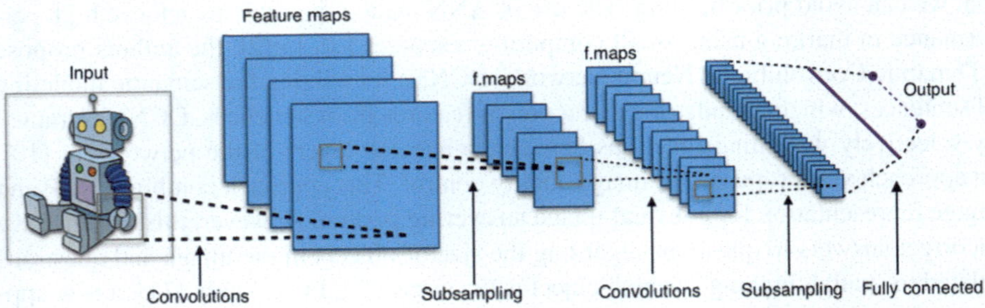

Fig. 1. Typical CNN architecture.

The network consists of many layers. After the initial layer (input image), the signal passes through a series of convolutional layers, in which the actual convolution and subsampling (pooling) alternate. Layer interleaving allows feature maps to be compiled from feature maps. The map decreases in size on each layer, but the number of channels increases. In practice, this means the ability to recognize complex feature hierarchies. Usually, after passing through several layers, the feature map degenerates into a vector or even a scalar, but there are hundreds of such features. At the CNN output, several layers of a fully connected neural network (perceptron) are additionally installed, at the input of which terminal feature maps are fed. In this network architecture, it is considered that information about the presence of the desired feature is more important than accurate knowledge of its coordinates. Therefore, from several neighboring neurons of the feature map, the maximum is selected and taken as one neuron of the condensed feature map of a lower dimension. Due to this operation, in addition to speeding up further calculations, the network becomes more invariant to the scale of the input image.

The convolution layer is the main block of CNN. The convolution layer includes its filter for each channel; the convolution kernel processes the previous layer fragment by fragment (summing the results of the element-wise product for each piece). The weights of the convolution kernel (small matrix) are unknown and are set during training. A feature of the convolutional layer is the relatively small number of parameters set during training. So, for example, if the original image has a dimension of 100 $\times$ 100 pixels in three channels (this means 30,000 input neurons), the convolutional layer uses filters with a 3 $\times$ 3-pixel kernel with an output of 6 tracks, then only 9 are determined during training. Weights of the kernel, however, for all combinations of channels, that is, 9 $\times$ 3 $\times$ 6 = 162; in this case, this layer requires finding only 162 parameters, which is significantly less than the number of necessary parameters fully connected neural network.

The scalar result of each convolution goes to the activation function, which is a kind of non-linear function. The activation layer is usually logically combined with the layer. Hyperbolic tangent ($f(x) = than(x)$), sigmoid ($f(x) = (1 + e^{-x})^{-1}$) or *ReLU* are used as the activation function, which can significantly speed up the learning process and at the same time simplify calculations. As of 2021, this function and its modifications Noisy ReLU, Leaky ReLU, and others are the most frequently used activation functions in CNN.

Thus, a key benefit of using CNNs architecture for NLP tasks is to apply a non-linear function to a sequence of words by moving the window over the sentence. This function, also called a "filter", converts the input (k-word window) into a d-dimensional vector of meaningful characteristics of the words in the window. Then the operation of combining vectors obtained from different channels is applied into one n-dimensional vector.

Currently, CNN models are used to solve such problems of NLP classification as semantic marking of roles [3], variety of question types [17], text categorization [6].

Also, note the study [18], in which the *mQA CNN*s model is used to visualize responses to questions about image content. This model consists of memory for retrieving the representation of the request, memory for processing the response context, and a response generation module that collects information from memory to respond. The answer can be in the form of a word, phrase, or sentence. In [14, 19], CNNs solve the problem of ambiguity in the semantic clustering of both short and long texts.

Thus, the advantages of using CNN for solving NLP problems are as follows:

- compared with a fully connected neural network, much fewer custom weights are required since one core of weights is used entirely for the entire image instead of looking for its weight coefficients for each pixel of the input image. It pushes the neural network during training to generalize the displayed information and not to memorize each displayed picture in a pixel-by-pixel manner in myriads of weight coefficients, as the perceptron does;
- parallelization of computations, and, consequently, the possibility of implementing algorithms for the operation and training of the network on graphic processors;
- relative stability to rotation and shift of the recognized image;
- learning using the classical backpropagation method.

The disadvantages of CNN include a large number of variable network parameters, namely: the number of layers, the dimension of the convolution kernel for each of the layers, the number of cores for each of the layers, the step of the kernel shift when processing a layer, the need for subsampling layers, the degree of their dimension reduction, the function by the decrease in dimension (choice of maximum, average, etc.), the transfer function of neurons, as well as the presence and parameters of the output fully connected neural network at the outcome of the convolutional one. All these parameters significantly affect the result but are chosen by researchers empirically. So far, there are no verified and working recommendations on which to build the optimal architecture of a convolutional neural network for a new task.

The authors propose to use the bioinspired algorithm to optimize CNN architecture.

2.3 Architecture Optimization of Convolutional Neural Networks by Evolutionary Algorithms

Let us look at a neuroevolutionary approach to solving the inherent problems of CNN.

Neuroevolution is a machine learning method that applies evolutionary algorithms to build and train neural networks [20]. In neuroevolution, the basic provisions of the theory of biological evolution are modeled: the processes of selection, mutation, and reproduction. The environment determines the behavior of agents. It is customary to call a set of agents a population. Such a population evolves by the selection rules and the target function set by the environment. Thus, each agent in the population is assigned a value for its suitability in the background. Only the most suitable species breed. Recombination and mutation allow agents to change and adapt to the environment.

In neuroevolution, a genotype is a representation of a neural network. In a direct coding scheme, the genotype is equivalent to the phenotype, neurons and connections are directly specified in the genotype. On the contrary, the rules and structures for creating a neural network are indicated in the scheme with indirect coding in the genotype. The neuroevolutionary process first initializes a set of these genomes and then applies them to the complex environment. Each genome is assigned a fitness score based on how well the neural network performs at the application. For example, this fitness indicator can be a function of the accuracy achieved in the task of processing linguistic information.

Evolutionary convolutional neural networks, in addition to learning, use evolution to optimize the CNN architecture. Neuroevolution is an agnostic process that does not provide specific topology or hyperparameters for the generated neural networks. A neural network evolves based on feedback on how effectively it works in a complex environment.

The proposed evolutionary algorithm works as follows. First, we initialize a random population of a small number of agents (no more than 5) and use the fitness function to obtain estimates for each agent in the population. Then we carry out a tournament selection between all agents and, with probability 1, apply the crossing over operator between the selected agents. The next generation retains the best agent from the previous generation (elitism strategy). Thus, there is a guarantee that information about a good CNN will not be lost.

The proposed algorithm uses several CNN layers: convolutional layer, maximum pooling (subsampling) layer, middle pooling layer, and fully connected layer. Positions in the list of layers in a chromosome contain the following information: the type of layer, the number of output maps, the number of neurons, and the size of the nucleus. The first step of the proposed algorithm is to create a population from a small number of CNNs satisfying the following constraints:

- CNN architecture must have at least three layers;
- the first layer must be convolutional;
- the output layer must be a fully connected layer with the number of neurons equal to the number of classes;
- in a convolutional layer, the number of filters (output feature maps) should not exceed a predetermined maximum number;
- the number of pooling layers must not exceed the specified limit.

The next step of the algorithm is to calculate the fitness function of each agent in the population, tournament selection, and the application of the crossover operator between

CNNs. The worst agent is replaced by the best one, and we need to check the algorithm's condition.

The proposed algorithm was tested on a local computer with an ASUS Cerberus GeForce® GTX 1050 Ti OC Edition 4 GB GDDR5 GPU, 8 GB RAM, and an Intel (R) Core (TM) i5-4670 processor. It took about 1 h to complete the execution of the algorithm. The popular TensorFlow framework was used to create and train CNN. To evaluate the algorithm, the publicly available datasets were used: MNIST [22] and MNIST-RB [23]. MNIST contains black and white handwritten letters and numbers with 50,000 training samples and 10,000 test samples. MNIST-RB is a modified version of MNIST because it tests how well CNN can summarize data containing non-problematic information. They all have 12,000 training samples and 50,000 test samples.

The evolutionary algorithm proposed for optimizing the architectures of convolutional neural networks has a clear advantage: the small population size that the algorithm uses for optimization. It is important because, due to the complexity of CNN, it takes a significant amount of time to train one network. Experimental results show that the proposed neuroevolutionary approach can quickly find an optimized CNN architecture for any given dataset. The algorithm allows you to find a CNN architecture with acceptable accuracy values.

3 Discussion and Future Research

Developing a machine learning system requires specialized knowledge in the subject area to create a classifier model and transform raw data into internal representation input data or vectors. It is called representation learning (RL), in which the model automatically feeds the raw data to determine the required representation.

In particular, representing words, phrases, sentences (statement or question) in natural language and their classification is essential for understanding the language. Deep learning approaches are like RL methods using multiple levels of symbolic processing [1].

These approaches can be combined with other DL models to improve the performance of NLP tasks. Feedforward neural networks can provide competitive performance in serial NLP problems. However, if the problem is related to the sparseness of the input data, it is more promising to solve it with the help of CNN. The CNN implementation is most used in NLP tasks such as sentence classification, text categorization, semantic clustering, and question answering. However, there is a problem with choosing the optimal convolutional neural network architecture for a new task. This problem can be solved by using the evolutionary algorithm proposed above, a distinctive feature of which is the small size of the CNNs population, which the algorithm uses for optimization.

The DL methods improve NLP tasks' performance due to modern processors. The DL methods use a multi-layer nonlinear filter system to extract features with transformations. Each subsequent layer receives the output of the previous layer as input. A deep learning system can combine supervised and unsupervised learning algorithms, with sample analysis representing unsupervised learning and classification is supervised learning. The deep learning system has several layers of revealing features or parameters of data presentation. In this case, the characteristics are organized hierarchically; the attributes of a higher level are derived from attributes of a lower level. During the learning process, layers are formed at several levels of representation, which correspond to different levels of abstraction; layers form a hierarchy of concepts.

The composition and number of specific nonlinear layers depend on the problem being solved. Schmidhuber highlights "very deep learning" when layers are more than ten.

We expect that RNN models dealing with sequences of inputs will significantly impact natural language understanding over the next few years, especially when dealing with text conversation, answering questions, and analyzing expressed opinions.

An overview of NLP tasks using DL approaches for semantic text analysis using ANN, CNN and RNN are presented in Table 1.

Table 1. Overview of NLP problems using DL-approaches for semantic text analysis.

Text comprehension tasks	ANNs	CNNs	RNNs
Classification of sentences		√	
Named entity recognition	√		
Text categorization		√	
Semantic role marking		√	
Semantic clustering		√	
Short text conversation	√		√
Answer to the question	√	√	√
Opinion analysis	√		√
Paraphrasing detection	√		
Documents classification	√		
Themes categorization		√	
Opinion prediction			√

4 Conclusion

The article analyzed deep learning-based approaches to natural language processing problems. Studies are presented covering various NLP tasks implemented by ANNs, CNNs, and RNNs. These architectures made it possible to solve many issues that previously did not lend themselves to practical solutions. For example, computer vision, machine translation, speech recognition, and natural language processing. Significant results were achieved in the following NLP tasks: sentence modeling, semantic role marking, named entity recognition, answering questions, text categorization, and machine translation.

The advantages of using CNN to solve NLP problems are as follows: a much smaller number of tunable weights in comparison with a fully connected neural network, the ability to implement algorithms for working and training the network on GPUs, the relative stability to rotation and shift of recognized objects, the ability to learn using the backpropagation method. The disadvantages of CNN include the following: many variable network parameters selected empirically by researchers. The authors propose an evolutionary algorithm for optimizing convolutional neural network architectures. The algorithm has the advantage of a small population size of networks. It turned out to be important because, due to the complexity of CNN, it takes a long time to train one network. The experimental results showed that the proposed neuroevolutionary approach could quickly find an optimized CNN architecture for a given dataset with acceptable accuracy.

The authors did not discuss the performance comparison of neural networks related to deep learning techniques used for natural language processing tasks. This should be comprehensively explored in future research.

Acknowledgements. The study was performed by the grant from the Russian Science Foundation № 22-21-00316, https://rscf.ru/project/22-21-00316/ in the Southern Federal University.

References

1. Ba, L., Caurana, R.: Do Deep Nets Really Need to be Deep? arXiv preprint arXiv:1312.6184 521(7553), 1–6 (2013)
2. Kalchbrenner, N., Grefenstette, E., Blunsom, P.: A convolutional neural network for modelling sentences. In Proc. of the 52nd Annual Meeting of the Association for Computational Linguistics, (ACL'2014), Baltimore, MD, USA, vol. 1, pp. 655–665 (2014)
3. Collobert, R., Weston, J., Bottou, M., Karlen, K., Kavukcuglu, P.: Kuksa: natural language processing from scratch. J. Mach. Learn. Res. **12**, 2493–2537 (2011)
4. Dos Santos, C.N., Guimarães, V.: Boosting named entity recognition with neural character Embeddings. ACL, (2014), 25–33 (2015)
5. Iyyer, M., Boyd-Graber, J., Claudino, L., Socher, R.: A neural network for factoid question answering over paragraphs. In: EMNLP (2014)
6. Johnson, R., Zhang, T.: Semi-supervised convolutional neural networks for text categorization via region embedding, 1–12 (2015)
7. Golosio, B., Cangelosi, A., Gamotina, O. Masala, G.L.: A cognitive neural architecture able to learn and communicate through natural language. PloS ONE, **10**(11) (2015)
8. Weston, J., America, N.E.C.L., Way, I.: A unified architecture for natural language processing: deep neural networks with multitask learning, 160–167 (2008)
9. Sun, Y., Lin, L., Tang, D., Yang, N., Ji, Z. Wang, X.: Modelling mention, context and entity with neural networks for entity disambiguation, 1333–1339 (2015)
10. Malinowski, M., Rohrbach, M., Fritz, M.: Ask your neurons: a neural based approach to answering questions about images. In: IEEE Int. Conf. on Computer Vision, pp. 1–9 (2015)
11. Yu, L., Hermann, K.M., Blunsom, P., Pulman, S.: Deep learning for answer sentence selection. In: NIPS Deep Learning Workshop, vol. 9 (2014)
12. Hermann, K.M., Blunsom, P.: Multilingual models for compositional distributional semantics. In: Proceedings of ACL (2014)
13. Shang, L., Lu, Z., Li, H.: Neural responding machine for short-text conversation. ACL 1577–1586 (2015)

14. Wang, P., et al.: Semantic clustering and convolutional neural network for short text categorization. Proceedings ACL **2015**, 352–357 (2015)
15. Dong, L., Wei, F., Tan, C., Tang, D., Zhou, M., Xu, K.: Adaptive recursive neural network for target-dependent twitter sentiment classification. ACL, 49–54 (2014)
16. LeCun Y., Bengio, Y.: Convolutional Networks for Images, Speech, and Time-Series, pp. 255–258. MIT Press (1995)
17. Kim, Y.: Convolutional neural networks for sentence classification. In: Proc. of the 2014 Conference on Empirical Methods in Natural Language Processing (EMNLP'2014), pp. 1746–1751 (2014)
18. Gao, H., Mao, J., Zhou, J., Huang, Z., Wang, L., Xu, W.: Are you talking to a machine? dataset and methods for multilingual image question answering. Arxiv, 1–10 (2015)
19. Shen, Y., He, X., Gao, J., Deng, L., Mesnil, G.: A latent semantic model with convolutional-pooling structure for information retrieval. In: Proc. of the 23rd ACM Int. Conf. on CIKM'2014, pp. 101–110 (2014)
20. Rodzin, S., Rodzina, O., Rodzina, L.: Neuroevolution: problems, algorithms, and experiments. In: Proc. of the 10th IEEE Int. Conf. on Application of Information and Communication Technologies (AICT'2016), 7991745 (2016)
21. Alshahrani, S., Kapetanios, E.: Are deep learning approaches suitable for natural language processing? In: Métais, E., Meziane, F., Saraee, M., Sugumaran, V., Vadera, S. (eds.) NLDB 2016. LNCS, vol. 9612, pp. 343–349. Springer, Cham (2016). https://doi.org/10.1007/978-3-319-41754-7_33
22. LeCun, Y., et al.: Back propagation applied to handwritten zip code recognition. Neural Comput. **1**(4), 541–551 (1989)
23. Larochelle, H., Erhan, D., Courville, A., Bergstra, J., Bengio, Y.: An empirical evaluation of deep architectures on problems with many factors of variation. ACM International Conf. Proc. Series **227**(2006), 473–480 (2007)

Forecasting Heat Production for a Large District Heating Network with NARX Neural Networks

Szabolcs Kováč(✉), Peter Střelec, Tibor Horák, German Michalčonok, and Pavel Važan

Institute of Applied Informatics, Automation and Mechatronics, Faculty of Materials Science and Technology in Trnava, Slovak University of Technology in Bratislava, Trnava, Slovakia
{szabolcs.kovac,peter.strelec,tibor.horak,german.michalconok, pavel.vazan}@stuba.sk

Abstract. Heat production planning is a challenging task that dispatchers have to solve each day because the production planning is based on their experience They are the ones who decide when to turn on/off boilers. District heating network operators must secure stable heat supply under different circumstances, of course, with regard to different economic, technical constraints. This task cannot be accomplished properly without knowing the future demand. In this research we propose a data driven approach, therefore the grid operator does not use historical data to create a forecast. The NARX (Nonlinear AutoRegressive Exogenous) neural network model was proposed to predict the future 24 h ahead heat demand. The created model can be implemented as a help tool for dispatchers.

Keywords: NARX · District heating · Prediction · Data analysis · Heat production planning

1 Introduction

We are currently witnessing notable changes in the energy sector in particular, the use of cogeneration units, renewable energy sources, the integration of energy storage units into energy infrastructure, and mainly the integration of Information and Communication Technologies (ICT) into the various stages of energy production and distribution. ICT is the key element toward the smart grid concept. Digitization at all stages of production and distribution plays an important role in production planning and optimization, which directly affects economic and environmental aspects. The consequence of ICT integration is a large amount of monitored and stored data, which opens a new research area. However, some energy producers still do not use the benefits of historical data which can be considered as a major drawback to meet the expectations of the smart grid concept. Therefore, it is essential to find new and effective approaches to model energy demand, which is a fundamental step for future development as well as for day-to-day operation.

To fulfill strict European Union emission restrictions, lowering heat price and increasing net income, it is necessary to predict. Generally, there exist two fundamental approaches to model the heat load. The first and well known heat demand modelling

R. Silhavy (Ed.): CSOC 2022, LNNS 502, pp. 131–139, 2022.
https://doi.org/10.1007/978-3-031-09076-9_12

approach is white box. In the white box approach, prior knowledge of the system is used, so the heat demand is modelled with detailed equations such as heat and mass transfer, etc. [1]. Modeling the whole district heating network (DHN), which supplies more than 200 thousand households, is very time-consuming and computationally consuming and requires an expert from the field of district heating. Some commercial solutions are available such as TRNSYS, Termis, but the annual license for these solutions is expensive especially when a large amount of data is collected each day. The second approach is based on historical data only. By applying suitable data pre-processing, data analysis techniques, it is possible to extract knowledge which represents the behaviour of the heating network, consumers, etc. By choosing an appropriate algorithm, the extracted features from the inputs with the combination of new data could easily be turned into prediction [2]. Also, the main advantage of black box (data-driven) modelling techniques is the fact that several algorithms are able to capture nonlinearity. Hence, DH is highly non-linear and past studies prove that data driven approaches outperform white box modelling.

1.1 Literature Overview

The prediction of heat demand is well studied research field. Although a relatively large number of research papers have been published, still exist no proven method which deals with the following challenges: size of the district heating grid, climate, socio-economic aspects, etc. Therefore, each research regarded as the Short-Term- Load-Forecasting (STLF) can be considered as unique. In this section, several data-driven methods will be presented which were used to predict the heat demand for small scale (buildings) up to big scale DH networks. Generally, data-driven methods could be divided into two main categories. The first well investigated category is a statistical approach. Several statistical methods for demand forecasting have been studied, such as linear regression [3, 4], autoregressive models such as the AutoRegressive Moving Average (ARMA) [5, 6], AutoRegressive Integrated Moving Average (ARIMA) [7, 8], and more complicated time series models. These models reflect linear dependencies between inputs which could be considered as a drawback hence heating demand is highly non-linear. These above mentioned lacks balance machine learning (ML) models. The number of ML models has been studied, e.g. Support Vector Machines (SVM) [9, 10], different Artificial Neural Networks (ANN) models [11, 12], fuzzy logic [13], neuro-fuzzy logic [14]. In this study, a data driven approach namely Nonlinear AutoRegressive with Exogenous (NARX) recurrent neural networks is presented to model the heat production for one day ahead (24 h). The main purpose of this research is to analyze the performance of the proposed NARX method for production planning purposes as a possible verification tool for production planning for DH dispatchers.

2 Theoretical Background

2.1 NARX Artificial Neural Network

ANNs are a group of learning models which have been inspired by the ability of the human brain to learn, memorize, and extract knowledge. Generally, the process of memorizing and mapping the input-output relationship is called learning. During learning, the synaptic weights are adjusted in terms to minimize the error function. ANNs can be divided into two main groups, namely feedforward (FFNN) and recurrent (RNN). FFNN consists of several layers of interconnected neurons (input-hidden-output), where the input signal is fed through the hidden neurons where the signal is processed and fed to the output layer neurons [15]. The main difference between FFNN and RNN architecture is that the RNNs use feedback connections and a delay module which provides a memory to the network which helps to map the nonlinearity more complexly. The mathematical description of the NARX model is following [16]:

$$y(k) = f(y(k-1), \ldots y(k-n), u(k-1), u(k-m)) \tag{1}$$

where *y(k)* and *u(k)* represent model inputs, respectively outputs, *k* – discrete time steps, *n* and *m* represents time lag, *f* represents a nonlinear function between inputs and outputs. To properly approximate the non-linear function *f* it is necessary to choose a robust learning algorithm. For our purposes, we chose widely used Levenberq-Marquardt algorithm which belongs to the group of the backpropagation algorithms. Compared to widely used gradient methods, this algorithm uses approximation of second order derivates which means no need to compute the Hessian matrix which affects computational requirements [17].

3 Model Development

3.1 Case Study

For this research paper, the district heating network in Bratislava, Slovakia is examined. The heat plant supplies more than 600 substations with a total piping length of 202 km. The input data used in this research are retrieved from the ProCop control system for the years of 2014–2018. Heat load (MW) from the sensors is recorded every 10 min which corresponds to 144 records/day. This time span contains various information from different heating seasons. Weather data (outdoor temperature in °C) is also provided by the district heating network operator. Our research aims to create a 24 h ahead prediction model for a winter heating season for the months of November to the end of March.

3.2 Data Pre-processing

The accuracy of prediction models is based on the quality of the raw data. For that reason, it is necessary to devote strict attention to the data pre-processing. Data from the sensor are often incomplete, noisy, and so on. Two basic pre-processing techniques were employed to deal with missing, outliers, and noisy data. First step is to detect outliers and replace them. Widely used method among data scientist is Z – score standardization. Data that were 3 standard deviations from the mean of the data could be considered as outliers. These marked data were replaced by linear interpolation. The next step was data smoothing. A simple moving average filter which is a basically a lowpass filter, was used with asymmetric window of 12 sample before and 12 sample after the current observation. It is important to mention that each intervention into the data set could lead to misleading information and therefore it is necessary to carefully modify the original data set. The corrected data set is depicted in Fig. 1.

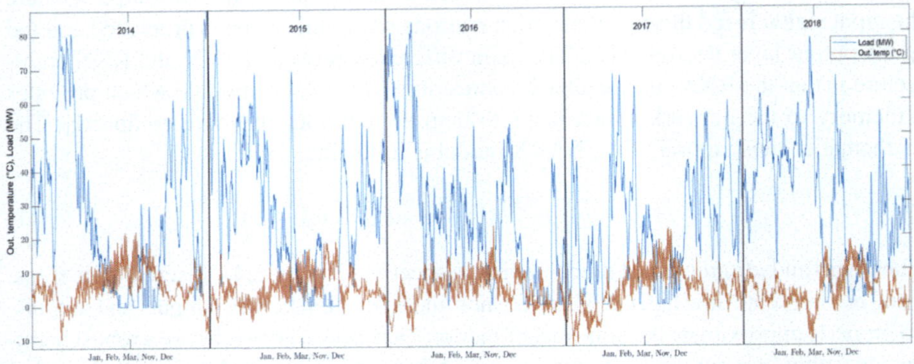

Fig. 1. Corrected data set for the winter heating season

3.3 Feature Selection and NARX Modelling

Generally, heat consumption is affected by many external and internal factors. To select the most influencing factors a correlation analysis was used. We found that the average outdoor temperature for the past 24 h has a higher correlation with the consumption compared to the outdoor temperature day before. To capture the dynamics of the consumption the historical consumption need to be considered. After correlation analysis we found out that average 24 h, respectively, 168 h consumption has higher correlation with the outdoor temperature compared to the consumption day, week before. Social components like hour of the day, day of the week and working/non-working day should not be omitted because the consumption is highly affect by the social behavior of the users connected to the grid. The following Table 1 shows the inputs for the proposed NARX model.

After preprocessing inputs, the next step is to scale data into dimensionless units. It allows unit comparison from different scales and it also prevents convergence problems (stuck in local minima). In our case, we used a min-max scaling so that the model inputs were scaled to the range [−1;1]. The proposed NARX model was trained with data from the 2014–2017 winter heating period. The data set was divided into three parts (70% training, 15% testing, 15% validation). The proposed model was further validated on the 2018 data set. The results obtained are presented in the next chapter.

Table 1. Selected inputs for the NARX model

Input no	Name (unit)	Predictor name
1	Previous day load at time h (MW)	$Q_{t\text{-}24h}$
2	Previous day moving average load (MW)	$Q_{avg\text{-}24h}$
3	Previous week moving average load (MW)	$Q_{avg\text{-}168h}$
4	Outdoor temperature (°C)	T_t
5	Previous day moving average outdoor temperature (°C)	$T_{avg\text{-}24h}$
6	Previous week moving average outdoor temperature (°C)	$T_{avg\text{-}168h}$
7	Hour of the day (h)	H
8	Day of the Week (–)	DoW
9	*Weekend	W

Iterative process was used for building a neural network model. Usually, several models are proposed to obtain acceptable results. First, the input and feedback delays must be determined. Autocorrelation and cross-correlation of inputs and desired targets were used to detect delays. At this stage, the initial number of hidden neurons was set to 10. The suitable range of input and feedback delays were determined by iterative approach. Increasing the range of input and feedback delays does not yield satisfactory results. After finding the suitable range of delays the next step was to find the appropriate number of hidden neurons. The initial number of hidden neurons (10) increased each training/testing/validation cycle by 2. The model with 20 hidden neurons produced the lowest error. The NARX model training consists of two phases. The first phase is called an open loop configuration where a single step ahead prediction is performed. The second phase where the multistep ahead prediction is performed is called closed loop configuration. Here, the actual outputs are fed back to the network to produce the prediction. The final NARX structure is presented in Fig. 2.

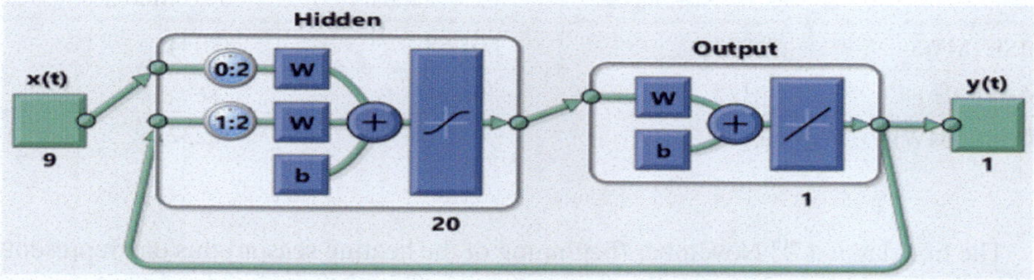

Fig. 2. NARX model (closed loop) implemented in MATLAB environment

4 Results

Experiments were performed on a PC with Intel i7 CPU @ 2.9 GHz and 16 GB of memory. MATLAB R2021b version was used to perform data pre-processing, model building, and evaluation.

4.1 Model Evaluation Criteria

Commonly used statistical metrics were used to measure the accuracy of the proposed method, namely: Mean Square Error (MSE), Mean Absolute Percentage Error (MAPE) and Root Mean Square Error (RMSE). MSE, RMSE measure the difference between the predicted and the real load and MAPE measure the relative percentage error.

$$MSE = \frac{1}{N}\sum\nolimits_{t=1}^{N}(\hat{y}_t - y_t)^2 \tag{2}$$

$$MAPE = \left[\frac{1}{N}\sum\nolimits_{t=1}^{N}\left(\frac{|\hat{y}_t - y_t|}{y_t}\right)\right] \cdot 100\% \tag{3}$$

$$RMSE = \sqrt{\frac{1}{N}\sum\nolimits_{t=1}^{N}(\hat{y}_t - y_t)^2} \tag{4}$$

where N stands for number of samples, $\hat{y}_t\hat{y}_t$ – predicted, $y_t y_t$ – measured heat load.

Three random days from the 2018 validation set were selected to evaluate the generalizability of the model. These days represent days with different heat demands. Table 2 presents the numerical results obtained between expected and predicted values. Figures 4, 5 and 6 depict the shapes of the expected and predicted day ahead heat demand with an hourly resolution.

Table 2. NARX model validation results

	17th November	25th January	12th March
MSE (MW)	0.232	0.068	0.054
MAPE (%)	4.125	0.431	∞
RMSE (MW)	0.479	0.261	0.232

The first day is 17th November (beginning of the heating season) this day represents a day with a mid heat demand. By visual checking (Fig. 3) we can clearly see that our proposed model captured the system behaviour properly. The MAPE error for this day was 4.125%.

The second experiment (Fig. 4) was performed on a day with a high heat demand. We can state that our model also captured systems dynamics properly also on a day with high demand. MAPE for this day was 0.431% which is extremely low.

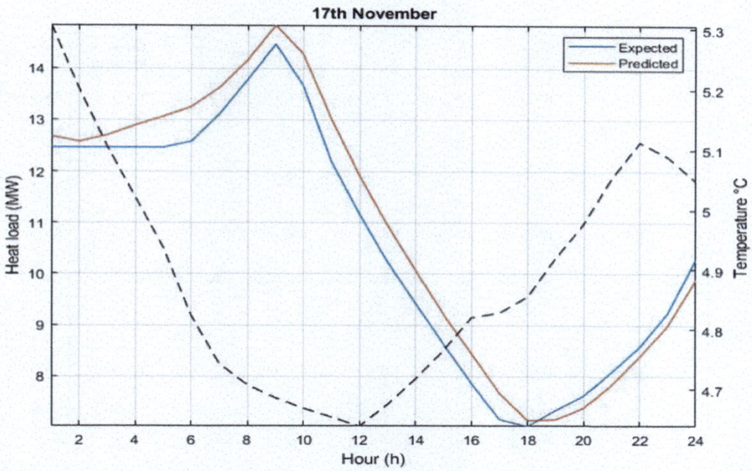

Fig. 3. Mid heat demand (Blue line - expected demand, Orange line – predicted, Black dashed line – outdoor temperature)

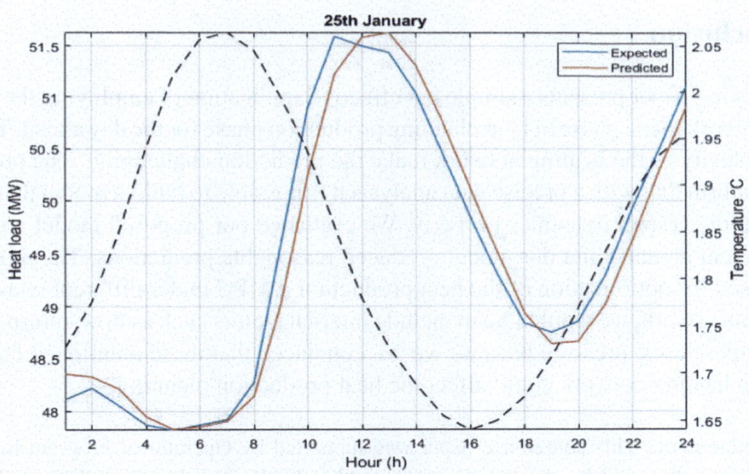

Fig. 4. High heat demand (Blue line - expected demand, Orange line – predicted, Black dashed line – outdoor temperature)

Interesting results were obtained from the last experiment, which represents a relatively warm day at the end of the heating season. On this day, the production was zero. The origin of this no production is unknown. It should be noted that examined heat source is mainly used during peaks, failures of the sources which have priority. We can see from this experiment that our prediction shows patterns like morning, evening peaks and fallback during the day. These patterns probably represents the diurnal pattern from the model inputs. It may also indicate that our model is too complex by many inputs. This is a good example of how necessary it is to employ MPC (Model Predictive Control) into production.

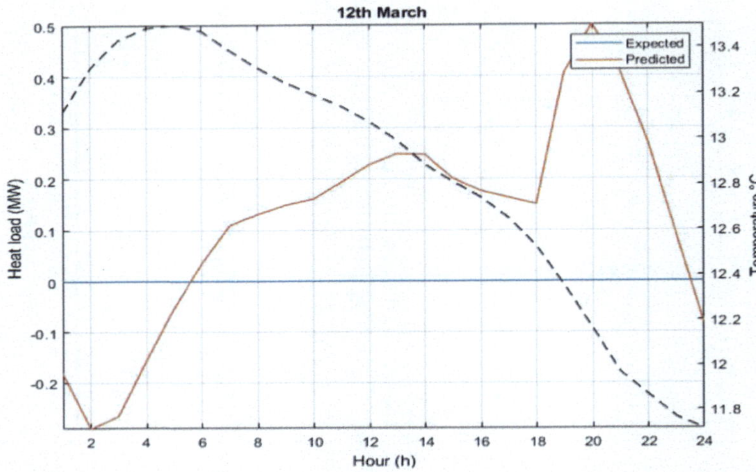

Fig. 5. Low heat demand (Blue line - expected demand, Orange line – predicted, Black dashed line – outdoor temperature)

5 Conclusion

The following paper presents a simple but effective application of employing the NARX neural network in the phase of heat planning production phase for the day ahead. The size and complexity of the heating network make the prediction challenging. The presented results confirm that with a precise data analysis it is possible to build a model that is able to capture the system dynamics properly. We evaluated our proposed model with days with different demand and our model produced reasonable predictions. The model can also be used for optimization of the heat production – MPC under different constraints.

For future work, we would like to include internal factors such as flow, return, supply water temperatures, pressure because we are convinced that these mentioned factors in such a big heating network could affect the heat production planning.

Acknowledgement. This conference paper was supported by Operational Program Integrated Infrastructure cofinanced by the ERDF, grant number ITMS 313011W988: "Research in the SANET network and possibilities of it further use and development", and Operational Program Integrated Infrastructure for the project: "Strategic research in the field of SMART monitoring, treatment and preventive protection against coronavirus (SARS-CoV-2)", Project no. 313011ASS8, co-financed by the European Regional Development Fund."

References

1. Fumo, N.: A review on the basics of building energy estimation. Renew. Sustain. Energy Rev. **31**, 53–60 (2014)
2. Montáns, F.J., et. al.: Data-driven modeling and learning in science and engineering. Comptes Rendus Mécanique 347(11), 845–855 (2019)

3. Rusovs, D., Jakovleva, L., Zentins, V., Baltputnis, K.: Heat load numerical prediction for district heating system operational control. Latv. J. Phys. Tech. Sci. **58**(3), 121–136 (2021)
4. Forrester, J.R.: Formulation of load prediction algorithm for a large commercial building. ASHRAE Trans. **90**, 536–551 (1984)
5. Chramcov, B., Balátě, J.: Time series analysis of heat demand. In: 22nd European Conference on Modelling and Simulation, pp. 3–6. ECMS, Nottingham (2008)
6. Pepescu, D., Ungureanu, F., Serban, E.: Simulation of consumption in district heating systems. In: WSEAS International Conference on Urban Rehabilitation and Sustainability, Environmental Problems and Development, pp. 50–55. WSEAS Press, Athens (2008)
7. Grosswindhagera, S., Voigt, A., Kozeka, M.: Online short-term forecast of system heat load in district heating networks. Tsp **1**(2), 1–8 (2011)
8. Fang, T., Lahdelma, R.: Evaluation of a multiple linear regression model and SARIMA model in forecasting heat demand for district heating system. Appl. Energy **179**, 544–552 (2016)
9. Yuan, J., Wang, C., Zhou, Z.: Study on refined control and prediction model of district heating station based on support vector machine. Energy **189**, (2019)
10. Protić, M., et al.: Forecasting of consumers heat load in district heating systems using the support vector machine with a discrete wavelet transform algorithm. Energy **87**, 343–351 (2015)
11. Golla, A., Geis, J., Loy, T., Staudt, P., Weinhardt, C.: An operational strategy for district heating networks: application of data-driven heat load forecasts. Energy Inform. **3**(1), 1–11 (2020). https://doi.org/10.1186/s42162-020-00125-5
12. Baltputnis, K., Petrichenko, R., Sauhats, A.: ANN-based city heat demand forecast. In: 2017 IEEE Manchester PowerTech, pp. 1–6. IEEE, Manchester (2017)
13. Ahmadi, S., Bevrani, H., Jannaty, H.: A fuzzy inference model for short-term load forecasting. In: Second Iranian Conference on Renewable Energy and Distributed Generation, pp. 39–44. IEEE, Tehran (2012)
14. Shamshirband, S., et al.: Heat load prediction in district heating systems with adaptive neuro-fuzzy method. Renew. Sustain. Energy Rev. **48**, 760–767 (2015)
15. Haykin, S.: Neural Networks: A Comprehensive Foundation. 2nd edn. Pearson, Delhi (1999)
16. Lin, T., et al.: Learning long-term dependencies in NARX recurrent neural networks. IEEE Trans. Neural Netw. **7**(6), 1329–1338 (1996)
17. Hagan, M.T., Menhaj, M.B.: Training feedforward networks with the Marquardt algorithm. IEEE Trans. Neural Netw. **5**(6), 989–993 (1994)

A Technique for Rapid Development of Declarative Knowledge Bases for Aircraft Diagnostics Based on Decision Tables

Aleksandr Yu. Yurin[1,2(✉)], Olga A. Nikolaychuk[1], Nikita O. Dorodnykh[1], and Yuri V. Kotlov[2]

[1] Matrosov Institute for System Dynamics and Control Theory, Siberian Branch of Russian Academy of Sciences (ISDCT SB RAS), Irkutsk, Russia
{iskander,nikoly}@icc.ru

[2] Moscow State Technical University of Civil Aviation, Irkutsk Branch (MSTUCA), Irkutsk, Russia

Abstract. Diagnostics of aircraft malfunctions remains an area of research that requires the use and creation of artificial intelligence methods and systems to support decision-making. The paper describes an approach to the rapid development of knowledge bases of intelligent aircraft fault diagnosis systems. At the same time, knowledge is presented in a declarative form as logical rules describing the relationships between external signs of malfunction, possible causes, and operations for their localization and elimination. The technology for prototyping rule-based expert systems based on transformations (PESoT) is used as a methodological basis. This technology implements some End-User Development principles, in particular, visual and wizard-based programming, and model transformations. By PESoT we developed domain models, transformed them to decision tables of a special type, filled them with information from troubleshooting manuals and fault accounting cards, transform filled tables to the elements of a knowledge base with consequent testing. The proposed methodology was also evaluated on a time criterion which showed its applicability for prototyping.

Keywords: Decision tables · Intelligent system · Knowledge base · Model transformation · Diagnostics of aircraft malfunctions

1 Introduction

One of the domain tasks that require the use of artificial intelligence methods and systems is the task of diagnosing aircraft malfunctions, which is especially important for new types of planes for which there are currently no representative failure statistics. One of these aircraft is an RRJ-95 (Sukhoi Superjet).

This paper is devoted to one of the aspects of the development of a decision support system for technical personnel in troubleshooting, namely "AviaTech Assistant", in particular, a technique for the rapid development of declarative knowledge bases for one of its subsystems.

R. Silhavy (Ed.): CSOC 2022, LNNS 502, pp. 140–149, 2022.
https://doi.org/10.1007/978-3-031-09076-9_13

The AviaTech Assistant is diagnostic software of the shell type that can be configured for a certain aircraft and its specific system. Within our project that was carried out in cooperation with Moscow State Technical University of Civil Aviation (MSTUCA) [1], the RRJ-95 power supply system is being considered.

The main purpose of the AviaTech Assistant is to support personnel in troubleshooting, including the formation of a list of candidate systems for failure, as well as an operating plan for the search, confirmation, elimination, and confirmation of the elimination of a malfunction or failure. The software consists of the following main modules: storage (database) containing information from operating manuals, maintenance manuals, troubleshooting manuals, and information about an aircraft; a case-based expert system for accumulating, searching, and storing information about failures and malfunctions; a rule-based expert system for defining potentially failed systems and the formation of plans of operations; the subsystem for generating reports; the domain-specific graphic user interface.

We used the original technology of prototyping expert systems based on transformations (PESoT) [2] to create declarative knowledge bases of an expert system. In our previous works [3, 4], we applied the formalism of event trees to form knowledge bases. Event trees are very popular among domain experts in the field of risk analysis [5, 6]. However, some difficulties of this formalism were identified during knowledge base engineering, related to the need to duplicate separate elements due to the impossibility of forming cycles when we describe scenarios of operations. To overcome this issue, we decided to use decision tables of a special type, which were tested earlier in solving other tasks [7, 8].

The paper is organized as follows: Sect. 2 briefly describes the proposed technique for creating declarative knowledge bases, while Sect. 3 presents our case study that illustrates the technique. Section 4 presents a discussion and concluding remarks.

2 Technique for Creating Declarative Knowledge Bases

The PESoT technology [2] implements some principles of End-User Development (EUD) [9, 10] in particular wizard-based and visual programming and model-driven development. These principles allow users of the software who are non-software developers to modify or extend separate software artifacts [11]. At the same time, important elements of model-driven development [12, 13] are various types of conceptual models, their meta-models, and model transformations [14]. PESoT provides the using certain types of conceptual models (class diagrams, concept maps, event trees, Ishikawa diagrams) as source information, transforming which the user can synthesize source codes or specifications for various platforms. In this case, the development of knowledge bases as components or modules of intelligent systems can be presented in the form of the following sequence of steps (Fig. 1).

The description of the steps (Fig. 1) is presented taking into account the use of decision tables as an intermediate way of representing knowledge aimed at non-programming end-users.

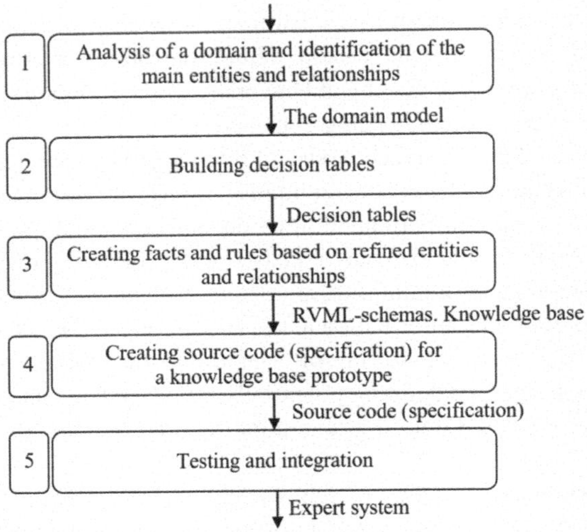

Fig. 1. The scheme of development of a knowledge base prototype for aircraft malfunction detection

At the first step, the domain is analyzed with the construction of detailed and generalizing conceptual models. Ontological models are used as detailing models, for the construction of which Protégé can be used. UML class diagrams were used as generalizing models.

At the second step, decision tables are constructed. This step includes two sub-steps: the formation of their structure (a list of tables with certain column headings); filling tables by specific data. For a first sub-step, class diagrams are mapped into logical rules and can be represented as RVML diagrams (Fig. 2). Then, based on RVML diagrams describing rule templates, decision tables of a certain structure are formed with their further filling. An example of mapping is shown in Fig. 2.

Decision tables [15] are the well-known formalism to describe logical rules that continues to be applied and developed [16]. It is a simple means for creating large knowledge bases and representing knowledge using widespread tabular editors, for example, Microsoft Excel.

In this paper, a specialized type of decision table was used (Fig. 2, block 3) enriched with additional semantics (or metadata), providing their transformation. The main features of this type: the ability to include a column with the names of rules ("RuleName"); the ability to indicate dependent columns with the "#" symbol; the ability to specify compound column names, including the name of the entity (or class name) and the name of its property, separated by the "::" string; no restrictions on the values in the cells, i.e. they can consist not only of a set of values {yes, no}, representing a degenerate form of truth tables, which allows one to use specific arbitrary values as cell values, and not only values indicating the presence or absence of certain property (component) in the rule structure.

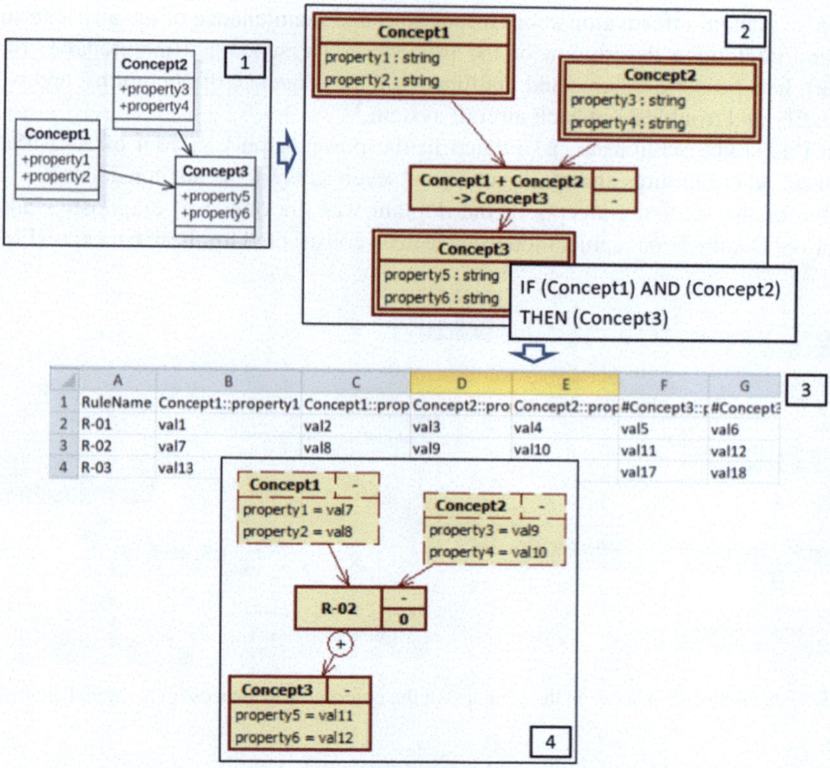

Fig. 2. An example of transformations at various steps of the technique: (1) the initial conceptual model; (2) a template for logical rules in the form of an RVML diagram; (3) a decision table based on the rule template; (4) a specific logical rule corresponding to a separate row of the decision table

The "Personal Knowledge Base Designer" (PKBD) [17] and its extension [18] were used as the main tool for implementing the proposed technique including the transformations described above. So, the obtained knowledge base structures are tested in the PKBD using an integrated CLIPS inference machine with the publication of the results through the GUI. The hid from the end-user generation of the CLIPS code is implemented, but code can be received on demand.

3 Case Study

The first step when applying the PESoT technology to the creation of the knowledge base for "AviaTech Assistant" is the analysis of the domain to identify the main domain concepts and relationships with subsequent building conceptual models.

The main sources of information were technical operation manuals (OM) and troubleshooting manuals (TM). The OMs describe information on maintenance, replacement, adjustment, inspection, and verification of the equipment of aircraft systems. The

TMs also contain information about inspections and maintenance of the airframe structure and contains a description of the procedures for scheduled maintenance of the aircraft; lists possible failures and malfunctions; the sequence of operations, and works to identify and troubleshoot each aircraft system.

In this study, section no. 24 related to the power supply system of RRJ-95 was analyzed, where failures and malfunctions of seven subsystems are considered.

One of the studied concepts of the domain was the object of diagnostics and its operation. Detailed conceptual models have been constructed for these concepts (Fig. 3).

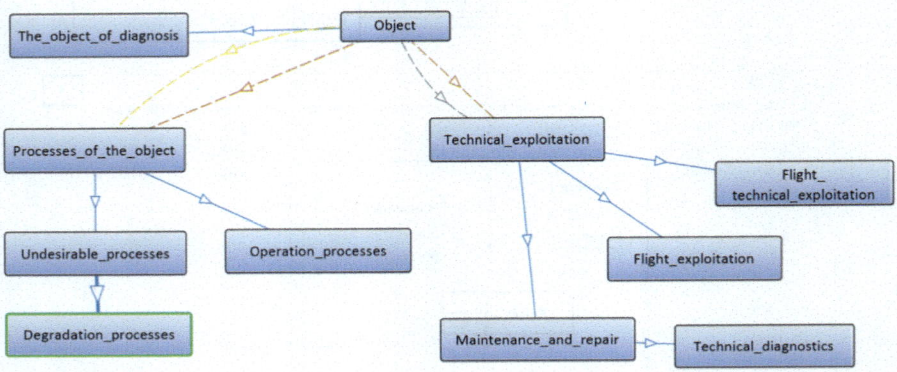

Fig. 3. The relationship between the concepts of the object and the process of technical diagnostics

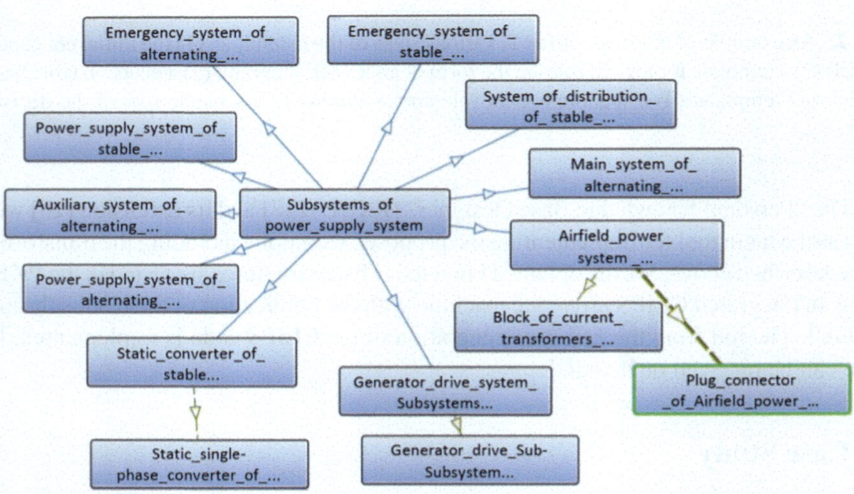

Fig. 4. A fragment of a conceptual model for the RRJ-95 power supply system

We used the ontological template [19] to clarify the object of diagnosis. This template reflects the hierarchical structure: "system – subsystem – sub-subsystem", and after its adaptation allows us to describe the mechanical system in the form of the following structure: "aircraft – aircraft systems – aircraft subsystems – aircraft sub-subsystems".

A hierarchy of the "general – specific" type was used to systematize information about the object.

Based on the "general – specific" hierarchy, the "part – whole" hierarchy for the structure of the aircraft was described (Fig. 4), a similar description was later carried out for operations (works) by various classifications, for example, a description of operations (works) according to the classification by a functional code.

The issue of detailing the properties of operations (works) and their relationships for separate systems was investigated too.

Currently, the conceptual model in the form of ontology contains 355 concepts and 1066 axioms. In the future, we plan to detail the conceptual model by covering other aircraft systems and subsystems.

For a more compact representation of main concepts and relationships, we have formed class diagrams, which were later used for automated prototyping of knowledge bases. Figure 5 shows one of the conceptual models describing an aircraft malfunction registration card. In accordance with this model, there is a certain aircraft of a certain type and with a registration number on which malfunctions are registered. Each malfunction is characterized by a date, a sign (external manifestation), a stage of detection, and a type of maintenance. In addition, the malfunction is associated with a certain system or subsystem of the aircraft, the consequences, and repair. Repair, in turn, is a certain set of operations or works, a detailed description of which is in regulatory documents (manuals for technical personal).

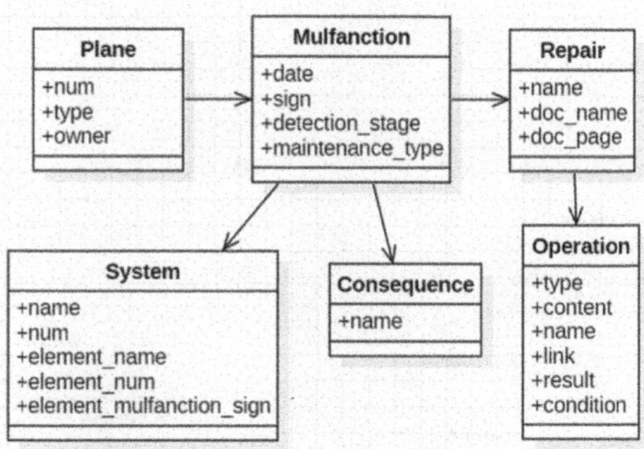

Fig. 5. The fragment of a conceptual model for an aircraft malfunction registration card in the form of a UML class diagram

Each relationship of the association type in the model can be interpreted as a causal relationship when transforming them into logical rules. As a result of the transformation of this model, a set of five rule templates was obtained, but after applying the aggregation procedure, only the following three remained:

```
IF (Plane) THEN (Malfunction)
IF (Malfunction) THEN (System) AND (Consequence) AND (Repair)
IF (Repair) THEN (Operation)
```

Based on the highlighted rules, decision tables of the following structure were formed:

```
RuleName,
plane::num, plane::type, #malfunction::date, #malfunction::sign,
#malfunction::detection-stage, #malfunction::maintenance_type
RuleName, malfunction::sign, malfunction::detection-stage, mal-
function::maintenance_type, #system::name, #system::num, #sys-
tem::element_name, #system::element_num, #consequence::name,
#repair::name, #repair::doc_name, #repair::doc_page
RuleName, repair::name, repair::doc_name, repair::doc_page,
#operation::type, #operation::name, #operation::content
```

Then the tables were filled in, a fragment of one of the tables is shown in Fig. 6.

1	RuleName	Mulfanctic	Mulfanction::sign	Mulfanction::dete	Mulfanctic	#System::name	#System::ı	#System::elemer	#System::ı	#System::ı	#Consequ
2	PMSCR-R-01		CAS MSG ELEC L GEN [	on land_start engi	DY	main AC system	24-22-00	generator contrc	6-X242	generator	no
3	PMSCR-R-02		EWD MSG APU GEN F/	on land_om	DY	power supply sy	24-00-00	APU generator	1-X242	periodic A	no
4	PMSCR-R-03		EWD MSG ELEC R GEN	on land_om	DY	AC generator dri	24-21-01	generator drive	16-X242	generator	no

Fig. 6. A fragment of a decision table describing aircraft malfunctions

The completed tables were imported into PKBD (Fig. 7) with subsequent testing (Fig. 8).

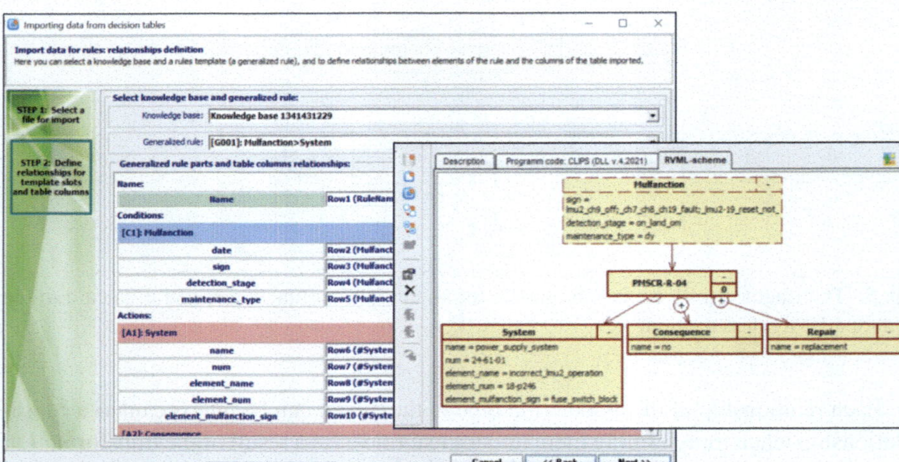

Fig. 7. Importing decision tables into PKBD: selecting source data for import and the result of the import in the form of the RVML diagram representing one of the rules

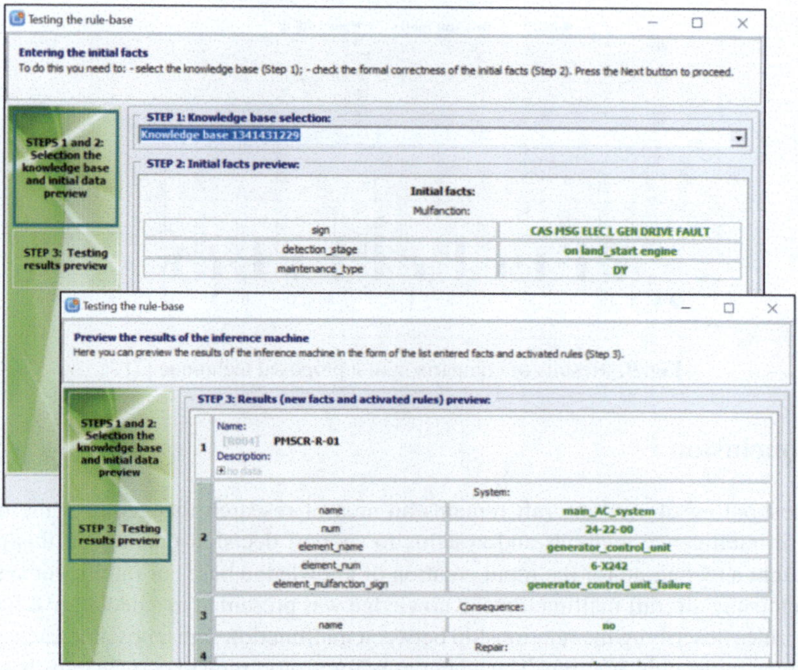

Fig. 8. Testing the knowledge base in PKBD: initial fact and results of logical inference

4 Discussion

In our project, several variants of knowledge bases were created for diagnostics of failures and malfunctions of the aircraft in the case of the RRJ-95 power supply system. The evaluation of the proposed technique and obtained knowledge bases were carried out from two points of view: 1) the effectiveness of creating a knowledge base; 2) the effectiveness of solving the troubleshooting task.

The effectiveness of creating knowledge bases was evaluated by comparing the time spent using the decision tables with hand-coding and visual and wizard-based programming in PKBD. The experiment with educational tasks is discussed in [7]. Briefly, these results are presented in Fig. 9, you can see that the use of decision tables can reduce the time spent compared to hand-coding (an average of 52%), however, it is less effective (an average of 38.8%) than to use PKBD visual and wizard-based programming. Though, as it was later determined, the use of decision tables is more preferable when a knowledge base contains more than 150 rules or the automated generation of decision tables is realized.

Currently, the evaluation of the effectiveness of the search and elimination of failures and malfunctions of the power supply system was carried out only on test examples, on which high scores were obtained. In the future, it is planned to try the system in real conditions based on the laboratories of MSTUCA and the maintenance service of IrAero Airlines [20].

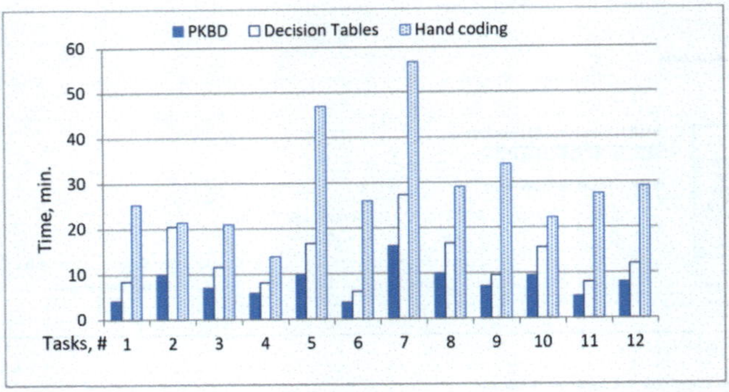

Fig. 9. Results of comparison of a proposed technique [7]

5 Conclusion

Troubleshooting of civil aircraft remains an area of research that requires the use of artificial intelligence methods and systems to support decision-making. In this paper, we present a technique for the rapid creation of knowledge bases of intelligent systems for diagnosing aircraft malfunctions. Knowledge was presented in a declarative form as logical rules describing the relationship between malfunction signs, possible causes, and operations for troubleshooting. The technology for prototyping expert systems based on transformations (PESoT) was used as a methodological basis and decision tables as the main mean for the representation of domain knowledge.

The evaluation of the approach showed its applicability in solving applied tasks and its usability for non-programming end-users, especially when describing domain knowledge.

Acknowledgement. The contribution of this work was supported by the Russian Science Foundation under Grant No. 22-21-00099.

References

1. Moscow State Technical University of Civil Aviation, Irkutsk Branch. http://www.if-mstuca.ru/. Last Accessed 17 Jan 2022
2. Yurin, A.Y., Dorodnykh, N.O., Nikolaychuk, O.A., Grishenko, M.A.: Designing rule-based expert systems with the aid of the model-driven development approach. Expert. Syst. **35**(5), 12291 (2018)
3. Kotlov, Y.V., Popov, V.M., Yurin, A.Y.: Towards designing knowledge bases for aircraft malfunctions diagnostics based on model trans-formations. J. Phys: Conf. Ser. **2060**, 012016 (2021)
4. Dorodnykh, N.O., Kotlov, Y.V., Nikolaychuk, O.A., Popov, V.M., Yurin, A.Y.: End-user development of knowledge bases for semi-automated formation of task cards. CEUR Workshop Proc. **2913**, 60–73 (2021)
5. Faisal, I.K., Abbasi, S.A.: Analytical simulation and PROFAT II: a new methodology and a computer automated tool for fault tree analysis in chemical process industries. J. Hazard. Mater. **75**(1), 23–56 (2001)

6. Kumamoto, H.: Fundamental Studies in Engineering, vol. 16, pp. 249–311 (1993)
7. Yurin, A.Y., Dorodnykh, N.O.: creating web decision-making modules on the basis of decision tables transformations. Commun. Comput. Inf. Sci. **1341**, 167–184 (2021)
8. Yurin, A.Y., Nikolaychuk, O.A., Dorodnykh, N.O.: Application of decision tables transformations for prototyping knowledge bases in the case of forest fire risk forecasting. CEUR Workshop Proc. **2984**, 34–41 (2021)
9. Barricelli, B.R., Cassano, F., Fogli, D., Piccinno, A.: End-user development, end-user programming and end-user software engineering: a systematic mapping study. J. Syst. Softw. **149**, 101–137 (2019)
10. Santos, M., Villela, M.L.B.: Characterizing end-user development solutions: a systematic literature review. Int. Conf. Hum.-Comput. Interact. 194–209 (2019)
11. Lieberman, H., Paternò, F., Klann, M., Wulf, V.: End-user development: an emerging paradigm. Hum.-Comput. Interact. Series **9**, 1–7 (2006)
12. Da Silva, A.R.: Model-driven engineering: a survey supported by the unified conceptual model. Comput. Lang. Syst. Struct. **43**, 139–155 (2015)
13. Cretu, L.G., Florin, D.: Model-Driven Engineering of Information Systems: Principles, Techniques, and Practice. Apple Academic Press (2014)
14. Mens, T., Gorp, P.V.: A taxonomy of model transformations. Electron. Notes Theor. Comput. Sci. **152**, 125–142 (2006)
15. Seagle, J.P., Duchessi, P.: Acquiring expert rules with the aid of decision tables. Eur. J. Oper. Res. **84**(1), 150–162 (1995)
16. Nalepa, G.J., Kluza, K.: UML representation for rule-based application models with XTT2-based business rules. Int. J. Softw. Eng. Knowl. Eng. **22**(4), 485–524 (2012)
17. Yurin, A.Y., Dorodnykh, N.O.: Personal knowledge base designer: software for expert systems prototyping. SoftwareX **11**, 100411 (2020)
18. Yurin, A.Y., Dorodnykh, N.O., Nikolaychuk, O.A.: Update (4.2020.0303) to "personal knowledge base designer: software for expert systems prototyping", (PII: S2352711019303334). SoftwareX **16**, 100825 (2021)
19. Zagorulko, Y.A., Borovikova, O.I.: Approach to the implementation of content patterns for the development of ontology of scientific subject domains. Syst. Inf. **12**, 27–40 (2018). https://doi.org/10.31144/si.2307-6410.2018.n12.p27-40(inRussian)
20. IrAero AirLines. https://iraero.ru/en/. Last Accessed 17 Jan 2022

Method for Adaptive Video Information Estimation in Aerospace Monitoring Systems

Valery Vladimirovich Vasilevsky(✉)

Moscow Aviation Institute (National Research University), 4, Volokolamsk highway, Moscow 125993, Russia

echinops777@rambler.ru

Abstract. The problem of increasing the efficiency of aerospace monitoring and automatic processing of video information under conditions of a priori uncertainty is investigated. To estimate the informative features of objects under observation in the contex of a priori uncertainty about the signal-noise environment, an approach is proposed that implements the procedures for obtaining a binary image and its clustering using adaptive minimax estimation of original multi-zone (grayscale) image elements. A description and testing of a computational algorithm for adaptive video information estimation in the problems of aerospace monitoring of observation objects is given. #CSOC1120.

Keywords: Aerospace monitoring system · Video information processing methods · Image model · Adaptive minimax estimation · Adaptive estimation algorithm

1 Introduction

1.1 Formulation of the Problem

One of the problems of using aerospace monitoring technologies in solving the problems of identifying and recognizing control objects is to increase the reliability and efficiency of processing video information delivered to consumers. One of the ways to solve it is related to the automation of stream processing of video information received by a photosensitive receiving device (FRD) of the corresponding spectral range [1].

Let us dwell on the task of automating the primary processing of video information of aerospace monitoring, search and recognition of scenery objects (anomalies), which, under the conditions of observation model parameters priori uncertainty, belongs to the class of statistical filtering problems.

The traditional approach to solving the problem is based on the use of algorithms for optimal processing of a multispectral-grayscale (color) image to highlight homogeneous region of anomalies (observed objects). Recognition of individual elements (segments) of the image frame, their structuring and linking into the corresponding sets (clusters) is the content of the task of primary processing of video information [2].

The problem of applying optimal image processing algorithms, in addition to their complexity, is the lack of complete priori information about models parameters and the

R. Silhavy (Ed.): CSOC 2022, LNNS 502, pp. 150–155, 2022.
https://doi.org/10.1007/978-3-031-09076-9_14

probabilistic characteristics of disturbances. In addition, optimal filtering and estimation methods are very sensitive to even slight deviations from the accepted assumptions and constraints under which they were obtained.

These circumstances were the reason for the study of image processing adaptive methods possibilities, which are based on the restoration of unknown stochastic parameters of the observation model [3].

Thus, the problem of automating the processing of multidimensional grayscale (color) images of the optical range when detecting objects (anomalies) of the Earth's surface is considered [1–3].

The aim of this work is to construct an adaptive minimax estimate of observation model parameters and develop an adaptive algorithm for processing a grayscale (color) image when receiving a binary data stream, taking into account the lack of monitoring conditions priori information awareness.

2 Image Model

Let the state of the current image be analyzed as a family $U_j\left\{u_{\bar{j}}, \bar{j} \in J\right\}$ random values of brightness u_j, which we represent as a random field (RF) changing in discrete time on a multidimensional lattice in discrete space and time

$$J = \left\{\bar{j} = (j_1, j_2, \ldots, j_n);\ j_l = \overline{1, M_l},\ l = 1, 2, \ldots, n\right\}$$

Since the RF varies with time, on a fixed time interval the family of quantities $\left\{u^t_{\bar{j}}, \bar{j} \in J_t\right\}$ will be called a row (column) of the current image.

The recurrent procedure for scanning and processing an image is as follows:

$$u_{\bar{j}} = U_{\bar{j}}\left(u_{\bar{l}}, \bar{\mathrm{l}} \in G_{\bar{j}}; \bar{\mathrm{j}}, \bar{\mathrm{l}} \in J_t\right), \tag{1}$$

where is the set of observed elements of the row (column) on the time interval;

$U_{\bar{j}}\left(u_{\bar{j}}\right), \bar{j}, \bar{l} \in J$ - brightness distribution functions $\left\{u_{\bar{j}}\right\}$.

For each discrete time interval, the state of RF of brightness distribution is determined based on the estimation of the values $\left\{u^t_{\bar{j}}, \bar{j} \in J_t, t \in T\right\}$ random variables and calculation of the estimating functional $Z_t(u)$ [4, 5].

We use an indefinitely stochastic model for observing the image current row (column), when measurements are carried out blocktype at time points and, at the same time, for different moments and observation errors can be correlated [4, 6]:

$$Z_t = X_t^* \rho + V_t, \quad \mathrm{t} = 1, 2, \ldots, \mathrm{T}, \tag{2}$$

where $Z_t \in R^r$ is the vector of the t- th observation of the brightness of the image row (column);

$$Z_t = col[\mathrm{Z}_1, \ldots, Z_T];$$

$X_t = \{X_{1t}, \ldots, X_{rt}\} \in R^{\rho \times r}$ non-random matrix of regressors - some set of volume r image elements, used to estimate the vector of the t -th observation;

$\rho = col[\rho_1, \rho_2] \in R^p$ composite indefinite-stochastic vector of observation model parameters;

$\rho_1 \in R^m$ - non-random uncertain parameter;

$\rho_2 \in R^l$ - random partially determined parameter;

$V_t \in R^r$ - vector random noise observation process with discrete time of the form:

$$\begin{cases} V_t = \sum_{k=1}^{q} B_k \cdot V_{t-k} + \Omega_t, \quad \text{t} = 1, 2, \ldots, \\ V_t = \Omega_t = 0, \quad \text{t} = 0, -1, \ldots. \end{cases} \tag{3}$$

The observation model (2) describes the r-dimensional autoregressive process of the q-th order and can be used in solving problems of estimating image model parameter.

Let us assume that in the observation model (3) the matrices are diagonal

$$B_k = diag[b_{k1}, \ldots, b_{k\mathrm{r}}],\ where\ b_i = col[b_{1\mathrm{i}}, \ldots, b_{\mathrm{ri}}];$$
$$\Omega_t = col[\omega_{1t}, \ldots, \omega_{\mathrm{rt}}],\ V_t = col[v_{1t}, \ldots, v_{\mathrm{rt}}].$$

If the parameters are known, then the minimax estimate of the parameters of the probabilistic observation model is determined by the relation:

$$\hat{\rho}_T = \left(\Phi_T^*(B) \cdot (I_T \otimes Q^{-1}) \cdot \Phi_T(B) + V_0^+\right)^{-1} \Phi_T^*(B) \cdot (I_T \otimes Q^{-1}) \cdot z_T(B) \tag{4}$$

where $\Phi_T(B) = col\left[U_1^*(B), \ldots, U_T^*(B)\right]$, $z_T(B) = col[z_1(B), \ldots, z_T(B)]$;

$\otimes$ - tensor product of matrices.

If the parameters $Q, B = \{B_k\}, V_0$ are unknown then there is an adaptive minimax estimate of the form:

$$\tilde{\rho}_T = \pi_T^{ad}(z_T), \quad \pi_T^{ad} \in \Pi, \tag{5}$$

where $z_T = col[Z_1, \ldots, Z_T]$, Π is the class of measurable transformations $R^n \to R^r$, $n = T \cdot r$;

π_T^{ad} - adaptive minimax estimation operator in the observation model (2) and (3).

At the same time, the adaptive estimate $\tilde{\rho}_T$ has an asymptotic estimation accuracy $\hat{\rho}_T(B, Q)$ and does not use a priori information about the parameters of the noise model $\{V_t\}$.

Estimators of probabilistic models' parameters with these properties are known as adaptive (asymptotically optimal), and the corresponding algorithms are known as adaptive estimation algorithms [3, 4, 6].

3 Computational Algorithm for Adaptive Estimation

The processing method is based on the use of two-dimensional data of the observed image when it is sequentially scanned along a row (column).

In the accepted assumptions and notations, the algorithm for constructing an adaptive estimate of image (row, column) observed section parameters contains the following computational procedures [4, 6, 7].

- The priori minimax estimate of the brightness distribution over an image row (column) is calculated:

$$\rho_T^a = \left(\sum\nolimits_{t=1}^{T} X_t \cdot X_t^*\right)^+ \cdot \sum\nolimits_{t=1}^{T} X_t \cdot Z_t = \Phi_T^+ \cdot z_T, \tag{6}$$

where $(\circ)^+$ is pseudoinverse Moore-Penrose matrix;
- It is found for each time $t = 1, \ldots, T$ moment of estimate of the random observation vector V_t:

$$\hat{V}_t = col\left[\hat{v}_{1\,t}, \ldots, \hat{v}_{r\,t}\right] = Z_t - X_t^* \cdot \rho_t^a. \tag{7}$$

- Estimates of auxiliary parameters are calculated $\{b_i\}_{i=1}^r$:

$$\hat{b}_i = \left(\sum\nolimits_{t=1}^{T} \hat{Y}_{i\,t} \cdot \hat{Y}_{i\,t}^*\right)^+ \cdot \sum\nolimits_{t=1}^{T} \hat{Y}_{i\,t} \cdot \hat{y}_{i\,t}, \quad \mathrm{i} = 1 \div \mathrm{r}, \tag{8}$$

where $\hat{Y}_{it} = col\left[\hat{y}_{i,t-1}, \ldots, \hat{y}_{i,t-q}\right]$.
- There is an estimate of the $\hat{Y}_T$ matrix $Q = cov\{\Omega_t, \Omega_t\}$:

$$\hat{Q}_T = T^{-1} \cdot \sum\nolimits_{t=1}^{T} \left(\hat{V}_t - \sum\nolimits_{k=1}^{q} \hat{B}_k \hat{\mathrm{V}}_{\mathrm{t-k}}\right)\left(\hat{V}_t - \sum\nolimits_{k=1}^{q} \hat{B}_k \hat{\mathrm{V}}_{\mathrm{t-k}}\right)^*, \tag{9}$$

where

$$\hat{B}_k = diag\left[\hat{b}_{k\,1}, \ldots, \hat{b}_{\mathrm{k\,r}}\right], \quad \mathrm{k} = 1 \div \mathrm{q}.$$

- An adaptive estimate of the vector $\tilde{\rho}_T$ of observation model parameters is constructed for the row (column) of the image:

$$\tilde{\rho}_T = W_T^+\left(\hat{B}, \hat{Q}_T\right) \cdot H_T\left(\hat{B}, \hat{\mathrm{Q}}_T\right), \tag{10}$$

where

$$\overline{W}_T\left(\hat{B}, \hat{\mathrm{Q}}_T\right) = \sum\nolimits_{t=1}^{T} U_t\left(\hat{B}\right) \cdot \hat{Q}_T^+ \cdot U_t^*\left(\hat{B}\right),$$
$$H_T\left(\hat{B}, \hat{\mathrm{Q}}_T\right) = \sum\nolimits_{t=1}^{T} U_t\left(\hat{B}\right) \cdot \hat{Q}_T^+ \cdot Z_t\left(\hat{B}\right),$$
$$W_T\left(\hat{B}, \hat{\mathrm{Q}}_T\right) = \overline{W}_T\left(\hat{B}, \hat{\mathrm{Q}}_T\right) + V_0^+, \quad \hat{\mathrm{B}} = \left[\hat{B}_1, \ldots, \hat{B}_q\right].$$

Estimate (10) is an adaptive minimax estimate, and algorithm (6), (7), (8), (9) and (10) is an adaptive minimax estimate algorithm.

An illustration of computational experiment results to study the applicability of the method of video information adaptive minimax processing when obtaining a binary image from the original grayscale image is shown in Fig. 1.

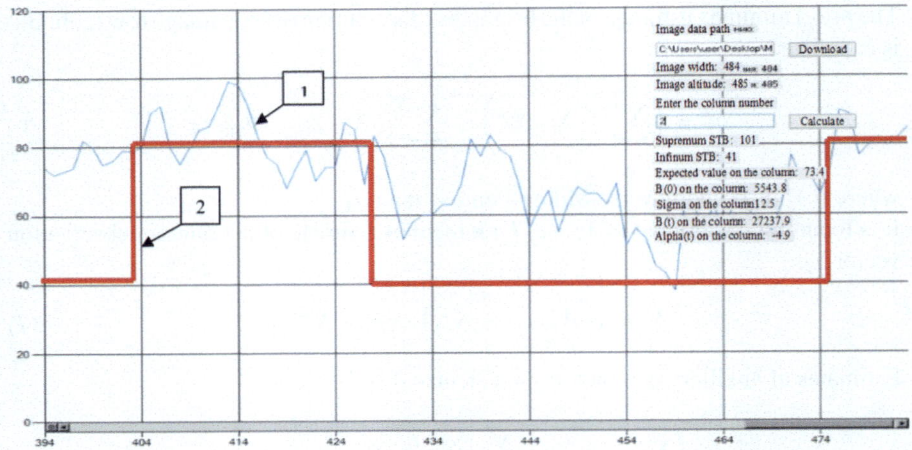

Fig. 1. Estimating the parameters of changes in grayscale image line (1) brightness and obtaining a binary data stream (2).

On Fig. 1, a continuous blue function (function 1) shows the brightness section of a real area image line, for which, in the case of traditional recognition scheme, it is almost impossible to assign decision boundaries in order to correctly separate pixels into object and background classes. The red discrete function (function 2) shows the binary representation of image line brightness, obtained using adaptive minimax estimation.

As follows from the test results, during automatic processing of video information using the adaptive estimation algorithm under conditions of a priori uncertain observation model parameters (brightness change dynamic range, data set size, threshold level, etc.), the pixels of the observed image can be correctly divided into classes 1 (background) and 2 (object), respectively.

The fields of computing module output file include the following information: 1) current image file; 2) data array on the brightness distribution in a row (column); 3) statistical characteristics of the image by row (column) estimate; 4) binary representation of an image row (column); 5) estimates array of image model parameters.

4 Conclusion

The paper proposes a solution to the problem of automatic processing and recognition of video information based on the use of an adaptive minimax image estimation algorithm.

The use of the algorithm for adaptive minimax estimation of multidimensional images shows the following possibilities for automating aerospace monitoring of objects on the Earth's surface:

- eliminating information redundancy when converting a grayscale (color) image into a binary data stream;
- reducing searching time for monitoring objects based on sequential image processing in rows (columns);
- improving the reliability of image recognition under conditions of a priori monitoring uncertainty when using an adaptive filtering and image estimation method.

References

1. Shovengerdt, R.A.: Remote Sensing. Models and Methods of Image Processing, p. 560. Technosfera, Moscow (2010)
2. Fukunaga, K.: Introduction to the Statistical Theory of Pattern Recognition, p. 368. Nauka. Main Edition of Physical and Mathematical Literature, Moscow (1979)
3. Fomin, V.N.: Recurrent Estimation and Adaptive Filtering, p. 288. Nauka, Moscow (1984)
4. Vasilevsky, V.V.: Adaptive minimax estimation of video information in the problems of aerospace monitoring. Semicond. Devices **2**(1(256)), 27–32 (2020)
5. Natanson, I.P.: Theory of Functions of a Real Variable, p. 480. Nauka, Moscow (1974)
6. Pugachev, V.S., Sinitsyn, I.N.: Theory of Stochastic Systems, p. 1000. Logos, Moscow (2000)
7. Ditkin, V.A., Prudnikov, N.P.: Integral Transformations and Operational Calculus, p. 544. Fizmatgiz, Moscow (1974)

Assessing the Disaster Tolerance of Digital Data

Alexander V. Solovyev(✉)

Federal Research Center "Computer Science and Control" of Russian Academy of Sciences, 44/2 Vavilova street, Moscow 119333, Russian Federation
soloviev@isa.ru

Abstract. In information system engineering it is often necessary to think over solutions that allow not only to ensure the security and reliability of digital data storage, but also to ensure the safety of digital data in the event of an external destabilizing effect, including a catastrophic one. The article proposes an assessment methodology and a mathematical model for assessing disaster tolerance. The disaster tolerance of digital data is understood as the ability to restore within a minimum period of time both the data itself and the operability of applications responsible for interpreting these data, as well as the operability of other software and hardware, without which the use of digital data is not possible. The methodology and mathematical model proposed in the article can be used to solve the problems of disaster tolerance of digital data of a wide class of information systems operating in the conditions of catastrophes and destabilizing factors. The proposed approach assumes the redundancy of software and hardware information systems, additional time spent in designing for compiling models, additional costs for storing the "history" of information systems functioning, describing destabilizing factors, etc. However, according to the author of the article, this is necessary and justified to ensure the safety of valuable digital data with a long storage period #CSOC1120.

Keywords: Information system engineering · Digital data · Long-term keeping · Stability · Disaster tolerance assessing

1 Introduction

In the article [1], the author analyzed the problems of disaster tolerance of digital data [2–7] in information systems, proposed a methodology for modeling the disaster tolerance of digital data during long-term storage [1]. The necessity of developing models for the disaster tolerance of digital data, checking and improving models, and carefully studying the results of modeling was proved. The need for constant monitoring of the disaster tolerance of digital data based on the created models is shown.

The disaster tolerance of digital data is understood as the ability to restore within a minimum period of time both the data itself and the operability of applications responsible for interpreting this data, as well as the operability of software and hardware, without which the use of digital data for its intended purpose is impossible.

R. Silhavy (Ed.): CSOC 2022, LNNS 502, pp. 156–163, 2022.
https://doi.org/10.1007/978-3-031-09076-9_15

The author's articles also show the importance of timely assessment of parametric disturbances in the software and hardware environment for storing digital data during their long-term storage [8] for the possibility of stabilizing storage parameters. In continuation of the study, this article proposes a methodology for developing a mathematical model for assessing the disaster tolerance of digital data to destabilizing effects, including those of a catastrophic nature.

Modern studies (see, for example, [10]) emphasize the importance of the digital data sustainability issues raised in the article, including cyber resilience, in the operation of applied systems operating under disaster conditions and destabilizing factors.

2 Methodology for Assessing the Disaster Tolerance of Digital Data

The methodology for assessing disaster tolerance is based on a scenario approach. That is, modeling the situation in which the impact of a catastrophic nature occurred, it is necessary to assess its impact on the functioning of the information system (IS) and determine the sufficiency of countermeasures (including redundancy tools) to compensate for the consequences of impacts and restore the IS to full capacity.

The factors that determine the scale and cost of countermeasures can be the following (see, for example, [9, 10]):

- value of digital data (including non-material)
- reliability indicators of IS
- performance indicators of IS
- availability of information resources of IS
- equipment and communication channels stability
- energy safety

Numerical values of indicators that identify potentially vulnerable elements of IS should be obtained on the basis of research or statistical analysis of the functioning of IS and stored in a separate secure database (DB). It should be possible to change this database when changing the composition of the IS software and hardware, changing the configurations of the IS, etc.

The author proposes the following sequence of steps of the methodology, which can be taken as a basis for compiling a methodology for assessing sustainability in the design and development of a specific information system:

1. Assessing the importance of IS elements

The importance of digital data storage media. At this step, the methodology can be given, for example, by the importance factor.

The coefficient of importance of each IS element is determined depending on the functions performed by the element, the value of information, potential vulnerability, etc. For example, given the types of data centers, consisting of various software and hardware, as well as equipment and communication channels. Considering this, we can introduce notations: let $\boldsymbol{b_1}$ correspond to the head data center, $\boldsymbol{b_{2j}}$ – to the j-th regional data center, b_{3k} – to the k-th district data center, etc. Thus, in the general case, we obtain the following relation:

$$1 >= b_1 > b_{2j} > b_{3k}. \quad (1)$$

The assignment of importance coefficients (by an expert method, to assess the value of digital data stored in IS elements, according to the information load of IS elements, etc.) to IS elements gives reason to believe that the damage caused to different elements as a result of the same or similar destabilizing effects.

2. Development of mathematical models of destabilizing factors (DF)

At this step of the methodology, the composition of the DFs that affect the digital data of the IS should be determined, they should be ranked according to the degree of damage, mathematical models of the DFs should be compiled.

In general, all DFs can be divided into those that can be affected by IS maintenance personnel, service centers, etc., and those that no one can influence (hurricane, tsunami, earthquake, etc.).

For DFs, which can be influenced in the direction of their reduction, mathematical models of their influence on IP are developed. The following are created: DF model, DF impact models on IP, DF counteraction models, DF consequences elimination models. For all DFs, it is necessary to establish what is the object of influence in each specific case (DPC, IS as a whole, a separate IS element: server, communication channel, etc.).

To model each DF, you need to determine:

- necessary statistical data
- the procedure and form of collecting statistical data, for example: survey, questioning, obtaining data from service centers, subsystems for monitoring the functioning of IS, etc.

Then it is necessary to collect the statistical data for a quantitative assessment or, in case of impossibility, a qualitative assessment of the degree of influence of the DF on the digital data of the IS.

For DF, the assessment of the impact of which on digital IS data is determined only qualitatively, 3 values can be entered: the minimum probability of occurrence ("optimistic" forecast), the maximum probability ("pessimistic" forecast), the highest probability (for example, based on the averaging of the two previous indicators). These values are used as restrictions that determine the potential state of the IC element.

In the case when it is not possible to obtain a quantitative assessment of the impact of the DF, it is permissible to perform the "acceptance" or "non-acceptance" of this DF on the basis of an expert assessment. Essentially, this means a logical condition: "DF affects the digital data" or "DF does not affect the digital data".

3. Selection of the main indicator for the digital data disaster tolerance model

The object of the sustainability assessment is the digital data of the IS as a whole. Therefore, for example, the coefficient of digital data availability, which characterizes the degree of data availability for the end users of the system, can be chosen as the main indicator. A number of "critical" digital data can be defined in the model, the lack of access to which (at least for some short time) leads to the impossibility of the IS to perform its functions.

Since one of the main indicators characterizing the effectiveness of the IS is timeliness, that is, the probability of the IS performing its functions in a time not exceeding the predetermined $T_D - P(T_{FUNC} \leq T_D)$, then the recovery time after a destabilizing impact T_{RDF} with an estimate of the probability of its implementation $P(T_{RDF} \leq T_{RDFd})$, based on the limitations and assumptions set out in [1]. $T_{RDFd.}$ is the predetermined recovery time after a destabilizing effect. An important assumption in modeling of disaster tolerance is that the redundancy tools will perform their functions in the event of a failure of the main IS assets.

Similar disaster tolerance assessment indicators should be evaluated for each element of the IS in which digital data is stored or which critically affect the safety of data.

4. Selection of private indicators of the disaster tolerance of digital data

Such indicators include, for example, the following indicators:

- energy safety
- fire safety
- terrorist security
- fault tolerance (see, for example, [9])
- climate resilience
- information security (see, for example, [10])
- efficiency (quality) of IS functioning

The composition of indicators can be supplemented and changed in relation to specific IS.

5. Development of models for assessing disaster tolerance indicators

In the general case, an IS can be organized according to a hierarchical principle [9]: head data center, regional data center, district data centers, communication channels center-region, region-district, etc. Therefore, the main indicator of the disaster tolerance of digital IS data is calculated based on the indicators of IS elements using weighting factors indicating the importance of the element (for example, by the number of users served, the disaster tolerance of the digital data of the element, the value of digital data, etc.). For the above formula (1), the importance of the central data center ($\boldsymbol{b_1}$) will be defined as 1, the importance of the regional data center ($\boldsymbol{b_{2j}}$) as, for example, the ratio of the data value of this data center to the total value of the IS data, etc. The main index of digital data resilience can be calculated as:

- static. For example, the coefficient of availability of IS digital data at an arbitrary point in time;

- dynamic. For example, as a coefficient of operational availability of IS digital data, which means availability not only at some arbitrary point in time, but also within a specified period of time after it.

6. Development of credibility and cost models for estimates of disaster tolerance indicators

Modeling the credibility and cost of assessments of disaster tolerance indicators must be carried out at all stages of the life cycle of creating information systems for long-term storage of digital data.

The credibility model should at least consist of the following models:

- confidence probability of the required accuracy (reliability) of the assessment of disaster tolerance indicators (the probability that within a given array of sustainability assessments there are no gross errors that lead to a violation of the required accuracy of the assessment);
- mean time to error (the ratio of the number of assessments made to the mathematical expectation of the number of errors that occur when performing assessments);
- the probability of correcting the disaster tolerance estimation error at a given time (the probability that the time spent on identifying and correcting an error during the assessment will not exceed a given one);
- average error correction time.

The value model should at least be composed of the following valuation models:

- the cost of developing mathematical models of disaster tolerance;
- the cost of assessments calculation;
- the cost of acquiring software and hardware for sustainability assessment;
- the cost of ownership of sustainability assessment tools;
- operation costs for a calendar period of time;
- cost of disposal of sustainability assessment tools.

3 Modeling the Reduction of the Influence of the DF on the Disaster Tolerance of Digital Data

For each type of DF, mathematical models of countermeasures should be developed, for example, based on collected statistics or analysis of the causes of DF.

The parameters of the countermeasure model should be:

- criticality of DF;
- the importance of digital data;
- cost of countermeasures;
- the probability of digital data recovery after the implementation of the DF;
- assessment of the "residual influence" of the DF in the implementation of countermeasures.

As a criterion for the applicability of a countermeasure in the model, the "efficiency/cost" ratio of the countermeasure can be used.

To obtain a more accurate assessment of the applicability of countermeasures, it is necessary to evaluate the cost of countermeasures and the reliability of their use according to the scheme similar to that given in Sect. 2, clause 6 of this article.

The analysis of the importance of digital data should be carried out according to the degree of influence of the implementation of the DF on the safety of data:

- according to the source of the greatest material losses;
- according to the maximum intangible losses;
- according to the most probable DF.

An analysis of the sustainability of digital data should identify the most vulnerable data that, under the influence of the DF and the consequences of their implementation, can be lost or distorted. If necessary, corrections are made to the disaster tolerance model in terms of the normalizing coefficients of the importance of digital data (see above formula (1)).

In general, when analyzing the impact of the DF on digital data, it can be noted that their vulnerability depends on the power (category) of the DF and the availability of means that neutralize negative factors: both the means that are part of the IS, and the ability to quickly ($P(T_{RDF} \leq T_{RDFd})$) eliminate the consequences of the implementation of the DF.

4 Mathematical Model of Digital Data Disaster Tolerance

Based on the developed methodology for modeling the disaster tolerance of digital data, a general mathematical model for assessing disaster tolerance is proposed. In the general case, the mathematical model of digital data disaster tolerance can be represented as:

$\boldsymbol{B = F(P_k, R_{nm}, R_m, P_{dl}, C, D, N)}$, where

$\boldsymbol{B}$ – integral indicator of digital data disaster tolerance;

$\boldsymbol{P_k}$ – the probability of implementing measures to counteract the disaster;

$\boldsymbol{R_{nm}}$ – assessment of non-material damage;

$\boldsymbol{R_m}$ – assessment of material damage;

$\boldsymbol{P_{dl}}$ – the probability of digital data loss or damage;

$\boldsymbol{C}$ – assessment of the cost of conducting disaster tolerance and countermeasure assessments;

$\boldsymbol{D}$ – reliability of the assessment of disaster tolerance and the implementation of countermeasures;

$\boldsymbol{N}$ – the function of the influence of available means and countermeasures to reduce the vulnerability of digital data (in the simplest case, it determines the presence of means and countermeasures associated with digital data in relation to each DF).

Based on private models of DF, counteraction to DF, a general model of digital data sustainability is created. B calculations are then carried out and strategies for improving the disaster tolerance of digital data are determined.

An important factor in the timely detection of disaster tolerance violations is the need for periodic assessment of the influence of DF. The developed methodology should allow quickly and efficiently assessing the impact of DF and determining the degree of disaster tolerance of digital data. The proposed methodology allows introducing new DF models, criteria for assessing the impact of DF, ways to collect the necessary data to assess the impact of DF, models of new countermeasures to neutralize the effect of DF on the disaster tolerance of digital data.

5 Possible Practical Application

The research topic covered in this article is extremely relevant in the era of rapid digitalization, since digital data are beginning to play a key role in all aspects of the functioning of both the economy and the state as a whole. Hence, their importance increases. Thus, it can be argued that the disaster tolerance of digital data is extremely important when designing an IS for the needs of the digital economy and its individual industries. And also during the operation of applied systems operating in the conditions of catastrophes and destabilizing factors (see, for example, [10]).

With long periods of storage of digital data, their resistance to DF should be ensured, since such data are of particular value. For IS, the operation of which is associated with the storage of valuable and especially valuable digital data, the loss of which can cause irreparable material and/or intangible (political, image, etc.) losses, it is necessary to pay great attention to the degree of protection of IS from destabilizing influences, including including catastrophic, not only natural, but also technogenic and anthropogenic. In particular, it is generally recognized that the main problems of creating and implementing information technologies in large organizational systems are associated with the influence of the human factor (see, for example, the author's study [6]). This once again emphasizes that the provision of disaster tolerance for digital data must be comprehensively thought out in order to avoid unaccounted risks on the one hand and unjustified costs on the other. It is necessary to develop a mathematical model for the disaster tolerance of digital data, its verification and improvement, and a thorough study of the simulation results. Constant monitoring of disaster tolerance based on the created models.

6 Conclusion

The methodological approach proposed in the article can be applied to solving the problems of disaster tolerance of digital data for a wide class of applied information systems. This approach assumes the redundancy of the IS software and hardware, additional time spent in designing for compiling models, additional costs for storing the "history" of the IS operation, describing the DF, etc. However, according to the author of the article, this is necessary and justified to ensure the safety of valuable digital data with a long storage period.

In further research, the author plans to carry out a series of works in order to develop an algorithmic apparatus for solving the problem of disaster tolerance of digital data. It is planned to test the methodology and algorithmic apparatus as part of the development of an electronic archive for long-term storage of digital data. Based on the results of the application, it is planned to formulate proposals for the implementation of the methodology and algorithmic apparatus in the form of software prototypes. It is also planned to conduct assessments of the cost and reliability of the results obtained and to correct the obtained DF models and countermeasures.

References

1. Bakanova, N.B., Solovyev, A.V.: Methodology for modeling the stability of digital data. Inf. Technol. Comput. Syst. **2**, 67–74 (2021)
2. Taylor, Z., Ranganathan, S.: Designing High Availability Systems: DFSS and Classical Reliability Techniques with Practical Real Life Examples. John Wiley & Sons, Inc., Hoboken, NJ, USA (2013)
3. Schmidt, K.: High Availability and Disaster Recovery: Concepts, Design, Implementation. Springer-Verlag, Berlin Heidelberg, Germany (2006). https://doi.org/10.1007/3-540-34582-5
4. Casti, J.: Connectivity, complexity, and catastrophe in large-scale systems Chichester etc. International Institute for Applied Systems Analysis, New York University (1979)
5. Poston, T., Stewart, I.: Catastrophe theory and its applications. Surveys and Reference Works in Mathematics, Pitman, London. (1978)
6. Solovyev, A.V.: Human reliability assessment in control systems. In: Radionov, A.A., Gasiyarov, V.R. (eds.) RusAutoCon 2020. LNEE, vol. 729, pp. 632–641. Springer, Cham (2021). https://doi.org/10.1007/978-3-030-71119-1_62
7. Akimova, G.P., Pashkina, E.V., Solovyev, A.V.: Situational analytical centers as a way to reduce the influence of the human factor on managerial decision-making in the operation of large information systems. Proc. ISA RAS **29**, 113–122 (2007)
8. Solovyev, A.V.: Long-term digital documents storage technology. In: Radionov, A.A., Karandaev, A.S. (eds.) RusAutoCon 2019. LNEE, vol. 641, pp. 901–911. Springer, Cham (2020). https://doi.org/10.1007/978-3-030-39225-3_97
9. Akimova, G.P., Tarkhanov, I.A., Solovyev, A.V.: Reliability Assessment Method for Geographically Distributed Information Systems. In: The IEEE 12th International Conference on Application of Information and Communication Technologies Proceedings, pp. 188–191, IEEE, Almaty, Kazakhstan (2018)
10. Budzko, V.I., Melnikov, D.A., Fomichev, V.M.: The information security and cyber resilience managing basics in the centralized information telecommunication systems of high availability. High Availab. Syst. **15**(1), 70–77 (2019)

Simulation of the Gaussian Distribution of a Random Variable Using a Standard Programm Rnd()

I. F. Spivak-Lavrov(✉), S. Y. Sharipov, T. Zh. Shugaeva, and F. N. Kaldybek

K Zhubanov Aktobe Regional University, Aktobe, Kazakhstan
spivakif@rambler.ru

Abstract. The paper considers three algorithms for modeling the normal distribution of a continuous random variable using a standard program Rnd() that generates random numbers: the integral equation method, the von Neumann selection-rejection method, and the Metropolis algorithm. Listings of programs written in the Visual Basic for Application (VBA) language, in which the indicated algorithms are implemented, are presented. The results of calculations obtained with the help of these programs are presented. The efficiency of using various algorithms is analyzed and conclusions are drawn about the expediency of their use in solving various problems using the Monte Carlo method #CSOC1120.

Keywords: Gaussian distribution · Metropolis algorithm · The von Neumann selection-rejection method

1 Introduction

Every modern software package has a standard program that generates random numbers uniformly distributed between zero and one [1–3]. With the help of such a program, the values of a continuous random variable are «played out», which take values on the interval [4] with equal probability. Each call to this program gives one such number with a certain number of decimal places. When implementing double precision on modern PC, the number of significant digits of the generator is usually 15.

In fact, there is no element of randomness in the random number generator (RNG) program. This program is deterministic, and the numbers it generates only seem to be random [5, 6]. Such numbers are called pseudo-random.

The principle of the program is as follows. When you first access the program, you must specify a certain number γ_0. With the help of a very specific sequence of actions, this number is converted into a new number

$$\gamma_1 = \Phi(\gamma_0), \tag{1}$$

where Φ is a function chosen in a certain way or a sequence of operations that converts γ_0 to γ_1. It defines the algorithm for generating random numbers. In turn, the number γ_1 serves as the basis for obtaining the next number according to the same recipe:

$$\gamma_2 = \Phi(\gamma_1). \tag{2}$$

R. Silhavy (Ed.): CSOC 2022, LNNS 502, pp. 164–175, 2022.
https://doi.org/10.1007/978-3-031-09076-9_16

Of course, the function Φ must be arranged so that all numbers $\gamma_1, \gamma_2, \ldots, \gamma_n$ satisfy the inequality: $0 \le \gamma_i \le 1$. They are the desired random numbers. Creating a good RNG is not an easy task. Usually, when creating an RNG, a linear congruent method is used [4, 7–10].

In MS Excel, the RNG is called using the standard function:

$$\gamma = Rnd(), \tag{3}$$

In order to get a random number x evenly distributed on the segment [a, b], you need to use a linear mapping:

$$x = a + (b - a)\gamma. \tag{4}$$

In order to model a continuous random variable with a probability distribution *f(x)* density using RNG, it is necessary to solve the following integral equation with respect to the upper limit of integration [4, 7]:

$$\int_{-\infty}^{x} f(\xi)d\xi = \gamma, \tag{5}$$

where γ is a number that is modeled by a standard program Rnd(). For each value γ a random variable x is found. In the case when the antiderivative function *F(x)* in the integral (5) can be found analytically, we get:

$$\int_{-\infty}^{x} f(\xi)d\xi = F(x) = \gamma, \tag{6}$$

In fact, it is an integral distribution function, which is equal to the probability of obtaining a value of a random variable less than or equal to x, and $F(-\infty) = 0$, and $F(\infty) = 1$. The task of modeling the distribution *f(x)* is simplified if it is possible to solve the equation for x, that is, to find the inverse function explicitly:

$$x = F^{-1}(\gamma). \tag{7}$$

However, solution (7) is not always possible. In this case, one has to look for other approaches, for example, in the case of a normal or Gaussian distribution. The present work is devoted to the discussion of this problem.

2 Gaussian Distribution and Its Properties

Very often, when solving various problems using the Monte Carlo method, one has to simulate a normal distribution [11, 12]. The normal or Gaussian probability density distribution of a continuous random variable is described by the following function:

$$f(x) = f_g(x) = \frac{1}{\sqrt{2\pi}\sigma} \exp\left(-\frac{(x-a)^2}{2\sigma^2}\right). \tag{8}$$

Here, σ is the standard deviation, a is the mean value of the random variable. Graphs of function (8) for $a = 2$ and three values $\sigma = 0.75$; 1.0; 1.25 are shown in Fig. 1.

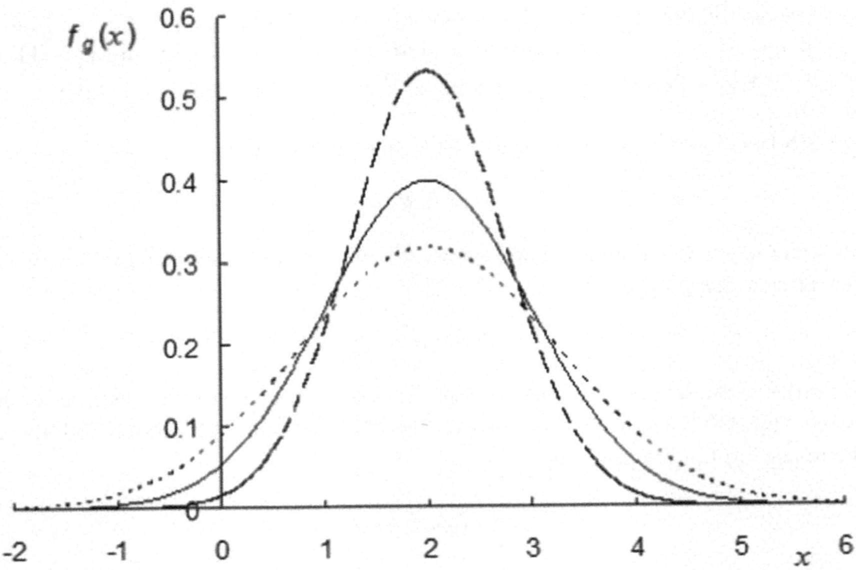

Fig. 1. Normal distribution for values $a = 2$and three values $\sigma = 0.75; 1.0; 1.25$.

First, the distribution is usually modeled for $\sigma = 1, a = 0$:

$$f_0(x) = \frac{1}{\sqrt{2\pi}} \exp\left(-\frac{x^2}{2}\right). \tag{9}$$

Then

$$f_g(x) = \frac{1}{\sigma} f_0\left(\frac{x-a}{\sigma}\right). \tag{10}$$

Most of the normal distribution points are located near the mean. The probability of a random variable x falling into the interval $x_1 \leq x \leq x_2$ is determined by the following integral:

$$P_{12} = P(x_1 \leq x \leq x_2) = \int_{x_1}^{x_2} f_g(x)dx. \tag{11}$$

Table 1 shows the values of the integral (11) for various integration limits.

Table 1. Values of the probability of a random variable falling into different intervals.

Interval	$a\text{-}\sigma \leq x \leq a+\sigma$	$a\text{-}2\sigma \leq x \leq a+2\sigma$	$a\text{-}3\sigma \leq x \leq a+3\sigma$	$a\text{-}4\sigma \leq x \leq a+4\sigma$
P_{12}	0.682690	0.954500	0.997300	0.999968

We also give two useful formulas that determine the average value of a random variable and its variance D:

$$a = \int_{-\infty}^{\infty} x f_g(x) dx, \tag{12}$$

$$D = \int_{-\infty}^{\infty} \left(x^2 - a^2\right) f_g(x) dx, \tag{13}$$

and
$\sigma = \sqrt{D}$.

3 Solution of the Integral Equation

Integral Eq. (6) can be reduced to solving the following differential equation:

$$\frac{dF}{dx} = f_0(x). \tag{14}$$

The algorithm for the numerical solution of Eq. (14) by the improved Euler method is implemented in the «Proba» program written in the Visual Basic for Application (VBA) language and implemented in the Excel environment. The «Proba» program calls two subroutines Sub Gauss(x0, hh, Rn, xk) and Sub Distrib(Mas, Dis, Knh, Nhh). The text of the programs is given below:

3.1 Program Listing «Proba»

```
Public Nm
Sub Proba()
'Returns a normally distributed random variable.
   Sheets("Norm").Select
   Dim Mx(1 To 1000000) As Double
   Dim Dis(1 To 2, 1 To 5000) As Double
   x0 = Cells(1, 2)
   Pi = Cells(2, 2)
   Nm = 1 / Sqr(2 * Pi)
   hh = Cells(4, 2)
   Nhh = Cells(5, 2)
   NN = Cells(6, 2)

   For i = 1 To NN
   Rn = Rnd()
   Call Gauss(x0, hh, Rn, xk)
   Mx(i) = xk
   Next i
   Call Distrib(Mx, Dis, NN, Nhh)
   For i = 1 To Nhh + 2
   Cells(9 + i, 1) = Dis(1, i)
   Cells(9 + i, 2) = Dis(2, i)
   Next i
End Sub

Sub Gauss(x0, hh, Rn, xk)
xx = x0
xx1 = x0 + hh / 2
yy = 0
yy1 = Nm * Exp(-xx1 * xx1 / 2)
AA:
yy = yy + yy1 * hh
If yy >= Rn Then
GoTo BB
End If
xx = xx + hh
xx1 = xx1 + hh
yy1 = Nm * Exp(-xx1 * xx1 / 2)
GoTo AA
BB:
xk = xx
End Sub
Sub Distrib(Mas, Dis, Knh, Nhh)
'Dim Mas(1 To 1000000) As Double
```

```
'Dim Dis(1 To 2, 1 To 5000) As Double
Max = Mas(1)
Min = Mas(1)
For i = 2 To Knh
If Mas(i) > Max Then
Max = Mas(i)
End If
Next i
For i = 2 To Knh
If Mas(i) < Min Then
Min = Mas(i)
End If
Next i
hh = (Max - Min) / Nhh
nn1 = 0
For i = 1 To Knh
If Mas(i) >= Min And Mas(i) < (Min + hh) Then
nn1 = nn1 + 1
End If
Next i
Dis(1, 1) = Min
Dis(2, 1) = 0
Dis(1, 2) = Min + hh / 2
Dis(2, 2) = nn1

For j = 2 To Nhh - 1
nn1 = 0
For i = 1 To Knh
If Mas(i) >= (Min + (j - 1) * hh) And Mas(i) < (Min + j * hh) Then
nn1 = nn1 + 1
End If
Next i
Dis(1, j + 1) = Min + (j - 1 / 2) * hh
Dis(2, j + 1) = nn1
Next j

nn1 = 0
For i = 1 To Knh
If Mas(i) >= Max - hh And Mas(i) <= Max Then
nn1 = nn1 + 1
End If
Next i

Dis(1, Nhh + 1) = Max - hh / 2
Dis(2, Nhh + 1) = nn1
```

```
Dis(1, Nhh + 2) = Max
Dis(2, Nhh + 2) = 0
End Sub
```

The subroutine Sub Gauss(x0, hh, Rn, xk) integrates Eqs. (14) by the improved Euler method from the initial value x0 with step hh for each value Rn = Rnd(), which is determined by the RNG, and outputs the value xk of a normally distributed random variable. From these values, an array Mx Mas is formed, which is then processed by the subroutine Sub Distrib(Mas, Dis, Knh, Nhh). The x0 and hh values are read from the Excel sheet ("Norm") and are subject to change.

Subroutine Sub Distrib(Mas, Dis, Knh, Nhh) processes the resulting one-dimensional array Mx Mas. Orders it by the values of a random variable, divides it into Nhh equal intervals, counts the number of points in each interval, and forms a two-dimensional array Dis(1 To 2, 1 To 5000). The first element of the array gives the number of points in each interval, and the second the coordinate of the center of the interval. Based on these data, a distribution graph is plotted: the values are plotted along the abscissa, and the corresponding number of points along the ordinate.

Initially, the values of the variables were chosen equal: x0 = −5, hh = 0.01, Nhh = 20. On Fig. 2 shows the distribution of the random variable obtained using our program «Proba» for the number of points NN = Knh = 10000 and 100000.

It can be seen from the figure that the given curves describe the normal distribution quite well. Although the shape of the curves obtained using the Proba program changes somewhat with different program launches, however, the error in all cases fits into the inverse root law and does not exceed a few percent.

4 Metropolis Method

Other methods for obtaining a non-uniformly distributed random variable with a known probability density *f(x)* are also possible. These are the «selection-rejection» method proposed by von Neumann and the Metropolis method proposed by Metropolis, Rosenblatt and Teller in 1953 [9, 10, 13, 14].

In the one-dimensional version of the Metropolis method, a «random walk» of points $\{x_i\}$ is modeled, the distribution of which, after a large number of steps, asymptotically tends to the distribution *f(x)*. A random walk is defined by specifying the probability of transition $w(x_{i\to}x_j)$ from one value x_i to another x_j for which the distribution of points $\{x_i\}$ converges to *f(x)*. It can be shown that for this it is neces sary to satisfy the condition of «detailed balance»:

$$f(x_i) = w(x_i \to x_j) = f(x_j) = w(x_j \to x_i). \tag{15}$$

The selection of points that satisfy condition (15) can be carried out using the Metropolis algorithm, which is implemented in VBA using the following subroutine:

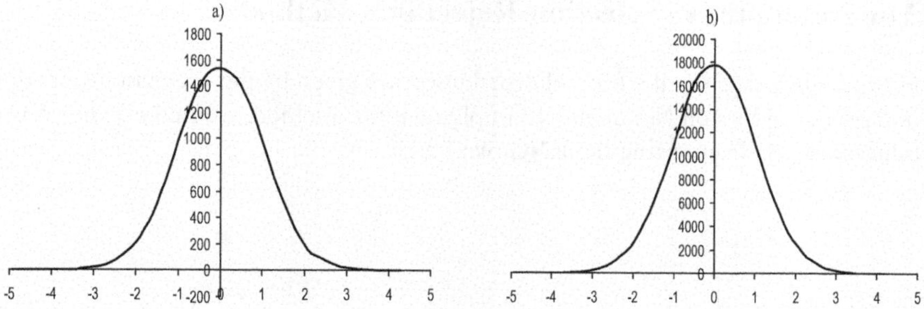

Fig. 2. Normal distribution obtained using the «Proba» program for NN = 10000 (a) and NN = 100000 (b).

```
Sub Metropolis(xn, xn1, delta, nm)
Rn = Rnd()
xt = xn + delta * (2 * Rn - 1)
ww = gaus(xt) / gaus(xn)
xn1 = xn
Rn1=Rnd()
If Rn1 <= ww Then
xn1 = xt
nm = nm + 1
End If
End Sub

Function gaus(x)
gaus = Exp (–x * x / 2)
End Function
```

The Metropolis subroutine moves from point x_n to point x_t with the maximum *delta* step. If this point satisfies condition (15), then it is selected in the distribution and the next step is made already from the point $x_{n+1} = x_t$, otherwise the point x_t is discarded. Approximately 1/3 of all considered values are accepted into the normal distribution array, which is a good result for this method [9].

The Metropolis method for a normal distribution gives acceptable results with a sufficiently large number of considered points. In this case, the array of selected points, just as for the Proba program, is processed by the Sub Distrib(Mas, Dis, Knh, Nhh) subroutine, which forms a two-dimensional array Dis(1 To 2, 1 To 5000) used to plot the Gaussian distribution. On Fig. 3 shows the results of the calculation of the Gaussian distribution, performed by the Metropolis method, for the number of points NN = Knh = 100000 and 1000000.

5 Von Neumann's «Selection-Rejection» Method

Good results in modeling the normal distribution are given by the «selection-rejection» method proposed by von Neumann. To implement this method, we find $c > \max f(x)$ on the interval [a, b]. Then, using the RNG, we find a

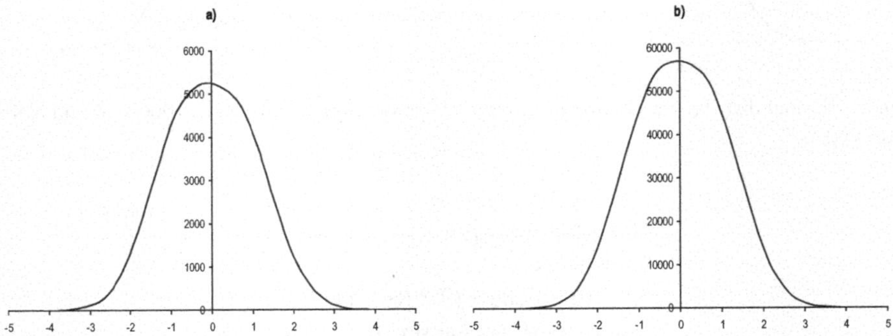

Fig. 3. The result of modeling the normal distribution using the Metropolis algorithm for NN = 10000 (a) and NN = 100000 (b).

random value x_i on the segment [a, b] and a random value y_i on the segment [0, c]. We calculate $f(x_i)$, if $f(x_i) > y_i$ then the point is selected in the distribution. The von Neumann algorithm is implemented using the Sub Gauss3() program, the listing of which is given below. The Gaussian distribution obtained with the help of this program after processing by the Sub Distrib(Mas, Dis, Knh, Nhh) subroutine for the number of points NN = Knh = 100000 and 1000000 is shown in Fig. 4.

5.1 Program Listing Gauss3

```
Public sigma, sigma1
Sub Gauss3()
'Von Neumann algorithm, normal distribution
    Dim Mx(1 To 2000000), Dis(1 To 2, 1 To 5000)

    Sheets("Norm").Select
    'Clearing a sheet of old data
    Range("A9:G16000").Select
    Selection.ClearContents
    Range("B1").Select
    'Data input
    NN = Cells(1, 2)
    Pi = Cells(2, 2)
    delta = Cells(3, 2)
    aa = Cells(4, 2)
    sigma = Cells(5, 2)
    Nhh = Cells(6, 2)
    nm = 0
    cc = 1

    For ii = 1 To NN
    xi = -delta + 2 * delta * Rnd()
    yi = Rnd()
    fxi = gaus(xi)
    If fxi - yi > 0 Then
    nm = nm + 1
    Mx(nm) = xi
    End If
    Next ii
    ss = 0
    For ii = 1 To nm
    ss = ss + Mx(ii) * Mx(ii)
    Next ii
    Cells(1, 3) = ss / nm
    For ii = 1 To nm
    Mx(ii) = aa + sigma * Mx(ii)
    Next ii
```

```
    Call Distrib(Mx, Dis, nm, Nhh)
    For ii = 1 To Nhh + 2
    Cells(ii + 9, 1) = Dis(1, ii)
    Cells(ii + 9, 2) = Dis(2, ii)
    Next ii
    Cells(7, 2) = nm
End Sub

Function gaus(x)
gaus = Exp(-x * x / 2)
End Function
```

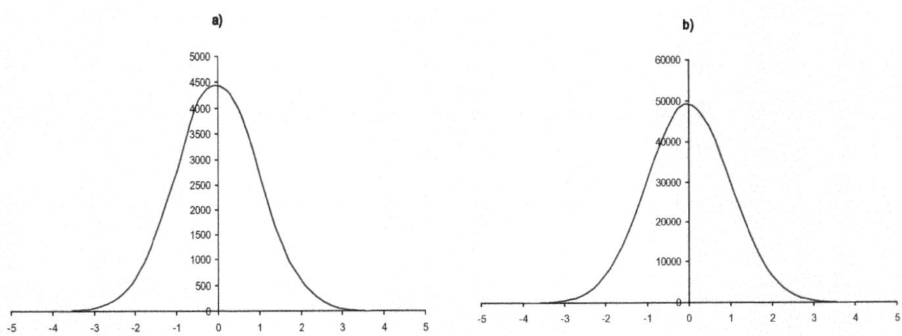

Fig. 4. The result of modeling the normal distribution using the algorithm von Neumann **for** NN = 100000 (a) and NN = 1000000 (b).

6 Conclusion

The paper considers three algorithms for modeling the normal distribution of a continuous random variable using a standard program that generates random numbers: the integral equation method ("Proba" program), the von Neumann selection-rejection method, and the Metropolis algorithm. Listings of programs written in the VBA language, in which the indicated algorithms are implemented, are presented. The results of calculations obtained with the help of these programs are presented. The presented programs differ in the speed of execution and the accuracy of reproducing the normal distribution. The Metropolis algorithm has the highest execution speed, followed by the von Neumann selection-rejection method, and the Proba program has the lowest execution speed. The Proba program selects all considered points in a normal distribution. While the Metropolis and von Neumann algorithms select only about 1/3 of the analyzed values into the distribution. Therefore, if you need to select a relatively small number of points that have a normal distribution, it is better to use the Proba program. When using large data arrays that have a normal distribution, it is better to use the Metropolis algorithms or the von Neumann selection-rejection method.

Acknowledgments. The work was carried out within the framework of a project with grant funding from the Science Committee of the Ministry of Education and Science of the Republic of Kazakhstan (IRN AR09258546).

References

1. Álvarez, R., Martínez, F., Zamora, A.: Improving the statistical qualities of pseudo random number generators. Symmetry **1**(2), 269 (2022)
2. Balakrishna, S.K., Shetty, S.M., Martis, J.E., Ramasamy, B.: Genetic algorithm-based pseudo random number generation for cloud security. In: Misra, S., Kumar Tyagi, A., Piuri, V., Garg, L. (eds.) Artificial Intelligence for Cloud and Edge Computing. IT, pp. 209–236. Springer, Cham (2022). https://doi.org/10.1007/978-3-030-80821-1_10
3. Ryabko, B.: Asymptotically most powerful tests for random number generators. J Stat. Planning Infer. **217**, 1–7 (2022)
4. Sobol, I.M.: Monte Carlo Method, p. 64. Nauka, Moscow (1968)
5. Fan, F., Wang, G.: Learning from pseudo-randomness with an artificial neural network-does god play pseudo-dice? IEEE Access **6**, 22987–22992 (2018)
6. Machicao, J., Bruno, O.M.: Improving the pseudo-randomness properties of chaotic maps using deep-zoom. Chaos **27**(5), 053116 (2017)
7. Sobol, I.M.: Numerical Monte Carlo Methods, p. 134. Nauka, Moscow (1973)
8. Efros, L.N.: Physics and Geometry of Disorder, p. 268. Nauka, Moscow (1982)
9. Gould, H., Tobochnik, Y.: Computer Modeling in Physics (in 2 parts). Part two. Mir, Moscow, p. 390 (1990)
10. Teylor, I.R.: An Introduction to Error Analysis. University Science Books, Oxford University Press (1982)
11. Fallahi, S., Skaug, H.J., Alendal, G.: A comparison of Monte Carlo sampling methods for metabolic network models. PLoS ONE **15**(7), e0235393 (2020)
12. Sudiyanto, A., Faradcy, A.A., Cahyadi, T.A., Poetranto, D.W.A., Suryono, H.: Analysis of loading and hauling equipment requirements with a comparison of Monte Carlo simulation methods, uniform distribution, normal distribution, and exponential distribution. AIP Conf. Proc. **2363**, 0061688 (2021)
13. Squires, G.L.: Practical Physics, 3rd edn. Cambridge University Press (1985)
14. Press, W.H., Flannery, B.P., Teukolsky, S.A., Vetterling, W.T.: Numerical Recipes. Cambridge University Press (1986)

Development of Information System for Monitoring the Rate of Digital Currency for Investing

Elena Butsenko(✉)

Ural State University of Economics, Ekaterinburg, Russia
evl@usue.ru

Abstract. The changes currently taking place in the world economy stimulate the emergence of new software tools for evaluating and analyzing financial investments. While investments are aimed at obtaining profit by the investor, they are not a guaranteed way to receive it. Different ways of investing provide different guarantees of income, but in all cases there is a risk that instead of profit, the investor will receive a loss. Therefore, the development of investment software that allows you to analyze and evaluate the growth or fall in the value of digital assets is an urgent issue. The computer system under consideration is designed to monitor and analyze the cryptocurrency exchange rate in real time. The purpose of the work is to describe the functional purpose and structural elements of the system for tracking changes in the cryptocurrency exchange rate to make a decision on its purchase/sale. The information system is written in the PyCharm integrated development environment in Python. The article presents the development technology of the system under consideration and shows a practical example of its work on the process of monitoring the bitcoin cryptocurrency. In the future, this system can be improved with additional functionality and a more flexible interface #CSOC1120.

Keywords: Exchange rate · Cryptocurrency · Digital currency · Investments · Monitoring · System development · Bitcoin · Python · Data parsing

1 Introduction

The exchange rate in one way or another affects all economic processes in the country. It determines the increase or decrease in business activity, the growth of investments in the country, and hence the possibility of successful employment of citizens. The exchange rate also affects the level of income of many enterprises, which subsequently affects salaries and other payments. However, most of all, its trends affect the rise in prices for imported goods and services. And this category also includes branded clothing, and household appliances, and gadgets, and cars of many brands, and other products [6].

As you can see, in one way or another, every person feels fluctuations in the exchange rate in his life. Those who track exchange rates can plan their expenses more accurately.

R. Silhavy (Ed.): CSOC 2022, LNNS 502, pp. 176–185, 2022.
https://doi.org/10.1007/978-3-031-09076-9_17

Even if a person is not engaged in entrepreneurial activity, he will extract the maximum benefit for himself from this or that economic situation.

The development of the digital economy and digital technologies has led to the emergence of new financial assets. The use of digital currency is increasingly gaining a place among the world's population, so much so that there are 2486 types of digital currency in history [8]. Buying/selling electronic currencies has become a very common investment service. The first major international online digital currency exchange services appeared back in 2014. Now, in terms of daily trading volume, this market has caught up with the trading volume of the stock exchange.

The popularity of cryptocurrencies is due to their volatility. The volatility is caused by the fact that the crypto market has a higher rate of return on capital and the terms of capital placement are shorter. And this market is still very young. The field of cryptocurrencies has not yet been fully regulated, since it has many fresh mechanics and an audience from various parts of the world. The price of assets is highly susceptible to investor sentiment, as a clear set of rules for the market has not been established.

In digital assets, the price is formed according to the classical model of the balance of supply and demand, since the number of the same bitcoins is limited, it is possible to form a price by adjusting the number of bitcoins in circulation. And the price in traditional markets is made up of many factors: forecasts of analytical agencies, financial performance of companies, ratings, government regulation, news background and other dependencies.

There is no definite set value of the asset at a particular moment in time on the crypto market, since different quotes are presented on the sites. This is due to the fact that the traditional stock or currency market is controlled by states and central banks, while crypto exchanges are not, since they are private companies and set the rules for trading on their sites.

The exchange rate is constantly in an unstable state. Tracking it manually is long and inconvenient. For investors, an important task is to continuously monitor changes in the rate of buying/selling cryptocurrency in order to make a deal in time with maximum profitability. Therefore, the development of a software tool for automatic tracking of the value of a cryptocurrency is an urgent issue, the solution of which will help to choose an effective investment option [3, 4].

We note the most recent most significant scientific achievements and work in this area.

In [11], the authors studied the rise in the price of bitcoin on the Binance cryptocurrency exchange from 2017 to 2021 using the least squares method. They revealed an underestimation of the equilibrium state of the empirical distribution of price data and the volume of daily trading in bitcoins along the channel of the right confidence interval. Also, based on economic modeling, the authors determined the forecast volume and price of bitcoin on this exchange.

The work [7] presents the development of a mobile application to simulate the use of cryptocurrencies. A model of a cryptocurrency trading and digital payment system, compatible with Android and iOS operating systems, is described. It is designed to assess the preferences of participants associated with various cryptocurrencies, according to the main characteristics (volatility, anonymity, trading speed and settlement time).

To develop the system, a technology for creating an external graphical interface is defined. The Progressive Web Application solution was chosen as the tool. And also the template of the internal software architecture of the mobile application is defined. In addition, the requirements for application development software are presented.

In the work [12], a study was made of the profitability and risk of investing in cryptocurrencies. For the analysis, a heteroscedastic model of profitability of 15 cryptocurrencies that had the highest value was used. The results showed that investments in most cryptocurrencies bring higher returns than investments in foreign currencies or the stock market. It is also noted that cryptocurrencies have a higher risk of loss and clustering of volatility or heteroscedasticity.

The study [9] presents the architecture of a bitcoin price monitoring system from 20 crypto markets. The services that provide currency price forecasts with the greatest accuracy have been studied. Required data retrieved using Web API. This data is then stored in a MySQL database. Then they can be continuously collected using PHP-Cronjob. And for the convenience of reading the collected data, Telegram Bot was used.

In [16], the author answers the important question of whether cryptocurrencies have intrinsic value. This study is essential for the economic classification of cryptocurrencies.

In [2], the authors propose to use machine learning to build a model for predicting prices for stocks and cryptocurrencies. This study uses available price, high price, low price, closing price, volume, and market capitalization, as well as the interdependencies of some cryptocurrencies, as the main factors. The authors conclude that the crypto market lacks firm regulatory structures and is unpredictable, making price prediction more difficult. The result of the analysis is the construction of a machine learning model for predicting the price of a cryptocurrency.

The work [5] explores the impact of distributed ledger technology (DLT) on monetary and financial activities in the relationship that bitcoin is a means of payment based on blockchain, a form of distributed ledger.

The work [15] identifies economic uncertainties that affect the change in the price of a cryptocurrency. To do this, stochastic trends were established in the cryptocurrency sphere and the main text uncertainty indices, and as a result of stochastic modeling, the following important point was observed. While the pricing dynamics of cryptocurrencies may be independent of global uncertainties, there remain some cases tied to global trends.

Real-time monitoring of bitcoin prices using the PHP programming language and reading the JSON format is discussed in [10]. Access to the API was provided by a specific exchange office. The program code and the results of the application of monitoring the movement of prices for bitcoin are presented.

Several deep machine learning methods that can be used to build cryptocurrency price predictions are presented in [13]. It also demonstrated the implementation of a machine learning model for predicting the price of bitcoin.

The work [1] reports on the particular relevance of the accuracy of calculating investment forecasts in crisis periods of economic development. The authors evaluate the performance of different types of machine learning algorithms for predicting the price of bitcoin when it is updated in a period of time from 5 min to 1 h. In their opinion, this will significantly reduce modeling errors. And with more accurate data, investors will make the best decisions for their investments.

So, real-time cryptocurrency rates allow you to quickly and efficiently invest your funds. At the same time, the cryptocurrency exchange rate is notable for its instability, so its constant monitoring is required to obtain benefits. Considering that the purchase of cryptocurrencies is currently a very risky investment, and after analyzing various scientific sources, it becomes obvious that it is necessary to develop an information system for monitoring the exchange rate of digital currencies in real time, which contributes to the main business goals - obtaining the greatest profit with the least risk.

2 Results

The development of an information system for monitoring the rate of digital currencies begins with a description of the general structural diagram of the future system. The general scheme of the information system being developed involves the following key actions:

- parse (collect and extract) data from the Google search results page;
- check changes in the exchange rate since the launch of the program;
- send an email in case of a strong exchange rate change.

The development of the software part of the system under consideration is carried out using the PyCharm integrated development environment. It implements all the functions laid down in the flowchart below.

Next, we will consider the algorithm for monitoring data on the price of bitcoin with a description of the main steps.

Step 0. Specify the data source in the selected software for collecting information about the bitcoin exchange rate in real time.
Step 1. Create a query for processing data received from the site.
Step 2. Read the code and find the bitcoin price values.
Step 3: Create a loop to send a request every 3 s.
Step 4. Evaluation of the new value of the price of bitcoin.
Step 5. Formation of a summary of data on the price of bitcoin.
Step 6. Determination of the parameter responsible for fluctuations in the bitcoin rate.
Step 7. Formation of notifications about the growth/fall of the bitcoin rate to the email.

The example of a flowchart in accordance with the data monitoring algorithm is shown in Fig. 1.

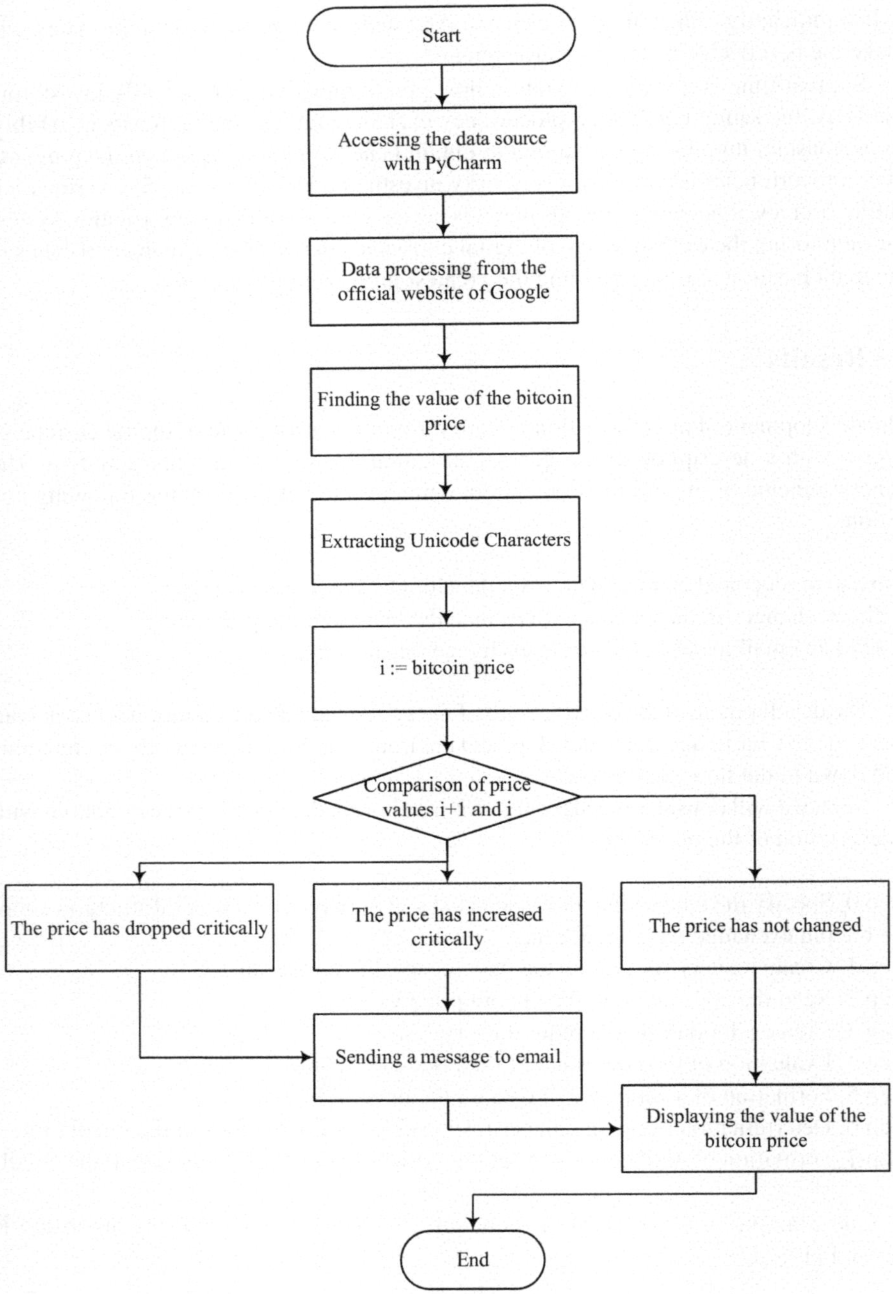

Fig. 1. Flowchart for monitoring the bitcoin price

Google.com has been chosen to parse the real-time bitcoin rate. This choice is justified by the general availability of the site, the lack of registration, unlike specialized services, the accuracy of the data, as well as the stability of the site and frequent updates of the cryptocurrency exchange rate.

Next, you need to make a selection through the site code to select the information we need. It is required to extract data from the line “<span class = “DFlfde SwHCTb” data-precision = “2” data-value = “38556.1”>38 556, 10 </span> ” (Fig. 2).

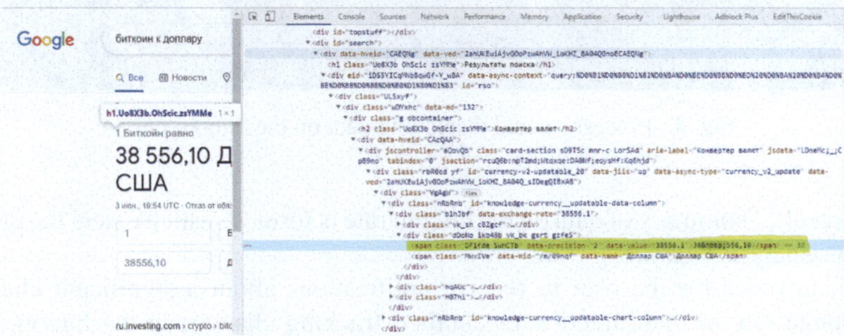

Fig. 2. Finding the right bitcoin exchange rate data

To get this data, a query is sent to the Google website, after which this code is read and the given string is searched (Fig. 3).

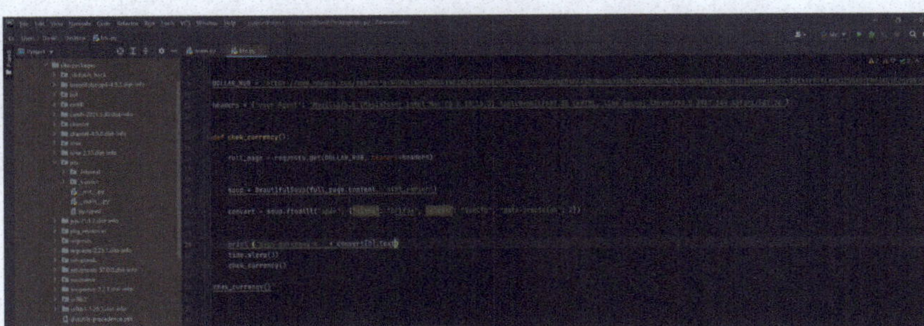

Fig. 3. Sending a query to retrieve data about the bitcoin rate

After that, information is collected and the numerical value of the bitcoin rate is displayed. And also, in order to receive all the data in a timely manner, a cycle was created that sends queries for data selection every 3 s (Fig. 4).

Fig. 4. Processing and displaying data on the bitcoin rate

As a result, a summary of data on the bitcoin rate is formed, which enters the program for further analysis every 3 s.

Then, in order for the user to receive notifications about a significant change in the exchange rate in one direction or another, tracking changes in the bitcoin rate is implemented in the form of a sequence of commands shown in Fig. 5.

```
# Проверка изменения валюты
def check_currency(self):
    currency = float((self.get_currency_price().replace(",", ".")).replace(u'\xa0','')
    if currency >= self.current_converted_price + self.difference:
        print("Курс сильно вырос, может пора что-то делать?")
        self.send_mail()
    elif currency <= self.current_converted_price - self.difference:
        print("Курс сильно упал, может пора что-то делать?")
        self.send_mail()
```

Fig. 5. Checking the change in the bitcoin rate

Now, with a significant change in the exchange rate, for which the user-settable *difference* parameter is used, a message is sent to the user's email using the SMTP network protocol (Fig. 6).

```
# Отправка почты через SMTP
def send_mail(self):
    server = smtplib.SMTP('smtp.gmail.com', 587)
    server.ehlo()
    server.starttls()
    server.ehlo()

    server.login('daniilkalmakov@gmail.com', 'password')

    subject = 'Currency mail'
    body = 'Currency has been changed!'
    message = f'Subject: {subject}\n{body}'

    server.sendmail(
        'От кого',
        'Кому',
        message
    )
    server.quit()
```

Fig. 6. Forming and sending a message to the user's email

The mail service gmail.com can be changed by the user to any other. Thus, when the cryptocurrency rate changes, a message is sent to the chat log, and this message is also sent to the user's email (Fig. 7).

```
Сейчас курс: 1 биткоина = 38638.1
Курс сильно вырос, может пора что-то делать?
```

Fig. 7. Sending cryptocurrency rate monitoring data to the user's email

The user receives the message shown in Fig. 8.

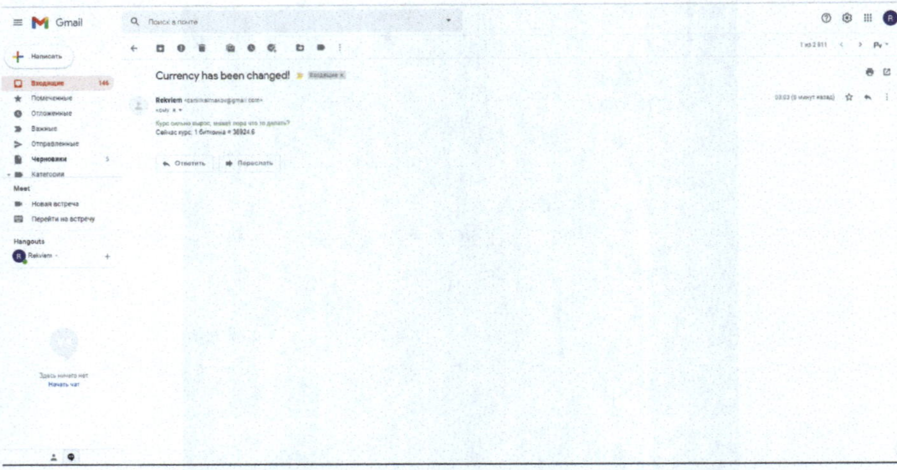

Fig. 8. Message from the user's email about a change in the bitcoin rate

Thus, the user will receive timely notifications about changes in the bitcoin exchange rate, which will allow him to timely and accurately use his financial assets in order to make a profit.

3 Conclusion

Real-time monitoring of the exchange rate is necessary for any investor who invests his savings in cryptocurrencies. This will help to profitably play on the jump in the exchange rate, avoiding mistakes in investing financial resources. The developed information system for monitoring the cryptocurrency exchange rate promptly receives up-to-date data on cryptocurrency quotes and monitors all changes in the currency price. It will be useful to everyone who works with exchange markets. The program interface is convenient and easy to use.

Further solutions to improve the automation of the information system for monitoring the exchange rate can be directed to the use of operational data from different sites for a deeper analysis and management of investors' money, as well as other additional functions for forecasting exchange rates [14].

Acknowledgments. This work was supported by the Russian Science Foundation, project no. 22-28-01868 «Development of an agent-based model of the network industrial complex in the context of digital transformation».

References

1. Akyildirim, E., Cepni, O., Corbet, S., et al.: Forecasting mid-price movement of Bitcoin futures using machine learning. Ann. Oper. Res. (2021). https://doi.org/10.1007/s10479-021-04205-x

2. Awotunde, J.B., Ogundokun, R.O., Jimoh, R.G., Misra, S., Aro, T.O.: Machine learning algorithm for cryptocurrencies price prediction. In: Misra, S., Kumar Tyagi, A. (eds.) Artificial Intelligence for Cyber Security: Methods, Issues and Possible Horizons or Opportunities. SCI, vol. 972, pp. 421–447. Springer, Cham (2021). https://doi.org/10.1007/978-3-030-72236-4_17
3. Butsenko, E.V.: Practical application of crowdfunding as an innovative financing mechanism (by the example of the development of rural territories of the region) (in Russian) Econ. Res. **3**, 2 (2014)
4. Butsenko, E.V.: The practice of using a network of economic and mathematical modeling of the process of investment planning. Tomsk State Univ. J. Econ. 1(33), pp. 147–158 (2016). (in Russian). https://doi.org/10.17223/19988648/33/12
5. Carvalho, C.E., Pires, D.A., Artioli, M., Oliveira, G.C.: Cryptocurrencies: technology, initiatives of banks and central banks, and regulatory challenges. Economia e Sociedade. **30**(2), 467–496 (2021). https://doi.org/10.1590/1982-3533.2021v30n2art08
6. Chuen, D.L.K., Teo, E.: The new money: the utility of cryptocurrencies and the need for a new monetary policy. In: Kaili, E., Psarrakis, D. (eds.) Disintermediation Economics, pp. 111–172. Springer, Cham (2021). https://doi.org/10.1007/978-3-030-65781-9_7
7. Fernandes, P.M.P.: Development of a mobile application and backend for the simulated use of cryptocurrencies (Doctoral dissertation, Universidade de Coimbra) (2021)
8. Fokri, W.N.I.W.M.: Classification of cryptocurrency: a review of the literature. Turk. J. Comput. Math. Educ. (TURCOMAT) **12**(5), 1353–1360 (2021). https://doi.org/10.17762/turcomat.v12i5.2027
9. Parlika, R., Atmaja, P.W.: Realtime monitoring of Bitcoin prices on several Cryptocurrency markets using Web API, Telegram Bot, MySQL Database, and PHP-Cronjob. In: 6th Information Technology International Seminar (ITIS), pp. 253–257 (2020). https://doi.org/10.1109/ITIS50118.2020.9321109
10. Parlika, R., Pratama A.: Use of the Web API as a basis for obtaining the latest data on bitcoin prices at 30 exchange places. IOP Conf. Ser.: Mater. Sci. Eng. **1125**, 012035 (2021)
11. Patashkova, Y., Niyazbekova, S., Kerimkhulle, S., Serikova, M., Troyanskaya, M.: Dynamics of Bitcoin trading on the Binance cryptocurrency exchange. Econ. Ann.-XXI, **187** (2021)
12. Putri, D.E., Ilham, R.N., Sinurat, M., Lilinesia, L., Saragih, M.M.S.: Analysis of potential and risks investing in financial instruments and digital cryptocurrency assets during the Covid-19 pandemic. J. SEKURITAS (Saham Ekon. Keuangan dan Investasi) **5**(1), 1–12 (2021)
13. Reddy, P.R.: Analysis of Bitcoin prices using deep learning techniques. Turk. J. Comput. Math. Educ. (TURCOMAT) **12**(6), 23–27 (2021)
14. Shorikov, A.F., Butsenko, E.V.: Network models for solving the problem of multicriterial adaptive optimization of investment projects control with several acceptable technologies. AIP Conference Proceedings. In: 9th International Conference for Promoting the Application of Mathematics in Technical and Natural Sciences, AMiTaNS 2017, p. 050013 (2017)
15. Sifat, I.: What Uncertainties Matter to Cryptocurrencies? SSRN. https://doi.org/10.2139/ssrn.3886238. Accessed 21 Dec 2021
16. Treiblmaier, H.: Do cryptocurrencies really have (no) intrinsic value? Electron Markets (2021). https://doi.org/10.1007/s12525-021-00491-2

Comparative Analysis of Machine Learning Methods for Assessing the Predictive Potential of Risk Factors for the Development of Cardiovascular Diseases

N. G. Plekhova[1], V. A. Nevzorova[1], I. N. Chernenko[1], L. G. Priseko[1](✉), and E. Yu. Shestopalov[1,2]

[1] Pacific State Medical University, 2 Ostryakova Avenue, Vladivostok 690002, Russia
pl_nat@hotmail.com

[2] Vladivostok Clinical Hospital No. 1, 22 Sadovaya Street, Vladivostok 690078, Russia

1 Introduction

The powerful development of modern technologies of laboratory diagnostics, including molecular diagnostics, offers a fairly large number of parameters for assessing the state of the body, the interpretation of which is difficult for doctors. From this perspective, machine learning methods can be useful as a means to determine predictively significant predictors among millions of phenotypic data points. Epidemiological studies in the field of creating prognostic models contain hundreds or thousands of variables and allow us to characterize the subclinical processes of the disease. The practical application of such models is justified by the identification of key risk factors for the development of CVD in order to timely apply preventive measures (for example, quitting smoking, statin therapy, blood pressure control) [1–6]. The forecast scales used for this purpose are based on data obtained as a result of large population studies, taking into account a set of firmly designated RF. Attempts to improve the predictive effectiveness of predictive scales consist in the allocation by expert communities of additional RF, Another principled approach to improving the prediction of CVD risks is the possibility of using artificial intelligence methods by creating a combination of complex mathematical models and algorithms in the presence of certain computing power conditions and improving the quality of databases [4].

The use of machine learning methods in processing large databases to identify new predictors and establish more complex and deep interactions between them creates other predictive possibilities in the field of a personalized approach to assessing CVD. When studying the possibilities of using machine learning methods, the attention of researchers is attracted by the resolution and reliability of using a particular method for conducting analytical calculations and constructing predictive models [2]. Objective: to evaluate the accuracy and reliability of machine learning-based models for solving problems of predicting the risks of cardiovascular diseases.

R. Silhavy (Ed.): CSOC 2022, LNNS 502, pp. 186–192, 2022.
https://doi.org/10.1007/978-3-031-09076-9_18

2 Materials and Methods

2.1 Data Source

The data of a single-stage anonymous survey of volunteers in Primorsky Krai were collected during the implementation of a multicenter observational study "Epidemiology of cardiovascular diseases in the regions of the Russian Federation, ESSE-RF". 2,131 people participated in the study on the basis of the written informed voluntary consent received.

2.2 Study Population

The data of volunteers aged 24 to 65 years at the time of the first examination included eight main baseline variables used in the established 10-year ACC/AHA risk prediction model [7]. Additionally, parameters were entered height, body weight, body mass index (BMI).

For the biochemical examination, following an overnight fast, blood samples were drawn into tubes and centrifuged the same day to separate serum, which was stored frozen for subsequent analysis. A venous blood sample was withdrawn on an empty stomach and parameters were deter-mined in certified laboratories using standard laboratory methods.

The starting date was set to January 1, 2014, which allowed monitoring of all patients in the cohort for 5 years. The end of the study period is defined as January 1, 2019. Persons with a history of cardiovascular diseases, hereditary lipid disorders, who were prescribed hypolipidemic drugs or outside the specified age range before or at the baseline date were excluded from the analysis.

2.3 Risk Factor Variables

In addition to the eight main risk variables for the development of CVD (see above) recommended by ACC/AHA 2013 [7], machine learning algorithms include indicators that can potentially be associated with CVD. These indicators were selected based on literature data, which indicates their potential as predictors of the risk of developing cardiovascular diseases [5–11]. For a general approach to working with missing values in machine learning, median imputation was used [12]. It was also hypothesized that the missing values of some clinical variables (for example, BMI and laboratory results) may indicate a perception of reduced relevance for some patients, given the incomplete recording of normal BMI values in primary health care medical records. In total, 26 variables were analyzed in machine learning models up to the basic level (with the exception of dummy variables for missing values).

2.4 Machine-Learning Algorithms

To compare machine-learning risk algorithms, the study population was split in the data set into a 'training' cohort in which the CVD risk algorithms were derived and a 'validation' cohort in which the algorithms were applied and tested. Using the randomization function 2 samples were formed: a training sample (80% of participants) and a test sample (20% of participants) to evaluate the effectiveness of the prognosis, which included data from patients with an established diagnosis of CVD. To create machine learning algorithms, the high-level Python 3.9.2 language (Python Software Foundation License) was used with the involvement of the Scilit-learn library, which represents a set of tools for data analysis. Four commonly used classes of machine-learning algorithms were utilised: logistic regression [13], random forest [14, 15], and neural networks [16, 17]. Replacement of missing data with an estimated value - imputation for all models was carried out using the missForest algorithm. The SCORE scale model is selected as the base model for comparison.

2.5 Statistical Analysis

A descriptive characteristic of the indicators of the studied population was presented. The accuracy of the machine learning prediction algorithms developed on the basis of the training cohort was evaluated using a 10-fold stratified cross-validation using a measure of the total area under the receiver operating characteristic curve (AUC-ROC). Using thresholds corresponding to a 10-year risk of CVD > 7.5%, as recommended by ACC/AHA [1], an analysis of binary classification of cases and non-cases in cohort validation was used to compare observed and expected forecasts. Statistical analysis evaluating the performance of the algorithm was performed using STATA 13 MP4.

3 Result

3.1 Data Extraction

In total, there were 2 500 volunteers in the database at the initial level (January 1, 2014) who met the eligibility criteria. After excluding 369 patients with coding errors (i.e. non-numeric entries for blood pressure/cholesterol) and extremely distant observations (> 5 SD from the mean), the analysis cohort consisted of 2131 volunteers. Then the cohort was randomly divided into a sample of 1 440 volunteers (80%) to create a machine learning algorithm, and the remaining sample was 360 volunteers to test it.

3.2 Study Population Characteristics

The average age of participants at the beginning of the study was 45.75 (11.7) years, males −874 (41%). During the follow-up period (5th-95th percentile: 3.44.7 years), 422 cases of CVD were detected in the age range of 60.2 ± 5.6 years for men and 61.1 ± 4.8 years for women. In the group of persons without arterial hypertension (AH, n = 1398), the presence of established CVD was noted in 13.38% (n = 187) of the subjects, whereas among persons with AH (n = 733) in 32.06% (n = 235).

3.3 Study Population Characteristics

The variables that make up the input data for machine learning models were trained using a cohort of 1,440 patients with 244 cases of CVD (35.1%) that developed during a 5-year follow-up period. As a first step to evaluate the im-portance of a variable in the training dataset, we used the training of a generated random forest. The second step was to evaluate the importance of displaying the parameter after training, for which their values were mixed and, again, the out-of-bag error was calculated. The importance of the parameter was estimated by averaging over all trees the difference in out-of-bag error indicators before and after mixing the values. Using the above permutation method allows us to estimate the decrease in the importance of each variable after the per-mutation of this variable for each decision tree.

In our work, we used a model of a multi-layer perceptron – the simplest kind of neural network. In such a model, the input signal propagates from layer to layer in a forward direction. The multilayer perceptron itself consisted of three main elements: an input layer including many neurons (20 variables), several layers of hidden computational neurons and one output layer. Training and optimization of the NS were carried out according to the adaptive estimation algorithm of the Adam moment, in 1000 epochs, the volume of data supplied at a time was 20 units. According to the diagram, the following variables make the greatest contribution to the implementation of the model: age, waist circumference, BMI, glucose, average heart rate.

3.4 Prediction Accuracy

The SCORE risk model served as the basis for comparison (AUC 0.728, 95% CI 0.723 ± 0.735). Sufficiently high values of the ROC AUC index −0.789 with a ratio to the base model SCORE of 8.98 indicate the greatest suitability of the neural network for predicting the selected diagnoses of CVD.

3.5 Classification Analysis

The basic SCORE model correctly predicted 1316 cases out of 2100 total cases, resulting in a sensitivity of 62.7%. The random forest algorithm resulted in a network increase of 191 cases of CVD compared to the base model, an increase in sensitivity to 65.3%. Neural networks worked best to improve the accuracy of forecasting, which showed an increase in the CVD to 1564 cases (sensitivity 74.5%) correctly predicted cases, respectively.

4 Discussion

Whereas, the advantage of multiple linear regression compared to a simple one is the use of several input variables in the model, which makes it possible to increase the proportion of the explained variance of the output variable and a hyperplane is used instead of a line [18]. This model is most often used in medicine, because of its ease of use and good accuracy. Using the least squares method, the indicators with the least im-portance were removed from the initial set of variables (26 indicators) and left with the greatest significance for the forecast (15 indicators). Based on the latest parameters, a regression forecast model was built to calculate the risk of developing CVD.

During the construction of the neural network, we used a multi-layer perceptron model, where the input signal propagates from layer to layer in a forward direction. In a series of preliminary computational experiments using the construction of predictive models (decision trees, multiple linear regression and neural networks), the best result was shown by the multilayer perceptron model, which is a model for the layered construction of a machine learning algorithm. The obtained indicators using ROC AUC testified to a high level of prediction of the CVD using machine learning methods. At the same time, the model of decision trees was preferable for applying the solution to the problem of determining the significance of risk factors in the development of the disease. The strength of the analysis is a fairly large amount of data during the regional stage "ESSE-RF, 2014–2019", in which a survey of the population of Primorsky Krai was conducted for 5 years. Focusing on specific CVD in conjunction with the initial examination data of conditionally healthy individuals who have not yet been diagnosed will increase the accuracy of predicting the risk of developing the disease. To improve the quality of models, additional development of various methods of preprocessing medical data is necessary.

The use of three machine learning algorithms in the study shows differences in the degree of importance of risk factors depending on the modeling method. This can help in the further study of various prognostic risk factors and the future development of new approaches to risk prediction and the algorithm for the development of CVD. Finally, the importance of missing values or lack of answers is often not evaluated using traditional tools for predicting the risk of CVD [7, 19]. Whereas our study demonstrates that missing values, in particular routine biometric variables such as BMI, are independent predictors of CVD.

5 Conclusion

The use of neural networks with a multi-layer perceptron, as one of the methods of layer-by-layer construction of a machine learning algorithm, is the most optimal for creating a prediction model and currently gives the most reliable result. A significant increase in the reliability of machine learning-based models is due to the fact that they use a larger number of predictors than when calculating on the SCORE scale. The use of a set of personal data obtained as a result of a survey, clinical and laboratory studies for machine learning in predicting diseases allows you to create individual therapeutic recommendations in a targeted manner.

References

1. Greenland, P., Alpert, J.S., Beller, G.A., Benjamin, E.J., Budoff, M.J., Fayad, Z.A., et al.: 2010 ACCF/AHA guideline for assessment of cardiovascular risk in asymptomatic adults: a re-port of the American college of cardiology foundation/American heart association task force on practice guidelines developed in collaboration with the American society of echocardiography, American society of nuclear cardiology, Society of atherosclerosis imaging and prevention, society for cardiovascular angiography and interventions, society of cardiovascular computed tomography, and society for cardiovascular magnetic resonance. J. Am. Coll. Cardiol. **56**(25), e50–e103 (2010). https://doi.org/10.1161/CIR.0b013e3182051b4c
2. Sjostrom, L., Lindroos, A.K., Peltonen, M., Torgerson, J., Bouchard, C., Carlsson, B., et al.: Life-style, diabetes, and cardiovascular risk factors 10 years after bariatric surgery. N. Engl. J. Med. **351**(26), 2683–2693 (2004). https://doi.org/10.1056/NEJMoa035622
3. Goldstein, B.A., Navar, A.M., Carter, R.E.: Moving beyond regression techniques in cardiovascular risk prediction: applying machine learning to address analytic challenges. Eur. Heart J. **38**, 1805–1814 (2017). https://doi.org/10.1093/eurheartj/ehw302
4. Piepoli, M.F., Hoes, A.W., Agewall, S.: European guidelines on cardiovascular disease prevention in clinical practice. Eur. Heart J. **37**(29), 2315–2381 (2016). https://doi.org/10.17863/CAM.17
5. Ambale-Venkatesh, B., et al.: Cardiovascular event prediction by machine learning: the multiethnic study of atherosclerosis. Circ. Res. **121**(9), 1092–1101 (2017). https://doi.org/10.1161/circresaha.117.311312
6. Weng, S.F., Reps, J., Kai, J., Garibaldi, J.M., Qureshi, N.: Can machine-learning improve cardio-vascular risk prediction using routine clinical data? PloS one **12**(4), e0174944 (2017). https://doi.org/10.1371/journal.pone.0174944
7. Goff, D.C., Lloyd-Jones, D.M., Bennett, G., Coady, S., D'Agostino, R.B., Gibbons, R., et al.: ACC/AHA guideline on the assessment of cardiovascular risk: a report of the American college of cardiology/American heart association task force on practice guidelines. Circulation **135**(11), 1–50 (2013)
8. Plekhova, N. G., et al.: Association of cardiovascular events and blood pressure and serum lipoprotein indicators based on functional data analysis as a personalized approach to the diagnosis. In: Silhavy, R., Silhavy, P., Prokopova, Z. (eds.) CoMeSySo 2020. AISC, vol. 1295, pp. 278–293. Springer, Cham (2020). https://doi.org/10.1007/978-3-030-63319-6_24
9. Emerging risk factors collaboration. C-reactive protein, fibrinogen, and cardiovascular disease prediction. N. Engl. J. Med. **367**(14), 1310±20 (2012). https://doi.org/10.1056/NEJMoa1107477, PMID: 23034020
10. Osborn DP, Hardoon S, Omar RZ, Holt RI, King M, Larsen J, et al. Cardiovascular risk prediction models for people with severe mental illness: results from the prediction and management of cardiovascular risk in people with severe mental illnesses (PRIMROSE) research program. JAMA Psychiatry **72**(2), 143±51 (2015). https://doi.org/10.1001/jamapsychiatry.2014.2133, PMID: 25536289
11. Wannamethee, S.G., Shaper, A.G., Perry, I.J.: Serum creatinine concentration and risk of cardiovascular disease: a possible marker for increased risk of stroke. Stroke J. Cereb. Circ. **28**(3), 557±63 (1997)
12. Batista GEAPA, Monard MC. An analysis of four missing data treatment methods for supervised learning. Appl. Artif. Intell. **17**(5–6), 519–33 (2003)
13. Hosmer, D.W., Lemeshow, S., Sturdivant, R.X.: Applied Logistic Regression, 3rd edn. Wiley, New Jersey (2013)
14. Breiman, L.: Random forests. Mach. Learn. **45**(1), 5±32 (2001)

15. Rigatti, S.J.: Random forest. J. Insurance Med. (New York, N.Y.), **47**(1), 31–39 (2017). https://doi.org/10.17849/insm-47-01-31-39.1
16. Hagan, M., Demuth, H., Beale, M., De Jesus, O.: Neural Network Design, 2nd edn. PWS Publishers, Boston (2014)
17. Uddin, S., Khan, A., Hossain, M.E., Moni, M.A.: Comparing different super-vised machine learning algorithms for disease prediction. BMC Med. Inform. Decis. Mak. **19**(1), 281 (2019). https://doi.org/10.1186/s12911-019-1004-8
18. Zahid, F.M., Heumann, C.: Multiple imputation with sequential penalized regression. Stat. Methods Med. Res. **28**(5), 1311–1327 (2019). https://doi.org/10.1177/0962280218755574
19. Quesada, J.A., et al.: Machine learning to predict cardiovascular risk. Int. J. Clin. Pract. **73**(10), e13389 (2019). https://doi.org/10.1111/ijcp.13389

Ontological Models as Components of the Decision Support Subsystem in Monitoring and Forecasting Systems of Hazardous Processes and Natural Phenomena of Coastal Infrastructure

I. B. Safronenkova[1(✉)], A. A. Rodina[1], and Y. E. Melnik[2]

[1] Federal Research Center The Southern Scientific Center of the Russian Academy of Sciences, 41, Chekhov Street, 344006 Rostov-on-Don, Russian Federation
safronenkova050788@yandex.ru

[2] Academy for Engineering and Technologies, Institute of Computer Technologies and Information Security, 44, Nekrasovsky Lane, GSP-17, 347922 Taganrog, Russian Federation

Abstract. The emergence of a new class of systems for monitoring and forecasting hazardous processes and natural phenomena, capable of taking into account information received from network users, actualizes the issues related to the collection, processing and transmission of large volumes of heterogeneous and semi-structured data from large geographically distributed territories. At the same time, one of the main problems is the integration and further analysis of such data in order to increase the level of decision-making automation. Ontological models make it possible to systematize and formalize semi-structured data, to obtain knowledge about the domain on their basis. This paper considers the architecture of a system for monitoring and forecasting hazardous natural processes, including an intellectual analysis module. It is proposed to use ontology models to determine the vertices of a cognitive map, on the basis of which further analysis will be carried out and a decision will be made. In this regard, the purpose of this study is to determine the vertices of the cognitive map of the aquatic ecosystem of the Ponto-Caspian region. To achieve this goal, it is necessary to develop domain ontologies that reflect knowledge about the domain of hazardous processes and natural phenomena of the coastal infrastructure of the Ponto-Caspian region, as well as knowledge taking into account information received from network users for a monitoring and forecasting system.

Keywords: Monitoring and forecasting systems · Ontological model · Cloud computing · Fog computing · Hazardous processes · Natural phenomena

1 Introduction

Over the past decades, scientists have noted a dramatically increased level of anthropogenic impact on the environment, which is primarily associated with the intensive

R. Silhavy (Ed.): CSOC 2022, LNNS 502, pp. 193–202, 2022.
https://doi.org/10.1007/978-3-031-09076-9_19

development of production and irrational consumption of natural resources [1, 2]. It entails an increase in the scale and frequency of natural hazards, various man-made processes and their consequences. The most vulnerable areas include coastal zones, where any natural disaster can cause a whole series of man-made accidents and lead to significant human casualties and material losses [3]. In this regard, the problems associated with the development and improvement of monitoring and forecasting systems in terms of collecting, processing and transferring large volumes of heterogeneous geographically distributed data, as well as identifying patterns of manifestation of natural disasters in coastal areas, are actualized. These problems do not lose their relevance even with the emergence of a new class of intelligent monitoring and forecasting systems capable of taking into account information from users when making a decision [4]. This is due to the fact that this information is also, as a rule, semi-structured, heterogeneous, which is collected from geographically distributed infrastructure.

Technologies of the "digital economy", such as the concepts of cloud and fog computing, distributed ledger technologies, solve most of the problems in terms of geographic distribution [5, 6]. However, the problems associated with data integration and their further analysis need to be addressed.

The use of intelligence in terms of data processing allows to increase the level of automation of the decision-making process and, thereby, to increase the efficiency of monitoring and forecasting systems. Currently, to solve problems associated with the processing of large heterogeneous and semi-structured data, including their integration, ontological models are widely used [7], which allow to formalize knowledge about the domain and data describing the domain.

In this study, it is proposed to use an ontological approach to knowledge representation in a new class of systems for monitoring and forecasting hazardous processes and natural phenomena of the coastal infrastructure of the Ponto-Caspian region, which allow taking information from users into account when making decisions based on cognitive analysis.

2 Review of Monitoring and Forecasting Systems Based on Digital Economy Technologies

Currently, monitoring and forecasting systems are used in a wide variety of areas of human activity. Modern information technologies make it possible to collect heterogeneous big data from geographically distributed areas, process it, carry out intelligent analysis and make decisions at an automated or automatic level. In view of this, at present, monitoring systems are used not only for the prevention, control and assessment of emergencies caused by natural or anthropogenic factors, such systems are also actively used in agriculture and forestry, land use, to control water resources, snow reserves and ice conditions [8].

Intellectualization of monitoring and forecasting systems implies the active use of domain models, processing input data, and its subsequent analysis, i.e. the transition from data to knowledge, which includes hidden dependencies and forecasting in the development of situations.

The analysis of open sources allows synthesizing a typical structure of an intelligent monitoring and forecasting system [8–10] and presenting it in the following form (see Fig. 1).

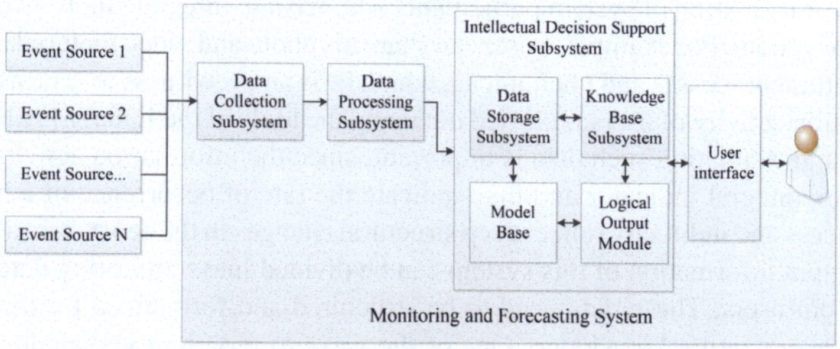

Fig. 1. The typical structure of a monitoring and forecasting system.

As seen from Fig. 1, the intellectual core of the system is the decision support subsystem, which includes the following blocks: database, knowledge base, model base, logical conclusion. Moreover, the forms of information presentation in the database, knowledge base and model base can be very diverse, as well as the methods used to support decision-making in the logical conclusion block.

For example, the work [11] proposes the concept of an intelligent system for monitoring and predicting hazardous processes and natural phenomena of the coastal infrastructure of the Ponto-Caspian region, including a decision support subsystem that monitors the social activity of Internet users. The enlarged architecture of the decision support subsystem is shown in Fig. 2.

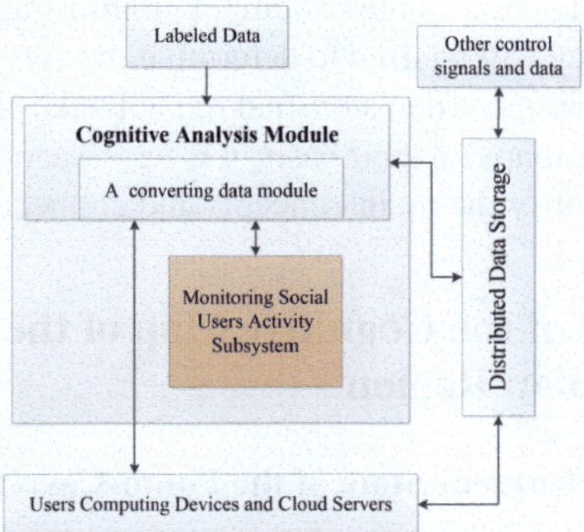

Fig. 2. The enlarged architecture of the decision support subsystem.

The subsystem includes a cognitive analysis module, which includes software for constructing cognitive maps and their analysis, realizes the specification of the system for a specific geographic region (in our case, the Ponto-Caspian). The module for converting data from monitoring natural phenomena into the format of a cognitive simulation model of the ecosystem is used for automated collection of data from various sources, including meteorological measurement data, data on the state of macrozoobenthos, and hydrological research data.

The module for monitoring the social activity of Internet users allows you to take into account the information of system participants who wish to integrate their own research data into the system. For example, a user can share his photo and video materials about the ongoing natural processes and phenomena, which he is an eyewitness to. The assessment of information activity of users is carried out using the tools of the IQbuzz [12] platform. Obtaining and evaluating such data is important, since the information activity of users is a complex integral indicator and may indicate the rate of occurrence of a hazardous natural process and indirectly reflect deep structural changes in the ecosystem as a whole. Thus, the input information of this system can be divided into monitoring data and data received from users. These data need to be structured and formalized for their further processing, i.e. gaining knowledge. One of the ways to represent knowledge is to use ontological models.

The formal model of the ONT ontology is as follows:

$$ONT = < C, P, R, A > \quad (1)$$

where C - a finite set of concepts (classes of entities) of the domain,

P – a finite set of properties of these concepts (classes);

R – a finite set of connections between concepts (classes);

A – a set of axioms, statements built from these concepts, their properties and connections between them.

Thus, an ontology is a system consisting of a set of concepts and a set of statements about these concepts, on the basis of which classes, objects, relations, functions and theories can be built [13].

It is assumed that the developed ontological models will be considered as structural components of the knowledge base of the system for monitoring and predicting hazardous processes, in particular, they will be used to determine the vertices of the cognitive map, on the basis of which modeling and decision-making will take place. To develop ontology models and build cognitive maps on their basis, it is necessary to comprehensively study the domain in order to identify the main concepts and connections between them.

3 The Development of the Cognitive Map of the Aquatic Ecosystem of the Ponto-Caspian Region

3.1 The Analysis of the Current State of the Ponto-Caspian Region

In recent years, coastal areas have come to be viewed as a single natural entity, consisting of many components and subject to a variety of extreme natural events that are a source of risks to human activities. However, it is important to note that since the beginning of the 21st century, not a single natural disaster, even for the small Azov-Black Sea basin, has been predicted [14]. In addition, in recent decades, previously unnoticed or rare natural phenomena began to develop in the Azov Sea basin, which may indicate both the onset of long-term dangerous changes that can lead to negative consequences for the population, and changes in the state of the ecosystem, which will not have significant impact on the economic state of the region and on people's lives:

1) Extreme ice events in the Azov Sea (extreme winter 2005–2006, which caused hummocking) [14].
2) Catastrophic floods in the Taganrog Bay and the Don delta on March 23–24, 2013 and September 24, 2014, when the hurricane advection of the Black Sea waters into the shallow Azov reservoir led to the influx of sea water with salinity up to 4–6% into the water supply system of the city of Azov [14].
3) A sharp decrease in the water level in the estuary area and delta of the Don River during extreme easterly winds, causing a decrease in the water level dangerous for navigation (October 2015 and 2016).
4) Development of thermal turbulence along the edge of the Don delta and on the Taganrog coastal area (July 2018), wind gusts up to 18–21 m/s were observed in the near-surface air layer [15].
5) Extreme surge in the Don River and the Azov Sea (November 15–26, 2019) For the first time in 20 years, the water in the Azov Sea retreated 100 m from the coastline. The drop in the level of the Don River in some areas exceeded 2 m. The air temperature these days dropped to −11 °C. Mass death of benthic invertebrates in the communities of the coastal zone was noted.

The occurrence of natural hazards and disasters in the coastal zone requires the pooling of complex efforts and the use of the most modern models in order to understand their nature, predict and reduce negative consequences.

Based on the methodology for monitoring estuarine zones in southern Russia and analyzing the current state of the Ponto-Caspian region, the main parameters have been identified that indicate the possible onset of dangerous phenomena. These include hydrobiological, hydrochemical, meteorological and other parameters.

3.2 Ontological Models Development

The Development of the Ontology Model of the Parameters of Hazardous Natural Phenomena for the Ponto-Caspian Region. The purpose of an ontology development is the need for integration, formalization, systematization and establishment of relationships between heterogeneous data belonging to the considered domain.

The development of this ontological model is carried out in accordance with the methodology described in the work of Natalya F. Noy and Deborah L. McGuinness [16], and comes down to the following stages:

- determination of the scope and scale of the ontology,
- consider options for reusing existing ontologies,
- enumeration of important terms in the ontology,
- definition of classes and class hierarchy,
- defining the properties of classes (slots); determination of restrictions on the values of properties (slots); creation of instances.

A complete and comprehensive analysis of the domain showed the need to highlight the following concepts that are included in the structure of the ontology for monitoring and predicting hazardous processes and phenomena: the state of the ecosystem and external factors. In terms of ontology, these concepts are reflected in two classes of one hierarchical level: Cl1 "Ecosystem_Condition", Cl2 "External_Factors". To describe the concept of the state of the ecosystem, it is necessary to introduce the concepts of basic parameters and objective function. In terms of ontology, these concepts will be subclasses of Cl1 "Ecosystem_Condition" and will be at the same hierarchical level: SubCl11 "Base_parameters", SubCl12 "Objective function". The objective function reflecting the state of the ecosystem can take two values, which is reflected in the instances of the developed ontology: Inst121 "transition_new_state" Inst122 "preservation_pre-existing_state". In turn, SubCl12 "Objective_function" has the following subclasses: SubCl 121 "Dangerous_hydrological_phenomena", SubCl 122 "Dangerous_meteorological_phenomena", SubCl 123 "Natural_fires". For the purpose of a detailed description of the domain, it is necessary to fill in each of the sub-subclasses (SubCl 121-SubCl 123) with specific examples. For example, SubCl 121 "Dangerous_hydrological_phenomena" includes such dangerous hydrological phenomena as inundation, flood, water_logging, which in the language of ontology are called instances of this class. SubCl 11 "Base_parameters" has the following instances: Inst 111 "disturbing", Inst 112 "indicator", Inst 113 "regulatory", Inst 114 "manager". Cl2 "External_Factors" has two subclasses: SubCl 21 "Hydrological", SubCl 22 "Meteorological". In turn, SubCl 21 "Hydrological" is divided into the following sub-subclasses: SubCl 211 "Hydrobiological", SubCl 212 "Hydrochemical". The following indicators can be considered as instances of SubCl 211 "Hydrobiological": Inst 2111 "fish_biomass", Inst 2112 "primary_production", Inst 2113 "area_distribution_ bottom_communities", Inst 2114 "area_flowering", Inst 2115 "average_annual_ benthos_biomassa". In SubCl 212 "Hydrochemical" the following instances are isolated: Inst 2121 "current_salinity", Inst 2122 "mineralization", Inst 2123 "nutrient_load", Inst 2124 "average_annual_salinity". In SubCl 22 "Meteorological" there are: Inst 221 "atmospheric_pressure", Inst 222 "wind_activity", Inst 223 "precipitation", Inst 224 "current_air_temperature". The hierarchical structure of the developed ontology for monitoring and forecasting hazardous processes and phenomena is shown in Fig. 3.

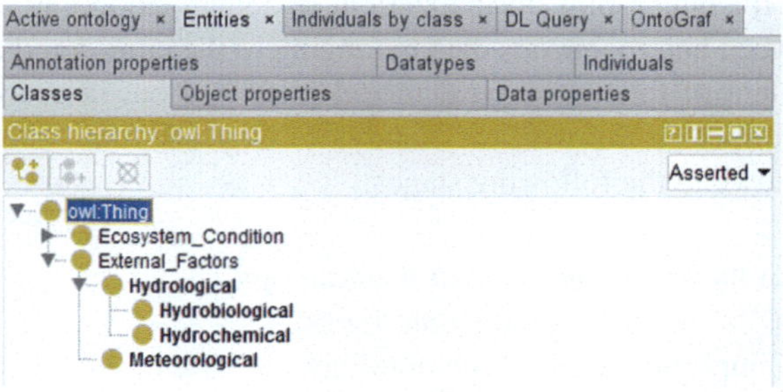

Fig. 3. The hierarchical structure of the developed ontology.

There is an example of filling the SubCl 212 "Hydrochemical" with instances (see Fig. 4).

Fig. 4. The list of instances of SubCl 212 "Hydrochemical".

The influence of biological monitoring parameters (including those related to the analysis of the behavior of benthos) in the form of a graph is shown. The proposed ontology also shows a relationship between the Cl2 "External_Factors" and the SubCl 11 "Base_parameters" through the "have" relation (see Fig. 5).

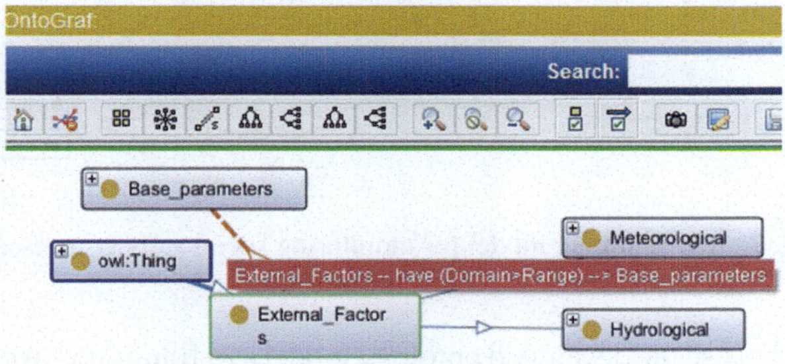

Fig. 5. The relationship between the Cl2 "External_Factors" and the SubCl 11 "Base_parameters".

The ontological model of the parameters of natural hazards for the Ponto-Caspian region is shown in Fig. 6.

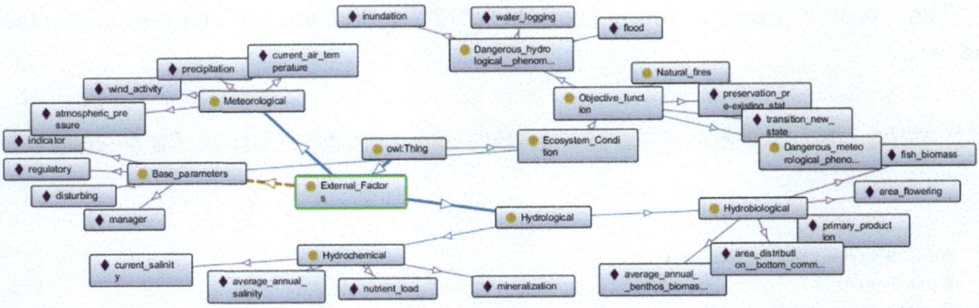

Fig. 6. The ontological model of the parameters of natural hazards for the Ponto-Caspian region.

The Development of the Ontology Model for Monitoring Social Activity of Users. The developed ontology model contains the following concepts: "User", "Message", "Type", "Authority", "Social_Networks". Each of the listed concepts contains a set of instances, and connections of relations among themselves, which are reflected in the ontology model (see Fig. 7).

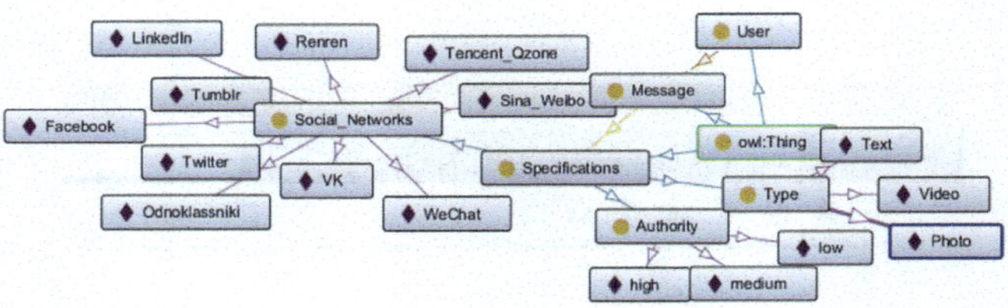

Fig. 7. The ontology model for monitoring social activity of users.

It is proposed to use the developed ontology model as a framework for the knowledge base in the prototype of the intelligent monitoring and forecasting system proposed in [11], including using these models to determine the vertices of the cognitive map of the aquatic ecosystem of the Ponto-Caspian region.

An example of a cognitive map of the aquatic ecosystem of the Ponto-Caspian region, reflecting the relationship and significance of parameters and processes to be monitored and predicted, taking into account the assessment of information activity of users of social networks regarding the onset of a hazardous natural phenomenon is shown in Fig. 8.

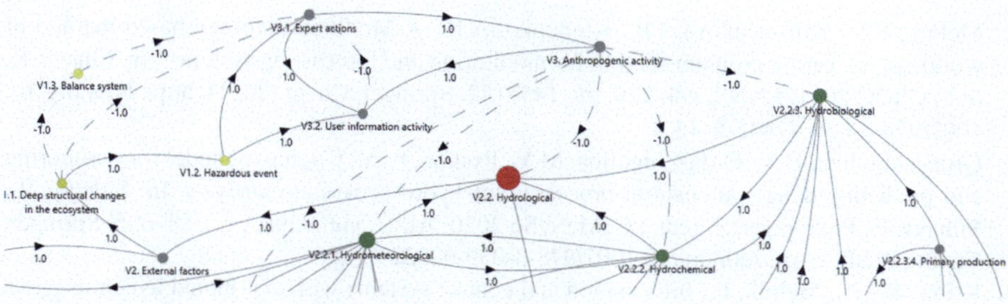

Fig. 8. An example of a cognitive map of the aquatic ecosystem of the Ponto-Caspian region.

When developing this cognitive model, an analysis of the current state of macrozoobenthos populations was carried out, as well as hydrobiological and hydrochemical monitoring of the aquatic ecosystem of the Ponto-Caspian region, an assessment of the information activity of users of social networks regarding the onset of a dangerous natural phenomenon was carried out.

4 Conclusion

In this paper, it is proposed to use ontological models as components of decision support subsystems in the monitoring and forecasting system of hazardous processes and natural phenomena of coastal infrastructure, capable of taking into account information from users to make a decision. An ontological model of the parameters of hazardous natural phenomena for the Ponto-Caspian region and an ontological model of monitoring the social activity of users have been developed. These models make it possible to formally describe the parameters that carry information about hazardous phenomena, as well as systematize external factors affecting the state of the ecosystem, as well as take into account user information in order to be able to monitor and predict hazardous phenomena and processes in the coastal zone. On the basis of the developed ontological models, the vertices of the cognitive map are determined, on the basis of which an intelligent analysis is carried out, which makes it possible to increase the level of automation in the decision-making procedure in the system for monitoring and predicting hazardous processes and natural phenomena of the coastal infrastructure of the Ponto-Caspian region.

Acknowledgments. This study is supported by the by the RFBR project 18-29-22046, 20-04-60485 and the GZ SSC RAS 122020100270-3.

References

1. Arihilam, N.H., Arihilam, E.: Impact and control of anthropogenic pollution on the ecosystem – a review. J. Biosci. Biotechnol. Discov. **4**, 54–59 (2019)
2. Chmielewski, J., Kusztal, P., Żeber-Dzikowska, I.: Anthropogenic: impact on the environment (case study). Ochrona Srodowiska i Zasobów Naturalnych. **29**, 30–37 (2018)

3. Melnik, E.V., Safronenkova, I.B., Klimenko, A.B.: A Modified ontology-based method of workload relocation problem solving for monitoring and forecasting systems. In: Silhavy, R. (ed.) CSOC 2021. LNNS, vol. 229, pp. 145–152. Springer, Cham (2021). https://doi.org/10.1007/978-3-030-77445-5_13
4. Orda-Zhigulina, D.V., Orda-Zhigulina, M.V., Rodina, A.A.: Cognitive model for monitoring and predicting dangerous natural processes for hydro ecosystem analysis. In: Silhavy, R., Silhavy, P., Prokopova, Z. (eds.) CoMeSySo 2020. AISC, vol. 1294, pp. 688–695. Springer, Cham (2020). https://doi.org/10.1007/978-3-030-63322-6_58
5. Klimenko, A., Melnik, E.: Information and control systems with distributed ledger usage: a reliability issue. In: Silhavy, R. (ed.) CSOC 2021. LNNS, vol. 229, pp. 133–144. Springer, Cham (2021). https://doi.org/10.1007/978-3-030-77445-5_12
6. Moura, J., Hutchison, D.: Fog computing systems: State of the art, research issues and future trends, with a focus on resilience. J. Netw. Comput. Appl. **169**, 102784 (2020)
7. Flotyński, J., Walczak, K.: Ontology-based representation and modelling of synthetic 3D content: a state-of-the-art review: ontology-based representation and modelling of synthetic 3D content. Comput. Graph. Forum **36**(8), 329–353 (2017)
8. Safiullin, R.N., Afanasyev, A.S., Rezchenko, V.V.: The concept of development of monitoring systems and management of intelligent technical complexes. J. Min. Inst. **237**, 322–330 (2019)
9. Abramov, N., Ardentov, Andrei, Emeljanova, Ju, Talalaev, A., Fralenko, Vitaly, Shishkin, O.: The architecture of the system for spacecraft state monitoring and forecasting. Program Syst. Theor. Appl. **6**, 85–99 (2015)
10. Sowa, J.F.: Architectures for intelligent systems. IBM Syst. J. **41**(3), 331–349 (2002)
11. Melnik, E.V., Orda-Zhigulina, M.V., Orda-Zhigulina, D.V.: Distributed library model based on distributed ledger technology for monitoring and diagnostics system. In: Silhavy, R. (ed.) CSOC 2021. LNNS, vol. 230, pp. 501–509. Springer, Cham (2021). https://doi.org/10.1007/978-3-030-77442-4_43
12. IQBuzz. https://startpack.ru/application/iqbuzz-smm. Accessed 03 Dec 2021
13. Gruber, T.: Ontology. In: Ling, L., Özsu, M.: Tamer (eds.) Encyclopedia of Database Systems. Springer, Verlag (2008). https://doi.org/10.1007/3-540-57697-5_13
14. Matishov, G.G., Matishov, D.G., Berdnikov, S.V., Yaitskaya, N.A.: Natural disasters in the Sea of Azov and Black Sea basin in the early 21st century. In: SSC RAS, Rostov-on-Don (2017)
15. Matishov, G.G., Valov, G.V., Kirillova, E.E.: Experience of registering squall wind in the don river delta-sea front. Sci. South Russ. **15**(1), 54–59 (2019)
16. Ontology development 101: a guide to creating your first ontology, ontology-tutorial-noy-mcguinness.pdf (stanford.edu). Accessed 03 Dec 2021

Implementation of Data Mining on a Secure Cloud Computing Over a Web API Using Supervised Machine Learning Algorithm

Tosin Ige[1(✉)] and Adewale Sikiru[2]

[1] Western Michigan University, Kalamazoo, MI 49008, USA
tosinolusola.ige@wmich.edu

[2] Virginia Technological University, Blacksburg, VA 24060, USA

Abstract. Ever since the era of internet had ushered in cloud computing, there had been increase in the demand for the unlimited data available through cloud computing for data analysis, pattern recognition and technology advancement. With this also bring the problem of scalability, efficiency and security threat.

This research paper focuses on how data can be dynamically mine in real time for pattern detection in a secure cloud computing environment using combination of decision tree algorithm and Random Forest over a web Application Programming Interface (API). We are able to successfully Implement data mining on cloud computing without any direct interaction with data warehouse and without any terminal involve by using combination of IBM Cloud storage facility, Amazing Web Service, Application Programming Interface and Window service along with a decision tree and Random Forest algorithm for our classifier. We were able to successfully bypass direct connection with the data warehouse and cloud terminal with 94% accuracy in our model.

Keywords: Cloud computing · Data warehouse · Data mining · Window service · Web API · Machine learning algorithm

1 Introduction

As millions of data are constantly being stored and retrieve through cloud computing on daily bases. With these also arises a new challenge on how to make use of the unlimited data available through data mining on cloud computing, and how it can be done in a highly secure, scalable and efficient manner to combat several security threats associated with cloud computing.

Economist needs to know how they can use readily available data in the cloud to predict consumers need and behavior, meteorologists need those data to make future weather forecast and predict or detect climatic change, government agency needs those data to make effective policy, police and intelligence agency needs data for background check and so on. All government and non-government parastatal depends on data from the cloud for one reason or the other. This further brings about the importance and urgency of mining data to detect previously unknown pattern in a scalable and efficient

R. Silhavy (Ed.): CSOC 2022, LNNS 502, pp. 203–210, 2022.
https://doi.org/10.1007/978-3-031-09076-9_20

manner through cloud computing to address daily needs with the guarantee of maximum security, authorization and authentication of the API user.

In this research, we made use of constantly updating real time data and without any connection to the cloud data warehouse for maximal data protection and dynamic predictive pattern with high accuracy. To achieve our objective, an application Programming Interface (API) and Background window service was developed to detect and fetch new and updated records from data warehouse, transformed to json and written to a file on the cloud. This ensures maximum security to the data as there is no direct connection to the data warehouse where we are getting data from. We used IBM Cloud object storage facility to host our csv file while the data warehouse is also on the cloud but from another channel entirely. Then we developed a restful API using Django python framework which was deployed to Amazon Web Services (AWS). Since our newly created API will be constantly pulling data from the data warehouse and writing directly to our CSV file which is on the IBM Cloud storage facility. The service can easily be overwhelmed, to prevent the service from being overwhelm, we developed a background service on a Microsoft console using Microsoft .NET Technology programming language.

Any record the web API misses or delayed in picking from the warehouse, the background service will pick it up. Hence, they complement each other and ensure efficiency so that the web API service is not overwhelmed. With this in place, we have a constantly updated and up to date data on our CSV file on the IBM cloud storage facility which is now our primary source of data. This ensure accessibility to updated records at every stage of our data mining activities, as there is an existing restful API and background window service that fetches data from the warehouse to the hosted file in cloud environment from which data will be mined from.

We also used combination of decision tree algorithm and Random Forest for classifier and also implement a graphic user interface (GUI) that automatically processed and displays newly discovered patterns in a graphical format on the screen at each and every update on new data.

We have four aims;

1. We don't want to connect or retrieve directly from the data warehouse
2. We needed access to real time data in an enabling environment on cloud
3. We want to shield the warehouse for maximum security and while also avoiding data lock.
4. We also want to automate the detection of any new pattern from the dataset and project to the screen with graphical illustration and analysis

To ensure, we achieve our fourth aim, we used a scheduler called Python schedule library so that the whole process is automatically initiated and repeated immediately a new pattern is detected.

2 Background Study

The importance of cloud computing cannot be overemphasize as it includes; unlimited storage, provisioning and updating, guaranteed privacy more secure (Ko et al. 2011).

Also, it is possible for users of cloud services to optimize server utilization, dynamic scalability, and minimize the development of new application life cycles (Al-Ruithe et al. 2019). In addition to the numerous benefits in cloud computing, there are also problems as a result of cloud outage since data storage is centralize in the cloud which can paralyze a company business (Gupta and Gupta 2014), also attack on the integrated cloud environment can cause loss of data and finance for both the service providers and subscribers (T.K and B 2016). There are also other risks associated with computing on cloud environment which includes the issues of threats to data security information confidentiality (Paquette et al. 2010, Wang 2011), and the possibility of information leakage and vulnerability (Inuwa 2015, Tchernykh et al. 2019).

Some of the problems associated with data mining on cloud computing are as a result of the existing method adopted in the industry and scientific world at large. Current cloud computation for data mining needed service provider to provide interface for user. The user does not need to bother about the infrastructure. It enables the retrieval of useful previously unknown data from integrated data warehouse, in such a way that users doesn't need to border about infrastructure, storage, or configuration and maintenance, the provider handles that. It is based on retrieving directly from the integrated data warehouse.

Ruxandra-Ştefania PETRE in his research work "Data mining in Cloud Computing", relies on extraction of previously unknown or meaningful pattern from unstructured or semi-structured data from the web sources. "The analysis steps in the Knowledge Discovery and Databases process" (Nodine, Ngu, Cassandra and Bohrer 2003). It listed three stages of research involving data warehouse which are staging, Integration, accessibility for the purpose of reporting and analysis. In the Review of Data Mining Techniques in Cloud Computing Database by Astha Pareek et al.

In addition to the problems associated with data mining on cloud computing in which majority of the problems are due to existing methods of data mining on cloud computing. The existing methods also creates a wide gap between data mining on cloud computing and application Programming Interface (API) which are yet to be closed. In the existing method we are yet to see an instance in which we call we only need to call an API endpoint, and then have everything done for us.

3 Research Methodology

For this research, we used data obtained from social network from github. To begin this research, we did four basic things;

1. Subscription to IBM Cloud Object Storage facility to host our CSV file
2. Setting up a data warehouse on cloud from another channel different from IBM
3. Developed a background window service using.NET Technology
4. Wrote a web service to be consume in the cloud using python
5. Scheduler using python scheduler library to trigger the web API at interval.

In order to prevent the web service from being overwhelm due to multiple calling of the endpoint at regular interval, we developed a background window service to support the restful API service. Both the web API and the window service are picking records from the integrated data warehouse and pushing to the CSV file on the IBM Object Cloud Storage facility. They automatically pick new records to the CSV, and if a record is modified, it will be picked and modified on the CSV as well. The essence of the scheduler which was written in python is to be calling the web API at regular interval to check and push from the integrated data warehouse to the CSV file.

Having set up our cloud environment and the necessary programs written, we proceeded by adopting the following data cleansing and preparation techniques;

Data Cleaning. We use preprocessing and cleaning methods to remove incomplete data that might cause system failure and also affect output prediction. Rows containing missing values where completely removed. We also use different methods to identify and remove noisy data, outliners, and other factors which can influence the output result

Data Reduction for Data Quality. In order to maintain data integrity, we needed to deal with all rows containing null value; hence we opted for python library tool called PyCaret. We have two options which is either to automatically fill all the null values or to remove any row(s) containing null values weighted. Having weighted the risk involves in both, we decided to remove any row with null or empty value from the data, and this was done using PyCaret python library.

Data Transformation. For us to make our data to acceptable format for easy data mining and pattern recognition, it needed to undergo some data transformation. To ensure data is fully transformed to acceptable format, we used Discretization, normalization, and data aggregation technique.

Data Mining. Having successfully set up our apparatus which includes IBM cloud object storage facility, running background window service, active web service, scheduler, and with the data being thoroughly pre-processed and transformed.

Unlike current data mining in cloud computing process in which, one needs to make direct connect to the data warehouse or directly call the csv file. We only called our restful API endpoint (Figure 1) which was developed and hosted on the cloud;

```
In [3]: data = pd.read_csv('http://cs4430.meritgbtech.com/customers/social')
        data.head()

Out[3]:
```

	User ID	Gender	Age	EstimatedSalary	Purchased
0	15624510	Male	19	19000	0
1	15810944	Male	35	20000	0
2	15668575	Female	26	43000	0
3	15603246	Female	27	57000	0
4	15804002	Male	19	76000	0

Fig. 1. Preview of first five rows of dataset.

This ensures a higher level of security and control over the data, the only thing that needed to be called is the endpoint of our API which automatically displays the data as seen in Figure 1 displaying the first five (5) records in the data using python library in panda (Fig. 2).

```
|--- feature_0 <= 0.63
|   |--- feature_1 <= 0.61
|   |   |--- feature_0 <= -0.16
|   |   |   |--- class: 0
|   |   |--- feature_0 >  -0.16
|   |   |   |--- feature_1 <= 0.40
|   |   |   |   |--- feature_1 <= -0.06
|   |   |   |   |   |--- class: 0
|   |--- feature_1 >  -0.06
|   |--- feature_1 <= 0.03
|   |   |--- class: 1
|   |   |   |   |   |--- feature_1 >  0.03
|   |   |   |   |   |   |--- feature_1 <= 0.26
|   |   |   |   |   |   |   |--- feature_1 <= 0.08
|   |   |   |   |   |   |   |   |--- feature_0 <= 0.14
|   |   |   |   |   |   |   |   |   |--- class: 0
|   |   |   |   |   |   |   |   |--- feature_0 >  0.14
|   |   |   |   |   |   |   |   |   |--- feature_0 <= 0.24
|   |   |   |   |   |   |   |   |   |   |--- feature_1 <= 0.06
|   |   |   |   |   |   |   |   |   |   |   |--- class: 1
|   |   |   |   |   |   |   |   |   |   |--- feature_1 >  0.06
|   |   |   |   |   |   |   |   |   |   |   |--- class: 0
|   |   |   |   |   |   |   |   |   |--- feature_0 >  0.24
|   |   |   |   |   |   |   |   |   |   |--- class: 0
|   |   |   |   |   |   |   |--- feature_1 >  0.08
|   |   |   |   |   |   |   |   |--- feature_0 <= 0.28
|   |   |   |   |   |   |   |   |   |--- feature_0 <= 0.14
|   |   |   |   |   |   |   |   |   |   |--- feature_1 <= 0.14
|   |   |   |   |   |   |   |   |   |   |   |--- class: 0

|   |   |   |   |   |   |   |   |   |   |--- feature_1 >  0.14
|   |   |   |   |   |   |   |   |   |   |   |---        truncatec
branch of depth 3
|   |   |--- feature_0 >  0.14
|   |   |--- class: 0
|   |   |   |   |   |   |   |   |--- feature_0 >  0.28
|   |   |   |   |   |   |   |   |   |--- feature_1 <= 0.13
|   |   |   |   |   |   |   |   |   |   |--- class: 1
|   |   |   |   |   |   |   |   |   |--- feature_1 >  0.13
|   |   |   |   |   |   |   |   |   |   |--- class: 0
|   |   |   |   |   |   |--- feature_1 >  0.26
|   |   |   |   |   |   |   |--- class: 0
|   |   |   |--- feature_1 >  0.40
|   |   |   |   |--- class: 1
|   |--- feature_1 >  0.61
|   |   |--- feature_0 <= -1.15
|   |   |   |--- class: 0
|   |   |--- feature_0 >  -1.15
|   |   |   |--- feature_1 <= 1.36
```

Fig. 2. Text representation of the feature algorithm from the decision tree

Formatted Decision tree algorithm was used to train the model having slatted the data into two equal part, half of it to train the model and the remaining half for testing of the model (Fig. 3);

```
feature_cols = ['Age','EstimatedSalary' ]
X = data.iloc[:,[2,3]].values y = data.iloc[:,4].values
#split the dataset into training and test
X_train, X_test, y_train, y_test = train_test_split(X,y,test_size = 0.25, random_state= 0) #perform feature scaling
from sklearn.preprocessing import StandardScaler sc_X = StandardScaler()
X_train = sc_X.fit_transform(X_train) X_test = sc_X.transform(X_test)
#fit the model in the decision tree classifier from sklearn.tree import DecisionTreeClassifier classifier = DecisionTreeClassifier()
classifier = classifier.fit(X_train,y_train)
```

Fig. 3. Code snippet for important feature selection and training of model.

We are able to obtain 90% accuracy on testing our model, after which we decide to optimize for more accuracy and better performance as seen in Figure 4.

```
# Create Decision Tree classifer object
classifier = DecisionTreeClassifier(criterion="entropy", max_depth=3)# Train Decision Tree Classifer
classifier = classifier.fit(X_train,y_train)#Predict the response for test dataset
y_pred = classifier.predict(X_test)# Model Accuracy, how often is the classifier correct? print("Accuracy:",accuracy_score(y_test,y_pred))
```

Fig. 4. Code snippet for our model accuracy and test performance

We are able to obtain accuracy of 94% after which we decided to visualize the performance of the model (Fig. 5).

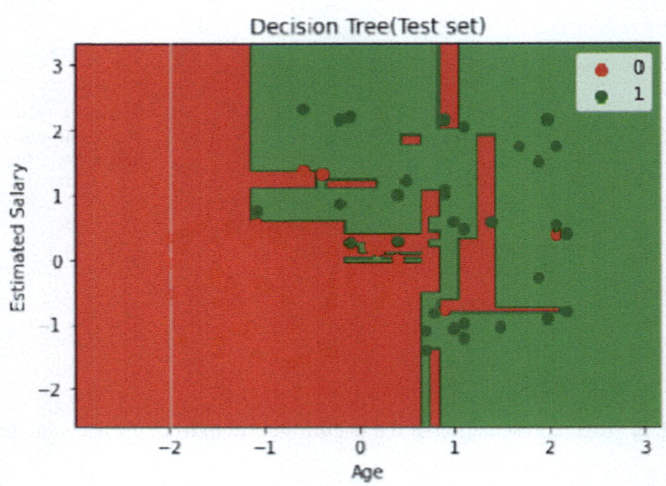

Fig. 5. Model visualization and performance evaluation.

We needed to know the level of fitting, because having an accuracy of 94% can be as a result of over fitting of the model with the training set of data, hence we decided to do two additional things; Firstly, we implemented Random Forest Algorithm for training the data again to check performance and then optimize, it gives an accuracy of 92%. Secondly, we fed another set of data in the format of the trained data but new to the model, the model performed very well with accuracy. So, we proceeded to measure the performance of the model using confusion matrix, classification report, and accuracy score which gives good indication of optimal performance. We are in a dilemma either to ensure about 100% accuracy or over fitting because in supervise machine learning, high accuracy of almost 100% can be as a result of over fitting of the model which we want to avoid by possible means. So, since we are able to feed the model with new set of data which had not been previously fed to the model for which it performed very well with high rate of accuracy. So, we gave priority to avoid over fitting of the model than the accuracy of the model since high accuracy for supervised learning can be as a result of over fitting of the model for which the model becomes less accurate or behaved weird when fed with unfamiliar data. The validation report can be seen in Figure 6.

```
---------------------------------------------------------------------
Mean Absolute Error: 0.09166666666666666
---------------------------------------------------------------------
Mean Squared Error: 0.09166666666666666
---------------------------------------------------------------------
Root Mean Squared Error: 0.30276503540974914
---------------------------------------------------------------------

 --------------------------------------------------------------------
 [[61  4]
  [ 7 48]]
 --------------------------------------------------------------------
              precision    recall  f1-score   support

           0       0.90      0.94      0.92        65
           1       0.92      0.87      0.90        55

    accuracy                           0.91       120
   macro avg       0.91      0.91      0.91       120
weighted avg       0.91      0.91      0.91       120
```

Fig. 6. Validation, cross validation and the visualization of mean square error (MSE).

4 Conclusion

In this applied research, we are able to achieve four goals;

1. Avoid direct interaction with integrated data warehouse
2. We are able to access data in a secure cloud computing environment
3. We are able to close the existing gap in data mining and web API, as all we needed to call is the endpoint of our web API. No need of terminal, or uniform resource locator (URL) of any csv is involved. It is simply over a web API.

4. As the patterns changes in the background, the algorithm automatically adjusts with accuracy of ninety-four (94)% after optimization.

We are able to implement data mining in a secure cloud computing environment with accuracy of ninety-four (94)% after optimization in our decision tree algorithm over web API. All that an AI engineer, data scientist, or machine learning engineer needs is just the endpoint of the API. This removes complexity while at the same time simplifying the whole process, it also add additional layer of security, and also remove unnecessary bottleneck as scientist and engineers will be able to concentrate more on optimizing their algorithm and model for optimal result since only API endpoint is what is needed to be called.

We hope that this will be de-facto standard in the data mining, machine learning, data science and other similar industry at large.

References

1. Frawley, W.J., Piatetsky-Shapiro, G., Matheus, C.J.: Knowledge discovery in databases: an overview. AI Mag. **13**(3), 57 (1992)
2. Wu, W., Gruenwald, L.: Research issues in mining multiple data streams. In: Proceedings of the First International Workshop on Novel Data Stream Pattern Mining Techniques, Stream KDD 2010, pp. 56–60. ACM, New York (2010)
3. Sarna, D.E.Y.: Implementing and Developing Cloud Computing Applications, CRC Press (2010). https://cwiki.apache.org/MAHOUT/k-means-clustering.html
4. Weiss, A.: Computing in the Clouds Networker (2007)
5. Wang, K., Xu, C., Liu, B.: Clustering transactions using large items. In: CIKM 1999: Proceedings of the Eighth International Conference on Information and Knowledge Management, pp. 483–490. ACM Press, New York (1999)
6. Berson, A., Smith, S., Thearling, K.: Building Data Mining Applications for CRM. McGraw-Hill, New York (2000)
7. Sitaram, D., Manjunath, G.: Moving to the Cloud: Developing Apps in the New World of Cloud Computing (2011)
8. Arias, T.: The Cloud Computing Handbook Everything You Need to Know about Cloud Computing (2012)
9. http://searchsqlserver.techtarget.com/definition/datamining
10. http://www.ijcaonline.org/volume15/number7/pxc3872623.pdf
11. http://www.waset.org/journals/waset/v39/v3972.pdf
12. http://www.estard.com/data_mining_marketing/data_mining_campaign.asp
13. http://dssresources.com/books/contents/berry97.html
14. https://www.marketingprofs.com/articles/2010/35%2067/the-nine-most-common-data-miningtechniques-usedin-predictive-analytics
15. Petre R.Ş.: Data Mining in Cloud Computing (2012). https://www.dbjournal.ro/archive/9/9_7.pdf
16. Pareek, A., Gupta, M.: Review of Data Mining Techniques in Cloud Computing Database (2012)

Lifting 2D Object Detection to 3D: Geometric Approach in Bird-Eye-View

Dmitriy Zhuravlev(✉)

T. Shevchenko National University, Kiev 01601, Ukraine
dzhuravlev@ukr.net
https://www.linkedin.com/in/dmitriy-viktorovich-zhuravlev/

Abstract. 3D object detection is fundamental for computer vision applications. Although motion is essentially linked with the orientation of the object, the trajectory of the object is not fully utilized in modern 3D object detectors. This paper proposes a method for object 3D bounding box estimation based on 2D tracking and Inverse Perspective Mapping. Unlike other solutions utilizing deep learning techniques for orientation estimation, the proposed method uses geometric constraints and object movements. Which causes the execution speed and errors of 3D localization to remain comparable to a pure 2D tracker.

Keywords: 3D object detection · Tracking

1 Introduction

Detecting 3D objects is one of the key components of computer vision. It recovers orientation, position, and the size of an object within a scene. While developed 2D detection algorithms are capable of processing large amounts of data in real-time [1,2], accurately detecting 3D objects remains challenging.

Special equipment-based methods, including LiDAR scanners [3] and RGB-D cameras [4] provide an approach for 3D direct object reconstruction, but the equipment is quite expensive.

Monocular camera-based methods [5], on the contrary, are attractive because of their cheaper cost and more flexible setup. However, the main disadvantage of using a monocular camera is that it only conducts observations in the plane of a two-dimensional image and is unable to directly measure the distance to an object. That is why it becomes important but still a challenge for cost-effective autonomous tracking and virtual reality.

While the recently developed 2D detection algorithms are capable of detecting and tracking large amounts of data in real time, accurate detection of 3D objects remains an open problem despite some recent promising findings. The difference between detecting 2D and 3D objects is significant. Solid shapes have three dimensions, center localization and orientation angles, whereas flat figures are limited by an axis-aligned rectangle with two dimensions.

R. Silhavy (Ed.): CSOC 2022, LNNS 502, pp. 211–225, 2022.
https://doi.org/10.1007/978-3-031-09076-9_21

The goal of this paper is to effectively infer the 3D bounding box on top of conventional 2D box object detection using the sequence of monocular images. We introduce the lifting pipeline that can be used with any 2D detection and tracking algorithms providing reliable results. The main contributions of this paper can be summarized as follows:

- a method for object orientation estimation via 2D boxes centers tracking;
- a geometric analysis of the relation between detection 2D box, orientation, and object sizes;
- a scheme for converting 2D tracking to 3D;
- a design of an end-to-end baseline architecture demonstrating reasonable performance and accuracy.

The remainder of this paper is organized as follows. In the next section, the review of popular 3D object detection techniques is given. Section 3 introduces a set of necessary concepts that will be used later in our method from Sect. 4. The results of real-world testing with a concluding summary are presented in Sect. 5.

2 Related Work

This section discusses the existing methods for 3D localization that intend to recreate the depth information from an RGB image.

Like most of the works related to 3D object detection from a single image, [5] proposed a method for regression of an observation angle and 3D object sizes from the cropped image patch enclosed by 2D detection box. Using the regressed dimensions and orientations of the 3D box and 2D detection box they try to minimize the re-projection error related to the initial 2D detection box constraints. Since each side of the 2D detection box can correspond to any of the eight corners of the 3D box, there are 8^4 configurations.

In the [12] the number of plausible configurations is reduced to 64. These configurations go through equations to find the vertices of a cuboid and the resulting solutions are ranked by approximation error or intersection over union (IoU) score between the tightest bounding box for the projected vertices of the fitted cuboid and the 2D detection rectangle.

To simplify object distance measurement, the Bird's Eye View (BEV) based architecture is proposed in [13]. Since all traffic participants are in contact with the ground, the distance can be estimated as detection of points of contact of the vehicle with the road. The proposed method involves two stages: motion cancellation and the detection of position, size, and orientation using convolutional neural networks. In this work, two types of detectors for rotated boxes prediction on BEV are proposed. The one-stage detectors have the advantage of fast operating speed because of direct prediction of the bounding box from the feature map grid without refinement. On the other hand, two-stage detectors predict the position and size of the object more precisely since the network predicts the box proposal and performs refinement in two steps. However, such defectors operate slower.

[14] extends the previous approach for uncalibrated traffic cameras using dual-view architecture. In this architecture, both the original view and BEV images are taken as input. The original view feature maps are warped to BEV and concatenated with BEV feature maps. The dual-view structure enables feature learning before warping, where the object shapes are regular and the knowledge propagation is easier but it still has the two-stage detectors issues.

In [15] the idea of learning moving object detection directly in BEV space is explored. The architecture based on two-stream RGB and optical flow fusion is proposed for object localization. However, the lengths of some objects are not captured perfectly and some of the objects are misclassified as false static objects, which shows that there is still room for improvement.

Moving object detection has received a lot of attention recently, especially in autonomous driving applications. Traffic information can be used as a signal for class-independent detection. Motion segmentation has been explored in [15] in order to predict object-moving in BEV using RGB image and optical flow.

All the mentioned methods study 3D object localization using different neural networks which not only get a 2D box but also extract additional information from the image to predict the distance and orientation. However, geometric constraints of moving objects are not explored enough. In this work, we identify real-world scenarios where high reconstruction accuracy is guaranteed via the geometric analysis and apply our algorithm to localize 3D object in real-time.

3 Background

3.1 Camera Model

The camera model is a basis for developing a computer vision application. In this article, the simplest model as shown in Fig. 1 with no distortion factor is used to describe the projection of 3D points onto an image plane.

Let us describe the correspondences between the camera, world, and picture coordinate systems. Let $p = (x_w, y_w, z_w)$ be a 3D point in the world coordinate system. It can be transformed into camera coordinates using the *extrinsic matrix* which consists of the rotation $R \in \mathbb{SO}(3)$ and translation $t \in \mathbb{R}^3$ between the two coordinate systems:

$$\begin{bmatrix} x_c \\ y_c \\ z_c \end{bmatrix} = R \begin{bmatrix} x_w \\ y_w \\ z_w \end{bmatrix} + t.$$

The new 3D point in the camera coordinate system is projected onto the image plane using the *intrinsic matrix* K which consists of internal camera parameters like the focal length, optical center, etc.:

$$\begin{bmatrix} u' \\ v' \\ w' \end{bmatrix} = K \begin{bmatrix} x_c \\ y_c \\ z_c \end{bmatrix},$$

where $u = \frac{u'}{w'}, \quad v = \frac{v'}{w'}$.

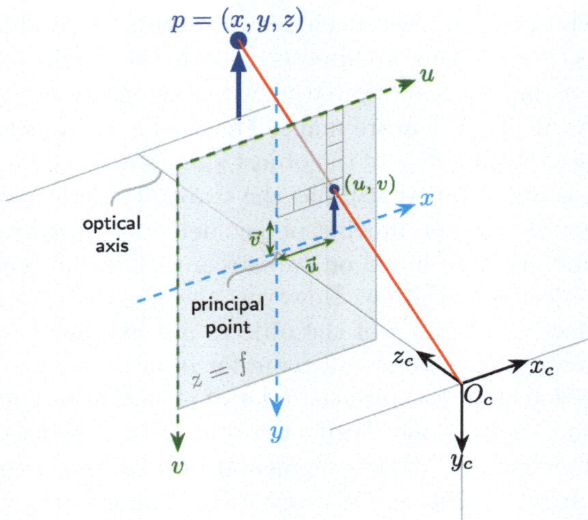

Fig. 1. Pinhole camera model.

The overall projection in homogeneous coordinates can be expressed by:

$$\begin{bmatrix} u' \\ v' \\ w' \end{bmatrix} = K \left(R \begin{bmatrix} x_w \\ y_w \\ z_w \end{bmatrix} + t \right) .$$

3.2 Inverse Perspective Mapping

The distance of the object from the camera and the observation angle should be taken into account for reconstruction the ground plane as if viewed from a camera placed above with its axis exactly vertical. *Inverse Perspective Mapping* (IPM) uses the frontal view to produce a *Bird's Eye View* (BEV) by removing perspective effects using a homography. The example of the IPM is shown in Fig. 2.

In general, the perspective transformation can be expressed as

$$p' = Gp \,, \tag{1}$$

where G can be expressed in terms of the intrinsic matrix K and the Euclidean *homography* matrix $H \in \mathbb{SL}(3)$ related to a planar viewed object. It should be noted that H depends on the rotation matrix R and the translation vector t corresponding to the rigid transformation between the real camera pose image and generated one.

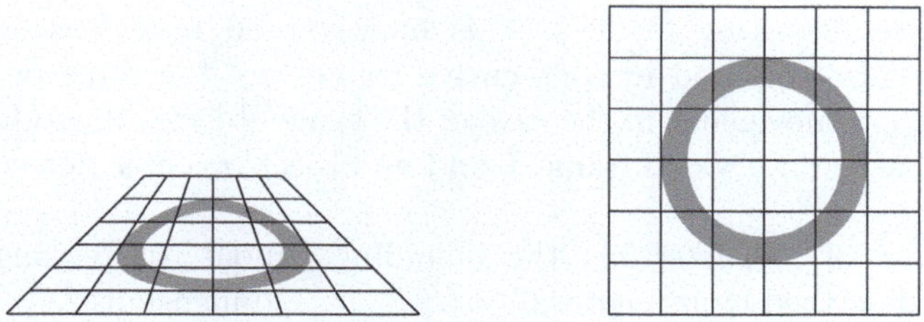

(a) Perspective view. (b) Top-down view.

Fig. 2. BEV image generated by IPM.

IPM can be conveniently computed automatically [7,8] or constructed using scene measurements, satellite images from publicly available map services (e.g. *Google Maps*).

3.3 Object Detection and Tracking

2D Object Detection is a common Computer Vision technique for identifying and locating objects of certain classes in the image. A 2D bounding box (rectangle) is used to describe the location of an object. It is determined by the rectangle center $C = (c_x, c_y)$, dimension $D = (d_x, d_y)$ with the width and height of the rectangle and type of the object T. The example of 2D detection using YOLO (You Only Look Once) algorithm is shown in Fig. 3.

Fig. 3. 2D object detection and classification with YOLOv3.

Multi-Object Tracking (MOT) [9] is mainly used in videos to determine whether objects recognized in consecutive frames are the same or not and to track the path of movement in the case of the same objects. It produces a set of trajectories based on a video frame f and an object o center detection as they move around.

In the case of *3D detection*, the bounding cuboid (or rectangular parallelepiped) is described by its center $C = (c_x, c_y, c_z)$, dimensions $D = (d_x, d_y, d_z)$, and orientation $R = (\varphi, \theta, \psi)$ parameterized by elevation, azimuth, and roll angles.

4 Proposed Method

Existing 2D detectors not only build a bounding box around the object but also allow it to be classified. For example, vehicles can be classified by body type. That makes it possible to get their average size as shown in Table 1. Moreover, using the classification by model [10], we get the exact dimensions of the object from https://www.automobiledimension.com. In the case of classification by model, the orientation will be determined exactly, and in other cases, with some error, which will be described below.

We will also use the directions of motion of the 2D detection box centers, that do not coincide with the movement of the object itself. However, a method for correction will be suggested.

The basic idea of the proposed method is to find object bottom box projection on the BEV using object sizes and moving direction. Further, it will be shown how, using the detected boxes and object sizes, to calculate the possible orientations and choose the one that best suits the movement of the object.

Table 1. Vehicle categories and their average dimensions.

Category	Vehicles included	Length (m)	Width (m)
Car	Car, jeep	3.72	1.44
Bus	Bus	10.1	2.43
Truck	Truck	7.5	2.35
LCV	Mini bus, vans	6.1	2.10
Tractor	Tractor, trailer	7.4	2.20
Three-wheeler	Three wheeler	3.2	1.40
Two-wheeler	Scooter, motorbike	1.87	0.64
Cycle	Bicycles	1.9	0.45

4.1 Model Assumptions

Our main task is to detect vehicles on the road and parking lot, but it is not limited to this application. Several assumptions have been made in order to build a

(a) 2D detection. (b) 3D detection.

Fig. 4. Vehicle detection variants.

mathematical model. We are most interested in the planar position of the objects and assume a vehicle is always on the (flat) ground. It allows assuming that only the azimuth (noted θ) matters and thus we do not predict the orientation R characterized by elevation and roll and we set $\varphi = 0$ and $\psi = 0$.

The camera position must be stationary. In another case, object moving vectors obtained from MOT have to be recalculated. And the vehicle being detected must be attached to the plane of the ground because only the points on the ground have the distance information.

In the context of our model, we will use a cuboid approximation of a 3D object which coincides with the definition of 3D detection as shown in Fig. 4. We assume that 3D detection perfectly fits into 2D detection. Therefore at least one vertex of the bottom box lies on the lower side of the 2D detection rectangle. Accordingly, we can see a similar situation in BEV as shown in Fig. 5.

Further, we will show how to find the projection of the lower face of the cuboid in BEV. This makes it possible to determine the orientation and center of the cuboid with known sizes. It should be noted that the orientation of the object is set up to the opposite since we do not geometrically distinguish between the front and back of a vehicle, which is appropriate in the case of cars since they can go both forward and backward, but not sidewards.

4.2 Bottom Box Determination

All calculations that are given in this section will refer to calculations in BEV. The domain of an IPM is an image with 2D objects detection.

Let the IPM image of an object detection rectangle will be non-degenerate quadrangle A, R, T, B with an inner rectangle whose three vertices lie on adjacent sides. Moreover, one of the vertices of the rectangle lies on the bottom side of the outer quadrangle as shown in Fig. 6. Our goal is to find touchpoints K, C, D, and angle φ. Further we will refer to such quadrangles as *bounding* ones. Under these conditions, we will prove the following statements.

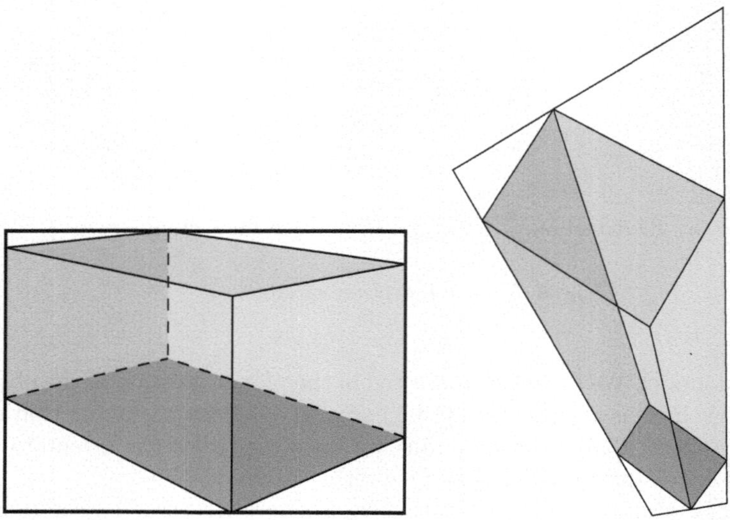

(a) Cube 2D detection. (b) IPM transformation.

Fig. 5. IPM of the 2D detection.

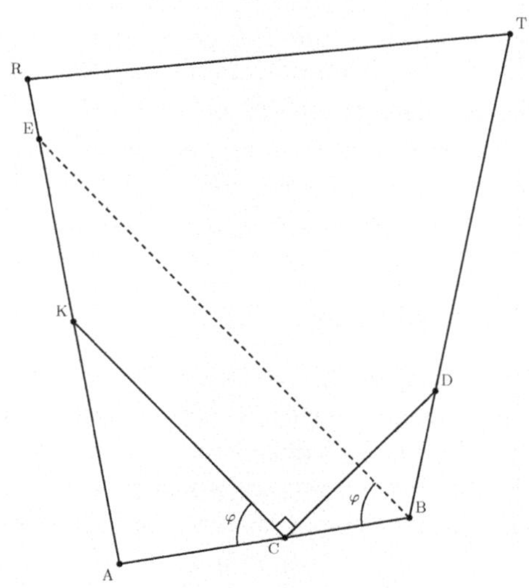

Fig. 6. Rectangle reconstruction by angle and sizes.

Proposition 1. *A rectangle with known dimensions and orientation can be reconstructed from bounding quadrangle.*

Proof. Let the rectangle be built on points K, C, and D with fixed sizes $KC = a$ and $CD = b$ as shown in Fig. 6. Let us draw a ray from B with the angle equal to the orientation. We denote E the intersection point of this ray with segment AR, and $\angle ABE = \varphi$, and let $EB = l$. According to the construction $\triangle KAC \backsim \triangle EAB$ and from the similarity we have $\frac{AC}{AB} = \frac{KC}{EB} = \frac{a}{l}$.

Hereof $AC = \frac{a}{l}(AC + CB)$ and $AC = \frac{a}{l-a}CB$.

Consequently $C = (C_x, C_y) = \left(\frac{(l-a)A_x + aB_x}{l}, \frac{(l-a)A_y + aB_y}{l}\right)$, which makes it possible to recover the rectangle.

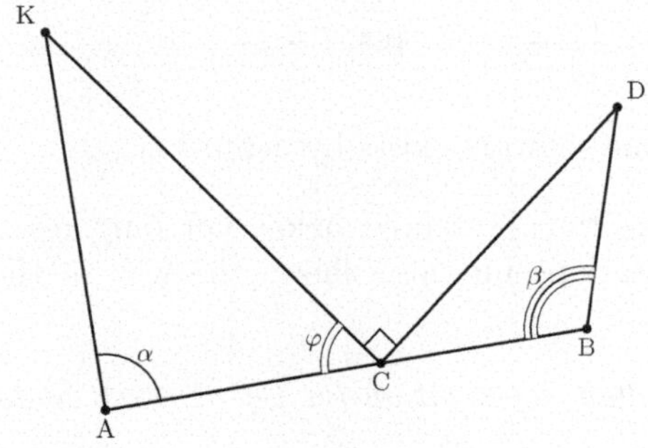

Fig. 7. Fixed sizes rectangle fitting.

Theorem 1. *There are at most 4 possibilities to fit a rectangle into a bounding quadrangle.*

Proof. Suppose we have bounding quadrangle with fixed $KC = a$, $CD = b$, $AB = d$, $\angle KAC = \alpha$, and $\angle CBD = \beta$ as shown in Fig. 7. To distinguish between different fitting possibilities let us assume that $a \geq b$. According to the sine rule

$$\frac{a}{\sin\alpha} = \frac{AC}{\sin(\alpha + \varphi)},$$

$$\frac{b}{\sin\beta} = \frac{BC}{\sin(\beta + \frac{\pi}{2} - \varphi)}.$$

As a result $AC + BC = \frac{a\sin(\alpha+\varphi)}{\sin\alpha} + \frac{b\cos(\beta-\varphi)}{\sin\beta} = d$
and consequently we obtain the following equation:

$$(a\sin\beta\sin\alpha + b\sin\alpha\cos\beta)\cos\varphi + \\ (a\sin\beta\cos\alpha + b\sin\alpha\sin\beta)\sin\varphi = d\sin\alpha\sin\beta. \quad (2)$$

Let us denote

$$
\begin{gathered}
e = a\sin\beta\sin\alpha + b\sin\alpha\cos\beta = \sin\alpha(a\sin\beta + b\cos\beta)\,,\\
f = a\sin\beta\cos\alpha + b\sin\alpha\sin\beta = \sin\beta(a\cos\alpha + b\sin\alpha)\,,\\
g = d\sin\alpha\sin\beta > 0\,.
\end{gathered}
$$

In terms of these notations, we get the equality $e\cos\varphi + f\sin\varphi = g$.
Further, if we denote $\dfrac{e}{\sqrt{e^2+f^2}} = \cos\theta$ and $\dfrac{f}{\sqrt{e^2+f^2}} = \sin\theta$,
where $0 \le \theta < 2\pi$, then we get that $\cos(\varphi - \theta) = \dfrac{g}{\sqrt{e^2+f^2}}$.
If $g^2 \le e^2 + f^2$ we obtain that

$$
\varphi = \theta \pm \psi + 2\pi n\,,
$$

where $\psi = \arccos\dfrac{g}{\sqrt{e^2+f^2}}$ and $0 \le \psi < \frac{\pi}{2}$.

Remark 1. At least one solution exists by definition.

Remark 2. The proposition presented below not only restricts the number of possible object orientations but also shows the way of the bottom rectangle reconstruction.

Corollary 1. *More than one solution for Eq. (2) exist under the conditions:*

$$
\begin{cases}
0 \le a + b\cot\beta \le d\,,\\
0 \le b + a\cot\alpha \le d\,,\\
(a + b\cot\beta)^2 + (b + a\cot\alpha)^2 > d^2\,.
\end{cases}
$$

Proof. Let us choose a coordinate system so that the axes coincide with the sides of rectangle K, C, D as shown in Fig. 8. We draw circles with chords KC and CD and related angles α and β. There is a solution with circles having the following centres and corresponding radiuses:

$$
\begin{aligned}
Q_1\left(-\frac{a}{2}\cot\alpha, \frac{a}{2}\right), R_1 &= \frac{a}{2\sin\alpha}\,;\\
Q_2\left(\frac{b}{2}, -\frac{b}{2}\cot\beta\right), R_2 &= \frac{b}{2\sin\beta}\,.
\end{aligned}
$$

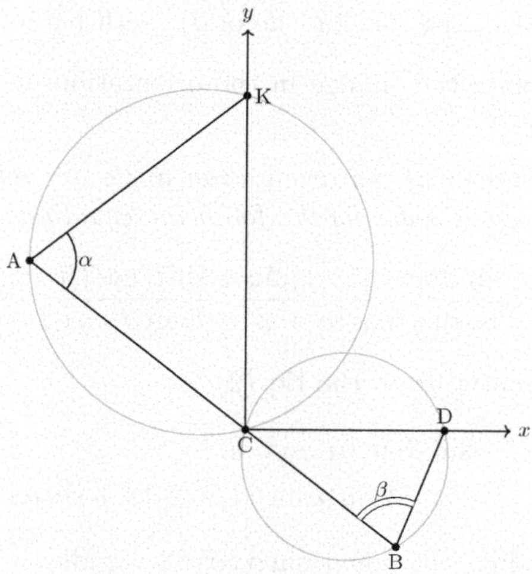

Fig. 8. The construction of the circles with fixed chords.

Which represented as equalities will have the form

$$\left(x + \frac{a}{2}\cot\alpha\right)^2 + \left(y - \frac{a}{2}\right)^2 = \left(\frac{a}{2\sin\alpha}\right)^2,$$
$$\left(x - \frac{b}{2}\right)^2 + \left(y + \frac{b}{2}\cot\beta\right)^2 = \left(\frac{b}{2\sin\beta}\right)^2;$$

or

$$x^2 + ax\cot\alpha + y^2 - ay = 0\,,$$
$$x^2 - bx + y^2 + by\cot\beta = 0\,.$$

Using the line equation $y = kx$ we will get

$$x^2 + ax\cot\alpha + k^2x^2 - akx = 0\,,$$
$$x^2 - bx + k^2x^2 + bkx\cot\beta = 0\,;$$

with the points of intersection of the circles

$$x_1 = \frac{ak - a\cot\alpha}{k^2+1}, \qquad y_1 = kx_1;$$
$$x_2 = \frac{b - bk\cot\beta}{k^2+1}, \qquad y_2 = kx_2.$$

Taking into account the fixed distance $AC + CB = d$:

$$k^2\left((a + b\cot\beta)^2 - d^2\right) - 2(b + a\cot\alpha)(a + b\cot\beta)k + (b + a\cot\alpha)^2 - d^2 = 0\,.$$

Therefore, two solutions exist for $(a + b\cot\beta)^2 + (b + a\cot\alpha)^2 > d^2$.

Let us try to evaluate the change in the orientation angle when the size of the rectangle changes.

Corollary 2. *The changes of the orientation angle are related to the changes of the sizes of the rectangle sides by the following equation:*

$$\Delta\varphi = -\frac{\sin\beta\sin(\alpha+\varphi)\Delta a + \sin\alpha\cos(\beta-\varphi)\Delta b}{a\sin\beta\cos(\alpha+\varphi) + b\sin\beta\sin(\beta-\varphi)} .$$

Proof. Take the differential from the Eq. (2)

$$a\sin\beta\cos(\alpha+\varphi)\mathrm{d}\varphi + \sin\beta\sin(\alpha+\varphi)\mathrm{d}a + b\sin\alpha\sin(\beta-\varphi)\mathrm{d}\varphi + \sin\alpha\cos(\beta-\varphi)\mathrm{d}b = 0 .$$

Hereof $\mathrm{d}\varphi = -\dfrac{\sin\beta\sin(\alpha+\varphi)\mathrm{d}a + \sin\alpha\cos(\beta-\varphi)\mathrm{d}b}{a\sin\beta\cos(\alpha+\varphi) + b\sin\beta\sin(\beta-\varphi)}$.

4.3 Algorithm of 3D Detection

Using the angle analysis described in the previous Sect. 4.2, we can build the Algorithm 1 for 3D detection. First of all, when the object appears on the scene under view, we set the vector $\vec{u}$ of the movement direction in BEV using well-known MOT techniques [9] and IPM. Due to the fact that a 2D detection center does not coincide with the center of a 3D object, this vector is not sufficient to correctly determine the orientation of the object.

Therefore, using Theorem 1 and Proposition 1, we can calculate no more than 4 possible object orientations R_i and related cuboid candidates P_i (usually only 2 according to Corollary 1). For each of them, we calculate the intersection over union (IoU) scores of the initial 2D box detection and P_i projection to choose the best solution minimizing the re-projection error (RPE) and proximity to the 2D motion vector $\vec{u}$:

$$R = \underset{R_i}{\operatorname{argmin}} \left(\gamma\|u\| \left| \langle \vec{u}, \overrightarrow{C_i K_i} \rangle \right| + (1-\gamma)\mathrm{RPE}(P_i) \right) ,$$

where $\gamma \in (0, 1]$ is configuration parameter.

It should be noted than we can aggregate the movements of several frames depending on the velocity of the object in order to equalize the random fluctuations of the 2D detector.

5 Experiments and Results

Experiments were captured on videos to test the functionality of the proposed method. All data is processed in real-time with a high degree of accuracy. Some results of vehicle detection and localization are shown in Fig. 9.

Algorithm 1. Lifting 2D detection to 3D.

Require: $\mathcal{V}$ ▷ video, 2D detector, IPM

Initialize $\mathcal{V} \leftarrow \{f \mid \text{ordered set of video frames}\}$

for each frame $f \in \mathcal{V}$ **do**

  $O^f \leftarrow \{(C, D, T) \mid \text{2D detected objects in the frame}\}$

  $V^f \leftarrow \left\{\overrightarrow{C_o^f C_o^{f-1}}\right\}$ ▷ track 2D objects

  for each object $o \in f$ **do**

    $\vec{u} \leftarrow \text{IPM}(\overrightarrow{C_o^{prev} C_o})$ ▷ rectangle movement direction in BEV

    $R \leftarrow \underset{R_i}{\text{argmin}} \left(\gamma \|\vec{u}\| \left|\langle \vec{u}, \overrightarrow{C_i K_i} \rangle\right| + (1-\gamma)\text{RPE}(P_i)\right)$

  end for

end for

(a) 2D YOLO detection with bottom projection estimation.

(b) Related IPM image.

Fig. 9. The visualization result of the proposed 3D object detection method.

The geometric 3d reconstruction depends entirely on the 2D detector, if the vehicle model and its 2D bounding box are detected with an error, this will impact 3D detection. Nevertheless, experimental results show that the proposed approach can provide faster and more accurate 3D object localization than the state-of-art methods.

6 Conclusion

In this paper, a novel monocular 3D object detection framework is introduced. Using the proposed bottom box projection reconstruction method on top of 2D tracking allows to improve the accuracy and performance of the 3D object detection. Moreover, the proposed method also is highly adaptable, that allows it to be applied on top of any 2D detection and tracking algorithms. The prospects of future research on non-static case with a moving camera will be considered.

References

1. Redmon, J., Farhadi, A.: Yolov3: An incremental improvement. arXiv preprint arXiv:1804.02767 (2018)
2. Fedorov, A., Nikolskaia, K., Ivanov, S., Shepelev, V., Minbaleev, A.: Traffic flow estimation with data from a video surveillance camera. J. Big Data **6**(1), 1–15 (2019)
3. Shi, S., Wang, X., Li, H.: PointRCNN: 3D object proposal generation and detection from point cloud. In: IEEE/CVF Conference on Computer Vision and Pattern Recognition (CVPR), pp. 770–779 (2019). https://doi.org/10.1109/CVPR.2019.00086
4. Wang, Y., Zell, A.: Yolo+FPN: 2D and 3D fused object detection with an RGB-D camera. In: 2020 25th International Conference on Pattern Recognition (ICPR), pp. 4657–4664 (2021). https://doi.org/10.1109/ICPR48806.2021.9413066
5. Mousavian, A., Anguelov, D., Flynn, J., Košecká, J.: 3D bounding box estimation using deep learning and geometry. In: IEEE Conference on Computer Vision and Pattern Recognition (CVPR), 5632–5640 (2017). https://doi.org/10.1109/CVPR.2017.597
6. Hartley, R., Zisserman, A.: Multiple view geometry in computer vision. Cambridge University Press (2003)
7. Zhou, J., Li, B.: Homography-based ground detection for a mobile robot platform using a single camera. In: Proceedings 2006 IEEE International Conference on Robotics and Automation 2006. ICRA 2006, pp. 4100–4105 (2006). https://doi.org/10.1109/ROBOT.2006.1642332
8. Jeong, J., Kim, A.: Adaptive inverse perspective mapping for lane map generation with SLAM. In: 2016 13th International Conference on Ubiquitous Robots and Ambient Intelligence (URAI), pp. 38–41 (2016). https://doi.org/10.1109/URAI.2016.7734016
9. Sharon, M., Latha, A.: Multi-object tracking based on Kalman filter by using multi-feature appearance model. In: International Conference on Inventive Research in Computing Applications (ICIRCA), pp. 544–549 (2018). https://doi.org/10.1109/ICIRCA.2018.8597365
10. Ghoreyshi, A., AkhavanPour, A., Bossaghzadeh, A.: Simultaneous vehicle detection and classification model based on deep YOLO networks. In: International Conference on Machine Vision and Image Processing (MVIP), pp. 1–6 (2020). https://doi.org/10.1109/MVIP49855.2020.9116922
11. Mauri, A., Khemmar, R., Decoux, B., Haddad, M., Boutteau, R.: Real-time 3D multi-object detection and localization based on deep learning for road and railway smart mobility. J. Imaging **7**, 145 (2021). https://doi.org/10.3390/jimaging7080145
12. Fang, J., Zhou, L., Liu, G.: 3D bounding box estimation for autonomous vehicles by cascaded geometric constraints and depurated 2D detections using 3D results. arXiv preprint arXiv:1909.01867 (2019)
13. Zhu, M., Zhang, S., Zhong, Y., Lu, P., Peng, H., Lenneman, J.: Monocular 3D vehicle detection using uncalibrated traffic cameras through homography. In: IEEE/RSJ International Conference on Intelligent Robots and Systems (IROS), pp. 3814–3821 (2021). https://doi.org/10.1109/IROS51168.2021.9636384

14. Kim, Y., Kum, D.: Deep learning based vehicle position and orientation estimation via inverse perspective mapping image. In: IEEE Intelligent Vehicles Symposium (IV), pp. 317–323 (2019). https://doi.org/10.1109/IVS.2019.8814050
15. Rashed, H., Essam, M., Mohamed, M., Sallab, A.E., Yogamani, S.: BEV-MODNet: monocular camera based bird's eye view moving object detection for autonomous driving. In: 2021 IEEE International Intelligent Transportation Systems Conference (ITSC), pp. 1503–1508 (2021). https://doi.org/10.1109/ITSC48978.2021.9564667

Handheld 3D Scanner Based on Intel RealSense Depth and Tracking Cameras

Stanislav Alexovič(✉), Milan Lacko, Ján Bačík, and Daniela Perduková

Department of Electrical Engineering and Mechatronics, Technical University of Košice, Košice, Slovakia
stanislav.alexovic@tuke.sk

Abstract. This paper describes design and construction of a 3D handheld scanner, which was based on Intel RealSense cameras. The goal was to create a device capable of 3D scanning of its environment and producing a 3D reconstruction of the scanned scene. The device contained its own battery with a single board computer, connected with the user's computer over a wireless network. This device should stand as a foundation for future research which will focus on the autonomous exploration of an unknown environment.

Keywords: 3D scanning · Visual SLAM

1 Introduction

3D scanning refers to the technology of acquiring visual and dimensional properties from an environment or an object from the real world and transferring this information into a digital form which can be processed or analysed using various algorithms [1]. This technology is a part of computer vision, a thriving field which has been gaining more and more attention from researchers and developers with various technological backgrounds.

There are multiple methods of acquiring a 3D reconstruction of a scene different mainly in their input sensors. The most common method in this field is stereo vision where two 2D cameras are used [2], much like the principle of depth sensing used by human beings. In stereo vision, both cameras scan the same object, each from a different angle. Since the distance between these cameras is known, algorithms can calculate the distance of the scanned object. Other approaches, like active stereo vision [3] or structured light principle [4], are very similar in their nature, because both of them make use of projected light to facilitate the scanning process. The former uses a random pattern projected on a scanned object which helps to determine depth on smooth or featureless objects, while the latter works with a comparison between a known light pattern and the scanned pattern which is deformed according to the shape of the scanned object. These deformations then serve as the main input for depth calculation. Other scanning methods are based on Time of Flight (ToF) factor [5], which calculates depth by measuring the time which transmitted light needs to hit scanned object and reflect back to a receiver.

R. Silhavy (Ed.): CSOC 2022, LNNS 502, pp. 226–235, 2022.
https://doi.org/10.1007/978-3-031-09076-9_22

The receiver in such scanning devices usually consists of CMOS chip which can measure depth of all pixels, making the information about the scene depth available all at once. There is one more device other than ToF cameras working on the same principle, a LiDAR [6]. This device emits a laser beam which illuminates a scene one point at a time by aiming always on different parts of the scene.

In this paper, the author describes a handheld 3D scanner built from affordable components. The main goal of this project is to test depth sensors and 3D reconstruction algorithm which will serve as foundation for future projects. One of them will be focused on an autonomous environment exploration algorithm [7] while in the second one, an autonomous Unmanned Aerial Vehicle (UAV) for autonomous exploration in GPS-denied environments [8] will be built and tested.

2 Hardware Components

Hardware design heavily depends on the primary requirement which is to build a 3D scanning device capable of reconstructing a previously scanned environment in 3D. These components will be reused in the following project, focusing on building an autonomous UAV, therefore this implies another set of constrains (Fig. 1):

- Low power consumption – All electric components will be powered by a local battery, therefore power efficiency might increase the operation time
- Large battery capacity – Local battery needs to have enough capacity to provide reasonable operation time
- Wireless communication – Communication with the user's computer needs to be accomplished wirelessly
- Computing power – Local computer which will host all internal logic must be capable of processing all algorithms required for 3D scene reconstruction

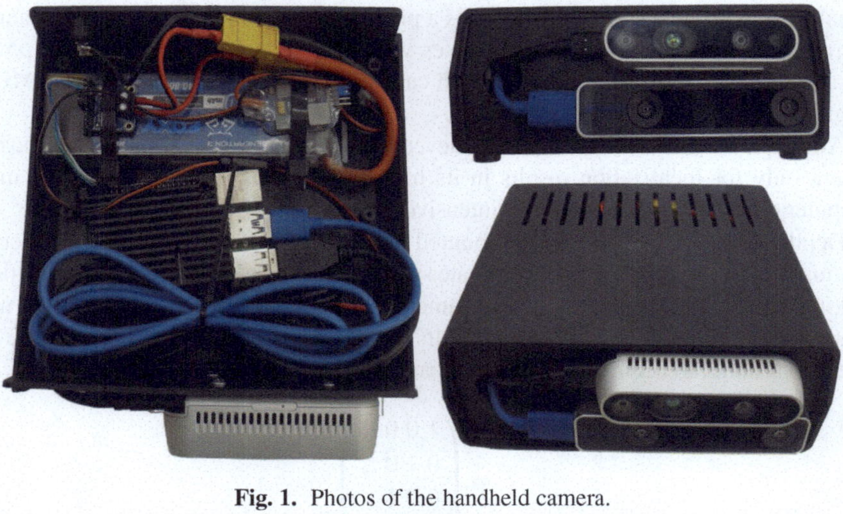

Fig. 1. Photos of the handheld camera.

2.1 Depth Camera

As far as depth sensing is concerned, Intel RealSense depth camera D435i was chosen. Even though this camera was developed mainly for the field of robotics its features fit the requirements for this project. It is small, light, power efficient, and can be used in both indoor and outdoor environments. Depth sensing technology in D435i is based on active stereo vision which uses Infra-Red (IR) projector and two IR cameras. The main technical specifications are listed in Table 1.

Another important feature of this camera is that it also contains Inertial Measurement Unit (IMU) which is essential for detecting its own rotation and movement. This is used in Visual Simultaneous Localization and Mapping (v-SLAM) algorithms. However, in this project a separate component for tracking its own movements was chosen for this purpose and will be described in the next sub-chapter.

Table 1. Intel RealSense D435i technical specifications

Depth Field of View (FOW)	87° × 58°
Depth output resolution	Up to 1280 × 720
Depth frame rate	Up to 90 fps
RGB sensor FOV (H × V)	69° × 42°
RGB frame resolution	1920 × 1080
RGB frame rate	30 fps

2.2 Tracking Camera

Tracking camera Intel RealSense T265 is a complement to the depth camera. It provides Visual SLAM on its own and its output is a precise localization of itself in the scanned environment. It contains two fisheye lenses with combined 163° FOV and low power consumption visual processing unit which can run V-SLAM algorithms. Also, IMU is integrated which is required by V-SLAM.

Although the same thing can be achieved with the first camera, using a dedicated camera only for localization results in its higher precision. On top of that, the main computer does not have to run CPU-intensive algorithms.

Depth and tracking cameras are mounted together so they move together. This combination requires specifying the coordinate system transformation between the position-tracking camera and the depth-sensing camera. Position of each voxel **v** is transformed with translation matrix **T** (1), which results in position shift (2). This will enable the handheld scanner to acquire precise localization of scanned point cloud.

$$T = \begin{pmatrix} 1\,0\,0\,t_x \\ 0\,1\,0\,t_y \\ 0\,0\,1\,t_z \\ 0\,0\,0\,1 \end{pmatrix} \tag{1}$$

$$\begin{pmatrix} 1 & 0 & 0 & t_x \\ 0 & 1 & 0 & t_y \\ 0 & 0 & 1 & t_z \\ 0 & 0 & 0 & 1 \end{pmatrix} \times \begin{pmatrix} v_x \\ v_y \\ v_z \\ 1 \end{pmatrix} = \begin{pmatrix} v_x + t_x \\ v_y + t_y \\ v_z + t_z \\ 1 \end{pmatrix} \tag{2}$$

2.3 Battery and Power Distribution

Handheld scanner must be able to operate on its own, this implies that it must have its own source of energy. For this purpose, 5000 mAh Li-Polymer 3S battery was used which is commonly present in RC models or small UAVs. Its 5000 mAh capacity can provide 11 h of idle running and 3, 5 h of active scanning (See Fig. 2).

3S battery specification refers to the count of cells used inside. Each cell's nominal voltage is 3.7 V, so the whole battery voltage is 11.1 V which refers to the nominal value and a minimal usable value. LiPo batteries have some drawbacks, one of them being that they should not be overcharged, which means not to exceed 4.2 V per cell or over discharged by going under 3.7 V per cell. Ignoring this might result in damaging the battery with permanent capacity lost or, in some cases, explosion of the cell and, possibly, fire. The second case, which is the worse one of the two, should be prevented by using a certified battery charger as prescribed by the manufacturer to prevent the battery connection to short. In light of the above, XT60 connectors which are good at preventing accidental short circuits, were used.

Battery Level Measurement

To prevent battery from being over discharged, a battery level measurement was implemented. It is based on Adafruit INA260 board which is capable of measuring voltage, current, and power consumption. It can measure voltage up to 36 V DC and current up to 15 A continuously. The principle of measurement is based on the shunt resistor with resistance of 2 mΩ. It is capable of high precision with resolution of 1.5 mA which is beyond the project needs. Internal logic is powered by 3.3 V or 5 V and measurement results are provided by I2C protocol. It even contains an alarm pin which can be programmed to trigger signal by reaching the predefined condition. This could be used to switch off the whole power supply with relay to prevent battery damage.

Despite the advantages stated above, a different solution was used in this project. When the mission computer is turned on it automatically starts a script which reads battery measurement data over the I2C protocol. If the level of voltage declines under nominal value of the battery, which is 11.1 V, mission computer is forced to shut down. While this is not a perfect solution, primarily because some components can drain the battery even in idle mode, it is sufficient for the needs of this project. It is unlikely that the handheld scanner will be running without a user and even if it happened, the user would notice this and turn it off with the main switch.

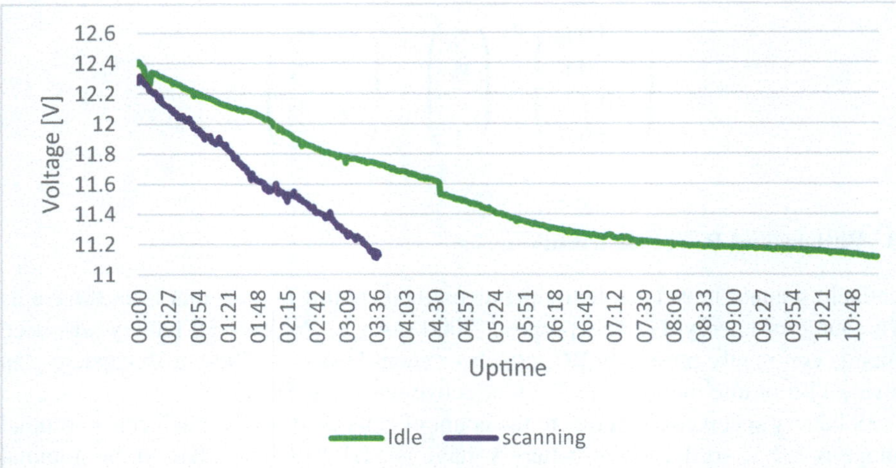

Fig. 2. Battery drain for idle and load mode

Converting Battery Voltage to 5 V

Batteries usually do not provide standard 5 V on their output. Also, in this case battery output voltage is derived from count of cell, where one cell provides 3.7 V and because it contains three cells, its output voltage is 11.1 V in total. As was already mentioned, this is only nominal value, albeit a fully charged battery provides 3 × 4.2 V or 12.6 V which is far beyond the required 5 V needed to power the mission computer and its peripheral sensors.

To convert battery output voltage to the required level, a Universal Battery Eliminator Circuit (UBEC) is used. Its job is also to stabilize the output level of voltage which, ideally, should stay the same regardless of the load. In this project, a 6 A UBEC is used which means that it can provide a stable output voltage level up to 6 A output current. However, in testing phase, this stability has proven itself only up to 3 A (See Fig. 3), only a half of what was declared by the manufacturer. Still, for standard operation it is more than what is needed because current from BEC never exceeded 1A.

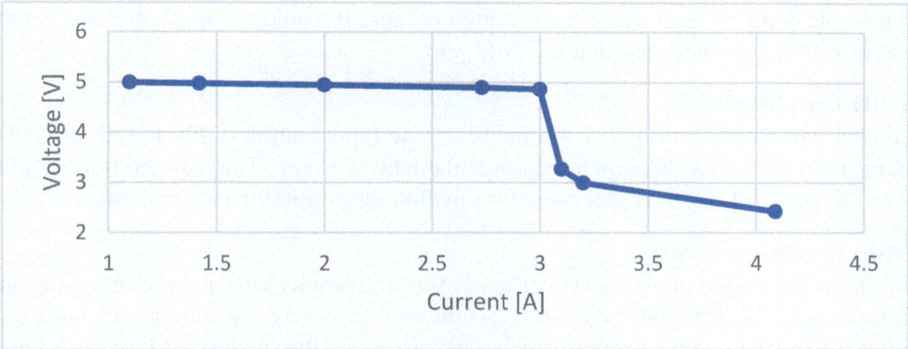

Fig. 3. Test of UBEC voltage stability

2.4 Mission Computer

Mission computer is the name for the main computer stored inside the handheld scanner. Raspberry Pi 4 model B was used in this project. This single-board computer provides sufficient computing power to run programs required for processing data from the depth camera Table 2.

Table 2. Raspberry Pi model B specifications

CPU	Quad core Cortex-A72 (ARM v8) 1.5 Ghz
RAM	8 GB
Wireless	2.4 GHz and 5.0 GHz IEEE 802.11ac
Bluetooth	5.0, BLE
Ethernet	1 Gbit
USB 2	2 ×
USB 3	2 ×

USB Devices

Raspberry Pi includes common ports used for connecting to peripheral devices. USB3 was used to connect the tracking camera and USB2 for the depth camera. The reason why a different USB version was used is that the tracking camera contains a bug which does not allow it to run immediately after the computer starts because it is detected as a different device. This can be solved simply by restarting the camera before use. It can be achieved with a tool called uhubclt which can restart power for a specific USB port, although in case of Raspberry Pi, only the whole USB hub can be restarted which results in restart of all USB ports for one version. Therefore, USB3 is automatically restarted after the computer starts to which only the tracking camera is connected. In contrast to this, the depth camera cannot be restarted because it would not work, and this is the reason why it must be connected to USB2 port.

GPIO Peripherals
Another computer-useful port is General Purpose Input Output (GPIO) used for communication with INA260 which measures the battery usage. The data are transmitted over I2C protocol which is processed by a python script running on a computer.

Wi-Fi Communication
The last required communication interface is Wi-Fi. It provides the main communication channel with the handheld scanner. Mission computer was preconfigured to always connect to a local wireless router which will assign to it the same IP address each time it connects. Therefore, mission computer is automatically accessible on a specific IP address after it is started (Fig. 4).

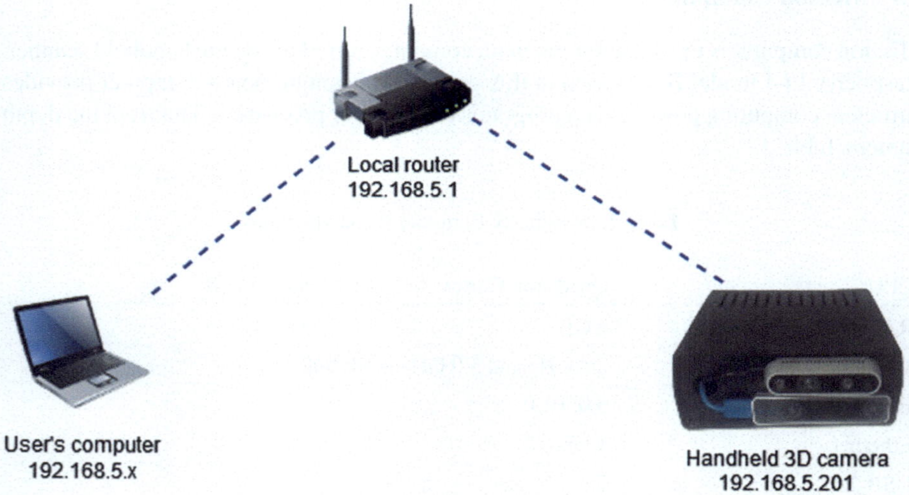

Fig. 4. Wi-Fi local network

3 Software Architecture

All logic in the handheld scanner is concentrated in the mission computer. As has already been mentioned, it is Raspberry Pi 4 which can run Linux operation system (OS). Ubuntu is one of the most popular Linux OS and here was used a version suitable for an embedded or a board computer called Ubuntu Mate 20.04. It is a big advantage to use a standard OS because it provides all tools which are commonly used for development, such as C++ compiler, Python interpreter, SSH and many command line interface tools.

The first configuration was done with a monitor, mouse, and a keyboard attached. Once networking was established over Wi-Fi, an SSH tool was used to connect to the mission computer from the user's computer and all next work was almost indistinguishable from working on a standard computer.

3.1 ROS

When it comes to the main software framework, Robot Operating System (ROS) was used which, in fact, is not so much an operating system as it is a set of robotics packages with standardized configuration and communication interfaces. One big advantage of this framework is that it can run programs called nodes independently of each other. This is achieved by a centralized broker called ROS core which interconnects nodes among themselves. Their communication is driven by a publish/subscribe pattern or a Remote Procedure Call (RPC). Nodes which are separately processed and written as either C++ executables or Python scripts can publish messages to topics and different nodes which are interested can subscribe to these topics and receive messages from other nodes. Nodes can even run on a different computer than the one used for visualization. The primary visualization tool called RViz runs on the user's computer and can receive updates from the handheld scanner and visualize the captured scene.

Access to camera data is also wrapped into nodes which are publishing depth, RGB and tracking information on specific topics. These topics are used by package called Real Time Appearance Based Mapping (RTAB-Map).

3.2 RTAB-Map

RTAB-Map is a Graph-Based SLAM approach based on an incremental appearance-based loop closure detector [9]. It can combine RGB-D, Stereo and Lidar to create a 3D point cloud representation of the scanned environment. It uses loop closure detectors for reducing error rate in generated graphs. In essence, loop closure is a method which can find previously scanned points and accordingly correct errors in a generated map graph.

The best way to present the feature of loop closure is to scan the premises inside a building and return to the starting point by taking a different route to that which was scanned. Already scanned environment is then detected and the current position aligned with the previously scanned environment. This alignment is back propagated to previously scanned parts.

RTAB-Map can operate in two regimes. One is mapping and one is localization. While mapping is used for scanning an unknown environment. The second regime, localization, is used for determining a position in a known (previously scanned) environment.

4 Testing the 3D Handheld Scanner

Testing was done in a house with good Wi-Fi coverage. Once handheld scanner was turned on, RViz on user's computer was launched, with environment variables which defined IP address of the handheld scanner where ROS core was launched. RViz can visualize also stream from the depth camera, which proved its usefulness for debugging whether camera is working. The main output from scanner is point cloud, a set of points located in 3D space. The point cloud was always extended when the scanner was obtaining data from a new area. However, the movements with scanner must be slow and smooth so that the mission computer would have enough time to process input depth and tracking stream from the cameras.

Also, the speed of visualization depended on Wi-Fi quality which varied according to the distance from the Wi-Fi router and the walls interfering with the Wi-Fi signal. Another tweak had to be done to reduce FPS of depth camera to only 6 FPS otherwise mission computer was unable to process such amount of input data.

After scanner was moved back to starting position a loop closure detector of RTAB-Map noticed previously scanned environment and corrected accumulated scanning error which resulted in slight movement of affected points in the point cloud. When scanning was done, data was still present inside the mission computer database which allowed post processing of the scanned data (Fig. 5).

Fig. 5. Example of a 3d scanned environment.

5 Conclusion

In this paper a 3D handheld scanner, build from accessible components, was described. The main goal was to do real time 3D scanning of environment with a standalone device. This was achieved with combination of depth and tracking cameras as main input sensors, mission computer which held all logic and processing power. And this was supported by own battery and power management circuits. Software part was based on ROS and RTAB-Map and for visualization tool was chosen RViz which was hosted on a separate computer and communicated with the scanner over Wi-Fi.

During the development of the handheld scanner, a significant amount of time was dedicated to the stable source of power for mission computer and its peripherals. This was not the first intention, but various problems related to voltage level influenced testing capacity of the battery and testing stability of power convertor. Protection against over discharge was included to protect battery and prolong its life.

Finally, testing proved that the handheld scanner could produce a real time 3D scene of the captured environment which could be used for various tasks, like reconstruction of a floor plan of a building.

Findings from this paper will be used in the following research which should be aimed at autonomous exploration of unknown environment. Handheld scanner will serve as a testing device for exploration algorithms which will propose new locations in captured environment to extend the already scanned space. Movement to this location will be done by the user carrying the handheld scanner. In future research, even a UAV which could even perform this movement fully autonomously, might be developed.

Acknowledgement. This work was supported by Slovak Research and Development Agency under project APVV-18-0436.

This work was supported by the Slovak Research and Development Agency under the Contract no. APVV-19-0210.

References

1. Ayaz, S.M., Khan, D., Kim, M.Y.: 3D handheld scanning based on multiview 3D registration using Kinect Sensing device. In: 2017 IEEE International Conference on Multisensor Fusion and Integration for Intelligent Systems (MFI), pp. 330–335 (2017). https://doi.org/10.1109/MFI.2017.8170450
2. Lee, C., Song, H., Choi, B., Ho, Y.-S.: 3D scene capturing using stereoscopic cameras and a time-of-flight camera. IEEE Trans. Consum. Electron. **57**(3), 1370–1376 (2011). https://doi.org/10.1109/TCE.2011.6018896
3. Shibata, M., Honma, T.: 3D object tracking on active stereo vision robot. In: 7th International Workshop on Advanced Motion Control. Proceedings (Cat. No.02TH8623), pp. 567–572 (2002). https://doi.org/10.1109/AMC.2002.1026983
4. Atif, M., Lee, S.: Adaptive pattern resolution for structured light 3D camera system. IEEE sens. 1–4 (2018). https://doi.org/10.1109/ICSENS.2018.8589640
5. Choi, O., Lee, S.: Wide range stereo time-of-flight camera. In: 2012 19th IEEE International Conference on Image Processing, pp. 557–560 (2012). https://doi.org/10.1109/ICIP.2012.6466920
6. Bastos, D., Monteiro, P.P., Oliveira, A.S.R., Drummond, M.V.: An overview of LiDAR requirements and techniques for autonomous driving. In: 2021 Telecoms Conference (ConfTELE), pp. 1–6 (2021). https://doi.org/10.1109/ConfTELE50222.2021.9435580
7. Yamauchi, B.: A frontier-based approach for autonomous exploration. In: Proceedings of the IEEE International Symposium on Computational Intelligence in Robotics and Automation pp. 146–151 (1997)
8. Zhu, C., Ding, R., Lin, M., Wu, Y.: A 3D frontier-based exploration tool for MAVs. In: 2015 IEEE 27th International Conference on Tools with Artificial Intelligence (ICTAI), pp. 348–352 (2015). https://doi.org/10.1109/ICTAI.2015.60
9. Labbé, M., Michaud, F.: RTAB-Map as an Open-source Lidar and visual SLAM library for large-scale and long-term online operation. J. Field Robot. **36**(2), 416–446 (2019)

Real-Time Intelligent Sensor Selection for Subsurface CO2 Flow and Fracture Monitoring

Klemens Katterbauer(✉), Abdallah Al Shehri, and Abdulaziz Al Qasim

Saudi Aramco, Dhahran 31311, Saudi Arabia
klemens.arabiya@gmail.com

Abstract. 4th Industrial Revolution (4IR) technologies have assumed critical importance in the oil and gas industry, enabling data analysis and automation at unprecedented levels. Formation evaluation and reservoir monitoring are crucial areas for optimizing reservoir production, maximizing sweep efficiency and characterizing the reservoirs. Automation, robotics and artificial intelligence (AI) have led to tremendous transformations in these areas, in particular in subsurface sensing. We present a novel 4IR inspired framework for the real-time sensor selection for subsurface pressure and temperature monitoring, as well as reservoir evaluation. The framework encompasses a deep learning technique for sensor data uncertainty estimation, which is then integrated into an integer programming framework for the optimal selection of sensors to monitor the CO2 penetration in the reservoir. The results are rather promising, showing that a relatively small numbers of sensors can be utilized to properly monitor the fractured reservoir structure.

Keywords: Artificial intelligence · CO2 · Sustainability · Sensor selection

1 Introduction

Sensing deep in the reservoir has always been a major objective to enhance reservoir formation understanding and optimize the recovery from the reservoir. In the early days of the oil and gas industry, determination of reservoir formation properties were based on assumed geological formations and structures encountered on the surface. Furthermore, retrieved rock cuttings assisted in getting a better understanding of the reservoir formation, however this information is limited to a small area and may not be representative of the reservoir formation as a whole or taking into account the heterogeneity in the reservoir. Another challenge for mature reservoirs is to determine the sweep efficiency in the reservoir, where besides production information and some surface reservoir monitoring, such as seismic or electromagnetics, there is no overall in-situ reservoir monitoring system available [1, 2]. As the reservoirs are dynamic, permanent monitoring of the reservoir is crucial to determine the saturation flow and the fracture channels. Hence, an in-situ monitoring of the reservoir becomes quintessential in order to overcome the existing challenges of limited information away from the wellbores.

R. Silhavy (Ed.): CSOC 2022, LNNS 502, pp. 236–246, 2022.
https://doi.org/10.1007/978-3-031-09076-9_23

The 4th industrial revolution (4IR) has become a major transformer of the upstream petroleum industry. Major advances were already achieved in enhancing production, performing real-time monitoring of wells and reservoirs and also forecast potential reservoir risks and workover requirements [3–5]. Several advances were also achieved in performing maintenance and installation operations remotely via the help of 4IR technology [6]. The main objective is to improve productivity and cost-effectiveness of the operations, as well as enhance safety. This allows to conduct maintenance in a much shorter time period and also allows to conduct the operations around the clock.

Enhancing production from and monitoring reservoirs are critical components for ensuring the effectiveness of CO2 storage and utilization operations and maintain its sustainability. For this, sensing is an essential area that allows to monitor the reservoir in real-time and investigate its evolution. Continuous sensing further allows monitor the behavior of a reservoir over time and forecast its future production potential. Conventional surface sensing covers an extensive area of the reservoir. However, the resolution and challenge connected to the multiple solutions of the inverse problem represent a significant problem. The challenge arises primarily from the lack of direct measurements and observations in the reservoir.

While surface sensing enables to cover an extensive area and deduce easier the correlations between different measurements, as well as the causes and effects, subsurface sensing operations are significantly more challenging. This is due to the lack of direct measurements and observations of the reservoir structure and formation, as well as challenge to place measurement equipment downhole [7, 8]. In order to overcome this challenge related to the lack of direct measurements, a more direct approach to sensing in the form of subsurface reservoir sensors is essential.

Miniaturized downhole sensors have been developed in recent years, allowing to achieve permanent downhole sensing that is both robust and efficient [8, 9]. Xu et al. [10] presented a temperature insensitive pressure sensor based on fiber-optics that has a size of only 125 μm. The authors demonstrated the ability to measure pressure levels over a significant range with minimal temperature effects, which may make these sensors applicable for downhole sensing. Similarly, Jia et al. [11] presented a fiber-optic FabryPerot gas refractive index sensor for high temperature applications. The miniaturized sensor allows to measure up to 800 °C, outlining the feasibility of high temperature permanent downhole monitoring with low power consumption.

Building upon these advances, commercial implementations of optical pressure and temperature sensors for downhole sensing were presented by Well-SENSE whose low-cost small and robust sensors can measure up to 20,000 psi and 300 °C [12].

Al Shehri et al. [9] presented innovative wireless sensors for the mapping of hydraulic fractures in subsurface reservoirs. The results outline the ability to accurately map fractures with a hybrid solution of electromagnetic and magnetic conduction wireless communication in order to overcome excessive path losses within the reservoir environment. Communication losses between the sensors represent a major challenge in addition to the power requirements of the sensors, requiring that there is sufficient proximity between the wireless sensors such that the data is adequately transmitted.

These advances outline the feasibility of downhole sensing in the reservoir with data transmission being conducted wirelessly. A major challenge arises from the powering of these downhole sensors in order to maximize the sensing duration in the downhole environment [14]. Given the connectedness of the reservoir as well as partial redundancy of the downhole sensors, not requiring all sensors to be operational at the same time, the objective is to maximize data acquisition while minimizing overall power consumption.

This conflict of interest between maximizing data acquisition and minimizing power consumption leads to the problem of selecting a number of sensors while achieving the target objective of the downhole sensing. These selection schemes can typically be classified in coverage schemes, target tracking and localization schemes, single mission assignment schemes and multiple missions assignment schemes [13]. Coverage schemes are selection schemes that ensure the sensing coverage of the location or the targets of interest, while target tracking and localization schemes focus on the selection of sensors for target tracking and localization purposes. The mission assignment schemes focus on the selection of sensors for a single or multiple missions that have to be accomplished.

The sensor selection problem is mathematically defined as given a set of sensors $S = \{S_1, \ldots, S_n\}$, then we need to select the best subset with k sensors that satisfy one or multiple missions. The challenge that arises from this problem is in most instances NP-complete, which implies that there is no polynomial-time algorithm for solving the problem. This represents a major challenge for real-time data interpretation and the optimization of the sensors, as there will be only approximative solutions available within an acceptable timeframe [16].

We will demonstrate a novel intelligent sensor selection framework for the optimization of sensor selection in real-time for CO2 flow and fracture monitoring. The objective of the framework is to maximize longevity of the operations while maintaining measurement accuracy and flow detection ability.

2 Method

We have developed an innovative real-time sensor utilization optimization framework that incorporates a deep learning driven optimization framework connected to a subsurface fracture network model. This forms then a crucial part of the sensor selection optimization problem that aims to optimize in real-time to minimize the number of sensors required in order to maintain sufficient data quality. This challenge is equivalent to maximize the longevity of the sensors deployed while maintaining sufficient reservoir coverage in order to limit the uncertainty in the multi-data interpretation.

The framework incorporates a deep learning approach for the sensor measurements combined with a fast iterative solver for real-time optimization of the sensor selection. The framework is outlined in **Error! Reference source not found** (Fig. 1).

First, a fracture-CO2 flow reservoir model is established using a connectivity and sensing data quality determination approach. The assumption is that the flow between the CO2 injecting wells is primarily within the fractures with only limited flow in the matrix structures. This is in line with conventional assumptions when utilizing discrete fracture network models, as well as observations on fractured carbonate reservoir rocks, where the flow is primarily in the fractures. The network flow model is then integrated into a deep learning framework for the sensor data estimation and the uncertainty in the estimates. The deep learning framework utilizes a feedforward network structure for

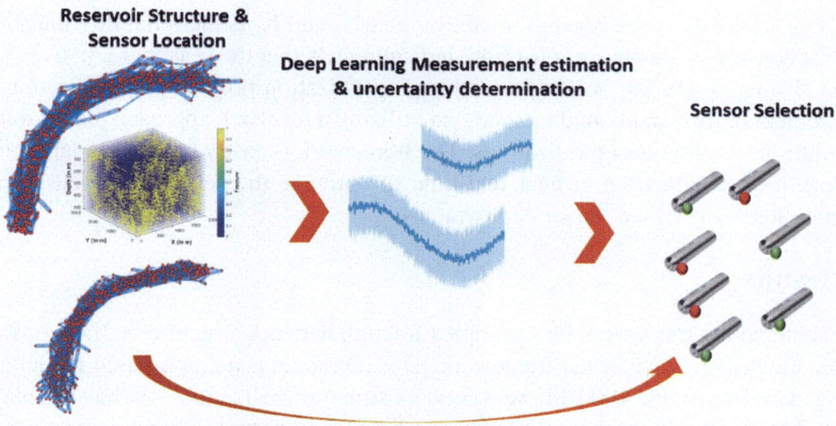

Fig. 1. Framework representation with the fracture network structure and the uncertainty estimates that are integrated into the sensor selection optimization.

determining from the sensor derived flow measurement data based on multiple potential scenarios in terms of the reservoir formation condition. The estimations relate to whether the sensors are close to the matrix or in the fracture, and what the water saturation in the vicinity of the sensor is. The main objective of the deep learning framework is to have a data-driven approach to the estimation of the fracture and water saturation in the vicinity of the sensor based on pressure and temperature measurements.

The sensor selection problem is then posed as an integer optimization problem as outlined below:

$$\min f'z$$

$$s.t.\, Cz > 0$$

$$Uz \leq b_u, \forall i \in N$$

$$z_i \in \{0, 1\}, \forall i \in N$$

The integer optimization problem is solved in real-time where the vector f is the cost function dependent power consumption over time of the sensors. For each update time step, the cost function is updated from the previous, implying that if the sensor i is operational, then f_i is gradually increasing, while for the inactive sensors, f_i may remain constant or is reduced in case the sensors can be recharged. The constraint $Cz > 0$ ensures that there is for each reservoir area at least one sensor that covers this area. The matrix C is the connectivity matrix between the sensors and the area, implying that $C_{ij} = 1$ if the j-th sensor covers the i-th area. This ensures that each area is covered, and that the sensor can connect and transfer data between each other.

The constraint $Uz \leq b_u$ implies that the data sensing reliability for each node is maintained, implying that the sensing uncertainty must be below a threshold value. The matrix or vector U is the sensing reliability matrix, and b_u is the reliability threshold. The last constraint is a binary constraint, indicating whether the sensor is active ($z_i = 1$) or inactive ($z_i = 0$). For solving the integer optimization problem, we utilized a fast and efficient branch and bound method, via utilizing a feedback approach incorporating the solutions of previous optimizations. The framework is easily scalable to larger flow network models, allowing in near real-time to optimize the selection of sensors and maintain longevity of the sensor deployments.

3 Results

We examined the framework on a complex fracture network structure in 2D in order to outline the performance of the framework. The 2D model is a graph-based model consisting of 500 nodes and 1000 different network structure realizations. We have displayed in Fig. 2 two examples of the different network realizations and connection between the fracture network nodes. The realizations illustrate the considerable difference between the connectedness of the fracture network which reflects the general challenge of monitoring and determining the fracture network structure and connectedness between the fractures.

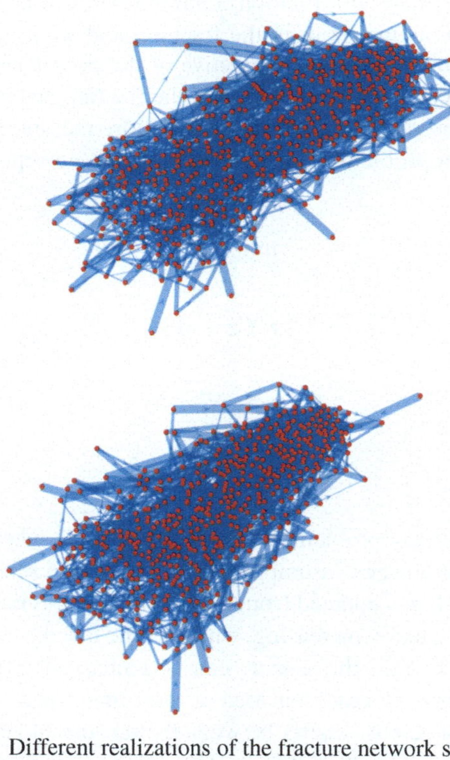

Fig. 2. Different realizations of the fracture network structure.

We then utilized a deep learning approach to estimate the uncertainty of the data based on the network structure. The data set was divided 75/15/15 into a training, validation and test dataset, and a fully connected feedforward neural network structure was used. For the optimization, we used a scaled conjugate gradient approach given the substantial size of the problem. The sensors record pressure and temperature data at each location, and for each of the sensors an interpreted uncertainty parameter is computed. The uncertainty parameter varies from 0 to 1.5, where a higher uncertainty parameter indicates stronger uncertainty in the measured data. The uncertainty measurement parameters are derived from multiple repeat measurements of the sensors that are then classified in terms of their accuracy and variation. The training, validation and testing results of the deep neural network are displayed in Fig. 3. The estimation results are rather strong, outlining overall accurate estimation of the sensor data uncertainty, with the larger number of data points for lower uncertainties only marginally affecting the estimation quality for higher uncertainties.

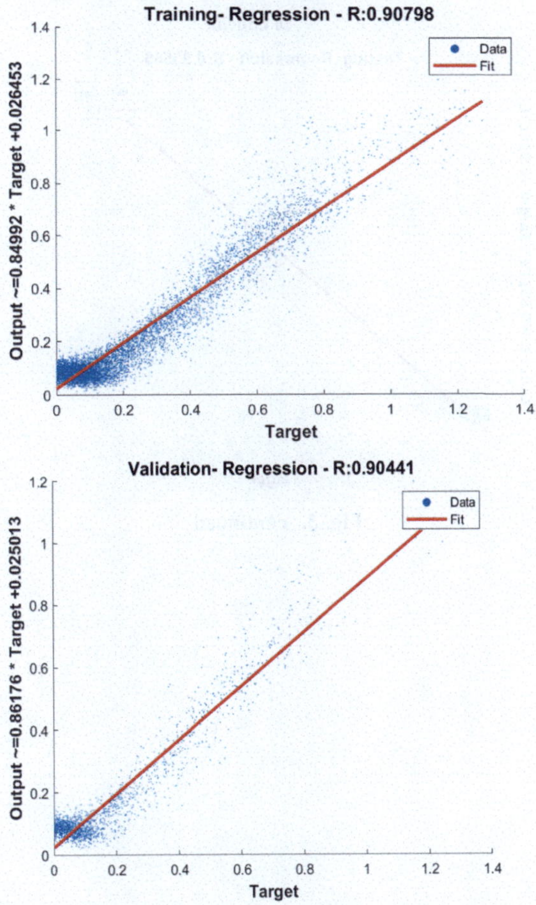

Fig. 3. Comparison of the neural network estimation of the data uncertainty.

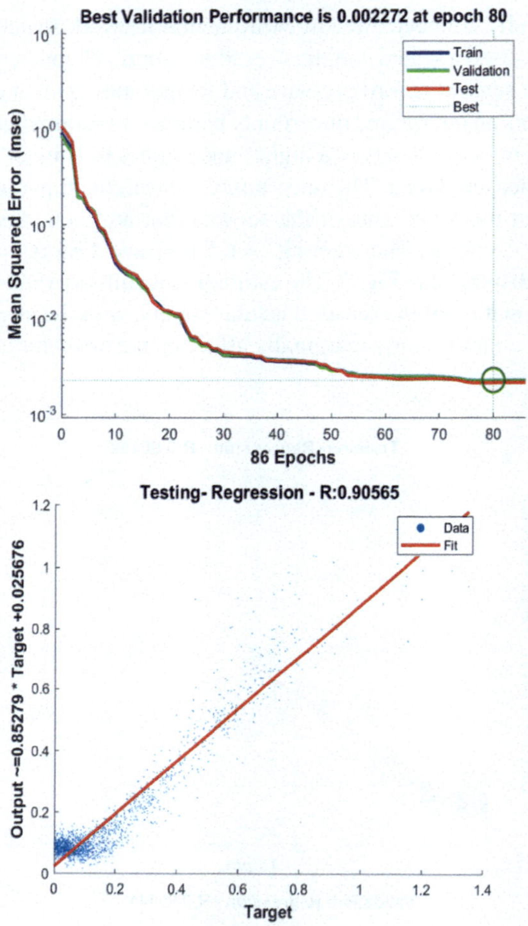

Fig. 3. continued

Utilizing the deep learning network model, we then solved the sensor selection problem in real-time under uncertainty. The uncertainty matrix U is updated in each simulation step to reflect the changing reservoir conditions as well as sensing parameters. The cost vector f for the sensors is increased in each step for the active sensor components, reflecting the power utilization of sensor and to penalize excessive usage of an individual sensor. In case the sensor is not anymore operational f_i (e.g. lack of power), then f_i was set to positive infinity.

The timeframe for the sensor optimization was from April 1st, 2019 until January 11th, 2020, where the sensors were optimized every 15 days.

The optimization results are displayed in Fig. 4, outlining the active sensors in green and the inactive in black.

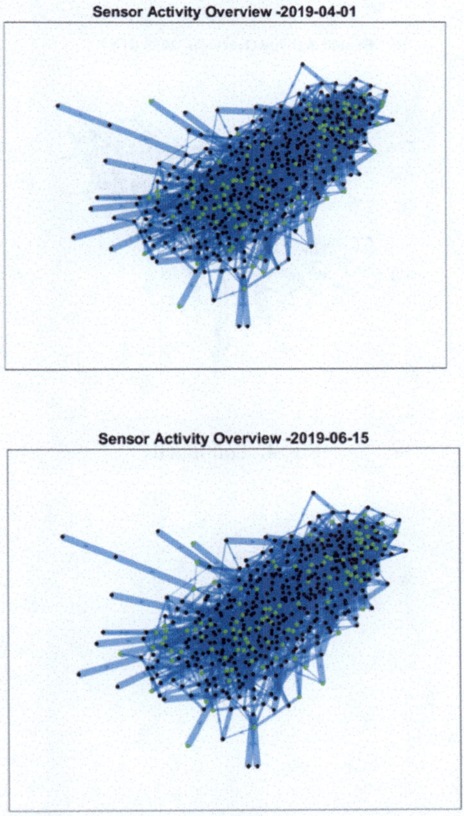

Fig. 4. Overview of the selected sensors for different time steps. In green are the active sensors and black indicates that a sensor is inactive.

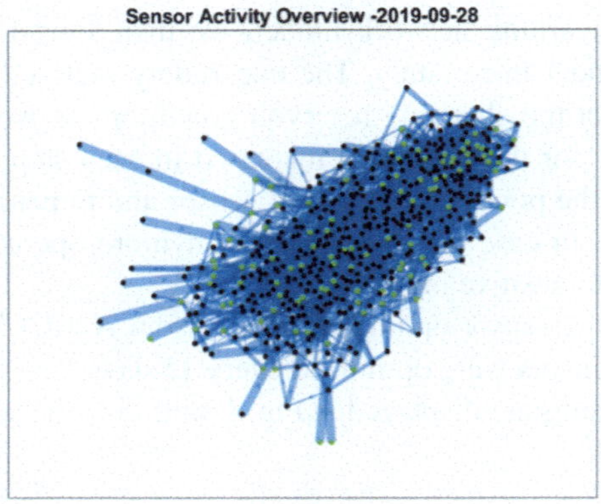

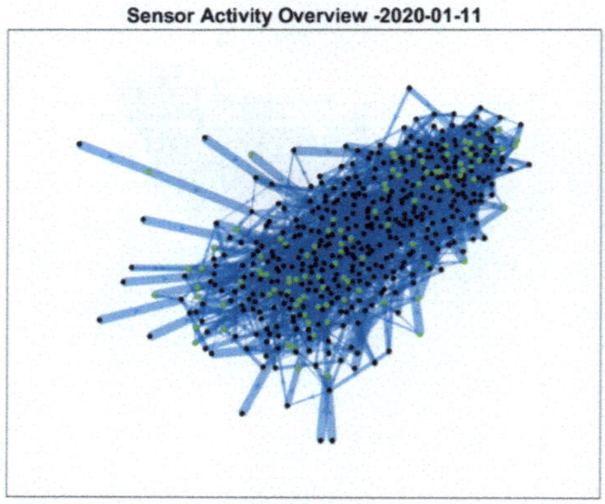

Fig. 4. continued

As observed there are certain sensor clusters that are active for longer durations indicating that these sensors are placed in crucial fracture intersection points as well as exhibit a low degree of measurement uncertainty. This is confirmed via a sensor utilization analysis for the 500 sensors in Fig. 5. The indication is that most sensor are rarely active, or solely active for a short period of time, while there are a few sensors that are heavily utilized and operational for more than 250 days out of 285 days.

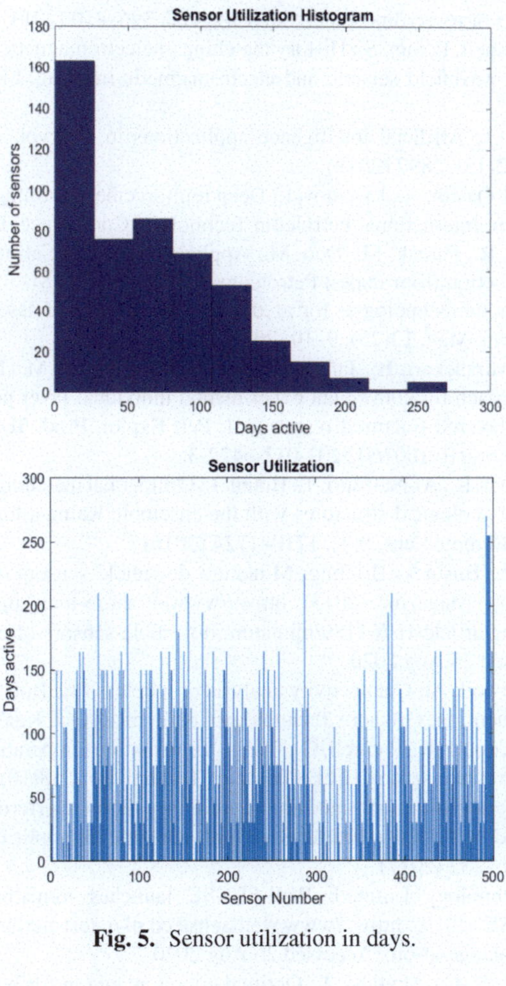

Fig. 5. Sensor utilization in days.

4 Conclusions

We presented a novel AI driven sensor selection framework for the optimal selection of subsurface pressure and temperature sensors in a fractured reservoir. The framework presents the ability to optimize the selection of sensors for subsurface sensing in real-time, thereby maximizing the overall coverage of the sensors for efficient waterfront tracking.

References

1. Katterbauer, K., Hoteit, I., Sun, S.: EMSE: synergizing EM and seismic data attributes for enhanced forecasts of reservoirs. J. Pet. Sci. Eng. **122**, 396–410 (2014)
2. Katterbauer, K., Hoteit, I., Sun, S.: History matching of electromagnetically heated reservoirs incorporating full-wavefield seismic and electromagnetic imaging. SPE J. **20**(5), 923–941 (2015)
3. Ertekin, T., Sun, Q.: Artificial intelligence applications in reservoir engineering: a status check. Energies **12**(15), 2897 (2019)
4. ,Miftakhov, R., Al-Qasim, A., Efremov, I.: Deep reinforcement learning: reservoir optimization from pixels. In: International Petroleum Technology Conference, Dhahran (2020)
5. Panja, P., Velasco, R., Pathak, M., Deo, M.: Application of artificial intelligence to forecast hydrocarbon production from shales. Petroleum **4**(1), 75–89 (2018)
6. Fumagali, S.: Robotic technologies for predictive maintenance of assets and infrastructure. IEEE Robot. Autom. Mag. **12**(25), 9–10 (2018)
7. Davarpanah, A., Mirshekari, B., Jafari Behbahani, T., Hemmati, M.: Integrated production logging tools approach for convenient experimental individual layer permeability measurements in a multi-layered fractured reservoir. J. Pet. Explor. Prod. Technol. **8**(3), 743–751 (2018). https://doi.org/10.1007/s13202-017-0422-3
8. Sana, F., Katterbauer, K., Al-Naffouri, T., Hoteit, I.: Orthogonal matching pursuit for enhanced recovery of sparse geological structures with the ensemble Kalman filter. IEEE J. Sel. Top. Appl. Earth Obs. Remote Sens. **9**(4), 1710–1724 (2016)
9. Offshore Magazine Business Briefing: Miniature downhole sensors offer improved shock resistance. Offshore Magazine (2013). https://www.offshore-mag.com/business-briefs/equipment-engineering/article/16761189/miniature-downhole-sensors-offer-improved-shock-resistance. Accessed 25 Aug 2020
10. Al Shehri, A., Shewoil, A.: Connectivity analysis of wireless FracBots network in hydraulic fractures environment. In: Offshore Technology Conference Asia, Kuala Lumpur (2020)
11. Xu, J., Wang, X., Cooper, K., Pickrell, G., Wang, A.: Miniature temperature-insensitive Fabry-Perot fiber optic pressure sensor. IEEE Photonics Technol. Lett. **18**(10), 1134–1136 (2006)
12. Jia, P., et al.: Temperature-compensated fiber-optic Fabry–Perot interferometric gas refractive-index sensor based on hollow silica tube for high-temperature application. Sens. Actuators B: Chem. **244**, 226–232 (2017)
13. Well-SENSE Technology Limited: Well-SENSE launches miniature optical gauge at OTC. Well-SENSE 2019. https://www.well-sense.co.uk/latest-news/well-sense-launches-miniature-optical-gauge-at-otc. Accessed 26 Aug 2020
14. Al Shehri, A., Lin, S.-C., Akyildiz, I.: Optimal energy planning for wireless self-contained sensor networks in oil reservoirs. In: 2017 IEEE International Conference on Communications (ICC) (2017)
15. Rowaihy, H., et al.: A survey of sensor selection schemes in wireless sensor networks. In: Proceedings of the SPIE, Unattended Ground, Sea, and Air Sensor Technologies and Applications IX, vol. 6562 (2007)
16. Yoo, T.-S., Lafortune, S.: NP-completeness of sensor selection problems arising in partially observed discrete-event systems. IEEE Trans. Autom. Control **47**(9), 1495–1499 (2002)

Predicting HIV Status Using Machine Learning Techniques and Bio-Behavioural Data from the Zimbabwe Population-Based HIV Impact Assessment (ZIMPHIA15-16)

Innocent Chingombe[1], Godfrey Musuka[1], Elliot Mbunge[2], Garikayi Chemhaka[3], Diego F. Cuadros[4], Grant Murewanhema[5], Simbarashe Chaputsira[1], John Batani[6], Benhildah Muchemwa[2], Munyaradzi P. Mapingure[1], and Tafadzwa Dzinamarira[1,7](✉)

[1] ICAP at Columbia University, Harare, Zimbabwe
td2581@cumc.columbia.edu
[2] Department of Computer Science, Faculty of Science and Engineering, University of Eswatini, Kwaluseni Campus, Kwaluseni, Eswatini
[3] Department of Statistics and Demography, Faculty of Social Sciences, University of Eswatini, Kwaluseni Campus, Kwaluseni, Eswatini
[4] Department of Geography and GIS, University of Cincinnati, Cincinnati, USA
[5] Unit of Obstetrics and Gynaecology, Department of Primary Health Care Sciences, Faculty of Medicine and Health Sciences, University of Zimbabwe, Harare, Zimbabwe
[6] Faculty of Engineering and Technology, Botho University, Maseru 100, Lesotho
[7] School of Health Systems and Public Health, University of Pretoria, Pretoria 0002, South Africa

Abstract. HIV and AIDS continue to be a significant public health concern globally, with about 36 million people currently living with the epidemic. Several HIV interventions have been implemented to intensify virus transmission prevention, screening, and diagnosis in sub-Saharan African countries, including Zimbabwe. HIV prevalence is substantially high in Zimbabwe despite the significant progress made in the previous years. As the country moves closer to attaining the epidemic control status, there is a need for targeted HIV interventions focusing on HIV risk individuals. Most current HIV interventions are based on evidence about specific sub-population groups, undermining the diversity of individual risk levels within such groups. Therefore, this study applied random forest classifier, support vector machine, and logistic regression to predict HIV status outcomes using Zimbabwe Population-Based HIV Impact Assessment data to identify high-risk individuals and develop targeted interventions based on risk. This study shows that logistic regression outperformed the random forest classifier and support vector machine with the prediction accuracy of 85%, recall of 98%, and F1-score of 92%. However, the random forest classifier has the highest precision of 87% compared to the other models. The support vector machine outperformed the random forest classifier in recall and F1-score metrics, with a recall of 96% and F1-score of 91%. Machine learning models can help identify individuals at high risk of contracting HIV and assist policymakers in developing targeted HIV prevention and screening strategies informed with socio-demographic and risk behavioural data. However, this study only used socio-demographic and behavioural predictors to predict HIV

R. Silhavy (Ed.): CSOC 2022, LNNS 502, pp. 247–258, 2022.
https://doi.org/10.1007/978-3-031-09076-9_24

status. There is a need to include other HIV clinical predictors to optimise HIV status prediction models better and further integrate them into real-world healthcare settings.

Keywords: HIV/AIDS · Prediction · Machine learning · ZIMPHIA · Logistic regression · Support vector machine · Random forest classifier

1 Introduction

Despite several decades of improved research and understanding of the human immunodeficiency virus (HIV), it continues to pose a significant public health challenge [1]. This is more so in sub-Saharan Africa (SSA), where most people living with HIV (PLWH) reside. Statistics from the Joint United Nations Programme on HIV/AIDS (UNAIDS) estimate the number of PLWH globally to be 37.7 million (30.2–45.1 million) [2]. Of these, an estimated 28.2 million were accessing highly active antiretroviral therapy (HAART) as of 30 June 2021 [3]. In 2020, 1.5 million (1.0–2.0 million) new infections were reported globally.

Despite a wide array of preventive behavioural and biomedical interventions [4], the numbers of incident infections remain too high, raising the need for more effort to eliminate new infections. Over the past four decades, SSA has been disproportionately affected by the HIV epidemic in terms of morbidity, mortality and socio-economic impacts on the population [5]. It is estimated that 67% of PLWH reside in SSA, and despite a key shift in the populations affected by new HIV infections, approximately 39% of incident infections in 2020 were in SSA [2]. Furthermore, 50% of all new HIV infections occur in young women and girls, with 4 200 incident cases in this population (15–24 years) weekly in SSA [3]. Several risk factors continue to put this population at greater risk than their male counterparts, calling for targeted interventions to reduce the burden among adolescent girls and young women [6].

Zimbabwe is among the most affected SSA countries, with an estimated HIV prevalence of 12.9% as of December 2020, corresponding to approximately 1.3 million adults [7]. Despite substantial progress in HIV care and treatment in the country, continued work to reduce new infections is needed significantly. As of 2019, 90% of PLWH in Zimbabwe were aware of their status, 94% of those diagnosed were on HAART, and of those diagnosed and on treatment, 86% were virally suppressed [8]. Strategies for controlling the HIV epidemic developed by the World Health Organization (WHO) and the UNAIDS have almost been universally adopted by SSA countries. However, they are at different stages of attaining the 95-95-95 goals. The 95-95-95 strategy is a crucial strategy for control [9]. The adoption of treatment as prevention (TaSP) for serodiscordant couples, test-and-treat policies and effective prevention of mother to child transmission guidelines coupled with the widespread availability of safer and effective HAART regimens have been significant steps towards the control of the epidemic in the country [10]. However, it is becoming more critical to understand HIV transmission dynamics, including high-rick communities suffering high burden of new infections and recency of infection. There has been a global scale of recency testing as an important strategy for public health stakeholders and policymakers to detect recently HIV-infected individuals [12], carry out real-time data analysis to understand transmission, design targeted programs to reduce/prevent incident infections in these vulnerable communities, and interrupt the chains of transmission with the ultimate goal of ending the HIV epidemic. This is especially important in countries with a burden of incident HIV infections, such as Zimbabwe, where effective interventions to reduce them remain critical significantly. Artificial intelligence techniques have gained increasing significance in HIV prevention in recent times and could substantially aid in improving the understanding of HIV transmission dynamics.

1.1 Contribution of the Study

Several HIV intervention efforts have been widely used for universal HIV screening [9]. However, due to HIV preventive measures and ART adherence [13], the HIV prevalence is substantially high, especially in SSA [14]. Therefore, there is a need to improve risk perception by integrating emerging technologies such as computational intelligence techniques for assessing, identifying and modelling HIV risk factors developing robust HIV diagnosis and prediction tools. Such tools are paramount, especially in clinical services, to identify people at-risk and effectively allocate and prioritise resources [15]. In addition, understanding the impact of HIV risk factors such as socio-demographic [11] and behavioural factors can tremendously assist policymakers in developing customised HIV interventions strategies to reduce HIV transmission and improve HIV early screening through coordinated efforts. Notably, machine learning (ML) models present unprecedented opportunities to develop HIV prediction tools for early HIV screening targeting recent infections. For instance, using electronic health data, machine learning tools were utilised to identify HIV predictors in SSA [16]. Machine learning algorithms can extract meaningful insights from huge amounts of data. Recently, machine learning models have been utilised in countries like Denmark, the United States of America and Eastern Africa to identify potential candidates for preexposure prophylaxis (PrEP) [17, 18]. A study conducted by Bao et al. [19] applied ML models to predict HIV or other sexually transmitted infections (STIs) diagnosis among men who have sex with men using clinical health records in Australia. However, there is limited literature on applying machine learning models to predict HIV status using Population-based HIV Impact Assessment (PHIA) data. Therefore, this study aimed to address this gap by applying and comparing the performance of machine learning models to predict HIV status using Zimbabwe PHIA data.

The remainder of this paper is structured as follows: Sect. 2 presents data sources, ethical clearance information, data preprocessing techniques and machine learning models. Discussion of results is presented in Sect. 3. Finally, the conclusion and future work are presented in Sect. 4.

2 Methodology

2.1 Data Sources and Ethical Considerations

The analysis is based on 20,577 individuals who consented to the face-to-face questionnaire and biomarker testing in this nationally representative household survey. The study received ethical approval from the Medical Research Council of Zimbabwe, the Research Council of Zimbabwe, the Centers for Disease Control and Prevention (CDC) Institutional Review Board (IRB), Columbia University IRB, and WESTAT IRB.

The methodology for the survey has been described elsewhere [20]. Briefly, ZIMPHIA 2015–2016, was a nationally-representative, household-based survey. The survey used 2-stage, cluster sampling design. In the first-stage sampling, 500 census enumeration areas (EAs) were selected using probability proportional to estimated size sampling based on the most recent census. The sample was stratified by place of residence (urban/rural) and subnational area i.e. provinces. In the second sampling stage, households were selected from each EA using an equal probability approach that allowed the number of households chosen to vary based on the total number of homes in the EA. This study explored the risk factors for HIV status in the 18795 females and males aged 15 and above years who had completed the survey and had ever had sex before the interview. The sample was reduced to 17130 by excluding missing cases on the selected risk factors for the study.

Several studies utilised socio-demographic, clinical data and behavioural characteristics to predict HIV infections, prognosis and mortality. These data have been also used to identify HIV predictors, HIV pre-exposure prophylaxis (PrEP) candidates, HIV risk estimation, response to HIV therapy, and early virological suppression in ART. For instance, Bisaso et al. [21] applied logistic regression to predict early virological suppression in ART for HIV patients using predictors such as sex, baseline age and body weight, tuberculosis (TB) disease status, ART regimen, baseline CD4 count and viral load count. A study conducted by Marcus et al. [17] applied machine learning to identify HIV PrEP candidates using electronic health record data. Also, Zhang et al. [22] applied an ensemble-based machine learning approach to predict HIV prognosis and mortality using demographics and clinical variables such as age, gender, race, ART, CD4, viral load, monocytes and granulocytes. Haas et al. [23] used ML to predict HIV risk estimation using diagnoses data, procedures, socio-demographic factors like gender, insurance type, and marital status, as well as the services. Notable studies including [24, 25] implemented machine learning to identify socio-behavioural predictors of HIV positivity and screening in Africa. Among other socio-behavioural predictors, age, gender, level of education, wealth quintile, age at the first experience of sex, number of sex partners, use of protection, excessive alcohol drinking, TB treatment, knowing HIV status, place of residence, marital status and circumcision status are frequently used HIV screening/infection predictors. In support, Mutai et al. [25] applied XGBoost algorithm to determine the association between HIV predictors aforementioned and HIV status in sub-Saharan Africa using the population-based HIV impact assessment (PHIA) dataset. Such predictors present tremendous opportunities to predict and model HIV status. For instance, Ahlström et al. [15] indicated of HIV status using logistic regression and nation-wide electronic registry data in Denmark. The nation-wide electronic registry dataset consists of socio-demographic data such as age, level of education, income level, sex, residency, marital status and migration information. Wray et al. [26] also used smartphone survey data and machine learning to identify situational and contextual risk factors for HIV risk behavior among men who have sex with men who are not on PrEP. Their study further alluded that identification of socio-behavioural HIV risk factors can enhance HIV screening especially in resource-constrained areas.

Table 1. Data coding description

Variables	Coding description
HIV status	HIV Positive (HIV+) - 1
	HIV Negative (HIV−) - 0
Age	15–29 - 1
	30–39 - 2
	40–49 - 3
	50+ - 4
Sex	Female - 0
	Male - 1
Ever married	Yes - 1
	No - 0
Education	Primary or less -1
	Secondary - 2
	Secondary+ - 3
Condom use	Yes - 1
	No - 0
HIV testing	Yes -1
	No - 0
Number of sexual partners	0 - 1
	1 - 2
	2 - 3
Ever forced sex	Yes - 1
	No - 0
Reside	Rural - 0
	Urban - 1
Province	Bulawayo - 0
	Manicaland - 1
	Mash Central - 2
	Mash East - 3
	Mash West - 4
	Mat North - 5
	Mat South - 6
	Midlands - 7
	Masvingo - 8
	Harare - 9

2.2 Data Preprocessing

The study utilised Zimbabwe Population-Based HIV Impact Assessment (ZIMPHIA) data. A nationally representative cross-sectional household-based survey was designed to assess HIV-related key health indicators [27]. We selected several established socio-demographic and behavioural factors linked to higher risk of HIV infection to predict HIV status. At first each socio-demographic and behavioural HIV predictor was tested for significance against the associated target class (HIV status) using the Chi-square test, as shown in Table 1. Since all risk factors are categorical Spearman rank correlation coefficients were computed and all variables were uncorrelated. Only 10 variables that were significant at p-value of 0.05 based on the Chi-square test in bivariate analyses were selected for the multivariable model. We excluded 5 variables namely; employment status, wealth status, alcohol use, sexual debut and whether or not respondents agree that men who are circumcised can have multiple sexual partners and not be at risk for HIV. After selecting significant HIV status variables, we applied multivariate analysis to determine further the association between HIV predictors features and HIV status to provide an intuitive understanding of key features in predicting HIV status in both sexes. Variance inflation factor (VIF) was used to check for multicollinearity in the multivariable model. One variable, stigma on PLWH, with a $VIF > 10$ was further removed from the study. The key selected socio-demographic and behavioural HIV predictors are shown in Table 1.

The key features (nominal and ordinal variables) include HIV status, age, sex, marital status, level of education, condom use, HIV testing, number of sexual partners, ever forced sex, residence and province. These HIV status predictors were encoded using one-hot encoding based on the information from the ZIMPHIA survey. The contributions of predicators risk to HIV status prediction using feature importance shows that province had the highest significant contribution (0.26), followed by age group (0.18), condom use (0.17), number of sex partners (0.08), level of education (0.08) and so on, as shown in Fig. 1.

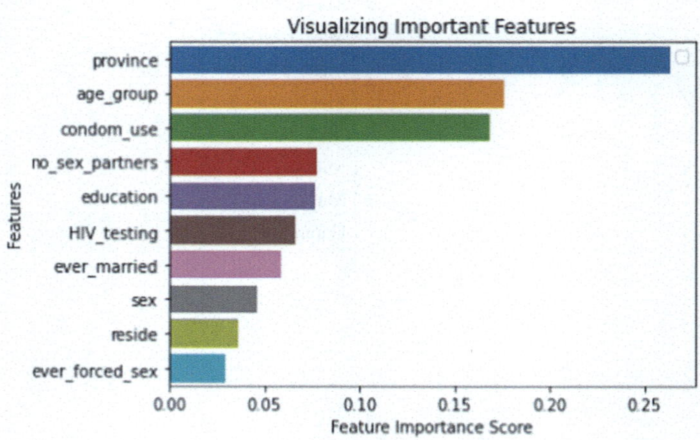

Fig. 1. HIV status predictors feature importance

After feature selection, we split up the dataset into training and testing using the ratio of 70% and 30%, respectively, as shown in Fig. 3. All rows with missing values were dropped. This resulted in 17130 adults (10332 females and 6798 males) in the final dataset. The Heatmap plot for Pearson's correlation coefficient between the selected significant predictors is shown in Fig. 2.

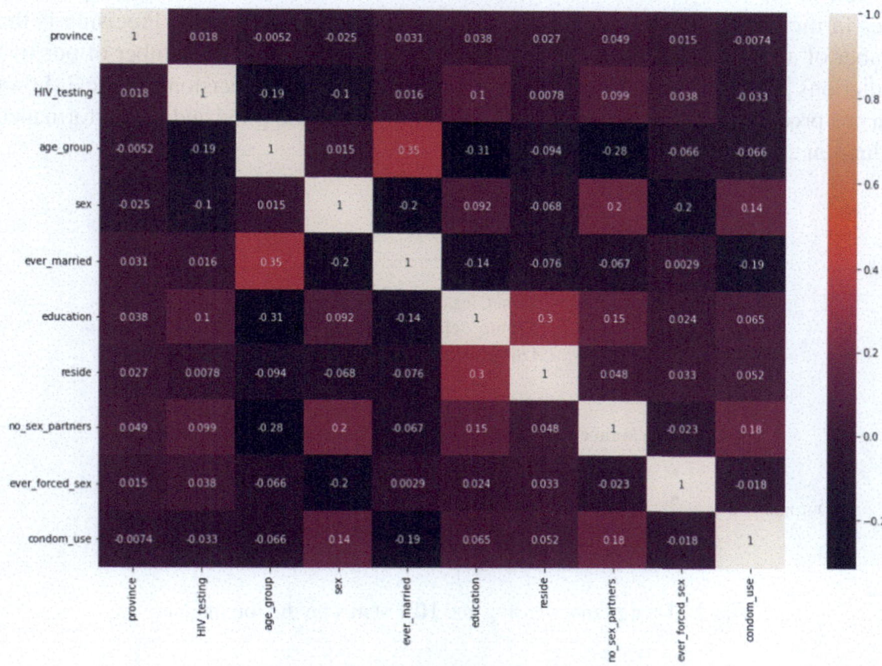

Fig. 2. Heatmap for HIV status predictors

After data pre-processing, the study implemented machine learning-based HIV status prediction models such as support vector machine (SVM), logistic regression and random forest classifier. SVM is a supervised learning machine learning algorithm used for classification and regression problems, albeit commonly used for classification [28]. Logistic regression is a supervised learning machine learning algorithm used for classification problems. Given multiple observations, this algorithm assigns them to a discrete set of categories or classes [29]. It uses the sigmoid function to return the probability of an observation falling into a specific class; thus, the range of output values of the sigmoid activation function is squashed between 0 and 1. The random forest classifier is an ensemble-based machine learning algorithm that comprises multiple individual decision trees [26, 30, 31]. Every tree in the forest predicts and votes, and the class that gets the majority vote is the model's overall prediction. The strengths of this algorithm include its ability to handle datasets that contain both continuous and categorical predictors, cope with datasets in which there are more variables than observations, and has high efficiency and accuracy [32, 33].

2.3 Performance Evaluation of HIV Status Prediction Models

The study used standardised performance evaluation metrics such as accuracy, precision, recall and F1-score to assess the performance of SVM, logistic regression and random forest classifier to predict HIV status using the ZIMPHIA dataset. Accuracy was calculated as the number of precise predictions dependent on the total number of predictions. A recall is the ratio of accurate positive predictions to the total number of positives cases in the dataset. This is also known as the true positive rate [34]. Precision is the number of accurate positive predictions as a percentage of the total number of positive predictions [35]. F1-score represents the average of recall and precision. Figure 3 shows data pre-processing steps and HIV status prediction models applied and the performance evaluation criteria.

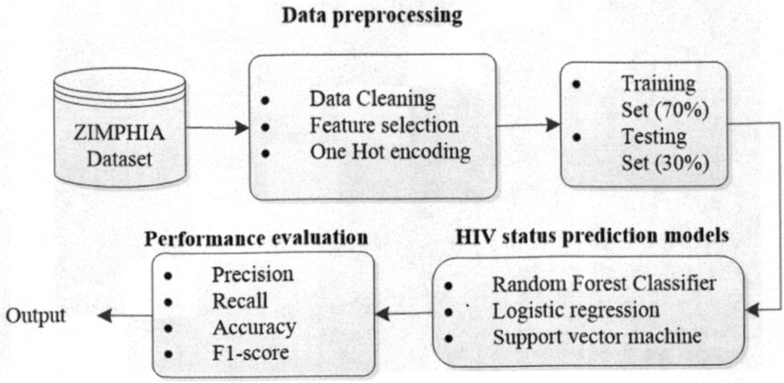

Fig. 3. Data pre-processing and HIV status prediction models

3 Results

Zimbabwe's annual overall HIV incidence among people aged 15 to 64 years is 0.47%, distributed as 0.33% and 0.6% among males and females, respectively [27]. This incidence rate corresponds to roughly 33 000 new HIV cases per year among people aged between 15 and 64 years inclusive. The HIV prevalence rate among people of the same age group (15 to 64 years) is 14.1%, distributed as 12% and 16% among males and females, respectively. Approximately 1 200 000 Zimbabwean residents aged between 15 and 64 years are HIV positive, with 72.9% of them claiming to know their HIV statuses. There are more (76.1%) females who know their HIV status than males (68.2%). Zimbabwe is doing dismally on this indicator since the target for awareness is 90%. Diagnosis continues to be the country's most significant challenge to achieving the 90–90–90 targets. Young people aged between 15 and 24 years constitute close to a fifth (18.9%) of the country's population. About 33% of people living with HIV in this age group is highly unlikely to be diagnosed (50.4%) than the general population of people aged 15 to 64 years (72.9%). Among the people who self-reported being HIV positive in the 15 to 24-year-old age group, almost 84% (83.7%) are on antiretroviral therapy (ART). Of these, 85.4% reported viral suppression, a result comparable to other adults. HIV prevalence among people of age 15 to 49 years was slightly different between rural (13.2%) and urban (13.6%) areas, with only a difference of 0.4%.

Geographically, HIV prevalence ranged between 10.2% (Manicaland Province) and 20.4% (Matebeleland South Province). Moreover, HIV prevalence was high (19.7%) in people without formal education and relatively lower (7.2%) in those with post-secondary school education. For the 15-to-64 years' age group, HIV prevalence by province was highest in Matebeleland South (21.7%), Matebeleland North (19.5%) and Bulawayo (17.9%), while the rest of the provinces have prevalence rates of below 15%.

The results in Table 2 show that logistic regression performed better than random forest classifier and support vector machine with the prediction accuracy of 85%, recall of 98% and F1-score of 92%. However, the random forest classifier has a high precision of 87% compared to other HIV status prediction models. In terms of recall and F1-score, the support vector machine performed better than the random forest classifier with a recall of 96% and F1-score of 91%.

Table 2. Performance of Machine learning based HIV status prediction models

ML based HIV status prediction model	Accuracy	Precision	Recall	F1-score
Random forest classifier	0.83	0.87	0.94	0.90
Logistic regression	0.85	0.86	0.98	0.92
Support vector machine	0.84	0.86	0.96	0.91

4 Conclusion

There is a need to adopt computational technologies like machine learning to predict individuals at high risk of HIV infection for targeted screening using socio-demographic and risk behavioral data. Notably, there is a rapid increase in surveys collecting individual data such as demographic characteristics, social history, laboratory tests, and various diseases. Such data could be utilised for differential diagnosis of HIV and other infectious diseases by integrating novel computational technologies. Here, we illustrated the usefullnes of machine learning algorithms combined with socio-demographic and behavioural factors from national representative samples like PHIA to predict HIV status. This timely information might become a key element for the implementation of micro-targeted strategies oriented towards the identification and disruption of recent infections. However, there is a need to include other HIV clinical predictors and optimise HIV status prediction models and further integrate them into some real-world healthcare settings to assist policymakers and healthcare professionals in HIV prevention, screening, and diagnosis.

Funding. This secondary data analysis research received no external funding. However, ZIMPHIA 2015–2016 was supported by the President's Emergency Plan for AIDS Relief (PEPFAR)

through the Centers for Disease Control and Prevention (CDC) to ICAP at Columbia University under the terms of cooperative agreement #U2GGH001226. The funder did not have any role in the study design, data collection and analysis, decision to publish, or preparation of the current manuscript.

Data Availability. The ZIMPHIA 2015 dataset is publicly accessible. Permission to access the dataset can be sought from the PHIA Project Document Manager (https://phia-data.icap.columbia.edu/).

References

1. Jones, J., Sullivan, P.S., Curran, J.W.: Progress in the HIV epidemic: identifying goals and measuring success. PLoS Med. 16 (2019). https://doi.org/10.1371/JOURNAL.PMED.1002729
2. Global HIV & AIDS statistics—Fact sheet|UNAIDS (n.d.) https://www.unaids.org/en/resources/fact-sheet. Accessed 4 Feb 2022
3. UNAIDS: Global AIDS Update 2018: Miles to Go: Closing Gaps Breaking Barrier Righting Injustices (2019)
4. Carlberg, R., Wolgast, E., Kristensson Hallström, I., Biru, M.: Caregiver and child factors predicting HIV status disclosure among children enrolled on ART: a cross-regional study in Addis Ababa and Oromia, Ethiopia **34**, 105–111 (2021). https://doi.org/10.1080/09540121.2021.1918622
5. Doat, A.R., Negarandeh, R., Hasanpour, M.: Disclosure of HIV status to children in sub-Saharan Africa: a systematic review. Medicina **55** (2019). https://doi.org/10.3390/MEDICINA55080433
6. Murewanhema, G.: HIV and sub-Saharan African women in the COVID-19 era and beyond. Int. J. Med. Rev. **8**, 74–79 (2021). https://doi.org/10.30491/IJMR.2020.247736.1142
7. National HIV Survey (ZIMPHIA 2020) Results Indicate Zimbabwe is on Track to Achieve HIV Epidemic Control by 2030 - U.S. Embassy in Zimbabwe (n.d.) https://zw.usembassy.gov/national-hiv-survey-zimphia-2020-results-indicate-zimbabwe-is-on-track-to-achieve-hiv-epidemic-control-by-2030/. Accessed 5 Feb 2022
8. Muchabaiwa, L., Mbonigaba, J.: Impact of the adolescent and youth sexual and reproductive health strategy on service utilisation and health outcomes in Zimbabwe. PLoS One **14** (2018). https://doi.org/10.1371/JOURNAL.PONE.0218588
9. Celum, C., Barnabas, R.: Reaching the 90-90-90 target: lessons from HIV self-testing. Lancet HIV **6**, e68–e69 (2019). https://doi.org/10.1016/s2352-3018(18)30289-3
10. Maman, D., et al.: Closer to 90-90-90. The cascade of care after 10 years of ART scale-up in rural Malawi: a population study. J. Int. AIDS Soc. **19** (2016). https://doi.org/10.7448/IAS.19.1.20673
11. Gonese, E., et al.: Comparison of HIV incidence in the Zimbabwe population-based HIV impact assessment survey (2015–2016) with modeled estimates: progress toward epidemic control **36**, 656–662 (2020). Https://HomeLiebertpubCom/Aid. https://doi.org/10.1089/aid.2020.0046
12. Conan, N., et al.: Successes and gaps in the HIV cascade of care of a high HIV prevalence setting in Zimbabwe: a population-based survey. J. Int. AIDS Soc. **23**, e25613 (2020). https://doi.org/10.1002/JIA2.25613
13. Cambiano, V., Miners, A., Phillips, A.: What do we know about the cost–effectiveness of HIV preexposure prophylaxis, and is it affordable? Curr. Opin. HIV AIDS **11**, 56–66 (2016). https://doi.org/10.1097/COH.0000000000000217

14. Branca, M., et al.: Factors predicting the persistence of genital human papillomavirus infections and PAP smear abnormality in HIV-positive and HIV-negative women during prospective follow-up. Int. J. STD AIDS **14**, 417–425 (2003). https://doi.org/10.1258/095646203765371321
15. Ahlström, M.G., Ronit, A., Omland, L.H., Vedel, S., Obel, N.: Algorithmic prediction of HIV status using nation-wide electronic registry data. EClinicalMedicine **17**, 100203 (2019). https://doi.org/10.1016/J.ECLINM.2019.10.016
16. Mutai, C.K., McSharry, P.E., Ngaruye, I., Musabanganji, E.: Use of machine learning techniques to identify HIV predictors for screening in sub-Saharan Africa. BMC Med. Res. Methodol. **21**, 1–11 (2021). https://doi.org/10.1186/S12874-021-01346-2
17. Marcus, J.L., Hurley, L.B., Krakower, D.S., Alexeeff, S., Silverberg, M.J., Volk, J.E.: Use of electronic health record data and machine learning to identify candidates for HIV pre-exposure prophylaxis: a modelling study. Lancet HIV **6**, e688–e695 (2019). https://doi.org/10.1016/S2352-3018(19)30137-7
18. Krakower, D.S., et al.: Development and validation of an automated HIV prediction algorithm to identify candidates for pre-exposure prophylaxis: a modelling study. Lancet HIV **6**, e696–e704 (2019). https://doi.org/10.1016/S2352-3018(19)30139-0
19. Bao, Y., et al.: Predicting the diagnosis of HIV and sexually transmitted infections among men who have sex with men using machine learning approaches. J. Infect. **82**, 48–59 (2021). https://doi.org/10.1016/J.JINF.2020.11.007
20. Sachathep, K., et al.: Population-based HIV impact assessments survey methods, response, and quality in Zimbabwe, Malawi, and Zambia. J. Acquir. Immune Defic. Syndr. **87**, S6-16 (2021). https://doi.org/10.1097/QAI.0000000000002710
21. Bisaso, K.R., Karungi, S.A., Kiragga, A., Mukonzo, J.K., Castelnuovo, B.: A comparative study of logistic regression based machine learning techniques for prediction of early virological suppression in antiretroviral initiating HIV patients. BMC Med. Inform. Decis. Mak. **18**, 1–10 (2018). https://doi.org/10.1186/S12911-018-0659-X/FIGURES/3
22. Zhang, X., et al.: Machine learning selected smoking-associated DNA methylation signatures that predict HIV prognosis and mortality. Clin. Epigenetics **10**, 1–15 (2018). https://doi.org/10.1186/S13148-018-0591-Z/FIGURES/5
23. Haas, O., Maier, A., Rothgang, E.: Machine learning-based HIV risk estimation using incidence rate ratios. Front Reprod. Heal. 96 (2021). https://doi.org/10.3389/FRPH.2021.756405
24. Orel, E., Esra, R., Estill, J., Marchand-Maillet, S., Merzouki, A., Keiser, O.: Machine learning to identify socio-behavioural predictors of HIV positivity in East and Southern Africa. MedRxiv (2020). https://doi.org/10.1101/2020.01.27.20018242
25. Mutai, C.K., McSharry, P.E., Ngaruye, I., Musabanganji, E.: Use of machine learning techniques to identify HIV predictors for screening in sub-Saharan Africa. BMC Med. Res. Methodol. **21**, 1–11 (2021). https://doi.org/10.1186/S12874-021-01346-2/TABLES/3
26. Wray, T.B., Luo, X., Ke, J., Pérez, A.E., Carr, D.J., Monti, P.M.: Using smartphone survey data and machine learning to identify situational and contextual risk factors for HIV risk behavior among men who have sex with men who are not on PrEP. Prev. Sci. **20**, 904–913 (2019). https://doi.org/10.1007/S11121-019-01019-Z/FIGURES/1
27. ZIMPHIA: Zimbabwe Population-Based HIV Impact Assessment (ZIMPHIA). Zimbabwe Ministry of Health Child Care, pp. 1–39 (2021)
28. Fashoto, S.G., Mbunge, E., Ogunleye, G., den Burg, J.V.: Implementation of machine learning for predicting maize crop yields using multiple linear regression and backward elimination. Malaysian J. Comput. (MJoC) **6**(1) 679-697 (2021)
29. Fashoto, S.G., Owolabi, O., Mbunge, E., Metfula, A.S.: Evaluating the performance of two hybrid feature selection model of machine learning for credit card fraud detection on classification and prediction methods. Adv. Appl. Sci. Technol. **2**, 70–87 (2019)

30. Macaulay, B.O., Aribisala, B.S., Akande, S.A., Akinnuwesi, B.A., Olabanjo, O.A.: Breast cancer risk prediction in African women using random forest classifier. Cancer Treat. Res. Commun. **28**, 100396 (2021). https://doi.org/10.1016/J.CTARC.2021.100396
31. Cutler, A., Cutler, D.R., Stevens, J.R.: Random forests. In: Zhang, C., Ma, Y. (eds.) Ensemble Machine Learning, pp. 157–175 (2012). Springer US, Boston. https://doi.org/10.1007/978-1-4419-9326-7_5
32. Mbunge, E., Muchemwa, B.: Deep learning and machine learning techniques for analyzing travelers' online reviews: a review. In: Gustavo, N., Pronto, J., Carvalho, L., Belo, M. (ed.) Optimizing Digital Solutions for Hyper-personalization in Tourism and Hospitality, pp. 20–39. IGI Global (2022). https://doi.org/10.4018/978-1-7998-8306-7.ch002
33. Mbunge, E., Fashoto, S.G., Bimha, H.: Prediction of box-office success: a review of trends and machine learning computational models. Int. J. Bus. Intell. Data Min. **20**, 192 (2022). https://doi.org/10.1504/IJBIDM.2022.120825
34. Govender, P., et al.: The application of machine learning to predict genetic relatedness using human mtDNA hypervariable region I sequence. PLoS One **17**, e0263790 (2022). https://doi.org/10.1371/JOURNAL.PONE.0263790
35. Mbunge, E., Simelane, S., Fashoto, S.G., Akinnuwesi, B., Metfula, A.S.: Application of deep learning and machine learning models to detect COVID-19 face masks - a review. Sustain. Oper. Comput. **2**, 235–245 (2021). https://doi.org/10.1016/J.SUSOC.2021.08.001

Evaluation of the Effectiveness of Investments in the Real Sector of the Russian Economy

Elena Nikolaevna Sochneva[1(✉)], Anna Andreevna Malakhova[1,2], Yurii Aleksandrovich Hegai[1], Tatiana Vasilievna Kiyan[3], Svetlana Petrovna Plotnikova[3], Aleksey Gennadevich Rusakov[3], Dmitry Valeryevitch Zyablikov[1], and Olga Valeryevna Starova[1]

[1] Siberian Federal University, Krasnoyarsk, Russia
Sochneva.e@inbox.ru
[2] Krasnoyarsk Medical State University, Krasnoyarsk, Russia
[3] Krasnoyarsk State Agrarian University, Krasnoyarsk, Russia

Abstract. The paper examines the methodological foundations for evaluating investment projects. The approaches to the assessment take into account the features of the modern Russian economy. The boundaries of the use of various discount factors for evaluating business projects are analyzed. The most effective method for calculating the discount factor for investment projects in Russia is substantiated. The authors draw conclusions based on the accumulated experience of analyzing business projects in the Russian economy over the past five years. The open data of the Moscow Interbank Currency Exchange are also used to calculate the discount factor by various methods. When analyzing the methods for calculating the discount factor, the W. Sharp method, the Fama and French method, the weighted average cost of capital method, the method of summing the inflation rate, normal profit and risk premium were used. The study revealed that the most acceptable method for the Russian economy is the method that takes into account the level of inflation, normal profit and risk premium. The authors present risk premium ranges for the Russian economy based on analyzed business projects over the past five years. The rationale for choosing the method for calculating the discount factor is due to the fact that other methods are more focused on the stock market, while in Russia the connection between the real sector and the stock market is not so obvious. The stock market is not as developed and does not affect the economy as it does abroad. Therefore, not all foreign models for estimating the discount factor can be adapted to the Russian economy. The conclusions are confirmed by calculations. The discount factor calculated using methods based on stock market indicators is too small and does not even cover the level of inflation in Russia. This casts doubt on the possibility of using it to evaluate investment projects in Russia. In Russia, inflation is growing due to the increase in production costs. The calculation of business project performance indicators should be adapted to such processes. Based on this, it is advisable not to discount investment flows at the comparison rate, but, on the contrary, to increase, that is, increase taking into account price increases.

Keywords: Investments · Discount factor · Valuation models · Modern Russian economy · Inflation · Risk premium · Normal profit · Efficiency

R. Silhavy (Ed.): CSOC 2022, LNNS 502, pp. 259–269, 2022.
https://doi.org/10.1007/978-3-031-09076-9_25

1 Introduction

The topic of investment is always relevant in all economic systems. Investment is the basis of the economy, a factor in its development. In different periods of time, the approaches to assessing the effectiveness of investments have changed. It is known that they are all based on the concept of discounting. Namely, thanks to discounting, it is possible to bring the value of future cash income to the current moment. In the literature, there are many approaches to assessing the discount rate.

The purpose of this work is to consider: how the real and financial sectors of the economy are related to each other, and in what manner, this should be reflected in the discount coefficient.

The research methodology will be based on a consistent analysis of existing approaches to assessing the discount rate (comparison rate), identifying the possibilities and boundaries of their use in the modern economy of Russia. There will also be presented the calculation of the discount rate by different methods with conclusions and proposals for the use of certain methods. In addition, the work will propose the authors' databases on the practice of applying the risk correction used in the calculation and justification of the discount rate.

2 Methods

So, the most famous methods for calculating the discount rate.

2.1 The Capital Asset Pricing Model (CAPM) Was Proposed by W. Sharp (1964)

$$r_i = r_f + \beta \cdot (r_m - r_f), \tag{1}$$

where r_f – is the profitability of a risk-free financial asset;

r_m – is the average market return in the financial market;

$\beta = \frac{\sigma_{im}}{\sigma_m^2}$ – this is a coefficient that reflects the riskiness of an investment in relation to the market, and shows the sensitivity of changes in stock returns to changes in market returns;

σ_{im} - is the standard deviation from the average market return on the financial market;

σ_m^2 - is the variance of the market yield.

Now, using formula (1), we calculate the discount coefficient based on the data of the Moscow Interbank Currency Exchange (MICEX).

So, as the average market return on the financial market (rm), it is logical to take the MICEX index, which as of July 31, 2021 was 14.67% [13].

As the yield on a risk-free financial asset, we take the yield on the deposit of Sberbank of the Russian Federation, the banking market also refers to the financial one). On 07/31/21 the average rate on Sberbank deposits is 3.5% [15].

Now let's calculate the coefficient β.

Since we are not evaluating the entire group, but only its individual values (sample), we use the standard deviation formula based on an unbiased variance estimate.

$$\sigma = \sqrt{\frac{\sum_{i=1}^{n}(P_i - P_a)}{n-1}}, \tag{2}$$

where P_a – is the average level of profitability of values in the sample;
P_i – is the value in the sample.

We take into account the profitability of the following investments in the financial market (Table 1).

Table 1. The profitability of financial instruments in Russia at 31.07.2021

№	Financial instrument	Profitability
1	*Oil and gas*	14?64%
2	*Power engineering*	−8,33%
3	*Telecommunications*	−4,86%
4	*Metals and mining*	12,85%
5	*Finance*	45,12%
6	*Chemistry and petrochemistry*	34,06%
7	*Construction companies*	50,19%
8	*Transport*	6,06%
9	*Corporate bonds*	−0,17%
10	*Information technology*	−3,67%

The percentage of profitability is given according to the MICEX data from the beginning of the year [13]. The sample included 10 financial assets selected by the authors for analysis.

Let's calculate the standard deviation (Table 2).

Table 2. Calculation of data to determine the standard deviation

№	$(P_i - P_a)$	$(P_i - P_a)^2$
1	(14,64% − 14,59%) = −0,19%	0,0361
2	(−8,33% − 14,59%) = −22,92%	525,3264
3	(−4,86% − 14,59%) = −19,45%	378,3025
4	(12,85% − 14,59%) = −1,74%	3,0276
5	(45,12% − 14,59%) = 30,53%	932,0809
6	(34,06% − 14,59%) = 19,47%	379,0809
7	(50,19% − 14,59%) = 35,6%	1267,36
8	(6,06% − 14,59%) = −8,53%	72,7609
9	(−0,17% − 14,59%) = −14,76	217,8576
10	(−3,67% − 14,59%) = −18,26%	333,4276
Total		**4109,2605**

$$\sigma = \sqrt{\frac{4109{,}2605}{10-1}} = 21{,}37$$

Based on this β:

$$\beta = \frac{21{,}37}{456{,}6769} = 0{,}0468$$

$$r_i = 3{,}5\% + 0{,}0468 \cdot (14{,}67\% - 3{,}5\%) = 4{,}02\%$$

So, the obtained discount coefficient (comparison rate) is too small and does not even cover the official inflation rate. Proceeding from this, it is impractical to carry out discounting on it. This size of the discount factor is due to the still weak development of the financial market in Russia. You can add the risk percentage to this small obtained indicator, and then we will get a different approach to estimating the discount rate.

For the purity of the experiment, we will recalculate this indicator according to the MICEX data for 04.08.2021.

We will include the same financial instruments in the portfolio. The MICEX index as of 04.08 was 15.71% [13] (Table 3).

Table 3. Profitability of financial instruments in Russia according to MICEX data as of 04.08.2021

№	Financial instrument	Profitability
1	*Oil and gas*	15,31%
2	*Power engineering*	−8,16%
3	*Telecommunications*	−4,28%
4	*Metals and mining*	14,06%
5	*Finance*	48,63%
6	*Chemistry and petrochemistry*	39,69%
7	*Construction companies*	53,00%
8	*Transport*	5,69%
9	*Corporate bonds*	−0,06%
10	*Information technology*	−4,72%

Let's calculate the standard deviation (Table 4).

$$P_a = [(15{,}31\% + (-8{,}16\%) + (-4{,}28\%) + 14{,}06\% + 48{,}63\%$$
$$+39{,}69\% + 53{,}00\% + 5{,}69\% + (-0{,}06\%) + (-4{,}72\%)]/10 = 15{,}92\%$$

$$\sigma = \sqrt{\frac{4819{,}77}{10-1}} = 23{,}14$$

Table 4. Calculation of standard deviation

№	$(P_i - P_a)$	$(P_i - P_a)^2$
1	(15,31% – 15,92%) = –0,61%	0,37
2	(–8,16% – 15,92%) = –24,08%	579,85
3	(–4,28% – 15,92%) = –20,20%	408,04
4	(14,06% – 15,92%) = –1,86%	3,46
5	(48,63% – 15,92%) = 32,71%	1069,94
6	(39,69% – 15,92%) = 24,49%	599,76
7	(53,00% – 15,92%) = 37,02%	1372,33
8	(5,69% – 15,92%) = –10,23%	104,65
9	(–0,06% – 15,92%) = –15,98	255,36
10	(–4,72% – 15,92%) = –20,64%	426,01
Total		**4819,77**

$$\beta = \frac{23,14}{535,49} = 0{,}0432$$

$r_i = 3{,}5\% + 0{,}0432 \cdot (15{,}71\% - 3{,}5\%) = 4{,}03\%.$

The result obtained differs little from the previous one, which indicates a consistently low value of the discount coefficient calculated using the CAPM model. This value does not even pay off the inflation rate.

2.2 Modified Model of W. Sharp's Method

The main difference between this method and the previous one is that the risk indicator (ru) is introduced into the calculations.

$$r_i = r_f + \beta \cdot \left(r_m - r_f\right) + r_u. \tag{3}$$

2.3 Calculation of the Discount Factor Based on the Estimated Inflation Rate, Normal Profit and Risk Premium

The advantage of this method is that it is suitable for any company, even one that does not issue securities. In the Russian practice of business design, this method is the most common.

The essence of the method is reduced to the calculation of the discount factor, based on their three components.

$$r = r_f + r_p + I, \tag{4}$$

where r_f – is normal profit, which in economics means income from the most risk-free type of investment (for example, income from a Sberbank deposit);

r_p – is the risk premium;

I – is the forecasted level of inflation.

In this technique, the most difficult is risk assessment.

There is the only methodology for assessing risks, enshrined in a regulatory document - Decree of the Government of the Russian Federation of November 22, 1997 N 1470 "On approval of the procedure for providing state guarantees on a competitive basis at the expense of the Development Budget of the Russian Federation and the Regulations on evaluating the effectiveness of investment projects when placing on competitive basis of centralized investment resources of the Development Budget of the Russian Federation" (Table 5).

Table 5. Risk assessment in accordance with Government Decree No. 1470

The magnitude of the risk	Project goal example	Risk adjustment
Small	Investments for the intensification of production	3–5%
Medium	Increasing product sales	8–10%
High	The risk of introducing a new type of product to the market	13–15%
Super high	Scientific research costs	18–20%

There are also several approaches to risk assessment from individual researchers. For example, a risk assessment from the Alt-Invest company is known [3].

There is also an approach of a group of Russian authors Vilensky P.L., Livshits V.N., Smolyak S.A. to calculate the aggregate risk indicator [1].

The method of using this method is that the investor sums up the estimates of all the factors that are present in his project. The maximum risk premium can be as high as 47%.

The scientist J. Honko proposed a method for estimating the risk premium depending on the purpose of the investment. In accordance with his approach, the investor chooses a goal and assigns a degree of risk to his project [2].

The conclusion for all the presented risk assessment methods is as follows: they have lost their relevance from the point of view of modern economic processes. Recently, business processes have undergone changes, including under the influence of the digitalization of the economy, and, accordingly, a new database of risk adjustment factors is required.

Thus, the previous modified model for determining the discount factor is difficult to implement due to the impossibility of determining the risk component.

2.4 Method of Weighted Average Cost of Capital (WACC) [4, 9, 11]

In the literature, this method is quite common, but in the practice of Russian investors it is rarely used to calculate the discount factor.

The formula for calculation as:

$$WACC = R_e \frac{E}{V} + R_d(1 - t)\frac{D}{V}, \tag{5}$$

where R_e – expected (required) return on equity;
R_d – is the expected (required) return on borrowed capital;
E – the amount of equity;
D – is the amount of borrowed capital;
V – is the total amount of capital;
t – income tax.

The limitations of using the method are that it is suitable only for an already operating enterprise that launches a business project, but cannot be used at the zero stage of the project. To apply this method, the enterprise must already know the ratio of equity and debt capital for the implementation of the project, and this is not always known immediately. Often the decision to use borrowed capital is made when the basic indicators for the project have already been calculated and the investor is thinking about whether he can implement this project on his own or with the involvement of borrowed capital.

In addition, the investor must pre-calculate the return on equity and borrowed capital. In Russia, this return is most often the same, because investors calculate the amount of profit for the entire invested capital, the investor is interested in the overall result, which is not tied to the ratio of equity and borrowed capital.

Let's calculate the discount factor using the WACC model.

Obviously, the values of Re and Rd will differ by industry. Let's define them at the level of 10% each, taking into account the crisis in the modern economy. As it was said for the Russian investor, the overall profitability of capital is important: own and borrowed.

Shares of own and borrowed capital are optimal at the level of 50/50.

Income tax in Russia is 20%.

Based on the foregoing:

$$\text{WACC} = 0{,}15 \cdot 0{,}5 + 0{,}15 \cdot (1 - 0{,}2) \cdot 0{,}5 = 0{,}135$$

That is, the discount factor will be 13.5%.

Another limitation of the model is that income tax is taken into account only when analyzing the return on debt capital, while in reality in Russia income tax will also have to be paid on equity income.

We propose to use the WACC formula, taking into account the agreed conditions for the Russian economy, in the following interpretation:

$$\text{WACC} = \text{R}_\text{i} \cdot (1 - \text{t}), \tag{6}$$

where R_i – is the expected real return on invested capital (taking into account inflation and other factors);

t – is the tax rate.

Then the discount factor for the same input data will be:

$$\text{WACC} = 15\% \cdot (1 - 0{,}2) = 12\%.$$

The component of the expected real value of return on invested capital is the normal profit, inflation and investment risk factors.

Thus, the WACC method, taking into account the peculiarities of the modern Russian economy, is reduced to the method that was described above under number 2.3.

3 Results

Practical advice to the investor based on the above is the following.

Evaluate the project using two discount factors, one of which will correspond to the minimum risk possible, the second - to the maximum. Thus, all real values will be between the two extreme values.

If we approach more precisely the measurement of risk, then we can establish the following. The authors have worked out a substantial base of business projects implemented in the Siberian Federal District (75 over the past three years).

The study showed that most investors use the discount rate estimation method, which is made up of the three components outlined in Method 3 above: inflation rate, normal return, and risk premium.

The analysis also showed that various authors use a discount rate in the range of 10 to 20%. Given that the inflation rate in Russia, according to Rosstat, is about 5%, the guaranteed income is about 3.5% (Sberbank deposit), or we use the guaranteed income rate that was used in a specific project, then we can assume that modern authors business projects include a risk component ranging from 1.5% to 11.5%.

Table 6 shows the calculation of the risk indicator for the business projects analyzed by the authors. The calculation was made according to the formula:

$$R_p = R_i - I - R_f, \tag{7}$$

where I - is the inflation rate;
Ri - is the discount rate;
Rf - is the minimum guaranteed income (normal profit).

Table 6. Risk indicators for analyzed projects

Project number	Risk indicator value,%	Project number	Risk indicator value, %	Project number	Risk indicator value, %
1	2,0	11	7,2	21	2,35
2	2,5	12	6,0	22	6,4
3	6,0	13	2,5	23	3,9
4	4,0	14	1,7	24	4,8
5	3,5	15	5,4	25	9,0

(continued)

Table 6. (*continued*)

Project number	Risk indicator value,%	Project number	Risk indicator value, %	Project number	Risk indicator value, %
6	7,0	16	3,35	26	11,5
7	1,5	17	2,25	27	7,8
8	6,5	18	9,8	28	6,6
9	8,0	19	10,8	29	3,4
10	4,2	20	11,2	30	2,2

4 Discussions

In any case, all the considered calculation methods are reduced to the sum of the minimum real income (normal profit) and the risk adjustment. In the Russian model, the inflationary component is singled out separately; in foreign models, it is included in the amount of the risk adjustment.

Now let's consider: what limit or what maximum value should be limited to the discount coefficient (comparison rate) accepted in the project.

It is believed that in order for the project to be recognized as a success, it is necessary that the discount rate adopted in the calculations is less than the internal rate of return (IRR) received for the project. Moreover, the larger the gap between the IRR and the accepted (and justified) discount factor in the calculations, the more effective the project.

This explanation is quite logical. Let's consider its essence.

The internal rate of return (IRR) is the discount rate at which the income flows are equal to the expenditure ones, that is, the discount rate corresponding to the breakeven point of the project, taking into account all the environmental factors mentioned above.

This means that if the discount factor calculated and used in the project should be less than the factor corresponding to the break-even point.

Now about the method of calculating the IRR indicator itself.

In the literature, the internal rate of return is calculated as follows.

IRR is the root of the equation that equates income and investment flows.

$$\sum_{i=1}^{n} \frac{R_i}{(1 + IRR)^i} = \sum_{j=1}^{m} I_j, \tag{9}$$

where I - investment flows;

Ri - income streams for the project;

i - is the number of the income stream;

n - is the number of income streams;

j - is the number of the investment stream;

m - is the number of investment flows.

The smallest positive root is used to solve the equation.

IRR shows the upper bound for borrowing. Often, when implementing a business project, there is a need to attract borrowed funds at a certain percentage. The maximum amount of this percentage is determined by the internal rate of return.

Now let's turn to the methodology for calculating cash flows in business design. Future income must be reduced by the discount factor, since income loses its value over time. The same amount today and a year later means completely different things to an investor. But costs increase over time, and, above all, due to cost inflation in Russia. Thus, the discount function is applicable to future income, while the accrual function is applicable to future costs. Both discounting and accrual should be carried out at compound interest.

Therefore, the formula for calculating IRR is transformed into the following form:

$$\sum\nolimits_{(i=1)}^{n} \frac{R_i}{((1+IRR)^i)} = \sum\nolimits_{(j=1)}^{m} \frac{(j=1)}{((1+IRR)^{(-j)})}, \tag{10}$$

Also the calculation of NPV from the traditional view:

$$NPV = \sum\nolimits_{(i=1)}^{n} \frac{R_i}{((1+r)^i)} - \sum\nolimits_{(j=1)}^{m} I_j, \tag{11}$$

where r - is the project discount factor.

Transforms into view:

$$NPV = \sum\nolimits_{i=1}^{n} \frac{R_i}{(1+r)^i} - \sum\nolimits_{j=1}^{m} \frac{I_j}{(1+r)^{-j}}, \tag{12}$$

5 Conclusion

1. Thus, we come to the conclusion that the discount coefficient in the modern Russian economy should be determined not by financial models (W. Sharp and other authors), but by means of methods that take into account the relationship between the financial and real sectors of the economy, and, most importantly, inflation rate. Thus, the most appropriate method for calculating the discount rate is to add the inflation rate, normal return, and risk premium.
2. The risk premium in business projects implemented in modern Russia ranges from 1.5% to 11.5%.
3. Taking into account the high inflation of costs in the Russian economy, for a real assessment of business projects, it is advisable to calculate investment flows over time by increasing the compound interest rate. This approach should be used when calculating NPV and IRR.

References

1. Vilensky, P.L., Livshits, V.N., Smolyak, S.A.: Assessment of the Effectiveness of Investment Projects: Theory and Practice: A Textbook, 4th edn., Rev. and add., 1103 p. DELO (2008)

2. Kasatov, A.D.: Development of Economic Management Methods for Integrated Corporate Structures in Industry: Investment Aspect. Publishing House. House "Economic newspaper", 324 p. (2010)
3. A set of teaching materials on the topic "The practice of commercial appraisal and examination of investment projects in industry." Alt-Invest LLC, Moscow, p. 71 (2006). https://rosinfostat.ru/inflyatsiya/
4. Tennent, J.: Cash Management: How not to be aground = Guide To Cash Management How To Avoid A Business Credit Crunch, 208 p. Alpina Publisher, Moscow (2014)
5. Sharpe, W., Alexander, G., Bailey, J.: Investments Publisher: Infra-M, 1035 p.
6. Fama, E.F., French, K.R.: The equity premium. J. Financ. **57**, 637–659 (2002). http://ssrn.com/abstract=309176
7. Fama, E.F., French, K.R.: The Capital Asset Pricing Model: Theory and Evidence (August 2003). CRSP Working Paper No. 550; Tuck Business School Working Paper No. 03-26. http://ssrn.com/abstract=440920. https://doi.org/10.2139/ssrn.440920
8. Fama, E.F., French, K.R.: The Corporate Cost of Capital and the Return on Corporate Investment (April 1998). CRSP Working Paper. http://ssrn.com/abstract=75999. https://doi.org/10.2139/ssrn.75999
9. Fernandes, N.: Finance for Executives: A Practical Guide for Managers. NPV Publishing (2014)
10. Carhart, M.M.: On persistence in mutual fund performance. J. Financ. **52**(1), 57–82 (1997). https://doi.org/10.1111/j.1540-6261.1997.tb03808.x.JSTOR2329556
11. Miles, J.A., Ezzell, J.R.: The weighted average cost of capital, perfect capital markets and project life: a clarification. J. Financ. Quant. Anal. **15**(3), 719–730 (1980)
12. Kenneth, R.: French Current Research Returns. http://mba.tuck.dartmouth.edu/pages/faculty/ken.french/data_library.html. Accessed 08 Apr 2021
13. Moscow Exchange Indices. Official site of the Moscow Interbank Currency Exchange. https://www.moex.com/ru/indices. Accessed 31 July 2021 and 04 Aug 2021
14. Inflation according to Rosstat data. Official site of Rosstat. https://rosinfostat.ru/inflyatsiya/. Accessed 08 Aug 2021
15. Deposits in Sberbank in 2021. The official website of Sberbank of the Russian Federation. www.sberbank.ru. Accessed 08 Aug 2021

Premature Birth Prediction Using Machine Learning Techniques

Kazi Rafat Haa Meem(✉), Sadia Islam, Ahmed Omar Salim Adnan, and Sifat Momen

Department of Electrical and Computer Engineering, North South University Bashundhara, Dhaka 1229, Bangladesh
{kazi.meem,sadia.islam5,ahmed.adnan01,sifat.momen}@northsouth.edu

Abstract. Premature babies are the most vulnerable to neonatal mortality, and their birth process is emotionally and physically painful for the mother and the family. The well-being of a premature baby can also be financially burdensome. This paper looks into the maternal factors associated with prematurity. The surveys took place in public maternity hospitals at the Western Brazilian Amazon. This research aims to predict whether a baby will be born prematurely using numerous distinct models. In order to serve this purpose, various machine learning classification algorithms (Decision tree, Naive Bayes, Random Forest, Extreme Gradient Boosting, and K-NearestNeighbors) were applied to the preprocessed data. The paper proposes models capable of predicting premature birth with an accuracy of about 80%. This research aids in developing a usable model that can detect premature births at an early stage, which will allow early treatment to prevent premature birth, substantially reducing child mortality and reducing the economic stress on families bearing a premature child.

Keywords: Preterm birth · Low birth weight · Life expectancy · Machine learning classifiers · Data analysis · Brazilian Amazon

1 Introduction

Preterm birth (also known as premature birth) is a cause of major concern and a global health problem at present, causing mass mortality and pregnancy problems. A preterm birth occurs before the gestational age of 37 weeks [9]. Preterm birth can be classified into four different categories according to its severity:

- Extreme preterm birth: Birth before 28 weeks of gestation
- Very preterm birth: Birth between 28 to 32 weeks of gestation
- Moderate preterm birth: Birth between 32 to 34 weeks of gestation
- Late Preterm Birth: Birth between 34 to 37 weeks of gestation [22]

R. Silhavy (Ed.): CSOC 2022, LNNS 502, pp. 270–284, 2022.
https://doi.org/10.1007/978-3-031-09076-9_26

Preterm birth is the principle cause of child mortality globally [26]. It is solely responsible for about 35% of total neonatal deaths worldwide in 2010 [2]. Along with its considerable negative effect on mortality, preterm birth also has an adverse impact on the health of a preterm baby if the baby survives. The total time of hospital admissions for infants born preterm are substantially more than term babies. Hospital duration for extreme and very preterm babies are 16 times than that of term babies [18]. Clinical studies have shown that preterm deliveries are majorly related to several health afflictions such as but not limited to long-term chronic Non-Communicable Diseases (NCDs) [4], cardiovascular diseases, and chronic lung diseases [10]. Preterm babies might also suffer from hearing and vision problem, such as Retinopathy of Prematurity (ROP) which is the principal cause of neonatal blindness worldwide [13]. To add to that, preterm babies who survive are also at a risk of suffering from neural and behavioural complications [6], and life-long disabilities. Preterm babies generally also tend to perform worse in academic and have less overall academic achievements [21]. There is a wide range of complex factors affecting babies to be born preterm [5], among which are mostly factors relating to maternal health. Preterm birth is affected by several factors, including, but not limited to, age of mother, history of preterm birth, chronic diseases etc. [20]. Preterm birth is a significant issue in several countries regardless of their economy [29].

Early detection of preterm birth could help prevent preterm birth or mitigate its effects. There are several preterm diagnostic methods in use at present. These methods such as obstetric interview, maternal features, and transvaginal ultrasound barely minimized the number of preterm births [29].

Recently machine learning has been found to be effective in providing early prediction to health related issues [12,17,19,25]. The purpose of this research lies in using machine learning techniques for early prediction of preterm birth at a clinically usable accuracy [5], using a very small dataset [24], that includes several necessary features and factors that affect a baby being born preterm. Multiple performance measures have been taken into account so that amount of false negatives and false positives can also be known as the wrong diagnosis for preterm birth can be fatal in some cases.

The background work related to preterm birth is discussed in Sect. 2. Section 3 describes the research methodologies followed for this research work. Section 4 contains experiments done and results obtained. Finally, in Sect. 5, we conclude the paper with information of future work we may carry out.

2 Literature Review

In recent times, there has been a significant increase in the usage of machine learning for clinical purposes, including preterm birth. Lee et al. [15] applied six machine learning algorithms to predict preterm birth: Logistic Regression, Decision Tree, Naïve Bayes, SVM (Support Vector Machine), Random Forest and ANN (Artificial Neural Network), and their study were conducted on a small group of 596 patients. They used the following features - diabetes mellitus, drinkers, IVF (In Vitro fertilization), hypertensive disorder, smoker, prior placenta previa, myomas, and adenomyosis. The accuracy they got from their model is –91.14% using ANN and 91.80% using Logistic Regression. Esty et al. [5] used R as their primary software to predict premature birth. In their work, they have applied C5. Decision Tree classifier and got specificity of 71.8% and sensitivity of 90.0%.

In another study, Rawashdeh and colleagues [23] proposed an intelligent system based on data mining techniques to predict premature birth for pregnant women with cervical cerclage. It was based on EHR (electronic health record) data, and applying KNN (K-Nearest Neighbours) and Random forest; they achieved an accuracy of 98%. Prema and Pushpalatha [20] proposed a solution to identify preterm risk factors of preterm birth based on two classifiers. The result shows that diabetes mellitus and gestational diabetes mellitus are major preterm birth risks. Koivu et al. [14] proposed a model which can predict the risk of stillbirth and preterm pregnancies using logistic regression and gradient boosting decision tree and the best performance they have achieved is 0.76 for (early stillbirth), 0.63 for (late stillbirth), and 0.64 for (preterm birth).In another work, Weber et al. [28] predicted nulliparous women preterm birth using a large dataset. LINDA K and colleagues [30] aimed to develop a system for predicting preterm labor risk. Applying some methods, they have achieved 53–88% accuracy in predicting preterm birth for 9419 women [29]. Frize et al. [7] used two methods to predict preterm birth in parous and nulliparous women. The first one is the ANN (artificial neural network)-we method, and the other one is the DT-ANN hybrid method. The classifier achieves for parous cases-sensitivity is 66%, specificity is 71%, and for nulliparous cases, sensitivity is 65%, specificity is 71%. Raja et al. [22] proposed a machine learning-based model to predict preterm birth in rural India. They used three classifiers decision tree, SVM, logistic regression, and the SVM classifier generates higher accuracy. The results of their proposed model are specificity = 0.783, sensitivity = 0.891, accuracy = 0.909. Ahadi et al. [1] used support vector machine to predict and classify the factors which are affecting preterm delivery. They collected the dataset from Tehran Milad Hospital. They have achieved 57% and 67% for logistic regression and support vector machine, respectively.

3 Methodology

The prime purpose of this paper is to develop machine learning models that will help predict preterm deliveries. The raw dataset [24] was pre-processed before applying machine learning techniques. Figure 1 shows a visual representation of the methodology adopted in this work.

Firstly, the raw dataset is carefully analyzed and insights of the dataset are completely understood. The dataset is then processed to remove unwanted data and features. Next, the dataset is split into train and test sets and feature selection techniques are then implemented to remove redundant and unnecessary features. The train set is then used to produce a model using different machine learning algorithms. The models are then optimized by tuning the

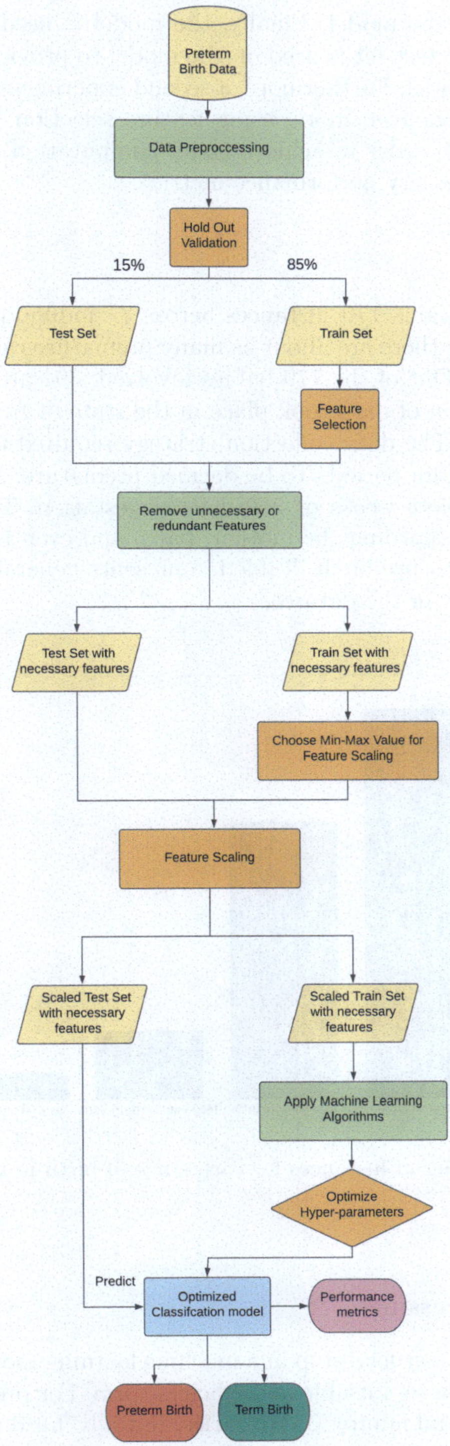

Fig. 1. Workflow of the research methodology

hyper-parameters of the model. Finally, the model is used to predict preterm or term birth and the test set is used on the model to provide performance metrics to analyze the model. Furthermore, a second experiment was done to reduce the number of features even more, using feature selection by information gain for comparison and in-order to achieve faster computation and maybe a better accuracy (or the necessary performance metric).

3.1 Dataset

The dataset [24] contains 779 instances across 77 individual attributes and is balanced, considering there are about as many premature cases (383) as controls or term cases (396). Out of the 779 babies involved, 394 are males, and 385 are females. The collection of data took place in the state of Acre between October 2016 and July 2017. The data collection strategy required interviews of women during their postpartum period. To be deemed premature, a birth had to occur in less than 37 complete weeks or 259 days of gestation. The dataset contains various information regarding the mother, baby, and even the family, and most of them are related to pre-birth. Table 1 represents general information about the attributes present in the dataset.

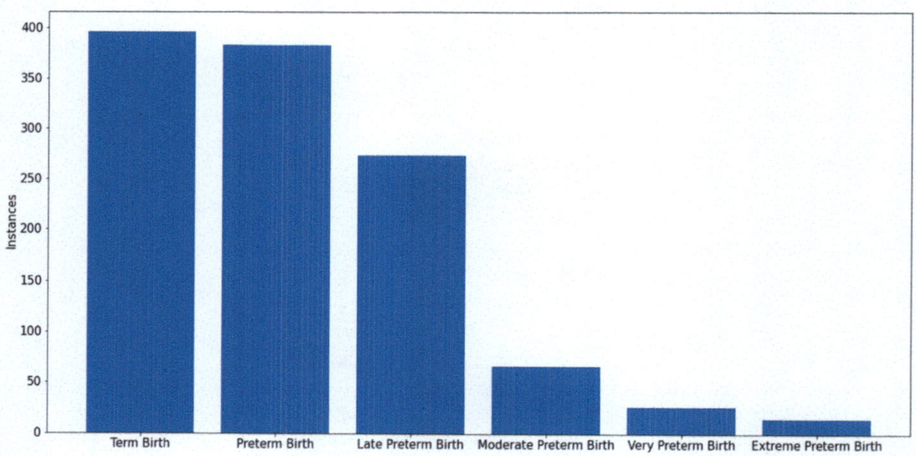

Fig. 2. Number of instances for each stage of birth in the dataset

3.2 Data Preprocessing

Data preprocessing is a crucial step in a machine learning model building process as it converts raw data to a usable and efficient form. For preprocessing we have used the pandas [16] and numpy [11] libraries. Initially, for data preprocessing all post-birth attributes were removed. All features that can be attained pre-birth

Table 1. Description of attributes of the dataset

Attribute	Data type	Description/Remarks
record_id	Integer	Identification number of cases
case_control	Integer	Whether the baby is born normal or premature
maternity	Integer	Whether the mother had any child previously
GA	Integer	Gestational age
mother's_age	Integer	Age of the mother
weight_baby_class	Integer	Weight of the baby after birth divided into 3 classes
resident_by_domicile	Integer	Area of permanent residence
water_piped	Integer	Whether the house of the patient is water piped
color_race	Integer	Color or race of the family
color_race/group	Integer	Color or race of the family (grouped)
Maternal_education	Integer	Education level of the mother
person_sustain_ education	Float	The education level of the person the family depends on
maternal_occupation	Integer	Occupation of the mother
Job	Float	Profession of the mothers
person_sustain_who	Integer	Which family member the family depends on
age_range_person_sustain	Float	Age range of the person the family depends on
family_structure	Integer	Structure of the family (Joined, Extended, Nuclear, Single Parent)
age_range_maternal	Integer	Age range of the mother represented in classes of numbers
BMI_classif	Float	Body mass index of the mother
weight_gain_class	Float	Weight gain of the mother during pregnancy divided into classes
abdominal_surgery_previous	Float	Whether the mother had any abdominal surgery before the birth of the baby
asthma_previous	Integer	Whether the mother had asthma before the birth of the baby
diabetes_previous	Integer	Whether the mother had diabetes before the birth of the baby
hypertension_previous	Integer	Whether the mother had any history of hypertension
Parity	Integer	Number of times the mother has reached a gestational age of 24weeks
premature_child_previous	Integer	Whether the mother had any premature child previously
child_underweight_previous	Integer	Whether the mother had any underweight baby born previously
death_child1	Integer	Number of dead child of the mother if any
periodontal_disease_previous	Integer	Whether the mother had any periodontal disease previously
pregnancy_planning	Float	Time taken for pregnancy planning
treatment_fertility	Float	Whether the mother had to undergo any treatment for fertility
City	Integer	The city the mother is from
interpregnancy_interval	Integer	Time between previous baby's birth and conception of the current baby's pregnancy.
abortion_previous	Integer	Whether the mother had any abortion previously
caesarean_previous	Float	Whether the mother had any child through caesarean previously
physical_activity	Integer	Whether the mother was involved with physical activities during pregnancy
Stress	Float	Level of stress the mother had to go through during pregnancy
physical_injury	Integer	Whether the mother was involved in any physical injury during pregnancy
vigorous_activity	Integer	Whether the mother was involved in any vigorous activity during pregnancy
tt_depression	Integer	Whether the mother had any history of terminal depression
prenatal_care	Integer	Whether the mother was under medical checkups throughout the pregnancy
marriage_duration	Integer	Duration of marriage of the father and mother
Smoking	Integer	Whether the mother has any habit of smoking
Mother's_reaction	Integer	Reaction of the mother after birth
Father's_reaction	Float	Reaction of the father after birth
Family_reaction	Float	Reaction of the family after birth
drinking_alcohol	Integer	Whether the mother has any habit of drinking
work_pregnancy	Integer	Whether the mother was involved in work during the pregnancy period
work_standing	Integer	Whether the mother's work involves standing for a long duration
physical_effort_work	Float	Amount of physical effort the mother had to do during pregnancy
Bleeding	Float	Whether bleeding occured during pregnancy
Anaemia	Integer	Whether the mother has anemia
vaginal_discharge	Float	Whether the mother had vaginal discharge issues
urinary_tract_infection	Float	Whether the mother has urinary tract infection
genitat_tract_infection	Integer	Whether the mother has any genital tract infection

(continued)

Table 1. (*continued*)

Attribute	Data type	Description/Remarks
abnormal_amniotic	Integer	Whether there was any abnormality in amniotic fluid
diabetes_gestational	Integer	Whether the mother develop high blood sugar (glucose) during pregnancy
Hypertension	Integer	Whether the mother has any history of hypertension
Hospitalization	Integer	Whether the mother was hospitalized during pregnancy
Syphilis	Integer	Whether the mother has syphilis
baby_sex	Integer	Whether the baby is male or female
type_gestation	Integer	Type of pregnancy (singlet or multiple pregnancy)
congenital_anomalies	Integer	Whether the baby was born with any congenital anomalies
Delivery	Integer	Type of delivery (caesarean or normal)
causes_of_premature_birth	Float	Reason which caused the premature
mothers_born_preterm	Integer	Whether the mother was born preterm herself
psychoactive_drugs	Integer	Whether the mother had to take any type of psychoactive drugs
abortion_attempt	Integer	Number of abortion attempts previously
occupation_person_sustain	Float	Occupation of the person the family depends on
place_of_residence	Integer	Current place of residence
household_income_BRL	Float	Household income in Brazilian currency
household_income	Float	Household income in USD
weight_gain _pregnancy	Float	Weight gain of the mother during pregnancy
birth_date	Date	Date of birth of the baby
weight_baby	Integer	Weight of the baby after birth
apgar1	Float	Appearance, Pulse, Grimace, Activity, and Respiration score of the baby after 1 min
apgar5	Float	Appearance, Pulse, Grimace, Activity, and Respiration score of the baby after 5 min

are kept. Exploratory Data Analysis (EDA) has been carried out on the dataset for a better understanding of the dataset and to gain insights about the dataset. Data analysis was done to get information on missing values, null values and to get statistical information about every attribute. The feature *Record_ID* has been removed as it is a unique identification number for the woman giving birth and is not useful for prediction. Duplicate columns are also removed such as *household_income_BRI* is same as *household_income_USD*. So, we dropped the attribute *household_income_BRI.* Null values in attribute *household_income_USD* have been replaced with the median of the attribute as these attributes cannot be null. The attribute *Caesarean_previous* represents if a woman had any previous births that were caesarean or not. However, the attribute has many missing values. After analyzing the dataset it is determined that the null values of the attribute,represents the woman who did not have any previous birth. So, they were replaced with zero that represented a woman who did not give any previous birth. Features with over forty missing values were removed because of the shortage of instances in the dataset. Finally, duplicate instances were dropped. As a result, the dataset was left with sixty-five individual attributes, including the attribute to be predicted.

The dataset was then divided into two parts: the train set and the test set. Splitting was done using the Hold-Out Validation technique to separate 85% of the data into the train set and the remaining 15% into the test set. Along with that, it was split using a fixed random state so that the same results could be repeated.

3.3 Feature Selection

Feature selection methods removes redundant and unnecessary features that can help improve computation and prediction accuracy [3]. The feature selection methods used, have only been implemented on the train set. The test set is used for analyzing the performance of the model as a result, it is kept completely unseen and not used in any model building processes. The attribute to be predicted was left out of these feature selection procedures as well. After applying feature selection, the features that were removed from the train set were also removed from the test set. For feature selection, the Pearson's correlation coefficient has been used to figure out the presence of any duplicate features. The Pearson's correlation coefficient compares two features and finds their similarity. The higher the coefficient, the stronger the positive correlation. The first feature, among a pair of features having a Pearson's correlation coefficient of .85 (85%) was removed. The train set and the test set were left with 63 features after the feature selection procedure was applied. After the Pearson's correlation was applied to the dataset it was further preprocessed using feature selection by mutual information. For mutual information all the features with a mutual information of zero with the feature to be predicted was removed. This left the dataset with 34 attributes. Similarly, to the previous feature selection method mutual information was used only on the train set and the features that were selected to be removed was removed from both the train and test set.

3.4 Classifier Algorithms and Hyper-parameter Optimization

After data preprocessing and feature selection, the data is now ready to be used to make a machine learning model. Five classifier algorithms were used to classify between preterm and term birth. Two levels of experiment have been conducted to detect the pregnancy outcomes. In the first experiment, the processed data was applied to the five different classifiers algorithms on a certain search space and via grid search cross-validation. From the grid search the optimum hyper-parameters were gathered; then the performance metrics of the models were collected. For the second experiment, the data was further processed using feature selection by information gain in-order to get better performance and faster computation. This data (processed by information gain) was put into the same five classifier algorithms on a search space from which the optimum hyper-parameters were known using a grid search cross-validation. Then performance metrics were collected for these five models as well. All the classifiers used are discussed as follows.

Decision Tree: The decision tree algorithm is a machine learning algorithm usually used for supervised learning. It generates a tree like structure from the root node to the leaf node that makes a decision based on the if-else statement [8]. It chooses the root node based on the lowest Gini index (1). The decision tree operates by using information gain and Gini and decides the best attribute to split. The decision tree can stop the splitting procedure by continuing splitting until the classes are completely homogenized or by early stoppage. The max-depth of the decision tree was chosen using grid search cross-validation on a broad search space of 1 to 80 tree depth. The optimum max-depth of the decision tree for the experiment without feature selection by information was 6 and the for the experiment with feature selection by information gain was 5.

$$Gini\ Index = 1 - \sum_{i=0}^{c-1} P_i(t)^2 \tag{1}$$

Random Forest: The random forest is a supervised ensemble machine learning algorithm that bases itself upon decision trees. The random forest prepares the T number of decision tree classifiers and then, via majority voting selects the appropriate class. For this ensemble algorithm, only the most vital hyper-parameters were taken into consideration for this classifier as it has several hyper-parameters. For this classifier a grid search was run across a wide search space and the optimum hyper-parameter were obtained. For the random forest classifier built on the first experiment had 200 decision trees, a max-depth of 5, a minimum leaf sample of 2, and entropy based splitting as its optimum hyper-parameters. The classifier built on the data for the second experiment had 320 decision trees, a max-depth of 6, a minimum leaf sample of 2 and entropy based splitting as its optimum hyper-parameters.

K-Nearest Neighbours: The K-Nearest Neighbours classifier is a machine learning algorithm used mainly for supervised learning. For every data to be predicted, the K-Nearest Neighbours algorithm maps all the instances onto an n-dimensional graph, where n is the number of features. For any instance to be predicted, the instance is put into the graph and the classifier then decides using majority voting what the instance will be classified as based on the K number of neighbours nearest to it. For our experiment, we used the Euclidean distance (2) to calculate the distance. The train data and the test data has to be scaled to a value between 0 and 1 so that all features have equal weight for which we used the Min-Max Scaling (3) procedure. For both the experiments the grid search cross-validation were applied on a large search space to obtain optimum values of K for both the experiments. The value of k by grid search for the first experiment was 27 and for the second experiment was 25.

$$d(x,y) = \sqrt{\sum_{i=1}^{n} (x_i - y_i)^2} \tag{2}$$

$$X_{scaled} = \frac{(X - X_{min})}{(X_{max} - X_{min})} \tag{3}$$

Extreme Gradient Boosting: Extreme Gradient Boosting (also known as XGboost) is a decision tree-based machine learning algorithm that uses the

Gradient Boosting algorithm, and the gradient boosting framework. It is used to create a robust classifier with the help of weak classifiers. Typically, multiple decision trees are used. Extreme Gradient Boosting is designed to be used to solve both regression and classification problems. Extreme Gradient Boosting can also be used to control data overfitting by tuning a series of parameters. As we are dealing with a small dataset, to avoid overfitting Extreme Gradient Boosting is a good choice. It is also capable of fine-tuning and algorithm enhancement. A grid search was applied across a search space for the Extreme Gradient Boosting algorithm to determine the optimum learning rate for both the experiments.

Naive Bayes: The Naive Bayes classifier takes a probabilistic approach for a classification task. The probability of a child being born preterm given a set of attributes can be figured out using Bayes Theorem. The assumption we make is that the prediction and the attributes are independent, which is why it is called Naive. [27]

$$P(y|x_1, ..., x_n) = \frac{P(x_1|y), ..., P(x_n|y)P(y)}{P(x_1), ..., P(x_n)} \tag{4}$$

4 Experiment and Results

As previously mentioned, there were two experiments conducted on the dataset to predict pregnancy outcomes. Firstly, machine learning algorithms were trained on the processed trained set containing 63 attributes and the feature to be predicted. For the second experiment, algorithms were trained on the same training set but after applying feature selection by information gain. The training set after feature selection by information gain contained 34 attributes. As features required now are less, collection and computation is supposed to be faster and easier. Using feature selection by information gain all the features that had a mutual information score of zero were found out. Attributes with a mutual information score of zero were removed. This is because the feature to be predicted has no relation to attributes with a mutual score of zero.

Each level of the experiment was run on the same five classifiers, i.e. Decision tree, Random Forest, K-Nearest Neighbours, Extreme Gradient Boosting(XGboost) and Naive Bayes. The performance measures calculated were training accuracy, testing accuracy, precision, recall(sensitivity), specificity, F1_score, and Area under the receiver operating characteristic curve(AUC_ROC).

4.1 Experimental Results Before Applying Feature Selection by Information Gain

For this experiment, the train set and test set contained 595 instances and 106 instances respectively. Each set also had 63 columns after preprocessing. Here, the data used was the data produced after applying the feature selection by the Pearson's correlation algorithm. The performance measures observed for each classification model are summarized in Table 2 and a visual comparison of the accuracy is shown in Fig. 3.

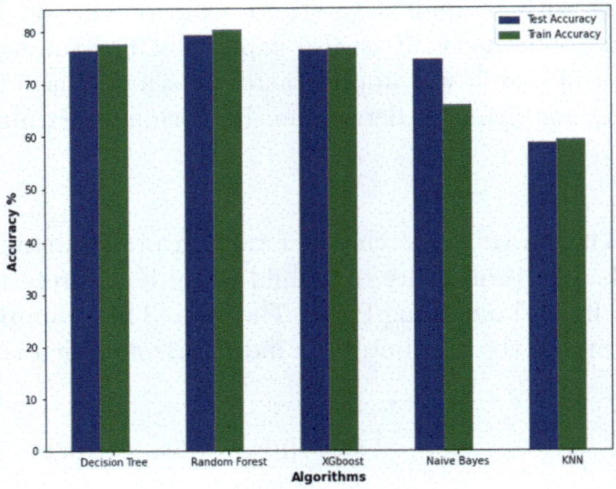

Fig. 3. Accuracy Score of different models trained with data before feature selection using mutual information

Table 2. Model performance on data before feature selection using information gain

Classifier name	Train accuracy	Test accuracy	Precision	Recall	Specificity	F1-score	AUC
KNN	63.53%	66.03%	0.43	0.50	0.63	0.46	0.57
Random forest	80.33%	79.25%	0.74	0.66	0.87	0.69	0.82
Decision tree	77.48%	76.42%	0.69	0.63	0.84	0.66	0.74
Naive bayes	65.71%	74.53%	0.67	0.58	0.84	0.62	0.77
XGBoost	76.64%	76.42%	0.70	0.61	0.85	0.65	0.77

4.2 Experimental Results After Applying Feature Selection by Information Gain

Data instances were the same for this experiment also, i.e. 595 and 106 for train set and test set respectively. However, after feature selection by information gain, the sets were left with 34 attributes. Here, the data used was the data produced after applying the feature selection by the Pearson's correlation algorithm and the mutual information technique. The experimental results of each classification model are listed in Table 3 and a visual of the accuracy metrics is shown in Fig. 4.

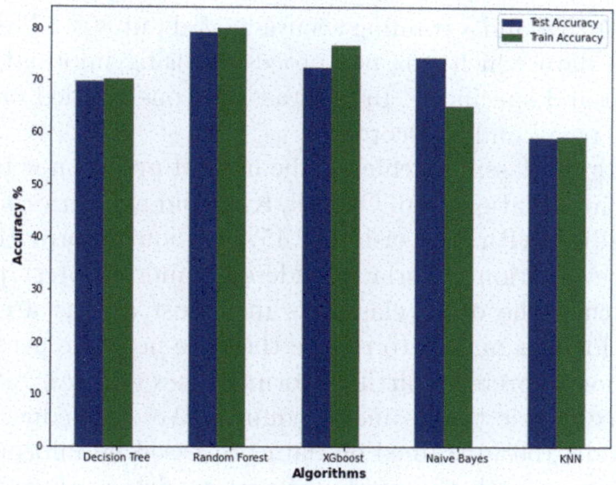

Fig. 4. Accuracy Score of different models trained with data after feature selection using mutual information

Table 3. Model performance on data after feature selection using information gain

Classifier name	Train accuracy	Test accuracy	Precision	Recall	Specificity	F1-score	AUC
KNN	59.16%	59.43%	0.45	0.55	0.62	0.49	0.57
Random forest	80.17%	79.25%	0.72	0.68	0.85	0.70	0.80
Decision tree	70.25%	69.81%	0.57	0.68	0.71	0.62	0.75
Naive bayes	65.21%	74.53%	0.68	0.55	0.85	0.61	0.81
XGBoost	76.64%	76.81%	0.70	0.61	0.85	0.65	0.77

4.3 Discussion

Tables 2 and 3 summarize the evaluation of results achieved through the classifiers. Evaluation of performances required assessment via accuracy, precision, recall, specificity, F1-score, and area under the ROC curve.

From the tables, the classifiers perform differently from each other and also perform differently based on the type of experiment. The K-Nearest Neighbours algorithm produces the lowest test and train accuracy. The decision tree based on the Gini index fairs quite well in both the experiment. Without feature selection using information gain, it got a training accuracy of 77.5% and testing accuracy of 76.42% compared to a decision tree model on further processed data, which got an accuracy of 70.25% for training and 69.81% for testing. The model is for predicting preterm birth; as a result, precision and recall are both critical, so a decision tree in the experiment with feature selection using information gain can have some advantages as it has a higher recall. Nonetheless, the random forest model for both experiments shows high performance in all regions compared to the other learning models. Both the random forest models are trained to have a test accuracy of 79.25% and a training accuracy of about 80%. The random forest model trained on data which was not processed using information gain shows a higher precision and specificity. In contrast, the one trained on the processed data has a better recall and F1 score.

The random forest classifier achieved the highest precision with a precision of 74% before using information gain. Overall, KNN did not perform well compared to the other classifiers with a precision of 45% without information gain.

With all the evaluation criteria considered, random forest performed considerably better than the other classifiers in almost all the areas in both the experiment. The focus is mainly to reduce the false-negative predictions so that the misclassification of preterm birth as normal cases reduces. As a result, recall is the performance metric that is most required. We can deduce that recall for random forest model (66%) trained on data processed by information gain performed best. However, both the random forest models work well enough to be used in real scenarios [5].

5 Conclusion

The birth of premature babies is a severe global issue. Several classifiers were put to the test to achieve viable performance. Work regarding preterm birth has a significant influence on life and the economy. This paper proposes machine learning approaches to predict preterm birth on a small datasets with an accuracy of 79.23%.

For future enhancement of the research, a more extensive dataset with more instances is pivotal. Deep neural networks and other deep learning techniques can be also be applied to achieve state-of-the-art performance.

References

1. Ahadi, B., et al.: Using support vector machines in predicting and classifying factors affecting preterm delivery (2016)
2. Blencowe, H., Cousens, S., Chou, D., Oestergaard, M., Say, L., Moller, A.B., Kinney, M., Lawn, J.: Born too soon: the global epidemiology of 15 million preterm births. Reprod. Health **10**(1), 1–14 (2013)
3. Chandrashekar, G., Sahin, F.: A survey on feature selection methods. Comput. Electr. Eng. **40**(1), 16–28 (2014)
4. Chehade, H., Simeoni, U., Guignard, J.P., Boubred, F.: Preterm birth: long term cardiovascular and renal consequences. Curr. Pediatr. Rev. **14**(4), 219–226 (2018)
5. Esty, A., Frize, M., Gilchrist, J., Bariciak, E.: Applying data preprocessing methods to predict premature birth. In: 2018 40th Annual International Conference of the IEEE Engineering in Medicine and Biology Society (EMBC), pp. 6096–6099. IEEE (2018)
6. Fergus, P., Cheung, P., Hussain, A., Al-Jumeily, D., Dobbins, C., Iram, S.: Prediction of preterm deliveries from ehg signals using machine learning. PLoS One **8**(10), e77154 (2013)

7. Frize, M., Yu, N., Weyand, S.: Effectiveness of a hybrid pattern classifier for medical applications. Int. J. Hybrid Intell. Syst. **8**(2), 71–79 (2011)
8. Géron, A.: Hands-on machine learning with Scikit-Learn, Keras, and TensorFlow: Concepts, tools, and techniques to build intelligent systems. O'Reilly Media (2019)
9. Goldenberg, R.L., Culhane, J.F., Iams, J.D., Romero, R.: Epidemiology and causes of preterm birth. The Lancet **371**(9606), 75–84 (2008)
10. Greenough, A.: Long term respiratory outcomes of very premature birth (< 32 weeks). Semin. Fetal Neonatal Med. **17**, 73–76 (2012)
11. Harris, C.R., Millman, K.J., van der Walt, S.J., Gommers, R., Virtanen, P., Cournapeau, D., Wieser, E., Taylor, J., Berg, S., Smith, N.J., et al.: Array programming with numpy. Nature **585**(7825), 357–362 (2020)
12. Hyland, S.L., Faltys, M., Hüser, M., Lyu, X., Gumbsch, T., Esteban, C., Bock, C., Horn, M., Moor, M., Rieck, B., et al.: Early prediction of circulatory failure in the intensive care unit using machine learning. Nat. Med. **26**(3), 364–373 (2020)
13. Kim, S.J., Port, A.D., Swan, R., Campbell, J.P., Chan, R.P., Chiang, M.F.: Retinopathy of prematurity: a review of risk factors and their clinical significance. Surv. Ophthalmol. **63**(5), 618–637 (2018)
14. Koivu, A., Sairanen, M.: Predicting risk of stillbirth and preterm pregnancies with machine learning. Health Inf. Sci. Syst. **8**(1), 1–12 (2020). https://doi.org/10.1007/s13755-020-00105-9
15. Lee, K.S., Ahn, K.H.: Artificial neural network analysis of spontaneous preterm labor and birth and its major determinants. J. Korean Med. Sci. **34**(16), e128 (2019)
16. McKinney, W., et al.: Data structures for statistical computing in python. In: Proceedings of the 9th Python in Science Conference, vol. 445, pp. 51–56. Austin, TX (2010)
17. Moor, M., Rieck, B., Horn, M., Jutzeler, C.R., Borgwardt, K.: Early prediction of sepsis in the icu using machine learning: a systematic review. Front. Med. **8**, 348 (2021)
18. Petrou, S., Mehta, Z., Hockley, C., Cook-Mozaffari, P., Henderson, J., Goldacre, M.: The impact of preterm birth on hospital inpatient admissions and costs during the first 5 years of life. Pediatrics **112**(6), 1290–1297 (2003)
19. Pranto, B., Mehnaz, S.M., Momen, S., Huq, S.M.: Prediction of diabetes using cost sensitive learning and oversampling techniques on bangladeshi and indian female patients. In: 2020 5th International Conference on Information Technology Research (ICITR), pp. 1–6. IEEE (2020)
20. Prema, N.S., Pushpalatha, M.P.: Machine learning approach for preterm birth prediction based on maternal chronic conditions. In: Sridhar, V., Padma, M.C., Rao, K.A.R. (eds.) Emerging Research in Electronics, Computer Science and Technology. LNEE, vol. 545, pp. 581–588. Springer, Singapore (2019). https://doi.org/10.1007/978-981-13-5802-9_52
21. Quigley, M.A., Poulsen, G., Boyle, E., Wolke, D., Field, D., Alfirevic, Z., Kurinczuk, J.J.: Early term and late preterm birth are associated with poorer school performance at age 5 years: a cohort study. Arch. Dis. Child. Fetal Neonatal Ed. **97**(3), F167–F173 (2012)
22. Raja, R., Mukherjee, I., Sarkar, B.K.: A machine learning-based prediction model for preterm birth in rural india. J. Healthc. Eng. **2021** (2021)
23. Rawashdeh, H., Awawdeh, S., Shannag, F., Henawi, E., Faris, H., Obeid, N., Hyett, J.: Intelligent system based on data mining techniques for prediction of preterm birth for women with cervical cerclage. Comput. Biol. Chem. **85**, 107233 (2020)

24. Santos, C.L., deMendonçaCosta, K.M., Dourado, J.E.C., deLima, S.B.G., Dotto, L.M.G., Schirmer, J.: Maternal factors associated with prematurity in public maternity hospitals at the brazilian western amazon. Midwifery **85**, 102670 (2020)
25. Shad, H.S., Jamal, Z., Ahmed, S.M.F., Momen, S., Mansoor, N.: Early prediction of chronic kidney disease using data mining techniques. In: Silhavy, R., Silhavy, P., Prokopova, Z. (eds.) CoMeSySo 2021. LNNS, vol. 231, pp. 947–957. Springer, Cham (2021). https://doi.org/10.1007/978-3-030-90321-3_79
26. Walani, S.R.: Global burden of preterm birth. Int. J. Gynecol. Obstet. **150**(1), 31–33 (2020)
27. Webb, G.I.: Naïve Bayes. In: Sammut, C., Webb, G.I. (eds.) Encyclopedia of Machine Learning, pp. 713–714. Springer, US, Boston, MA (2010). https://doi.org/10.1007/978-0-387-30164-8_576
28. Weber, A., Darmstadt, G.L., Gruber, S., Foeller, M.E., Carmichael, S.L., Stevenson, D.K., Shaw, G.M.: Application of machine-learning to predict early spontaneous preterm birth among nulliparous non-hispanic black and white women. Ann. Epidemiol. **28**(11), 783–789 (2018)
29. Włodarczyk, T., Płotka, S., Szczepański, T., Rokita, P., Sochacki-Wójcicka, N., Wójcicki, J., Lipa, M., Trzciński, T.: Machine learning methods for preterm birth prediction: a review. Electronics **10**(5), 586 (2021)
30. Woolery, L.K., Grzymala-Busse, J.: Machine learning for an expert system to predict preterm birth risk. J. Amer. Med. Inf. Assoc. **1**(6), 439–446 (1994)

Machine Learning Approach to Predicting Attrition Among Employees at Work

Sudipta Bhatta(✉), Isfaf Uz Zaman, Nuzhat Raisa, Shazzadul Islam Fahim, and Sifat Momen

Department of Electrical and Computer Engineering, North South University, Block-B, Bashundhara, Dhaka 1229, Bangladesh
{sudipta.bhatta,isfaf.zaman,nuzhat.raisa, shazzadul.fahim,sifat.momen}@northsouth.edu

Abstract. Employee attrition for any organization is a major problem. The rate of quitting jobs may cause the loss of talented employees as well as the loss of time and money to any organization. Machine learning techniques have been found to be effective in assisting organizations and decision-makers in minimizing employee attrition. In this paper, we report a machine learning based approach applied on IBM Watson dataset to predicting employee attrition. Several classifiers such as ZeroR, Decision Tree, K-nearest neighbor (KNN), Support Vector Machine (SVM), Random Forest as well as Voting Classifier were applied on the dataset. Hyperparameter optimization was applied to further improve the results. Our best model leveraging an ensemble technique with a Voting classifier demonstrates that the employee attrition model can achieve a high AUC (Area Under ROC Curve) score of 0.89, including competitive metrics on other measures like precision, recall, F1-score, and accuracy having a score of 0.90, 0.88, 0.85 and 0.88 respectively.

Keywords: Employee attrition · Machine learning · Ensemble learning · Evaluation measures

1 Introduction

Employees are considered as valuable assets for an organization. Their contributions effectively help to build the reputation of an organization and ensure its long-term success. Having a strong workforce enhances productivity, cost reductions, and profitability of a company, which can not be attainable without human resources. Inability for an organization to retain employees can result an organization to incur losses in the long term [10]. This phenomena, commonly known as Employee Attrition [4], is an event in an organization where employees leave the organization due to factors including personal or professional reasons, inconvenient work environment, extended office hours, and poor salary. This intentional act of abandonment by the employees is also known as voluntary attrition.

R. Silhavy (Ed.): CSOC 2022, LNNS 502, pp. 285–294, 2022.
https://doi.org/10.1007/978-3-031-09076-9_27

According to the ADP Research Institute, the average monthly attrition rate in the United States is 5% [10]. To predict voluntary attrition, samples were taken from 1900 firms with at least 1000 employees. The research was conducted on a total of 7 million employees from the year 2015 to 2016. In these companies, approximately (60–70)% attrition was predicted to be voluntary. For some situations, attrition can lead to finding better talent, but high attrition rates lead to business liability.

In this paper, machine learning algorithms have been applied to predict employee attrition that can help an organization to make faster management decisions by enhancing effective strategies with internal-external affairs. The dataset used was the HR Analytics Employee Attrition and Performance dataset [13] that was created by IBM data scientists in 2017 and comprise of information such as employee income, satisfaction, work environment, and other demographics.

The remainder of this paper is structured as follows: related works have been discussed in Sect. 2 followed by the methodology adopted is discussed in Sect. 3. Experimental results are given in Sect. 4. Finally, the paper is concluded in Sect. 5 with the best result and discussion about future work.

2 Literature Review

Alhashmi et al. [5] applied the decision tree on the IBM employee dataset [13] to find the factors that can affect an employee to leave the organization.

Using a Random forest classifier, Joseph and colleagues [7] predicted the attrition rate of the employee so that the organizations can have a better plan for their employees. Their method achieves a competitive accuracy of 86.00%.

Alduayj and colleagues [4] also used the IBM HR Dataset [13] for the employee attrition prediction task. They leveraged the adaptive synthetic oversampling (ADASYN) algorithm to address the class imbalance problem and applied classification algorithms in the new oversampled dataset. The best F1-score of 0.93 was achieved using the K-nearest neighbor classifier.

Subhashini and Gopinath [12] performed their experiments using a set of machine learning algorithms with different feature selection methods. Their best accuracy of 89.15% was achieved using the Gradient Boosting Tree (GBT) method with the Chi-Square Feature Selection technique on the IBM employee dataset [13].

Vignesh and colleagues [14] explored the reasons, challenges and solutions for employee attrition with their descriptive research. The main objective of the authors was to help HR managers to reduce the employee attrition rate through precautions. The attrition scenario in India and the reason from different perspectives were briefly discussed.

Skorikov et al. [11] predicted absenteeism using a dataset produced by a courier company in Brazil comprised of the absence information of their employees [1]. Multiple machine learning models were tested with and without the

SMOTE technique as their dataset also suffered from the class imbalance problem. The best accuracy was 92.3% achieved by the KNN algorithm with the Chebyshev distance metric.

3 Methodology

Figure 1 shows the methodology used in this work. Firstly, the IBM HR Dataset was collected and following that, it was pre-processed. Then the dataset was split into train and test set in the ratio of 80% and 20% respectively. Machine learning algorithms have been applied on the train set which creates the fitted models. After that, the hyperparameters of the models were tuned using grid search. These models were consequently evaluated on an unseen test set from which the best model was finally determined.

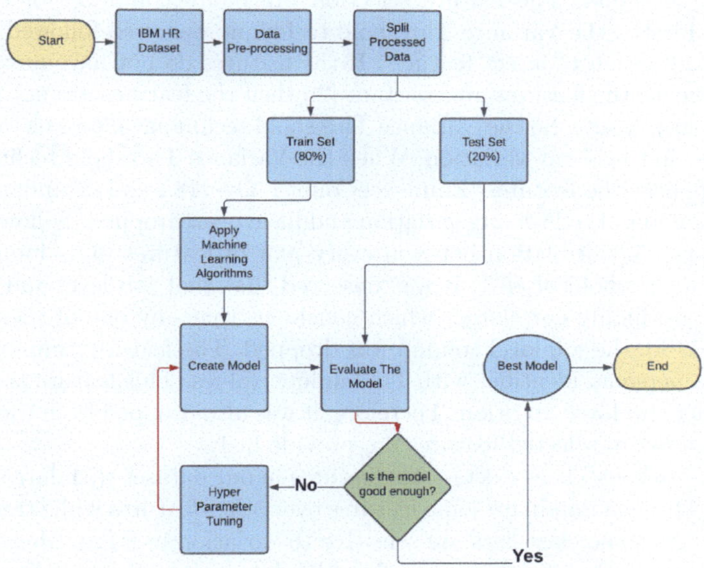

Fig. 1. High-level workflow of the methodology.

3.1 Dataset Acquisition

The dataset used in this research contains 1470 instances representing information of all employees and 35 features including the target class. While examining the gender feature, it was observed that there are 60% males and 40% females in the dataset. Some other key features for our experiments were job involvement and job satisfaction. On average, about 28.29% of the employees with low job involvement or low job satisfaction resulted in an attrition. It was also found that about 31.25% of employees with bad work-life balance left the company when compared to 17.65% of attrited employees who had the best work-life balance.

3.2 Data Pre-processing

After the data acquisition process, the dataset is pre-processed. The following pre-processing techniques for our initial dataset have been used.

1. *Data Encoding:* The features 'Department', 'Education', 'Gender', 'WorkLifeBalance' etc. were encoded because they were the categorical features in the dataset. Machine learning algorithms to work on them, the numerical representation of these categorical features is necessary. Therefore, Label Encoding was used to create numerical representation of these categorical features. After the feature selection phase, categorical features were converted to separate columns in terms of 0 and 1 to train different models. The dimensions were expanded based on each distinct value of every original column in our dataset. Pandas Get Dummies was used to accomplish this task.
2. *Feature Selection:* The feature selection procedure comprises of two techniques. Firstly, the Variance Threshold technique was used followed by Pearson's Correlation to select features. If the features do not have any variance which means the features are constant [2], then the features are not very useful for our purpose. So the Variance Threshold technique allows us to find out features that lack any variation. When the Variance Threshold technique has been applied, the features 'EmployeeCount', 'Over18', and 'Standard-Hours' have been found to lack any variation and have been dropped. Following that, the Pearson's Correlation between every pair of features [9] is found. Upon setting a threshold of 85%, it was observed that that job level and monthly income are highly correlated, which concludes that any one of them can be dropped. So, the job level column was dropped. The feature 'EmployeeNumber' is a personal identifier with 1470 unique values. This feature is unusable to predict employee attrition. Therefore, it was also dropped from the dataset. The final list of selected features are provide in 1.
3. *Feature Scaling:* There exists some features in our dataset with large range of values. This can dominate some features over other features while training the dataset. As some classifiers are sensitive to variation in large values between the features, it is crucial to normalize the data and thus, the *min-max normalization* technique [8] was used in our work. The *min-max normalization* was implemented using Eq. 1, which rescales the data converting all large values into the range between 0 and 1. A list of scaled features is provided in Table 2.

$$X_{scaled} = \frac{X - X_{min}}{X_{max} - X_{min}} \tag{1}$$

Table 1. List of the top 29 selected features.

1. Age	16. Job involvement
2. Business travel	17. Job role
3. Daily rate	18. Job satisfaction
4. Department	19. Marital status
5. Distance from home	20. Monthly income
6. Education	21. Monthly rate
7. Education field	22. Number of companies worked
8. Environment satisfaction	23. Overtime
9. Gender	24. Percent salary hike
10. Hourly rate	25. Performance rating
11. Relationship satisfaction	26. Stock option level
12. Total working years	27. Years in current role
13. Training times last year	28. Years since last promotion
14. Work life balance	29. Years with current manager
15. Years at company	

Table 2. List of 12 scaled features

1. Age	7. Percent salary hike
2. Daily rate	8. Total working years
3. Distance from home	9. Years at company
4. Hourly rate	10. Years in current role
5. Monthly income	11. Years with current manager
6. Monthly rate	12. Years since last promotion

3.3 Splitting Dataset

Splitting a dataset into training and testing partitions are crucial for the machine learning process. The training data is used to train machine learning models and the test data is used to evaluate the trained models. During the training phase, a model will be incapable of viewing the test data, enabling us to estimate how the model reacts to new samples [6,9]. In our work, the dataset was separated into the train set consisting of 80% of the instances and test set consisting of 20% of the instances using stratified train-test split. Stratified train test split means splitting the dataset into the same proportions of data in each output class to avoid underfitting or overfitting [3].

3.4 Machine Learning Classifiers

In this research, the data were fed to different supervised classification and ensemble learning algorithms. After completing the pre-processing stage, the trained models were utilized to predict on the unseen data (test set). ZeroR, Decision Tree, K-Nearest Neighbor, Support Vector Machine, Random Forest and Voting Classifier algorithms were evaluated. After initial evaluation, hyperparameter tuning was further performed for each classifier. Below brief descriptions of the classifiers implemented in our work are provided.

1. *ZeroR:* ZeroR is a baseline classifier that ignores all the features and predicts based on the majority class. It has no predictability power, but it can be treated as a benchmark to compare performance of other classification methods.
2. *Decision Tree:* A decision tree is a tree-like structure that represents a set of rules that the model has learned [8]. A decision tree is a very useful machine learning algorithm with high interpretability. Each node is selected based on high information gain by measuring impurity such as gini index and entropy [6]. It was observed that max depth = 2 is the best for the IBM employee dataset [13] and avoids overfitting. For Information Gain, Gini has been used to split the data. The formula for Gini has been given in Eq. 2 where $p_i(t)$ is the frequency of class i at node t, and c is the total number of classes. Figure 2 shows the tree generated by decision tree algorithm.

$$Gini = 1 - \sum_{i=0}^{c-1} p_i(t)^2 \tag{2}$$

3. *K Nearest Neighbor Classifier (KNN):* K nearest neighbor uses maximum voting mechanism of the k nearest neighbors of the test query to make the prediction of the test query [8]. For this dataset, $k = 9$ was found to be the best fit. The distance of neighbors can be calculated using distance metrics such as manhattan distance (Eq. 3) and euclidean distance. Hyperparameter tuning reveals that manhattan distance is a better suited distance metric for this dataset.

$$|A - B| = \sum_{i=1}^{d} |a_i - b_i| \tag{3}$$

4. *Support Vector Machine:* Support vector machine (SVM) is a versatile machine learning classifier [6]. With the help of grid search cross-validation, the best hyperparameters were indentified.
5. *Random Forest:* Random forest is a machine learning algorithm that trains many decision trees and averages the prediction. It is built on top of many decision trees. Therefore, the random forest is also an example of an ensemble learning model [6]. In our research, a total of 600 decision trees were used to train the model suggested by grid search cross-validation during hyperparameter optimization.
6. *Voting Classifier:* Combining the predictions of each classifier and predicting the class with the most votes is one ensemble technique to improve the classification [6]. As illustrated in Fig. 3, Soft Voting has been used which selects the highest class probability averaged by each classifier.

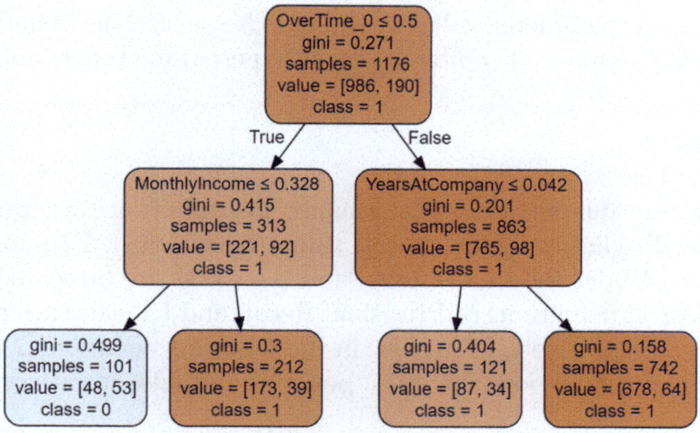

Fig. 2. Decision tree showing the number of samples, gini, values and their repective class for each partitioned node.

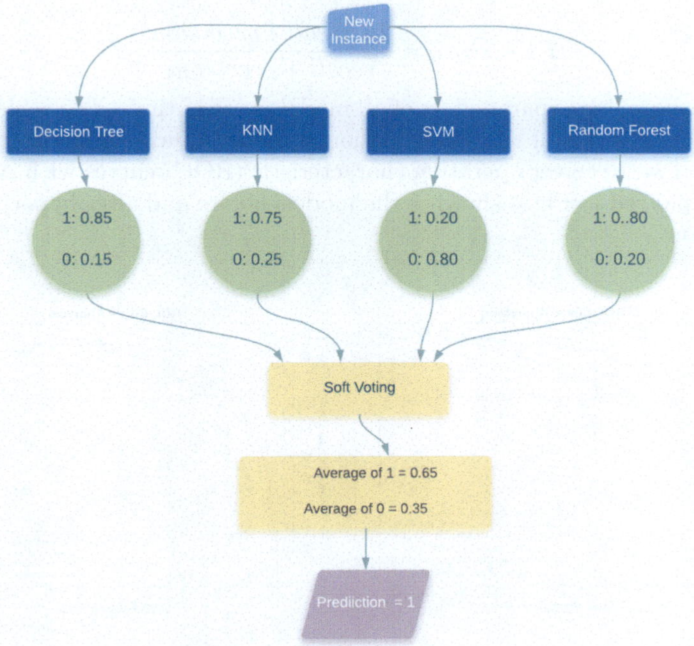

Fig. 3. Soft voting classifier for prediction

3.5 Hyperparameter Tuning

Hyperparameters are configurations that are defined by a developer for a dataset to control the machine learning process. The grid search cross-validation technique has been used to perform hyperparameter tuning in our work. In the case of grid search cross-validation, a set of best hyperparameters is found by evaluating every hyperparameter value in the search space. This technique is very helpful when the number of combinations of hyperparameters is not very high.

4 Results

The results are evaluated using performance measures such as precision (also called Positive Predictive Value), recall value (Sensitivity), F1-score, accuracy and AUC (Area Under the ROC Curve) and the scores are presented in Table 3. The mathematical definitions of Precision, Recall and F1-score are presented in Eqs. 4, 5, and 6 respectively. Moreover, in the following equations TP, FP, FN refers to number of true positives, false positive, and false negatives.

$$precision = \frac{TP}{TP + FP} \tag{4}$$

$$recall = \frac{TP}{TP + FN} \tag{5}$$

$$F1 - score = \frac{2 * recall * precision}{recall + precision} \tag{6}$$

Furthermore, hyperparameters of all models were optimized to achieve better performance in terms of the aforementioned metrics and summarize the scores in Table 4. Two Receiver operating characteristic (ROC) curves with AUC score have been plotted in Fig. 4 showing the models before and after hyperparameter optimization.

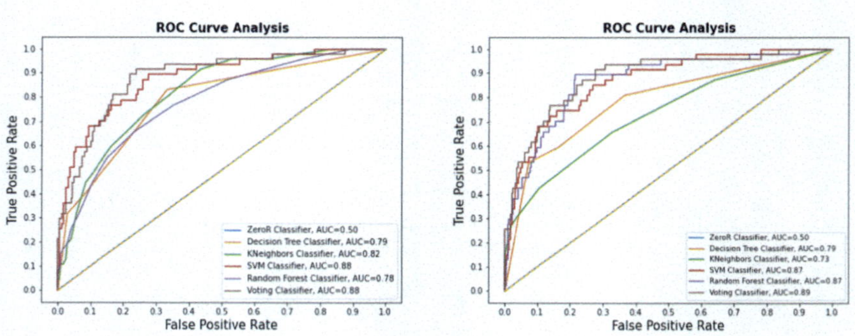

(a) Before Hyperparameter Optimization **(b) After Hyperparameter Optimization**

Fig. 4. ROC and AUC Plot for different models.

It is evident that in each case of Table 4, the F1 score using precision and recall validated remarkable results, satisfying the purpose of this research. It demonstrates that the voting algorithm outperforms all other algorithms in predicting employee attrition because the value of AUC is 0.89 which is a significant improvement.

Table 3. Results prior to hyperparameter optimization

Classifier Name	Precision	Recall	F1-score	Accuracy	AUC
ZeroR	0.71	0.84	0.77	0.84	0.5
Decision tree	0.85	0.87	0.85	0.87	0.79
KNN	0.88	0.86	0.81	0.86	0.82
SVM	0.87	0.88	0.87	0.88	0.88
Random forest	0.83	0.85	0.82	0.85	0.78
Voting	**0.89**	**0.87**	**0.83**	**0.87**	**0.88**

Table 4. Results after hyperparameter optimization

Classifier Name	Precision	Recall	F1-score	Accuracy	AUC
ZeroR	0.71	0.84	0.77	0.84	0.5
Decision tree	0.86	0.86	0.86	0.86	0.79
KNN	0.88	0.86	0.82	0.86	0.73
SVM	0.87	0.88	0.86	0.88	0.87
Random forest	0.85	0.86	0.82	0.86	0.87
Voting	**0.90**	**0.88**	**0.85**	**0.88**	**0.89**

5 Conclusion

A high attrition rate of employees is undesirable for any organization or company. Losing a highly qualified employee is a loss for any company, particularly those companies that invest in their employees. Moreover, finding a replacement remains difficult, costworthy, and time-consuming. In order to reduce the attrition rate of any organization, this research paper presents experimental approaches to predict the departure rate of employees using the historical data kept in human resources (HR) departments. After using the variance threshold technique and Pearson's correlation technique to select important data, five supervised machine learning algorithms consisting of ZeroR, Decision tree, KNN, SVM and Random Forest and one ensemble learning algorithm (Voting Classifier) were compared with performance measure techniques (F-measure and AUC score) because of the imbalanced dataset. It has been found that all classifiers acquired varying degrees of results - however the Voting classifier achieved the best results. In future, we plan to use explainable AI techniques (XAI) to better understand how these models are making the predictions.

References

1. Absenteeism at work data set. https://archive.ics.uci.edu/ml/datasets/Absenteeism+at+work?fbclid=IwAR1Nv_i1jNKAbaQhjph51L6AFudIeL9X_kNm02V0ZKi_XH9QuyqX3TBofcc

2. Sklearn feature selection variancethreshold. https://scikit-learn.org/stable/modules/generated/sklearn.feature_selection.VarianceThreshold.html
3. Sklearn stratified train-test split. https://scikit-learn.org/stable/modules/generated/sklearn.model_selection.train_test_split.html
4. Alduayj, S.S., Rajpoot, K.: Predicting employee attrition using machine learning. In: 2018 International Conference on Innovations in Information Technology (IIT), pp. 93–98. IEEE (2018)
5. Alhashmi, S.M.: Towards understanding employee attrition using a decision tree approach. In: 2019 International Conference on Digitization (ICD), pp. 44–47. IEEE (2019)
6. Gron, A.: Hands-On Machine Learning with Scikit-Learn and TensorFlow: Concepts, Tools, and Techniques to Build Intelligent Systems, 1st edn. O'Reilly Media Inc, Sebastopol (2017)
7. Joseph, R., Udupa, S., Jangale, S., Kotkar, K., Pawar, P.: Employee attrition using machine learning and depression analysis. In: 2021 5th International Conference on Intelligent Computing and Control Systems (ICICCS), pp. 1000–1005. IEEE (2021)
8. Mitchell, T.M.: Machine Learning. McGraw-Hill, New York (1997)
9. Raschka, S.: Python Machine Learning. Packt Publishing Ltd, Birmingham (2015)
10. Research, A.: Adp.com. https://www.adp.com/tools-and-resources/adp-research-institute/research-and-trends/-/media/RI/pdf/ADP_PredictingTurnOver_White_Paper.ashx
11. Skorikov, M., et al.: Prediction of absenteeism at work using data mining techniques. In: 2020 5th International Conference on Information Technology Research (ICITR), pp. 1–6 (2020). https://doi.org/10.1109/ICITR51448.2020.9310913
12. Subhashini, M., Gopinath, R.: Employee attrition prediction in industry using machine learning techniques (2021)
13. Swaminathan, S., Hagarty, R.: Data science process pipeline to solve employee attrition (2019). https://developer.ibm.com/patterns/data-science-life-cycle-in-action-to-solve-employee-attrition-problem/
14. Vignesh, S.G., Sarojini, V., Vetrivel, S.: Employee attrition and employee retention-challenges & suggestions. In: International Conference On Economic Transformation with Inclusive Growth (2018)

Optimization Method for Reconstruction of the Polyhedron 3D Model by Its Labeled Projection

Vladimir A. Klyachin(✉) and Elena G. Grigorieva

Volgograd State University, University Avenue, 100, Volgograd 400062, Russia
{klyachin.va,e_grigoreva}@volsu.ru

Abstract. The article deals with the problem of the reconstruction an object from its image of projection. The task is reduced to 3D reconstruction in space of a finite set of points from its central projection. In article we use parameterized polyhedrons as a considered objects. Simple mathematical reasoning allows us to conclude that with a small motion in the space of a finite set of points, its central projection remains unique. This provides a theoretical basis for developing an algorithms for the above reconstruction. Note that if in addition we admit scaling the model we have not uniqueness. Therefor the some restrictions on the spatial position of 3D model are introduced. The article proposes an algorithm for reconstructing a finite set of points in space based on minimizing the Hausdorff distance function between a given flat set and an auxiliary finite set of a 3D model of an object. To minimize this function, a combined method for the numerical calculation of the minimum point is proposed, in which each subsequent method uses the calculated point as the initial one. In addition, the article proposes an algorithm for choosing the starting point of the iterative minimization algorithm, based on the known spatial dimensions of the object under study.

Keywords: Hausdorff distance · 3D model · Minimizing a function

1 Preliminaries

1.1 Introduction

It is well known one of the urgent problems in the construction of computer vision systems is the problem of determining the spatial orientation and spatial position of an object from a photograph. In particular, this task is especially important for autonomous driving systems, where the positioning of nearby vehicles is a key issue for autonomous vehicles in an urban environment. In this regard, for example, in 2019, the Baidu Robotics and Autonomous Driving Laboratory, together with Peking University, set an appropriate task for the Kaggle community (https://www.kaggle.com/c/pku-autonomous-driving) and provided more than 60,000 copies of three-dimensional cars marked from 5,277 real images. It should be noted that attempts to solve similar problems and not only for cars as objects have been made earlier. So, we should note the work [1], which presents

R. Silhavy (Ed.): CSOC 2022, LNNS 502, pp. 295–306, 2022.
https://doi.org/10.1007/978-3-031-09076-9_28

one of the methods for assessing the positions and sizes of 3D boxes covering a car in space by its image in a photograph. The same techniques were used in works [2, 3]. Also, a similar problem of determining the 3D position of objects was solved in [4], in which modified convolutional neural networks and specially prepared object masks with downsampling were used. In [5], a method based on frame-by-frame tracking of an object was used, and the training model was built on the basis of recurrent LSTM neural networks. We also note the work [6], in which for the integral reconstruction of a three-dimensional scene its description in the form of a grammar is proposed, and the authors construct the result of the reconstruction in the form of a special graph. It should be noted that similar reconstruction tasks arise, for example, when restoring a three-dimensional model of a person's face and posture of a person's body from a photograph (see, for example, [7, 8]). Analysis of these works and works cited in them shows that neural networks are most often used to solve problems of spatial image analysis. This is understood because convolutional neural networks can automatically detect features in images and determine their significance based on the dataset used in training. Certainly, the approach based on the use of neural networks deserves attention, but it requires a very large set of data for training. Therefore, it seems to us, it would be useful to be able to carry out direct calculations of spatial data based solely on the characteristic features of the images of the objects themselves. So, for example, in [9] it was shown that the orientation of the plane of a circle in space can be obtained from the characteristics of an elliptical region, which is a projection of the circle. In [10], an algorithm for reconstructing the profile of the visible part of the surface of revolution from its projection was obtained and described.

In this paper, it is proposed the reconstruction method of an object on the basis of an approach that can be called geometric regression: a given finite set is approximated by projections of the holes of the 3D model of the object, minimizing the Hausdorff distance.

1.2 Hausdorff Distance

The Hausdorff distance is determined as follows. For any closed sets A, $B \subset R^n$, the quantity

$$h(A, B) = max\ min\ |x - y|,$$

where the minimum is taken over all points y from the set B, and the maximum is taken over all points x of the set A. The Hausdorff distance $dist(A, B)$ is then determined by the formula

$$dist(A, B) = max\{h(A, B), h(B, A)\}.$$

Geometrically, the inequality $dist(A, B) < \varepsilon$ means that each point of the set A is located at a distance not exceeding ε from some point of the set B and vice versa. So that the Hausdorff distance is zero if and only if the sets coincide. A number of works [11–14] should be noted, in which the Hausdorff distance is used to solve various kinds of practical problems.

1.3 Reconstruction 3D Object by Its Projection

The problem of determining the spatial position of an object from a photograph can be reduced to the mathematical problem of reconstructing a spatial object from its projection. We consider images as a result of the central projection $\pi_C : R^3 \to R^2$, which is given by the equations

$$X = d \cdot x/z, Y = d \cdot y/z, for\ all(x, y, z), z \neq 0, d > 0,$$

where d is a focal length of the camera. Let P be some finite subset of R^3. The problem is to determine the set P from its projection $\pi_C(P)$ in assumption that P belongs some family of polyhedrons. This family can be obtained as a set of images some transformations group – translations, rotations, scaling. The possibility of such reconstruction can theoretically be substantiated by referring, for example, to the results of work [15]. From these works, we can conclude that for a small motion of the set P in R^3, its projection $\pi_C(P)$ remains unique. Below we describe the algorithm for this reconstruction. We note but P can be reconstructed up to scaling relative to the center of projection. Therefor we need to introduce some restrictions to the polyhedron model. In particular, we can assume that some vertex or geometric center of P must be on some plane. Other way is to assume that the P has some fixed size in case when we consider, for example, shapes of parallelepiped.

We introduce some families of polyhedrons by the following way. The polyhedron P from some family F is defined by set of parameters $z = (z_1, \ldots, z_m)$. For example, The family of cubes in R^3 can be described by parameters (x_0, y_0, z_0, a, b, c) such that cube Q has vertexes

$$(x_0, y_0, z_0), (x_0 + a, y_0, z_0), (x_0 + a, y_0 + b, z_0), (x_0, y_0 + b, z_0),$$
$$(x_0, y_0, z_0 + c), (x_0 + a, y_0, z_0 + c), (x_0 + a, y_0 + b, z_0 + c), (x_0, y_0 + b, z_0 + c).$$

Also we can add parameter φ as a value of the angle rotation of the cube around axis Oz. In this case Q has the following vertexes:

$$(x_0, y_0, z_0), (x_0 + a \cdot \cos\varphi, y_0 + a \cdot \sin\varphi, z_0),$$
$$(x_0 + a \cdot \underline{\cos\varphi} - b \cdot \sin\varphi, y_0 + a \cdot \sin\varphi + b \cdot \cos\varphi, z_0), (x_0 - b \cdot \sin\varphi, y_0 + b \cdot \cos\varphi, z_0),$$
$$(x_0, y_0, z_0 + c), (x_0 + a \cdot \cos\varphi, y_0 + a \cdot \sin\varphi, z_0 + c),$$
$$(x_0 + a \cdot \cos\varphi - b \cdot \sin\varphi, y_0 + a \cdot \sin\varphi + b \cdot \cos\varphi, z_0 + c), (x_0 - b \cdot \sin\varphi, y_0 + b \cdot \cos\varphi, z_0 + c).$$

Another example of parameterized polyhedron family is drawn on the Fig. 1.

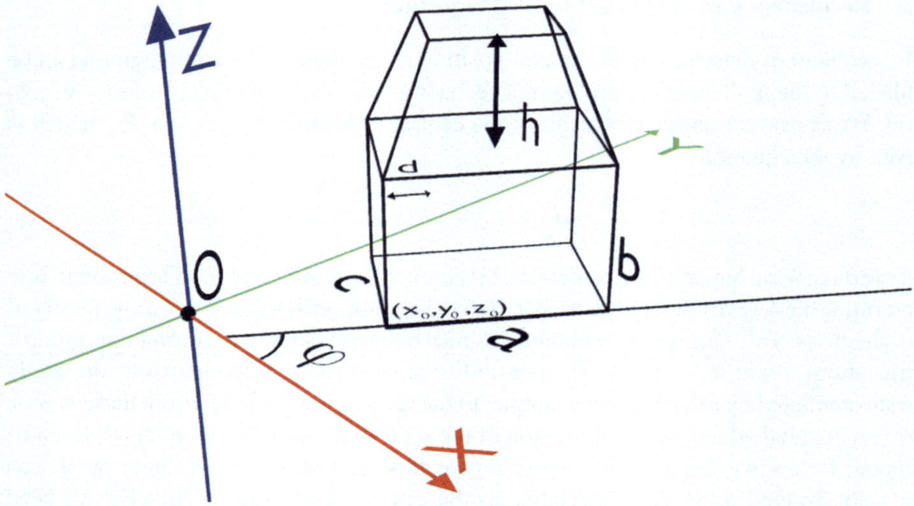

Fig. 1. The parameterized 3D model.

In this case the parameters are $(x_0, y_0, z_0, a, b, c, d, \varphi)$ and the vertices of the 3D model are given by formulas:

$$(x_0, y_0, z_0), (x_0 + a \cdot cos\varphi, y_0 + a \cdot \sin\varphi, z_0),$$
$$(x_0 + a \cdot \cos\varphi - b \cdot \sin\varphi, y_0 + a \cdot \sin\varphi + b \cdot \cos\varphi, z_0), (x_0 - b \cdot \sin\varphi, y_0 + b \cdot \cos\varphi, z_0),$$
$$(x_0, y_0, z_0 + c), (x_0 + a \cdot \cos\varphi, y_0 + a \cdot \sin\varphi, z_0 + c),$$
$$(x_0 + a \cdot \cos\varphi - b \cdot \sin\varphi, y_0 + a \cdot \sin\varphi + b \cdot \cos\varphi, z_0 + c), (x_0 - b \cdot \sin\varphi, y_0 + b \cdot \cos\varphi, z_0 + c),$$
$$(x_0 + d \cdot \cos\varphi, y_0 - d \cdot \sin\varphi, z_0 + c + h), (x_0 + a \cdot \cos\varphi - d \cdot \cos\varphi, y_0 + a \cdot \sin\varphi - d \cdot \sin\varphi, z_0 + c + h),$$
$$(x_0 + a \cdot \cos\varphi - b \cdot \sin\varphi - d \cdot \cos\varphi, y_0 + a \cdot \sin\varphi + b \cdot \cos\varphi + d \cdot \sin\varphi, z_0 + c + h),$$
$$(x_0 - b \cdot \sin\varphi - d \cdot \cos\varphi, y_0 + b \cdot \cos\varphi + d \cdot \sin\varphi, z_0 + c + h).$$

2 Algorithm

2.1 The Labeled Projections

We define the labeled projection of the polyhedrons by the following way. We suppose that there are some polyhedrons $G_1, G_2, \ldots, G_N$ in R^3 and each polyhedron belongs some family F_j. The labeled projection is a set of tuples (X_i, Y_i, F_j, N_k), where N_k is a number of polyhedron, F_j is a its family, and (X_i, Y_i) – coordinates projection its vertexes. On the Fig. 2 the labeled cubes marked by the colored points (X_i, Y_i) of the separated polyhedron are painted by own color.

Fig. 2. The labeled image of cubes.

In this example we restrict positions of cubes on the horizontal plane $z = z_0 = const$. We are aimed to calculate parameters of each cube on the given image by the drawn points. The proposed algorithm treats the labeled image of polyhedron the arbitrary known family.

2.2 Algorithm in General

The main idea is as follows. For the object under study, it is necessary to create a 3D model (polyhedron) and select its parameters gives its position in such a way that its projection would bring the characteristic points of the object in the image as close as possible. We choose the Hausdorff distance as a measure of approximation. The calculation of this value does not require the comparison of points of the model and points of the investigated object. And the theoretical premises described in the previous section give strong grounds to believe that by minimizing the Hausdorff distance, we will be able to correctly calculate (estimate) the needed parameters of the object. Thus, the problem is reduced to the problem of minimizing the function

$$L(z) = dist(A(z), B) \to min,$$

here z is vector $z = (z_1, \ldots, z_m)$ of parameters of the set P in family $\{P(z), z = (z_1, \ldots, z_m)\}$ of polyhedrons, B is a set of the labeled points, and $A(z)$ is a projection $A(z) = \pi_C(P(z))$.

In addition for uniqueness reconstruction we need to introduce some restrictions on the parameters. For example, we can admit that all objects is situated on the horizontal plane. In this case we choose $z_0 = const$. See Fig. 3.

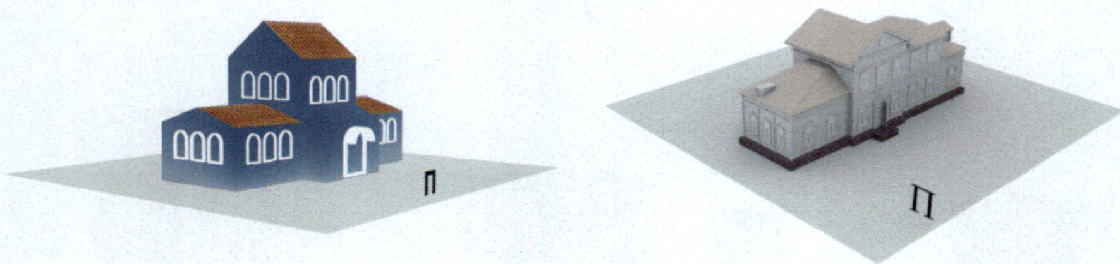

Fig. 3. 3D models of objects on the plane Π.

The block diagram of the described algorithm is shown in Fig. 4. Note that the found z value will contain all the required sought-for values of the parameters of the object under study. The calculated parameters z give ability to reconstruct the 3D model.

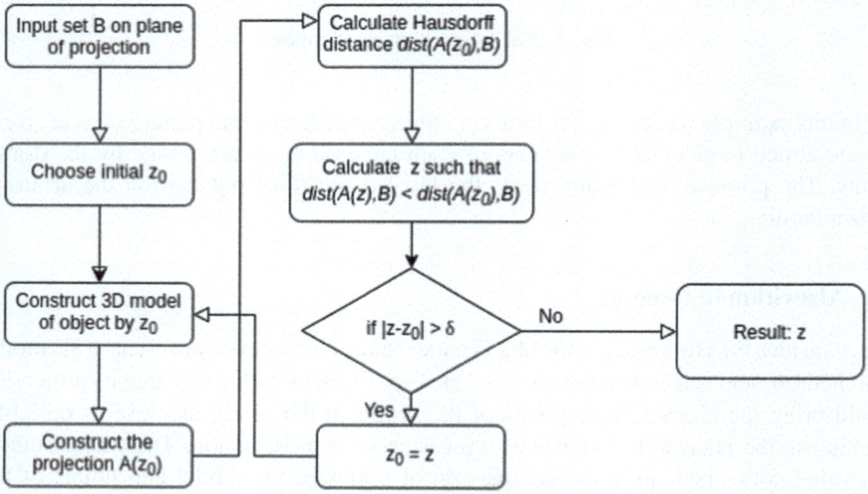

Fig. 4. Flowchart of the algorithm.

Below, in Fig. 5, results of minimizing the function $L(z)$ are shown. Points of the sets B are marked in blue, edges of the set $A(z)$ are marked in other colors.

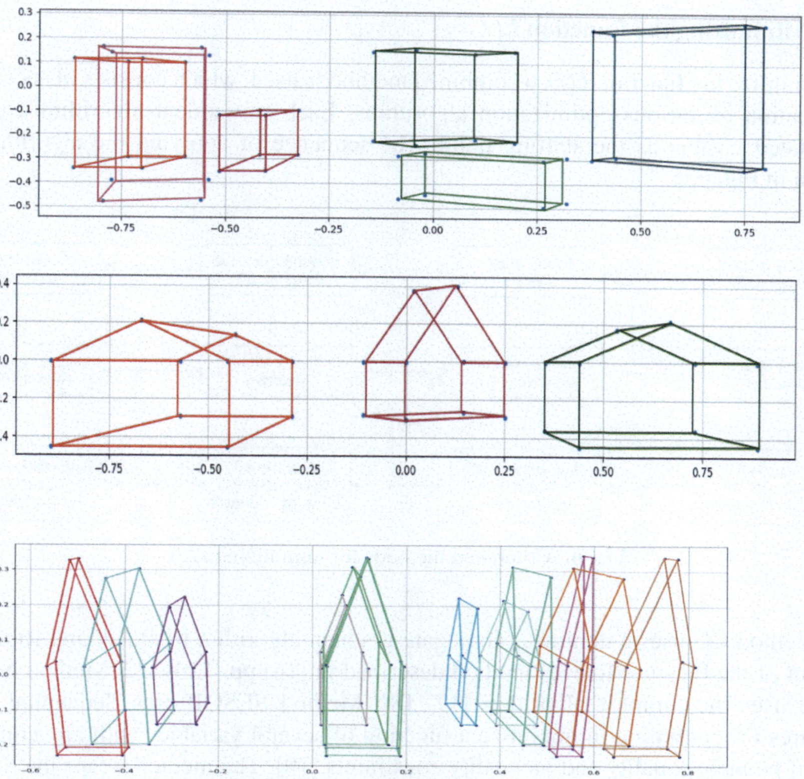

Fig. 5. The results of minimizing Hausdorff distance

2.3 Algorithm for Choosing the Initial Value z_0

Let us give a brief description of the algorithm for calculating the initial value of the parameter $z = (z_1, \ldots, z_m)$. We assume that z contains coordinates (x_0, y_0, z_0) of the some chosen vertex of polyhedron or its geometric center. The coordinates of the geometric center of the set B are taken as the quantities (x_0, y_0). The value z_0 is calculated using the f value of the camera focal length by the formula

$$z_0 = f \cdot diam(A)/diam(B) + diam(A)/2,$$

where $diam(X)$ is a diameter of the set X. We denote by (X_i, Y_i) the labeled points on image. Other initial parameters are calculated within of the type of 3D model. In particular, for 3D model on the Fig. 1. we put

$$\begin{aligned} b &= |(Y5 - Y0) \cdot z0)|/f \\ \alpha &= Y5/Y2 \\ a &= |z0| \cdot \sqrt{((\alpha - 1)^2 + (\alpha \cdot X_2 - X_3)^2} \\ d &= a/2 \\ \varphi &= arcsin(z_0 \cdot (Y5 - Y2)/(Y2 \cdot a)). \end{aligned}$$

2.4 Minimizing the Function *L(Z)*

To minimize the function $L(z)$, a combined method is used, which consists of sequential application of various optimization algorithms. Each subsequent algorithm uses the calculated z-value as the starting point. The sequence of applying the algorithms is shown in Fig. 6.

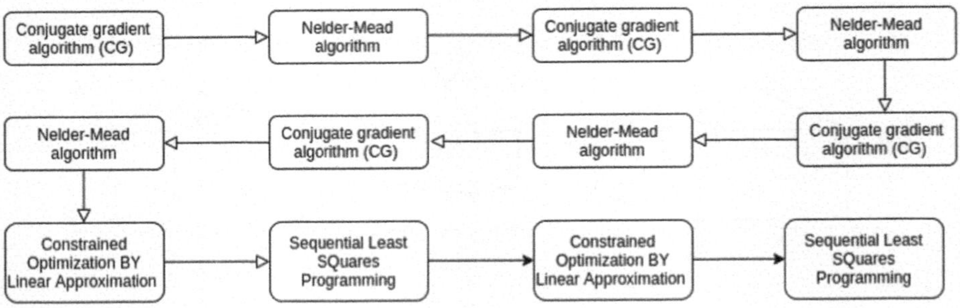

Fig. 6. Combined methods for minimizing $L(z)$

Method CG uses a nonlinear conjugate gradient algorithm by Polak and Ribiere, a variant of the Fletcher-Reeves method described in [16] pp. 120–122. Method Nelder-Mead uses the Simplex algorithm [17, 18]. Method SLSQP uses Sequential Least SQuares Programming to minimize a function of several variables with any combination of bounds, equality and inequality constraints [19]. The method wraps the SLSQP Optimization subroutine originally implemented by Dieter Kraft [20].

The calculated z-value can be used for reconstruction the 3D model from its projection. We just construct the mesh of the model by its information about faces and calculating the vertexes by the parameters vector z. We note that in [21] we considered the case where 3D model can't be scaled and proposed algorithm for calculating the spatial position and orientation of the object by its projection.

Now we consider visualization of some results. At first, we input labeled image with polyhedrons (see Fig. 7).

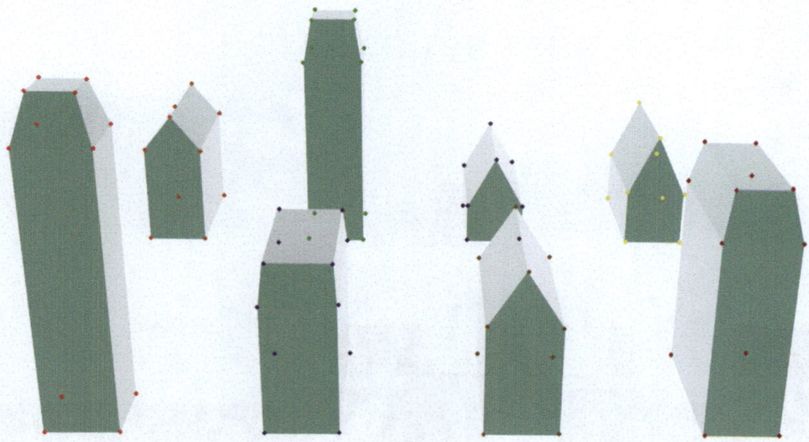

Fig. 7. The image with the labeled polyhedrons

Below on the Fig. 8 we can see the result of comparison the calculated 3D models using labeled image and the given 3D models.

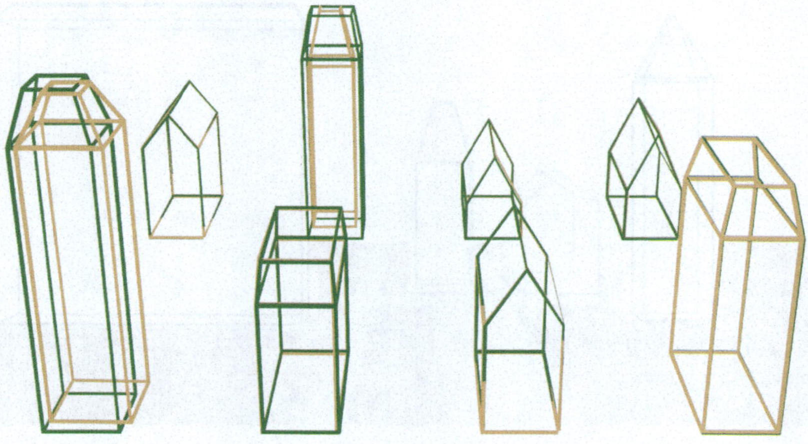

Fig. 8. Comparison the calculated 3D models (green colored) and the given 3D models (brown colored)

Now we consider some 3D scene with models of buildings (Figs. 9 and 10):

Fig. 9. The labeled image of 3D scene.

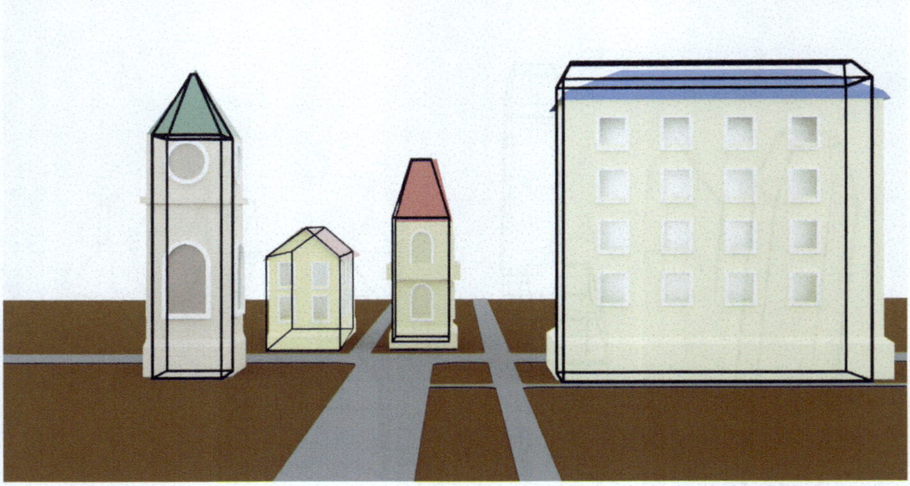

Fig. 10. The result of 3D reconstruction of the 3D scene by the labeled its image.

3 Conclusion and Future Work

The results obtained allow us to make the following conclusions. Using Hausdorff distance as a measure of difference of finite sets on image can be useful for problem of calculations the parameters of the object and as a consequence we obtain spatial information about object. We plan to use the obtained results for design the machine learning model which will be used for solution the problem of reconstruction 3D scenes by one or more its images.

Acknowledgments. This work was supported by the Ministry of Education and Science of Russia (the project "Development of Virtual 3D Reconstruction of Historical Objects Technique", scientific theme code 2019-0920, project number in the research management system FZUU-0633-2020-0004).

References

1. Mousavian, A., Anguelov, D., Flynn, J., Kosecka, J.: 3D bounding box estimation using deep learning and geometry. In: 2017 IEEE Conference on Computer Vision and Pattern Recognition (CVPR) (2017). https://doi.org/10.1109/cvpr.2017.597
2. Gordeev, A.Y., Klyachin, V.A.: Determination of the spatial position of cars on the road using data from a camera or DVR. In: Popkova, E.G., Sergi, B.S. (eds.) ISC 2020. LNNS, vol. 155, pp. 172–180. Springer, Cham (2021). https://doi.org/10.1007/978-3-030-59126-7_20
3. Alexey, G., Klyachin, V., Eldar, K., Driaba, A.: Autonomous mobile robot with AI based on Jetson Nano. In: Arai, K., Kapoor, S., Bhatia, R. (eds.) FTC 2020. AISC, vol. 1288, pp. 190–204. Springer, Cham (2021). https://doi.org/10.1007/978-3-030-63128-4_15
4. Ren, M., Pokrovsky, A., Yang, B., Urtasun, R.: SBNet: Sparse Blocks Network for Fast Inference. arXiv:1801.02108 (2018)
5. Hu, H., et al.: Joint Monocular 3D Vehicle Detection and Tracking (2018). arXiv:1811.10742v3
6. Huang, S., Qi, S., Zhu, Y., Xiao, Y., Xu, Y., Zhu, S.-C.: Holistic 3D scene parsing and reconstruction from a single RGB image. In: Ferrari, V., Hebert, M., Sminchisescu, C., Weiss, Y. (eds.) ECCV 2018. LNCS, vol. 11211, pp. 194–211. Springer, Cham (2018). https://doi.org/10.1007/978-3-030-01234-2_12
7. Jackson, A.S., et al.: Large pose 3D face reconstruction from a single image via direct volumetric CNN regression. In: 2017 IEEE International Conference on Computer Vision (ICCV), pp. 1031–1039. IEEE (2017)
8. Ferkova, Z., Urbanova, P., Cerny, D., Zuzi, M., Matula, P.: Age and gender-based human face reconstruction from single frontal image. Multimed. Tools Appl. **79**, 1–26 (2018)
9. Klyachin, V.A., Grigorieva, E.G.: Algorithm of automatic determining of camera orientation parameters based on characteristic elements on image. Tendencii razvitiya nauki I obrazovaniya **45**(6), 10–20 (2018)
10. Klyachin, V.A., Grigorieva, E.G.: A 3D reconstruction algorithm of a surface of revolution from its projection. Sib. Zh. Ind. Mat. **23**(1), 84–92 (2020)
11. Liu, P., et al.: Deep learning to segment pelvic bones: large-scale CT datasets and baseline models. arXiv:2012.08721
12. Huang, P., Lin, L., Xue, F., Xu, K., Cohen-Or, D., Huang, H.: Hausdorff Point Convolution with Geometric Priors. arXiv:2012.13118
13. Dubuisson, M.-P., Jain, A.K.: A modified Hausdorff distance for object matching. In: Proceedings of 12th International Conference on Pattern Recognition, vol. 1, pp. 566–568 (1994)
14. Engelmann, F., Rematas, K., Leibe, B., Ferrari, V.: From Points to Multi-object 3D Reconstruction (2020). arXiv preprint arXiv:2012.11575
15. Klyachin, A.A., Klyachin, V.A.: Existence and uniqueness theorems of inverse projection geometry problems for 3D reconstruction by photograph. Chebyshevskiy sbornik **21**(4), 117–128
16. Nocedal, J., Wright, S.J.: Numerical Optimization. Springer, New York (2006). https://doi.org/10.1007/978-0-387-73301-2_5

17. Nelder, J.A., Mead, R.: A simplex method for function minimization. Comput. J. **7**, 308–313 (1965)
18. Wright, M.H.: Direct search methods: once scorned, now respectable. In: Griffiths, D.F., Watson, G.A. (eds.) Numerical Analysis 1995: Proceedings of the 1995 Dundee Biennial Conference in Numerical Analysis, pp. 191–208. Addison Wesley Longman, Harlow (1996)
19. Kraft, D.: A software package for sequential quadratic programming. Technical report, DFVLR-FB 88-28, DLR German Aerospace Center –Institute for Flight Mechanics, Koln, Germany (1988)
20. Powell, M.J.D.: A direct search optimization method that models the objective and constraint functions by linear interpolation. In: Gomez, S., Hennart, J.-P. (eds.) Advances in Optimization and Numerical Analysis, pp. 51–67. Kluwer Academic, Dordrecht (1994)
21. Klyachin, A.A., Klyachin, V.A.: Geometric fitting in problem of spatial position of object determination by image. In: Arai, K. (ed.) FTC 2021. LNNS, vol. 359, pp. 210–217. Springer, Cham (2022). https://doi.org/10.1007/978-3-030-89880-9_16

Exclusion of Parcel's Borderline Pixels and Related Impact on Crop Classification Accuracy

Ján Mojžiš(✉)

Institute of Informatics, Slovak Academy of Sciences, Bratislava, Slovakia
jan.mojzis@savba.sk

Abstract. Crop classification with the use of remote sensing satellite images is a broadly discussed topic in the literature. Much less attention is paid to borderlines of farmer's parcels and the fact, that these pixels may contain mixed vegetation content due to decreased spatial resolution of the satellite images, comparing to images obtained with the use of Unmanned Aerial Vehicles (UAV). In this paper an accuracy of the crop classification is improved by exclusion of borderline pixels of parcels. For the sunflower by 12 and sugar beet by 10 percentual points. Rules for the exclusion of these pixels are presented. Sentinel 2 level 2 satellite images are used to evaluate the proposed solution. Random Forest and Support Vector Machines machine learning models are used for crop classification. There is a comparison of the crop classification accuracy results with and without the borderline pixels use.

Keywords: Remote sensing · Support vector machine · Random forest · Crop classification · Aggregation · Sentinel 2 · Buffer size · Parcel · Borderline

1 Introduction

Remote sensing is used in the field of Earth's atmosphere and surface monitoring. In order to monitor Earth's land surface, the Surface Reflectance (SR), also called Bottom of Atmosphere (BOA) or Level 2, satellite images are used. These types of images are atmospherically corrected for the scattering and absorbing effects of atmospheric gases and aerosols [1].

Crop monitoring is important for food supply analysis and overall crop and vegetation condition [2]. Images of the land can be obtained either from the satellite or by using aerial device with mounted camera (Unmanned Aerial Vehicles (UAV)). Using satellite images for crop monitoring is less expensive than using UAV and revisits are periodical (for Sentinel 2 since March 2017 revisit frequency is on five days basis) [3]. Most limiting factor here is the occurrence of the clouds, where landscape pixels are covered in clouds, thus no useful land data could be observed.

For UAV an orthorectification and mosaicking stages are used to create orthophotomap (larger area image is created from many small area images). In comparison to

R. Silhavy (Ed.): CSOC 2022, LNNS 502, pp. 307–319, 2022.
https://doi.org/10.1007/978-3-031-09076-9_29

Satellite images, UAV could be used during cloudy weather (but not during rain). This is an advantage comparing to Satellite imaging. UAV offers better spatial resolution and control during crop monitoring and imaging due to flight planning, but due to various camera issues, a recommended overlapping in adjacent images should be followed [4, 5] and there may be other issues with camera, like missing rolling shutter support for Parrot Sequoia camera type in OpenDroneMap software used for orthophotomap creation [6].

Traditional procedures in crop monitoring (either by using images from satellite or UAV) includes calculation of Normalized Difference Vegetation Index (NDVI), which is a normalized index suited for chlorophyl reflectance and vegetation condition, used to monitor vegetation health in general. Classification or recognition of crop types is also commonly used in the EU to prevent overpayments and unjustified payments to farmers. In the EU paying agencies provide direct payments to farmers. These payments are based on various eligibility rules. An average of 80% of field (farmer's parcels) inspections are performed using remote sensing [3].

A crop classification using satellite or UAV images can be performed with the use of image features – pixels after preprocessing. A buffering technique is used in order to analyze the land surface using remote sensing. A buffer (a vector shape mask) is created with a specified size and shape (shape of the parcel) and then used as a mask to denote the parcel. Pixels, located in this parcel are processed and classified. Machine Learning (ML) models are used to classify crop pixels in image of the parcel. A per-parcel approach here is when individual pixels of a given parcel are not used, instead, individual pixels are statistically precomputed by aggregating functions and the results of these calculations are used as features for ML models to classify the crops. Classification is done for a whole parcel instead of individual pixels, located in this parcel. Another possibility would be to use an object-oriented approach with the use of segmentation. After segmentation, these segments are further processed and analyzed with the use of ML [7].

In this paper a proposal of per parcel classification is presented. Pixels of each parcel are preprocessed with aggregating functions (Sect. 4.1) and new features are constructed. A special focus for the results of classification accuracy is on the borderline pixels comparison: inclusion and exclusion of these borderline pixels. Two ML models are used: Random Forest (RF) and Support Vector Machines (SVM). The rest of this paper is organized as follows. Related works, Borderline pixels detection, Datasets, Results and Discussion.

2 Related Works

Automatic crop classification (with the use of Machine Learning (ML) models) is a well discussed topic in the literature [2, 7–13]. Badhwar [8] in 1984 used Landsat images to classify corn and soybean crops. He used Landsat satellite image acquisitions. The crop proportion estimated using this technique yielded results only slightly inferior to those obtainable when full season acquisition data were available. It is thus possible to use reduced temporal range in order to successfully classify the crops.

Siachalou et al. [9] uses Hidden Markov models to classify croplands in a Mediterranean area. Class labels (or crop types) were gathered from LPIS, an EU integrated system for croplands administration. LPIS is used in the form of georeferenced vector shapes (polygons) to mask relevant areas of croplands to classify. A pixel-based classification is used, and it is stated, that there are extended classification errors along the parcel's borderline. A suggestion to overcome this drawback is proposed, which consists of the incorporation of vegetation dynamics and the implementation of image enhancement techniques.

Zhang et al. [10] discuss a detection of land cover changes using CNN-based classification. Beijing-2 satellite images from 2017 are used along with land cover GIS maps. He focusses on the improvement of the land cover classification performance for very high resolution (VHR, 0.5 m pixel size) images.

Su et al. [11] states, that due to decreased spatial resolution, a pixel would have more mixed ground information. This would result it over- or underestimating the degree of vegetation. Pixels near borderline thus may represent mixed vegetation content (trees, bushes, crop) or even parcel's access roads. Johnson states, that various buffer sizes. In his work, a buffer is a circular shape, and it is used as a mask to denote pixels for processing. Pixels laying inside this shape are processed and analyzed. Change in a buffer size can lead to variability in the results. He investigated the impacts of buffer size and sensor spatial resolution on identified associations between NDVI and health outcomes. And the identified health outcomes were similar across sensors of different spatial resolutions, but a mean trend was identified in bigger buffer size being associated with greater health outcome.

Kurucu [12] propose the use of Rapid Eye and SPOT satellite images to monitor the crop pattern in parcel level and perform the farmer statement checks (to verify their statements in the Farmer Registration System). Although, no machine learning was used, the verification was carried out by GIS laboratory experts employed in Soil Department of Agricultural Faculty of Ege University. Satellite images, of the relevant crop parcels were reviewed and statements of the farmers checked.

After the launch of Sentinel satellite in 2015, there is an effort to combine Landsat and Sentinel images to obtain more dense set of observations. Multimodal dataset (Landsat and Sentinel images) crop classification is proposed by Johnson and Mueller [2]. Johnson uses Random Forest model to improve the classification accuracy by combining Landsat and Sentinel using sophisticated methodology, that included composite images. Class labels are used from Cropland Data Layer (CDL).

An aggregated form of data representation is widely proposed in the literature [7, 9, 12, 14, 15]. An aggregated data form can help to obtain more robust representation when comparing to individual pixels crop representation of a given parcel. Additionally, it can lead to classification improvement in the case of use of pixels of decreased resolution [11] and borderline pixels classification errors [9]. Blaschke et al. [7] states, that per-parcel classification (aggregated form) is positively contributing to the classification of remote sensing imagery of high to medium geometric resolution. In this paper, Sentinel 2 where highest resolution of 10 m is used, which is a medium geometric resolution (Beijing-2 satellite images having 1 m [10] and MODIS satellite having 250 m [13]). Aggregation reduces variability of individual pixels and can be important for crop classification on smaller areas or parcels.

For the extent of this paper, individual pixels are not used, instead, statistical computations are used (aggregation) on pixels and the results of these functions are used as features for ML in classification.

3 Borderline Pixels Detection

A preprocessing stage, detecting borderline pixels is performed. In this stage, pixels laying at parcel's borderline path are detected. Two different settings are used to detect these borderline pixels. First is, the evaluation by a central point of the pixel. If this central point lies outside of the parcel, then this pixel is not in intersection with parcel's borderline path and is excluded from the further processing and from the dataset. Second setting use four corner points of the pixel. Only if all these four points are located inside parcel's polygon, then this pixel is considered as intersecting pixel and is included, otherwise this pixel is excluded from the processing and dataset.

3.1 Evaluation E_1 by a Central Point

Let x_0, y_0 be geospatial coordinates of the upper left corner of the image, p_w, p_h width and height of each pixel in the image in meters.

For each pixel p and its position x, y, $x \geq x_0$ and $y \geq y_0$ a coordinates of a central point $c_{(x,\, y)}$ can be calculated using the formulas:

$$o_{c(x)} = ((p_x \times p_w) + x_0) + \frac{p_w}{2} \quad (1)$$

$$o_{c(y)} = \left(\left(p_y \times p_h\right) + y_0\right) + \frac{p_h}{2} \quad (2)$$

where $o_{c(x)}$ and $o_{c(y)}$ are x and y spatial coordinates of the central point o_c of the pixel p. With the use of these equations, we find the half of the pixel size, plus the spatial shift of the upper left corner pixel of the image. This shift is necessary in order to correctly calculate geospatial coordinates for the points of the pixels in the image. When the spatial coordinates x, y of the point o_c are known, one can evaluate if a given central point o_c of the pixel p is located within a given geometry G of the crop parcel:

$$o_{c(x,\, y)} \text{ within } G \quad (3)$$

where G is a two-dimensional geometrical shape defined by an ordered pair (x, y) of points p such that $G = \{p_1, p_2, \ldots, p_n\}$ where n is the number of points and x, y are spatial coordinates in two-dimensional space. Function within evaluates whether a point $o_{c(x,\, y)}$ is located in geometry G[1]. For extent of this paper, this evaluation is denoted as E_1.

[1] https://docs.djangoproject.com/en/4.0/ref/contrib/gis/geos/.

3.2 Evaluation E_2 by the Four Corner Points

In order to exclude borderline pixels of the parcel's geometry *G*, where anyone of the pixel's four corners lie outside geometry *G*, we first calculate spatial coordinates of the four corner points of the given pixel *p*, namely top left $o_{tl(x,\ y)}$, top right $o_{trl(x,\ y)}$, bottom right $o_{br(x,\ y)}$ and bottom left $o_{bl(x,\ y)}$ corner points of the pixel *p*. Their spatial coordinates *x*, *y* are calculated using the following equations, for leftmost *x* and topmost *y* ($o_{tl(x,\ y)}$):

$$o_{tl(x)} = (p_x \times p_w) + x_0 \tag{4}$$

$$o_{tl(y)} = (p_y \times p_h) + y_0 \tag{5}$$

For rightmost *x* and bottommost *y* ($o_{br(x,\ y)}$):

$$o_{br(x)} = ((p_x \times p_w) + x_0) + p_w \tag{6}$$

$$o_{br(y)} = ((p_y \times p_h) + y_0) + p_h \tag{7}$$

Remaining two points top right (o_{tr}(x, y)) and bottom left (o_{bl}(x, y)) are combinations of Eqs. (4), (7) and (6), (5). Using Eqs. (4) to (7) to calculate four corner points and then applying Eq. (3) on all four corner points, we can exclude borderline pixels of the parcel if these corner points lie outside of the geometry *G*. In other words, only pixels having all their four corner points located within a given geometry *G*, ale approved for further processing. Pixels p that lies within geometry *G* construct a vector *V* of pixels p such that $V = (p_1, \dots p_n)$ where *n* is the number of pixels of given geometry *G* that are located within this geometry, because four corner points of these pixels are located within this geometry. This evaluation we denote as E_2.

3.3 Visualization of the Principle of E_1 and E_2

The goal of E_2 is a) to exclude the parcel's borderline pixels and b) pixels that are laying around a borderline according to calculations. E_1 is here for a comparison and to evaluate the purpose of E_2, regarding the impact on the accuracy of the crop classification, because it is possible that, due to decreased spatial resolution (10 m), borderline pixels represent other objects than the crop of the parcel, for example bushes, trees or parcel's access roads or even a public road. Inclusion of these pixels may contaminate the dataset and therefore cause loss of accuracy of given parcel's crop classification. Even if the features used are not directly raw pixels, instead, an aggregated features of the pixels are used, a negative impact on the accuracy could still be present.

Figure 1 displays a sample parcel as seen on RGB satellite image (a) and the same parcel using Sentinel 2 band, pixel size is 10 m (b). Next two images ((c) for E_1 and (d) for E_2) show the difference of pixel selection. Area around a borderline is not very useful for crop classification. In this area there can be a parcel's access road (used by a harvester), bushes, trees or other types of objects or even other types of crops. Therefore, the usage of this area's pixels can lower the AOI crop classification accuracy.

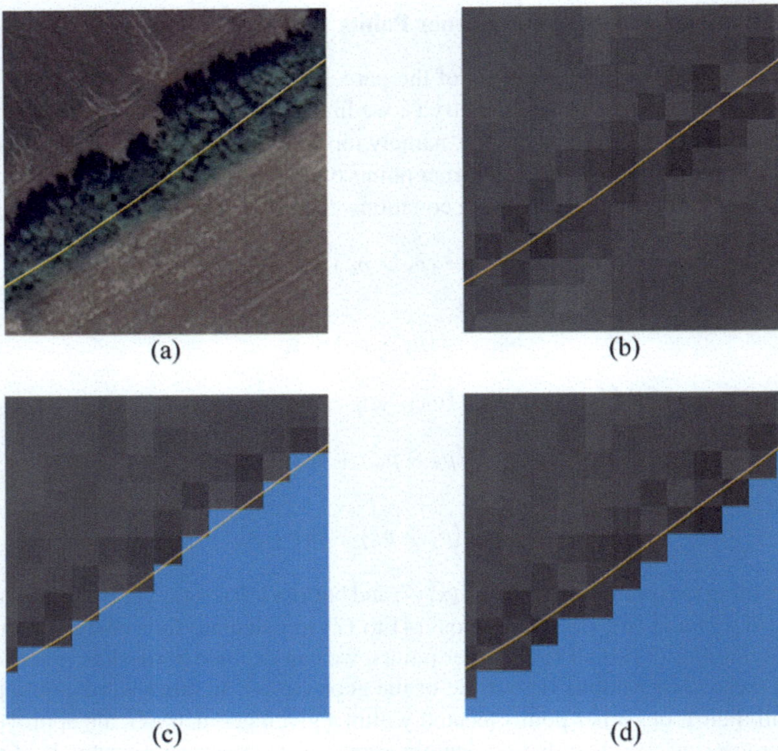

Fig. 1. An illustration of proposed evaluations in practice. From top left (a) Google Maps satellite image of the area with a yellow line denoting the parcel (extent of geometry G) located in the bottom half. There are some trees or bushes visible near or at the parcel's borderline path. Top right (b) the same parcel displayed using a Sentinel 2 SR satellite image (monthly median of Band 8 for April 2018). Bottom left (c) the same as (b), but parcel's pixels are blue masked according to E_1. Bottom right (d) pixels blue masked according to E_2.

4 Dataset

A single-type satellite images are used, Sentinel 2, level 2 (surface reflectance). Location is a southwestern part of Slovakia called "Podunajská nížina". Temporal period is from April 2018 to August 2018 and cloud occupied pixels are partially purged by using monthly temporal compositing, where for each month a single composite image is created by using an aggregation (median) on temporal pixel basis. The resulting data consists of five images representing five months, each image having 12 bands. All 12 bands are used in classification. Crop class labels in the georeferenced vector polygonal form are obtained with the help of National Agriculture and Food Centre[2]. These crop labels denote crop types for parcels.

Original class distribution was strongly imbalanced with a strong representation of winter wheat, corn and grassland (Table 1). Potato was underrepresented with only seven

[2] https://www.vupop.sk/eng/index.php.

parcels. A preprocessing was performed. The idea here was to split large enough, but underrepresented crop parcels into more sub-parcels in order to compensate for the crop's underrepresentation. Parcels with a strong representation (crop classes with a count of parcels >550) were also treated. A threshold at 550 was chosen and crops with more parcels than this threshold were randomly selected and purged out of the dataset. Table 2 displays the class distribution after this preprocessing. However, a potato class is still underrepresented and the results in the classification output for this crop, for the extent of this paper, are ignored.

Table 1. Original crop distribution in dataset.

Crop	Train (80%)	Test (20%)	Total
Corn	944	237	1181
Grassland	965	242	1207
Potato	5	2	7
Rapeseed	530	132	662
Soybean	68	17	85
Spring barley	372	93	465
Sugar beet	89	22	111
Sunflower	388	96	484
Winter wheat	1344	336	1680

Table 2. Crop distribution in dataset after preprocessing.

Crop	Train (80%)	Test (20%)	Total
Corn	440	110	550
Grassland	440	110	550
Potato	17	5	22
Rapeseed	440	110	550
Soybean	440	110	550
Spring barley	372	93	465
Sugar beet	440	110	550
Sunflower	387	97	484
Winter wheat	440	110	550

A vector V is created for each parcel's pixels p and satellite band in the form $V = (p_1, p_2, p_n)$, where n is the number of pixels for a given parcel (according to E_1 or E_2) and satellite image band. A spatial aggregation is then performed for each vector V.

4.1 Feature Collection

The following aggregating functions are performed: minimum, maximum, first, second and third quartile and arithmetic mean. Resulting numbers of these aggregating functions are then used as features for machine learning. All these features are calculated for each one of 12 bands. Additional features include the area of the parcel's geometry G_a calculated in square meters and pixel-based difference vectors D_v and $|D_v|$ constructed with the following equations:

$$D_v = p_1 - p_2, p_2 - p_3, \ldots, p_{n-1} - p_n, \forall p \in V \tag{8}$$

$$|D_v| = (|p_1 - p_2|, |p_2 - p_3|, \ldots, |p_{n-1} - p_n|), \forall p \in V \tag{9}$$

5 Results

Following tables present the results of the crop classification accuracy for rules E1 (Table 3 and Table 4) and E2 (Table 5 and Table 6). There is also a comparison of the results in Table 7 and Table 8 by subtraction $E_2 - E_1$ results in the table cells. Two tables are given for Random Forest and SVM respectively.

5.1 Results for E_1

Table 3. Classify results (accuracy by crop and month) of Random Forest model. Pixel processing according to E_1. Best results are highlighted. Note that potato class results should be ignored due to its underrepresentation in this dataset.

Crop	April	May	June	July	August	Max
Corn	50.4	62.8	**90.4**	**91.3**	88.9	91.3
Grassland	**92.2**	85.3	**93.9**	89.2	83.3	93.9
Potato	100	100	100	100	100	100
Rapeseed	**97.2**	**98.1**	93.6	83.5	66.3	98.1
Soybean	86.7	90.6	**97.3**	**98.2**	**96.5**	98.2
Spring barley	**83**	74.2	64.8	65.3	38.2	83
Sugar beet	88.2	86.1	**97.3**	**97.3**	**99.1**	99.1
Sunflower	53.3	64	**94.3**	**93.3**	79.2	94.3
Winter wheat	**92.6**	**90.7**	72.1	69.2	45.9	92.6

Table 4. Classify results of SVM model.

Crop	April	May	June	July	August	Max
Corn	34.5	41.3	**97.1**	90.4	86.3	97.1
Grassland	**90.8**	77.3	87.9	81.1	87.4	90.8
Potato	0	0	0	0	0	0
Rapeseed	**96.3**	**98.1**	**96.1**	82.5	55	98.1
Soybean	52.6	55.6	**98.1**	**99**	95.1	99
Spring barley	**75.7**	68.5	57.1	67.1	35.8	75.7
Sugar beet	47.1	55	**98.2**	**96.3**	91.4	98.2
Sunflower	31.4	62.3	**88.2**	83.3	79.4	88.2
Winter wheat	**95.2**	91.6	68.3	61.8	39	95.2

5.2 Results for E_2

Table 6. Classify results of SVM model.

Crop	April	May	June	July	August	MAX
Corn	40.8	41.4	**97.2**	89	84.7	97.2
Grassland	**90.3**	79.6	89	85.2	88.6	90.3
Potato	0	0	0	0	0	0
Rapeseed	94.5	**98.1**	95.2	83.8	50.5	98.1
Soybean	55.3	56.3	**99**	**98.1**	92.2	99
Spring barley	**80**	68.8	58.1	63.4	36	80
Sugar beet	57.1	61.5	**97.3**	**95.3**	89.4	97.3
Sunflower	43.1	54.4	**89.4**	83.2	76.8	89.4
Winter wheat	**96.1**	92.5	68.6	60	41.1	96.1

Table 5. Classify results (accuracy by crop) of Random Forest model. Only pixels satisfying our rule E_2 are included. Best results are highlighted. Note that Potato class results should be ignored due to that Potato class having too few samples.

Crop	April	May	June	July	August	MAX
Corn	50.8	64.1	**93.6**	**91.3**	89.5	93.6
Grassland	**90.5**	89.4	**92.2**	87.5	85.3	92.2
Potato	100	100	100	100	100	100
Rapeseed	**97.2**	**99**	93.5	82	61.5	99
Soybean	92.4	90.4	**98.2**	**97.3**	**96.5**	98.2
Spring barley	**82.3**	74	62.1	71.9	37.3	82.3
Sugar beet	89.2	82.5	**98.2**	**98.2**	**98.2**	98.2
Sunflower	53.3	61.1	**94.4**	91.3	77.6	94.4
Winter wheat	**90.7**	**91.6**	74.6	67	40.6	91.6

5.3 A Comparison of E_1 and E_2 Results

Table 7. Difference Δ of Random Forest crop classification accuracies between E_1 and E_2 pixel processing.

Crop	Δ April	Δ May	Δ June	Δ July	Δ August
Corn	+0	+1	+3	+0	+1
Grassland	-2	+4	-2	-2	+2
Potato	+0	+0	+0	+0	+0
Rapeseed	+0	+1	-0	-2	-5
Soybean	+6	-0	+1	-1	+0
Spring barley	-1	-0	-3	+7	-1
Sugar beet	+1	-4	+1	+1	-1
Sunflower	+0	-3	+0	-2	-2
Winter wheat	-2	+1	+3	-2	-5
Σ monthly	+3	+0	+3	-1	-11

Table 8. Difference Δ of SVM crop classification accuracies between E_1 and E_2 pixel processing.

Crop	Δ April	Δ May	Δ June	Δ July	Δ August
Corn	+6	+0	+0	-1	-2
Grassland	-1	+2	+1	+4	+1
Potato	+0	+0	+0	+0	+0
Rapeseed	-2	+0	-1	+1	-5
Soybean	+3	+1	+1	-1	-3
Spring barley	+4	+0	+1	-4	+0
Sugar beet	+10	+6	-1	-1	-2
Sunflower	+12	-8	+1	-0	-3
Winter wheat	+1	+1	+0	-2	+2
Σ monthly	+34	+3	+3	-4	-10

6 Discussion

A direct comparison of E_1 and E_2 based on a simple difference Δ of the resulting accuracies of classification with the use of RF and SVM reveals only a few notable improvements for E_2 and only for SVM.

For RF, the best classification improvement is for soybean crop in April (+6 pp) and spring barley in July (+7 pp). A negative impact on the classification accuracy can be observed for the rapeseed and winter wheat in August (both −5).

For SVM (Table 8), sugar beet and sunflower are among two most positively affected by the change of the preprocessing from E_1 to E_2. There is a notable (more than 10 percentual points) increase in the classification accuracy, but only in April. However, a negative change in the classification accuracy is for the sunflower in May (−8 percentual points (pp)). Another notable negative change is for the rapeseed and month August (−5 pp).

Direct comparison of RF (Table 7) and SVM (Table 8) results reveals, that the improvement for sugar beet and sunflower (in April) for SVM model is not observed for RF model. Relevant table (Table 7) crop cells of RF are unaffected. However, a negative change is the same for the rapeseed in August (−5). A completely different behavior between SVM and RF is for the sugar beet in May (−4 and +6) and spring barley in July (+7 and −4) respectively. A monthly sum (Σ monthly in the Crop column) of differences reveals that, for SVM and RF, the best is April and worst August. The cumulative (sum per month) improvement in the classification accuracy of RF is not discovered, but the increased cumulative loss is still present in August. In that case the worst are rapeseed and winter wheat (−5 percentual points each). A loss in the classification accuracy is the same for SVM and RF for the rapeseed crop in August.

Based on these results it would be interesting to evaluate the impact on accuracy if the distance from the borderline is further increased. Except for notable improvement in SVM model for sugar beet and sunflower (April), the reason of reduced classification accuracy improvements in E_2 pixel processing could be that an aggregation of the pixels is used. If the classification is performed on a pixel-wise basis (e. g. individual pixels are preserved for the classification), there can be a further increase or loss in the classification accuracy, when E_2 pixel processing is used.

Acknowledgement. This work was supported by following grants. The Scientific National Grand Agency under the contract no. VEGA 2/0155/19 and the Slovak Research and Development Agency under the Contract no. APVV-19-0220.

References

1. Landsat Collection 2 Level-2 Science Products. https://www.usgs.gov/landsat-missions/landsat-collection-2-level-2-science-products. Accessed 23 Feb 2020
2. Johnson, D.M., Mueller, R.: Pre-and within-season crop type classification trained with archival land cover information. Remote Sens. Environ. **264**, 112576. https://doi.org/10.1016/j.rse.2021.112576
3. Using new imaging technologies to monitor the Common Agricultural Policy: steady progress overall, but slower for climate and environment monitoring. https://op.europa.eu/webpub/eca/special-reports/new-tech-in-agri-monitoring-4-2020/en/. 23 Feb 2022
4. Flight preparation – How to plan your flight mission for best Digital Orthophoto. https://www.3dsurvey.si/upload/files/002%202D%20Orthophoto%20instructions_v002.pdf. Accessed 23 Feb 2020
5. OpenDroneMap, Tutorials. https://docs.opendronemap.org/tutorials/. Accessed 23 Feb 2020
6. Problems with output from Parrot Sequoia RGB. https://community.opendronemap.org/t/problems-with-output-from-parrot-sequoia-rgb/10216/30. Accessed 23 Feb 2022
7. Blaschke, T., Lang, S., Lorup, E., Strobl, J., Zeil, P.: Object-oriented image processing in an integrated GIS/remote sensing environment and perspectives for environmental applications. Environ. Inf., Plan. Polit. Public **2**, 555–570
8. Badhwar, G.B.: Automatic corn-soybean classification using Landsat MSS data. II. Early season crop proportion estimation. Remote Sens. Environ. **14**, 31–37. https://doi.org/10.1016/0034-4257(84)90005-1
9. Siachalou, S., Mallinis, G., Tsakiri-Strati, M.: A hidden Markov models approach for crop classification: linking crop phenology to time series of multi-sensor remote sensing data. Remote Sens. **7**, 3633–3650. https://doi.org/10.3390/rs70403633
10. Zhang, C., Wei, S., Ji, S., Lu, M.: Detecting large-scale urban land cover changes from very high resolution remote sensing images using CNN-based classification. ISPRS Int. J. Geo-Inf. **8**, 189. https://doi.org/10.3390/ijgi8040189
11. Su, J.G., Dadvand, P., Nieuwenhuijsen, M.J., Bartoll, X., Jerrett, M.: Associations of green space metrics with health and behavior outcomes at different buffer sizes and remote sensing sensor resolutions. Environ. Int. **126**, 162–170. https://doi.org/10.1016/j.envint.2019.02.008
12. Kurucu, Y., Erden, H.: Control of farmer statements integrated in national farmer register system by remote sensing data. In: Fourth International Conference on Agro-Geoinformatics (Agro-Geoinformatics), pp. 97–100. IEEE (2015)
13. Wardlow, B.D., Egbert, S.L., Kastens, J.H.: Analysis of time-series MODIS 250 m vegetation index data for crop classification in the US Central Great Plains. Remote Sens. Environ. **108**, 290–310. https://doi.org/10.1016/j.rse.2006.11.021

14. Kenyeres, M., Kenyeres, J.: Average consensus over mobile wireless sensor networks: weight matrix guaranteeing convergence without reconfiguration of edge weights. Sensors **20**, 3677. https://doi.org/10.3390/s20133677
15. Kenyeres, M., Kenyeres, J.: Distributed mechanism for detecting average consensus with maximum-degree weights in bipartite regular graphs. Mathematics **9**, 3020. https://doi.org/10.3390/math9233020

Ensembles of Artificial Neural Networks for Smart Grids Stability Prediction

Dorin Moldovan(✉)

Independent Researcher, Cluj-Napoca, Romania
dorin.i.moldovan@gmail.com

Abstract. The prediction of the stability of the smart grids in demand-response contexts is a challenging research problem as the stability is highly influenced by the nature of the energy prosumers. The prediction complexity is addressed in this article using an ensemble of deep learning models. The main contributions of the article are: (1) a review of the literature approaches for the prediction of the smart grids stability, (2) the development of 5 Artificial Neural Network (ANN) models for smart grids stability prediction, (3) the proposal and the comparison of 4 ANN ensembles based on standard heuristics and bio-inspired heuristics, and (4) the comparison of the obtained results with the literature results. The best prediction results were obtained by the Grey Wolf Optimizer (GWO) based ensemble, as follows: 0.950 precision, 0.945 recall, 0.962 accuracy, and 0.947 F1Score.

Keywords: Artificial neural network · Stability · Ensembles · Smart grids · Prediction

1 Introduction

The electricity is expected to contribute up to 40% of the final energy consumption increase in the context of an energy driven economy [1]. Three challenges regarding the electrical grid paradigm, according to the Paris UN climate conference agreement, are: the need to reduce the carbon emissions, the need to decentralize the production of the electrical energy, and the need to increase the electrification of the transportation. The smart grids are a proposed solution to overcome the mentioned challenges.

From an architectural point of view, the smart grids have three types of layers [2]: (1) the top level is represented by the electrical network infrastructure, (2) the middle level considers primary and secondary substations which provide metering and aggregation of information, and (3) the bottom level is the one where the smart meters are installed and where they provide contractual operations and metering.

Smart grids are modernized grids which enable bidirectional flows of energy and which use two-way control and communication capabilities that lead to new applications and functionalities [3]. The main functions of the smart grids [4] are

R. Silhavy (Ed.): CSOC 2022, LNNS 502, pp. 320–336, 2022.
https://doi.org/10.1007/978-3-031-09076-9_30

the analysis of the data, the monitoring, and the control. The Internet of Things (IoT) sensors are installed in different stages of the smart grids in order to track and to manage various network statistics.

The advancement of the smart grids [5] enable the energy consumers to schedule the appliances in order to respond to a variety of demand-response programs. Two of the direct effects of the smart scheduling of the household appliances are the save of money in the case of the customers and the ability of the utility company to reduce the carbon emissions.

The efficiency and the stability of the electric grids are improved by the implementation of the real time response-demand programs and by the renewable energy sources (RES) integration [6]. Moreover, the transition from the traditional grids to the smart grids is needed in order to reduce the fossil fuels use. One solution to reduce the emission of the gases is the replacement of the existing power plants with sources of renewable energy.

In the energy management research field [7], the demand response management has become a research hotspot due to the smart grids' cyber-physical fusion properties [8,9]. Moreover, the utilization of the changes in the usage of energy in order to cope with electricity prices which vary in time has as main objective the reduction of the costs or of other types of interests, as can be seen in [10].

The continuous monitoring of the stability in the case of the power systems/smart grids is essential in order to maintain the stability both under normal and abnormal conditions when the systems are characterized by disturbances [11]. The smart grids stability is highly influenced by the fluctuating nature of the sources of renewable energy, as can be seen in [12], where the authors employ a deep learning model in order to improve the generalized Smart Grid Stability Prediction (SGSP).

The main contributions of the article are presented below:

(1) a critical review of the ensembles of ANNs for classification problems and of the literature approaches for smart grids stability prediction;
(2) the proposal of 5 ANN models for the prediction of the smart grids stability;
(3) the proposal of 4 ANN ensembles based on: the average heuristic, Particle Swarm Optimization (PSO) [13], Grey Wolf Optimizer (GWO) [14], and Chicken Swarm Optimization (CSO) [15,16];
(4) the comparison of the obtained results with the results from previous research studies;

The article has the following structure: Sect. 2 presents the research background, Sect. 3 presents the high level view of the smart grids model considered in the experiments, Sect. 4 presents the method based on ANNs for the prediction of the stability of the smart grids, Sect. 5 presents the experimental results, and Sect. 6 presents the conclusions.

2 Research Background

2.1 Ensembles of ANNs for Classification Problems

The authors of [17] considered the application of an ensemble of Recurrent Neural Networks (RNNs) for the detection of the IoT cyber attacks. Their approach integrated a set of Long Short-Term Memory (LSTM) models. The models were merged using a Decision Tree (DT) in order to obtain an aggregated output in the final stage.

The authors of [18] considered an ensemble of RNNs for the forecasting of the wind speed. Different architectures were considered in order to build the RNNs such as the LSTM, the Gated Recurrent Unit (GRU), and the dropout layer. The proposed model was compared to other soft computing models, and the results showed the superiority and the feasibility of the proposed model.

In [19], the authors considered an ensemble of ANNs in order to forecast the solar photovoltaic (PV) power. In that approach, the data was modelled in 15-minute averages time steps and the forecasts were obtained ahead from 15 to 120 min. The range achieved by the forecast skills was from 9.79% to 23.75%.

Finally, the approach presented in [20] considered the application of ensembles of ANNs for the robust time series forecasting, using 4 different datasets as experimental support.

2.2 Literature Approaches for Smart Grids Stability Prediction

The approach presented in [21] considered the application of machine learning techniques for the classification of the smart grids stability, using a method which consisted of 3 steps: (1) features extraction, (2) features selection, and (3) classification. The features were extracted using tsfresh [22] from Python. The features were selected using Multivariate Adaptive Regression Splines (MARS) [23], a binary version of PSO, and a binary version of Kangaroo Mob Optimization (KMO) [24]. Finally, the four different classifiers that were applied were: Random Forest (RF), Gradient Boosted Trees (GBT), Logistic Regression (LR), and Multilayer Perceptron Classifier (MPC).

The approach presented in [25] considered the application of a modified binary version of Horse Optimization Algorithm (HOA) for features selection in order to classify the stability of a smart grid. Similar to the approach from [21], this approach had three steps. The features were extracted using the tsfresh library, the processed features were then selected using the modified binary version of HOA, and the smart grid stability was classified using a RF classification algorithm. The results obtained using the modified binary version of HOA were compared to the results obtained using modified binary versions of CSO and Cat Swarm Optimization Algorithm (CSOA) [26].

Another approach, the one from [27], used an improved version of KMO for the classification of the smart grids stability. Compared to the approaches from [21] and [25], that approach was different as it did not consider a features extraction step and a features selection step. The smart grids data was classified using LR such that the tolerance parameter (tol) and the regularization parameter (reg) were tuned using the Improved Kangaroo Mob Optimization (IKMO).

In [28], the authors considered an ensemble based method. The flowchart was as follows: the loaded data was processed using feature engineering, the data was divided in training and testing sets, the proposed Stacking Ensemble Classifier (SEC) consisted of Categorical Boosting (Catboost) and eXtreme Gradient Boosting Machine (XGBoost), and the applied metaclassifier for the generating of the forecasting results was the Light Gradient Boosting Machine (LGBM). The proposed SEC results were compared to the results obtained by LGBM, Catboost, XGBoost, Extra Trees (ET), Multilayer Perceptron (MLP), RF, and Support Vector Classifier (SVC).

The authors of [29] applied the Voting Margin (VM) scheme based on k-Nearest Neighbor (kNN) on several datasets with two classes, one of them being the Smart Grid Stability Data. That approach considered all 10000 elements, 13 features, and a value of k equal to 19. The obtained accuracy was 94.49%.

3 Smart Grids Model

The smart grids model considered in this article is based on the principles and the assumptions presented in [30] and [31]. A few assumptions are the following:

(1) the stability of the smart grids is influenced positively by the presence of a consumer that reacts slowly;
(2) the smart grids are destabilized when there are delays greater than a threshold which is equal to 8.7 s;
(3) the smart grids are unstable when the delay value between a price change is 8 s and the averaging time necessary to measure the price signal is 2 s;

The final equation of the smart grids dynamics [30], in terms of the j and k participants of the smart grids, is:

$$\frac{d^2\Theta_j}{dt^2} = P_j - \alpha_j \times \frac{d\Theta_j}{dt} + \sum_{k=1}^{N} K_{jk} \times sin\,(\Theta_k - \Theta_j) - \\ -\frac{\gamma_j}{T_j} \times [\Theta_j(t-\tau_j) - \Theta_j(t-\tau_j-T_j)] \quad (1)$$

such that the elements of the final equation have the definitions:

- Θ_j = generator/prosumer j's rotor angle;
- t = time;
- P_j = produced/consumed mechanical power;
- α_j = participant j's damping constant;
- N = participants number;
- K_{jk} = coupling strength value;

- Θ_k = participant k's rotor angle;
- γ_j = a value proportional to the elasticity's price;
- T_j = time of averaging;
- τ_j = reaction time;

4 Method Based on ANNs Ensembles for Smart Grids Stability Prediction

Next are presented the steps for the prediction of the smart grids stability, considering the Smart Grids Stability Data from the UCI machine learning repository [32] as reference.

4.1 Smart Grids Data Preprocessing

The data is preprocessed as follows: the first row of the table is removed since it contains the names of the columns, the column **stab** is removed, the elements of the column **stabf** are converted to numerical values such that the value **unstable** is converted to 0 and the value **stable** is converted to 1, and the data is normalized.

4.2 Cross Validation

The data is shuffled and then divided in 5 folds of data with equal numbers of samples. Cross validation is performed 5 times such that each time the testing data is represented by one fold of data and the training data is represented by the other 4 folds of data. The final evaluation results are averages of the results obtained for each fold of data.

4.3 ANNs Ensemble High-Level View

Table 1 presents the structure of the ANNs ensemble for the classification of the smart grids stability. The ANNs are adapted after the ones presented in [33].

Table 1. ANNs ensemble structure

Number of nodes	ANN1	ANN2	ANN3	ANN4	ANN5
Input layer	12	12	12	12	12
1st hidden layer	24	24	24	24	24
2nd hidden layer	24	36	48	48	48
3rd hidden layer	24	24	24	48	36
4th hidden layer	12	12	12	12	12
Output layer	1	1	1	1	1

All of these five ANNs have the following structure: one input layer, four hidden layers, and one output layer. The number of input nodes is equal to 12 in all five cases because this is the number of features for each smart grids data sample. The number of outputs is equal to 1 because the prediction is a numerical value which can be either 0 (unstable) or 1 (stable). The kernel initializer is 'uniform' for all ANNs. The activation is 'relu' for all layers, with the exception of the output layers for which the activation is 'sigmoid'. The ANNs are compiled using the 'adam' optimizer, the 'binary_crossentropy' as loss, and the accuracy for the metrics. The number of epochs was set to 200 for all five ANNs.

A different weight is assigned to each ANN such that the final prediction of the ensemble is computed considering the average of the predictions of each ANN.

First, the average of the predictions is computed using the formula:

$$p = \sum_{i=1}^{5} \frac{w_i}{w_1 + w_2 + w_3 + w_4 + w_5} \times p_i \tag{2}$$

where:

- w_1, w_2, w_3, w_4, w_5 are the weights assigned to ANN1, ANN2, ANN3, ANN4, and ANN5, respectively;
- p_1, p_2, p_3, p_4, p_5 are the predictions returned by ANN1, ANN2, ANN3, ANN4, and ANN5, respectively;
- p is the average of the predictions;

Second, the final prediction of the ANNs ensemble is computed using the formula:

$$p_{final} = \begin{cases} 0 \text{ if } p < \frac{1}{2} \\ 1 \text{ otherwise} \end{cases} \tag{3}$$

4.4 Proposed Ensembles

This section describes the two types of ensembles considered in the paper. The first type is a classical one and is based on the average heuristic. The second type considers bio-inspired heuristics in order to determine the optimal values of the weights for the predictions returned by each ANN.

Average Ensemble. In this case, all weights are equal. In particular, since the total number of weights is equal to 5, the following relation is true:

$$w_1 = w_2 = w_3 = w_4 = w_5 = \frac{1}{5} \tag{4}$$

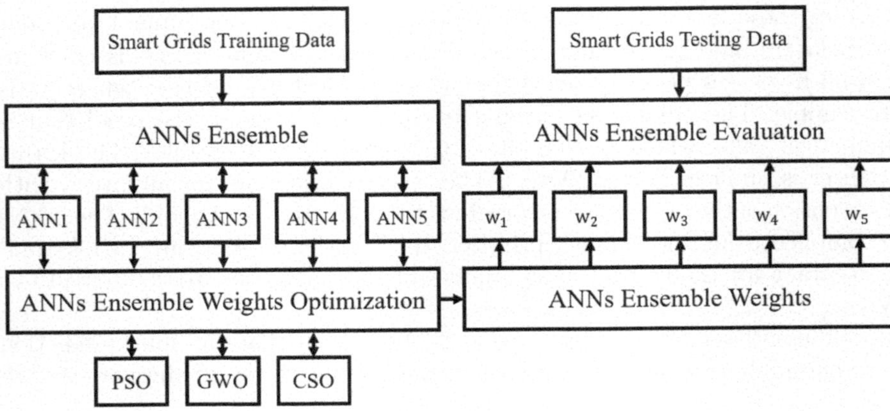

Fig. 1. High-level view of the method based on bio-inspired heuristics for ANNs ensemble weights optimization

Bio-Inspired Heuristics Ensembles. The three types of bio-inspired heuristics considered in this article are the ones based on PSO, GWO, and CSO. Figure 1 presents the high-level view of the method based on bio-inspired heuristics for the optimization of the weights of the ANNs ensemble. The proposed method is adapted after the methods presented in [34] and [35]:

The input of the method is represented by the **Smart Grids Training Data**. This data is processed by the **ANNs Ensemble** which consists of **ANN1**, **ANN2**, **ANN3**, **ANN4**, and **ANN5**, respectively. The three bio-inspired heuristics considered in the **ANNs Ensemble Weights Optimization** phase are **PSO**, **GWO**, and **CSO**.

In all three cases, the positions of the individuals are restricted to take values from a range of values $[X_{min}, X_{max}]$ such that X_{min} represents the minimum possible value for the positions and X_{max} represents the maximum possible value for the positions.

Next are presented the steps for computing the **ANNs Ensemble Weights**:

Step 1. The initial weights $(W_1, W_2, W_3, W_4, W_5)$ are set to the values of the positions of the individuals such that they correspond to ANN1, ANN2, ANN3, ANN4, and ANN5, respectively.

Step 2. For each initial weight, the value X_{max} is added, obtaining the new intermediary weights:

$$W_i' = W_i + X_{max} \tag{5}$$

where $i \in \{1, 2, 3, 4, 5\}$.

Step 3. If $\sum_{i=1}^{5} W_i' = 0$, then the weights w_i where $i \in \{1, 2, 3, 4, 5\}$ are all set to the value $\frac{1}{5}$. Otherwise, the weights are computed using the formula:

$$w_i = \frac{W_i'}{\sum_{i=1}^{5} W_i'} \tag{6}$$

where $i \in \{1, 2, 3, 4, 5\}$.

Using the resulted weights w_1, w_2, w_3, w_4, and w_5, the predictions of the ensemble are evaluated using the equation (3).

The objective function used in the **ANNs Ensemble Evaluation** phase considers the formula:

$$OF(X) = \frac{1}{2 \times N} \times \sum_{i=1}^{N} (p_{final,i} - z_i)^2 \tag{7}$$

such that N is the number of samples and for each $i \in \{1, ..., N\}$, $p_{final,i}$ is the prediction returned by the ensemble and z_i is the actual label of the sample from the **Smart Grids Testing Data**.

4.5 Evaluation

Figure 2 presents the confusion matrix for the evaluation of the RNNs ensembles based method considered in the prediction of the smart grids stability in terms of: True Positives (TP), False Positives (FP), False Negatives (FN), and True Negatives (TN).

	Actual Stable	Actual Unstable
Predicted Stable	TP	FP
Predicted Unstable	FN	TN

Fig. 2. Confusion matrix for the evaluation of the RNNs ensembles for the prediction of the smart grids stability

The evaluation metrics are combinations of the elements of the confusion matrix [17]:

(1) *Precision*: A metric which indicates the number of smart grids stability data samples which were correctly classified:

$$Precision = \frac{TP}{TP + FP} \tag{8}$$

(2) *Recall*: The ratio of the smart grids stability data samples which were correctly classified:

$$Recall = \frac{TP}{TP + FN} \tag{9}$$

(3) *Accuracy*: The number of data samples which were correctly identified by a classifier divided by the total number of data samples:

$$Accuracy = \frac{TP + TN}{TP + FP + TN + FN} \tag{10}$$

(4) *F1Score*: The harmonic mean between the *Precision* and the *Recall*:

$$F1Score = 2 \times \frac{Precision \times Recall}{Precision + Recall} \tag{11}$$

5 Experimental Results

The experiments were performed on a machine with the properties: 64-bit Operating System, 8.00 GB installed RAM, and Intel(R) Core(TM) i7-7500U CPU @ 2.70GHz processor.

5.1 Smart Grids Data Description

The data which is used in the experiments is the Smart Grid Stability Data from the UCI Machine Learning Repository [32]. The dataset consists of 10000 data samples and each data sample is characterized by 14 attributes. The dataset is not characterized by missing values.

Table 2 presents a detailed description of the attributes of the experimental dataset.

Table 2. Smart grid stability data attributes detailed description

Attribute	Detailed Description	Range
tau1	producer reaction time	$[0.5s, 10.0s]$
tau2	consumer 1 reaction time	
tau3	consumer 2 reaction time	
tau4	consumer 3 reaction time	
p1	producer nominal power	$[-0.5s^{-2}, -2.0s^{-2}]$
p2	consumer 1 nominal power	
p3	consumer 2 nominal power	
p4	consumer 3 nominal power	
g1	producer gamma coefficient	$[0.05s^{-1}, 1.0s^{-1}]$
g2	consumer 1 gamma coefficient	
g3	consumer 2 gamma coefficient	
g4	consumer 3 gamma coefficient	
stab	characteristic equation root's maximal real part	a real value
stabf	system stability label	{stable, unstable}

The characteristic equation root is the equation (1) presented in more details in Sect. 3 and in the original article [30].

The two goal attributes are **stab** and **stabf**. The other 12 attributes are predictive with the exception of p1 which is non-predictive and which is computed as follows:

$$p1 = |p2 + p3 + p4| \tag{12}$$

Figure 3 presents the four-nodes star motif smart grids design, adapted after the electricity flow simulated in [30], for a configuration in which the smart grid is stable:

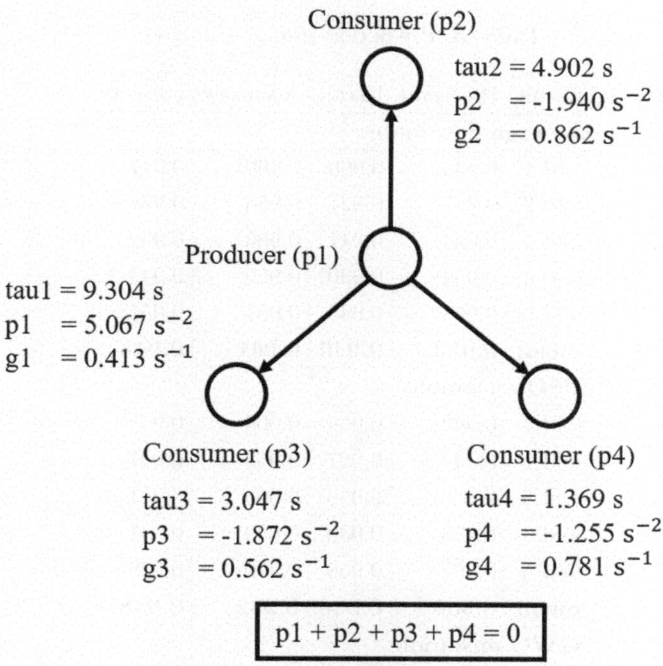

Fig. 3. Four-node star motif smart grids design (adapted after [30])

5.2 Bio-inspired Heuristics Configuration Parameters

The configuration parameters of the bio-inspired heuristics considered in the experiments are presented below:

- **PSO**: 30 particles, 50 iterations, positions range ($[X_{min}, X_{max}]$) = $[-20, 20]$, 5 dimensions, inertia value (w) = 0.900, velocities range ($[V_{min}, V_{max}]$) = $[-12, 12]$, cognitive component constant (c_1) = 2, social component constant (c_2) = 2;
- **GWO**: 30 wolves, 50 iterations, 5 dimensions, positions range ($[X_{min}, X_{max}]$) = $[-20, 20]$;

- **CSO**: 30 chickens, 50 iterations, 5 dimensions, hierarchy update frequency $(G) = 10$, flight range $([FL_{min}, FL_{max}]) = [0.5, 0.9]$, constant for avoiding division by 0 $(\epsilon) = 0.000000001$, roosters percentage $(RP) = 20\%$, hens percentage $(HP) = 60\%$, chicks percentage $(CP) = 20\%$;

5.3 Prediction Results

Table 3 presents the prediction results for each fold and the means of these results.

Table 3. Prediction results summary

Result	Precision	Recall	Accuracy	F1Score
average ensemble				
fold 1	0.934	0.960	0.960	0.947
fold 2	0.947	0.927	0.955	0.937
fold 3	0.964	0.941	0.966	0.952
fold 4	0.954	0.930	0.960	0.942
fold 5	0.968	0.943	0.967	0.955
mean	**0.953**	**0.940**	**0.961**	**0.946**
PSO ensemble				
fold 1	0.942	0.959	0.963	0.951
fold 2	0.947	0.927	0.955	0.937
fold 3	0.946	0.956	0.965	0.951
fold 4	0.953	0.930	0.959	0.941
fold 5	0.962	0.958	0.970	0.960
mean	**0.950**	**0.946**	**0.962**	**0.948**
GWO ensemble				
fold 1	0.942	0.959	0.963	0.951
fold 2	0.947	0.927	0.955	0.937
fold 3	0.946	0.956	0.965	0.951
fold 4	0.953	0.930	0.959	0.941
fold 5	0.964	0.954	0.970	0.959
mean	**0.950**	**0.945**	**0.962**	**0.947**
CSO ensemble				
fold 1	0.934	0.960	0.960	0.947
fold 2	0.947	0.927	0.955	0.937
fold 3	0.964	0.941	0.966	0.952
fold 4	0.954	0.930	0.960	0.942
fold 5	0.968	0.943	0.967	0.955
mean	**0.953**	**0.940**	**0.961**	**0.946**

As can be seen in the table, the PSO and the GWO ensembles returned the best mean results. The values for the precision and accuracy were 0.950 and 0.962, respectively. However, GWO performed slightly better because in the case of the fold 5 it returned a better precision value, a value equal to 0.964. That difference can also be seen in the mean values for recall and F1Score. These values were equal to 0.946 and 0.948 in the case of PSO, and they were equal to 0.945 and 0.947 in the case of GWO.

Another observation is that the average ensemble and the CSO ensemble performed quite similarly, returning the following mean values: 0.953 precision, 0.940 recall, 0.961 accuracy, and 0.946 F1Score.

Table 4 presents the weights returned by each bio-inspired heuristic.

Table 4. Bio-inspired heuristics weights summary

Weight	w_1	w_2	w_3	w_4	w_5
PSO ensemble					
fold 1	0.070	0.212	0.134	0.331	0.253
fold 2	0.145	0.169	0.222	0.254	0.210
fold 3	0.145	0.196	0.386	0.171	0.102
fold 4	0.073	0.121	0.258	0.206	0.342
fold 5	0.0	0.230	0.184	0.275	0.311
GWO ensemble					
fold 1	0.068	0.211	0.136	0.334	0.251
fold 2	0.146	0.173	0.221	0.253	0.207
fold 3	0.151	0.205	0.359	0.180	0.105
fold 4	0.076	0.119	0.259	0.202	0.344
fold 5	0.005	0.225	0.186	0.274	0.310
CSO ensemble					
fold 1	0.199	0.200	0.200	0.200	0.201
fold 2	0.199	0.200	0.200	0.201	0.200
fold 3	0.198	0.198	0.208	0.198	0.198
fold 4	0.199	0.200	0.199	0.200	0.202
fold 5	0.200	0.200	0.200	0.200	0.200

Even though PSO and GWO returned similar results, the weights were different. In the case of CSO, the weights were approximately equal to 0.200 for all five folds.

5.4 Comparison with Literature Results

Table 5 presents a comparison of the best results obtained using the approaches proposed in this article with the previous research results and with the results obtained in other research studies. However, the comparison is not 100% accurate because the dataset was processed differently by each approach and different cross validation configurations were applied.

Table 5. Literature results comparison

Approach	Method	Accuracy
This approach	Average ensemble	0.961
	PSO ensemble	**0.962**
	GWO ensemble	**0.962**
	CSO ensemble	0.961
Moldovan et al., 2021 [36]	12-6-6-6-2	**0.868**
Massaoudi et al., 2021 [28]	LGBM	0.960
	Catboost	0.983
	XGBoost	0.930
	ET	0.955
	MLP	0.972
	RF	0.944
	SVC	0.820
	proposed model	**0.989**
Moldovan, 2021 [27]	Default LR	0.816
	IKMO LR	**0.818**
Moldovan, 2020 [25]	FRESH DBHOA RF	0.903
	FRESH DBCSO RF	0.904
	FRESH DBCSOA RF	**0.905**
Breviglieri, 2020 [33]	12-24-12-1	0.950
	12-24-24-12-1	**0.958**
Liu et al., 2019 [29]	Voting margin scheme	**0.944**
Moldovan et al., 2019 [21]	FRESH all MPC	0.932
	FRESH MARS MPC	0.919
	FRESH BPSOFS MPC	**0.938**
	FRESH BKMOFS MPC	0.915

The current approach performed better than the approach from [36] where a 12-6-6-6-2 LSTM RNN architecture with 30 epochs was used.

The comparison with the method presented in [28] is not accurate, because the authors of that method applied a 10 fold cross validation. The accuracy obtained in this approach was much better compared to the accuracy obtained by the methods based on LGBM, XGBoost, ET, RF, and SVC. However, the methods based on Catboost, MLP, and the proposed model were better.

Compared to the method from [27], the results of this method were better than the results returned by the default LR and by the IKMO LR methods.

The methods presented in [25] which extracted the features using the FRESH algorithm from the tsfresh library, selected the features using Discrete Binary Horse Optimization Algorithm (DBHOA), Discrete Binary Chicken Swarm Optimization (DBCSO), and Discrete Binary Cat Swarm Optimization Algorithm (DBCSOA), and classified the data using RF, were also not better.

The results are comparable to the ones from [33] where a 10 fold cross validation was used. The table presents the results of the best configurations: 12-24-12-1 architecture with 20 epochs and 12-24-24-12-1 architecture with 50 epochs.

Regarding the method based on voting margin scheme proposed in [29], where a 70%:30% ratio between training data and testing data was used, the comparison is also not 100% accurate. The top accuracy of that method was 0.944, which was lower than the accuracy returned by the proposed ensembles.

Finally, the proposed method is better than the methods presented in [21] where the features were extracted using FRESH, various features selection approaches were considered such as all features, MARS, Binary Particle Swarm Optimization Features Selection (BPSOFS), and Binary Kangaroo Mob Optimization Features Selection (BKMOFS), and the data was classified using MPC.

6 Conclusions

This article presented a method based on ensembles of ANNs for the prediction of the stability of the smart grids. The four types of ensembles that were considered were the average ensemble and the ones based on PSO, GWO, and CSO. The obtained results were compared to the results from previous research studies and to the results obtained by other researchers when the same experimental dataset was used. The comparison shows that the obtained results are performing and comparable to the ones obtained by other researchers.

References

1. Martins, J., Strasser, T.I., Sanduleac, M.: Guest editorial: smart meters in the smart grid of the future. IEEE Trans. Ind. Inf. **18**(1), 653–655 (2022). https://doi.org/10.1109/TII.2021.3094182
2. D1.1 architecture and design for use cases (2015). http://www.segrid.eu/wp-content/uploads/2014/11/SEGRID-report-deliverable-1.1-EU-Public-format-V1-01.04.2015.pdf
3. Song, E.Y., FitzPatrick, G.J., Lee, K.B., Griffor, E.: A methodology for modeling interoperability of smart sensors in smart grids. IEEE Trans. Smart Grid **13**(1), 555–563 (2022). https://doi.org/10.1109/TSG.2021.3124490
4. Ghosh, S.: Neuro-fuzzy-based IOT assisted power monitoring system for smart grid. IEEE Access. **9**, 168587–168599 (2021). https://doi.org/10.1109/ACCESS.2021.3137812
5. Rehmana, A.U., et al.: An efficient energy management in smart grid considering demand response program and renewable energy sources. IEEE Access **9**, 148821–148844 (2021). https://doi.org/10.1109/ACCESS.2021.3124557

6. Khan, H.W., et al.: Intelligent optimization framework for efficient demand-side management in renewable energy integrated smart grid. IEEE Access **9**, 124235–124252 (2021). https://doi.org/10.1109/ACCESS.2021.3109136
7. Wan, Y., Qin, J., Yu, X., Yang, T., Kang, Y.: Price-based residential demand response management in smart grids: a reinforcement learning-based approach. IEEE/CAA J. Automatica Sinica **9**(1), 123–134 (2022). https://doi.org/10.1109/JAS.2021.1004287
8. Deng, R., Yang, Z., Chow, M.-Y., Chen, J.: A survey on demand response in smart grids: mathematical models and approaches. IEEE Trans. Ind. Inf. **11**(3), 570–582 (2015). https://doi.org/10.1109/TII.2015.2414719
9. Ding, L., Wang, L.Y., Yin, G.Y., Zheng, W.X., Han, Q.-L.: Distributed energy management for smart grids with an event-triggered communication scheme. IEEE Trans. Control Syst. Technol. **27**(5), 1950–1961 (2019). https://doi.org/10.1109/TCST.2018.2842208
10. Hao, H., Corbin, C.D., Kalsi, K., Pratt, R.G.: Transactive control of commercial buildings for demand response. IEEE Trans. Power Syst. **32**(1), 774–783 (2017). https://doi.org/10.1109/TPWRS.2016.2559485
11. Venayagamoorthy, G.K.: Intelligent sense-making for smart grid stability. In: 2011 IEEE Power and Energy Society General Meeting, pp. 1–3, Detroit, MI, USA (2011). https://doi.org/10.1109/PES.2011.6039876
12. Massaoudi, M., Abu-Rub, H., Refaat, S.S., Chihi, I., Oueslati, F.S.: Accurate smart-grid stability forecasting based on deep learning: point and interval estimation method. In: 2021 IEEE Kansas Power and Energy Conference (KPEC), pp. 1–6, Manhattan, KS, USA (2021). https://doi.org/10.1109/KPEC51835.2021.9446196
13. Kennedy, J., Eberhart, R.: Particle swarm optimization. In: Proceedings of ICNN 1995 - International Conference on Neural Networks, vol. 4, pp. 1942–1948, Perth, WA, Australia (1995). https://doi.org/10.1109/ICNN.1995.488968
14. Mirjalili, S., Mirjalili, S.M., Lewis, A.: Grey wolf optimizer. Adv. Eng. Softw. **46**, 46–61 (2014). https://doi.org/10.1016/j.advengsoft.2013.12.007
15. Meng, X., Liu, Yu., Gao, X., Zhang, H.: A new bio-inspired algorithm: chicken swarm optimization. In: Tan, Y., Shi, Y., Coello, C.A.C. (eds.) ICSI 2014. LNCS, vol. 8794, pp. 86–94. Springer, Cham (2014). https://doi.org/10.1007/978-3-319-11857-4_10
16. Moldovan, D., Slowik, A.: Chicken swarm optimization. In: Swarm intelligence algorithms. A tutorial. Slowik, A. (eds.), pp. 71–84. Taylor & Francis Group (CRC Press), Boca Raton, USA (2020). https://doi.org/10.1201/9780429422614-6
17. Saharkhizan, M., Azmoodeh, A., Dehghantanha, A., Choo, K.-K.R., Parizi, R.M.: An ensemble of deep recurrent neural networks for detecting IoT cyber attacks using network traffic. IEEE Internet Things J. **7**(9), 8852–8859 (2020). https://doi.org/10.1109/JIOT.2020.2996425
18. Cheng, L., et al.: Ensemble recurrent neural network based probabilistic wind speed forecasting approach. Energies **11**(8), 1958 (2018). https://doi.org/10.3390/en11081958
19. de Paiva, G.M., Pimentel, S.P., Marra, E.G., de Alvarenga, B.P., Mussetta, M., Leva, S.: Intra-day forecasting of building-integrated PV systems for power systems operation using ANN ensemble. In: 2019 IEEE Milan PowerTech, pp. 1–5, Milan, Italy (2019). https://doi.org/10.1109/PTC.2019.8810480

20. Krstanovic, S., Paulheim, H.: Ensembles of recurrent neural networks for robust time series forecasting. In: Bramer, M., Petridis, M. (eds.) SGAI 2017. LNCS (LNAI), vol. 10630, pp. 34–46. Springer, Cham (2017). https://doi.org/10.1007/978-3-319-71078-5_3
21. Moldovan, D., Salomie, I.: Detection of sources of instability in smart grids using machine learning techniques. In: 2019 IEEE 15th International Conference on Intelligent Computer Communication and Processing (ICCP), pp. 175–182, Cluj-Napoca, Romania (2019). https://doi.org/10.1109/ICCP48234.2019.8959649
22. Christ, M., Braun, N., Neuffer, J., Kempa-Liehr, A.W.: Time series feature extraction on basis of scalable hypothesis tests (tsfresh - a python package). Neurocomputing **307**, 72–77 (2018). https://doi.org/10.1016/j.neucom.2018.03.067
23. Kartal Koc, E., Bozdogan, H.: Model selection in multivariate adaptive regression splines (MARS) using information complexity as the fitness function. Mach. Learn. **101**(1), 35–58 (2014). https://doi.org/10.1007/s10994-014-5440-5
24. Moldovan, D., Anghel, I., Cioara, T., Salomie, I., Chifu, V., Pop, C.: Kangaroo mob heuristic for optimizing features selection in learning the daily living activities of people with Alzheimer's. In: 2019 22nd International Conference on Control Systems and Computer Science (CSCS), pp. 236–243, Bucharest, Romania (2019). https://doi.org/10.1109/CSCS.2019.00046
25. Moldovan, D.: Horse optimization algorithm: a novel bio-inspired algorithm for solving global optimization problems. In: Silhavy, R. (ed.) CSOC 2020. AISC, vol. 1225, pp. 195–209. Springer, Cham (2020). https://doi.org/10.1007/978-3-030-51971-1_16
26. Chu, S.-C., Tsai, P., Pan, J.-S.: Cat swarm optimization. In: Yang, Q., Webb, G. (eds.) PRICAI 2006. LNCS (LNAI), vol. 4099, pp. 854–858. Springer, Heidelberg (2006). https://doi.org/10.1007/978-3-540-36668-3_94
27. Moldovan, D.: Improved kangaroo mob optimization and logistic regression for smart grid stability classification. In: Silhavy, R. (ed.) CSOC 2021. LNNS, vol. 229, pp. 469–487. Springer, Cham (2021). https://doi.org/10.1007/978-3-030-77445-5_44
28. Massaoudi, M., Abu-Rub, H., Refaat, S.S., Chihi, I., Oueslati, F.S.: An effective ensemble learning approach-based grid stability assessment and classification. In: 2021 IEEE Kansas Power and Energy Conference (KPEC), pp. 1–6, Manhattan, KS, USA (2021). https://doi.org/10.1109/KPEC51835.2021.9446197
29. Liu, S., Reviriego, P., Hernandez, J.A., Lombardi, F.: Voting margin: a scheme for error-tolerant k nearest neighbors classifiers for machine learning. IEEE Trans. Emerg. Top. Comput. **9**(4), 2089–2098 (2019). https://doi.org/10.1109/TETC.2019.2963268
30. Arzamasov, V., Bohm, K., Jochem, P.: Towards concise models of grid stability. In: 2018 IEEE International Conference on Communications, Control, and Computing Technologies for Smart Grids (SmartGridComm), pp. 1–6, Aalborg, Denmark (2018). https://doi.org/10.1109/SmartGridComm.2018.8587498
31. Schäfer, B., Grabow, C., Auer, S., Kurths, J., Witthaut, D., Timme, M.: Taming instabilities in power grid networks by decentralized control. Eur. Phys. J. Spec. Top. **225**(3), 569–582 (2016). https://doi.org/10.1140/epjst/e2015-50136-y
32. Dua, D., Graff, C.: UCI Machine learning repository. University of California, School of Information and Computer Science, Irvine, CA (2019). http://archive.ics.uci.edu/ml
33. Predicting Smart Grid Stability with Deep Learning, Release 2.0 - June, 2020, Paulo Breviglieri (2020). https://www.kaggle.com/pcbreviglieri/predicting-smart-grid-stability-with-deep-learning/notebook

34. Moldovan, D., Anghel, I., Cioara, T., Salomie, I.: Particle swarm optimization based deep learning ensemble for manufacturing processes. In: 2020 IEEE 16th International Conference on Intelligent Computer Communication and Processing (ICCP), pp. 563–570, Cluj-Napoca, Romania (2020). https://doi.org/10.1109/ICCP51029.2020.9266269
35. Koohestani, A., Abdar, M., Khosravi, A., Nahavandi, S., Koohestani, M.: Integration of ensemble and evolutionary machine learning algorithms for monitoring diver behavior using physiological signals. IEEE Access **7**, 98971–98992 (2019). https://doi.org/10.1109/ACCESS.2019.2926444
36. Moldovan, D.V., Salomie, I.: Ph.D thesis - Abstract - machine learning for heterogeneous distributed data streams (2021). https://doctorat.utcluj.ro/theses/view/ylP5A3KnopVG23dpY7utdym5lm61FQ2mKUsf8ivF.pdf

Extracting Relations from NER-Tagged Sentences for Ontology Learning

Pavel Lomov[1], Marina Malozemova[1(✉)], and Maxim Shishaev[1,2]

[1] Institute for Informatics and Mathematical Modeling, Kola Science Centre of the Russian Academy of Sciences, Fersman St. 24A, Apatity, Murmansk 184209, Russia
{lomov,malozemova,shishaev}@iimm.ru

[2] Apatity Branch of Murmansk Arctic State University, Lesnaya St. 29, Apatity, Murmansk 184209, Russia
http://www.iimm.ru

Abstract. This paper is a continuation of the research on training and using a neural-network language model for ontology learning. The paper deals with the problem of analyzing domain natural language texts and extracting possible relations between concepts from them. A brief overview of existing approaches to solving this problem is presented. A procedure for extracting relations between concepts based on the analysis of the sentence syntax dependency tree is proposed. The results of evaluating the efficiency of extracting relations using this procedure are considered.

Keywords: Relation extraction · Ontology learning · NLP

1 Introduction

Today the task of extracting relations from unstructured text data is quite urgent in solving various applied problems of natural language processing (NLP). In particular, the relation extraction task (RE) involves the detection of relations between entities in text data. These relations can be represented as "triples" *(subject, relation, object)*, which can be extracted using approaches that involve the use of lexico-syntactic patterns, automatic or manually created rules (heuristics), as well as machine learning methods (neural networks).

Relation extraction is one of the key stages in the ontology development process. An ontology is a conceptual domain model shared by a certain group of agents (people, organizations, information systems, etc.) [1]. Ontology development is considered in context of the ontology learning process, which includes steps such as extracting taxonomic and non-taxonomic relations from text. The extraction of taxonomic relations allows building the main hierarchy of concepts found in the domain texts, and the extraction of non-taxonomic relations allows reflecting the domain-specific relations between them.

R. Silhavy (Ed.): CSOC 2022, LNNS 502, pp. 337–344, 2022.
https://doi.org/10.1007/978-3-031-09076-9_31

In this paper, we consider the problem of extracting relations from the domain texts in order to further add them to the ontology. In this case, the extraction is carried out from those sentences in which, using the previously proposed technology [2,3] possible ontology concepts were detected. Therefore, the subsequent use of the resulting set of sentences with labels indicating the position of the found concepts is considered here.

The remaining part of the paper is structured as follows. Section 2 discusses in detail the past experience of studying the ontology learning problem. Section 3 provides an overview of existing approaches to extracting relations from unstructured text. Section 4 presents the proposed procedure for extracting relations between concepts based on the sentence syntax dependency tree. Section 5 discusses the results of experimental evaluation of the effectiveness of the procedure. The conclusion contains conclusions and directions for further work.

2 Background

In the previous work [2], a technology was proposed that involves the analysis of an ontology to form an initial list of its concepts, the collection and analysis of texts related to the domain of the ontology, with the formation of a training set of marked sentences as a result. The label assigned to the sentence, in this case, contained the boundaries of the concept found in it and the category of the concept.

Further, this set was used to train a neural-network language model focused on solving the problem of named entity recognition (NER). The model was subsequently used to extract new concepts from texts - candidates for adding to the ontology.

In the work [3], steps were added to the technology to "clarify" the concepts and augment the resulting set. The clarification step assumed the inclusion of some syntactically related words of the sentence into the concept. For example, clarifying the concept "аэропорт" ("*airport*") to "Международный аэро порт Шереметьево" ("*Sheremetyevo International Airport*") in the sentence "Международный аэропорт Шереметьево является крупнейшим в России." ("*Sheremetyevo International Airport is the largest one in Russia.*"). This made it possible to adjust the boundaries of the concept in sentences of the training set more precisely and thereby correctly set the context of its use, which had a positive effect on model training.

Augmentation, in turn, assumed the generation of new samples - marked sentences by replacing some combinations of words from the context of the concept with alternative ones proposed by a pre-trained language BERT model for the Russian language. It also gave some (insignificant in comparison with the clarification of concepts) improvement in the performance of the model.

In this paper, we will consider the problem of extracting relations from marked sentences obtained as a result of applying an already trained model to a set of texts. It is assumed that in such sentences there already exists a concept relevant to the initial ontology, and it is required to find another concept related to it by some relation.

3 Related Work

Extracting relations from text is a sub-task of NLP focused on identifying relations between pairs of entities in unstructured text data. Early work in this area used rule-based approaches (e.g. Hearst patterns [4] and [5]) as well as feature-based models. For example, Mintz et al. [6] considered lexical features in the model, such as: the sequence of words between two entities and their part-of-speech (POS) tags; a marker indicating which entity appears first in the sentence; the number of tokens to the left of the first entity and the number of tokens to the right of the second entity, as well as their part-of-speech tags. The authors also extracted the dependency path between two entities as a syntactic feature, as well as the named entity types for the two entities.

In [7] to extract relations the following features were taken into account: head tokens of two entities, tokens of two entities, tokens between two entities, their POS tags, the order of two entities, the distance between them, and the Brown cluster [8] for each token.

In recent studies, the relation extraction problem is solved using neural-network models based on the Transformer [9] architecture. These works can be divided into two groups: the first group is focused on extracting relations within a sentence [10–12], and the second group - on extracting relations within a document [13,14]. For example, [10] presents a BERT-based CASREL framework that allows to identify all possible triples of tokens - (subject, relation, object) in a sentence, where some of these triples may include the same entities. At the first stage, all possible subject tokens in a sentence are identified. At the second stage, using the taggers trained for each type of relation and the identified subject tokens, all possible relations and the corresponding object tokens are identified.

In [11], in the sentence passed to the input of the pre-trained BERT model the positions of two entities between which it is necessary to determine the relation are marked with special tokens. BERT considers these positions, as well as the context of the sentence, so it more precisely predicts the relation between entities. And in this work [12], the authors generate training datasets containing relation statements, which are sentences where entities are replaced by the [BLANK] token. The BERT model takes a pair of such relation statements containing the same masked entities as input, and builds their similar vector representations of relations as output.

In addition to the Transformer architecture, it has recently become common practice to use the architecture based on graph convolutional networks (GCN). The GraphRel model [15] allows extracting not only relations, but also entities in two stages. At the first stage, using bidirectional RNN and GCN networks, sequential and local features of words are extracted, taking into account which relations are predicted for each pair of words, as well as their types. At the second stage, for each predicted relation, full relational graphs are built, to which GCNs are sequentially applied. As a result of this stage, sufficient features of words are extracted, using which a more accurate classification of entities and relations is performed.

Examples of work considering extracting relations within a document are [13,14]. Han and Wang [13] also apply the BERT model for this purpose. To more precisely describe each mention of an entity in the text, the authors suggest masking entities with two special tokens, one of which is the entity type, and the second one indicates the number of the entity in the document (for example, "[LOC] Russia [MASK_1]"). This representation of the input data provides a more accurate definition of the types of relations between entities, and also allows to solve the coreference problem by linking words with the same token in different sentences.

Sahy et al. [14] use a graph convolutional network (GCN) to extract relations within a document. The authors consider each token in the document as a node in the graph, and to create links between them, they use relations from the sentence dependency tree, coreference edges (to connect document tokens that refer to the same entity), adjacency edges of words and sentences (to keep the sequence information) and self-node edges (to take into account the node information itself).

Another example of solving the relation extraction problem is to consider it as a question answering problem [16]. At the first step, using templates of questions about the type of entity, the entity-subject of the relation from the initial sentence is extracted. In the second step, a question is generated using relation templates where this entity-subject is placed. As a result of answering this question, the relation and the entity-object of the relation are extracted.

Ding et al. [17] propose prototypes for relation extraction. Prototypes are representations (embeddings) in the feature space that highlight the most important semantics of relations between entities in sentences. Prototypes are the centers of clusters representing a set of sentences expressing the same relation. According to the authors, this approach allows to identify meaningful, interpretable prototypes for the final classification of relations.

4 Proposed Procedure for Extracting Relations Between Concepts Based on a Sentence Syntax Dependency Tree

In this work, the extraction of relations for their subsequent addition to the ontology is carried out within separate sentences from the domain texts. In this case, only those sentences are considered in which the model trained using the technology described in the previous work [3] found concepts corresponding to the domain of the populated ontology.

Thus, it is assumed that the words representing the relation and the concept that it relates are in the same sentence. The algorithm is based on the hypothesis that the relation between concepts can be expressed through:

- a predicate represented by a verb that is the root of the sentence syntax dependency tree;
- a syntactic relation between words within noun phrases. For example, in the noun phrase *"Sussex police spokesman"* one can distinguish the relation between the pairs of words: *"police"* - *"spokesman"* and *"Sussex"* - *"police"*.

Taking into account this hypothesis, the analysis of sentences in order to detect relations in them is carried out as follows:

1. For each sentence, using a pre-trained language model from the spaCy [18] library, a syntax dependency tree is formed.
2. Next, the tree is traversed, starting from its top, and n-grams are formed by combining a word from the current level of the tree and related words from child levels. This is how various options for breaking up the noun phrases of the sentence into fragments corresponding to possible concepts were presented. At this stage of the study, unigrams and bigrams, that is, combinations of one and two words, were considered. For example, the noun phrase *"Sussex police spokesman"* was represented as n-grams: *"spokesman"*, *"police spokesman"*, *"police"*, *"Sussex police"*.
3. Among the formed n-grams, a search is made for possible domain concepts, i.e. relation participants. To identify them, the Word2Vec model [19] is first trained on a large set of domain texts. It is used to evaluate the proximity of the vectors of the obtained n-grams and the vectors of the initial list of ontology concepts. If the average distance between the generated n-gram vector and ontology word vectors does not exceed the threshold value, then the n-gram is considered as a domain concept. For example, if the bigram *"police spokesman"* has the proximity above the threshold, but the unigram *"spokesman"* does not, then it means that in the domain texts the concept *"spokesman"* is less common than *"police spokesman"* and therefore it is not necessary to consider the concept *"spokesman"* as a separate one. The threshold value is calculated for each sentence separately - as the average value of the proximity of all its n-grams to ontology concepts. Thus, at this step, possible related concepts represented by words within the same noun phrase are defined.
4. If the sentence contains more than one n-gram, defined in the previous step as domain concepts, then it is assumed that there is a relation between each pair of them. In this case, if n-grams do not overlap each other and are connected in the syntax dependency tree through other words of the sentence, then the set of intermediate words ordered in accordance with the sentence is considered as some context of the intended relation, indicating its meaning.
5. The following corrective procedures are applied to a set of relations derived from a single sentence:
   - if two relations have an n-gram as one concept, in which the main word is the same verb, then they are combined into one relation according to this concept. The n-gram with a verb goes into the context of this relation. For example, the relations *(Rosneft company, bought, context:*{}*)* and *(In*

2011, bought, context:{}) will be merged into one relation with the new context *(Rosneft company, In 2011, context:{bought})*.
- if the relation includes n-grams in which the main word is the same noun, then this relation is interpreted as taxonomic (is-kind-of). In this case, the n-gram with fewer words is assigned as the intended parent. For example - *(Oil Company, Company, context:is_kind_of, parent: Company, child: Oil Company)*.

Thus, as a result of applying this procedure, a set of relations is formed in the form of a pair of n-grams and sometimes its context. Next, the resulting set is presented to the ontology developer for the final decision on adding the detected relations to the ontology.

5 Evaluation

To evaluate the effectiveness of the proposed procedure for extracting relations, two experiments were carried out, for which two test sets were manually formed, containing sentences and reference results of extracting relations from them. The first set was used to evaluate the extraction of relations of an arbitrary type. It includes 500 samples, where the sample is a pair *(sentence, relation)*. The second test set was formed to evaluate the extraction of only taxonomic relations (is-kind-of), since they are especially important in the context of ontology learning - they allow to build a hierarchy of domain concepts. Its size was 75 thousand samples, similar in structure to the first test set.

Thus, the results of the procedure were compared with the reference results of the test sets, and estimates of precision and fullness were calculated. The following results were obtained:

Experiment 1. Evaluation of extracting relations of an arbitrary type from sentences of the 1st test set:

- precision = 0.016;
- fullness = 0.052.

Experiment 2. Evaluation of extracting taxonomic relations (is-kind-of) from sentences of the 2nd test set:

- precision = 0.128;
- fullness = 0.207.

The low precision estimates obtained as a result of the first experiment indicate the presence in the resulting set of a large number of ontology-irrelevant concepts in the extracted relations. This is because the model used to filter them in the next Word2Vec step misses a significant number of n-grams that are not domain concepts.

Improving the performance of the model probably requires finer tuning of its metaparameters during training. First of all, it is the size of the dictionary, which defines the minimum number of detections of a certain n-gram in the analyzed texts in order for it to be taken into account by the model. That is, if an n-gram is not in the dictionary, then its proximity to the initial concepts of the ontology is not evaluated and it will not be in a relation. Thus, increasing the value of this metaparameter can help reduce the number of n-grams that are not related to the domain of the ontology and, accordingly, the relations containing them.

The results of the detection of taxonomic relations were higher. This indicates the expediency of using heuristics to clarify the types of relations important for the formation of the basic ontology structure (is-kind-of, part-of, depend-on).

The proposed procedure for extracting relations can be used as an additional step in the previously developed ontology population technology, as well as for forming a hierarchy of concepts in the ontology learning process.

6 Conclusion

In this paper, we consider the problem of extracting relations from domain texts for their possible addition to the existing ontology. As a solution, a procedure is proposed that is supposed to be used as extension of the previously proposed technology for using neural-network models for learning ontologies. The procedure is focused on the search for relations in sentences in which the possible domain concepts (candidates for inclusion in the ontology) were found at the previous stages of the technology.

In the context of this procedure, the syntax dependency tree of the sentence is analyzed in order to identify noun phrases and consider their possible interpretations in the form of pairs of domain concepts related by a relation. In this case, the definition of words within the noun phrase representing probable domain concepts is made by assessing the proximity of various combinations of words (n-grams) to the initial set of ontology concepts in the vector space. The latter is represented by a separate Word2Vec model trained on a set of texts relevant to the ontology domain.

The obtained results of the experiments indicate the need, first of all, to improve the filtering of the formed n-grams. To do this, it is supposed to analyze the influence of various values of metaparameters of the Word2Vec model on the result obtained, as well as to consider the possibility of replacing and/or supplementing it with an additional model (e.g. binary classifier) to obtain a more accurate proximity estimation during the filtering.

References

1. Gruber, T.R.: A translation approach to portable ontology specifications. Knowl. Acquisition **5**(2), 199–220 (1993)
2. Lomov, P., Malozemova, M., Shishaev, M.: Training and application of neural-network language model for ontology population. In: Silhavy, R., Silhavy, P., Prokopova, Z. (eds.) CoMeSySo 2020. AISC, vol. 1295, pp. 919–926. Springer, Cham (2020). https://doi.org/10.1007/978-3-030-63319-6_85

3. Lomov, P., Malozemova, M., Shishaev, M.: Data augmentation in training neural-network language model for ontology population. In: Silhavy, R., Silhavy, P., Prokopova, Z. (eds.) CoMeSySo 2021. LNNS, vol. 231, pp. 669–679. Springer, Cham (2021). https://doi.org/10.1007/978-3-030-90321-3_55
4. Hearst, M.A.: Automated discovery of wordnet relations. In WordNet: An Electronic Lexical Database, pp. 26. MIT press edition, Cambridge (1998)
5. Garcia, M., Gamallo, P.: A weakly-supervised rule-based approach for relation extraction, p. 10 (2011)
6. Mintz, M., Bills, S., Snow, R., Jurafsky, D.: Distant supervision for relation extraction without labeled data. In: Proceedings of the Joint Conference of the 47th Annual Meeting of the ACL and the 4th International Joint Conference on Natural Language Processing of the AFNLP, pp. 1003–1011. Association for Computational Linguistics, Suntec, Singapore (2009)
7. Ren, X., et al.: CoType: joint extraction of typed entities and relations with knowledge bases. In: Proceedings of the 26th International Conference on World Wide Web, pp. 1015–1024. International World Wide Web Conferences Steering Committee, Perth Australia (2017)
8. Implementation of the brown word clustering algorithm. https://github.com/percyliang/brown-cluster
9. Devlin, J., Chang, M.W., Lee, K., Toutanova, K.: BERT: pre-training of deep bidirectional transformers for language understanding (2018). http://arxiv.org/abs/1810.04805 version: 1
10. Wei, Z., Su, J., Wang, Y., Tian, Y., Chang, Y.A.: Novel cascade binary tagging framework for relational triple extraction. In: Proceedings of the 58th Annual Meeting of the Association for Computational Linguistics, pp. 1476–1488. Association for Computational Linguistics (2020)
11. Wu, S., He, Y.: Enriching pre-trained language model with entity information for relation classification (2019). http://arxiv.org/abs/1905.08284
12. Soares, L.B., FitzGerald, N., Ling, J., Kwiatkowski, T.: Matching the blanks: distributional similarity for relation learning. In: Proceedings of the 57th Annual Meeting of the Association for Computational Linguistics, pp. 2895–2905. Association for Computational Linguistics, Florence, Italy (2019)
13. Han, X., Wang, L.: A novel document-level relation extraction method based on BERT and entity information. IEEE Access **8**, 96912–96919 (2020)
14. Sahu, S.K., Christopoulou, F., Miwa, M., Ananiadou, S.: Inter-sentence relation extraction with document-level graph convolutional neural network. In: Proceedings of the 57th Annual Meeting of the Association for Computational Linguistics, pp. 4309–4316. Association for Computational Linguistics, Florence, Italy (2019)
15. Fu, T.J., Li, P.H., Ma, W.Y.: GraphRel: modeling text as relational graphs for joint entity and relation extraction. In: Proceedings of the 57th Annual Meeting of the Association for Computational Linguistics, pp. 1409–1418. Association for Computational Linguistics, Florence, Italy (2019)
16. Li, X., et al.: Entity-relation extraction as multi-turn question answering. In: Proceedings of the 57th Annual Meeting of the Association for Computational Linguistics, pp. 1340–1350. Association for Computational Linguistics, Florence, Italy (2019)
17. Ding, N., et al.: Prototypical representation learning for relation extraction, p. 16 (2021)
18. Russian spaCy models. https://spacy.io/models/ru#ru_core_news_sm
19. Word2Vec model. https://radimrehurek.com/gensim/auto_examples/tutorials/run_word2vec.html#word2vec-model

Investment Portfolio Synthesis Problem: Evidence from the Russian Stock Market

Anna Andreevna Malakhova[1,3](✉), Elena Nikolaevna Sochneva[1], Grigory Borisovich Dobretsov[2], Olga Valeryevna Starova[1], Dmitry Valeryevitch Zyablikov[1], Evgeny Evgenievitch Savchenko[3], Svetlana Anatolyevna Yarkova[3], Svetlana Pavlovna Dudina[4], and Dmitry Ivanovitch Kravtsov[1]

[1] Siberian Federal University, Krasnoyarsk, Russia
malahovaanna@mail.ru, amalakhova@sfu-kras.ru
[2] Research Institute of the Federal Penitentiary Service of Russia, Moscow, Russia
[3] Krasnoyarsk Institute of Railway Transport - Branch of the Irkutsk State Transport University, Krasnoyarsk, Russia
[4] Reshetnev Siberian State University of Science and Technology, Krasnoyarsk, Russia

Abstract. The emerging markets present interest for international investors due to their growth potential and resulting market returns, offering an attractive risk-return trade-off relatively to those of the developed markets. Considering the increased interest in investing in the EMEA (Eastern Europe, Middle East and Africa) financial emerging markets, it is a topical issue to develop the methods and approaches to investment techniques consistent with the market under consideration. The present state of the art demonstrates a strong likelihood that classical portfolio analysis approaches applied to the developed financial markets fail to perform the excellency in the emerging markets due to the market illiquidity, the model factors instability and information inequities. The present paper is the attempt to develop the widely used optimization techniques for investment portfolio composed of the assets of the highly volatile emerging market, namely the Russian stock market. The investment portfolio is considered under the premise that the asset classes have been selected but the values of asset returns and variances are random. The sensitivity of the traditional Markowitz Model to market uncertainties is described by the 'conservative', 'nominal' and 'optimistic' formulations of the problem. The effect of introducing the weight-coefficient constraint is studied. Alternative models for investment portfolio synthesizing based on the decision theory criteria and on the 'downside' risk measures are proposed. The applicability of the developed model is demonstrated by considering an illustrative example on the basis of stocks of one of the leading EMEA segments – the Russian stock market – traded via the Moscow Stock Exchange. Portfolio efficiency evaluation is performed on the basis of the IMOEX index as a benchmark. #CSOC1120.

Keywords: Asset allocation · Investment portfolio optimization · Markowitz model · Decision making · Minimum-variance set · Portfolio performance evaluation

R. Silhavy (Ed.): CSOC 2022, LNNS 502, pp. 345–361, 2022.
https://doi.org/10.1007/978-3-031-09076-9_32

1 Introduction

The emerging markets present interest for international investors due to their growth potential and resulting market returns, offering an attractive risk-return trade-off relatively to those of the developed markets. The present state of the art of investment portfolio analysis reveals such features of emerging markets as information inequities, market illiquidity, poor investor protection, increased vulnerability and integration with global markets that complicate the financial investment process. Therefore, despite economic, financial and political reforms, issues related to financial transparency and political instability are still powerful obstacles to investments in these markets [2, 3].

Historically, emerging markets have offered high rates of returns and high risk (or volatility), along with low return correlations with developed markets. In theory, passive including emerging markets equities in portfolios could give investor higher returns and a reasonable level of risk because of the low correlations. Such approach to portfolio optimization is implemented into the traditional formulation of the Markowitz portfolio model [6].

Nevertheless, the recent fundamental researches on the emerging markets reveal that the correlation of emerging markets with developed markets has increased over time, making the low-correlation argument for including emerging markets less attractive [7]. The overall market volatility in developed markets has increased and their expected returns have decreased substantially. Consequently, the growth potential and resulting potential returns of emerging markets may now offer an attractive risk-return trade-off relatively to developed markets. Due to the named reasons, the investors' flows to the emerging markets continue to grow. In this context the decision to invest in emerging market depends on the level of risk tolerance.

Considering the increased interest in investing in the EMEA emerging markets, it is a topical issue to develop the methods and approaches of investment techniques under condition of the increased market volatility, asymmetric return distributions, information inequity. Some of the classical investment approaches accepted in the international practice may not work efficiently in the emerging market. A key assumption of various models is that the model factors are stable over time. To invest in assets traded on financial emerging markets, portfolio managers need to accommodate departures from traditional investment practices and frameworks [3].

The following four distinct possible problems associated with usage of mean-variance optimization algorithms for emerging market investment process are not well considered my practitioners. They are:

I) The expected returns on the investment assets may not be normally distributed, calling into question the validity of variance as a measure of dispersion of future returns.
II) The usual mean-variance utility function is not an adequate representation of the true preferences of the investor (or their agent).
III) Informational inputs to the algorithm do not adequately express the investor's beliefs about future events affecting asset returns and the probabilities of those events.

IV) The optimization algorithm takes the informational inputs that are forecasts of future events as parameters of known probability distributions. Estimation error is simply ignored [1, 3, 4, 8].

In comparison with the other emerging markets in Eastern Europe, the Middle East and Africa (EMEA emerging markets), Russia holds the leading positions in foreign investment flows. According to the Morgan Stanley Capital International assessment, the weight of investment in Russia reaches over 10% of total global emerging markets (GEM) investments [11].

The aim of this paper is to develop the widely used optimization techniques for investment portfolio composed of the assets of the highly volatile emerging market on the example of the Russian stock market. The paper is organized as follows. In the second chapter of the paper, we formulate the Markowitz approach under parametric uncertainty. We describe the market uncertainties (asset returns and volatilities) by defining 'nominal', 'conservative', and 'optimistic' models of the market. The weight coefficient constraint is introduced. The modified models for investment portfolio synthesizing based on the decision theory criteria and on the introduced 'downside' risk measure are proposed. In the third chapter experimental study of the developed model is performed on the illustrative example of stocks of the Russian leading companies traded via the Moscow Stock Exchange. The sensitivity of the Markowitz model to parametric uncertainty is evaluated. The effect of introducing constraints on the weights of assets and on the level of risk is studied. The results of portfolio synthesizing based on decision theory criteria and on the introduced 'downside' risk measure are presented. The portfolio performance is assessed using the efficiency evaluation criterion based on the IMOEX (Moscow Stock Exchange) index as a benchmark. The obtained results are discussed in the fourth section. The main findings of the research are summarized in the conclusion.

2 Methods

2.1 The Markowitz Model for Portfolio Optimization Under Parametric Uncertainty

The Markowitz model reveals the trade-off between return and variance of the portfolio. The model is based on the principle of 'effective diversification' and allows constructing an efficient frontier on which the optimal portfolios are located, that is, the efficient set provides the investor with the best combinations of portfolio risk and return. This approach is known as 'mean-variance' optimization [1, 4, 6].

Initial premises of the classical Markowitz problem are as follows [1]:

1. Investors evaluate investment portfolios based on expected returns and standard deviations over the holding period.
2. Investors are never jaded, that is, other conditions being equal, they prefer a portfolio with a higher expected return.
3. Investors are rational, that is, other conditions being equal, they prefer a portfolio with less risk.

4. The portfolio consists only of risky assets, the risk measure of which is the standard deviation of return.
5. Investors assume a normal probability distribution of the expected return on assets. The portfolio return aggregates the return on the securities included in it.
6. Private assets are infinitely divisible, that is, an investor can buy part of a share.
7. Taxes and transaction costs are immaterial.

The mean return of a portfolio of n assets is obtained as:

$$R_p = \sum_{i=1}^{n} w_i \cdot \overline{R}_i, \tag{1}$$

where w_i is the weight of the asset i in the portfolio, $\overline{R}_i$ is the mean return on the asset i.

The variance of a portfolio return is defined as [4]:

$$\sigma_p^2 = \sum_{ij} w_i \cdot w_j \cdot Cov_{ij} \tag{2}$$

where w_i and w_j are the weights of the assets i and j in the portfolio, respectively, Cov_{ij} is the covariance between two assets i and j.

The points of the efficient frontier are obtained by solving two optimization problems that yield the same result: to minimize (maximize) the variance (the mean return) of (on) the portfolio under fixed value R (σ^2) of the mean return (variance). That is [5, 9]:

$$J_{Markowitz} = \sum_{ij} w_i \cdot w_j \cdot Cov_{ij} \rightarrow min \tag{3}$$

subject to

$$\begin{aligned} &\sum_i w_i \cdot \overline{R_i} = R \\ &\sum_i w_i = 1 \\ &w_i \geq 0,\ i = \overline{1, n} \end{aligned} \tag{4}$$

or

$$J_{Markowitz} = \sum w_i \cdot \overline{R}_i \rightarrow max \tag{5}$$

subject to

$$\begin{aligned} &\sum_{ij} w_i \cdot w_j \cdot Cov_{ij} = \sigma^2,\ i = \overline{1, n} \\ &\sum_i w_i = 1,\ i = \overline{1, n} \\ &w_i \geq 0,\ i = \overline{1, n} \end{aligned} \tag{6}$$

The solution of the problem yields the optimal weight coefficients w_i^* for the assets in the portfolio.

While all mentioned problems contribute to the difficulty of using optimization algorithms, we will try to optimize the portfolio considering investors' changing expectations about future events: the nominal, optimistic and pessimistic cases [10].

To evaluate the sensitivity of the optimal solution we consider 'conservative' and 'optimistic' formulations of the problem. We obtain the 'conservative' optimal solution w_{icons}^*, which provides the investor with the guaranteed values of return and variance, and represents the worst possible result. Similarly, the 'optimistic' optimal solution w_{iopt}^* is determined which yields a value of return impossible to exceed, and the least possible variances as well. For 'conservative' ('optimistic') formulation of the Markowitz problem we assume $\overline{R}_i = \overline{R}_{imin}$ ($\overline{R}_i = \overline{R}_{imax}$) and $\sigma_i^2 = \sigma_i^2{}_{max}$ ($\sigma_i^2 = \sigma_i^2{}_{min}$).

2.2 Introducing the Weight Coefficient Constraint

Investment practitioners who use mean-variance optimization techniques for portfolio construction are often disappointed in the result. The most widely noted complaint is a lack of robustness on the optimization results. Small changes in informational inputs can have a seemingly overly dramatic impact on the resulting suggested portfolio. Some investors are reluctant to put optimization into practice because they regard optimizers as error maximizers that can cause inaccurate inputs to be translated into potentially even more inaccurate portfolio weights. The critique of optimization may be technically correct: optimizer inputs, since they are statistical estimates, are necessarily inexact. There is no way to make the precise estimates that would be needed for absolute confidence in the outputs of an optimizer. The applied optimization technique may imply the essential shift in the asset weight in the portfolio and may be conducive to extreme weights of the assets in the portfolio undesirable for a risk-averse investor. Introducing the weight coefficient constraint in the mean-variance optimization model can be the approach to resolve the mentioned problem.

We introduce the constraints on the weights of individual assets by modifying the Markowitz model (3, 4) as follows:

$$J_{Markowitz} = \sum_{ij} w_i \cdot w_j \cdot Cov_{ij} \rightarrow min \tag{7}$$

subject to

$$\begin{aligned} &\sum_i w_i \cdot \overline{R_i} = R \\ &\sum_i w_i = 1 \\ &w_i \le w_{lim} \\ &w_i \ge 0,\ i = \overline{1, n} \end{aligned} \tag{8}$$

where w_{lim} is the constraint on the weight coefficient.

Given the fact that market equilibrium weights remain a question under discussion for the Russian Stock market, we propose to fix the weight constraints at different levels in order to see whether it will affect the optimal set of portfolios.

2.3 Optimization Models Based on Decision Theory Approach and on the 'Downside' Risk Measure

As it has been discussed above the investor's aim is to maximize the expected portfolio return and to keep small the portfolio risk. In terms of the portfolio risk measured by the variance and using the criteria of the Decision Making Theory [5] we formulate the following formulations of $(R_p - k \cdot \sigma_p^2)$- and $(R_p - k \cdot \sigma_p)$-maximizing problems:

$$J_{\left(R_p - k\sigma_p^2\right)} = \sum w_i \cdot \overline{R}_i - k \cdot \sum w_i^2 \cdot \sigma_i^2 \rightarrow max \tag{9}$$

subject to

$$\begin{gathered} \sum\nolimits_i w_i = 1 \\ w_i \geq 0,\ i = \overline{1, n} \end{gathered} \tag{10}$$

and

$$J_{(R_p - k\sigma_p)} = \sum w_i \cdot \overline{R}_i - k \cdot \sqrt{\sum w_i^2 \cdot \sigma_i^2} \rightarrow max \tag{11}$$

subject to

$$\begin{gathered} \sum\nolimits_i w_i = 1 \\ w_i \geq 0,\ i = \overline{1, n} \end{gathered} \tag{12}$$

where k is the risk aversion coefficient.

In this subsection we introduce a risk-measure, based on the downside mean (we call it 'low-mean' return). The idea of downside risk measure is not new. Among the most popular downside measures it is necessary to mention a widely-used VaR (Value-at-Risk)-measure, Conditional VaR (sometimes called 'shortfall'), downside dispersion [1, 9].

The peculiarity of our approach is that the introduced risk measure considers the gap between the mean and the low-mean ('downside') return. Thus, by minimizing $(R_p - R_{plow})$-value for the desired level of the portfolio mean we 'pull' the low-mean return to the desired level.

We formulate the optimization model as a risk-minimizing problem: to minimize the difference of mean and 'low-mean' for a fixed value of mean-return:

$$J_{(R_p - R_{plow})} = R_p - R_{plow} \rightarrow min \tag{13}$$

subject to

$$\begin{gathered} \sum\nolimits_i w_i \cdot \overline{R_i} = R \\ \sum_i w_i = 1 \\ w_i \geq 0,\ i = \overline{1, n} \end{gathered} \tag{14}$$

2.4 Portfolio Efficiency Criteria

There are different indicators to evaluate the efficiency of investment portfolio. The main approaches to evaluate portfolio performance are the following [2, 3]:

I) The measures of portfolio performance which enable to evaluate whether the synthesized portfolio has outperformed the benchmark portfolio. For this purpose, market indices are used (for example, S&P 500, DJIA), which evaluate the mean-market return.

II) The measures of portfolio performance which take into consideration both – the risk and the return of the portfolio (Sharpe Index, Treynor Index, Jensen Index).

III) Performance attribution analysis which enables to reveal the factors influencing portfolio performance: variability in markets, trading activity, earnings-price ratio (EPR), book-price ratio (BPR), earnings variation (EVR) etc. One of the most common used is BARRA's performance analysis factor model.

In Russian traders' practice it is commonly used to compare the portfolio mean-return with the return of the market. An investment portfolio is considered to be efficient if its return is not less than the return of the market.

We consider the return on the IMOEX index - the official Moscow Exchange benchmark - to be the market return. The returns on investment portfolios R_{pt} synthesized by using different models can be compared with the market return to evaluate their efficiency R_{It}. Figure 1 demonstrates the approach to portfolio efficiency evaluation relatively to the market index.

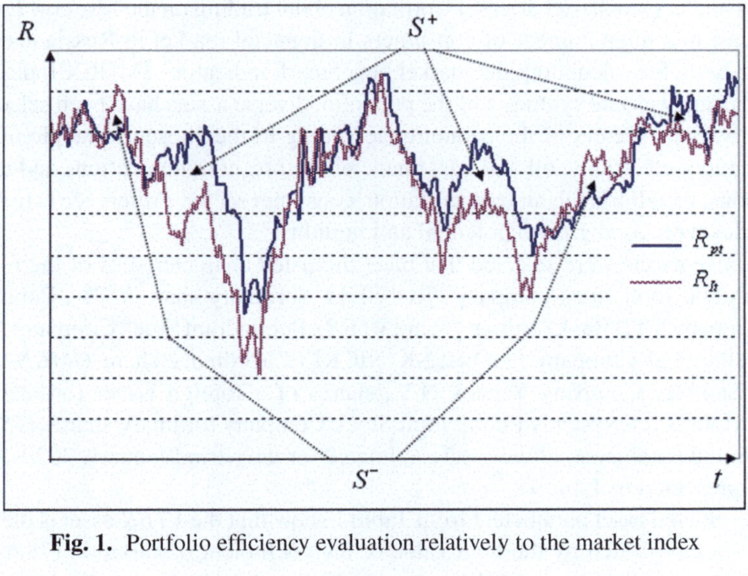

Fig. 1. Portfolio efficiency evaluation relatively to the market index

To evaluate the efficiency of the selected portfolio against the market we propose two criteria, termed 'portfolio quality' criteria:

$$PQ_1 = \frac{S_1}{S_2} \tag{15}$$

$$PQ_2 = \frac{S_1}{S_1 + S_2}, \tag{16}$$

where S_1 is the total 'upper' ('positive') area, S_2 is the total 'lower' ('negative') area in Fig. 1.

PQ_1 criterion shows the reward (prime) per one unit of loss related to the mean-market return. PQ_2-criterion shows the weight of reward (prime) in the sum of positive and negative deviations from the mean-market return.

3 Results

The 'Moscow Exchange' was chosen as the information basis for conducting this part of the experimental study as the oldest organizer and the leading trading platform for the securities market in Russia [10]. A crucial factor in the choice of this trading platform was the open access to information on trading for the entire historical period of its existence including the former 'Russian Trading System', which had been affiliated further to 'Moscow Exchange', namely the period of 1995–2021. The prices that are formed in Moscow Exchange are a generally recognized guideline for investors who transact with Russian shares and depositary receipts on other Russian stock exchanges, on European and US stock exchanges. Information about trading on the Moscow Exchange is broadcast to a huge number of consumers in financial market in Russia and abroad and is the basis for calculating the market benchmark indicator - IMOEX Index [12].

For the analysis and synthesis of the portfolio, diverse assets have been selected that represent various sectors of the economy according to the chosen principle of branch diversification: electricity, oil and gas, financial sectors, communications and telecommunications, metallurgical, air transportation, consumer sector. Preference is to be given to securities with good growth potential and liquidity.

Thus, five assets were selected that meet the listed characteristics of the following issuers: Public Joint stock company "ROSSETI" (ordinary share *RSTI*), Public Joint-Stock Company VTB Bank (ordinary share *VTBR*), Public Joint Stock Company "Mining and Metallurgical Company "NORILSK NICKEL" (ordinary share *GMKN*), Public Limited Liability Company Yandex N.V., shares of a foreign issuer (ordinary share *YNDX*), Mobile TeleSystems Public Joint Stock Company (ordinary share *MTSS*) [12].

The initial parameters of the assets estimated for the period January, 2020- January, 2021 are presented in Table 1.

The estimated asset parameters from Table 1 show that the *VTBR*-asset is the riskiest one, as is demonstrated by the high value of the coefficient of variation (78.6%). The least variable are *MTSS*- and *GMKN*-assets (6.62% and 9.46%, respectively).

Table 1. Statistic parameters for the Markowitz model

Asset	Asset parameters			
	$\overline{R}_i$	σ_i	σ_i^2	Coefficient of variation
GMKN	1,023189	0,096822	0,009375	0,094628
MTSS	1,006183	0,066658	0,004443	0,066249
YNDX	1,062315	0,107592	0,011576	0,10128
VTBR	1,058802	0,832589	0,693204	0,78635
RSTI	1,03565	0,151427	0,02293	0,146215

Variance-covariance matrix is given in Table 2.

Table 2. Covariance matrix for the selected assets

Asset	Cov_{ij}				
	GMKN	MTSS	YNDX	VTBR	RSTI
GMKN	0,00933704	0,00413593	0,00232696	0,0115345	0,00986401
MTSS	0,004135931	0,00442554	0,00393745	0,0097648	0,00733078
YNDX	0,00232696	0,00393745	0,01152967	0,0071513	0,00528334
VTBR	0,011534529	0,00976478	0,00715132	0,6904312	0,02447735
RSTI	0,009864008	0,00733078	0,00528334	0,0244773	0,02283856

We solve the asset allocation problem as a variance-minimizing problem for the parameters of the Table 1, using (3), (4). The arbitrary values R are chosen using a 0.5% step. The solution of the problem yields the optimal weight coefficients $w^*_{iMarkowitz}$ for efficient frontier portfolios, presented in Table 3.

Table 3. Portfolio parameters and weight distribution for optimal portfolios $w^*_{iMarkowitz}$

Portfolio return	Standard deviation	Weights of the assets				
		GMKN	MTSS	YNDX	VTBR	RSTI
1,0623	0,107376	0,00000100	0,00000000	1,00000100	0,00000000	0,00000000
1,0600	0,102399	0,05880663	0,00000000	0,93705812	0,00413525	0,00000000
1,0550	0,093020	0,18680083	0,00000000	0,81132568	0,00187348	0,00000000
1,0500	0,085690	0,31473468	0,00000000	0,68526532	0,00000000	0,00000000

(*continued*)

Table 3. (*continued*)

Portfolio return	Standard deviation	Weights of the assets				
		GMKN	MTSS	YNDX	VTBR	RSTI
1,0450	0,080875	0,39074852	0,03610852	0,57314296	0,00000000	0,00000000
1,0400	0,077089	0,34301222	0,15845785	0,49852993	0,00000000	0,00000000
1,0350	0,073762	0,29527592	0,28080716	0,42391692	0,00000000	0,00000000
1,0300	0,070958	0,24753961	0,40315651	0,34930388	0,00000000	0,00000000
1,0250	0,068743	0,19980331	0,52550584	0,27469085	0,00000000	0,00000000
1,0200	0,067173	0,15206700	0,64785517	0,20007782	0,00000000	0,00000000
1,0150	0,066295	0,10433069	0,77020453	0,12546478	0,00000000	0,00000000
1,0010	0,066136	0,05660396	0,89252827	0,05086677	0,00000000	0,00000000

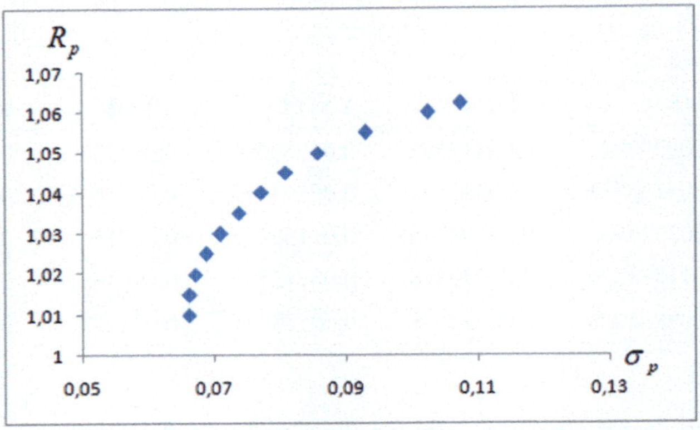

Fig. 2. Efficient frontier for the Markowitz model

The minimum-variance set for the Markowitz model is shown in Fig. 2.

We evaluated minimum and maximum values of returns and variances, assuming $\overline{R}_i = \overline{R}_{ikmin}$ ($\overline{R}_i = \overline{R}_{ikmax}$) and $\sigma_i^2 = \sigma_{ik\ max}^2$ ($\sigma_i^2 = \sigma_{ik\ min}^2$), $k = \overline{1, 12}$. The results are shown in Table 4.

Table 2 provides us with interval values of $\overline{R}_i$ and σ_i^2 to describe the market parametric uncertainty.

For 'conservative' ('optimistic') formulation of the Markowitz problem we assume $\overline{R}_i = \overline{R}_{imin}$ ($\overline{R}_i = \overline{R}_{imax}$) and $\sigma_i^2 = \sigma_{i\ max}^2$ ($\sigma_i^2 = \sigma_{i\ min}^2$) (see Table 2). Thus, we solve the problem for these new parameters. The solution of the problem yields the optimal weight coefficients w_{icons}^* and w_{iopt}^* for 'conservative' and 'optimistic' portfolios.

Table 4. The market parametric uncertainty estimators

Asset	Asset parameters			
	$\overline{R}_{imin}$	$\overline{R}_{imax}$	σ^2_{imin}	σ^2_{imax}
MTSS	0.897688	1.083603	0.000617	0.013937
GMKN	0.81476	1.102451	0.000816	0.016183
YNDX	0.90562	1.097415	0.000429	0.008696
VTBR	0.782302	1.147571	0.000459	0.009378
RSTI	0.953155	1.1218	0.000884	0.01877

To verify the applicability of the solutions w^*_{inom}, w^*_{icons} and w^*_{iopt} we calculated $R_p(w^*_{inom})$, $R_p(w^*_{icons})$, $R_p\left(w^*_{iopt}\right)$ and $\sigma^2_p(w^*_{inom})$, $\sigma^2_p(w^*_{icons})$, $\sigma^2_p\left(w^*_{iopt}\right)$ using the moving monthly-mean return, which is determined daily during one year. The mean – standard deviation diagrams for 'conservative', 'optimistic' and 'nominal' cases are presented in Fig. 3.

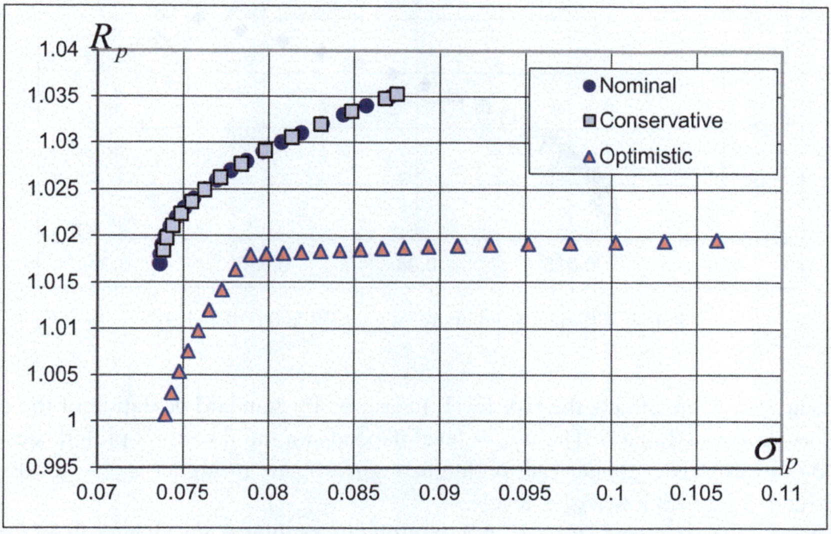

Fig. 3. Mean - standard deviation diagrams for 'conservative', 'optimistic' and 'nominal' formulations of the problem

The interval values for 'conservative', 'optimistic' and 'nominal' portfolio parameters are given in Table 5.

It can be found out that the 'conservative' and 'realistic' frontiers are similar, while the 'optimistic' frontier provides the worst results in real time.

In the next step of the experimental study, we solved the optimization problem for various levels of weight-coefficient constraint. Given the fact that market equilibrium

Table 5. Portfolio parameters for 'conservative', 'optimistic' and 'nominal' portfolio parameters

Portfolio scenario	Portfolio return	Standard deviation
Conservative	[1.01842; 1.035253]	[0.073849; 0.08504]
Nominal	[1.017; 1.035001]	[0.073672; 0.087141]
Optimistic	[1.000708; 1.019597]	[0.073973; 0.106178]

weights remain a question under discussion for the Russian stock market, we propose to fix the weight constraints at different levels ($w_{lim} = 0.5, 0.6, 0.7.$) in order to test whether it affects the optimal set of portfolios.

The efficient frontiers for $w_{lim} = 0.5, 0.6, 0.7$, and $w_{lim} = 1.0$ (constraint-free frontier) are shown in Fig. 4. The 0.5-frontier is included into the 0.6- and 0.7-frontier and they all are included into the constraint-free frontier.

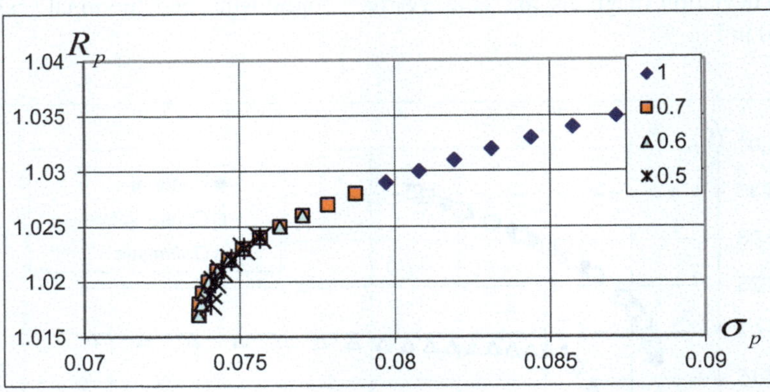

Fig. 4. Efficient frontiers for $w_{lim} = [0.5; 0.6; 0.7; 1.0]$.

Each w_{lim}-value affects the risk-level, measured by standard deviation of the portfolio, as shown in Table 6. The $w_{lim}-$ level depends on the investor's preferences and attitude towards risk: greater risk preference corresponds to higher $w_{lim}-$ levels and risk-aversion implies low $w_{lim}-$ levels.

The $(R_p - k \cdot \sigma_p^2)$- and $(R_p - k \cdot \sigma_p)$-maximizing problems have been solved for the parameters given in Table 1. The k-values have been set within the intervals [0, 185] and [0, 13.4] for the $(R_p - k \cdot \sigma_p^2)$- and $(R_p - k \cdot \sigma_p)$-maximizing problems respectively, which yield nonnegative values of the objective functions. For the corresponding k, weight distributions $w^*_{(R-k\sigma^2)}$, $w^*_{(R-k\sigma)}$ were obtained.

The dependence of the parameters of the portfolio and the objective utility function on the parameter k is analyzed. The dependences of the utility functions $(R_p - k\sigma_p^2)$, $(R_p - k\sigma_p)$ on the k coefficient are shown in Fig. 5 and Fig. 6, respectively.

According to the results of the experimental study, the objective function decreases from k. We conclude, that the coefficient k can be regarded as the coefficient of risk

Table 6. Portfolio standard deviation and w_{lim}-level

$w_{\lim}$	σ_p
0.5	0.075648
0.6	0.077017
0.7	0.078735
1.0	0.087141

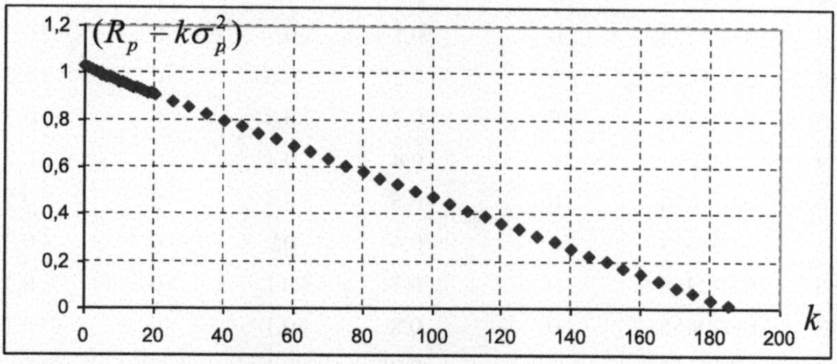

Fig. 5. Dependence of the objective $(R_p - k\sigma_p^2)$- function on k coefficient.

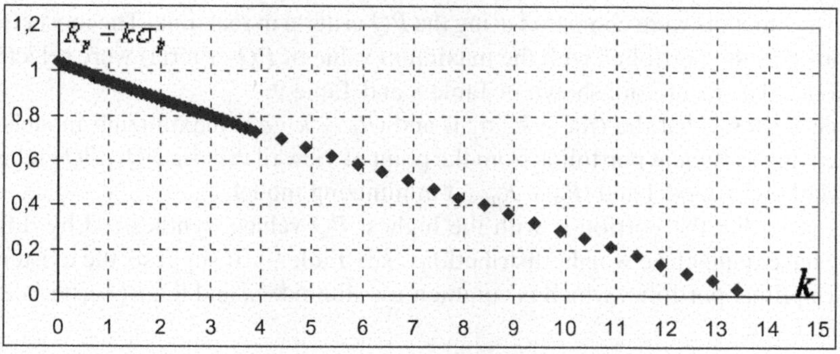

Fig. 6. Dependence of the objective $(R_p - k\sigma_p)$- function on k coefficient.

aversion. The portfolio return, the variance (standard deviation) and the value of the objective function for different k are shown in Fig. 5 and Fig. 6.

Note that the obtained weight distribution $w^*_{i(R-k\sigma^2)}$ is close to $w^*_{i(R-k\sigma)}$ and they both are close to the Markowitz nominal weight distribution.

The optimal portfolios on the efficient frontier of the minimum-$(R_p - R_{plow})$ set are presented in Table 7.

Table 7. Portfolio parameters and weight distribution for optimal portfolios $w^*_{i(R-Rlow)}$

Portfolio return	Portfolio risk $(R_p - R_{plow})$	Weights of the assets				
		GMKN	MTSS	YNDX	VTBR	RSTI
1,061964	0,062055	0	0	0,9	0,1	0
1,051369	0,047273	0	0,1	0,7	0	0,2
1,045756	0,041942	0	0,2	0,6	0	0,2
1,037476	0,036426	0	0,3	0,4	0	0,3
1,034529	0,03377	0	0,4	0,4	0	0,2
1,028916	0,027876	0	0,5	0,3	0	0,2
1,023303	0,024553	0	0,6	0,2	0	0,2
1,020356	0,023308	0	0,7	0,2	0	0,1
1,014743	0,020455	0	0,8	0,1	0	0,1
1,011796	0,020427	0	0,9	0,1	0	0

Comparing the results of the $(R_p - R_{plow})$-minimizing problem and $(R_p - k \cdot \sigma_p^2)$- and $(R_p - k \cdot \sigma_p)$-maximizing problems we conclude that the weight distribution $w^*_{i(R-Rlow)}$ is close to $w^*_{i(R-k\sigma^2)}$, $w^*_{i(R-k\sigma)}$.

The portfolios synthesized by the Markowitz model, $(R_p - R_{plow})$-, $(R_p - k \cdot \sigma_p^2)$- and $(R_p - k \cdot \sigma_p)$-models were compared using the *PQ*-criteria in real time. The most efficient portfolios (i.e. the portfolios with the maximum value of *PQ*-criteria) were selected for each model. The results are shown in Table 8 and Table 9.

Table 8 shows that the $(R_p - k \cdot \sigma_p^2)$- and $(R_p - k \cdot \sigma_p)$-maximizing models have produced more efficient portfolios from the point of view of criteria PQ_1, PQ_2 relatively to the Markowitz model and $(R_p - R_{plow})$-minimizing model.

It is seen that the portfolios with the highest *PQ*-values, synthesized by different models, have quite close weight distributions (see Table 9). It suggests the existence of a (quasi)optimal portfolio, which is common for all models, and the *PQ*-criteria enable us to find it out.

Table 8. Portfolio parameters for the most efficient portfolios

Formulation of the model	Portfolio return	Standard deviation	PQ_1	PQ_2
$w^*_{i(R-k\sigma^2)}$	1.0292265	0.079989	4.436742	0.816066
$w^*_{i(R-k\sigma)}$	1.029405	0.080135	4.43535	0.816019
$w^*_{iMarkowitz}$	1.029001	0.815976	4.434059	0.815976
$w^*_{i(R-Rlow)}$	1.029692	0.815846	4.430234	0.815846

Table 9. Optimal weight distributions for the most efficient portfolios

Formulation of the model	Weights of the assets				
	GMKN	MTSS	YNDX	VTBR	RSTI
$w^*_{i(R-k\sigma^2)}$	0	0	0.268855	0	0.731145
$w^*_{i(R-k\sigma)}$	0	0	0.262637	0	0.737363
$w^*_{iMarkowitz}$	0	0	0.280555	0	0.719445
$w^*_{i(R-Rlow)}$	0	0	0.250000	0	0.750000

4 Discussions

The differences of the optimal solutions w^*_{inom}, w^*_{icons} and w^*_{iopt} confirm that the Markowitz model is sensitive to parametric uncertainty under the considered financial market conditions.

It has been found out that the 'conservative' portfolios are equally efficient as the 'nominal' portfolios and their weight distributions are close. Under parametric uncertainty 'conservative' portfolios provide the guaranteed values of return and variance. The 'optimistic' portfolios have not proved their applicability in real time.

A modification of the Markowitz model is proposed which consists of imposing constraints on weights of assets to take into consideration the investors' risk preferences measured by standard deviation. The intuitive position of stock exchange traders to introduce constraints on the weight of assets may be seen as a risk-level limitation.

In this paper we have proposed alternative approaches to synthesize investment portfolios based of the Decision Making Theory: the $(R_p - R_{plow})$-minimizing, and the $(R_p - k \cdot \sigma_p^2)$- and $(R_p - k \cdot \sigma_p)$-maximizing models.

The weight distribution of $(R_p - k \cdot \sigma_p^2)$-model is close to the weight distribution of $(R_p - k \cdot \sigma_p)$-model and they both are close to the Markowitz weight distribution. It may be explained by the low values of the covariances of the assets, the ignorance of which yields similar results.

In the $(R_p - k \cdot \sigma_p^2)$- and $(R_p - k \cdot \sigma_p)$-models the parameter k may be considered as a risk-aversion coefficient: higher values of k correspond to lower values of the expected portfolio return and of the standard deviation.

The difference between the portfolio mean return and 'low-mean' return may be considered as a downside risk-measure. The $(R_p - R_{plow})$- risk measure is especially suitable for the returns with the left-skewed distribution.

The 'portfolio quality' criteria have been proposed to compare the portfolio efficiency with that of the market. The classical Markowitz approach and the introduced models are compared using the proposed criteria for portfolio performance and efficiency evaluation on the basis of the IMOEX Index as a benchmark. According to the *PQ*-criterion all models under discussion (the Markowitz model, the $(R_p - R_{plow})$-minimizing model, the $(R_p - k \cdot \sigma_p^2)$- and $(R_p - k \cdot \sigma_p)$-maximizing models) have shown a greater efficiency than the market (IMOEX Index). The 'portfolio quality' criteria may be used to compare the efficient frontiers of different models, as well.

It has been shown that the $(R_p - k \cdot \sigma_p^2)$- and $(R_p - k \cdot \sigma_p)$-maximizing models have produced more efficient portfolios than the Markowitz model and the $(R_p - R_{plow})$-minimizing model from the point of view of criteria PQ_1, PQ_2 for the market under consideration. A possible explanation may be found in the different approaches to define the objective function. The Markowitz model and $(R_p - R_{plow})$-minimizing model are risk-minimizing (or mean return-maximizing) problem, while in the objective functions $(R_p - k \cdot \sigma_p^2)$ and $(R_p - k \cdot \sigma_p)$ of the proposed models we consider simultaneously both risk and return on the portfolio.

The introduced *PQ*-criteria enable the selection of the optimal portfolio on the efficient frontier, as well as the comparison of portfolios synthesized using different models. To the best of our knowledge both PQ_1 and PQ_2 criteria yield the maximum value for the close weight distribution w_i^*. It suggests the existence of a (quasi)optimal portfolio, which is common for all models, and the *PQ*-criteria enable us to find it out. Such a portfolio can be considered as an equilibrium point of diverse portfolio optimization formulations. Further investigation is needed to explain this fact.

The obtained weight distributions for the alternative risk measures turned out to be closer to each other than to the weight distributions for the traditional risk measures.

Note that portfolios with downside risk measures produced better results in real time than portfolios with traditional risk assessment: higher mean return values were obtained at a lower risk. This finding confirms the need of using the 'left-side' measures for risk assessments under financial market conditions.

5 Conclusion

In this paper a mean-variance optimization technique is modified for a portfolio composed of the assets of a volatile emerging market. An asset allocation problem is studied under the premise that the asset classes in investment portfolio are selected but the values of asset returns and variances are random. The sensitivity of the Markowitz Model to market uncertainties is described by 'conservative', 'nominal' and 'optimistic' formulations of the problem. Under discussed conditions the 'conservative' portfolios provide the guaranteed values of return and variance. The 'optimistic' portfolios have not proved their applicability in real time.

The approaches to synthesize investment portfolios based on the Decision Making Theory have been proposed: the $(R_p - k \cdot \sigma_p^2)$- and $(R_p - k \cdot \sigma_p)$-maximizing models.

The difference between the portfolio mean return and 'low-mean' return may be considered as a downside risk-measure. The $(R_p - R_{plow})$- risk measure is especially suitable for assets with the left-skewed return distribution.

The performance of the proposed models has been compared with the commonly used Markowitz model in real time. It has been shown that the $(R_p - k \cdot \sigma_p^2)$- and $(R_p - k \cdot \sigma_p)$-maximizing models have produced more efficient portfolios than Markowitz model from the point of view of the introduced *PQ*-criteria for the market under consideration.

The introduced *PQ*-criteria enable the selection of the optimal portfolio on the efficient frontier, as well as the comparison of portfolios synthesized using different models. To the best of our knowledge both PQ_1 and PQ_2 criteria yield the maximum value for the close weight distribution w_i^*. It suggests the existence of a (quasi)optimal portfolio, which is common for all models, and the *PQ*-criteria enable us to find it out. Further investigation is needed to explain this fact.

To the best of our knowledge that portfolios with downside risk measures produced better results in real time than portfolios with traditional risk assessment. These findings confirm the validity of application and the need for further development and research of alternative measures for risk assessment under financial market conditions.

References

1. Sharpe, W.F., Alexander, G.J., Beiley, J.V.: Investments. Prentice Hall International Inc., Upper Saddle River (1995)
2. Bodie, Z., Kane, A., Marcus, A.J.: Essentials of Investments. McGraw-Hill, New York (2001)
3. Fabozzi, F.J.: Quantitative strategies: multi-asset. J. Portfolio Manag. Multi-Asset Special Issue **46**(6), 1–3 (2020)
4. Luenberger, D.G.: Investment science. Oxford University Press, New York (1998)
5. Taha, H.A.: Operations Research: An Introduction, 10th edn. Global Edition. University of Arkansas, Fayetteville (2017)
6. Markowitz, H.M.: Portfolio selection. J. Financ. **7**(1), 77–91 (1952)
7. Breuer, W., Gürtler, M.: Performance evaluation and preferences beyond mean-variance, Swiss society for financial market research. Finan. Mark. Portfolio Manag. **17**(2), 213–233 (2003)
8. Chen, Z.-P., Zhao, C.-E.: Is the MV efficient portfolio really that sensitive to estimation errors. Asia-Pac. J. Oper. Res. **19**, 149–168 (2002)
9. Malakhova, A.A., et al.: Portfolio optimization model for asset allocation problem based on alternative risk measures. In: Silhavy, R. (ed.) CSOC 2021. LNNS, vol. 229, pp. 673–693. Springer, Cham (2021). https://doi.org/10.1007/978-3-030-77445-5_61
10. Malakhova, A.A., et al.: Mean-variance portfolio optimization under parametric uncertainty. In: Silhavy, R., Silhavy, P., Prokopova, Z. (eds.) CoMeSySo 2020. AISC, vol. 1295, pp. 797–812. Springer, Cham (2020). https://doi.org/10.1007/978-3-030-63319-6_74
11. Morgan Stanley Capital International. https://www.msci.com/. Accessed 11 Jan 2022
12. Moscow Exchange. https://www.moex.com/. Accessed 20 Nov 2021

Finding Optimal Paths of All Lengths in Directed Grid Graphs

Abdullah N. Arslan(✉)

Department of Computer Science, Texas A & M University – Commerce, Commerce, TX 75428, USA
abdullah.arslan@tamuc.edu
http://faculty.tamuc.edu/aarslan

Abstract. We study a problem of finding optimal paths of all lengths on a directed edge-weighted grid graph. Variations of this problem appear in different contexts using similar frame-works. We propose sequential and parallel algorithms for it. Our algorithms have applications for *k-vertex problem* in one dimension, and *edit distance* problem in two dimensions. They offer fast solutions for special cases of these problems.

Keywords: Path optimization problem · Grid graph · Parallel algorithm · Dynamic programming · Suffix tree · Path graph

1 Introduction

A *directed path graph* (also known as *directed linear graph*) G is a graph whose vertices can be listed in a linear order $t_1, t_2, \ldots, t_n$ such that its edges are (t_i, t_{i+1}) for all $i \in [1, n-1]$. We represent a path graph G by its *vertex sequence* $t_1 t_2 \ldots t_n$ such that each edge (t_i, t_{i+1}) has a given weight $d[i, i+1]$.

We define a *directed grid graph* G in two dimensions as a graph whose vertices can be embedded into an $m \times n$ grid ($m, n \geq 1$) such that in this embedding, for each vertex (i, j), $((i, j-1), (i, j))$ is an edge iff $(i, j-1)$ is a vertex; and $((i-1, j), (i, j))$ is an edge iff $(i-1, j)$ is a vertex. In other words, each row and each column of a grid graph is a path graph. At the same time, a path graph is a special case of a grid graph in one dimension. From this perspective, for a grid graph, we can have a recursive definition which can be extended to dimensions larger than two.

We focus on grid graphs in one and two dimensions because these graphs have been used for important applications in the literature. We consider the problem of obtaining optimal paths of all lengths, where the *length of a path* is defined as the number of vertices on it.

We present output-size sensitive sequential and parallel algorithms for this problem for grid graphs in one dimension. We also propose a parallel algorithm for this problem for grid graphs in two dimensions.

The outline of this paper is the following: In the remaining part of this section (in Subsect. 1.1) we give the necessary definitions related to the problem statements in this work. In Sect. 2 we explain the methods we use in developing solutions for these problems. We present our algorithms and their complexity results

R. Silhavy (Ed.): CSOC 2022, LNNS 502, pp. 362–380, 2022.
https://doi.org/10.1007/978-3-031-09076-9_33

in Sect. 3. We remark the advantages of these algorithms, and contributions of this work in Sect. 4.

1.1 Definitions

Let G be a *directed grid graph* G in two dimensions whose vertices are embedded into a grid such that there is a one-to-one correspondence between lattice-points (i.e. positions in two dimensions) in this grid and vertices in G; and all edges to each vertex are the edges from its *predecessors*, where a predecessor of a vertex v differs from v in only one coordinate in which its value is smaller by one (e.g. for vertex (i, j), vertices $(i-1, j)$ and $(i, j-1)$ are its predecessors, and vertex $(i-1, j-1)$ is not). A grid graph in one dimension is a *directed path graph.*

A *directed path* in graph G is represented by a sequence of vertices $v_1 v_2 \dots v_k$ of *length* k, where each (v_i, v_{i+1}) is an edge. Let $c[k]$ denote the optimum total edge-weights over all paths of length k in G. Also, let $d[v_j, v_{j+1}]$ denote the weight of an edge (v_j, v_{j+1}) in G.

Figure 1 illustrates an example directed grid graph in two dimensions. For clarity, no edge weights are included in the figure.

We define the following optimization problem which we call *Optimal Paths of All Lengths in Directed Grid Graph in r Dimensions (ADG(r))*:

Definition 1. *[Problem ADG(r)]: Given a directed grid graph G in* $r \geq 1$ *dimensions, compute the optimum distance* $c[k]$ *for all possible path-lengths k in G.*

First, we study Problem $ADG(1)$.

Let $s = t_1 t_2 \dots t_n$ be a vertex-sequence for a directed weighted path graph. We refer to any subsequence $t_j t_{j+1} \dots t_{j+k-1}$ of length k as a *k-sequence* of s starting at position j, where $j \in [1, n-k+1]$.

For a given vertex-sequence $s = t_1 t_2 \dots t_n$ with distances $d[j, j+1]$ for all $j \in [1, n-1]$, and for a given $k \in [2, n]$, the *distance of the k-sequence* starting at vertex $i \in [1, n-k+1]$ is $\sum_{j=i}^{i+k-2} d[j, j+1]$. For $k = 1$, the 1-vertex sequence starting at vertex $i \in [1, n]$ is $s = t_i$, and $c[1] = 0$. Let $c[k]$ denote the optimum k-sequence with minimum (or maximum distance, based on the objective) obtained over all k-sequences in s for $k \in [2, n]$. More formally (without loss of generality, let the objective be minimization),

$$\begin{aligned} &c[1] = 0, \\ &c[k] = \min_{i \in [1, n-k+1]} \textstyle\sum_{j=i}^{i+k-2} d[j, j+1] \text{ for all } k \in [2, n] \ . \end{aligned} \tag{1}$$

In this case, Problem $ADG(1)$ can be rewritten as the following problem specialized for one dimension: Given a directed path graph G represented by vertex-sequence $t_1 t_2 \dots t_n$, for all $k \in [2, n]$, compute the optimum distance $c[k]$ of k-vertex directed sub-graph G_k starting at vertex t_i and ending at vertex t_{i+k-1} in G over all $i \in [1, n-k+1]$. That is,

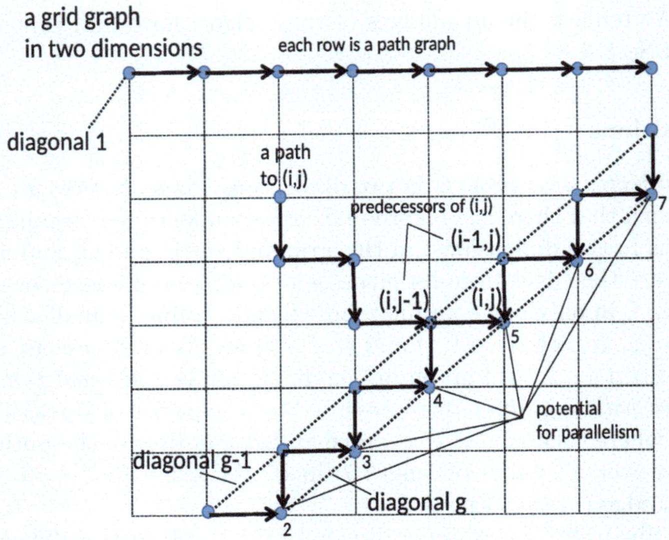

Fig. 1. An example grid graph in two dimensions. Each row (also each column) is a path graph (a grid graph in one dimension). Each vertex (i, j) has two predecessors $(i-1, j)$ and $(i, j-1)$. The figure also includes a directed path that ends at vertex (i, j). The diagonal that contains vertex (i, j) is named by integer $g = i + j - 1$. Two diagonals $g - 1$ and g are shown. Computations in each vertex (i, j) on diagonal g depends on the computed values in its predecessor vertices $(i-1, j)$ and $(i, j-1)$ on diagonal $g - 1$. Computations for all vertices in the same diagonal can potentially be done in parallel.

Definition 2. *[Problem ADG(1)]: Given a vertex-sequence $t_1 t_2 \dots t_n$ of n vertices with distances $d[j, j+1]$ between t_j and t_{j+1} for all $j \in [1, n-1]$, compute optimum k-sequence distance $c[k]$ defined in Eq. 1 for all $k \in [1, n]$.*

We note that Problem $ADG(1)$ can be a minimization or maximization problem. If the objective is maximization, we can negate the distances and solve the resulting minimization problem. In this case, optimum distances are the maximum, not the minimum distances. Therefore, we use both types of problems (minimization and maximization problems) interchangeably without reiterating this relation in each case.

We list several real-life applications for $ADG(1)$. For each application in this list, an instance of Problem $ADG(1)$ can be created. Consider, for example, a sequence $t_1 t_2 \dots t_n$ of stoppage points for which the distances between all consecutive two stoppage points are also given such that $d[j, j+1]$ is the total cost (including the energy costs, toll-road charges, parking fees) of travelling from stoppage point t_j to t_{j+1} for all $j \in [1, n-1]$. For trip organizers, the following query can be of interest: on a planned travel path (i.e. a sequence of stoppage points), what are the minimum costs of k consecutive stoppage points (i.e. k-sequence distances) for all k in $[2, n]$? The organizers can use these results in identifying best stoppage points sequence such that for all k-sequence of stoppage points, the costs are minimal assuming that the numbers of (paying) passengers are very similar over all trips of k-stoppage points (i.e. assuming that numbers of passengers are nearly independent of k) for all k in $[2, n]$.

For another instance for Problem $ADG(1)$, we list a problem in the domain of work-load analysis. For all $j \in [1, n-1]$, let $d[j, j+1]$ denote the number of database queries processed in hours interval $[t_j, t_{j+1})$. Given a sequence of hours (at which reports are generated) in ascending order $s = t_1 < t_2 < \ldots < t_n$ along with query numbers $d[j, j+1]$ for all $j \in [1, n-1]$, find the maximum number of processed database queries $c[k]$ (of k-sequences) for all $k \in [2, n]$. Finding these would be useful information for selecting the sequence of reporting hours such that reporting is done when the largest changes are expected within the last k reports (for all $k \in [2, n]$) assuming that database queries repeat similar hourly frequency-patterns over time (within the day everyday).

Problem $ADG(1)$ is related to a problem known in the literature as the *k-vertex sets* problem and studied mostly by researchers in computational geometry. The problem aims to find sets of k vertices based on topological criteria used for selecting vertices in the underlying number of dimensions. For example, the following is a variant of this problem in two dimensions [8,11]: Given a set P of n points in the plane, and a parameter k, compute smallest area/perimeter axis-aligned rectangle that contains k points of P. A number of other variations of this problem have been studied in the literature [1,5,8,9,11,12]. They use different topological criteria for selecting k vertices (points) such as diameter, variance, rectilinear diameter or perimeter [1], convex hull perimeter [9,12] or area [12]. These problems have found a number of applications in computational geometry (e.g. clustering [15], approximating Euclidean Travelling Sales-Person Problem [3]). Recent results in most studied variants in two dimensions can be found in [5]. The variant in one dimension has roots in [1]. This variant is stated in [13] as the following: "[David Eppstein] *Given a sorted list of* n *real numbers, find the smallest interval containing exactly* k *elements, for all values of* k *between* 1 *and* n." We call this problem $ASKPI$ (*All Shortest k-Point Intervals*). The statement in [13] continues with the following related open question: "*There is a simple dynamic programming algorithm that solves this problem in quadratic time, but the best known lower bound is only linear. Either describe a faster algorithm, or prove a bigger lower bound in some reasonable model of computation.*" We describe such a dynamic programming solution in Sect. 2. This would yield a $\Theta(n^2)$-time algorithm. It is easy to see that a natural lower bound for the time-complexity of Problem $ASKPI$ is linear (i.e. $\Omega(n)$).

A very recent work (Lemma 4.0.1 in [10]) describes a solution for finding the smallest interval containing k points in $O(n + (n-k)\log(n-k))$ time using $O(n)$ space for a given set P of n points in $\mathcal{R}^1$ and a given integer k such that $1 \leq k \leq n$. We note that the problem solved here is not for all k, but for a given single k. If we solve $ASKPI$ by calling this algorithm for all $k \in [1, n]$, the resulting time complexity would not be better than $O(n^2)$.

Reductions between two computational problems are studied to find these problems' relative complexities with respect to each other [6]. We show that $ASKPI$ reduces to $ADG(1)$ in linear time. To see this, consider a given ordered sequence $p_1 < p_2 < \ldots p_n$ of n real numbers in ascending order. From this sequence, we construct a vertex-sequence $t_1 t_2 \ldots t_n$ that represents a unique path graph G, and we set edge weights $d[j, j+1] = p_{j+1} - p_j$ for all edges (t_j, t_{j+1}), where $j \in [1, n-1]$. Clearly, for all $k \in [2, n]$, $c[k]$ calculated for the constructed instance of Problem $ADG(1)$ is the minimum distance for k-vertex sequence (i.e. smallest k-point interval) for the inputted instance of $ASKPI$ (trivially, $c[1] = 0$ is also the smallest 1-point interval length). Constructing the corresponding instance of $ADG(1)$ takes time $O(n)$. Solving $ASKPI$ by performing this reduction and then solving the resulting problem $ADG(1)$ takes the same asymptotic time as the time required for solving $ADG(1)$.

The synergy between optimization problems that arise in real-world applications and a wide range of algorithmic methods used for them has been very well discussed in the literature (see for example [16]). Our objective is to provide additional evidence for this synergy. While we list very practical real-world problems for Problem $ADG(1)$, we also propose solutions for it by following two very different approaches. One uses an efficient data structure originally employed by algorithms for very different computational problems. The other one uses parallelism.

We note that problems $ADG(1)$ and $ASKPI$ seem to be very similar in definitions. However, $ADG(1)$ has one important feature that makes it more general than $ASKPI$. It is that in $ASKPI$ the edge-weights (distances) are defined in a metric space since it is presented as a computational geometry problem. This requirement is dropped in $ADG(1)$. Therefore, $ADG(1)$ has more potential applications. $ADG(1)$ is more suitable for the applications we list in which the edge-weights are not distance metrics in any geometry. On an additional note, the algorithms we propose solve the more general problem $ADG(1)$, and therefore they are applicable to $ASKPI$, too. Developing an algorithm for $ASKPI$ asymptotically faster than $O(n^2)$ has been posed as a computational challenge by the research community. Our algorithms achieve this goal for special cases or by using parallel architectures.

Next, we study Problem $ADG(2)$, which is the case of the problem in Definition 1 with the number of dimensions $r = 2$. That is, it can be stated as the following:

Definition 3. *[Problem $ADG(2)$]: Given a directed grid graph G in two dimensions, compute the optimum distance $c[k]$ for all possible path-lengths k in G.*

In a directed grid graph in two dimensions, each vertex has two predecessors connected from at most one horizontal and at most one vertical arcs. Paths are formed by horizontal and vertical arcs. Figure 1 has a directed grid graph in which a directed path that ends in vertex (i, j) is shown.

In a grid graph in two dimensions, the length of a given path is the number of vertices, and its total weight is defined as the sum of edge-weights on this path.

For an example application of Problem *ADG*(2), suppose that we would like to compare two strings $X = x_1 \dots x_{m-1}$ and $Y = y_1 \dots y_{n-1}$, where $m, n > 1$. For this purpose, we consider editing $\#X$ into $\#Y$ by performing insert, delete, and substitute operations, where $\#$ is a character not included in the alphabet for X, Y, and all editing costs on $\#$ are zero. All possible edits between two strings can be analyzed on an *edit graph* [14]. If substitutions are not allowed or not advantageous, then the edit graph becomes exactly a directed weighted grid graph in two dimensions. In this graph, each horizontal edge corresponds to an insert operation on Y, and each vertical edge corresponds to a delete operation on X. Each of these operations has a given cost (i.e. edge-weight). Each position (i, j) in the graph corresponds to the length-i prefix string of $\#X$, and length-j prefix string of $\#Y$, where $\#$ is placed at position $(1, 1)$. For $1 \leq i' \leq i < m$, $1 \leq j' \leq j < n$, a path starting at vertex (i', j') and ending at vertex (i, j) defines a transformation of substring $x_{i'} \dots x_i$ in $\#X$ to $y_{j'} \dots y_j$ in $\#Y$. The total weight of this path is the editing cost of substring $x_{i'} \dots x_i$ into substring $y_{j'} \dots y_j$. We also note that the length of this path is $|i - i'| + |j - j'| + 1$ since the path length is the number of vertices on it.

Classical *local sequence alignment* problems [4,14,18] aim to find local similarities in two strings. They are related to a problem of finding optimal paths of all different path lengths, which can be found by solving Problem *ADG*(2). The results obtained would reveal local similarities for substrings of all different sizes.

2 Methods

The dynamic programming approach offers solutions to Problem *ADG*(r) in r dimensions. More formally, a dynamic programming description for Problem *ADG*(r) can be defined as the following: Let G be a grid graph in r dimensions. Each vertex v has at most r predecessors. Let $q[v, k]$ denote the optimum total weight of a path over all possible paths of length k ending at vertex v. For all vertices v in G,

$$\begin{aligned} q[v,1] &= 0, \\ q[v,k] &= \min\{q[u,k-1] + d[u,v] \mid u \text{ is a predecessor of } v \text{ in } G\}. \end{aligned} \tag{2}$$

Then for all k, $1 \leq k \leq$ maximum possible path-length in G,

$$c[k] = \min\{q[v,k] \mid v \text{ is a vertex in } G\}. \tag{3}$$

The general dynamic programming formulae in Eq. 2 and the solution in Eq. 3 can be rewritten more specifically for problems *ADG*(1) and *ADG*(2) to obtain additional insights into these problems. This requires redefining expressions for variables q and c as needed.

We first consider Problem *ADG*(1). Let $p_1, p_2, \dots, p_n$ be n real numbers in ascending order. Let $q[i, k]$ be the interval length containing k consecutive elements for all $k \in [1, n]$ ending at element p_i. We initialize $q[i, 1] = 0$ for all $i \in$

$[1, n]$. For all k in $[2, n]$, $i \in [k, n]$, we calculate $q[i, k]$ as $q[i, k-1] + d[i-1, i]$. Let $c[k]$ be the shortest interval length over all intervals containing k (consecutive) elements in $p_1, p_2, \ldots, p_n$. Clearly, $c[1] = 0$; and for all $k \in [2, n]$, we can compute $c[k]$ by finding the minimum $q[i, k]$ for all $i \in [2, n]$ over all possible interval lengths (i.e. for given $i \in [2, n]$, for all k such that $2 \leq i + k - 1 \leq n$).

The following is a dynamic programming solution for $ADG(1)$:

$$\begin{aligned} q[i,1] &= 0 \qquad \text{for all } i, 1 \leq i \leq n, \\ q[i,k] &= \min\nolimits_{1 \leq i \leq n-1,\ 2 \leq i+k-1 \leq n} \{q[i, k-1] + d[i-1, i]\} \\ &\quad \text{for all } k, 2 \leq k \leq n. \end{aligned} \tag{4}$$

Then, for all $k \in [1, n]$,

$$c[k] = \min_{1 \leq i \leq n,\ 1 \leq i+k-1 \leq n} \{q[i, k]\} \tag{5}$$

Equations 4 and 5 would yield a $\Theta(n^2)$-time algorithm.

The dynamic programming solution in Eqs. 2 and 3 can be translated to the following formula for $ADG(2)$: For a grid graph G embedded in a lattice of size $m \times n$, where (without loss of generality) $m \leq n$,

$$\begin{aligned} q[i,j,1] &= 0 \text{ for all i,j, } 1 \leq i \leq m, 1 \leq j \leq n; \\ q[i,j,k] &= \min \left\{ \begin{array}{l} q[i, j-1, k-1] + d[(i, j-1), (i, j)], \\ q[i-1, j, k-1] + d[(i-1, j), (i, j)] \end{array} \right\} \\ &\text{for all i,j,k, } 2 \leq i \leq m, 2 \leq j \leq n, 2 \leq k \leq i + j - 1. \end{aligned} \tag{6}$$

Then, for all $k \in [1, m + n - 1]$,

$$c[k] = \min_{1 \leq i \leq m, 1 \leq j \leq n, 1 \leq k \leq i+j-1} \{q[i, j, k]\}. \tag{7}$$

Equations 6 and 7 would yield a $\Theta(n^3)$-time algorithm.

For a special case of Problem $ADG(1)$ in which the edge-weights can be mapped to a fixed finite set, we propose an algorithm that uses a *suffix tree*. A suffix tree for a given sequence s is a labeled rooted tree in which each suffix of s is spelled out in exactly one branch [7,14,19]. Suffix trees are very efficient data structures that are used in algorithms in many different areas from data compression (e.g. [17]) to computational biology (e.g. [14]). For a comprehensive coverage of them, see [7,14]. In Fig. 2, a suffix tree created for string $ABBCABB$ is shown.

We also propose parallel algorithms for both problems $ADG(1)$ and $ADG(2)$. Our algorithms assume a parallel random-access machine (PRAM) with concurrent read and exclusive write capability (CREW) for the same memory location on shared memory. All processors can concurrently read from the same location. When a processor writes to a memory location (an element of an array), no other processors can concurrently access the same element-location for read or write. Our parallel algorithm for $ADG(1)$ exploits the parallelism based on a simple observation we make. Our parallel algorithm for $ADG(2)$ computes the values in the two dimensional grid in the order of diagonals. Diagonal by diagonal computation in two dimensional grid or matrix structures has been used in the literature for other parallel algorithms to guarantee availability of required elements when computing new elements [20]. We take this one step further. In order to guarantee exclusive access for writes to shared memory locations, every processor counts in a loop, array-element-index in cyclic order such that each processor starts at a different index, and waits for each other in synchronization points. By this way, each processor writes to a different array-index at any given time, eliminating the race condition.

3 Results

3.1 Algorithms for Problem *ADG*(1)

We propose two algorithms for Problem *ADG*(1). The first algorithm performs well for a special case of the problem. The second algorithm is a parallel algorithm that performs well for the general problem. Without loss of generality, we assume that the objective is minimization, therefore, the algorithms find k-sequences with the minimum distances.

For a special case of the problem, we assume that there is a constant number of distinct distances $d[j, j+1]$ for all $j \in [1, n-1]$. Each distinct distance can be represented by a symbol in an alphabet Σ. We clarify this on an example. Let an instance of Problem *ADG*(1) contain a vertex-sequence $t_1t_2t_3t_4t_5t_6t_7t_8$ of a path graph with the edge weights (distances) $d[1,2] = d[5,6] = 2$, $d[2,3] = d[3,4] = d[6,7] = d[7,8] = 1$, and $d[4,5] = 1.5$. We create a *distance sequence* $s = s_2s_3 \dots s_8$ such that each s_{i+1} is an element in Σ that corresponds to distance (edge-weight) $d[i, i+1]$. The distance-sequence created from the above example instance is $ABBCABB$, where $\Sigma = \{A, B, C\}$, and the distances $f(s_i)$ for symbols s_i in Σ are the following: $f(A) = 2$, $f(B) = 1$, and $f(C) = 1.5$. Figure 2, part (a) illustrates the distance sequence created from the given sequence of elements. For this case of the problem, we propose constructing and using a suffix tree. Figure 2 includes a suffix tree for the above described example instance. All suffixes of the distance sequence are listed (note that $\$ \notin \Sigma$ is appended to the original sequence s as the end marker) in part (b); and the resulting suffix tree is shown in part (c) of Fig. 2.

We present Algorithm 1 for Problem *ADG*(1) with a number of distinct distances bounded from above by a constant. Algorithm 1 uses a constant-size alphabet Σ whose elements are used for representing distinct distances. The algorithm takes as input a sequence $t_1t_2 \dots t_n$ of n elements and $n-1$ distances. These distances are $d[j, j+1]$ for all consecutive elements t_j and t_{j+1}, $j \in [1, n-1]$. The algorithm assumes that the distinct distances $d[j, j+1]$ are mapped to an alphabet Σ whose size $|\Sigma|$ is bounded by a constant. The algorithm is based on depth-first traversal on a suffix tree during which optimum k-sequence distances are updated. In line 1, if n is 1, then $c[1]$ is zero, and the algorithm returns. Otherwise, it continues. In line 2, the algorithm runs through the entire sequence and computes the n-sequence distance $c[n]$ as the

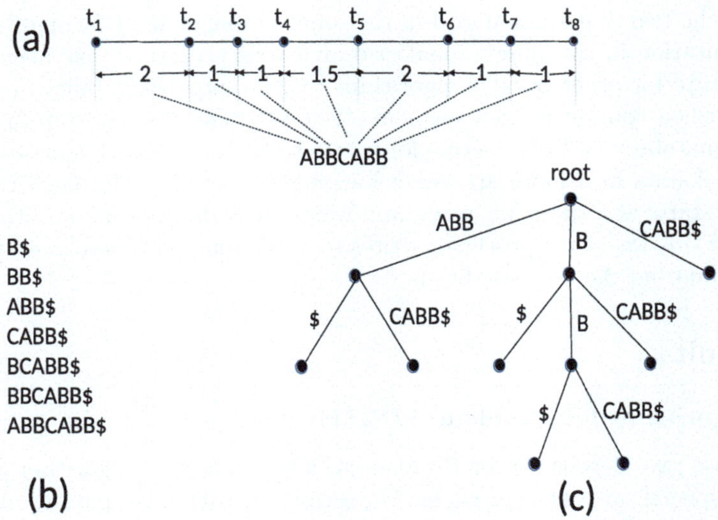

Fig. 2. An example instance of Problem $ADG(1)$: (a) the generated distance sequence; (b) list of all suffixes with end marker \$; (c) the constructed suffix tree.

sum $\sum_{j=1}^{n-1} d[j, j+1]$. In line 3 for all $k \in [2, n-1]$, it initializes all k-sequence distances $c[k]$ to the absolute value $|c[n]|$ of $c[n]$ as this value is the upper bound for all $c[k]$. These values are updated to correct values later in the algorithm. In line 4, the algorithm scans all distances $d[j, j+1]$ for all $j \in [1, n-1]$, and associates each next distinct distance with the next unused symbol in alphabet Σ. It creates a distance sequence $s_2 \ldots s_n$ from the given sequence $t_1 t_2 \ldots t_n$ of n elements and from distances $d[j, j+1]$ for all $j \in [1, n-1]$ using the alphabet Σ. Let $f(s)$ denote the distance associated with symbol $s \in \Sigma$. Line 5 creates a suffix tree T from the distance sequence $s_2 \ldots s_n$ using \$ as the end marker. For any node w, let x_w and y_w denote respectively, the total string length and total calculated distance from the root node to node w. More formally, if the concatenation of labels from the root to w is $s'_1 s'_2 \ldots s'_m$ (the end marker \$ is not included) for some $m \geq 1$, then $x_w = m+1$, and $y_w = \sum_{i=1}^{m} f(s'_i)$, where $f(s'_i)$ is the associated distance for symbol $s'_i \in \Sigma$. If the label of the edge from its parent to w is \$ only, node w is not processed as x_w and y_w are not defined in this case. For the root node $root$, $x_{root} = 1$ and $y_{root} = 0$. Lines 6 – 8 perform a depth-first traversal on T. The edge-labels are read and used during the traversal. Let $e_{u,v}$ be the label of the currently visited edge from vertex u to v excluding the end marker \$ (if it appears in $e_{u,v}$). Also, let $p[k]$ be a temporary distance calculated at a position at offset k from the root while processing symbols on edges. If $e_{u,v} = s_{x_u+1} s_{x_u+2} \ldots s_{x_v}$, for all $k \in [x_u+1, x_v]$, the distances $p[k]$ on this edge are calculated as the following: $p[x_u] = y_u$; $p[k] = p[k-1] + f(s_k)$ for all $k \in [x_u+1, x_v]$. During these calculations for all $k \in [x_u+1, x_v]$, if $p[k] < c[k]$ then $c[k]$ is updated to $p[k]$. When the depth-first traversal back-track to the ancestor node u, the depth-first traversal can proceed with another child node

Algorithm 1: Suffix tree based algorithm for a special case of Problem $ADG(1)$.

Data: An alphabet Σ of constant size.
Input: A sequence $t_1 t_2 \ldots t_n$ of n vertices with distances $d[j, j+1]$ between t_j and t_{j+1} for all $j \in [1, n-1]$.
Result: Minimum k-sequence distances $c[k]$ for all k in $[1, n]$.

1 If $(n == 1)$ return $c[1] = 0$
2 Calculate $c[n]$ as the sum $\sum_{j=1}^{n-1} d[j, j+1]$
3 For all $k \in [2, n]$, initialize $c[k] = |c[n]|$
4 For all $k \in [1, n-1]$, associate each distinct distance $d[j, j+1]$ between t_j and t_{j+1} with a unique element in Σ; let this symbol be s_{j+1}. Create a distance sequence $s_2 \ldots s_n$ from these symbols.
5 Create a suffix tree T from sequence $s_2 \ldots s_n$ using \$ as the end marker
6 **for** *all edges* (u, v) *in depth-first traversal in* T **do**
7   process the label $s_{u,v}$ one symbol at a time and update running $c[k]$ for the corresponding k on every symbol visited
8 **end**
9 return array c

of u. Stored values for variables x_u, y_u at ancestor node u help. They do not need to be recalculated. They are re-used in calculating new values when the edge-label string to the next child is processed. We also note that the end marker \$ does not contribute to the calculated distances.

We trace Algorithm 1 on the $ADG(1)$ problem-instance shown in Fig. 2. Figure 3 illustrates the intermediate results as the depth-first traversal progresses. For clarity, we did not name the nodes except for the root node. We also note that x, y are local to each node, and therefore, we drop the node-subscript in x, y. The values in $p[k]$ are the values calculated on the currently visited edges. The optimum $c[k]$ values are updated as the depth-first traversal progresses. When the traversal is completed, array c has the optimum value for all $k \in [1, n]$ ($n = 8$ in this example).

In Algorithm 1, line 1 takes constant time. Lines 2 and 3 take $O(n)$ time. For line 4, we propose a loop in which the index j runs from 1 to $n-1$. For every $d[j, j+1]$, in the list of used symbols of Σ, the associated distances are compared with $d[j, j+1]$; if $d[j, j+1]$ is found, the corresponding symbol is used, otherwise, a new symbol of Σ is added to this list, and it is associated with $d[j, j+1]$. Line 4 takes $O(n)$ time since the alphabet size is constant. The suffix tree construction in line 5 can be done by a parallel algorithm (e.g. [2]), or by a sequential algorithm (e.g. [19]) in the literature. Since our algorithm is sequential, we propose implementing line 5 in $O(n)$ time by using Ukkonen's linear-time suffix tree construction algorithm [19]. The total time taken in lines 6–8 is linear in the total string-length of the edge-labels. The total string-length of edges in a suffix tree T constructed from a string of length n can be anywhere between $\Theta(n)$ and $\Theta(n^2)$ [14]. The other lines of the algorithm take constant time. Therefore,

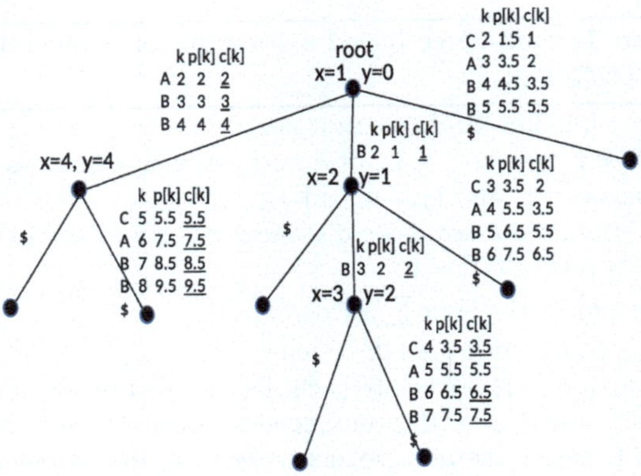

Fig. 3. Tracing Algorithm 1 on the $ADG(1)$ problem-instance in Fig. 2. The edge-labels are shown vertically. The x and y values are local for every node, and not shown for leaf nodes since they are neither defined (if the last edge-label is $) nor reused. Values of k, $p[k]$, and $c[k]$ are shown after processing each symbol of Σ on edge-labels during the depth-first traversal. When changes occur in $c[k]$, new values are underlined.

the time complexity of Algorithm 1 for this special case of Problem $ADG(1)$ is output-size sensitive. It is $O(|T|)$, where $|T|$ denotes the total edge-label lengths in suffix tree T constructed from the given instance of the problem with constant-size alphabet, and $|T|$ is between $\Theta(n)$ and $\Theta(n^2)$.

Our second algorithm is Algorithm 2, which is a parallel algorithm for the general case of Problem $ADG(1)$. That is, this algorithm makes no assumption for the problem parameters. It is designed for a concurrent read exclusive write (CREW) parallel random-access machine ($PRAM$).

In Algorithm 2, iterations of the loop in lines 1–12 are done in parallel using n processors having local variable p, accessing shared memory for concurrent read for array d, and for concurrent write for array c. Each $c[k]$ is exclusively accessed by processor P_k. Each iteration is handled by a separate processor. More precisely, processor P_k handles iteration k that calculates $c[k]$ by using its local variable p. In line 2, P_1 calculates and returns $c[1] = 0$; and processor P_k calculates the k-sequence distance $\sum_{j=1}^{j+k-1} d[j, j+1]$ starting at vertex 1 by adding the first $k-1$ distances, and initializes $c[k]$ to this sum. In line 3, p is set to this same calculated value for $c[k]$ in line 2 which serves as an upper bound for $c[k]$. The for-loop in lines 4–11 slides the starting position of the window for k-sequence by one position to the right in each iteration. The current k-sequence (i.e. the current window of k elements) is formed from the previous window by removing one element and adding another element as shown in consecutive lines in the last column in Fig. 4. The distance of the current k-sequence that starts at position j is p (which is also denoted for illustrative purposes by $d[j, j+k-1]$ in the comment line 5 of the algorithm). The running p in the current iteration is updated by adding $d[j+k-2, j+k-1]$ and subtracting $d[j-1, j]$. This is implemented in line 6. If the current k-sequence distance p is smaller than

Algorithm 2: Parallel algorithm for the general $ADG(1)$ problem.

Input: A vertex-sequence $t_1 t_2 \dots t_n$ of n elements with distances $d[j, j+1]$ between t_j and t_{j+1} for all $j \in [1, n-1]$.
Result: Minimum k-sequence distances $c[k]$ for all k in $[1, n]$.

1 **for** *all* $k \in [1, n]$ *in ascending order of* k *in parallel* **do**
2   If ($k == 1$) return $c[k] = 0$; else calculate $c[k] = \sum_{j=1}^{j+k-1} d[j, j+1]$
3   Set $p = c[k]$ /* p is local for each processor */
4   **for** *all* $j \in [2, n-k+1]$ *is ascending order of* j **do**
5   /* $p = d[j, j+k-1]$ */
6         $p = p + d[j+k-2, j+k-1] - d[j-1, j]$
7   **if** $p < c[k]$ **then**
8             $c[k] = p$
9   **end**
10   **end**
11 **end**
12 return /* $c[k]$ for all $k \in [1, n]$ are available in shared memory */

$c[k]$, then $c[k]$ is updated to p. When the loop in lines 4–11 terminates, $c[k]$ has the running minimum over all k-sequences. It is easy to see that this "sliding-window" method calculates the correct value for $c[k]$ for all $k \in [2, n]$. Processor P_k writes $c[k]$ in line 12. We note that for $k = 1$, the algorithm calculates and returns $c[1] = 0$ in line 2.

For the correctness of Algorithm 2, we note that for every $k \in [2, n]$ in lines 4–11, distance calculations for k-sequences appearing in two consecutive lines (in two consecutive sliding windows) include all the same distances except for two. In the next window, the distance between the first two elements in the previous window is subtracted from, and the distance between the last two elements in the next window is added to the running distance as shown in column three in Fig. 4.

In each of the n parallel processors, the loop in lines 4–11 takes $O(n)$ time; line 2 takes $O(k)$ ($k \leq n$) time; and all other lines take constant time. That is, each of the n processors takes $O(n)$ time. Therefore, the entire algorithm uses n processors and solves the problem in $O(n)$ time using CREW PRAM.

We remark that in our algorithms for Problem $ADG(1)$, for every $k \in [1, n]$, if an optimal k-sequence along with $c[k]$ is desired, then we can maintain k-sequence beginning index for an optimal k-sequence through the iterations by updating this information whenever $c[k]$ changes, and return it along with $c[k]$. This would not change the complexity of our algorithms.

3.2 A Parallel Algorithm for $ADG(2)$

We present Algorithm 3 for Problem $ADG(2)$. It is a parallel algorithm that implements the dynamic programming-based solution in Eqs. 6 and 7. Algo-

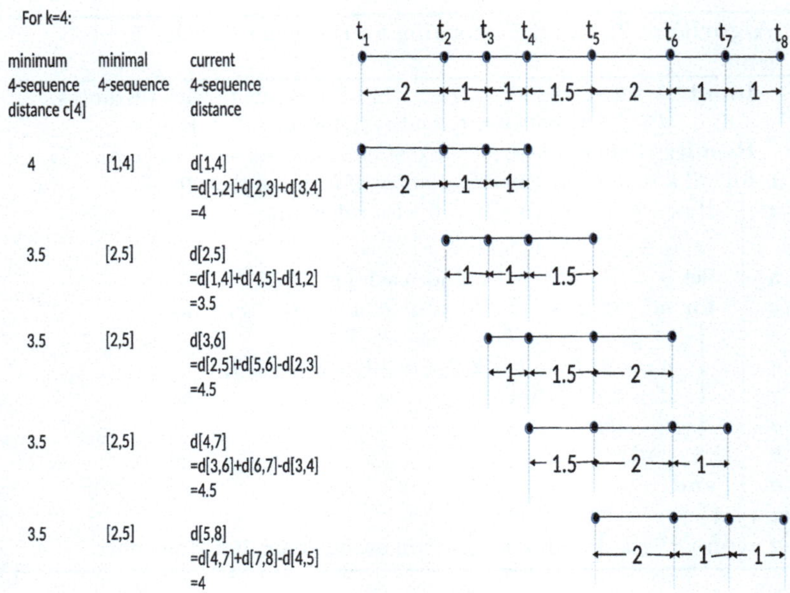

Fig. 4. Illustration of Algorithm 2 for Problem $ADG(1)$. The work of processor P_4 for $k = 4$ is shown (it executes lines 2–11 in the algorithm, and returns $c[4]$ in line 12). Each consecutive two rows summarize computations in one iteration of the loop in lines 4–11. In column three, for all $j \in [2, n-k+1]$, $d[j, j+k-1]$ denotes the distance of the k-sequence starting with vertex t_j and ending at vertex t_{j+k-1} whose positions are shown in column two in the current iteration. Current k-sequence distances $p = d[j, j+k-1]$ are calculated by using the previous value of p, and distances $d[j+k-2, j+k-1]$, $d[j-1, j]$ given in the problem instance. Distances of k-sequences $d[j, j+k-1]$ for $k > 2$ in column two are not needed separately, however, they are shown to simplify the explanation of calculations in column three. The running minimum k-sequence distance $c[k]$ is shown in column one after every iteration. Processor P_4 calculates and writes $c[4] = 3.5$ in this case.

rithm 3 computes $q[i, j, k]$ in diagonal order, where the index of the diagonal passing through vertex (i, j) is expressed as $i + j - 1$. In these calculations, only the current and previous diagonals are needed and saved in memory at any given time. More precisely, the diagonal entries (i, j) are calculated in the current diagonal g by using the computed values for predecessors $(i-1, j)$, $(i, j-1)$ in diagonal $g - 1$, and weights of edges $((i-1, j), (i, j))$, $((i, j-1), (i, j))$ that connect these predecessors to (i, j) in diagonal g. This dependency description of calculations allows for potential parallelism if they are done in ascending order of diagonal indices as shown in Fig. 1, and with some additional care as illustrated in Fig. 5. After the values in the current diagonal g are computed, the current diagonal g becomes the previous diagonal (with index $g-1$) in the next iteration. The calculated new values are collected in array $current[j]$ while needed values from the previous diagonal are found in arrays $prev_{up}[j]$ and $prev_{left}[j]$. Each of these is a linear array of integers of size $m+n-1$ stored locally in a processor j assigned for each column j. However, for clarity of presentation, we show them as two dimensional arrays $current[j,*], prev_{up}[j,*]$, and $prev_{left}[j,*]$, respectively. At vertex (i,j), for all k, $current[j,k]$ corresponds to $q[i,j,k]$, $prev_{up}[j,k]$ corresponds to $q[i-1,j,k]$, and $prev_{left}[j,k]$ corresponds to $q[i,j-1,k]$. For the next iteration, at each vertex (i,j), $prev_{up}[j,*]$ and $prev_{left}[j+1,*]$ will be updated from the values in $current[j,*]$. In Fig. 6, part (a) illustrates how $q[i,j,k]$ in Eq. 6 is calculated at vertex (i,j) for k. Part (b) shows the vertices $(i,j+1)$ and $(i+1,j)$ to which this calculated value is copied. This value plays a different role in these vertices as shown in part (b) of the figure. After every iteration on a new diagonal, improved values encountered in $current[j,k]$ (if any) are used to update the optimum $c[k]$ values for each k. Only one processor writes to $c[k]$ for any given k at any given time.

In more details, Algorithm 3 receives as input three integers m, n, and real distance array d, where expression $m \times n$ specifies the size of the directed grid graph in two dimensions; distances $d[(i-1,j),(i,j)]$ and $d[(i,j-1),(i,j)]$ are, respectively, the weights of the edges $((i-1,j),(i,j))$ (if exists) and $((i,j-1),(i,j))$ (if exists). Without loss of generality, we assume that $m \leq n$ (otherwise, we can transpose the input graph G). We note that at the boundary, some of these vertices do not exist (e.g. vertices $(0,1)$, $(1,0)$).

In line 1 of Algorithm 3, the minimum distance $c[1]$ for path length 1 is set to 0 (the distance obtained by a path of a single vertex and no edges); and for all other possible path lengths k in $[2, m+n-1]$, $c[k]$ is initialized to ∞ (i.e. maximum positive numerical value in the executing machine). In lines 2–4, all optimum distances of path length 1 at column j is set to $c[j,1] = 0$ for all $j \in [1,n]$. The first diagonal contains only vertex $(1,1)$. In lines 5–34, $c[k]$ for all k in $[2, m+n-1]$ are calculated and returned in line 35. The loop in lines 5–34 considers all diagonals g in $[2, m+n-1]$ in ascending order. For this loop, preparations are done to do calculations for all vertices (i,j) in diagonal g in parallel. For this purpose, in lines 6–11, calculated optimum distances from array $current$ are copied to arrays $prev_{up}$ and $prev_{left}$. Array $current$ has the optimum values calculated for the vertices in the current diagonal. These vertices are predecessors of the vertices in the next diagonal.

Diagonal and column numbers identify a unique vertex including its row number. The row numbers are used in detecting the borders of the grid. For the vertex at column j in the next diagonal, $current[j,k]$ in the current diagonal is needed as $prev_{up}[j,k]$ which is the optimum distance of a path of length k ending at the predecessor at the same column, but on the preceding row. For the vertex at column $j+1$ in the next diagonal, $current[j,k]$ in the current diagonal is needed as $prev_{left}[j+1,k]$ which is the optimum distance $q[i,j,k]$ of a path of length k ending at the same row, but on the preceding column. Therefore, arrays are copied by the assignment statements $prev_{up}[j,k] = prev_{left}[j+1,k] = current[j,k]$ in lines 7–10. These are also summarized in part (b) in Fig. 6. Each processor does them for a separate column j in parallel. They all wait in line 12 for all involved processors to complete. Calculations for the vertex at column 1

Algorithm 3: Parallel algorithm for Problem $ADG(2)$.

Input: Integers m, n for a grid graph G in two dimensions assumed embedded into an $m \times n$ grid. That is, vertices are (i, j), $1 \le i \le m$, $1 \le j \le n$. Input array d specifies the distances such that $d[(i', j'), (i, j)]$ is the weight of edge $((i', j'), (i, j))$, where $(i'j')$ is a predecessor of vertex (i, j).

Result: Minimum distances $c[k]$ of path-lengths k for all k in $[1, m + n - 1]$.

```
set c[1] = 0; and c[k] = ∞ for all k in [2, m + n + 1]
for all columns j in [1, n] do
    set current[j, 1] = 0
end
for all diagonals g in [2, m + n − 1] in ascending order of g do
    for all columns j in [1, min{g, n}] in parallel do
        prev_up[j, k] = current[j, k] for all k in [1, g − 1]
        if j < n then
            prev_left[j + 1, k] = current[j, k] for all k in [1, g − 1]
        end
    end
    wait until all processors reach here
    for all columns j in [1, min{g, n}] in parallel do
        set i = g + 1 − j; if i > m then set i = m;
        set k_start = j + 1; if k_start > g then set k_start = 2;
        for all k making a complete cycle of size g in increments of
          one from k = k_start until k = j by skipping k = 1 do
            if (j == 1) then
                current[1, k] = prev_up[1, k − 1] + d[(i − 1, 1), (i, 1)]
            else if (i == 1) then
                current[g, k] =
                  prev_left[g − 1, k − 1] + d[(1, g − 1), (1, g)]
            end
            else
                set
                  current[j, k] = prev_up[j, k − 1] + d[(i − 1, j), (i, j)]
                if
                  prev_left[j, k − 1] + d[(i, j − 1), (i, j)] < current[j, k]
                  then
                    set current[j, k] =
                      prev_left[j, k − 1] + d[(i, j − 1), (i, j)]
                end
            end
            if current[j, k] < c[k] then
                c[k] = current[j, k]
            end
            wait until all processors reach here
        end
    end
end
return /* c[k] for all k in [1, m + n − 1] are available in shared
memory */
```

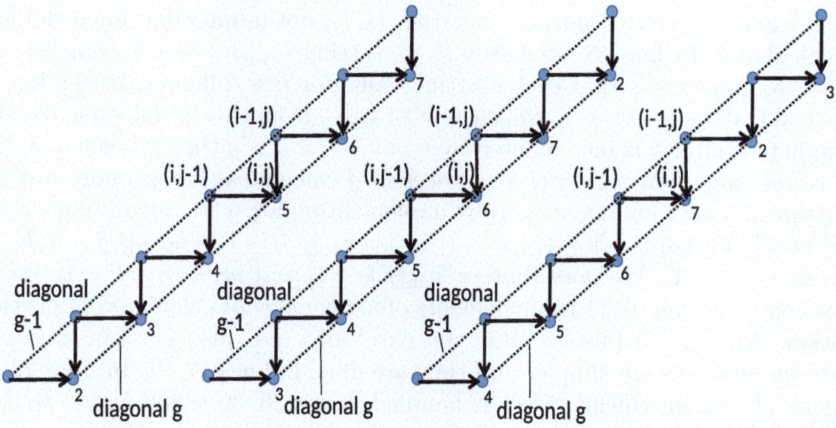

Fig. 5. An example grid graph in two dimensions. Calculations proceed diagonal by diagonal. At each vertex, counting is done cyclically starting with a different initial value. Three consecutive iterations on diagonal g are shown.

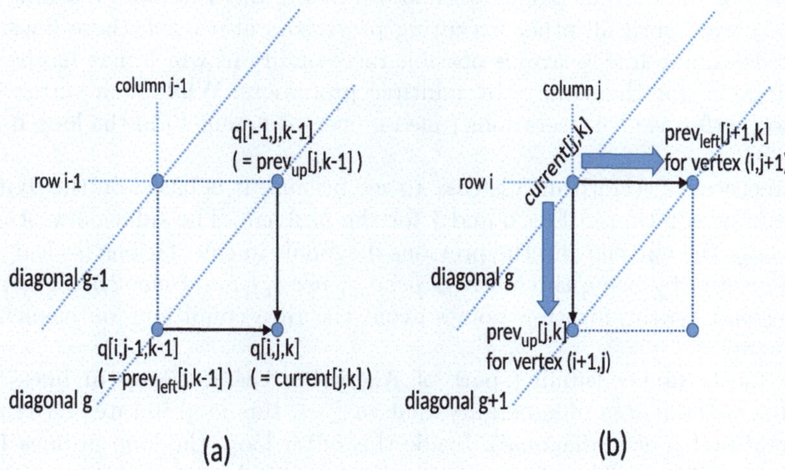

Fig. 6. Diagonal by diagonal dynamic programming calculations for Problem $ADG(2)$ in parallel: (a) calculating Eq. 6; (b) copying calculated values as preparation of the next diagonal addressed in the next iteration.

or row 1 in the next diagonal are handled in a special way in later steps (in lines 17–21). For each diagonal g, calculations for each column j are done in parallel by using $O(n)$ processors. For the given diagonal g and column j, the row number i is obtained for vertex (i, j) as shown in line 14. For each vertex (i, j), the optimum distances obtained by optimal paths ending at vertex (i, j) are calculated from the optimum values at its predecessors by using Eq. 6. This is done with some care to avoid the race condition caused by concurrent writes and reads for $c[k]$

for the same k. For this purpose, at each (i, j), optimum value for a different k is calculated. In line 15, at vertex (i, j), setting $k_{start} = j + 1$ ($k_{start} = 2$ if $k_{start} > g$) guarantees that each starting value for k is different. In the loop in lines 16–32, at each vertex, a cyclic count of size g from this initial value to $k = j$ is completed (after k is incremented by $k = k+1$, if $k > g$ then k is set to 2). For each count-value k at vertex (i, j), Processor j calculates the optimum distance for an optimal path ending at (i, j) by implementing Eq. 6 for calculating $q[i, j, k]$ in $current[j, k]$, using $q[i-1, j, k-1]$ in $prev_{up}[j, k-1]$, and $q[i, j-1, k-1]$ in $prev_{left}[j, k-1]$. We note that $prev_{up}[j, k-1]$ and $prev_{left}[j, k-1]$ are the values copied in lines 6–11 from elements of array $current$ calculated in previous iteration. When $j = 1$ (also $i = d$ in this case), or $i = 1$ (also $j = g$ in this case), these computations are simpler and they are done in lines 17–21. In lines 22–27, the general case in which $i, j > 1$ is handled. Lines 28–30 test if $current[j, k]$ is less than $c[k]$; and if so, $c[k]$ is updated in line 29. Each of the $O(n)$ processors executes the loop in lines 13–33 in parallel. At any given time, each processor can update $c[k]$ for a different k. This is guaranteed by the synchronization points in lines 12 and 31, and by cyclic counting starting with a different initial value in vertices of the current diagonal as shown in Fig. 5. Processors reaching lines 12 and 31 wait until all other executing processors also reach these lines. The synchronization in line 31 avoids possible race conditions which may happen due to writing $c[k]$ for the same k by multiple processors. While at a vertex, each processor performs $O(n)$ iterations (one for every different k) in the loop in lines 16–32.

Correctness of Algorithm 3 is easy to see because it is based on the dynamic programming solution in Eqs. 6 and 7 for the problem. The sufficiency of maintaining only the current and the previous diagonals in calculations is clear. This is implemented by using three arrays $prev_{up}$, $prev_{left}$, and $current$. The cyclic counting and synchronization points avoid the race conditions on elements of shared array c.

The most time-consuming part of Algorithm 3 is the loop in lines 5–34. Assuming without loss of generality that $m \leq n$, this loop iterates $O(n)$ times (one iteration for each diagonal). Inside this outer loop, the loop in lines 16–32 iterates $O(n)$ times. Therefore, the total time complexity of the algorithm is $O(n^2)$ using $O(n)$ processors. The algorithm also requires $O(n^2)$ total additional space due to arrays $prev_{up}$, $prev_{left}$, and $current$.

We note that in Algorithm 3, for every $k \in [2, m+n-1]$, if beginning and ending positions of an optimal path of length k along with $c[k]$ are desired, then we can carry these two positions along with the value of $c[k]$. This requires that the elements of arrays $current$, $prev_{up}$, and $prev_{left}$ include these two positions for each optimal distance value. Assignments to elements in these arrays, and changes in $c[k]$ in line 29 are done by also updating information about optimal paths. By this way, beginning and ending positions of an optimal path can be obtained in addition to $c[k]$ for all k. This would not change the complexity of the resulting algorithm. That is, all the described work can be done in time $O(n^2)$ using $O(n)$ processors and $O(n^2)$ additional space. If the actual optimal paths are also desired for any k, after finding the beginning position (i', j') and ending position (i, j) for an optimal path of length k (as described above), $O(n)$ processors can reconstruct a shortest path from (i', j') to (i, j) in parallel in additional time $O(n)$. If this is done for all k, then the required total additional time and space would be $O(n^2)$.

4 Discussions

We define a directed weighted grid graph as a graph that can be embedded into an r dimensional grid such that each vertex has an edge from its predecessor in each of the r dimensions. We define an optimization problem $ADG(r)$ that asks for optimum distances for all path-lengths in an r-dimensional grid graph for $r \geq 1$. We consider two special cases of this problem in one and two dimensions.

A directed grid graph in one dimension is a path graph. Problem $ADG(1)$ has practical applications one of which is the k-point sets problem (i.e., the problem of finding shortest k-point intervals in a set of n points on an axis). We present fast algorithms for $ADG(1)$. The first algorithm runs in linear time for a special case of this problem. The second algorithm solves the general case in linear time using linear number of processors on a CREW PRAM. This is an attempt in developing a sub-quadratic-time algorithm for this problem, which is presented as a computational challenge in the literature. Our proposed algorithms meet this challenge for a special case of the problem or by using a parallel architecture with a linear number of processors.

Problem $ADG(2)$ has applications for finding optimum local similarities (more precisely local distances between substrings of all total lengths in given strings). We present a parallel algorithm that assume CREW PRAM and runs in time $O(n)$ using $O(n^2)$ total additional space. This algorithm computes entries in diagonal order in parallel, and uses cyclic counting of array indices to avoid race conditions involving writes for the same shared memory locations.

Our definition of directed grid graph is not constrained to one or two dimensions only. It also includes larger dimensions. Problem $ADG(r)$ can have applications for dimensions r larger than two. In addition, the general ideas in the dynamic programming solution for $ADG(2)$, and our parallel algorithm for $ADG(2)$ can be used in developing solutions for Problem $ADG(r)$ with $r > 2$. These are topics for future work.

References

1. Aggarwal, A., Imai, H., Katoh, N., Suri, S.: Finding k points with minimum diameter and related problems. J. Algorithms **12**, 38–56 (1991)
2. Apostolico, A., Iliopoulos, C.S., Landau, G.M., Schieber, B., Vishkin, U.: Parallel construction of a suffix tree with applications. Algorithmica **3**(1), 347–365 (1988). https://doi.org/10.1007/BF01762122
3. Arora, S.: Polynomial time approximation schemes for Euclidean traveling salesman and other geometric problems. J. ACM **45**(5), 753–782 (1998). https://doi.org/10.1145/290179.290180

4. Arslan, A.N., Eğecioğlu, Ö.: Dynamic programming based approximation algorithms for sequence alignment with constraints. INFORMS J. Comput. **16**(4), 441–458 (2004). https://doi.org/10.1287/ijoc.1040.0097
5. Chan, T.M., Har-Peled, S.: Smallest k-enclosing rectangle revisited. Discret. Comput. Geom. **66**, 769–791 (2021)
6. Cormen, T.H., Leiserson, C.E., Rivest, R.L., Stein, C.: Introduction to Algorithms, MIT Press, Cambridge (2001). ISBN 978-0-262-03293-3
7. Crochemore, M., Hancart, C., Lecroq, T.: Algorithms on Strings, Cambridge University Press, (2007). ISBN 978-0-521-84899-2
8. Das, S., Goswami, P.P., Nandya, S.C.: Smallest k-point enclosing rectangle and square of arbitrary orientation. Inf. Process. Lett. **94**(6), 259–266 (2005). https://doi.org/10.1016/j.ipl.2005.02.013
9. Dobkin, D.P., Drysdale, R.L., Guibas, L.J.: Finding smallest polygons. In: Preparata, F.P. (eds.) Computational Geometry. Advances in Computing Research, vol. 1, pp. 181–214. JAI Press (1983)
10. Ehrencron, T.M.: Data structures for point enclosing problems on the circle and the square. Master thesis, ICA-5651808, Computing Science, Utrecht University (2021). https://dspace.library.uu.nl/handle/1874/404951
11. Eppstein, D., Erickson, J.: Iterated nearest neighbors and finding minimal polytopes. Discret. Comput. Geom. **11**, 321–350 (1994)
12. Eppstein, D., Overmars, M., Rote, G., Woeginger, G.: Finding minimum area k-gons. Discret. Comput. Geom. **7**, 45–58 (1992)
13. Erickson, J.: Open Algorithmic Problems (2021). https://jeffe.cs.illinois.edu/open/algo.html. Accessed 13 Feb 2021
14. Gusfield, G.: Algorithms on Strings, Trees and Sequences: Computer Science and Computational Biology. Cambridge University Press (1999). ISBN 0-521-58519-8
15. Hansen, B., Jaumard, B.: Cluster analysis and mathematical programming. Math. Program. **79**, 191–215 (1997)
16. Melab, N., Gmys, J., Korosec, P., Chakroun, I.: Synergy between parallel computing, optimization and simulation. J. Comput. Sci. **44**, 101168 (2020)
17. Na, J.C., Apostolico, A., Iliopoulos, C.S., Park, K.: Truncated suffix trees and their application to data compression. Theor. Comput. Sci. **304**(1–3), 87–101 (2003)
18. Smith, T.F., Waterman, M.S.: Identification of common molecular subsequences. J. Mol. Biol. **147**(1), 195–197, PMID 7265238 (1981). https://doi.org/10.1016/0022-2836(81)90087-5
19. Ukkonen, E.: On-line construction of suffix trees. Algorithmica **14**(3), 249–260 (1995). https://doi.org/10.1007/BF01206331
20. Xia, Z., et al.: A review of parallel implementations for the Smith-Waterman algorithm. Interdisc. Sci. Comput. Life Sci. (2021). https://doi.org/10.1007/s12539-021-00473-0

A Reinforcement Learning-Based Approach for Detection Zero-Day Malware Attacks on IoT System

Quoc-Dung Ngo(✉) and Quoc-Huu Nguyen

Posts and Telecommunications Institute of Technology, Hanoi 10000, Vietnam
dungnq@ptit.edu.vn, huu.nqh@gmail.com

Abstract. The rapid growth of Internet of Thing devices (IoT) carries various smart services, however the malicious attacks have been increase quickly either. Besides, the evolution of IoT Botnet malware in both quantity and sophistication. These malwares are detected using machine learning (ML) have been attended over recent years and received better detection performance. There have been many studies proposed IoT botnet detection based on supervised learning using static analysis and dynamic analysis. Although ML-based detectors have shown improve detection performance, they are sensitive with the malware evolution and sophistication such as new IoT botnet variants. In this work, we proposed a robust IoT botnet detection model, that utilize supervised learning combine with adversarial learning using system call feature, to detect zero-day attacks. Experiments on dataset demonstrate the ability of detect Iot Botnet as well as zero-day attack. The proposed model archives 88,94% accuracy on relatively known dataset and highly increase the ability of detect new variants of IoT Botnet.

Keywords: IoT Botnet · Dynamic analysis · System calls · Reinforcement learning · Adversarial learning

1 Introduction

The advancement of the Fourth Industrial Revolution carries various renovations in many industries in the world, especially Internet of thing (IoT) is one of the core technologies. Over last decade, IoT has been a promising technology over several researches and applications not only in industrial but also daily life. The developed potential of IoT is very huge due to its effect to improve life's quality via smart services such as smart home, smart factory, self-drive car, health cares, and others. The IoT devices including electronic devices, sensors, cameras,... are connected allowing them to exchange information and have communication ability. Recently, IoT devices are widely applied and become cheaper, so it's very easy to reach smart services. Predicted by Cisco [1], 29 billion IoT devices will be used by the end of 2023.

With the significant development of IoT devices in the market to meet the requirement of smart-services. The characteristic of IoT devices are resource limitation, low compute ability, low memory because IoT devices are demanded to adapt to various environments.

R. Silhavy (Ed.): CSOC 2022, LNNS 502, pp. 381–394, 2022.
https://doi.org/10.1007/978-3-031-09076-9_34

It leads to many IoT devices being manufactured without any security policy [2]. The missing mechanism and standard security have created many vulnerabilities exploited. Once hacker has the control of massive number IoT devices, it would be extremely huge attacks. The common scenario is hacker uses controlled IoT device to become part of IoT botnet [3]. In this way, hacker can create and extend IoT botnet, then botnet will be utilized to different attacks in the internet space. The consequences of IoT botnet attack are extremely damaging. Over recent years, one of the largest distributed denial of service (DDoS) attack using IoT botnet to Dyn, the DNS provider company in 2016 [4]. This attack used a giant botnet against Dyn, included tens of millions of devices—IP cameras, routers, printers and other smart gadgets from the Internet of Things. They flooded Dyn's site with requests claimed 1.2 terabits per second. The IoT botnet gaining popularity due to its irrefutable impact, is called Mirai [5]. IoT devices are heterogeneous, detecting IoT vulnerability as well as IoT Botnet becomes difficult. Therefore, detecting IoT Botnet is a necessary approach.

Detection of IoT botnet is a crucial task while malware keeps evolving to hide its malicious activities. To address IoT botnet detection, a lot of studies have been conducted with various machine learning algorithms [6–8]. There are categorized by three major types of malware feature analysis consist of static, dynamic and hybrid analysis. The static analysis extracts feature by analyzing the source code and related file in the system without run them such as printable string information (PSI), function length frequency (FLF), operation codes (opcodes) [9–11]. On other hand, dynamic analysis extracts features that require the running application such as requested permissions, application programming interface (API) calls, inter-component communication (ICC) features, system calls [12–14]. Several papers proposed a hybrid approach that uses both static and dynamic features to improve malicious activities detection [15, 16].

Even though many works on IoT instruction detection using several machine learning algorithms and get high accuracy, adversarial attack and zero-day attack recently are problems with almost machine learning algorithms. There are only limited works that have reported adversarial sample crafting and against as well as zero-day in IoT [17], especially IoT botnet detection.

In this paper, we propose a zero-day attack detection approach based on reinforcement learning and the dynamic analysis is utilized by extracting system calls. Our model uses Q-learning to generate adversarial samples and improve the intrusion detection system by early detecting zero-day attack. The main contributions of the paper include:

(1) Proposing a reinforcement learning approach in the generation of the adversarial sample;
(2) Experimenting the proposed method on the real-world dataset and zero-day attack dataset.

The remainder of this paper is organized as follows: In Sect. 2, we summarize the literature in the related area. Section 3 explains the static and dynamic features used in our proposed method for malware detection. Next, in Sect. 4, with a description of the dataset along with the evaluation of the experimental results. This section also provides a comparison of the results with existing works, followed by the conclusion in Sect. 5.

2 Related Works

In this section, we briefly discuss the works on IoT botnet detection as well as adversarial crafting and reinforcement learning approach studies in malware detection. In recent works, a lot of studies in analysis method: static analysis, dynamic analysis or both analysis to ensure the security of IoT devices. Nguyen et al. [18] has proposed an effort using hybrid analysis method. The integration mechanism, the pretreatment process, and the classification based on the integration features which has printable string information graph (PSI graph) and system call graph (SCG). The classifier used include Decision Tree (DT), k-Nearest Neighbor (KNN), SVM (RBF kernel), and Random Forest (RF). This approach was tested on a data set of 6520 samples dataset consisting of 4644 IoT botnet samples and 1876 IoT benign samples and reaching 99.18% accuracy. However, this method is limited by using obfuscation techniques and the sandbox supports only five types of architecture. For similar studies in detecting IoT botnet malware, Ngo et al. [19] proposed an approach using reinforcement learning model to generate adversarial samples then training the PSI-graph machine learning classifier. This model was evaluated by a data set with 10010 PSI graph samples including 6165 IoT botnet samples and 3845 benign samples. The classifier reaches 95.13% accuracy and from 40% to 94.10% after trained by add more 4083 adversarial samples into original data set. Even though, the limitation of this approach is inherent from static analysis methods, which cannot disassembly complex and sophisticated obfuscation samples as well as cannot catch the trigger-point of IoT botnet. Khoda et al. [20] presented a method to robust the classifier using deep neural network by crafting adversarial samples. The clean data set was extracted either static or dynamic features. Permissions, API calls, and ICC features (activities, android intents, content providers and broadcast receivers) were extracted as static features while System calls of during the runtime were logged as dynamic features. The subset of adversarial samples is combined with the original samples methods to retrain the classifier. Rather than selecting random adversarial samples for retraining, this work proposed two techniques for selecting adversarial samples. The data set was used for this study including 3000 malwares and 5000 benign applications. The accuracy of this method reaches 98.3% with clean data and increase from 71% to 86.08% after 65% adversarial samples was selected for retraining. This work may open the approach for selecting the adversarial samples but the experiment only for industrial IoT application and it is questionable to perform with other IoT malware datasets.

An another adversarial sample generation work was done by Liu et al. [21] proposed a model named testing framework for learning-based Android malware detection systems (TLAMD) for against adversarial attack. Firstly, the author attempts to train five different classifier models, including logistic regression (LR), DT, and fully connected neural network (NN). The adversarial samples were generated by use genetic algorithms to search for the minimum perturbation value that causes the detection model to be misjudged. The data set was used is DREBIN dataset with a total of 123,453 sample data for Android applications, including 5,560 malicious samples and contains as many as 545,333 behavioral features. The experiment of this framework showed the success rate close to 100% to make adversarial samples by using static analysis of dataset. The limitation of this work is black-box approach need frequent model requests and only for IoT Android malware. The similar work was done by Song et al. [22] with framework named multi-armed bandit (MAB-Malware) to adversarial sample generation. This framework has two modules including the Binary Rewriter and the Action Minimizer. The Binary Rewriter is reinforcement learning model to generate the adversarial sample with input is three categories of features (hash-based signatures, rule-based signatures, data distribution based features) and action space relatively each features. The output of Binary Rewriter is Evasive Samples are handled by Action Minimizer to remove redundant actions while keep the evasion. The dataset was used including 5000 samples Windows PE binaries and five target classifiers. Results show it has over 74%–97% evasion rate for two state-of-the-art machine learning detectors (EMBER, MalConv) and over 32%–48% evasion rate for commercial anti-virus in a pure black-box setting. However, the experiment dataset was relatively small and the only consist of malware and benign samples on Window OS. In the same respective, Abusnainay et al. [23] proposed approach by using static analysis including eight representations of malware binary. In this study, four machine learning algorithms were used to evaluated in the malware classification phase, namely LR, RF, Deep Neural Network (DNN), Convolutional Neural Network (CNN) for classifier. Adversarial samples generation phase was compared among six types. The dataset was collected consists of 3,000 malware samples of three families—Gafgyt, Mirai, and Tsunami and addition to 2,295 benign samples. The experiment obtained the most robust classifier, providing a stable 98.96% baseline accuracy. This works was performing with several methods to generate adversarial samples but its static and not flexible if adversarial attack model develops.

In the dynamic analysis with network traffic approach, Alauthman et al. [24] proposed a sophisticated traffic reduction mechanism, integrated with a reinforcement learning technique. The model was proposed includes five phases. First phase is network traffic capture by Jpcap [25] and packet reduction by separate network traffic into time-windows. The second phase is feature extraction which extract the connection features and represent the state in time-window. Next, malicious activity would be detected using neural network. The last is improving botnet detection by using reinforcement learning with the Partially Observable Markov Decision Process (POMDP) context. This work used three datasets include non-malicious and malicious network traffic and reach detection rate of 98.3%. Although this approach can reduce network traffic and get high detection rate, however it may not perform well if bot masters explore using rootkits.

We also discuss some works applying reinforcement learning to intrusion detection and classification. An approach using reinforcement learning for malware classification was proposed by Caminero et al. [26]. The author provide an intrusion detection framework named adversarial environment reinforcement learning (AE-RL) which using Deep-Q network to address the classification and adversarial crafting problem. The model uses two agents are classifier agent and environment agent corresponding purpose classifier and selector new features for randomly generate adversarial samples. Two datasets were used in the field of intrusion detection: NSL-KDD and AWID. The NSL-KDD dataset has 125973 training samples and 22544 test samples, containing 41 features while AWID dataset contains 154 features with 1.795,574 and 575,642 samples for the training and test datasets. The accuracy of this method can reach greater 80% even though using huge dataset. Despite of high performance of this method, it was only evaluated by normal malware and necessary to experiment how it performs in

IoT environment. On other hand, the reinforcement learning via Deep-Q Network algorithm was used by Sethi et al. [17] to classification malware based on network traffic feature vectors. The model named Multi-Agent Reinforcement Learning (MARL) was proposed by using two types of agent in network. The learning agent equipped in a router for collecting network data, then label node into non-intrusion or intrusion class. The central agent gets information from learning agent and feedback the reward based on the learning agent's action. Moreover, author robust the model by use the JSMA method to generate various adversarial samples for adversarial attack detection and attention-based for indexing learning agent. This work used NSL-KDD and CICIDS2017 dataset. NSL-KDD has total 1,48,517 records, including attacks and non-attacks, each record of the dataset contains 41 features. CICIDS2017 contains 28,30,743 records, each record of the dataset has 79 features. This approach obtained high accuracy in case against adversarial attack from 84.2% to 96.5% with CICIDS2017 dataset and from 86.8% to 94.2% with NSL-KDD dataset. This experiment also highlights the applicability of the proposed system under zero-day attack scenarios by adding more samples into test dataset. Nonetheless, this method used network traffic for dynamic analysis, it could be difficult if applied in the huge network due to large of mount traffic data should be handled.

Through analysis of the above studies, there are a lot of studies have been made approaching adversarial machine learning and reinforcement learning to improve IoT instruction detection. However, it is still quite less studies in zero-day malware attack on IoT system.

3 The Proposed Method

In this section, we present the proposed IoT Botnet intrusion detection model including the detail of data preprocessing phase, the classifier mechanism based on the features vector and a reinforcement learning model to zero-day attack detection improvement.

3.1 Overview

The proposed model includes five major components: System Calls Collector (SCC), Feature Vectors Extractor (FVE), IoT Botnet Detector (IBD), Adversarial Samples Generator (ASG), Reinforcement Learning Agent (RLA). The overview architecture is illustrated in Fig. 1. The operation of model is separated to two phases:

(1) The first phase is machine learning classification phase which the IBD will be trained by using different machine learning classifiers. Firstly, dynamic analysis is used to process the dataset. The Extensible Linking Format (ELF) files are handled by SCC using a sandbox environment, consequently we got the system calls graph are result of Sandbox. Those system calls are analyzed by FVE using graph2vec algorithm [27]. Afterward, graphical data of SCG is embedded into vector with fix dimension. Last step of this phase, the IBD uses several machine learning algorithms to classify this data is benign or malware.

(2) The second phase is classification improvement, the IBD is evolved by reinforcement learning to detect the zero-day attack. Initially, RLA uses reinforcement learning to find the best way to create the new structure system call graph. Then the ASG

combines the default malware system call graph with new system call graph from RLA. To ensure malicious properties of malware, ASG adds the new call graph from RLA into first node of default malware call graph to make the adversarial SCG. Next, the adversarial SCG is analyzed and extract feature vectors as same as first phase. Afterward, the IBD executes classify of the adversarial sample. Consequently, the result of IBD is the reward return the RLA, the result can be positive if this adversarial sample can defeat IBA and otherwise. Finally, the evaluation of model by using test dataset with zero-day samples to evaluate effeciency and accuracy. The detail processing of model is presented in the next section of paper.

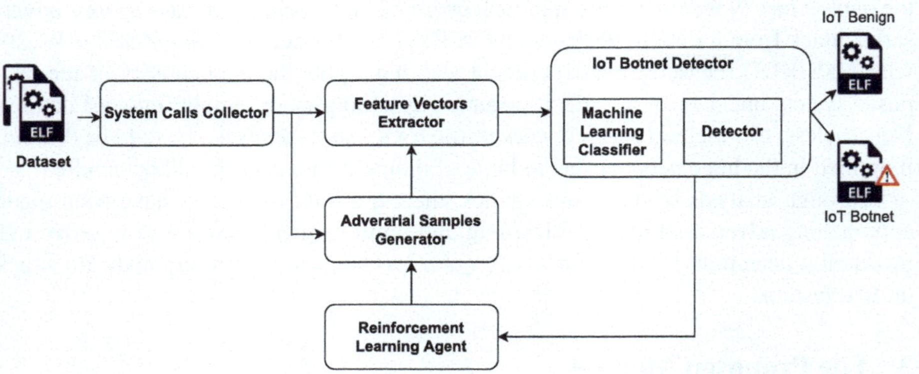

Fig. 1. Overview of proposed method

3.2 System Call Graph Collecting

The dynamic analysis approach is used to tracking the malicious behavior of IoT Botnet and collect the system call logs. Initially, the ELF files are executed in sandbox environment that has properties similar with IoT device. And the output is record of the system call data carrying the system's call information of build-in system operation as well as the call order. This system call is record of the interaction of IoT botnet with malicious behaviors such as try to scan port, brute force attack, attempt the connection to C&C server, detect other devices via open various service ports. To ensure the best condition of IoT Botnet, the sandbox require multi-architecture support ability, provide a dynamic library, C&C server demands emulator. Therefore, we used V-Sandbox [28] to collect malicious behavior information. System calls log files then are handled to generate SCG. We define the SCG as follows:

Definition 1: *SCG is a directed graph define as $G_{SCG}(V, E)$ where:*

- *V is the vertices set composed system calls with process id, name and argument.*
- *E is the edges set which each edge connected from vertex v_i to vertex v_j of the graph reflecting the relationship of two system calls. With loops, it counts as only one edge.*

Next step, The SCG is extracted through the system call data by looking at the call order of the input sample from V-Sandbox. The SCG includes vertices labeled with process id (PID), name, argument and edges are the connection between vertices. With the system calls log file contain the order of system call, the list of system calls will be generated corresponding the vertices and edges of SCG. Due to the property of "fork()" system call, IoT Botnet can generate child process to perform another task such as scan devices, open the connection with C&C server, therefore if the "fork" or "clone" system call belong to above list, a branch of graph will be generated by a recursive procedure. This branch of graph with child process can completely analyze to detect IoT Botnet behavior. The Fig. 2 illustrates the malware SCG start from execve() system call.

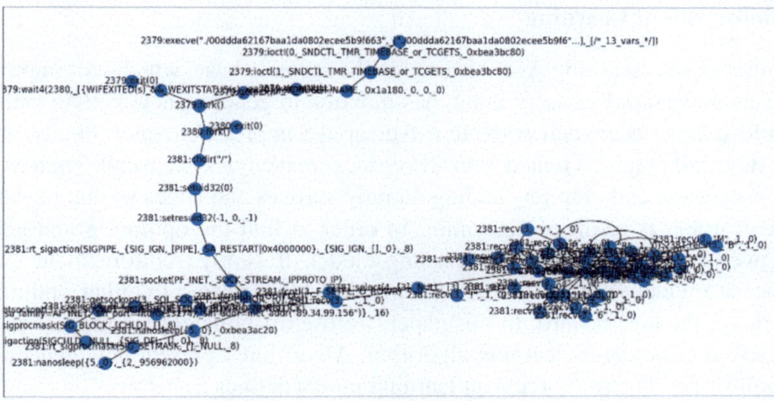

Fig. 2. System call graph of malware ELF file

3.3 Extract the Feature Vectors of System Call Graph

To adaption with the input of machine learning classifiers, the SCG data is present as a feature vector with fix-length not only because extract the feature and data normalization but also reduce computing time and cost. We used the graph embedding technique to solve the problem that each SCG carries the property of each input ELF files with different number of vertices and edges. The whole graph is converted into a feature vector, then it is fed to machine learning classifiers. In this paper, we use Graph2vec framework [27] proposed by Narayanan et al. The author considers the entire graph as a document and the rooted sub-graphs around every node in the graph as a word, processing the graph by extending Skipgram method for text (Doc2vec). The graph embedding is learned in an unsupervised learning method. Hence, they could be used for any task such as classification, clustering, and event for supervised machine learning methods. In the first phase, the initial system call graphs created by SCC are extracted to feature vector and then become the input of classifiers. On other hand, in the second phase, the FVE will extract feature vectors from the adversarial samples containing modified system call graphs and put them into classifiers once again. The feature extraction is illustrated as Fig. 3.

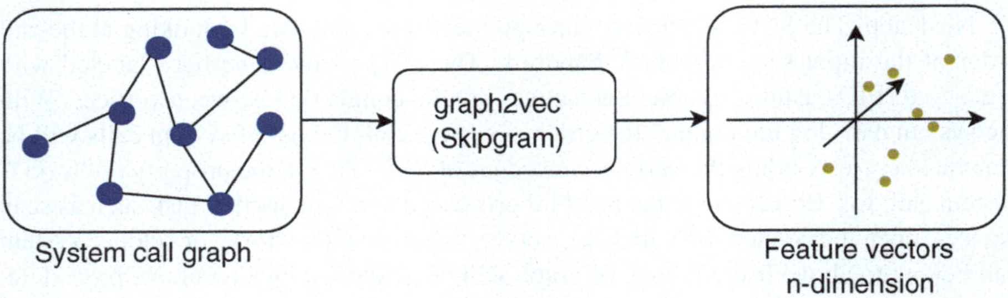

Fig. 3. Feature vector extraction

3.4 Reinforcement Learning

The Reinforcement Learning Agent is used in the second phase, which will improve the classifier by adversarial samples input, has function to generate new system call graph and then feed new system call graph to Adversarial Samples Generator. Firstly, the new initial system call graph is created with all vertices are benign system call. Then we try to modify this system call graph by adding dummy vertices and edges so that new benign system call graph is result of algorithm. In order to find the optimal graph generate method, we use the reinforcement learning model. Recently, reinforcement learning techniques are widely used to resolve the problems that related to either optimize the method or get the best reward. In this paper, we use off-policy Q-learning as a simple but effective reinforcement learning algorithm. Algorithm 1 present how generate new system call graph. The reinforcement learning model defines as below:

Environment contains the IBD that has responsibility return the reward to agent. The interact of environment to agent by give high positive reward when agent do some actions that make IBD classify wrong result and otherwise punish agent with a negative reward.

Action is the method agent choose to modify structure of graph by add a number of edges into graph. The action set is $A = \{a_1, a_2, a_3, \ldots, a_n\}$ corresponding add a number of edges into vertex a.

State is the vertices set of initial system call graph. The set of state is $S = \{x_0, x_1, \ldots, x_N\}$ where N is number of vertex. Agent moves to new state after execute an action. The state is considered to be a terminated state when the new system call can fool the classifier or exceed the number of step.

Reward r is number that agent received when take an action. If IBD classifies wrong result with an agent's action, it is a properly action and agent will get positive reward ($r = 1000$). Otherwise, if agent moves each step and the IBD classifies right result, a punishment is set with $r = -1$.

Algorithm 1: Reinforcement learning agent algorithm

Input: *The dummy benign SCG* G_{Dummy_SCG} (V, E)
Output: *The new benign SCG* G_{RL_SCG} (V, E)
Initialize *SCG* G_{Dummy_SCG} (V, E) *with N vertices and M edges, step = 0;*
Initialize *ε, γ, α (greedy factor, discount factor, learning rate);*
E ← Number of episodes;
for i in range 1 to E do:
  while step < 10 do:
    Select randomly a vertex j;
    Set current state as x_j*;*
    Select the action depends on greedy method;
      If IBD classifies a wrong result:
        Get reward r=1000 from IBD;
        Break;
      else reward r = -1;
    Calculate Q_value in state x_j *and update Q_table using Bellman's equation:*
    *Qnew(s, a) = r + γ*Qold(s, a);*
    step = step +1;
  end
end

The purpose of reinforcement learning model is make the misclassification of IBD by generating new dummy benign system call graph. This system call graph is combined with default malware system call graph in first phase and then extract to feature vector, finally fed to classifier.

3.5 Adversarial Sample Generation

According to malicious behavior of IoT Botnet, hacker may add some noise signals in malware by several ways with purpose fool IDS system. The noise signals are the benign behaviors, therefore malicious actions cannot detect by IDS.

Depend on this above approach, we modify the malware system call graph by adding benign system call. Because benign activities, the additional information of new system call graph will not change the malicious property of malware. Firstly, the new benign system call is generated by RLA. Next, ASG merges above system call graph with default malware system call graph from dataset. The combination of two system call by adding "fork" in the first vertex of malware system call graph, this fork system call generates a child process which is benign system call, as consider to noise signals while keep original malware system call graph structure. Finally, this new system call graph is adversarial sample and the malicious behavior of malware is not changed.

3.6 Classify by Machine Learning Algorithms

In this section, we present how we apply machine learning classifiers to detect IoT Botnet. After dataset was pre-processing by FVE, the input of classifiers are fix-length feature vectors. Finally, normalized dataset is fed to classical machine learning classifiers for training and testing. In this paper, we use popular machine learning algorithms such as Naive Bayes (NB), SVM, KNN, Decision Tree, Random Forest.

The classifiers were trained by two phases to improve the zero-day attack detection ability. Firstly, the default dataset is fed to classifiers, the known IoT Botnet of the default dataset then can be detected properly by classifiers. Lastly, the default dataset, which is slightly modified by RLA and ASG but keep the malicious behavior, is fed to classifiers. The adversarial dataset can improve the zero-day attack detection.

4 Experimental and Evaluation

This section presents the experiments to validate the effeciency and accuracy of proposed method. To evaluate the ability to detect zero-day IoT Botnet, in addition of conducting experiments with the proposed method, we also conduct the same experiment with two phases mentioned in 3.1 with unknown dataset. Afterward, we compare corresponding results and evaluate zero-day IoT Botnet detection ability of model. All experiments were performed on the same dataset, with a source-code written in Python and running on a PC configuration: Intel Xeon CPU E5–2650 v2 @ 2.60GHz × 4 with 8 GB RAM, Ubuntu 16.04 LTS. In addition, all source code was implemented using Python 3.7.1 and the library for ML task was scikit-learn [29]. The classifiers used include NB, DT, KNN, SVM (RBF kernel), and RF. The dataset is divided into a training set and testing set with a 7:3 ratio, in which the training set uses tenfold cross-validation during parameter refinement.

In order to verify the ability to detect new variant of IoT Botnet, we use two scenarios to evaluate the proposed model. The first scenario evaluates the machine learning classifier without adversarial learning meanwhile the second scenario evaluates the proposed model within adversarial learning. Then we compare the result to show the robust of proposed model in new IoT Botnet variant detection ability.

4.1 Data Description

Our dataset contains 1107 samples including 507 malware and 600 benign samples based on multi-architectures of microprocessor. IoT malware dataset is a set of 507 IoT Botnet malware samples, randomly selected from IoTPOT [30], a telnet-based honeypot which is now extended to other services. To determine if a file is malicious, we uploaded the samples on VirusTotal [30] and gathered the scan results corresponding to each of the malware. For sample labeling, each of the IoT malware, which is detected by at least one of the antivirus software scanners listed in VirusTotal, is considers as IoT Botnet and otherwise, the file is considered as benign. The IoT Botnet samples are collected belong to various variants such as Mirai, Bashlite, and others. This data set is separated to 2 sub-datasets to train and evaluate the efficiency of proposed model. The first sub-dataset includes 400 IoT Botnet malware samples and 600 benign sample while the second sub-dataset includes 107 newer IoT Botnet malware samples reported by VirusTotal. The second sub-dataset is considered as un-known malware and the first sub-dataset is considered as known malware. Consequently, the first sub-dataset is also divided into training set and testing set corresponding to 7:3 ratio. Meanwhile the second sub-dataset is used only for testing set for verification the ability to detect new IoT Botnet malware variants of proposed model.

4.2 Evaluation Criteria

The following criteria used for evaluating the effeciency and accuracy of the proposed method:

- True positive (TP): indicates that a malware sample was detected correctly and labeled as malware.
- True negative (TN): indicates that a benign sample was correctly detected and labeled as benign.
- False positive (FP): indicates that a benign sample was wrong and labeled as malicious.
- False negative (FN): indicates a malware sample was wrong and labeled as benign.

Based on the above criteria, we measure the following performance metrics:

- Accuracy = (TP + TN)/(TP + FP + TN + FN)
- TPR = TP/ (TP + FN)
- FPR = FP/ (FP + TN)
- Precision = TP/(TP + FP)
- F1 = 2*Precision*TPR/(Precision + TPR)

4.3 Results and Evaluation

In the first experiment scenario, the result in Table 1 show the ability of classification and ability to against zero-day attack. Before applying adversarial learning, the model has best performance with the Random Forest classification algorithm ($ACC = 0.8059$). Besides, the false positive rate is low *($FPR = 0.0111$)* at of the model show effeciency of machine learning algorithm to detect known IoT Botnet variants. With unknown malware, this classifier got low result with only detects *39/107* unknown IoT Botnet variants.

Table 1. Evalation results in first scenario.

Machine learning classifier	ACC	TPR	FPR	Precision	F1-score	AUC	Detect new malwares
NB	0.7125	0.5154	0.0389	0.9435	0.6667	0.8325	24/107
DT	0.7641	0.6256	0.0611	0.9281	0.7474	0.7822	34/107
KNN	0.7764	0.6035	0.0056	0.9928	0.7507	0.8895	27/107
SVM	0.7985	0.6564	0.0222	0.9739	0.7842	0.9527	37/107
RF	0.8059	0.6608	0.0111	0.9868	0.7916	0.9313	39/107

In order to verify the robust of proposed model using adversarial learning, we conducted the second experiment scenario with the same test dataset with first scenario. The result show as Table 2 has proven the better performance compared with first scenario. Using the same classifier with first scenario, the *ACC* result is increase to *0.855*, the FPR is still low ($FPR = 0.0444$) and TPR is increase ($TPR = 0.7753$). On other hand, the *detect new malwares rate* highly increase (*66/107*) of this model show the outstanding increase performance when detect new variants of IoT Botnet malware. In the second experiment scenario, the best classifier is SVM with $ACC = 0.8894$ and *detect new malwares rate* = *73/107*.

Table 2. Evalation results in second scenario.

Machine learning classifier	ACC	TPR	FPR	Precision	F1-score	AUC	Detect new malwares
NB	0.7101	0.5198	0.05	0.9291	0.6667	0.8649	25/107
DT	0.8378	0.7665	0.0722	0.9305	0.8406	0.8471	63/107
KNN	0.8673	0.7797	0.0222	0.9779	0.8676	0.9358	65/107
SVM	0.8894	0.8282	0.0333	0.9691	0.8931	0.9738	73/107
RF	0.8550	0.7753	0.0444	0.9565	0.8564	0.9549	66/107

Table 3 presents the comparison result between two scenarios. The comparison shows clearly the robust of proposed model after performing adversarial learning. Over the experiment scenarios, it is confirmed that the efficiency of adversarial learning as well as reinforcement learning to detect new variants of IoT botnet malware when co-operate with popular machine learning classification algorithm.

Table 3. Comparison result without adversarial learning and within adversarial learning

Metric	Scenario 1 (RF)	Scenario 2 (RF)
ACC	**0.8059**	**0.8550**
TPR	0.6608	0.7753
FPR	0.0111	0.0444
Precision	0.9868	0.9565
F1-Score	0.7916	0.8564
AUC	0.9313	0.9549
Detect new malwares	**39/107**	**66/107**

5 Conclusion

In this paper, we propose a robust machine learning classifier model to against zero-day IoT Botnet attack. Dynamic analysis based on sequentially structure of system calls is used to analyze and exploit IoT Botnet malware. The system calls are vectorized and transform to several representations as adversarial samples, then input for train a malware detection model using popular machine learning algorithms. We evaluated the performance of proposed model with real state-of-art dataset include 1107 samples. With two different scenarios, the adversarial learning is shown the robustness performance compared with original machine learning classifier in the context of zero-day attack. The experiments prove our method is effective, consequently it opens the detection method approach for our future work, which captures the unknown malware. However, the proposed model is still limited the structure of system call graph in adversarial sample, therefore sensitive with malware variant has complexible system call graph. In addition, the sandbox currently supports limited architectures. So that, our future work will continue to make proposed method adapt to more novel variant of IoT botnet. Simultaneously, we will perform more experiment with more dataset to evaluate the comprehensive IoT botnet detection model.

Acknowledgments. This work has been supported by Research Project: VAST01.06/21-22, Vietnam Academy of Science and Technology and funded by the Ministry of Science and Technology, Vietnam grant number KC-4.0-05/19-25.

References

1. Cisco Annual Internet Report - Cisco Annual Internet Report (2018–2023) White Paper. In: Cisco. https://www.cisco.com/c/en/us/solutions/collateral/executive-perspectives/annual-internet-report/white-paper-c11-741490.html. Accessed 23 Feb 2022
2. Kumar, S.A., Vealey, T., Srivastava, H.: Security in Internet of Things: challenges, solutions and future directions. In: 2016 49th Hawaii International Conference on System Sciences (HICSS), pp. 5772–5781. IEEE (2016)
3. Dange, S., Chatterjee, M.: IoT botnet: the largest threat to the IoT network. In: Jain, L., Tsihrintzis, G., Balas, V., Sharma, D. (eds.) Data Communication and Networks, vol. 1049, pp. 137–157. Springer, Cham (2020). https://doi.org/10.1007/978-981-15-0132-6_10
4. IoT role in Dyn cyberattack|Kaspersky official blog. https://www.kaspersky.com/blog/attack-on-dyn-explained/13325/. Accessed 23 Feb 2022
5. Kolias, C., Kambourakis, G., Stavrou, A., Voas, J.: DDoS in the IoT: Mirai and other botnets. Computer **50**, 80–84 (2017)
6. Thanh Vu, S.N., Stege, M., El-Habr, P.I., et al.: A survey on botnets: incentives, evolution, detection and current trends. Future Internet **13**, 198 (2021)
7. Ngo, Q.-D., Nguyen, H.-T., Le, V.-H., Nguyen, D.-H.: A survey of IoT malware and detection methods based on static features. ICT Express **6**, 280–286 (2020)
8. Kumar, A., Kuppusamy, K.S., Aghila, G.: A learning model to detect maliciousness of portable executable using integrated feature set. J. King Saud Univ. Comput. Inf. Sci. **31**, 252–265 (2019)
9. Nguyen, H.-T., Ngo, Q.-D., Le, V.-H.: A novel graph-based approach for IoT botnet detection. Int. J. Inf. Secur. **19**(5), 567–577 (2019). https://doi.org/10.1007/s10207-019-00475-6

10. Alasmary, H., Khormali, A., Anwar, A., et al.: Analyzing and detecting emerging Internet of Things malware: a graph-based approach. IEEE Internet Things J. **6**, 8977–8988 (2019)
11. Trajanovski, T., Zhang, N.: An automated and comprehensive framework for IoT botnet detection and analysis (IoT-BDA). IEEE Access **9**, 124360–124383 (2021)
12. Jeon, J., Park, J.H., Jeong, Y.-S.: Dynamic analysis for IoT malware detection with convolution neural network model. IEEE Access **8**, 96899–96911 (2020)
13. Le, H.-V., Ngo, Q.-D., Le, V.-H.: IoT botnet detection using system call graphs and one-class CNN classification. Int. J. Innov. Technol. Explor. Eng. **8**, 937–942 (2019)
14. Tong, F., Yan, Z.: A hybrid approach of mobile malware detection in Android. J. Parallel Distrib. Comput. **103**, 22–31 (2017)
15. Yen Y.S., Chen, Z.W., Guo, Y.R., Chen, M.C.: Integration of static and dynamic analysis for malware family classification with composite neural network. https://arxiv.org/abs/1912.11249 (2019)
16. Sethi, K., Madhav, Y.V., Kumar, R., Bera, P.: Attention based multi-agent intrusion detection systems using reinforcement learning. J. Inf. Secur. Appl. **61**, 102923 (2021)
17. Ngo, Q.-D., Nguyen, H.-T., Tran, H.-A., Nguyen, D.-H.: IoT botnet detection based on the integration of static and dynamic vector features. In: 2020 IEEE Eighth International Conference on Communications and Electronics (ICCE), pp 540–545. IEEE (2021)
18. Ngo, Q.-D., Nguyen, H.-T., Nguyen, V.-D., et al.: Adversarial attack and defense on graph-based IoT botnet detection approach. In: 2021 International Conference on Electrical, Communication, and Computer Engineering (ICECCE), pp. 1–6. IEEE (2021)
19. Khoda, M.E., Imam, T., Kamruzzaman, J., et al.: Robust malware defense in industrial IoT applications using machine learning with selective adversarial samples. IEEE Trans. Ind. Appl. **56**, 4415–4424 (2019)
20. Liu, X., Du, X., Zhang, X., et al.: Adversarial samples on android malware detection systems for IoT systems. Sensors **19**, 974 (2019)
21. Song, W., Li, X., Afroz, S., et al.: MAB-malware: a reinforcement learning framework for attacking static malware classifiers. arXiv preprint https://arxiv.org/abs/2003.03100 (2020)
22. Abusnaina, A., Anwar, A., Alshamrani, S., et al.: ML-based IoT malware detection under adversarial settings: a systematic evaluation. arXiv preprint https://arxiv.org/abs/2108.13373 (2021)
23. Alauthman, M., Aslam, N., Al-Kasassbeh, M., et al.: An efficient reinforcement learning-based botnet detection approach. J. Netw. Comput. Appl. **150**, 102479 (2020)
24. Zihao, S., Hui, W.: Network data packet capture and protocol analysis on JPCAP-based. In: 2009 International Conference on Information Management, Innovation Management and Industrial Engineering, pp 329–332. IEEE (2009)
25. Caminero, G., Lopez-Martin, M., Carro, B.: Adversarial environment reinforcement learning algorithm for intrusion detection. Comput. Netw. **159**, 96–109 (2019)
26. Narayanan, A., Chandramohan, M., Venkatesan, R., et al.: graph2vec: learning distributed representations of graphs. arXiv preprint https://arxiv.org/abs/1707.05005 (2017)
27. Le, H.-V., Ngo, Q.-D.: V-sandbox for dynamic analysis IoT botnet. IEEE Access **8**, 145768–145786 (2020)
28. scikit-learn: machine learning in Python — scikit-learn 1.0.2 documentation. https://scikit-learn.org/stable/. Accessed 24 Feb 2022
29. Pa, Y.M.P., Suzuki, S., Yoshioka, K., et al.: IoTPOT: analysing the rise of IoT compromises. In: Proceedings of the 9th USENIX Conference on Offensive Technologies. USENIX Association, USA, p. 9 (2015)
30. VirusTotal - Home. https://www.virustotal.com/gui/home/upload. Accessed 24 Feb 2022

Performance Analysis of an Ontology Model Enabling Interoperability of Artificial Intelligence Agents

Andrei Tara[1,2(✉)], Nicolae Taban[2], and Hjalmar Turesson[3]

[1] Lucian Blaga University of Sibiu, Victoriei 10, 550024 Sibiu, Romania
research@andreitara.com

[2] Openfabric Network SRL, Iosif Velceanu 16, 550057 Sibiu, Romania
nicu@openfabric.ai

[3] York University, Keele Street, Toronto 4700, Canada
hturesson@schulich.yorku.ca

Abstract. In the context of current information systems, the necessity of processing vast amounts of information enforces a significant paradigm shift. The monolithic data formats are gradually replaced with semantic models enclosing light, layered, and easily replicable structures. Advances represented by blockchain, decentralized data storage and replication along with semantic models open new possibilities to combine Artificial Intelligence (AI) algorithms in multi-agent, cross-organizational systems. By taking a semantic model approach, the communication and coordination between agents can be abstracted by the ontology concepts used to define the data models. This paper explores the feasibility of utilizing ontology-based semantic data models as communication interfaces among multiple AI agents. Specifically, it focuses on the storage, retrieval, and replication part of this process in a decentralized medium.

Keywords: Decentralized AI · Distributed ontologies · Blockchain

1 Introduction

Initially, AI technology gained popularity as a way to mimic human intelligence. Since then, it has seen growing applicability in various domains. The current AI landscape is centred around the development of standalone, monolithic and disconnected AI models suited for operating under constrained conditions [4] and particular use-cases [11]. While this approach has its clear-cut advantages, such as reducing the prototyping time, and simplifying the overall process of understanding complex models [12], it also has significant drawbacks, limiting the AI applicability to narrow-scoped use-cases. Overcoming this challenge demands a paradigm shift from the traditional monolithic data structures to distributed and collaborative semantic models, where multiple AI agents can interconnect and operate as a network of nodes capable of addressing more complex problems.

This paper explores the feasibility of multi-agent semantic interoperability by utilising a shared ontology model, the practicality, limitations, and performance

R. Silhavy (Ed.): CSOC 2022, LNNS 502, pp. 395–406, 2022.
https://doi.org/10.1007/978-3-031-09076-9_35

of this approach. It provides an experimentation environment in which decentralized technologies and blockchain solutions are analyzed. These technologies were chosen due to their inherent properties of transparency [20], trustworthiness, redundancy [16], fault tolerance [6] and tamper-resistance [13].

Pioneering research in decentralized AI was conducted in [7,8] from a fundamental perspective. These studies considered the case of autonomous homogeneous agents interacting in simulated environments [18] with a limited set of rules and interactions. Other papers extrapolated the biological behaviour of swarms of animals like ants, (which solve problems cooperatively) [9] stressing the importance of the environment [1] acting as a medium of message transmission. Subsequent studies [19,26] assigned different roles and actions to each agent as well as heterogeneous characteristics and explored the consensus possibility [25] for first-order [24] and second-order [21] multi-agent systems.

Significant advancements in designing multi-agent systems were introduced in [23] where ontologies were used to gain additional support for interoperability, reusability, and development activities (such as the agent reasoning process and overall system analysis).

As blockchain and distributed ledger technologies started to gain in popularity, they enabled a collaborative approach [10] where agents use blockchain as a communication medium. Approaches such as DInEMMo [17] facilitated access to the models to all paying participants in a marketplace, enabling simple search and retrieval. Other approaches such as [14] solve the communication medium problem by storing the models in smart contracts, but this limits continual updates and collaborative model building.

In this paper, the analysis is focused on exploring the performance of different DLT systems acting as communication pipelines linking heterogeneous AI agents into emergent networks. In order to achieve this goal, a change in perspective is required, namely fragmenting the monolithic ontology design into multiple layers. Each layer addresses a different aspect of the communication and entails a precise function [22]. We perform a comprehensive empirical study utilizing the Interplanetary Filesystem (IPFS) [3], Hyperledger Fabric [2] and Tendermint Cosmos [15].

2 Proposed Model

Shifting the paradigm from monolithic AI-capable data processing applications to collaborative AI agents demands that some preconditions be met. First of all, agents must be able to communicate their messages and to interpret the messages they receive. This level of cooperation is achieved only if data is wrapped in a common data format, which can be described via ontology concepts. Another precondition is that agents must be connected together in order to perform emergent tasks that compete with their traditional counterparts. Therefore, a model for interconnecting agents has to be provided. Last but not least, agents must operate on a platform that facilitates their communication. The rest of this section is dedicated to discussing these topics.

2.1 Data Format

Interoperability can be achieved if the messages that are sent among agents adhere to a flexible and abstract model. The model is represented by ontology concepts whose details are presented here. These concepts are designed to model complex messages and are described below. The architecture of a concept is composed of ten layers, each encoding an aspect of the message:

- Structural layer: enumerates all the properties which compose the concept, along with their (raw or complex) data types
- Connection layer: provides information about the locations of the complex data types present in the structural layer
- Encoding layer: specifies encoding formats for the concept's properties
- Defaults layer: provides implicit values for the properties in the structural layer
- Validation layer: verifies that the values stored in the concept's properties conform to their definition
- Restriction layer: performs validation across subsets of a concept's properties
- Naming layer: provides user-friendly terms for the properties
- Instruction layer: offers guidance about a concept's purpose and the manner in which it should be populated
- Versioning layer: contains information about the current version of a concept and requests for structural modifications
- Template layer: provides a series of views for the concept, for use-cases in which not all properties are necessary

The first six layers are machine-oriented and require specification if messages are to be interpreted correctly by agents. The last four layers are user-oriented and optional. They serve the purpose of adding context to the message, allowing it to be displayed and interpreted by the end-user (e.g. a message that contains the results of processing).

2.2 Agent Interaction

Figure 1 depicts the algorithm composition process. An AI is composed of algorithms (Al 1, Al 2 up to Al N - top dashed rectangle) where the manifests are composed of Input and Output ontology concepts. The algorithms are connected via the ontology concepts (A, B, C, D, E and F) (middle dashed rectangle). The composition of algorithms making up the AI is specified in the bottom layer (bottom rectangle). The decentralized storage holds the actual concepts, as the middle layer in Fig. 1 shows. The bottom rectangle in Fig. 1 shows the actual implementation of a composite AI, which is programmed using stubs. The stubs offer access to the methods of the subcomponent AIs and invoke their methods by using remote procedure calls to the algorithms, which are instantiated in execution environments.

The first step towards implementing the platform described in the Environment section below is to ensure that the ontology concepts can be stored

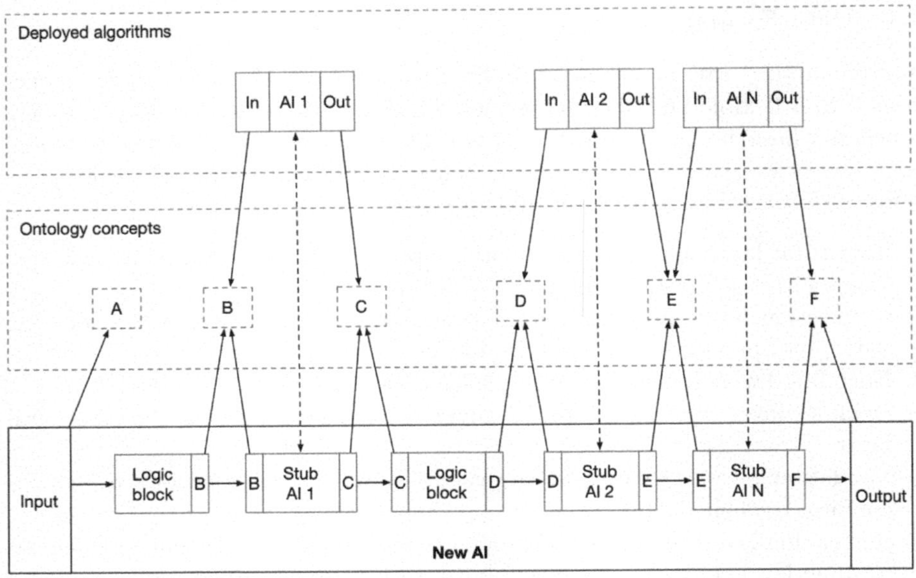

Fig. 1. Algorithm composition

efficiently in a decentralized environment. That is why the main focus of this paper is to evaluate the efficiency of storing ontology concepts on decentralized storage solutions which differ in the approach they take to storing assets. That is why we have chosen to perform our evaluations on Hyperledger Fabric, a permissioned blockchain that employs multiple types of specialised nodes, Tendermint Cosmos, a permissionless blockchain and IPFS, a peer-to-peer solution which is specifically designed for file storage.

2.3 Environment

Figure 2 presents the architectural overview of the proposed environment. Conceptually, this architecture contains all the necessary building blocks for creating a medium in which agents can be connected to perform tasks. Below, a brief description of the components is given.

The **Clients** represent the components through which data and algorithms can be introduced in the environment. They are responsible for issuing commands (e.g. starting algorithms), providing datasets and obtaining results from Coordinators.

A **Coordinator** acts as the principal component of the system. It is a software component that has multiple responsibilities:

- it receives high-level commands from Clients and transforms them into instructions for the components that run under its supervision
- it monitors the state of the components under its supervision (e.g. checking the responsiveness of AI algorithms and querying the application state)

- it generates transactions that will be sent to the Blockchain State Machine when it receives events from the components under its supervision (e.g. the successful execution of an algorithm)
- it validates incoming transactions received from the State Machines of other Coordinators and passes them to its State Machine
- it informs the Execution Environment Manager of what algorithm instances to create or stop
- it persists and fetches assets by communicating with the Decentralized Storage layer.

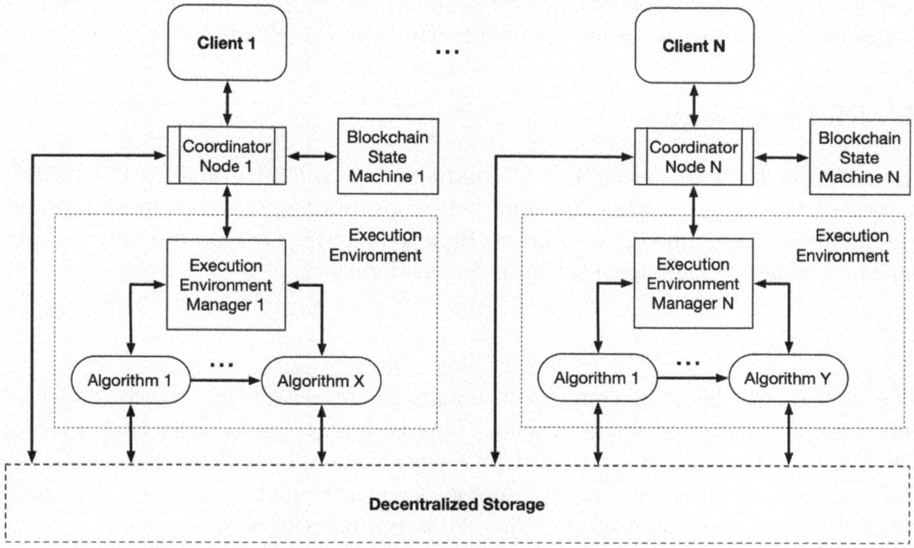

Fig. 2. Architectural overview

Maintaining the state of the application is accomplished via the **Blockchain State Machine**. The application's primary goal is to allow the decentralized execution of interoperable AI agents. Therefore, the Blockchain State Machine must store both the locations of the algorithms as well as their input and output concept definitions in the decentralized storage. Also, its state must contain information about the status of the AI agents, the locations of their required assets in the decentralized storage, as well as audit information that ensures traceability and transparency.

Allocating resources for the AI agents is done by the **Execution Environment Manager**. When instantiating agents, it must first create execution environments (e.g. Virtual Machines, Docker Containers, etc.), fetch the binaries and their input/output definitions from the decentralized storage and inject them into the execution environments. Furthermore, this module should monitor the states of the algorithms and report them to the Coordinator. Finally, it should return the results of the algorithms' execution to the Coordinator.

The **Decentralized Storage** holds all the assets that are necessary for instantiating and executing AI algorithms. Therefore, the decentralized storage contains the binaries for the AI agents as well as the input and output definitions, where the latter allows the algorithms to interpret the data they receive and format their results. Algorithm cooperation is accomplished by maintaining the format of their messages in ontology concepts.

Algorithms are intelligent artifacts that can be connected to perform emergent tasks. Their binaries are kept in the decentralized storage and are instantiated in execution environments. AI agents can use the results of other agents to complete their tasks, which leads to the concept of a mesh of AI algorithms that work together to solve a complex problem. They can either receive properly formatted data from other agents or directly from the client.

3 Methodology

As mentioned above, research was conducted in order to observe the performance of storing and retrieving ontology concepts some of the most popular decentralization technologies. This section defines the metrics and the configuration parameters that were taken into consideration.

3.1 Metrics

The goal of the benchmarking experiments is to measure the performance of the three decentralized environments. This is done by measuring two types of latencies: **write latency** and **read latency**. These metrics are used to observe the behaviour of different network topologies when trying to manipulate ontology concepts as the ones presented in the Datas Format subsection.

3.2 Configuration Types

Two types of configurations were applied, namely *network topology configurations* and *concept size variations*. Network size varies in our experiments between two, four, and eight nodes, and the topology for the chosen technologies differs as well.

Topology Configurations: The Tendermint network is composed of only one type of node, which is instantiated for the appropriate number of times, according to the network size. However, the nodes were operated in two different modes, namely Commit and Sync. Further details about them will be given in the Results section.

The IPFS topology was tested in two modes. The first topology only includes a network of IPFS daemons (regular peers), which transmit information among themselves only when it is requested. The second topology includes a Cluster network alongside the Daemon network, and each peer in the Daemon network has one correspondent in the Clustering network. Having a clustering network brings the benefit of disseminating the files received by one member of the Daemon network to all the peers in the same network, by emitting signals in the Clustering network and thus improving replication and fault-tolerance.

Hyperledger Fabric, being a permissioned blockchain, requires multiple types of nodes so that an operational network can be formed. Every configuration contains one orderer node that receives signed transactions from the peers, packages them into blocks and appends the blocks to the blockchain. The components that were varied in this topology are the peer nodes, which are responsible for validating, executing and signing the transactions. They were grouped into organizations of two and each organization also contained one Certificate Authority node, that is responsible for generating certificates that attest the identity and trustworthiness of the nodes in the organization.

Concept Sizes: The concept sizes are varied identically irrespective of the tested technology, from 10^1 properties per concept to 10^5 properties per concept. In absolute terms, the concept sizes are presented in Table 1. The sizes are approximated due to the fact that property names were assigned randomly using the Faker Java library and the names vary in size.

Table 1. Concept sizes in relation to the number of properties

Properties per concept	Concept size	Unit of measure
10^1	~41.4	KB
10^2	~54.5	KB
10^3	~186.1	KB
10^4	~1.5	MB
10^5	~14.6	MB

Out of the ten layers described in the Data Format section, the experiments vary the number of properties in the structural layer. At the same time, the naming layer is varied in accordance with the structural layer, because each property at the structural layer is given a new name that can be displayed to the user. Concretely, each concept is composed of ten files - one for each layer - plus a meta file which identifies the concept and its author. These files are the ones that are persisted to the decentralized storage for each concept.

Test Suite and Hardware Configuration: For the benchmarking experiments, a suite of tests, whose goal is to persist and read ontology concepts, was developed using the Java Microbenchmark Harness tool. The experiments adhere to the guidelines presented in [5] and each experiment consisted of five warm-up iterations of ten seconds each, followed by ten thirty-second measurement iterations. The experiments were run on a machine leveraging an Intel 7700HQ CPU, 24 GB of RAM and 1 TB of SSD.

4 Results

Figure 3 depicts the read and write latencies of the IPFS networks as functions of concept sizes and the number of peers. The figure also shows the two types

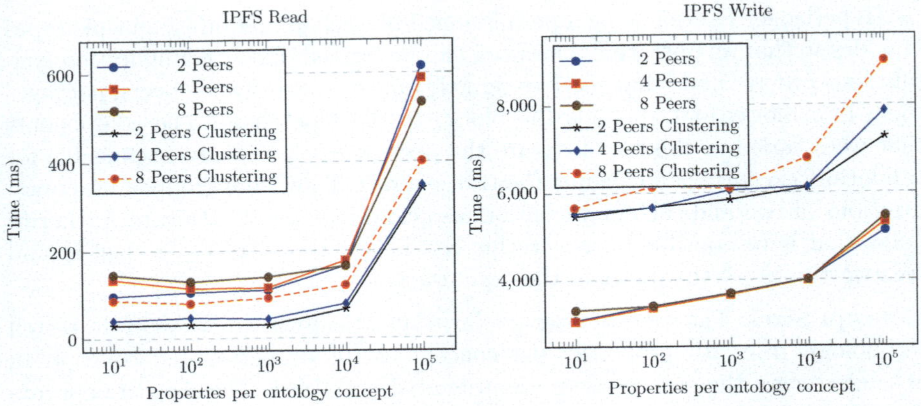

Fig. 3. IPFS results

of topologies described in Subsect. 3.2 (Daemon network vs Daemon network + Clustering network). The advantage of the composite topology makes itself evident in the case of the read operations. In this configuration, the Daemon nodes serve the content faster because they were prompted by the Clustering peers to replicate the ontology concepts amongst themselves. The differences between read latencies among configurations of the same topology stem from the concept sizes and the inter-node data propagation times (because, in the benchmarks, data was written on one node and requested on a distinct one each time).

However, the disadvantage of the composite topology becomes apparent in the case of the write operations. The differences in write latencies for the topology that only contains the Daemon network are insignificant at each concept size. On the other hand, the write latencies depend on the network size in the case of the composite topology because the concepts are propagated across the whole network.

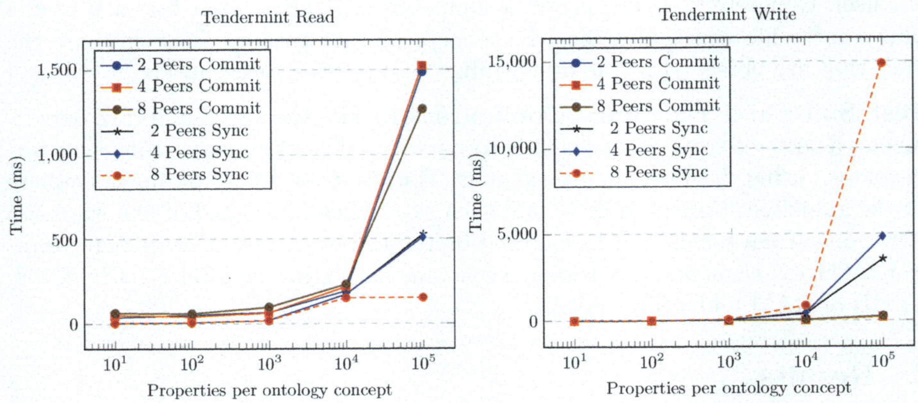

Fig. 4. Tendermint results

The results of the Tendermint experiments are outlined in Fig. 4. This figure also depicts two modes of operating the Tendermint nodes. Depending on the endpoint through which the transaction is passed to the node, it reports the success of the operation at different stages of the consensus protocol. Therefore, the Commit operating mode reports an operation as being successful when the transactions reaches the commit stage of the consensus protocol. The Sync operation mode reports a transaction as being successful at an earlier stage in the consensus. This allows for parallel processing of transactions but introduces issues as well. These issues are visible among the Synchronous Tendermint Read results, where apparently the 100.000 property concepts are read faster from an 8-peer network than from the 4-peer and 2-peer equivalents. This, however is actually a side-effect of the fact that the nodes cannot reach consensus when writing the block contents, and when the read is performed an empty response is returned. This failure in reaching consensus causes the nodes to timeout and disconnect from the network, a behaviour that acts as a clear limitation of the protocol. However, as long as the transactions are accepted into blocks, the Sync Strategy reduces the read and write latencies, as the figures indicate.

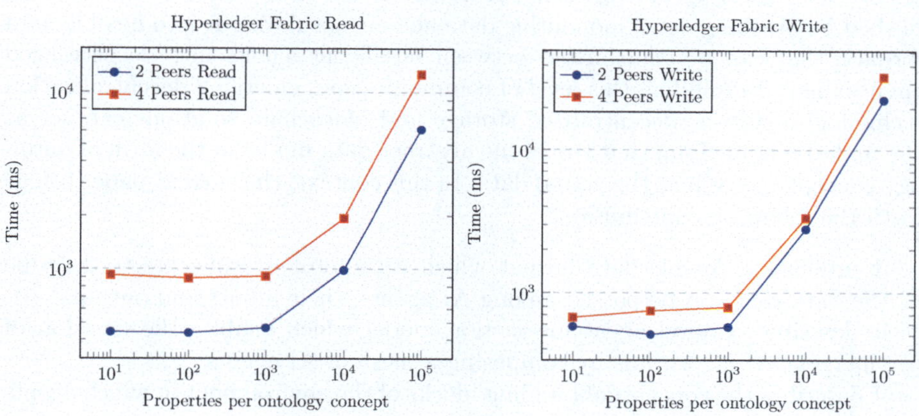

Fig. 5. Hyperledger Fabric results

The last set of results, those of the Hyperledger Fabric experiments, is depicted in Fig. 5. This set includes only the 2-peer and 4-peer network configurations because, as described in Subsect. 3.2, the Hyperledger blockchain is composed of multiple types of nodes, and they are resource-intensive. The nodes in the 8-peer Hyperledger network proved to be highly CPU and memory bound. While the memory bottleneck was alleviated by adding a swap partition, the network depleted all the CPU threads. This behaviour was not observed in the IPFS and Tendermint networks of similar sizes.

The results from the Hyperledger Fabric experiments demonstrate a clear dependence on networks size, where both the read and write speeds decrease as the number of peer nodes increases. In an 8-peer network (including the associated orderer and certificate authority nodes), this dependence becomes prohibitive, and the resource demands too high for the hardware platform used in the experiment. Thus, those results are omitted in Fig. 5.

Overall, the experiments indicate that the ontology concepts will be handled most efficiently if they are persisted on a decentralized storage that utilizes the IPFS substrate. Although the read latencies for IPFS and Tendermint are comparable, the write latency tips the balance in favour of IPFS. The experiments show that Hyperledger Fabric is limited to smaller networks and ontology concepts with fewer properties and that more complex configurations become too resource-intensive.

5 Conclusion

Present-day data-processing systems face an ever-increasing amount of information that has to be manipulated, interpreted and turned into insightful knowledge. However, the sheer quantity of data requires us to rethink the manner in which it is processed. By using lightweight semantic data models, multiple cooperative intelligent agents can collaborate in solving such tasks. This is accomplished by breaking down monolithic data and encapsulating it into flexible data formats that can be transmitted between intelligent agents in a decentralized environment. In fulfilling this level of communication, emerging decentralization technologies such as decentralized storage and blockchain solutions can act as the underlying medium that stores the abstract data model in the form of ontology concepts, as well as the actual data. In this context, the current paper brings forth the following contributions:

- it presents a flexible data format which uses ontology concepts to describe the data that can be passed among AI agents, their inputs and outputs
- it describes a novel agent interaction model which enables the creation of emergent AI algorithms by composing a mesh of AI agents
- it describes the conceptual building blocks of the environment in which agents can cooperate
- it evaluates the performance of the three widespread decentralization technologies (IPFS, Tendermint, Hyperledger Fabric) from the perspective of storing and retrieving ontology concepts

This area of research leaves plenty of room for exploration, and, as further development, the interconnection of AI agents will be evaluated from a performance standpoint. Furthermore, a study of the interactions between the components of the decentralized environment will be performed.

References

1. Admane, L., Benatchba, K., Koudil, M., Siad, L., Maziz, S.: AntPart: an algorithm for the unsupervised classification problem using ants. Appl. Math. Comput. **180**(1), 16–28 (2006)

2. Androulaki, E., et al.: Hyperledger fabric: a distributed operating system for permissioned blockchains. In: Proceedings of the Thirteenth EuroSys Conference, pp. 1–15 (2018)
3. Benet, J.: IPFS-content addressed, versioned, P2P file system. arXiv preprint arXiv:1407.3561 (2014)
4. Brayford, D., Vallecorsa, S., Atanasov, A., Baruffa, F., Riviera, W.: Deploying AI frameworks on secure HPC systems with containers. In: 2019 IEEE High Performance Extreme Computing Conference (HPEC), pp. 1–6. IEEE (2019)
5. Costa, D., Bezemer, C.P., Leitner, P., Andrzejak, A.: What's wrong with my benchmark results? Studying bad practices in JMH benchmarks. IEEE Trans. Softw. Eng. **47**(7), 1452–1467 (2019)
6. Crosby, M., Pattanayak, P., Verma, S., Kalyanaraman, V., et al.: Blockchain technology: beyond bitcoin. Appl. Innov. **2**(6–10), 71 (2016)
7. Demazeau, Y., Müller, J.P.: Decentralized AI, 2nd edn. Elsevier (1991)
8. Flores-Mendez, R.A.: Towards a standardization of multi-agent system framework. XRDS: Crossroads ACM Mag. Students **5**(4), 18–24 (1999)
9. Hamada, M.: An environment for simulating multi-agents based on ants behavior. In: Proceedings of the 7th Conference on 7th WSEAS International Conference on Systems Theory and Scientific Computation, vol. 7, pp. 294–299. Citeseer (2007)
10. Harris, J.D., Waggoner, B.: Decentralized and collaborative AI on blockchain. In: 2019 IEEE International Conference on Blockchain (Blockchain), pp. 368–375. IEEE (2019)
11. Höb, M., Kranzlmüller, D.: Enabling EASEY deployment of containerized applications for future HPC systems. In: Krzhizhanovskaya, V.V., et al. (eds.) ICCS 2020. LNCS, vol. 12137, pp. 206–219. Springer, Cham (2020). https://doi.org/10.1007/978-3-030-50371-0_15
12. John, M.M., Holmström Olsson, H., Bosch, J.: Architecting AI deployment: a systematic review of state-of-the-art and state-of-practice literature. In: Klotins, E., Wnuk, K. (eds.) ICSOB 2020. LNBIP, vol. 407, pp. 14–29. Springer, Cham (2021). https://doi.org/10.1007/978-3-030-67292-8_2
13. Kogure, J., Kamakura, K., Shima, T., Kubo, T.: Blockchain technology for next generation ICT. Fujitsu Sci. Tech. J **53**(5), 56–61 (2017)
14. Kurtulmus, A.B., Daniel, K.: Trustless machine learning contracts; evaluating and exchanging machine learning models on the Ethereum blockchain. arXiv preprint arXiv:1802.10185 (2018)
15. Kwon, J., Buchman, E.: Cosmos whitepaper (2019)
16. Liu, D., Zhang, Y., Jia, D., Zhang, Q., Zhao, X., Rong, H.: Toward secure distributed data storage with error locating in blockchain enabled edge computing. Comput. Stan. Interfaces **79**, 103560 (2022)
17. Marathe, A., Narayanan, K., Gupta, A., Manoj, P.: DInEMMo: decentralized incentivization for enterprise marketplace models. In: 2018 IEEE 25th International Conference on High Performance Computing Workshops (HiPCW), pp. 95–100. IEEE (2018)
18. Pipattanasomporn, M., Feroze, H., Rahman, S.: Multi-agent systems in a distributed smart grid: design and implementation. In: 2009 IEEE/PES Power Systems Conference and Exposition, pp. 1–8. IEEE (2009)
19. Ren, W., Beard, R.W.: Distributed Consensus in Multi-vehicle Cooperative Control, vol. 27. Springer, Heidelberg (2008)
20. Rizal Batubara, F., Ubacht, J., Janssen, M.: Unraveling transparency and accountability in blockchain. In: Proceedings of the 20th Annual International Conference on Digital Government Research, pp. 204–213 (2019)

21. Sun, Y.G., Wang, L.: Consensus problems in networks of agents with double-integrator dynamics and time-varying delays. Int. J. Control **82**(10), 1937–1945 (2009)
22. Tara, A., Butean, A., Zamfirescu, C., Learney, R.: An ontology model for interoperability and multi-organization data exchange. In: Silhavy, R. (ed.) CSOC 2020. AISC, vol. 1225, pp. 284–296. Springer, Cham (2020). https://doi.org/10.1007/978-3-030-51971-1_23
23. Tran, Q.N.N., Low, G.: MOBMAS: a methodology for ontology-based multi-agent systems development. Inf. Softw. **50**(7–8), 697–722 (2008)
24. Vicsek, T., Czirók, A., Ben-Jacob, E., Cohen, I.: Novel type of phase transition in a system of self-driven particles. Phys. Rev. Lett. **75**(6), 1226 (1995)
25. Wang, L., Xiao, F.: Finite-time consensus problems for networks of dynamic agents. IEEE Trans. Autom. Control **55**(4), 950–955 (2010)
26. Zheng, Y., Zhu, Y., Wang, L.: Consensus of heterogeneous multi-agent systems. IET Control Theor. Appl. **5**(16), 1881–1888 (2011)

Preterm Birth Prediction of Pregnant Women in Post Conization Period Using Machine Learning Techniques

Mian Ahmed Jamiul Bari, Mohammad Imtiaz Faisal, Mahmud Hasan(✉), Labiba Islam, Md. Sabbir Hossain, and Sifat Momen

Department of Electrical and Computer Engineering, North South University, Bashundhara, Dhaka 1229, Bangladesh
{mian.ahmed,imtiaz.faisal03,mahmud.hasan03,labiba.islam,sabbir.hossain11, sifat.momen}@northsouth.edu

Abstract. Pregnant women who underwent excisional surgeries (conization) for cervical intraepithelial neoplasia (CIN) display high risks of preterm birth. It is crucial to predict the risks of preterm birth amongst women in their post conization periods as this has severe consequences in terms of the cost as well as the health of the mother and the baby. This paper presents a preterm birth prediction system using machine learning approaches which will allow to evaluate the risk of a preterm birth. The dataset used in this work consisted of longitudinal cervical length (CL) of different gestational periods from 725 pregnant women undergoing surveillance programs in three clinics at London University Hospitals. Several machine learning algorithms were applied to make the prediction. Model based on decision tree achieved the highest accuracy (99.3%) on the test dataset.

Keywords: Cervical intraepithelial neoplasia · Machine learning · Preterm birth · Post conization · Supervised learning

1 Introduction

A baby born without completing a full 40 weeks of gestation period (born before 37 weeks) is considered to be premature baby and the term of the delivery is called preterm birth [10]. The severity of preterm birth depends on how early the delivery occurs. According to WHO, approximately 15 million preterm deliveries happen every year and about 1 million babies die due to the consequences of preterm birth. Furthermore, children who survive preterm birth face a multitude of health related issues. One particular condition that amplifies the risk of preterm birth is Cervical intraepithelial neoplasia (CIN) or Cervical dysplasia. It is regarded as a precancerous condition where abnormal cells grow on the epithelial tissues of the cervix [10]. Several excisional procedures (i.e. cone biopsy or conization procedures) such as CKC, LEEP and laser conization are

R. Silhavy (Ed.): CSOC 2022, LNNS 502, pp. 407–416, 2022.
https://doi.org/10.1007/978-3-031-09076-9_36

considered as effective treatments to counter CIN [11]. However, these conization procedures leaves a negative impact on a woman's pregnancy and greatly increases the risk of preterm birth. As early as 1938, the fact that conization may result in a negative effect on future pregnancy and cause preterm deliveries was recognized [11].

The risk factors of potential preterm birth delivery are numerous which make predicting preterm birth highly challenging. Several studies indicate that exisional surgeries considerably increases the risk factors that contribute towards preterm birth [12,13]. A study reported measurement of the cervical length (CL) and a short mid-trimester cervix which has a length less than 25mm is considered a strong predictor that may cause preterm birth to happen spontaneously [19]. Kindinger et al. [9] developed a traige model that performs risk categorization of preterm birth for post conization pregnancies into high and low using attributes such as longitudinal cervical length (CL). Tests like CL screening test, ultrasound screening, different forms of progesterone, indomethacin therapy, cervical cerclage and constantly under supervision of medical staff are high-cost clinical practices which are commonly used to prevent or limit the risks of preterm birth to a safe threshold [1]. On the contrary, the ability to predicting preterm birth using machine learning (ML) approaches will be more cost-effective. Recently, ML based models have proven to be highly competent in detecting various diseases (see e.g. [3,6,14,20] and [16]). This paper proposes an approach that inducts machine learning techniques in this vast field of study of predicting preterm birth of women in post conization period. In this paper, we present a system that will intelligently predict term or preterm birth.

The dataset that was used to train the model was collected from [7]. The proposed solution and methodologies have been adequately described in Sect. 3 with visual representations, data collection methods, data pre-processing followed by the application of several machine learning algorithms. The developed model has been trained with several suitable machine learning algorithms. The detailed study regarding the evaluation of the model based on its generalization capabilities and accuracy is discussed in Sect. 4 for better understanding.

2 Related Works

Many investigations and studies have been published on the elevated risk factors for preterm birth in women with excisional surgery (conization) for Cervical Intraepithelial Neoplasia (CIN) prior to pregnancy. Based on these studies, examinations are being conducted to either prevent preterm birth or predict the possible risk factors for preterm birth of women in their post-conization period.

From May 1996 till April 2009, Nam and colleagues [13] conducted studies by reviewing the medical data of all Korean patients who had cervical conization and delivered healthy singleton infants in a Korean hospital. Potential risk factors for preterm birth were observed to be the type of excisional surgery performed and second trimester cervix lengths at different gestational periods. Other research have backed it up as well [12].

Kindinger et al. [9] supervsied a retrospective analysis in the preterm surveillance clinics from January 2004 to January 2014, situated in three London University clinics that specializes in caring pregnant women of post conizaiton phase. Their research also discovered that women who had formerly undergone excisional surgery had an elevated risk of premature birth. Furthermore, they revealed that targeted cervical cerclage can minimize the risk.

From January 2013 to December 2014, Miyakoshi and colleagues [12] exploited the perinatal records of Japanese women to do a retrospective research of a group of singleton pregnancies. Their findings revealed that pre-pregnancy excisional cervical operation is linked to a higher risk of preterm delivery. Japanese cohorts have corroborated it as well [2,8,15,17].

Crane et al. [5] selected pregnant women admitted in a healthcare center situated in Canada with singleton pregnancies between June 2001 and June 2004 for a prospective cohort study. Women undergone surgical treatments for CIN such as CKC, LEEP, or cryotherapy previously and with singleton pregnancies were incorporated in the study based on the criteria. It was observed that women with past experience of conization surgeries had their cervix length reduced when related to groups of women with no experience of conization. Furthermore, LEEP and CKC were linked to spontaneous preterm delivery, with a cut-off of < 3.0cm being the best suitable for predicting spontaneous preterm delivery in the LEEP category at <37 weeks.

As can be seen, statistical analysis has yielded a number of impressive results when it comes to identifying major risk variables in premature births. However, what sets this study apart from others is that it seeks to predict a desired outcome, namely whether a pregnant woman in post-conization would have a term or preterm birth. Machine learning techniques will be used to make this prediction. According to our knowledge, no previous research based on this subject using a machine learning approach has been undertaken.

3 Proposed Work

This paper proposes a machine learning solution that develops a prediction model using several machine learning algorithms that predicts and classifies the risk into either high or low preterm birth. The work is based on women who have had excisional surgery (conization) done in their past. The dataset used in this work was collected from an open access dataset repository called Dryad and was created by Kindinger et al. [7]. Data of 725 pregnant women undergoing preterm surveillance program for their post conization period at three clinics of London University Hospitals were collected over a period of ten years [7]. Features of this dataset comprises longitudinal cervical length, age, ethnicity, gestational age at cl meaning the cervical length at different gestational weeks, gestation at spontaneous delivery. The dataset consisted of a target label that led us to use supervised machine learning techniques. The target label of this dataset is whether the birth is preterm or not. The methodology adopted in this work is shown in Fig. 1.

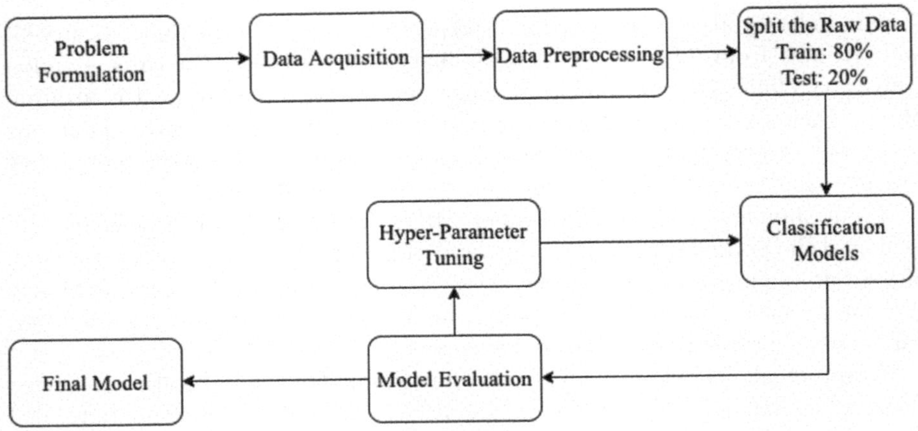

Fig. 1. Schematic diagram.

3.1 Problem Formulation

This paper addresses a severe problem of women with past experience of conization surgeries have higher risks of delivering preterm birth. A solution of developing an effective machine learning model is introduced in this paper which can be used to predict if a pregnant woman in her post-conization phase will deliver a preterm delivery or not.

3.2 Dataset Acquisition

The dataset was compiled by collecting data of women in their post conization periods undergoing treatments at three preterm surveillance clinics in London for ten years.

3.3 Data Preprocessing

Data preprocessing was applied so that machine learning algorithms can use the translated form of raw data efficiently.

Preprocessing of Predictor Attributes: Categorial attributes like ethnicity and cerclage were converted into numerical values so that they can be used for training the model. We filled the missing values of age for the age column using interpolation.

Preprocessing of the Class Attribute: Class attribute is the one that we wish to predict. We encoded the preterm variable from categorical to numerical values. We used 0 to represent term entries and 1 to represent Preterm entries.

Class Imbalance Issue: Our target variable had a class imbalance issue. There were 613 term entries and only 71 preterm entries. To overcome this problem Synthetic Minority Over-sampling Technique (SMOTE) [4] was applied. Using this technique, we synthetically increased the number of the minority Preterm entries and made it equal to the majority Term entries - thus we managed to get a balanced dataset.

3.4 Dataset Splitting

The original dataset was divided into two portions: a train set consisting 80% of the data, and a test set consisting 20% of the data respectively. After data pre-processing, there remained 684 instances where 547 of instances remained in the training set, and 137 instances remained in the test set.

3.5 Classifiers Used

Supervised classification algorithms were applied on the dataset to detect whether the baby will be born preterm or not. In this work, six different classification algorithms were used.

Zero R: Zero R is a baseline classifier that predicts the majority class [18]. In spite of the fact that there is no predictability ability in Zero R, it is helpful for deciding as a benchmark for other classification techniques.

Logistic Regression: Logistic regression, a binary classification algorithm that uses linear regression with modified cost function is used. This is followed by a sigmoid function which enables it to make binary classification.

Naive Bayes Classifier: The Naive Bayes classifier makes a prediction based on the probability, presuming that the predictors are independent.

Support Vector Machine: Support Vector Machine, SVM, is a machine learning algorithm that can address classification as well as regression problems. This technique falls under the umbrella of supervised learning.

K-Nearest Neighbor: K-Nearest Neighbor uses majority voting amongst k nearest neighbors of a test query to make the prediction of the query data.

Decision Tree: Decision trees use greedy approaches to split nodes such that the gain is maximized. In this work the measure of impurity is determined using gini index.

$$Gini\text{-}Index = 1 - \Sigma_{i=0}^{c-1} p_i(t)^2 \tag{1}$$

where $p_i(t)$ is the probability of class i at node t and c is the number of classes.

4 Results and Discussion

To train the model, six classification algorithms were applied on the train set. The accuracies of six classification algorithms are shown in Table 1 and 2 after performing stratified k - fold cross validation on training data which consisted a total number of 547 instances (80% of the original dataset).

Table 1. Train and test accuracy before smote. (The best results are highlighted with bold font)

Algorithms	Training accuracy	Test accuracy
Zero R	89.54%	89.78%
Logistic Regression	90.8%	91.9%
Naïve Bayes	95.06%	89.05%
SVM	99.82%	97.81%
K-Nearest Neighbor	94.70%	96.35%
Decision Tree	**99.82%**	**99.81%**

Table 2. Train and test accuracy affter smote. (The best results are highlighted with bold font)

Algorithms	Training accuracy	Test accuracy
Zero R	50%	92.70%
Logistic Regression	99.63%	98.5%
Naïve Bayes	93.31%	88.3%
SVM	100%	97.8%
K-Nearest Neighbor	98.14%	92.70%
Decision Tree	**99.9%**	**99.3%**

The comparison between the accuracies of the algorithms before and after applying SMOTE are represented in Fig. 2 and 3. Besides accuracy, other performance metrics like TPR (True Positive Rate), FPR (False Positive Rate), precision (Positive Predictive Value), recall (Sensitivity), F-Measure, and AUC (Area Under ROC curve) were also evaluated.

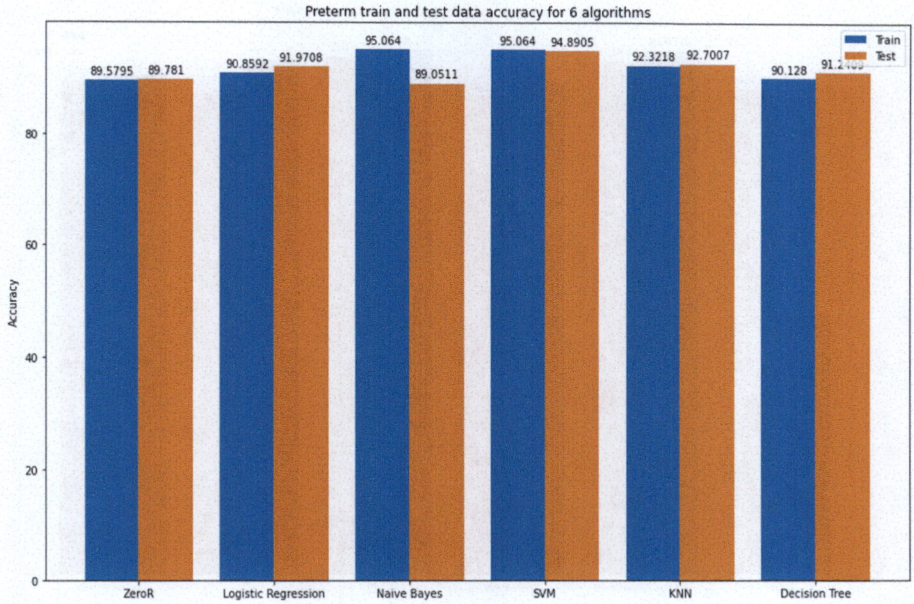

Fig. 2. Comparison between algorithms before SMOTE.

Table 3. Other performance metrics on test set.

Algorithms	FP rate	Precision	Recall	F-Measure	Area under ROC
Zero R	0.091	0	0	0	0.500
Logistic Regression	0	1	0.984	0.992	1
Naïve Bayes	0.500	0.959	0.929	0.944	0.954
SVM	0	1	0.984	0.992	0.992
K-Nearest Neighbor	0.700	0.947	0.984	0.965	0.857
Decision Tree	0	1	0.976	0.988	0.859

The empirical error rate for a classifier is given by Eq. 2. This equation is then used to determine the accuracy.

$$Error\text{-}Rate = \delta(h(x), y) = \begin{cases} 1, & y \neq h(x) \\ 0, & y = h(x) \end{cases} \tag{2}$$

Equation 3 is used to determine False Positive rate. False Positive rate is denoted by α.

$$\alpha = \frac{FP}{TN + FP} \tag{3}$$

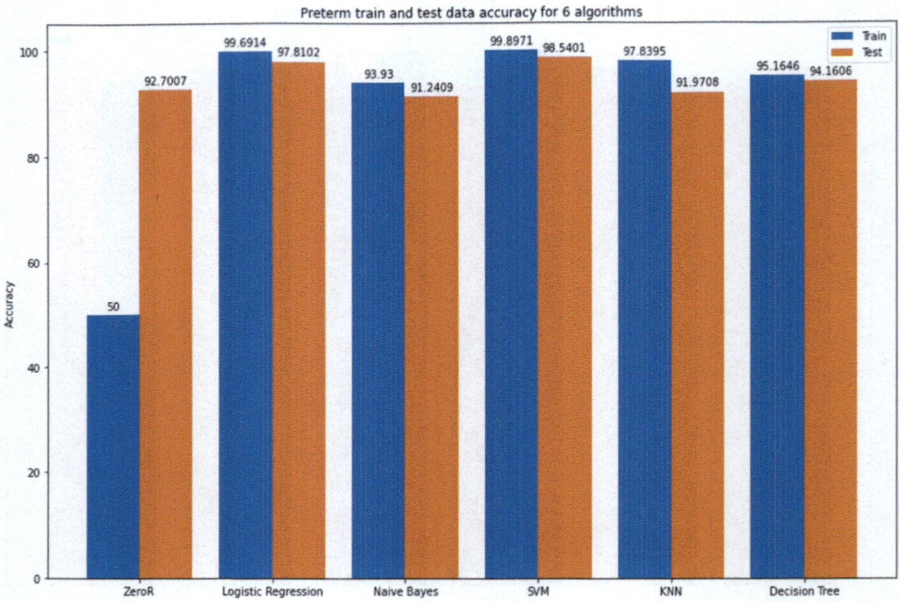

Fig. 3. Comparison between algorithms after SMOTE.

This equation is used to determine Precision.

$$Precision = \frac{TP}{TP + FP} \tag{4}$$

Equation 5 is used to determine Recall which is also called true positive rate.

$$Recall = \frac{TP}{TP + FN} \tag{5}$$

The following equation is used to determine F-Measure.

$$F\text{-}Measure = \frac{2 \times recall \times precision}{precision + recall} \tag{6}$$

5 Conclusion and Future Work

This paper presents a machine learning model that predicts term or preterm birth delivery of pregnant women who have ungergone conization surgeries. It is hoped that, this model will serve its purpose and avoid unnecessary high cost treatments and ensure preventive measures to overcome preterm delivery. The dataset was trained with six machine learning algorithms (including that of Decision Tree, KNN, Logistic Regression, SVM, Naïve Bayes, and ZeroR).

The results of these algorithms have varied in accuracy. The Decision Tree algorithm was found to attain the best accuracy score with 99.3% accuracy on the test data.

In future, we intend to use explainable AI techniques to better comprehend the models' prediction.

References

1. Berghella, V.: Novel developments on cervical length screening and progesterone for preventing preterm birth. BJOG: Int. J. Obstet. Gynaecol. **116**(2), 182–187 (2009)
2. Bevis, K.S., Biggio, J.R.: Cervical conization and the risk of preterm delivery. Am. J. Obstet. Gynecol. **205**(1), 19–27 (2011)
3. Callahan, A., Shah, N.H.: Machine learning in healthcare. In: Key Advances in Clinical Informatics, pp. 279–291. Elsevier (2017)
4. Chawla, N.V., Bowyer, K.W., Hall, L.O., Kegelmeyer, W.P.: Smote: synthetic minority over-sampling technique. 0J. Artif. Intell. Res. **16**, 321–357 (2002)
5. Crane, J.M., Delaney, T., Hutchens, D.: Transvaginal ultrasonography in the prediction of preterm birth after treatment for cervical intraepithelial neoplasia. Obstet. Gynecol. **107**(1), 37–44 (2006)
6. Fuhad, K., et al.: Deep learning based automatic malaria parasite detection from blood smear and its smartphone based application. Diagnostics **10**(5), 329 (2020)
7. Iams, J.D., et al.: The length of the cervix and the risk of spontaneous premature delivery. New Engl. J. Med. **334**(9), 567–573 (1996)
8. Jakobsson, M., Gissler, M., Sainio, S., Paavonen, J., Tapper, A.M.: Preterm delivery after surgical treatment for cervical intraepithelial neoplasia. Obstet. Gynecol. **109**(2), 309–313 (2007)
9. Kindinger, L.M., et al.: Preterm birth prevention post-conization: a model of cervical length screening with targeted cerclage. PLoS One **11**(11), e0163793 (2016)
10. LLC, M.: Preterm birth (2018). https://www.who.int/news-room/fact-sheets/detail/preterm-birth
11. Miller, N.F., Todd, O.E.: Conization of the cervix. Surg. Gynecol. Obstet. **67**, 265–70 (1938)
12. Miyakoshi, K., Itakura, A., Abe, T., Kondoh, E., Terao, Y., Tabata, T., Hamada, H., Tanaka, K., Tanaka, M., Kanayama, N., et al.: Risk of preterm birth after the excisional surgery for cervical lesions: a propensity-score matching study in Japan. J. Maternal-Fetal Neonatal Med. **34**(6), 845–851 (2021)
13. Nam, K.H., Kwon, J.Y., Kim, Y.H., Park, Y.W.: Pregnancy outcome after cervical conization: risk factors for preterm delivery and the efficacy of prophylactic cerclage. J. Gynecol. Oncol. **21**(4), 225–229 (2010)
14. Pranto, B., Mehnaz, S., Mahid, E.B., Sadman, I.M., Rahman, A., Momen, S., et al.: Evaluating machine learning methods for predicting diabetes among female patients in Bangladesh. Information **11**(8), 374 (2020)
15. Sadler, L., Saftlas, A., Wang, W., Exeter, M., Whittaker, J., McCowan, L.: Treatment for cervical intraepithelial neoplasia and risk of preterm delivery. Jama **291**(17), 2100–2106 (2004)
16. Shad, H.S., Jamal, Z., Ahmed, S.M.F., Momen, S., Mansoor, N.: Early prediction of chronic kidney disease using data mining techniques. In: Silhavy, R., Silhavy, P., Prokopova, Z. (eds.) CoMeSySo 2021. LNNS, vol. 231, pp. 947–957. Springer, Cham (2021). https://doi.org/10.1007/978-3-030-90321-3_79

17. Shanbhag, S., Clark, H., Timmaraju, V., Bhattacharya, S., Cruickshank, M.: Pregnancy outcome after treatment for cervical intraepithelial neoplasia. Obstet. Gynecol. **114**(4), 727–735 (2009)
18. Wahbeh, A.H., Al-Radaideh, Q.A., Al-Kabi, M.N., Al-Shawakfa, E.M.: A comparison study between data mining tools over some classification methods. Int. J. Adv. Comput. Sci. Appl. **8**(2), 18–26 (2011)
19. Wang, L.: Value of serial cervical length measurement in prediction of spontaneous preterm birth in post-conization pregnancy without short mid-trimester cervix. Sci. Rep. **8**(1), 1–5 (2018)
20. Wiens, J., Shenoy, E.S.: Machine learning for healthcare: on the verge of a major shift in healthcare epidemiology. Clin. Infect. Dis. **66**(1), 149–153 (2018)

The Use of a Robotic Module for Monitoring and Diagnostic Systems for Various Applications

E. V. Melnik, M. V. Orda-Zhigulina(✉), and D. V. Orda-Zhigulina

Southern Scientific Center of the Russian Academy of Sciences, st. Chehova, 41, Rostov-on-Don 344006, Russia
jigulina@mail.ru

Abstract. This article describes the solution of scientific and technical problems that arise when integrating a robotic module into monitoring and diagnostic systems when solving problems of various directions and varying complexity, which ensures an increase in the efficiency of processing and transmission of information, an increase in reliability and survivability, and an expansion of interaction opportunities ("openness") monitoring systems through the use of "digital economy" technologies ("fog" and "edge" computing, distributed registry, big data) and resource reconfiguration.

Keywords: Processing and storage systems · Decentralized decision support system · Blockchain · Robotics · Hazardous phenomena

1 Introduction

Currently, the tasks of monitoring and diagnostics for various fields of activity exist in various fields from industrial production and health care, security, to observations of natural hazards and the prevention of their consequences. The emergence of new intelligent sensors leads to an increase in the volume of information requiring processing in real time. In addition, when monitoring and diagnosing in various areas of human activity, not only the volume and speed of new information requiring processing is growing, but the initial data themselves are heterogeneous and diverse. There is also a limit of information that a person can understand and process in a certain period of time (adaptive level of perception of an expert). All this leads to changes in the traditional methods of operation of centralized monitoring and diagnostic systems and requires not only automation and data processing and analysis processes, but also the intellectualization of information and organizational processes, the development and implementation of new methods and technologies for decision support based on artificial intelligence [1], as well as the creation of new principles for building systems.

R. Silhavy (Ed.): CSOC 2022, LNNS 502, pp. 417–421, 2022.
https://doi.org/10.1007/978-3-031-09076-9_37

Therefore, new monitoring and diagnostic systems should be highly versatile in order to be used in various areas of human activity. Thus, it is necessary to ensure an increase in the efficiency of processing and transmission of information, an increase in reliability and survivability, an expansion of the interaction possibilities ("openness") of such systems through the use of technologies of the "digital economy" ("fog" and "edge" computing, distributed registry, big data) and resource reconfiguration.

To solve the above scientific and technical problems, the authors developed and then integrated prototypes of the cognitive analysis module, the robotic complex and the distributed data storage module. These component prototypes have been combined within the decision support subsystem, which is one of the subsystems in monitoring and forecasting for various applications. Such integration of the system components was carried out on the basis of the developed set of methods for intelligent data processing for monitoring and diagnostic systems for various fields of activity, as well as the methodology previously proposed by the authors [2–7] for building highly reliable reconfigurable systems based on "digital economy" technologies for information processing, providing access, storage and transmission of data based on a distributed registry, the method of information integration, the method of robustly stable control of the movement of a group of mobile robots in solving monitoring problems by multi-robotic complexes of the methodology.

When integrating various components of the monitoring system, it was possible to combine various participants within one system without their hierarchical subordination and the possibility of automated assessment of the likelihood of occurrence of hazardous processes or phenomena from various areas of human activity, for example, the likelihood of occurrence of hazardous natural processes and phenomena.

The integration of the developed components (cognitive analysis module, multi-robotic complex and distributed data storage) was carried out in the form of a decision-making subsystem, the structure of which is shown in Fig. 1.

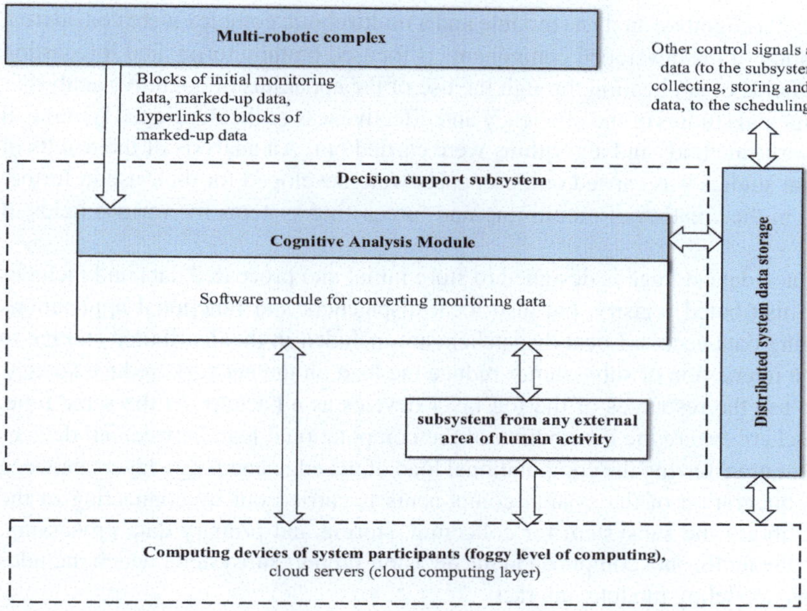

Fig. 1. Scheme of integration of models of components of the monitoring and forecasting system for various areas of human activity.

2 Integration of the Model of the Component of the Monitoring and Diagnostics System - A Multi-robotic Complex

The authors developed an "integrated approach" to create systems for monitoring and diagnosing various fields of application [12]. Within the framework of this approach, a mobile robotic complex is used, which is a group of mobile robots (UAV robotic unmanned aerial vehicles) as terminal devices (sensors) for collecting information, as well as UAVs with separate sensors or analyzers on board, photo and video capabilities.

Therefore, when integrating a mobile robotic complex into systems for monitoring and diagnosing various areas of application, the urgent task of managing a group of UAVs was solved, namely: determining the number of UAV subgroups needed to scan areas; distribution of scanning sub-areas between UAV subgroups; distribution of UAVs by subgroups; distribution of the UAV over the scanning bands. To solve the problem of scanning a set of areas using a UAV, a formal statement and solution of the problem was shown, which is considered in detail here [4, 14].

3 Conclusion

The developed system components were integrated into the prototype of the monitoring and forecasting system model for various fields of application based on digital economy technologies. The principles of joint operation of a prototype of a distributed data warehouse model implemented using distributed registry technology and its interaction with a prototype of a cognitive analysis module and a multirobotic complex were considered. The integration of the developed components is focused on monitoring and forecasting for various fields of application through the use of the apparatus of cognitive analysis.

Experimental studies of the efficiency and effectiveness of the developed (within all stages of work) methods and algorithms were carried out. An analysis of the results of experimental studies was carried out, proposals were developed for their use in further studies and in the creation of monitoring and forecasting systems for various fields of application.

Distributed data storage is designed to store initial and processed data and includes not only a distributed registry, but also local dispatchers and functional applications for organizing data access. Local dispatchers are included in the distributed storage to organize the interaction of subsystems, reduce the load on the network, reduce latency, and also to use the resources of the fog layer devices as a backup. At the same time, local dispatchers ensure the distribution of the computational load between all devices of the system, transferring the computational load of the edge and foggy layers between the layers. Integration of the system components is carried out by connecting to the distributed library the subsystem for collecting, storing and primary data processing, with which the multirobot complex and the decision support subsystem, which includes the cognitive modeling module, interact.

The cognitive analysis module allows you to automate the process of monitoring and forecasting based on cognitive maps. This module solves the problem of analyzing the behavior of systems in various fields of application, and its results can become the basis for creating an intelligent decision support system in terms of predicting and evaluating the dynamics of changes in the behavior of the system to ensure timely adoption of the necessary management decisions. Cognitive analysis is carried out on the basis of the parameters of the state of the observed, social activity of the population and expert assessments based on such technologies of the digital economy as data mining (cognitive modeling), distributed registry, fog and cloud computing, wireless technologies and the Internet of things. All operations for collecting, storing, processing data and automating management decisions are implemented using functional applications and are processed and stored in the corresponding sections of the distributed registry.

A multi-robotic complex implemented on the basis of digital economy technologies, in particular, blockchain and distributed registry technologies, allows solving the problem of simultaneously scanning several sufficiently large areas of space by a heterogeneous group of UAVs, which is the basis of MRTK SMD in a time not exceeding a given one. At the same time, the developed methods and algorithms for solving the problem of distributing scanning areas between a group of mobile robots, which represent a group of UAVs, are simple and have low computational complexity, which allows their implementation by on-board control devices of individual UAVs of the group, united by communication channels into a computer network. An approach to autonomous trajectory planning is also proposed, which allows minimizing trajectory intersections and thereby reducing the risk of UAV collisions with each other during operation. All operations for the collection, storage, processing of UAV data are also processed and stored in sections of the distributed registry.

The application of the obtained scientific results makes it possible to develop distributed monitoring and diagnostic systems without centralizing the control function, to implement advanced scalability, to provide increased fault tolerance and to increase the reliability of monitoring and diagnostic systems.

The obtained scientific results cover a wide range of types of monitoring and diagnostic systems, which makes it possible to use them in various fields of activity.

The scientific novelty of the results obtained lies in the complex application of such technologies of the digital economy as a distributed registry, data mining, wireless communication technologies, "fog and cloud" computing in monitoring and diagnostic systems for various fields of application.

The study was carried out with the financial support of GZ No. 122020100270-3.

References

1. Orda-Zhigulina, M.V., Melnik, E.V., Ivanov, D.Y., Rodina, A.A., Orda-Zhigulina, D.V.: Combined method of monitoring and predicting of hazardous phenomena. In: Silhavy, R. (ed.) Software Engineering Methods in Intelligent Algorithms, vol. 984, pp. 55–61. Springer, Cham (2019). https://doi.org/10.1007/978-3-030-19807-7_6
2. Orda-Zhigulina, D.V., Orda-Zhigulina, M.V., Rodina, A.A.: Cognitive model for monitoring and predicting dangerous natural processes for hydro ecosystem analysis. In: Silhavy, R., Silhavy, P., Prokopova, Z. (eds.) Software Engineering Perspectives in Intelligent Systems, vol. 1294, pp. 688–695. Springer, Cham (2020). https://doi.org/10.1007/978-3-030-63322-6_58
3. Melnik, E.V., Bulysheva, N.I., Orda-Zhigulina, M.V., Orda-Zhigulina, D.V.: Component of decision support subsystem for monitoring and predicting of hazardous processes at the base of analysis of macro zoobenthos communities of azov sea. In: Silhavy, R., Silhavy, P., Prokopova, Z. (eds.) Software Engineering Perspectives in Intelligent Systems, vol. 1294, pp. 676–687. Springer, Cham (2020). https://doi.org/10.1007/978-3-030-63322-6_57
4. Melnik, E.V., Orda-Zhigulina, M.V., Orda-Zhigulina, D.V., Ivanov, D.Y., Rodina, A.A.: Fog computing in new approach for monitoring of hazardous phenomena. J. Phys. Conf. Ser. **1333**(7), 072016 (2019)
5. Melnik, E.V., Orda-Zhigulina, M.V., Orda-Zhigulina, D.V., Ivanov, D.Y., Rodina, A.A.: Primenenie tehnologij cifrovoj ekonomiki pri razrabotke sredstv monitoringa i prognozirovanija opasnyh processov i obespechenija bezopasnosti naselenija i beregovoj infrastruktury. In: Zakonomernosti formirovanija i vozdejstvija morskih, atmosfernyh opasnyh javlenij i katastrof na pribrezhnuju zonu RF v uslovijah global'nyh klimaticheskih i industrial'nyh vyzovov ("Opasnye javlenija"), pp. 289–291. Izdatelstvo SSC RAS, Rostov-on-Don (2019)
6. A new approach to consensus: Swirlds hashgraph. http://sammantics.com/blog/2016/7/27/hashgraph-consensus. Accessed 02 Sep 2020
7. Hedera hashgraph. Explained. https://medium.com/datadriveninvestor/hedera-hashgraph-explained-c5d8ce4730a6. Accessed 02 Sep 2020
8. Hedera hashgraph, consensus, and scalability. https://medium.com/@saratechnologiesinc/hedera-hashgraph-consensus-and-scalability-2315133a3e33. Accessed 02 Sep 2020
9. LeMahieu, C.: Nano: a feeless distributed cryptocurrency network. https://content.nano.org/whitepaper/Nano_Whitepaper_en.pdf. Accessed 25 Jan 2021

Scenario and Architecture for Intelligent Decision Support Based on User Digital Life

Alexander Smirnov and Tatiana Levashova(✉)

St. Petersburg Federal Research Center of the Russian Academy of Sciences, 39, 14th line, St. Petersburg 199178, Russia
{smir,tatiana.levashova}@iias.spb.su

Abstract. Human digital traces as a source of personal information are becoming increasingly popular in personalized systems. Such traces collected for a user organize the user digital life. The paper proposes a scenario of intelligent decision support based on user digital life and service-oriented architecture of a decision support system that provide recommendations based on context-sensitive user types. A user ontology represents the user types and infers a type of the user based on the information identified in the user's decision-making activities contained in the digital life model of this user. Recommended decisions are proposed based on the information about the user preferences and the kinds of decisions that the users of the same type as the active user make. DSS behaviour is demonstrated by providing recommendation to a user that searches for a book.

Keywords: Decision support · Digital traces · User digital life model · User type · Ontology classification · Type-sensitive recommendations

1 Introduction

The present reality supposes intensive online human activities during which users leave hundreds of digital traces. Each trace is a source of data/information/knowledge. Among many other things, digital traces contains information that explicitly and/or latently characterises the user and the user's behaviour.

Numerous decision support systems (DSSs) including recommender ones, consider user digital traces as an information source. For instance, it is a source for recommending preventive and predictive decisions [2–4]. The most commonly analyzed traces contained in this kind of source is the traces left by user communications in social networks [5]. For example, a system adopting a user-centric recommendation model [6] analyses personal digital traces left in social networks and e-mail traces, reveals from them users' interests, and provides personalized news and local event recommendations for their utility and societal importance. The research on meta-analysis on the use of social media investigates predictive power of digital traces over personality traits [7]. The purposes of the predictions cover various domains including recommender systems. Personality prediction based on an analysis of users' activities in social networks is the aim of the method that integrates text, image, and users' meta features from Twitter and Instagram

R. Silhavy (Ed.): CSOC 2022, LNNS 502, pp. 422–433, 2022.
https://doi.org/10.1007/978-3-031-09076-9_38

[11]. User clustering techniques refer to digital traces to generate multidimensional feature description of products to provide personalize offers to retail customers, mainly with online purchases [8].

Opportunities that digital traces provide for decision support systems motivated their usage for the identification in it a user-specific information based on which context-aware decisions can be recommended. A conceptual framework of intelligent decision support based on a user digital life model [1] embodies this idea. This framework addresses the concept of user digital life, which is a collection of user's digital traces. In the framework, the user digital life model is a structured representation of weekly systematized and poorly curated content of user digital traces related to the decision-making process of the user. The framework is purposed to recommend to the user a decision that the users sharing their preferences and decision-making behaviour with this user would make in the user context.

The goal of the present paper is to propose a decision support scenario developed in accordance with the conceptual framework of intelligent decision support based on a user digital life model and to offer architecture of a DSS implementing this framework.

The proposed decision support scenario considers situations that the DSS addresses from the time instant when the user starts decision-making until a recommendation has been provided. The architecture of the DSS adopts the concept of service-oriented architecture. It comprises a set of services to implement the expected functionality of the DSS. The service interactions while supporting the user with a book recommendation demonstrate the DSS behaviour according to the scenario.

The rest of the paper is as follows. Section 2 outlines the conceptual framework of intelligent decision support based on user digital life and proposes the decision support scenario based on this model. Section 3 presents the architecture of a DSS that implements the conceptual framework. Section 4 shows implementation of the scenario for a book search problem. Main concluding remarks are discussed in the Conclusion.

2 Scenario of Intelligent Decision Support Based on User Digital Life

The Section introduces the conceptual framework of intelligent decision support based on user digital life model [1] in accordance with which the decision support scenario is designed and describes this scenario.

2.1 Conceptual Framework of Intelligent Decision Support Based on User Digital Life

The conceptual framework for intelligent decision support based on user digital life (Fig. 1) is intended to recommend decisions that the user would made in the current situation (context). The main components of this framework are user digital traces, user profile, user digital life model, user segment, user ontology, and context [1]. The user digital life model is the main source of information for the rest of the components.

User profile is a set user characteristics that can be used to create a descriptive portrait of an individual and to identify one. *User digital traces* is a set of records fixing

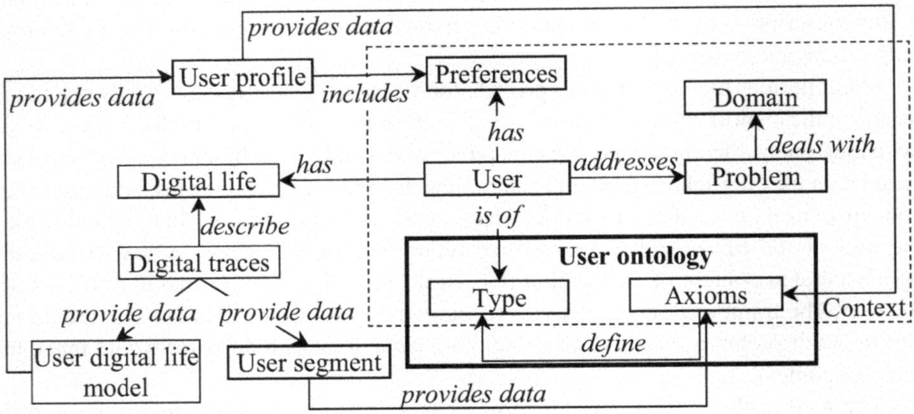

Fig. 1. Conceptual framework for intelligent decision support based on user digital life.

information on the user activities including decision-making. *User digital life model* is a structured representation of a part of the content of user digital traces, which carries information related to the decision-making process of the user. *User segment* is a group of users with common needs and behavioural reactions when making decisions. *User ontology* is a multi-aspect user model, which formalizes knowledge to classify a user into a user type, i.e. into a category of users distinguished by common preferences and decision-making behaviour. *Context* is any information that characterizes the situation of the user in the decision-making process. In the conceptual framework, context comprises the user identifying information and the information on the user preferences, the user type, the problem requiring a decision, and the knowledge domain that this problem deals with. When the user requests the DSS with a problem requiring a decision or when the DSS finds out that the user needs a recommendation, the system infers the user type and recommends a decision based on the knowledge about the kinds of decisions customary for the users of this type.

The framework components are modelled using the set-theoretic approach.

User profile (*UP*):

$$UP = (User\_ID,\ \ P\_out, P\_in(C)), P\_in(C) = DM\_Type(C) \bigcup P_c(C) \bigcup Pr(C), \tag{1}$$

where *User_ID* is the unique user identifier, *P_out* is the set of context-independent user characteristics; *P_in* is the set of context-sensitive user characteristics in the context $C(T)$; $DM\_Type(C)$ is the user type in the context $C(T)$; $Pr(C)$ is the set of user preferences in the context $C(T)$; $P_c(C)$ is the set of context-sensitive user characteristics other than the user type and the user preferences (e.g., the user location, local time, etc.); T is the period of existence of the context C.

User digital life model (DL):

$$DL = (User\_ID, Problem(t_0, t_n), Domain(t_0, t_n), \{Action(t_a^-, t_a^+)\}, Decision(t_n), R_1, R_2, R_3), R_1 \in Problem \times Domain, R_2 \in Problem \times Decision, R_3 \in Action(t_a^-, t_a^+) \times Problem \quad (2)$$

where *Problem* is the kind of the decision-making problem ($Problem(t_0, t_n)$ means that the user addresses the problem *Problem* in the interval (t_0, t_n)); t_0 is the time instant when the user starts decision-making; t_n is the time instant when the user has made a decision; *Domain* is the knowledge domain that the problem *Problem* deals with ($Domain(t_0, t_n)$ means that the domain knowledge is dealt with in the interval (t_0, t_n)); $Action(t_a^-, t_a^+)$ is the action carried out in the interval t_a^-, t_a^+ $(t_0 \le t_a^-,\ t_a^- < t_a^+,\ t_a^+ < t_n)$; $Decision(t_n)$ is the decision made at the time instant t_n.

User segment (*S*):

$$S = (Domain, Consumer\_Type, Var, R_4, R_5), R_4 \in Domain \times Var, R_5 \in Var \times Consumer\_Type \quad (3)$$

where *Consumer_Type* is the type of users sharing a common behavioural pattern when choosing a decision, *Var* is the set of behavioural variables providing data to the behavioural pattern.

User ontology (O_U):

$$O_U = (Cl, Rel, A),\ Cl = Cl_o \bigcup Type,\ A = A_o \bigcup A_{DM\_Type},$$

where *Cl* is the set of ontology classes, *Rel* is the set of class relationships ($Rel \to Cl \times Cl$), *Type* is the class that represents the user types, $Cl_o = Cl \backslash Type$, *A* is the set of ontology axioms, A_{DM_Type} is the set of axioms that define the membership of the class *Type* by a user, $A_o = A \backslash A_{DM\_Type}$.

Two classes are direct subclasses of the class *Type*. They are decision maker type (*DM_Type*) and consumer type (*Consumer_Type*). The class *DM_Type* represents types of users based on their characteristics (*P_in*) that influence the decision-making style. This class classifies the users into spontaneous, rational, inert, risky, and cautious decision makers. The class *Consumer_Type* classifies the users as consumers of the recommended decisions based on the behavioural variables (*Var*); it is supposed that the users' behavioural patterns when they are thinking of either accept a decision or do not reflect the behavioural patterns of these users as consumers. Progressives, consolidators, always-hurrying, traditionalists, and security-concerned consumers are represented in this class. Since the user characteristics influence a user type and some of these characteristics are context-sensitive, a user type becomes a context-sensitive. That is, the same user in different contexts can be classified into different user types.

Context (C):

$$
\begin{aligned}
&C(T) = (user\_ID,\, user\_type(T),\, domain(T),\, problem(T),\, Pr_u(T),\, R_6),\\
&user\_ID \subset User\_ID,\ user\_type(T) = (dm,\, consumer),\ dm \subset DM\_Type,\\
&consumer \subset Consumer\_Type,\ domain(T) \subset Domain,\ problem(T) \subset Problem,\\
&Pr_u(T) \subseteq Pr,\ R_6 \in domain(T) \times Pr,
\end{aligned} \tag{4}
$$

where *user_ID* is the unique user identifier; *user_type(T)* is the user type in the context *C(T)*; *problem(T)* is the problem for that the user is making a decision in the context *C(T)*; *domain(T)* is the knowledge domain that the problem *problem(T)* deals with; $Pr_u(T)$ is the set of user preferences in the context *C(T)*; $T = (t_0, t_n)$.

2.2 Decision Support Scenario

The scenario of intelligent decision support based on user digital life (Fig. 2) considers situations that the DSS addresses on the interval from the time instant when the user starts decision-making until a recommendation has been provided. The considered situations correspond to the scenario stages. For each stage, its purpose, inputs and results (outputs) are defined.

Stage A. A problem that the user faces (decision-making problem) initiates the operation of the DSS. The information on the unique user identifier *user_ID*, the time t when the user performs the actions causing the system reaction, these actions $action(t^-, t^+)$, $action(t^-, t^+) \subseteq Action(t_a^-, t_a^+)$ (the very first action in time is $action(t_0, t^+)$), the decision-making problem *problem(t)*, and the domain *domain(t)* is identified in the user digital traces and becomes represented in the user digital life model. This model serves as the input information source for the stage A. At this stage, a formal problem model is built to the DSS would be able to solve this problem. For simple problems, an ontology representation language can be used to model them. For complex problems, where means of such a language is not sufficient any other formalisms should be used. The main requirement for the problem formalization is that the DSS must "understand" it. The output information of this stage is the kind of the problem (*problem(t), t* $\subset$ *T*) that the user faces, the domain of knowledge that this problem deals with (*domain(t)*), the problem model (*PM*), and the function *M*: *problem(t)* $\rightarrow$ *PM* that assigns the kind of the problem its formal model.

Stage B. At this stage, the user ontology gets values of the user properties based on which the ontology can infer a user type (*Type*). User properties include user's characteristics and behaviour variables. The main sources of information about these properties are the user profile representing and the user digital life model. Additionally, a decision mining service (not shown in the conceptual framework) reveals user preferences based on an analysis of the decisions that the user has made [9, 10]. Information about such decisions this service gets from the user digital life model. The values for the user properties (including the preferences) fully or partly instantiate the ontology axioms and rules that define the user types. The fully instantiated axioms and rules become assertions (axioms and rules describing individuals [11]). The set of the assertions and the user preferences $Pr_u(t)$ ($Pr_u(t) \subset P\_in(t)$) are the output of the stage B.

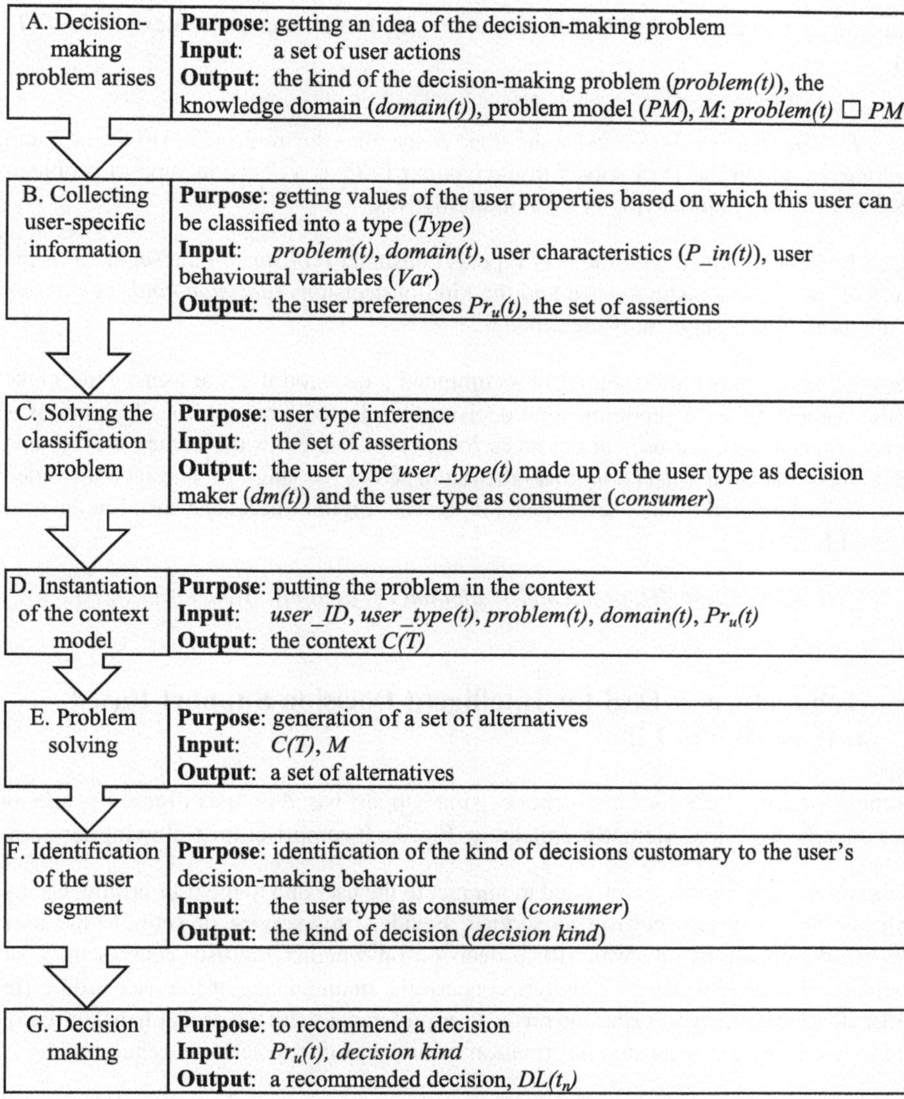

Fig. 2. Scenario of intelligent decision support based on user digital life.

Stage C. Based on the set of the assertions that describe the user, the ontology solves the classification problem and derives the user type (*user_type(t)*) made up of the user type as decision maker (*dm(t)*) and the user type as consumer (*consumer*).

Stage D. The information on the user identifier (*user_ID*), the user type (*user_type(t)*), the kind of the problem that the user addresses (*problem(t)*), the knowledge domain

(*domain(t)*), and the set of user preferences ($Pr_u(t)$) instantiates the context model *C(T)* (4).

Stage E. The function *M* defined at the stage A specifies the model (*PM*) of the problem *problem(t),* which the DSS solves in the context *C(T)* as a decision support problem. The result of problem solving is a set of alternatives.

Stage F. In accordance with the user type as consumer (*consumer*) the segment of the users of this type is distinguished and the kind of decisions (*decision kind*) customary to the users of this segment is identified.

Stage G. The final stage is aimed to recommend a decision that the user would make in the context *C(T).* A recommended decision is selected from the set of alternatives in accordance with the user preferences $Pr_u(t)$ represented in the context model and the kind of decisions (*decision kind*) distinguished at the stage E. The recommended decision is delivered to the user, appears in the user digital traces, and saved in the user digital life model:

$$(DL(t_n) = (user\_ID, problem(T), domain(T), \{action(t_n)\}, decision(t_n))).$$

3 Architecture of DSS for Intelligent Decision Support Based on User Digital Life

Architecture of a DSS for intelligent decision support based on user digital life adopts the concept of service-oriented architecture (Fig. 3). It comprises the following services.

User Agent. The service is intended to interact to the user and to the user profile. User's requests to the system get to this service. Besides, this service presents to the user the recommendations and waits for a decision on whether the user accepts them or does not. Additional service activities concern the maintenance of the user profile (in particular, introducing information on the user preferences and the decision maker type), and transmitting the requested information from the profile to services-requestors.

Digital life service transmits the information structured in the user digital life model to services-requestors, and discoveries in the actions represented in this model problem-specific information (e.g., variables, explicitly formulated criteria, etc.). As well, this service discovers information that can be used to determine user properties other than preferences and processes it.

Problem formalization service is aimed at building a model of the decision-making problem. It interacts to the digital life service on the kind of this problem, problem parameters, and other problem-specific information. Additionally, it interacts to the problem solving service with the reference to the model built.

Decision-mining service is purposed to reveal user preferences based on the decisions that this user make. It interacts with the user agent and the ontology service to transmit them these preferences.

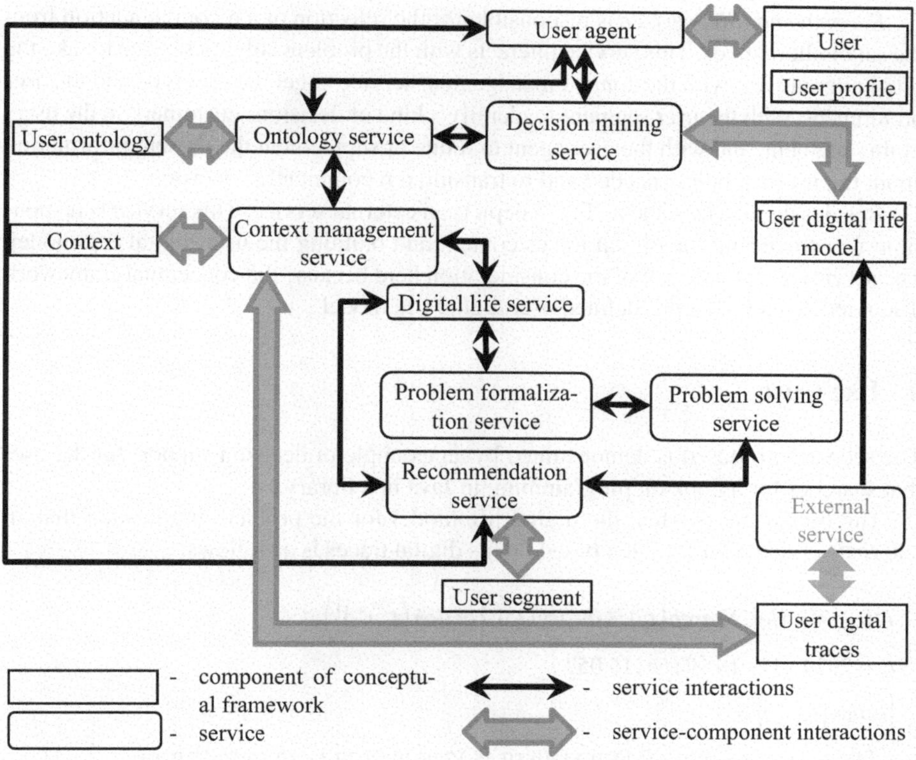

Fig. 3. Architecture of DSS for intelligent decision support based on user digital life.

Ontology service is responsible for the user type identification. This service interacts with the decision mining service and the user agent to get the user preferences, with the user agent to get from the profile the context-sensitive user characteristics determining the user type, and with the decision life service to get the user properties other than the preferences and the characteristics from the profile. Additional service actions address the creation of a set of assertions that describe the user, transmitting them to the ontology, and receiving the result of the ontology inference. Finally, the ontology service interacts with the context management service and the user agent and transmits them the user type.

Context management service in case of proactive decision support is responsible for monitoring the user digital traces and for the realization that the user needs support. This service keeps the context model up to date. It interacts with the ontology service and the digital life service to take the model parameters and with the recommendation service to transmit the user type and preferences.

Problem solving service is intended to solve the problem that the user addresses as a decision support problem. It interacts with the problem formalization service to take the reference to the problem model and with the recommendation service to transmit the set of problem solutions (alternatives).

Recommendation service is responsible for the selection of a recommendation from the set of alternatives. This service interacts with the problem solving service to take the set of alternatives, with the context management service to get the user type and the user preferences, with the user segment to identify a kind of decisions customary to the users of this segment, and with the user agent to refine or supplement the present information about the user when it is needed and to transmit a recommended decision.

Besides the services above, Fig. 3 depicts an external service. This service is responsible for structuring the digital traces content and building the user digital life model. The service actions are out of the consideration here because the conceptual framework discussed deals with a pre-defined user digital life model.

4 Use Case

The scenario proposed is demonstrated by an example of decision support for the user that searches for a book on programming in Java in a library.

The user name is Alex; the digital life model for the problem in question that an external service built for Alex based on his digital traces is as follows:

$$DL_{Alex} = \left(Alex, Search\left(t_0\right), Library\left(t_0\right), \left\{action\left(t_0, t^+\right)\right\}\right),$$

$$t_0 = 2020-11-19\ 19:55:16.057\ ,$$

$$\left\{action\left(t_0, t^+\right)\right\} =$$

$$= \left\{\begin{array}{l} Press.MenuItem.Search\left(\text{2020-11-19 19:55:17.0648, 2020-11-19 19:55:17.926}\right), \\ Chose.Option.Title\left(2020-11-19\ 19:55:16.057, \text{2020-11-19 19:55:18.873}\right), \\ Enter.Title.\text{"Programming in Java"}\left(\text{2020-11-19 19:55:19.203, 2020-11-19 19:55:19.936}\right) \end{array}\right\}.$$

The model above represents that the user identified as Alex (user_ID = Alex) addresses the problem of searching (*problem*$(t) =$ *Search*(t_O)) in the library (*domain*$(t) =$ *Library*(t_O)) a book entitled "Programming in Java".The problem formalization service builds an ontology-based model for the problem under consideration (Fig. 4). OntoGraph plug-in for Protégé ontology management tool [12] is used to provide the visualization of this model.

The preferences of Alex coming from his profile for the problem of book searching declare that the preferable language for Alex is English.

The digital life service and the decision mining service reveal from the actions and decisions represented in the digital life model of Alex values for the properties used in the ontology axioms that define the user types. Table 1 presents the properties that the user ontology uses in the type definitions and values that the services revealed for Alex. The services analyze not only the actions and decisions represented in the model DL_{Alex} for the given time (for the time interval $[t_O, t^+]$ this model does not contain a decision), but all the actions and decisions represented in the model of Alex' digital life that concern the problem of books searching. In accordance with the property values, the ontology classifies Alex as a spontaneous decision maker and as an always-hurrying consumer.

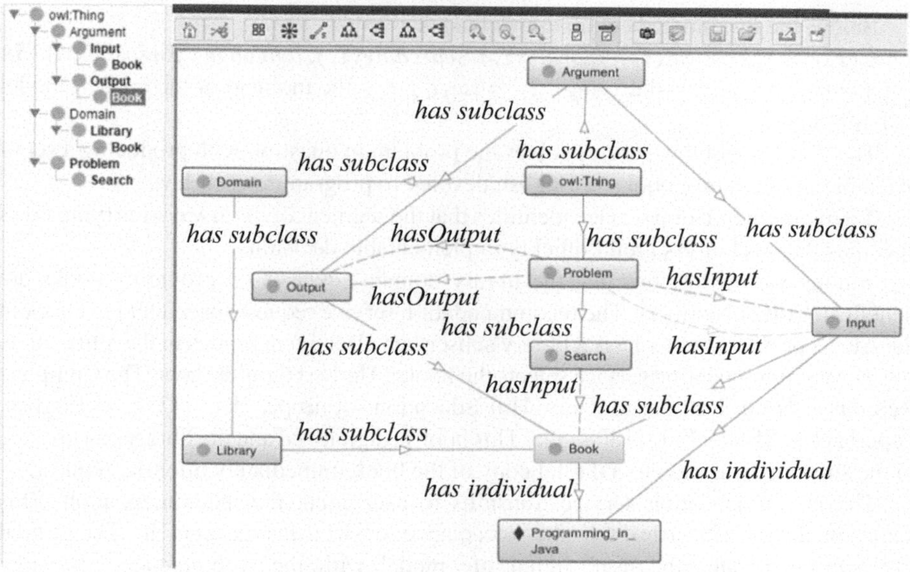

Fig. 4. Ontology-based problem model.

Table 1. User type inference based on type definitions in ontology.

User type property in ontology	Value/function in user digital life model	User type inferred
Decision maker type		
Decision-making time	Low	Spontaneous
Preference for an individual or collective decision-making style	One-two	
Complexity of decision-making procedure	Simple	
Criterion	Maximizing rapidity of getting benefit	
Consumer type		
Time spent in the Internet	Moderate	Always-hurrying
Degree of Internet services consuming	Medium	
Degree of interest to innovations	Low	
Level of loyalty to organizations services of which the user consumes	High	
Preferable communication means	Written	
Criterion	Maximizing own benefits, minimizing the decision implementation time	

Below, the instantiated context model (4) is given.

$C(T_c) = (Alex, SA(T_c), Library(T_c), searching(T_c), language : English(T_c))$, SA = (spontaneous, always-hurrying), $T_c = [t_0, t_c)$, t_c – the moment of the context model instantiation.

The problem solving service solves the problem in question with producing a set of available in the library books in English devoted to programming in Java.

The recommendation service identifies that the segment of the always-hurrying users specifies that such users prefer quickly implementable decisions.

Among various options that the library domain suggests on providing books are benefits for the subscribers. The recommendation service requests user agent to check in the Alex's profile whether he is a library subscriber. User agent replies in the affirmative.

The recommendation service selects the book: "Herbert Schildt, Java: The Complete Reference, Ninth Edition, McGraw-Hill Education – Europe, 2014, 1312 p. (English, Paperback), ISBN: 9780071808552." This book is digitalized and the library can provide to the subscribers access to a digital copy of the book immediately after the request.

The recommendation service transmits to user agent the recommendation. This agent interacts with Alex for the acceptance of it. Alex accepts it. The digital life service updates the user digital life model with the recommended decision: $DL_{Alex} = (Alex, Search(t_0, t_n), Library(t_0, t_n), \{action(t_0, t_n)\},$ "9780071808552"), $decision(t_n) =$ "9780071808552"

5 Conclusion

The paper offers a scenario and architecture for intelligent decision support based on user digital life. The scenario allows DSSs to recommend to a user a decision that users of the same type as this user would make in the given context. The recommendations are suggested based on the information on the user specifics that the user digital traces contain. The service-oriented architecture comprises a set of services to implement the expected functionality of DSSs that operate in accordance with the proposed scenario. The service interactions while supporting the user with a book recommendation demonstrate the DSS behaviour.

An arguable remark on the recommended decisions is that these decisions are comfortable for the given types of users. Speaking about efficiency of them, they may not be such. As a compromise, based on the premise that rational users have a tendency to make efficient decisions, the decisions recommended for the users whose type differs from the rational one can be offered to compare with the decisions that would be supposed the rational users make.

Acknowledgements. The conceptual framework for intelligent decision support based on user digital life (Sections 2.1, 4) is due to the grant from RFBR no. 20-07-00490, the scenario implementing this framework and the architecture of DSS (Sections 2.2, 3, 4) are due to the grant from RFBR no. 20–07-00455, user digital life model created from unstructured digital traces (Sections 2.1, 4) is due to State Research (project no. FFZF-2022-0005).

References

1. Smirnov, A., Levashova, T.: Intelligent decision support based on user digital life model: principles and conceptual framework. In: 2020 27th Conference of Open Innovations Association (FRUCT), pp. 247–252. IEEE (2020). https://doi.org/10.23919/FRUCT49677.2020.9211055
2. Araujo, T., Helberger, N., Kruikemeier, S., de Vreese, C.H.: In AI we trust? Perceptions about automated decision-making by artificial intelligence. AI Soc. **35**(3), 611–623 (2020). https://doi.org/10.1007/s00146-019-00931-w
3. Han, M.L., Kwak, B.I., Kim, H.K.: CBR-based decision support methodology for cybercrime investigation: focused on the data-driven website defacement analysis. Secur. Commun. Netw. **2019**, Article ID 1901548 (2019). https://doi.org/10.1155/2019/1901548
4. Surendro, K.: Predictive analytics for predicting customer behavior. In: 2019 International Conference of Artificial Intelligence and Information Technology (ICAIIT), pp. 230–233. IEEE (2019). https://doi.org/10.1109/ICAIIT.2019.8834571
5. Suhaim, A.B., Berri, J.: Context-aware recommender systems for social networks: review, challenges and opportunities. IEEE Access **9**, 57440–57463 (2021). https://doi.org/10.1109/ACCESS.2021.3072165
6. Hsieh, C.-K., Yang, L., Wei, H., Naaman, M., Estrin, D.: Immersive recommendation. In: Proceedings of the 25th International Conference on World Wide Web, pp. 51–62. International World Wide Web Conferences Steering Committee, Republic and Canton of Geneva, Switzerland (2016). https://doi.org/10.1145/2872427.2883006
7. Azucar, D., Marengo, D., Settanni, M.: Predicting the Big 5 personality traits from digital footprints on social media: a meta-analysis. Pers. Individ. Differ. **124**, 150–159 (2018). https://doi.org/10.1016/j.paid.2017.12.018
8. Zhuravlev, Y., Dokukin, A., Senko, O., Stefanovsky, D., Saenko, I., Korolev, N.: Comparison of two approaches to recommender systems with anonymous purchase data. In: Thalheim, B.K., Sychev, A.V., Makhortov, S.D. (eds.) Supplementary Proceedings of the XXII International Conference on Data Analytics and Management in Data Intensive Domains (DAMDID/RCDL 2020), pp. 114–120. CEUR Workshop Proceedings (2020)
9. Smirnov, A., Pashkin, M., Levashova, T., Shilov, N., Kashevnik, A.: Role-based decision mining for multiagent emergency response management. In: Gorodetsky, V., Zhang, C., Skormin, V.A., Cao, L. (eds.) AIS-ADM 2007. LNCS (LNAI), vol. 4476, pp. 178–191. Springer, Heidelberg (2007). https://doi.org/10.1007/978-3-540-72839-9_15
10. Smirnov, A., Shilov, N.: Ontology matching in collaborative recommendation system for PLM. Int. J. Prod. Lifecycle Manag. **6**, 322–338 (2013). https://doi.org/10.1504/IJPLM.2013.063210
11. Glimm, B., Horrocks, I., Motik, B., Shearer, R., Stoilos, G.: A novel approach to ontology classification. J. Web Semant. **14**, 84–101 (2012). https://doi.org/10.1016/j.websem.2011.12.007
12. Protégé. https://protege.stanford.edu/. Accessed 21 Feb 2022

Analyzing Embedded AIoT Devices for Deep Learning Purposes

Roman Budjac[1](✉), Martin Barton[1], Peter Schreiber[1], and Martin Skovajsa[2]

[1] Faculty of Materials Science and Technology in Trnava, Institute of Applied Informatics, Automation and Mechatronics, Slovak University of Technology in Bratislava, Bratislava, Slovakia

{roman.budjac,martin.barton,peter.schreiber}@stuba.sk

[2] Research Centre, University of Zilina, Univerzitna 8215/1, 010 26 Zilina, Slovakia

martin.skovajsa@uniza.sk

Abstract. In this article, we focused on the artificial intelligence of things (AIoT) embedded devices that are characterized by higher computing power. These devices are designed directly to work with the implementation of artificial intelligence - deep learning. We mentioned the importance of deep learning applications and their current use in various industry applications. They also meet their role as IoT devices. Therefore, they can be included in a separate AIoT category. In the article, we described the role of IoT in Industry 4.0 and we also focused on AIoT devices and their use in industry. The aim of the article was to analyze and verify Jetson equipment from the point of view of heat dissipation during operation time and to visually inspect printed circuit boards for the installation of these equipment, taking into account the thermal requirements for the design of electrical equipment where Jetson embedded devices will be physically installed.

Keywords: Deep learning · Resource-limited devices · AIoT · Embedded devices

1 Introduction

The fourth industrial revolution Industry 4.0 is still a big chapter in scientific articles. Even with an established strategy for the implementation of Industry 4.0 technologies, companies are moving towards digitalization very slowly. Digitalization will be an increasingly important milestone in the near future for all companies, otherwise companies will be in a danger of collapse due to the growing demands of digitized companies [1]. In other words, if companies want to remain competitive, they will have to digitize their business processes. For this reason, it is necessary to approach digitization responsibly and prepare the ground for the gradual implementation of principles and technologies.

Industry 4.0 has its general fundamental pillars. These include Artificial Intelligence, IoT, Big Data and Analytics, Simulation, Additive Manufacturing, Horizontal, and Vertical System Integration, Cloud Computing, Cybersecurity, Autonomous Robots, and

R. Silhavy (Ed.): CSOC 2022, LNNS 502, pp. 434–448, 2022.
https://doi.org/10.1007/978-3-031-09076-9_39

Augmented Reality. Each of these pillars has its own technologies and is more or less demanding in terms of the implementation of rebuilding existing corporate structures, human resources, and finance. That requests the question – Which pillars need to be implemented to bring the companies to a level close to Industry 4.0? The answer is complicated and studies do not have a clear answer, but the most frequently mentioned main pillars are Big Data and Analytics, IoT and Artificial Intelligence [2–5]. Horizontal and Vertical System Integration is also mentioned if the companies have several non-communicating systems [6]. Probably the most essential building block in the whole structure of the Industry 4.0 concept are IoT devices. Without them, it is not possible to collect process data. Therefore, the implementation of these devices is crucial in the industry.

The applications of IoT devices are varied in different domains. IoT devices, based on the environment in which are applied, can be divided into many subsets. The main idea behind IoT is to interconnect embedded devices that measure or collect data from the industrial system in the field devices layer of the automation pyramid. These data are transferred through the layers of this pyramid individually from mentioned field layer to supervisory layers factory plant schedules. For example, critical production data needed for analysis for predictive maintenance needs. One of these subsets is Artificial Intelligence of Things or known as AIoT. The expectation of the AIoT devices in the fields of application is to forward an AI driven solution to the edge. It is a well- known fact that in the field of deep learning (DL) application all models have a large number of parameters. Training sequence in general leaves room for redundant neuronal connections in the models. Despite this, it is not an issue in the case when we are not limited to utilizing enough computer power resources. From a different point of view, in the use case we want to implement AI decision making into the field layer of the pyramid, we can quickly realize that based on the sequence mentioned before, it is not possible to add AI-driven decision making into the level of IoT device. A lack of computer power mainly causes this possibility.

The IoT devices are focusing on lower power consumption devices towards some compactness is required, and they are used as embedded devices in deployment. From our perspective, the IoT devices have many subsets from the computational power point of view. From the lowest power consumption, for instance, microcontrollers where IoT systems are optimized for the lowest possible power consumption, for instance, using 1-wire solutions with parasitic induction to reduce the energy requirements of peripheral sensors [7, 8]. Another type of devices we can present are in the middle computer power category. These devices, for example raspberry pi and its industrial modifications for example, UniPi embedded devices [9]. Moreover, the third category device specialized device manufactured specially for deploying artificial intelligence models also called for inference. During design of these devices, emphasis is placed for GPU-based computational power. Hardware architecture is oriented on parallel computing, and the number of GPU cores is significant.

Such a designed device has the capability to compete in the field of implementing deep learning algorithms.

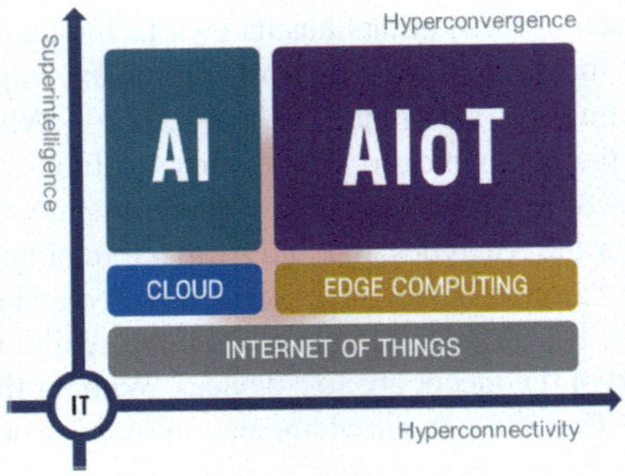

Fig. 1. Components of the artificial intelligence of things [10].

In Fig. 1, there is a graph showing IoT, AI, and AIoT in terms of intelligence and connectivity. AIoTs are above standard IoT devices in terms of intelligence and AI in terms of connectivity.

Deep learning has become an integral part of machine learning and its applications extend to all areas that can be defined as data-oriented and it is possible to quantify a certain explicit type of data that characterizes the area of application. Subsequently, it is possible to group this data into different areas according to preferences and collect these data.

Depending on the area of application, the research can deal mainly with three areas from the perspective of Deep Learning, namely: classification of data into categories of interest, prediction of data based on historical data. Such cases can be, for example, data processing in the manufacturing industry. In this area, data processing technology for predictive maintenance of machines and production equipment is implemented. The aim of which is to minimize the costs of unexpected failures. Unexpected production interruptions increase the costs of eliminating the failure and have the potential to disrupt the entire production process, and the resulting costs will be several times higher. For example, implemented deep learning fault diagnosis methods of rotating machinery based on vibration signal analysis. To prevent with expectation higher precision in fault classification than in traditional methods using manual feature extraction [11].

Another place to implement deep learning in the industry is to identify unwanted product defects. One of the most common cases of DL use is in optical quality control, where high-speed cameras in the quality control process check for defects in products. For example, identification of defects on the steel surface [12] or detecting defects in products made of glass [13]. The same is with detecting welding faults during the process of joining material made of steel [14]. Automatic optical inspection is also used in the case of PCB manufacturing, where automatic defect classification is applied for sander, dirty, sold mask unclean, normal and real defects [15]. The possibility of DL application also extends to areas outside the production process. In the case of construction, DL is applicable to classifying multi-lines bridge defects on bridge, which helps to classify minor defects challenging to identify even for human analysis [16].

In the area of transport infrastructure, research is underway to automatically detect the condition of the road surface, where the software has the task of identifying problem areas of the road. Subsequent analysis will determine preventive maintenance on individual sections of the road in order to prevent property damage [17, 18]. From another point of view, deep learning models can be used for classifying road users and traffic signs [19]. This information is especially important in the field of driver assistants installed in cars and autonomous vehicles, which must be able to recognize their surroundings to make the right decisions in the environment in where they are located. Deep neural networks have also proved effective in the use of agriculture to support so-called smart farming, which aims to support food production, positively affect the environmental impact of farming, processed food safety [20], where deep learning can be used in cases such as image recognition. For example, we are talking about automatic recognition of decreased fruit and vegetables grade [21]. Another is a pest detection with DNN to reduce the amount of pesticide based on precision agriculture policy [22].

The benefit of using deep neural networks is that such a system based on deep architecture can, based on the observation of the simulated system obtained in the form of a training and test set, is able to learn this DL system to imitate a real system. Such a system, when implemented in real time and in a real environment, needs to be maintained and provided with new data from the real environment in order to be able to respond to changes that may occur in this environment.

2 Materials and Methods

2.1 Nvidia Jetson AIoT Devices

Nvidia has provided graphic programming units for decades and it is now the world's largest semiconductor company. With other competitors in the semiconductor segment it is becoming more and more attractive for these companies to expand artificial intelligence chips to business [23]. Not only by bringing to the market some graphics cards and graphics chips, as well as special embedded devices. These devices are specially designed for artificial intelligence tasks. The main task are computer vision tasks due to the high hardware performance. Another possible implementation is to use these devices as a typical IoT device partially. By the word typical we mean to use Jetson as a field level device. In such a case, they would be used to measure physical quantities and ensure their transmission over the network to the higher accounting management of the production plant. This approach, however, does not exploit the full potential of AIoT devices. Although we do not consider this to be a suitable solution, such use is also possible and fully functional. It has even several advantages. The main advantage may be the power reserve, which we can fill so that one device will replace tens of IoT in some cases. For example, in the case of collecting and transmitting data from sensors between data nodes of the internal network or using the Internet in cases where devices and sensors are located outside the cable network.

Nvidia Jetson

Nvidia Jetson Nano is high performance device equipment designed to support the implementation of neural network models and their deployment in a real environment. Nvidia Jetson Nano is part of Nvidia Jetson's lowest performance product line and consequently the cheapest one. A quad-core ARM Cortex-A57 processor powers the

Jetson board with a frequency of 1.43 GHz utilizing the Armv8-A architecture used [24]. The graphics processor used in the Nano board is a 128-core based on second generation Maxwell architecture and GPU processor board GM20B with boost clock up to 921 MHz. Computing capability is calculated as 5,3 [25]. The Jetson Nano B01 version operates with the system that shares 4 GB LPDDR4 special memory which provides less energy consumption and fast data transmission. This memory is especially convenient for embedded devices, mobile devices, wearables, etc. As we mentioned before, the powerful hardware of Jetson Nano prepared him for artificial intelligence tasks. From a software perspective, Jetson runs on Jetson Linux. It includes Linux Kernel 4.9, bootloader, utilities and it is based on Ubuntu 18.04. The Jetson Linux Driver package also supports all Jetson Jetson Nano series [26]. Board supports UHS-1 microSD card 16 GB memory or larger. NVMe memory disk. All technical data for the Jetson Nano 4 GB are listed in Table 1.

Table 1. Nvidia Jetson Nano hardware specification [27].

GPU	128-core Maxwell
CPU	Quad-core ARM A57, 1.43 GHz
Memory	4 GB 64-bit LPDDR4 25.6 GB/s
Storage	microSD+
Video encoder Video decoder	4K 30, 4x 1080p 30, 9x 720p 30 (H.264/H.265) 4K 60, 2x 4K 30, 8x 1080p 30, 18x 720p 30
Camera	1x MIPI CSI-2 DPHY lanes
Connectivity	Gigabit Ethernet, M.2 Key E
Display	HDMI 2.0 and eDP 1.4
USB	4x USB 3.0, USB 2.0 Micro-B
Other periferals	GPIO, I2C, I2S, SPI, UART
Device dimension	100 mm × 80 mm × 29 mm

The Jetson Nano board consists of two parts. The first part is the board consisting of main computational parts. And the second one is board composed of I/O sets which includes USBs, 2-lane MIPI CSI, SD card slots, UART, I2C, SPI etc.

In professional industry application, it is possible to use Jetson as a System On Module device. Such a solution allows using own designed carrier board or buying compatible industry-oriented boards [28].

Nvidia Jetson Xavier

Nvidia Jetson Xavier NX is the second member of the Nvidia Jetson product family. It is in the same product line as Jetson Nano, however, it is equipped with more powerful hardware. Xavier NX is powered by 6 core ARM CPU Nvidia Carmel processor cores with operating frequency up to 1.9 GHz. As a graphic processor, Nvidia Volta architecture is used. The most important feature of Volta graphic architecture is introduced Volta

Streaming Multiprocessor which was design to perform deep learning tasks better. The major performance difference is focused on reducing cache latency, which brings the ability to program Tensor Cores. This includes also an acceleration in matrix calculation which are basic mathematic operations during calculating neural network tasks. The GPU consists of 384 Nvidia Cuda cores and 48 programmable Tensor cores. Like the Jetson Nano, the Jetson Xavier NX consists of two parts: the carrier and the main board. The difference in cooling between the two devices is visible at glance. The Xavier board is significantly more powerful and passive cooling is insufficient. This is the reason that active cooling is used. As a result, the higher temperatures, it is also advisable to use self-cooling by a combination of a passive cooler and active warm air extraction. By reducing the heating of the carrier and the reliable heat dissipation from the main board, it is possible to ensure optimal operation of the device. Powerful cooling is also essential from the point of view of the inference deep learning model (Table 2).

Table 2. Jetson Xavier NX specifications [23].

GPU	384 NVIDIA CUDA cores and 48 Tensor cores
CPU	6-core NVIDIA Carmel ARMv8.2 64-bit CPU 6 MB L2 + 4 MB L3
Memory	8 GB 128-bit LPDDR4x 51.2 GB/s
Storage	16 GB eMMC 5.1
Video encoder	2x464 MP/sec (HEVC) 2x 4K 30 (HEVC) 6x 1080p 60 (HEVC) 14x 1080p 30 (HEVC)
Video decoder	2x690 MP/sec (HEVC) 2x 4K 60 (HEVC) 4x 4K 30 (HEVC) 12x 1080p 60 (HEVC) 32x 1080p 30 (HEVC) 16x 1080p 30 (H.264)
Power	10 W–15 W
Camera	Up to 6 cameras
Connectivity	10/100/1000 BASE-T Ethernet 1 USB 3.1 (10 Gbps) l 3x USB 2.0
Display	HDMI 2.0 and eDP 1.4
USB	4x USB 3.0, USB 2.0 Micro-B
Other IOs	1x SDIO/2x SPI/3x UART/2x I2S/4x I2C/1x CAN/GPIOs
Device periferals	45 mm × 69.6 mm 260 pin SO-DIMM connector

3 Boards Visual Inspection

The main goal of visual inspection is to ensure that the electronics equipment is properly set up. The modern visual inspection gives advanced possibilities to detect PCB production errors before sending the product to the customer. The highest standards of semiconductor components manufacturers are a matter of course today.

The printed circuit board is assembled according to the technical documentation of the manufacturer and surface mounted technology is utilized. The first factor to be checked is that soldered components and verify their location according to the product datasheet. In this section, we have done basic visual board inspection to verify that the equipment complies with the technical documentation. These concerns arose because during the operating time of the Jetson equipment we encountered several problems of both software and hardware origin. Based on Putera et al. we recognized 14 types of PCB defects [29].

For the purposes of this paper, we selected certain defects on which we focused our attention. The reason for this approach is that the devices we used in this paper assume a high standard of production and a high rate of detection of defects in production. Therefore, our goal is not to detect manufacturing errors in the production of PCBs, but to verify by a simple visual inspection whether the equipment that we will continue to use on our other projects is entirely in order after in terms of the hardware.

Another method of visual inspection is the Automatic Optical Inspection (AOI). Such a visual check of the security system is the evaluation of defects and problems on the PCBs automatically and allows to significant automate of the entire production of PCBs. In the case of using automatic optical inspection, the defect categories are determined, which the system identifies automatically. At present, deep learning by applying neural networks is also gaining prominence in this industrial area. However, there is no general model and therefore, each AOI solution using deep learning is different and has different categories, which the algorithm classifies as defects [30].

These categories always depend on the optical inspection methodology that is applied, and it is often the case that the individual motors overlap, and their combinations are applied in order to achieve better control. The issue of Automatic Inspection is beyond the scope of this proposal. One of the impressive and very accurate methods of PCB inspection is X-ray defect identification. The applied X-rays have the capability to detect defects that are not on the surface of the soldering material, but on its internal structure. The advantage is that it can also show defects in multiple layers [31].

We distinguish between two-dimensional and three-dimensional X-ray tomography. A typical application is to classify plated through a hole defect, missing ball, bridge, offset, etc. [32, 33].

4 Experiment and Results

Table 3 consists of defect types that we verified to see if this type of error was found on the Jetson Nano and Jetson Xavier boards. We focus on two main types of defects: Soldering defects and Component defects. We used a microscope with 1200x magnification for our visual inspection. The inspection was successful, and we did not notice any serious defects on the printed circuit board or any of the devices mentioned above. The result of the test scenarios can be found in Table 3.

Figure 2 and Fig. 3 depict some of the images recorded by a microscope to verify the individual components and soldered spots on the Nvidia Jetson Nano PCB. There were no places on the printed circuit boards where there were no components or disturbing paths between the components. There were also no places where there was a lack of solder or joints where the solder was completely missing. In the next part of the visual inspection, we monitored the defects on the components. Again, we did not find any

Table 3. Test scenarios.

	Defect type	Nvidia Jetson Nano 4 GB	Nvidia Jetson Xavier
Soldering defects	Components placed on PCB	✓	✓
	Open circuits	✓	✓
	Solder bridges	✓	✓
	Solder quality	✓	✓
	Insufficient solder	✓	✓
	Solder void	✓	✓
	Excess solder	✓	✓
Component defects	Lifted lead	✓	✓
	Missing component	✓	✓
	Misaligned or misplaced component	✓	✓

significant errors on any of the boards. There were no gaps on the board where no component was missing, or no component was misplaced in such a way as to break the conductive connection in the circuit.

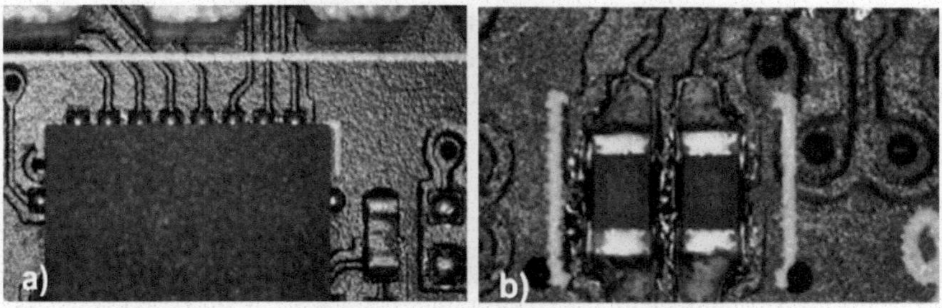

Fig. 2. NVIDIA Jetson visual inspection a) solder quality, b) excess solder.

Thermal management is one of the most critical aspects when designing a printed circuit board. The thermal analysis of the circuit board has the task of displaying heat transfer scenarios when the device is loaded with electrical current. Three phenomena occur in the analysis of heat transfer on the PCB. The heat conduction, heat convection, and thermal radiation. It is necessary to verify the thermal properties of the circuit by simulation and also experimentally, since heat transfer on the board does not only occur by direct contact of the components mounted on the board, despite the fact that heat also continues without direct contact between the components [34]. Subsequently, the entire board emits heat, which thermally affects the space around it. This heat emission must then be taken into account when designing electrical cabinets made of PVC material,

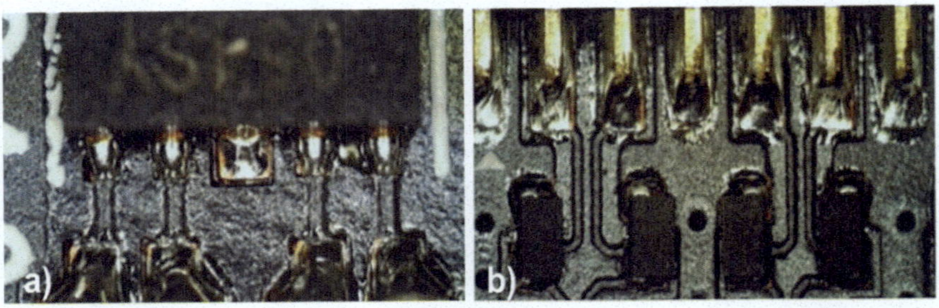

Fig. 3. NVIDIA Xavier Visual Inspection – a) insufficient solder, b) misaligned or misplaced component

which has excellent thermal insulation properties and does not dissipate heat out of the cabinet, but on the contrary retains it. This issue also seems to be important for devices that concentrate higher computing power and are part of a more complex system stored in a PVC box, which should meet the protection against external influences.

4.1 Proposal of Experiments

The environment has a stable temperature in the range of 20,8 to 21,3 °C. The first type of measurement was 24 h long. This measurement aimed to determine the surface temperature of the Jetson Nano and Jetson Xavier. The measurement of surface temperature and thermal heat dissipation across the board took place in the Device Ready mode. The Device Ready system is considered to be the state when the system is booted and waiting for the script to run. Only the operating parts of the operating system are running, and no other application is running. The measurement period was set at every 6 h.

We measured temperature in Ready Mode in two places:

- On peripherals (Temperature 1)
- Cooler (Temperature 2)

The second parameter measured was the electrical current flowing into the device from the electrical supply network. The Table 4 consists of three columns: mode, Temperature 1, and Temperature 2.

Table 4. Temperature measurement Jetson Nano

Ready mode	Temperature 1	Temperature 2
6 h	27 °C	19 °C
12 h	32 °C	16 °C
18 h	33 °C	18 °C
24 h	35 °C	21 °C

In Table 5, we measured electrical current and calculated the power consumption of the device in each defined time period. The highest value measured by the amperemeter was 0,56 A in the Device Ready mode. The measured electrical voltage was stable at value 5,1 V. The calculated power consumption of Nano device was from 2,51 W to 2,85 W.

Table 5. Electrical current measurements - Jetson Nano

Ready mode	Electrical current	Power consumption
0 h	0,49 A	2,51 W
6 h	0,50 A	2,55 W
12 h	0,52 A	2,65 W
18 h	0,51 A	2,61 W
24 h	0,56 A	2,85 W

We used a thermal camera to measure thermal scattering. The following images show the heat dissipation on the plate during each measurement. From the images, we can conclude that the largest heat dissipation is around the main board and then the heat spreads the most towards the I/O module. The plug-in peripherals for connecting the SD card module are also subjected to thermal resistance.

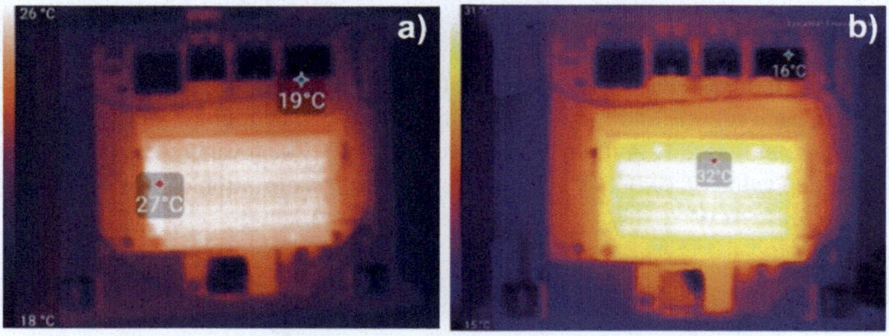

Fig. 4. Heat dissipation on the printed circuit board - Jetson Nano a) 6 h b) 12 h

In Fig. 4 are shown thermal images and we can see temperatures measured in the different time periods. These time periods are defined in the table above. We can clearly see that temperatures were rising slowly during periods and in the last two measurement periods stabilized from 27–35 °C. It is observable that thermal energy from main module is spreading to SD slot and affecting the temperature of these module (Fig. 5).

The second measurement also took place in Device Ready mode, but with the Jetson Xavier NX. This means that the device was turned on, but no targeted calculations or scripts were performed outside the operating system. We recorded the individual measurements in Table 6. In this case, the measurement period was the same as when

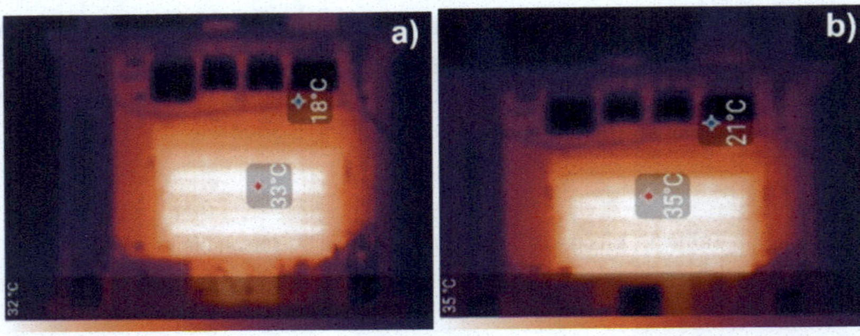

Fig. 5. Heat dissipation on the printed circuit board - Jetson Nano a) 12 h b) 24 h

measuring the Jetson Nano heat emission in Device Ready mode. The highest value of temperature in the Device Ready mode was 45 °C.

Table 6. Temperature measurement Jetson Xavier

Ready mode	Temperature 1	Temperature 2
6 h	26 °C	19 °C
12 h	32 °C	16 °C
18 h	41 °C	18 °C
24 h	45 °C	21 °C

In Table 7, there is measured electrical current and also calculated power consumption of device in each regime and in the defined time period. The highest value measured by the amperemeter was 0,41 A in the Device Ready mode. The measured electrical voltage was stable at a value of 19.0 V. The calculated power consumption of Xavier device was from 5,7 W to 7,79 W.

Table 7. Electrical current measurements - Jetson Xavier

Ready mode	Electrical current	Power consumption
0 h	0,30 A	5,7 W
6 h	0,35 A	6,65 W
12 h	0,36 A	6,84 W
18 h	0,40 A	7,60 W
24 h	0,41 A	7,79 W

Figure 6 and Fig. 7 show thermal images, and we can see temperatures measured in the different time periods. These periods are defined in the table above. We can see temperatures were rising slowly during periods and in the last two measurement periods stabilized around 40–45 °C. It is observable thermal energy from the main module is spreading to MPI CSI-2 and affecting the temperature of these modules.

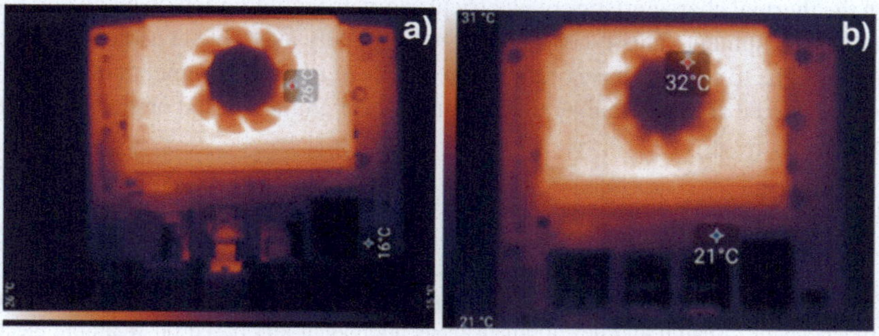

Fig. 6. Heat dissipation on the printed circuit board - Jetson Xavier a) 6 h b) 12 h

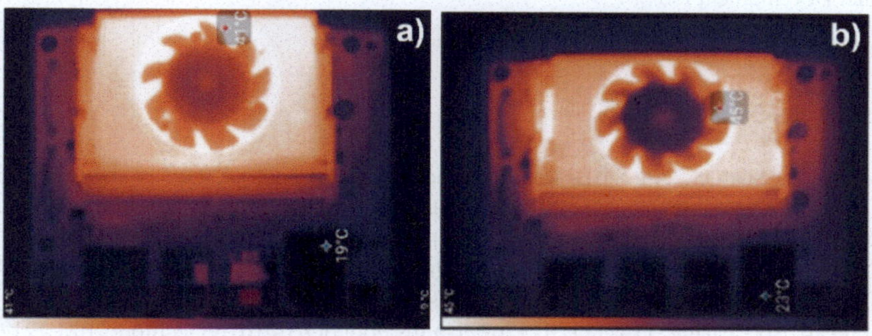

Fig. 7. Heat dissipation on the printed circuit board - Jetson Xavier a) 12 h b) 24 h

5 Future Work

In the future, we expect to continue experimenting with Jetson Nano and Jetson Xavier NX modules. Based on the circuit model, we want to make a model and then perform thermal analysis in some simulation software and compare and evaluate these results with experimental measurements of thermal emission of individual boards. The findings are then used in the technological design of the thermal parameters of the electrical boxes, which will include a Jetson device.

6 Conclusion

In this proposal, we have devoted the need for the Industry 4.0 concept and the importance of IoT equipment. Gradually, we came to the description of a subset of IoT devices specialized in the field of artificial intelligence, the so-called AIoT devices. Subsequently, we described the printed circuit board inspection methods that we used for our experimental measurements. The performed experiments yielded data on temperatures measured in Device Ready mode. We also monitored the heat transfer across the printed circuit board. These measurements were made to identify temperatures around Jetson equipment to design the thermal characteristics of electrical cabinets including Jetson Nano and Jetson Xavier NX equipment. This type of data is not officially available. The next step in our research will be to include the experimental results of our measurements in the thermal analysis and to be able to simulate the amount of heat emitted based on the power of the device.

Acknowledgements. This publication was supported by the project of the project Young Scientist 1360 (2021–2022): "Applying machine learning at industry environment" supported by internal grant Young Scientist at Slovak University of Technology in Bratislava.

This paper was supported under the project of Operational Programme Integrated Infrastructure: Independent research and development of technological kits based on wearable electronics products, as tools for raising hygienic standards in a society ex-posed to the virus causing the COVID-19 disease, ITMS2014+ code 313011ASK8. The project is co-funding by European Regional Development Fund.

References

1. Menon, S., Shah, S.: Are SMEs Ready for industry 4.0 technologies: an exploratory study of I 4.0 technological impacts. In: 2020 International Conference on Computation, Automation and Knowledge Management (ICCAKM), pp. 203–208 (2020)
2. Di Vaio, A., Hassan, R., Alavoine, C.: Data intelligence and analytics: a bibliometric analysis of human–Artificial intelligence in public sector decision-making effectiveness. Technol. Forecast. Soc. Chang. **174**, 121201 (2022). https://doi.org/10.1016/j.techfore.2021.121201
3. Aoun, A., Ilinca, A., Ghandour, M., Ibrahim, H.: A review of Industry 4.0 characteristics and challenges, with potential improvements using blockchain technology. Comput. Ind. Eng. **162**, 107746 (2021). https://doi.org/10.1016/j.cie.2021.107746
4. Karnik, N., Bora, U., Bhadri, K., et al.: A comprehensive study on current and future trends towards the characteristics and enablers of industry 4.0. J. Ind. Inf. Integr. 100294 (2021). https://doi.org/10.1016/j.jii.2021.100294
5. Calderón, R.R., Izquierdo, R.B.: Machines for industry 4.0 in higher education. In: 2020 IEEE World Conference on Engineering Education (EDUNINE), pp. 1–4 (2020)
6. Powell, D., Morgan, R., Howe, G.: lean first ... then digitalize: a standard approach for industry 4.0 implementation in SMEs. In: Dolgui, A., Bernard, A., Lemoine, D., von Cieminski, G., Romero, D. (eds.) APMS 2021. IAICT, vol. 631, pp. 31–39. Springer, Cham (2021). https://doi.org/10.1007/978-3-030-85902-2_4
7. Dudak, J., Gaspar, G., Michalconok, G.: Extension of 1-wire measuring system SenSys. In: Proceedings of 15th International Conference MECHATRONIKA, pp. 1–4 (2012)
8. Dudak, J., Sladek, I., Gaspar, G.: Proposal and implementation of universal and non-volatile 1-wire module. In: 2016 17th International Conference on Mechatronics - Mechatronika (ME), pp. 1–5 (2016)
9. Unipi. https://www.unipi.technology/. Accessed 24 Feb 2022
10. AIoT.concepts.e.png (Obrázok PNG, 1385 × 1080 bodov). https://upload.wikimedia.org/wikipedia/commons/2/23/AIoT.concepts.e.png. Accessed 23 Feb 2022

11. Xia, M., Li, T., Xu, L., et al.: Fault diagnosis for rotating machinery using multiple sensors and convolutional neural networks. IEEE/ASME Trans. Mechatron. **23**, 101–110 (2018). https://doi.org/10.1109/TMECH.2017.2728371
12. Yunhui, Y., Kechen, S., Zhitao, X., Xuehui, F.: The strip steel surface defects classification method based on weak classifier adaptive enhancement. In: 2011 Third International Conference on Measuring Technology and Mechatronics Automation, pp. 958–961 (2011)
13. Bao, Y., Qibing, Z., Min, H.: Image identification of glass defects based on non-negative matrix factorization and sparse representation classification. In: 2012 24th Chinese Control and Decision Conference (CCDC), pp. 3225–3229 (2012)
14. Kan, Y.C., Kalkan, H.: Automatic detection and classification of laser welding defects. In: 2021 Innovations in Intelligent Systems and Applications Conference (ASYU), pp. 1–5 (2021)
15. Deng, Y.-S., Luo, A.-C., Dai, M.-J.: Building an automatic defect verification system using deep neural network for PCB defect classification. In: 2018 4th International Conference on Frontiers of Signal Processing (ICFSP), pp. 145–149 (2018)
16. Lin, B.-S., Cheng, J.-S., Liao, H.-C., et al.: Improvement of multi-lines bridge defect classification by hierarchical architecture in artificial intelligence automatic defect classification. In: 2020 International Symposium on Semiconductor Manufacturing (ISSM), pp. 1–4 (2020)
17. Menegazzo, J., von Wangenheim, A.: Multi-contextual and multi-aspect analysis for road surface type classification through inertial sensors and deep learning. In: 2020 X Brazilian Symposium on Computing Systems Engineering (SBESC), pp. 1–8 (2020)
18. Mandal, V., Uong, L., Adu-Gyamfi, Y.: Automated road crack detection using deep convolutional neural networks. In: 2018 IEEE International Conference on Big Data (Big Data), pp. 5212–5215 (2018)
19. Yang, Y., Luo, H., Xu, H., Wu, F.: Towards real-time traffic sign detection and classification. IEEE Trans. Intell. Transp. Syst. **17**, 2022–2031 (2016). https://doi.org/10.1109/TITS.2015.2482461
20. Liu, J., Xu, L., Cao, X., et al.: Review on the architectures and applications of deep learning in agriculture. In: 2020 7th International Conference on Information Science and Control Engineering (ICISCE), pp. 1234–1240 (2020)
21. Mohapatra, D., Choudhury, B., Sabat, B.: An automated system for fruit gradation and aberration localisation using deep learning. In: 2021 7th International Conference on Advanced Computing and Communication Systems (ICACCS), pp. 6–10 (2021)
22. Albanese, A., Nardello, M., Brunelli, D.: Automated pest detection with DNN on the edge for precision agriculture. IEEE J. Emerg. Sel. Top. Circ. Syst. **11**, 458–467 (2021). https://doi.org/10.1109/JETCAS.2021.3101740
23. NVIDIA Xavier - Description. https://developer.ridgerun.com/wiki/index.php?title=Xavier/Processors/GPU/Description. Accessed 24 Feb 2022
24. Cortex-A57. https://developer.arm.com/Processors/Cortex-A57. Accessed 24 Feb 2022
25. NVIDIA GM20B GPU Specs. In: TechPowerUp. https://www.techpowerup.com/gpu-specs/nvidia-gm20b.g819. Accessed 24 Feb 2022
26. Jetson Linux. In: NVIDIA Developer (2015). https://developer.nvidia.com/embedded/linux-tegra. Accessed 24 Feb 2022
27. NVIDIA Jetson NANO Developer Kit
28. Quark Carrier for NVIDIA® Jetson Nano™. In: Connect Tech Inc. https://connecttech.com/product/quark-carrier-nvidia-jetson-nano/. Accessed 24 Feb 2022
29. Putera, S.H.I., Dzafaruddin, S.F., Mohamad, M.: MATLAB based defect detection and classification of printed circuit board. In: 2012 Second International Conference on Digital Information and Communication Technology and it's Applications (DICTAP), pp. 115–119 (2012)

30. Anitha, D.B., Rao, M.: A survey on defect detection in bare PCB and assembled PCB using image processing techniques. In: 2017 International Conference on Wireless Communications, Signal Processing and Networking (WiSPNET), pp. 39–43 (2017)
31. Doi, H., Suzuki, Y., Hara, Y., et al.: Real-time X-ray inspection of 3-D defects in circuit board patterns. In: Proceedings of IEEE International Conference on Computer Vision, pp. 575–582 (1995)
32. Huang, X., Zhu, S., Huang, X., et al.: Detection of plated through hole defects in printed circuit board with X-ray. In: 2015 16th International Conference on Electronic Packaging Technology (ICEPT), pp. 1296–1301 (2015)
33. Ma, J.-Q., Kong, F.-H., Ma, P.-J., Su, X.-H.: Detection of defects at BGA solder joints by using X-ray imaging. In: 2005 International Conference on Machine Learning and Cybernetics, vol. 8, pp. 5139–5143 (2005)
34. Chen, Y.: On the four types of weight functions for spatial contiguity matrix. Lett. Spat. Resour. Sci. **5**, 65–72 (2012). https://doi.org/10.1007/s12076-011-0076-6

Assessment and Rehabilitation of Low Back Pain (LBP) Using Artificial Intelligence and Machine Learning – A Review

Manvendra Singh(✉) and Abhijit Sharma

Department of CSE, National Institute of Technology, Durgapur, India
ms.19cs1109@phd.nitdgp.ac.in, abhijit.sharma@cse.nitdgp.ac.in

Abstract. Low Back Pain (LBP) influences millions of individuals worldwide at some point in their lives, and it can cause a serious impact on physical, mental, and emotional health for an extended period. People dealing with LBP may have trouble concentrating or find themselves unable to perform daily activities or work obligations. Proper clinical assessment and rehabilitation of LBP are thus important for performance and overall long-term quality of life. During the last couple of decades, artificial intelligence and machine learning (AI/ML) algorithms have increasingly been acquired for medical diagnosis, prediction and classification of diseases. Nowadays, AI has improved the treatment, prevention, and prognosis associated with LBP due to the growth in clinical data. Recent research trends related to the assessment and rehabilitation of low back pain using AI/ML have been reviewed and presented in this paper.

Keywords: Low Back Pain classification · Assessment · Rehabilitation · Artificial intelligence · Machine learning · Clinical decision support systems

1 Introduction

The human body is susceptible to disease during its lifespan, which requires the patient to go to the hospital. Low Back Pain (LBP) impacts about 80% of the world's population at some point in their lives [1], and it can cause severe discomfort for an extended period. When an individual experiences low back pain, they are at risk of decreased functionality. Low back pain is a significant reason for nonattendance [2] due to the severity of the pain, and the employees have to take early leave from work. Pain in the lower back restricts occupational activities, becoming a significant reason for early retirement and income poverty [3]. Its financial liability is represented by the high expenses of healthcare expenditure and declined income [4]. Many factors can cause lower back pain [5]. Work at the job, lifestyle changes, and working environments can lead to lower back pain. LBP can be triggered by lifting heavy objects or performing painstaking tasks. Accidents or injuries can also generate low back pain strain. Irritation, ligaments sprain, or damage in the discs, joints and bones of a human body also produces low back pain. This condition occurs when muscles in the spine become inflamed, causing them to stiffen and weaken, reducing the overall quality of life.

R. Silhavy (Ed.): CSOC 2022, LNNS 502, pp. 449–461, 2022.
https://doi.org/10.1007/978-3-031-09076-9_40

Acute back pain develops unexpectedly and is often a consequence of an identifiable injury. It can resolve within days, but it may last up to six weeks. On the other hand, chronic pain develops more gradually over time and may become progressively worse. There are multiple factors involved in low back pain. Each element has a condition that can generate many underlying problems of varying levels of severity. Low back pain is classified into three types: mechanical back pain, non-mechanical back pain, and referred pain from internal organs [6]. Any remarkable change in the mechanical structure of the lower back, severe pain is experienced and compressed nerve roots, non-specific musculoskeletal strain, herniated disc, degenerative disc or joint disease, and a ruptured vertebral subject to mechanical back pain. Swelling around the flexure of the back, any type of tumours and infection are the causes of non-mechanical back pain. Specific lower back pain, which falls under the category of non-mechanical back pain, is also coming under mechanical lower back pain because it is due to the movement of the spinal cord. Some lower back pain comes under the referred pain category. Such pain is caused by complications in internal organs such as kidney infections, gallbladder disease, kidney stones, aortic aneurysm, and others. The part of non-specific LBP [7] in the diagnostic Classification of LBP is about 90%, where the cause of LBP is not clearly visible, and treatment is focused on reducing pain and its outcome.

Diagnosis and treatment of LBP in time is essential since any type of low back pain gives rise to a severe problem if it is late. Thus, recognising the method of diagnosis and treatment of LBP becomes very important. Guidelines for specifying clinical practice [8] to detect red flags, know the patient's disease history and neurological testing for the radicular syndrome, and perform imaging and psychosocial assessment for serious pathology testing are necessary. Clinical guidelines for treating acute low back pain patients recommend the patient to return to normal activities, use nonsteroidal anti-inflammatory drugs (NSAIDs), avoid prolonged bed rest and recommend weak opioids for a modest period. Patients with chronic low back pain are prescribed NSAIDs and antidepressants, exercise therapy, and psychosocial interventions as per clinical guidelines. A referral is recommended in certain suspected cases with specific pathologies or radiculopathy or when the patient does not improve after four weeks of treatment.

During the last couple of decades, artificial intelligence (AI) and machine learning (ML) algorithms have increasingly been acquired for medical diagnosis, prediction, and Classification of diseases. Nowadays, AI has improved the treatment, prevention, and prognosis associated with LBP. It plays a valuable role in clinical practice due to the growth in clinical data used to train and develop good AI models. Improvements leading to the latest AI models are due to the outbreak of deep learning. It has enhanced the computing capabilities, leading to a more autonomous and accurate model. Through this paper, we have reviewed the existing literature on the application area of AI/ML algorithms on the diagnosis and treatment of LBP to illustrate the state of the art of such technology and its potential applications.

In the remainder of this paper, we present the complete overview of AI/ML-based approaches for the assessments and rehabilitation of Low Back Pain. Section 2 begins with a detailed discussion of related evaluations and rehabilitation of Low Back Pain (LBP) using AI/ML. Finally, we conclude the paper in Sect. 3.

2 Review Work

AI/ML models are broadly used to extract high-level abstract features applied to understanding and processing lower back-pain data. AI/ML model allows performing a plethora of tasks that may advance the clinical exercise, such as automation of detection and localisation of lumbar structure with segmentation. Besides, it will enable extracting a set of features from the image that can be used as an input for different machine learning algorithms to provide decision support to the clinician or, in other cases, directly suggest the most appropriate diagnosis.

In this section, w have discussed various existing AI/ML-based methods for the assessments and rehabilitation of Low Back Pain and provided a qualitative comparison of them, please refer to Table 1.

2.1 Pain Assessment

Classifying patients into different treatment groups leads to improved healing outcomes. However, Clinical decision-making is, as of now, generally based on qualitative questionnaires. Recent advances in machine learning techniques and wearable sensors allows clinicians to capture and analyse patients' movements during daily life activities. This facilitates a cost-effective and improved assessment tool that can be used for prescribing appropriate treatment.

This section reviews [9–12] machine learning and sensor-based pain assessment methods, classifying the patients into different groups.

2.1.1 Motion Sensor Based

The authors in [9] have investigated the potential of using an efficient motion-capture sensor like the Inertial Measurement Unit (IMU) and involve placing the IMU on the sternum of ninety-nine patients and allowing the patients to act as many trunk flexion/extension activities as quickly in the sagittal plane within a 14-s period. In addition to the gained trunk kinematic data, a Wii Balance Board was used to record the centre of pressure data. The author found that the machine learning methods could distinguish non-specific low back pain patients into high risk and low-medium risk patients with an accuracy of $>75\%$.

2.1.2 Physical and Muscle Activity Information Based

The work [10] is based on an existing EmoPain dataset containing movements from thirty participants carrying out physical activity, described by two features: full-body motion capture and muscle activity. The classification experiments were performed using four different methods: decision trees, nearest neighbour, linear discriminant analysis, and Random Forest. The pain recognition task of the proposed method classifies the pain into three Healthy categories, Low-level pain and High-level pain.

2.1.3 Magnetic Resonance Imaging (MRI) Based Classification of Disc

[11], the purpose of this paper is to determine whether a machine-learning model can classify fssures and pain based on magnetic resonance imaging (MRI) data with great accuracy and precision. This study classified concurrent pain-response with pain-positive discograms performed in 86 discs with pressure-controlled discography. CT-discograms are used to classify fssures and discs with fssures extending into the external annulus, machine-learning algorithms trained on the extracted MRI markers and the disc classified as painful or non-painful.

2.1.4 NLBP Symptomatic Muscles Recognition Method (EasiSMR) Locate the Symptomatic Muscles

Non-specific low back pain (NLBP) is originated in the human body due to anonymous pathoanatomical reasons and treatment centred on the doctor's experience. To help the doctor and give precise patient treatment, it is vital to locate symptomatic muscles. In [12], the author proposed an NLBP Symptomatic Muscles Recognition method (EasiSMR) to locate the symptomatic muscles precisely. EasiSMR is a machine learning model for low back pain that enables the recognition of the symptomatic muscles through an isometric exercise. EasiSMR has two main contributions to enhancing recognition performance; first, the Multi-Muscles Fusion (MMF) method, which uses muscle compensation data. Second, it extracts new features for isothermic exercise, which suggests behaviours of muscle firing stage. Results show 86.16% average accuracy for recognition for six low back pain muscles.

2.2 Diagnosising LBP Based Classification

Low back pain is a widespread symptom and the leading cause of disability worldwide. Based on the prediction and identification of low back pain, we have reviewed [13–17] LBP identification methods that classify the patients into different groups. Machine learning algorithms are applied on the sensor and clinical notes data identification of lower back pain.

2.2.1 LBP Analysis by Free-Text Clinical Notes

Clinical notes are used to describe systematic learning of treatment approaches management, and billing guidelines options for acute LBP. Free-text clinical notes, free text (ICD-10), is used to track and record patients' health, and it is helpful to recognise low back pain using a machine learning algorithm. The authors in [13] used Free-text clinical notes (ICD-10) to assess the prospect of automatically differentiating acute LBP episodes. Many supervised and unsupervised approaches are compared for automated identification of LBP. ConvNet gives the best results with an area under the receiver operating characteristic curve of 0.98 and an F score of 0.70.

2.2.2 Classification Based on Kinematics Data

In [14], the author described a deep learning approach to recognise the low back pain in populations from healthy controls with human balance and body sway performance during standing trials using their kinematic data. Models trained from motion sensors' kinematics data outperformed a model trained with centre of pressure (CoP) trajectories. Results show that a deep learning network could solve the above classification problem with good precision and recall performance.

2.2.3 Classification Based on RMI Data

The author used lumbar magnetic resonance imaging (MRI) [15] to estimate LBP. In a Specific MRI, the texture of T2-weighted MR images is analysed using texture analysis and machine learning. Machine learning (logistic regression) algorithm is used to classify the texture feature and dichotomous pain variable based on the questionnaire. The data is classified into symptomatic and asymptomatic by applying principal component analysis and logistic regression classifiers to textural feature extraction. Applying texture analysis on the two lowest intervertebral discs (L4-L5 and L5-S1) give an accuracy of 83%, a specificity of 83%, a sensitivity of 82%, a negative predictive value of 94%.

2.2.4 Classification Based on Objective Functional Testing

In paper [16], the authors evaluated the range of diagnostic information obtained from the 5R-STS test as it may vary between patients with the performance of the 5R-STS with or without disc herniation (LDH), lumbar spinal stenosis (LSS) or low-grade spondylolisthesis or with chronic low back pain (CLBP). Based on the patient's physical examination, history and imaging, the patient has been classified in the gold standard diagnostic categories and adjusted. Crude comparisons of 5R-STS performance are performed between the three diagnostic types. A machine learning algorithm has been trained to classify patients into three categories, which uses the patient's age, gender, weight, height and 5R-STS test time for this Classification. Here, there is a variation in the performance of the 5R-STS among patients with LDH, LSS and discogenic CLB. Thus, a simple combination of objective functional impairment and machine learning algorithms can help in early diagnosis to accurately classify patients with back or leg pain.

2.2.5 Classification Based Ranking Degree of Importance

This study [17] aims to manage low back pain on a day-to-day basis and to identify whether a person is abnormal or normal using collected physical spine data of 381 patients with 12 parameters. This study also identifies the degree of importance of each parameter used in the Classification and ranks those parameters accordingly.

2.3 Decision Making on LBP Treatment

A general practitioner or medical specialist provides patients with low back pain (LBP) some referral advice as needed. Making decisions on treatment is a very complex problem for the medical profession practitioner. Conflicting advice and treatments may have adverse costs for optimal recovery. A clinical decision support system (CDSS) may support physicians in selecting appropriate therapies for patients with LBP. With the correct data, integration methods, and personnel in place, machine learning can advance clinical decision support and help providers deliver optimal care.

In this section, essential decision-making methods [18–20] have been reviewed where the authors contributed to integrating machine-learning platforms into the clinical decision support systems.

2.3.1 Designing a Clinical Decision Support System

The author in [18] has designed clinical decision systems that help patients with LBP for self-referral to primary care. The author further explored whether a supervised machine learning-based classification model, i.e., 1. random forest, 2. decision tree, and 3. boosted tree, is capable of assisting clinical decision support systems and compared the performance of models. Generate these supervised machine learning models, 70% of the training dataset is used, and 30% of the training dataset is used for total accuracy, as well as kappa and sensitivity, specification, and precision, to measure the performance of the model during validation. Accuracies found in Boosted Tree are the best in all three machine learning algorithms for the training dataset and test dataset. They are best for predicting referral guidance with fair accuracy, i.e., kappa between 0.2 and 0.4.

2.3.2 Decision Making on Conservative and Operative Therapy

The paper [19] shows whether deep learning techniques with lumbar disc herniation patient data predict the need for surgical or conservative treatment after six months of treatment. An algorithm is developed to make correct predictions of 6 randomly selected test patients using ten-fold cross-validation. The result shows that 100% prediction accuracy can be achieved by dividing the ODI scale into 12 sections. After six months of therapy, predictions made with the individually predicted and actual ODIs yield a maximum absolute difference of only 3.4%.

2.3.3 Multi-modal Biomarkers to Identify LBP and HC

The author in [20] describes the structure and reorganisation of the human brain and how morphological changes in cerebral cortical thickness (CT) and rest-state functional connectivity (rsFC) can be used as potential biomarkers to guide LBP treatment. Structural MRI scans, rsFC and self-reported clinical scores find CT differences in LBP patients concerning HC. Cortical thickness trains the Support Vector Machine (SVM) model to differentiate the LBP and healthy controls (HC).

2.4 Adherence Monitoring

Low back pain (LBP) is a significant cause of disability, and its treatment is complex. As a recurrent condition, prevention is crucial. Home-based exercise is an effective way to treat LBP, but adherence and proper exercise become very important to get rid of LBP. It is problematic to monitor manually by medical professionals and therapists. The concept and technology that can be included are garment stretch capacitive sensor array, wearable T-shirt inductive sensor, exoskeleton supports, computer vision-based systems, accelerometers or IMU. Applying Machine learning (ML) to rehabilitation and adherence monitoring is a solution for people with LBP if it adequately monitors adherence performance while providing patient guidance. Thus, in this section, reviewed papers [21–26] that have used AI/ML techniques for adherence and exercise monitoring.

2.4.1 Fuzzy System Based Postural Therapies

This study describes spinal column posture with three angular motions to measure spinal angular activity to prevent low back pain and reduce permanent damage based on postural control and maintenance of postural stability. The author [22] has presented a wearable device used to measure the three spinal sections using an inertial unit with a triaxial accelerometer, gyroscopes and magnetometers and device. The device uses fuzzy systems to adjust the back posture, which controls the vibration unit and integrates the absolute and relative orientation obtained by the device sensor. The back posture can be estimated in real-time. Measurement of spinal angular momentum to determine vibration response in real-time the posture of the spinal column with three angular motions is described 1. flexion 2. tilt 3. flexion. The device has three internal states that determine the vibration response in real-time. 1) State 0: The posture is correct. 2) State 1: The pose is incorrect. 3) State 2: The pose is incorrect, and the vibration unit activates.

2.4.2 sEMG Topography Applied to Identify the LBP Patients

In [21], the topography feature parameters are used for both symmetrical and asymmetrical trunk movement, which provides an innovative tool with great accuracy to predict the prognosis of functional restoration rehabilitation in LBP patients. These tools can help clinicians to recommend an appropriate medication to patients with LBP to achieve a 12-week rehabilitation program. To achieve the proper functional restoration rehabilitation, two groups are created a "responding" group another "nonresponding" group of patients. To differentiate the groups, and Support Vector Machine (SVM)-based classifier is used to identify patients responding to functional restoration rehabilitation, and this will help health workers to improve the efficiency of LBP rehabilitation's feature parameters following rehabilitation in the "responding" group showed a significant difference ($P < 0.05$) with the one in the "nonresponding" group.

2.4.3 Deep Neural Network (DNN) Approach to Detect Lumbar-Pelvic Movements (LPMs)

In [23], the author presents an approach for LPM detection based on Deep Neural Network (DNN) and includes lateral flexion, flexion, rotation and extension, locally on-device. The data collected from the dual IMUs sensor system is clinically validated. An enhanced adapted CNN -LSTM for additional context information processing was introduced, using domain adaptation to perform additional context information processing through feature augmentation and a global temporal-wise attention mechanism. The results established that adding domain adaptation techniques and attention mechanisms can meaningfully advance the performance of CNN-LSTM on their dataset. Moreover, the proposed model is suitable for processing multiple IMU sensory data. The presented method has a capable real-time device detection performance, including efficient power consumption rate (in this case, good battery life for 8 h of monitoring) and acceptable LPM detection accuracy on feedback from medical experts.

2.4.4 Deep LSTM and CNN Models Using MEx Datasets [22].

In this paper [22], the author presents the multi-model heterogeneous sensor "MEx" data set and three machines learning k-Nearest Neighbours (k-NN) and Support Vector Machine (SVM).algorithms that benchmark measures with comparative studies. Also, introduce representations for shallow and deep architecture learning with MEx datasets. MEx is a collection of the sensory dataset for seven exercises, collected with four sensors which are publicly available at the UCI Machine Learning Repository1. In general, Multi-Layer Perceptron (MLP) classifiers achieved the best performance across different feature representation methods compared to k-Nearest Neighbours (k-NN) and Support Vector Machine (SVM).

2.4.5 Classifying Exercise Movements as CPE or TCM

In [25] this paper, the author has proposed a classification model that detects the compensatory movement as typical compensatory movement (TCM) and a clinically prescribed exercise (CPE) of an autonomously performed LBP exercise, i.e. (a) Prone-Rocking, (b) Bird-Dog and (c) Bent-over Rowing. A binary decision tree is used to classify these autonomously performed exercise movements into CPE or TCM. This study shows that three out of two tests for three exercises by one sensor is capable of monitoring exercise-based rehabilitation and prevention. The accuracy of Classification in CPE and TCM uses one sensor three (head) and for Bird-Dog, Prone-Rocking and Rowing 97.7%, 98.3% and 76.4%, respectively, and it is also capable of detecting specified TCMs. Thus, it is possible to conduct rehabilitation and prevention related to LBP in a community setting by preventing worsening activities and providing guidance to patients.

2.4.6 Wearable Sensing System for Posture Estimation

Posture estimation is used in many types of body activities. Wearable devices prove to be very useful in the measurement of posture. Some of these Wearable sensing devices are used in LBP therapy to control posture. In [26] paper, the author has done real-time 3D posture estimation using a wearable sensing device for lower back pain patients. In this approach, three components have been used: Arduino-based sensing device, second Wearable Wrist Watch Device, and third computer vision unit for real-time 3D posture estimation. These devices use BLE technology for communication and IMU sensor, which is used to display the position and orientation of the body. Measurement. The motion sensor used in this is compact, low power and cost-effective and these sensing systems are calibrated with experimental data and cross-validated real-world datasets. MediaPipe and YOLO framework is used posture estimation, and DNN (Deep Neural Network) is proposed for the prediction model. In this way, computer vision systems for human posture are effectively estimated using a machine learning algorithm greater than 97%.

Table 1. Qualitative comparison summarizes the main characteristics of the study

Author/year of publication	Data source or senser	Dataset	AI/ML algorithm	Outcome	Result/accuracy
LBP classification on pain location and severity					
Abdollahi, M et al. [9], (2020)	IMU and STarT Back Screening Tool (SBST)	Ninety-four patients	SVM, MLP	Classification into High and Low-medium	Accuracy SVM-75% MLP-60%
F. Haider et al. [10], (2020)	sEMG	EmoPain dataset	SVM, LSTM	Pain recognition: 1) 0: Healthy 2) 1: Low-Level pain 3) 2: High Movement behaviour: 1) 0: Not protective 2) 1: protective	Averaged FScore of 58.08%
Kerstin, L et al. [11], (2021)	Magnetic resonance imaging (MRI), PCD, CT	30 LBP patients	Random forest	1. Classify discs with fissures extending into the outer annulus or not 2. Classify discs as painful or non-painful	Result: Fissure model: 0.97 Pain model:.069
N. Wang et al. [12], (2018)	sEMG Isometric Exercise	Six sEMG NLBP datasets	Random forest	Recognize the symptomatic muscles for NLBP	Average accuracy 86.16%
Diagnosising LBP based classification					
Miotto R et al. [13], (2020)	Free text	17,409 clinical notes,10th revision (ICD-10) code (M54.5)	Convolutional neural network	Automatically identify acute LBP	ConvNet led to a good separation between acute low back pain clinical notes
Hu, B., Kim et al. [14], (2018)	C7 Sensor	Forty-four chronic LBP	Deep neural networks, LSTM	Differentiate LBP patients	Precision 97.2% and recall up to 97.2%
Ketola J.H.J et al. [15], (2020)	T2w MRI	Symptomatic (N = 110, 21%), Asymptomatic (N = 408, 79%)	Logistic regression	Classification intervertebral disc	Specificity 83%, sensitivity 82%, the negative predictive value of 94%, the precision of 56%, and AUC of 0.91
Staartjes et al. [16], (2020)	History, physical examination, and imaging. Crude and adjusted comparisons of 5R-STS	262 patients, 188 = LDH, 59 = LSS with or without spondylolisthesis, and 15 = CLBP	Linear regression	5R-STSperformance with LDH, LSS, and discogenic CLBP	Classification accuracy 96.2%
Gaonkar et al. [17] (2017)	Pain data consisting of 12 parameters	381 patients with 12 parameters	Random forest classifier, SVM classifier, KNN	Classification of lower back pain into two classes abnormal or normal	k-NN 85.48 RF = 87.09 Decision Tree 90.29, SVM = 85.4838

(*continued*)

Table 1. *(continued)*

Author/year of publication	Data source or senser	Dataset	AI/ML algorithm	Outcome	Result/accuracy
LBP classification based on decision making on treatment					
Oude Nijewemeet al. [18], (2018)	Referral advise during a vignette study [26]. NO sensor	1288 cases with 65% GP, 33% physio, 2% self-care cases	1. Decision tree 2. Random Forest 3. Boosted tree	CDSS for self-referral	1.decision tree: 70%, 2. random forest: 69%, 3. boosted tree:72%
Wirries et al. [19], A(2021)	MOS 36-Item Short Form Survey (SF-36) Oswestry Disability Index (ODI), Depression Scale (HADS)	60 patients: 1. Male-31, female-29 surgical therapy: 13, conservative therapy: 27	LR, ElasticNet Nearest neighbour Decision trees Stochastic gradient descent regressor, Support vector regressor	Conservative vs. Operative therapy Target variable: ODI after 6 months	Accuracy 87.6%
B. Lamichhane et al. [20], (2020)	MRI	MRI and fMRI data	SVM	CT and rsFC may act as potential biomarkers for LBP to guide therapy	Average classification accuracy of 74.51%, AUC = 0.787 (95% CI: 0.66–0.91)
Classification on adherence monitoring					
A. Rodriguez et al. [22], (2021)	IMUs	50 measurements at random orientations,	DCM algorithm	Assist patients with LBP for postural therapy	Result obtaining an rms deviation $\leq$1.24°
[21], Jiang, N, et al. (2017)	(SEMG) topography	30 Patients	SVM	Identify the patient is responding to functional restoration Rehabilitation "responding" group, "nonresponding" group	Accuracy 96.67%, sensitivity 100%, specificity 3.75%, AUC 0.8925
[23], Zhang, et al. (2021)	IMUs	Data collected from the ViMove system	CNN-LSTM	Identify symptoms of LBP as well as improve therapy	CNN-LSTM Single IMU accuracy 83.07 Dual IMU accuracy 85.81
[24], Wijekoon, A et al. (2020)	1.Obbrec Astra Depth Camera 2, Sensing Tex Pressure Mat 3 and Activity AX3 3-Axis, Logging Accelerometer	Seven exercises, 30 users and four sensors	KNN, SVM, NLP	End-to-End digital intervention for exercise adherence monitoring	Deep LSTM and CNN models achieve best performances for 4 sensors 90.2%, 63.4%, 87.2% and 74.1%
[25], Sellmann, A et al. (2022)	3 Sensors (One Head, Two Upper back Sensor)	Thirty volunteers for LBP rehabilitation exercises (24 sets (2 categories × 4 exercises × 3 sets)	Binary decision trees	Distinguish CPE from TCMM in Bird-Dog, Prone-Rocking and Rowing	Accuracy for Bird-Dog Prone-Rocking Rowing i.e. 97.7%, 98.3%, 76.4%

(continued)

Table 1. (*continued*)

Author/year of publication	Data source or senser	Dataset	AI/ML algorithm	Outcome	Result/accuracy
[26], A. Seth et al. (2021)	Wearable inertial measurement units (IMU) sensors	Standing Straight, moving (Right, Left), Tilted (Right, left) sample	Posture estimation, MediaPipe and YOLO	Posture estimation compared to manual observation	Accuracy standing 0.89, Moving right 0.96, moving left 0.98, tilted right 0.84, tilted left 0.84

3 Conclusion

In this paper, we have reviewed and presented research papers on the assessment and rehabilitation of low back pain using artificial intelligence and machine learning. Selected papers were classified according to their applicability in assessing and rehabilitating low back pain. We have presented papers for Low back pain identification, Pain assessment, Decision Making on Treatment, and Adherence monitoring. Finally, a comparative study amongst these works is briefed which will definitely help to have an overview of Artificial Intelligence and Machine learning-based strategies for Assessment and Rehabilitation of Low Back Pain.

References

1. Ganesan, S., Acharya, A.S., Chauhan, R., Acharya, S.: Prevalence and risk factors for low back pain in 1,355 young adults: a cross-sectional study. Asian Spine J. **11**(4), 610–617 (2017). https://doi.org/10.4184/asj.2017.11.4.610
2. Serranheira, F., Sousa-Uva, M., Heranz, F., Kovacs, F., Sousa-Uva, A.: Low back pain (LBP), work and absenteeism. Work (Reading Mass.) **65**(2), 463–469 (2020). https://doi.org/10.3233/WOR-203073
3. Wong, A.Y.L., Karppinen, J., Samartzis, D.: Low back pain in older adults: risk factors, management options and future directions. Scoliosis Spinal Disord. **12**, 14 (2017). https://doi.org/10.1186/s13013-017-0121-3
4. Olafsson, G., Jonsson, E., Fritzell, P., Hägg, O., Borgström, F.: Cost of low back pain: results from a national register study in Sweden. Eur. Spine J. **27**(11), 2875–2881 (2018). https://doi.org/10.1007/s00586-018-5742-6
5. Wong, A.Y., Karppinen, J., Samartzis, D.: Low back pain in older adults: risk factors, management options and future directions. Scoliosis **12**, 14 (2017). https://doi.org/10.1186/s13013-017-0121-3
6. Allegri, M., Montella, S., Salici, F., et al.: Mechanisms of low back pain: a guide for diagnosis and therapy. F1000Res. **5** (2016). F1000 Faculty Rev-1530. https://doi.org/10.12688/f1000research.8105.2
7. Bardin, L.D., King, P., Maher, C.G.: Diagnostic triage for low back pain: a practical approach for primary care. Med. J. Aust. **206**(6), 268–273 (2017)
8. Oliveira, C.B., et al.: Clinical practice guidelines for the management of non-specific low back pain in primary care: an updated overview. Eur. Spine J. **27**(11), 2791–2803 (2018). https://doi.org/10.1007/s00586-018-5673-2

9. Abdollahi, M., et al.: Using a motion sensor to categorize nonspecific low back pain patients: a machine learning approach. Sens. (Basel Switz.) **20**(12), 3600 (2020). https://doi.org/10.3390/s20123600
10. Haider, F., Albert, P., Luz, S.: Automatic recognition of low-back chronic pain level and protective movement behaviour using physical and muscle activity information. In: 2020 15th IEEE International Conference on Automatic Face and Gesture Recognition (FG 2020), pp. 834–838 (2020). https://doi.org/10.1109/FG47880.2020.00065
11. Kerstin, L., Hanna, H., Helena, B.: Identification of potentially painful disc fissures in magnetic resonance images using machine-learning modelling. Eur. Spine J.: Off. Publ. Eur. Spine Soc. Eur. Spinal Deformity Soc. Eur. Sect. Cervical Spine Res. Soc. (2021). https://doi.org/10.1007/s00586-021-07066-x
12. Wang, N., Xiao, J., Cui, L.: EasiSMR: recognizing non-specific low back pain symptomatic muscles using multi-muscles fusion based machine learning. In: 2018 IEEE 4th International Conference on Computer and Communications (ICCC), pp. 2002–2006 (2018). https://doi.org/10.1109/CompComm.2018.8780772
13. Miotto, R., et al.: Identifying acute low back pain episodes in primary care practice from clinical notes: observational study. JMIR Med Inform. **8**(2), e16878 (2020). https://doi.org/10.2196/16878. PMID: 32130159; PMCID: PMC7068466
14. Hu, B., Kim, C., Ning, X., Xu, X.: Using a deep learning network to recognise low back pain in static standing. Ergonomics **61**(10), 1374–1381 (2018). https://doi.org/10.1080/00140139.2018.1481230
15. Ketola, J.H.J., Inkinen, S.I., Karppinen, J., Niinimaki, J., Tervonen, O., Nieminen, M.T.: T2-weighted magnetic resonance imaging texture as predictor of low back pain: a texture analysis-based classification pipeline to symptomatic and asymptomatic cases. J. Orthop. Res. **39**, 2428–2438 (2020). https://doi.org/10.1002/jor.24973.-DOI-PubMed
16. Staartjes, V.E., Quddusi, A., Klukowska, A.M., Schröder, M.L.: Initial classification of low back and leg pain based on objective functional testing: a pilot study of machine learning applied to diagnostics. Eur. Spine J. **29**(7), 1702–1708 (2020). https://doi.org/10.1007/s00586-020-06343-5
17. Gaonkar, A., Kulkarni, R., Caytiles, R., Iyenger, N.C.S.N.: Classification of lower back pain disorder using multiple machine learning techniques and identifying degree of importance of each parameter. Int. J. Adv. Sci. Technol. **105**, 11–24 (2017). https://doi.org/10.14257/ijast.2017.105.02
18. Oude Nijeweme-d'Hollosy, W., van Velsen, L., Poel, M., et al.: Evaluation of three machine learning models for self-referral decision support on low back pain in primary care. Int. J. Med. Inform. **110**, 31–41 (2018)
19. Wirries, A., Geiger, F., Hammad, A., Oberkircher, L., Blümcke, I., Jabari, S.: Artificial intelligence facilitates decision-making in the treatment of lumbar disc herniations. Eur. Spine J. **30**(8), 2176–2184 (2020). https://doi.org/10.1007/s00586-020-06613-2
20. Lamichhane, B., et al.: Multi-modal biomarkers of low back pain: a machine learning approach. NeuroImage. Clin. **29** (2021). Article 102530. https://doi.org/10.1016/j.nicl.2020.102530
21. Jiang, N., Luk, K.D., Hu, Y.: A machine learning-based surface electromyography topography evaluation for prognostic prediction of functional restoration rehabilitation in chronic low back pain. Spine **42**(21), 1635–1642 (2017). https://doi.org/10.1097/BRS.0000000000002159
22. Rodriguez, A., Rabuñal, J.R., Pazos, A., Rodríguez Sotillo, A., Ezquerra, N.: Wearable postural control system for low back pain therapy. IEEE Trans. Instr. Meas. **70**, 1–10 (2021). Art no. 4003510. https://doi.org/10.1109/TIM.2021.3057935
23. Zhang, Y., Delir Haghighi, P., Burstein, F., Yao, L., Cicuttini, F.: On-device lumbar-pelvic movement detection using dual-IMU: a DNN-based approach. IEEE Access. **9**, 62241–62254 (2021). https://doi.org/10.1109/ACCESS.2021.3074755

24. Wijekoon, A., Wiratunga, N., Cooper, K., Bach, K.: Learning to recognise exercises for the self-management of low back pain. In Barták, R. and Bell, E. (eds.) Proceedings of the 33rd International Florida Artificial Intelligence Research Society (FLAIRS) 2020 Conference (FLAIRS-33), 17–20 May 2020, Miami Beach, USA, pp. 347–352. AAAI Press, Palo Alto (2020)
25. Sellmann, A., et al.: Detection of typical compensatory movements during autonomously performed exercises preventing low back pain (LBP). Sensors **22**, 111 (2022). https://doi.org/10.3390/s22010111
26. Seth, A., James, A., Mukhopadhyay, S.: Wearable sensing system to perform realtime 3D posture estimation for lower back healthcare. In: 2021 IEEE International Symposium on Robotic and Sensors Environments (ROSE), pp. 1–7 (2021). https://doi.org/10.1109/ROSE52750.2021.9611778

Controllers Synthesis for Computer Research of Dynamic Conveyor Belt Model Using Intelligent Algorithms

Olga N. Masina[1], Olga V. Druzhinina[2], and Alexey A. Petrov[1](✉)

[1] Bunin Yelets State University, Yelets, Russia
olga121@inbox.ru, xeal91@yandex.ru

[2] Federal Research Center "Computer Science and Control" of Russian Academy of Sciences, Moscow, Russia
ovdruzh@mail.ru

Abstract. This paper is devoted to the synthesis and analysis of such a controlled model of a belt conveyor with switching modes of operation, which takes into account the axial resistance and the dynamic change in the angle between the horizontal plane and the plane of the conveyor belt. The problems of constructing a neural network controller and a PID controller for studying the model are considered. The solution of optimal control problems with the use of constructed controllers is proposed. The results of computer experiments on the study of trajectory dynamics are presented. The implementation of the proposed algorithms for constructing trajectories and intelligent control is carried out using the Python language and the corresponding mathematical libraries. The synthesis of control laws for stabilizing the elevation angle of the conveyor belt is carried out taking into account the variable mass of the loads, as well as taking into account the change in the center of mass of the conveyor. Qualitative effects that arise due to axial resistance are studied. The questions of the adaptability of neural network control and PID control in relation to a linear increase in axial resistance are considered. A comparative analysis of computational experiments with the use of various types of regulators is carried out. The obtained results can be applied to the design of new control systems for production lines, as well as in the stabilization of dynamic controlled models. The considered method of estimating the influence of axial resistance on the trajectory dynamics can be used in the study of various generalized models of conveyor transport control. The developed software is aimed at being used as part of a software package for modeling conveyor transport systems with intelligent control #CSOC1120.

Keywords: Intelligent control algorithms · Switching dynamic models · Mathematical modeling · Stabilization · Synthesis of controllers · Conveyor belt · Axial resistance · Optimal control · Software

R. Silhavy (Ed.): CSOC 2022, LNNS 502, pp. 462–473, 2022.
https://doi.org/10.1007/978-3-031-09076-9_41

1 Introduction

Mathematical modeling of conveyor transport systems is an actual scientific direction related to the design, implementation and automation of new production lines [1–6]. The modern problems of automation of conveyor belt transport include, in particular, stabilization of the linear load on the conveyor belt, stabilization of the traction factor of the conveyor, obtaining parameters for optimal control of conveyor systems. The solving of these problems requires the development of adequate mathematical models using methods of control theory, optimization theory and artificial intelligence.

The construction of predictive and control logic components of technical systems is associated with the development of artificial intelligence algorithms. In this direction, the use of artificial neural networks is effective. The study of various technical control systems using artificial intelligence methods is carried out, for example, in [7–12]. The solving of the problems of modeling conveyor transport systems using artificial intelligence methods is contained in [13–18].

Modeling and stabilization of systems using neural network control and PID control is an urgent area of research [7,19–23]. Combining PID controllers with intelligent controllers is used in modeling various technical systems [2–22]. For modeling conveyor transport systems, PID controllers and intelligent controllers are considered in [14,17,18,24] and in other papers. Modern machine learning methods can be used to adjust the parameters of the controllers [25,26].

An actual direction is the creation of problem-oriented software packages for modeling conveyor transport systems with intelligent control. When creating software packages, the Python high-level programming language and the corresponding mathematical libraries are effective tools [27,28].

In this paper, the controllers synthesis problems for computer research of a dynamic model of a belt conveyor with intelligent control are considered. The purpose of this work is to develop an approach to modeling conveyor transport systems with a dynamically variable conveyor angle.

In Sect. 2, such a conveyor model with switching modes of operation is synthesized, which takes into account the axial resistance and the dynamic change in the angle between the horizontal plane and the plane of the conveyor belt. In Sect. 3, the solving of optimal control problems based on a neural network controller and a PID controller is proposed. The results of computer experiments for the trajectory dynamics study are presented. A discussion of the results and promising areas of research are contained in Sect. 4.

2 Generalized Model of a Belt Conveyor with Intelligent Control

We study such a model of a single-drive belt conveyor with a dynamic angle of elevation, which is described by differential equations of the form

$$\begin{aligned}
&\dot{x}_0 = x_1,\\
&\dot{x}_1 = \frac{u_1(t) - kx_1 - m_0 g \sin(\alpha_0)}{m},\\
&\dot{\alpha}_0 = \alpha_1,\\
&\dot{\alpha}_1 = \frac{u_2(t) - l\alpha_1}{cm\varepsilon^2} - \frac{g\cos(\alpha_0)}{\varepsilon},\\
&u_1, u_2 \in U,\ m \in M,\ \varepsilon \in E,
\end{aligned} \tag{1}$$

where α_0 is the elevation angle of the conveyor relative to the zero position, α_1 is the angular speed of the conveyor, g is the acceleration of gravity, x_0 is the conveyor belt movement, x_1 is the linear speed of the conveyor belt, m is the full conveyor mass, m_0 is the own mass of the conveyor belt, $u_1(t)$ is the conveyor traction control function, $u_2(t)$ is the belt elevation angle control function, ε is the position of the conveyor mass center relative to the lower roller, c is the coefficient that determines the of the conveyor inertia moment, k is the rolling friction coefficient, l is axial resistance coefficient. The sets M, E, U include all possible values of the total mass of loads, mass center and controls, respectively. Changes in the modes of operation in model (1) correspond to the moments of choosing the values of m and ε according to the given rules. System (1) belongs to the class of switching systems [29,30].

To simulate the impact of loaded and unloaded elements, we accept the following conditions

$$m = m_0 + \hat{m},$$

where $\hat{m}$ is the total mass of loads on the belt. In addition, we will assume that there is such a function f (operator for switching the mass of loads) for which $t \xrightarrow{f} \hat{m}\ \ \forall t \geq 0$.

It should be noted that while switching the mass of loads, the law of momentum conservation is taken into account, according to which we have

$$\lim_{t \to t_0} mx(t) = mx(t_0)\ \ \forall m, t.$$

We consider the following formulation of the optimal control problem for the model (1). It is necessary (by the aid of the controls u_1, u_2) to implement the transient response and stabilize the phase state of the system (1) near the target point $E_1 = (\hat{x}_1, \hat{\alpha}_0, \hat{\alpha}_1)$. It should be noted that when solving the optimal problem, one should take into account the invariance with respect to the initial conditions. We consider the minimization problem based on following optimality criterion

$$\lim_{t_n \to \infty} \frac{1}{t_n} \int_0^{t_n} \|E_1 - X\| dt,$$

where $t \in (0, t_n)$, $X = (x_1(t), \alpha_0(t), \alpha_1(t))$.

To solve the optimal control problem, software has been developed in Python 3 using the mathematical libraries SciPy and Numpy. Stabilization of the linear speed of motion for the mathematical model of the conveyor is carried out using control based on the sliding mode. Conveyor elevation angle control is implemented on the basis of three types of controllers: a fuzzy logic controller, a controller based on artificial neural networks and a PID controller. The description of the technique and the results of computer modeling without taking into account the axial resistance are given in [17,18].

Next, we study the influence of various types of controllers (for $l \neq 0$) on the dynamics of the system (1). We use a neural network controller designed on the basis of a reinforcement machine learning algorithm. In addition, we use a PID controller to control the elevation angle of the conveyor.

3 Synthesis of Controllers and Computer Study of the Model

The structure of the neural network controller is as follows. We use a feedforward perceptron that has two inputs, eight neurons in one hidden layer, and one output (with linear gain). The controller inputs take the mismatch error value and the value of this value error derivative. These values are determined in the process of numerical solving of the system (1). The neurons in the hidden and output layers have a tangential activation function. We use a reinforcement learning algorithm based on differential evolution [31,32]. The controller scheme and the approach to adjusting the weight coefficients are described in [17,18]. The results of computational experiments for $l = 50$ are shown in Fig. 1, 2, 3 and 4.

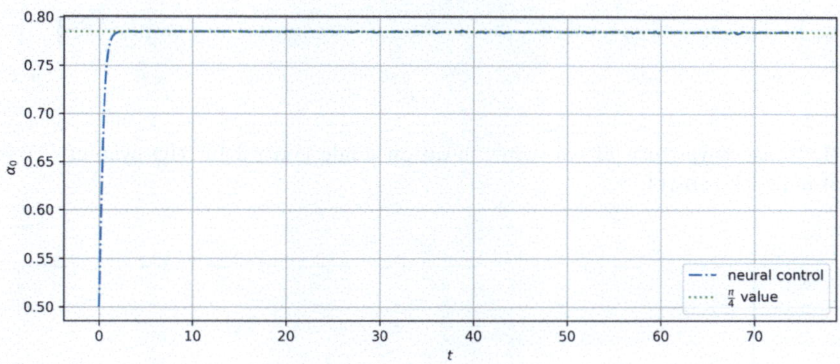

Fig. 1. Angular position of the conveyor for the model (1) with neural network control.

According to Fig. 1, 2, 3 and 4, insertion of the linear axial resistance does not lead to significant changes in the qualitative behavior of the system (1) compared to $l = 0$. Nevertheless, a number of changes can be noted regarding the quantitative indicators of axial stability. The first change is a longer transient response settling time (compared to $l = 0$). This change is consistent with the physical sense of the axial resistance. The second important quantitative change is the increase in quantitative indicators stability of the conveyor angular position.

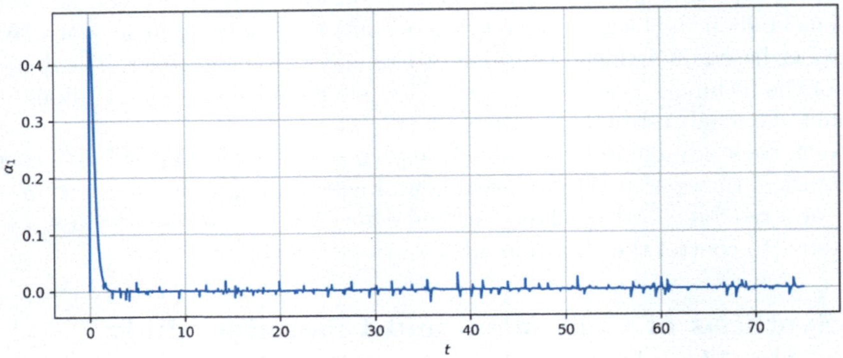

Fig. 2. Conveyor angular speed for the model (1) with neural network control.

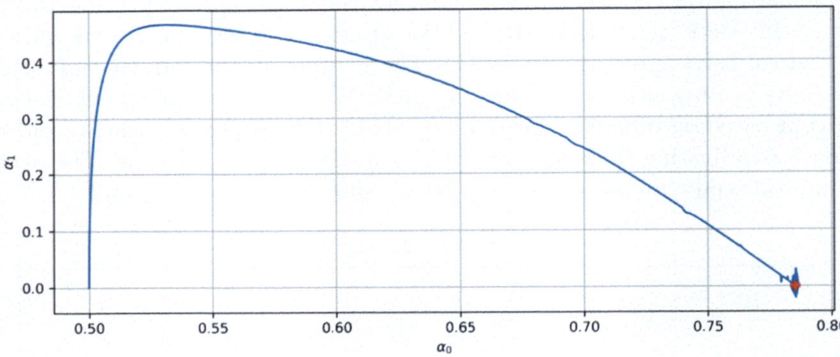

Fig. 3. Phase trajectory of conveyor elevation angle control for the system (1) with neural network control.

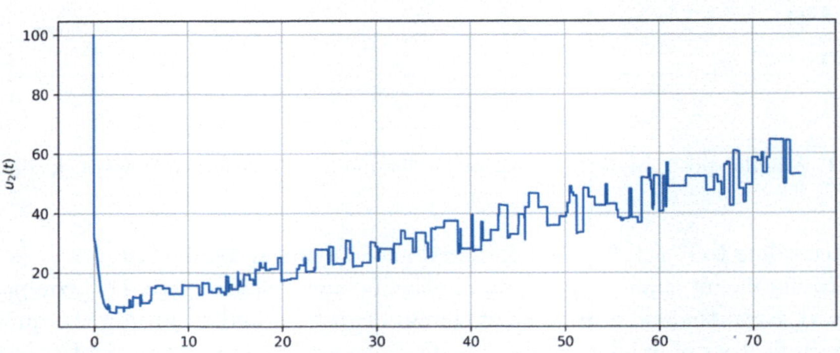

Fig. 4. Control based on a neural network algorithm.

For numerical interpretation of the stability of the conveyor angular position we introduce the following indicator

$$s = \int_{t_1}^{t_2} \|E_1 - X\| dt,$$

where $t_1 < t < t_2$ is conveyor motion time. It should be noted that a decrease in the indicator s corresponds to an increase in the stabilization level of the system (1). We carry out three computational experiments with the values $t_1 = 35$, $t_2 = 75$ for a quantitative assessment of the stabilization of the elevation angle in the system (1). The results of these computational experiments are shown in Fig. 5.

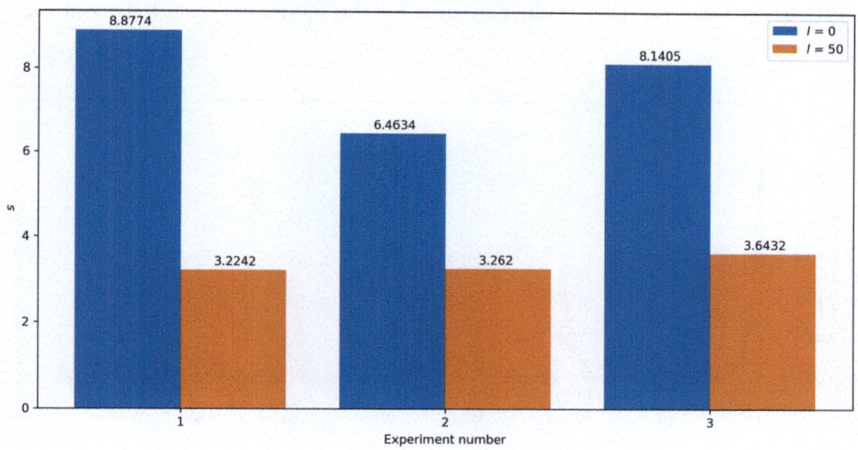

Fig. 5. Stabilization of the elevation angle in the system (1) with linear axial resistance (while using neural network controller).

The obtained results indicate that the insertion of axial resistance increases the stability of the elevation angle in system (1) for a given parameters set. The absolute scatter of s is 27% and 12% for $l = 0$ and $l = 50$, respectively.

The results of computational experiments using the PID controller are shown in Fig. 6, 7, 8 and 9. The coefficients for tuning the PID controller are obtained using a machine learning algorithm based on differential evolution [32].

According to the results of computational experiments presented in Fig. 6, 7, 8 and 9, the PID controller shows a longer transient response settling time compared to the neural network controller. In general, the trajectory dynamics are similar to the results obtained on the basis of the neural network controller. A comparative analysis of the controllers used demonstrates a lower stabilization efficiency based on the PID controller compared to the neural network controller. The results of a quantitative assessment of the stabilization of the elevation angle in the system (1) with control based on the PID controller are shown in Fig. 10.

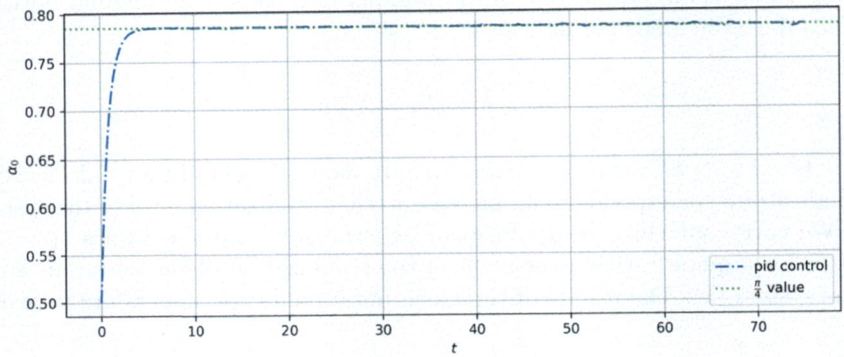

Fig. 6. Angular position of the conveyor for the model (1) with PID control.

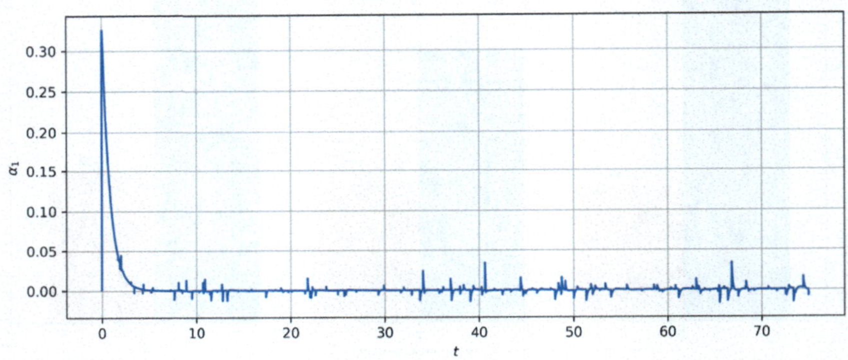

Fig. 7. Conveyor angular speed for the model (1) with PID control.

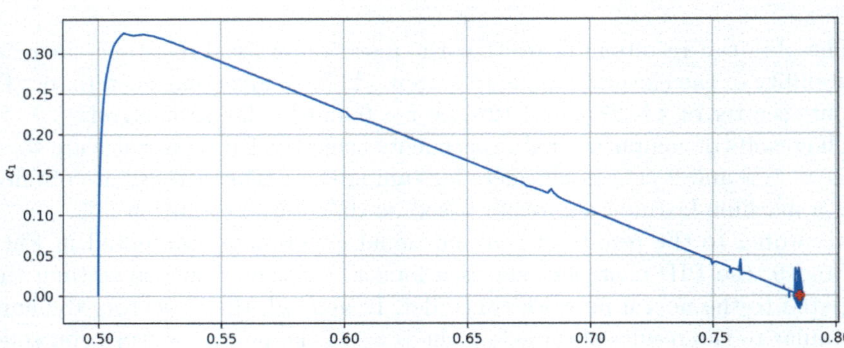

Fig. 8. Phase trajectory of conveyor elevation angle control for the system (1) with PID control.

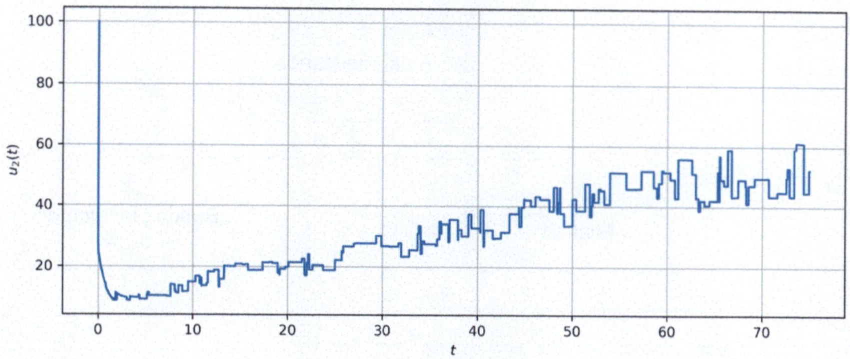

Fig. 9. Control based on a PID algorithm.

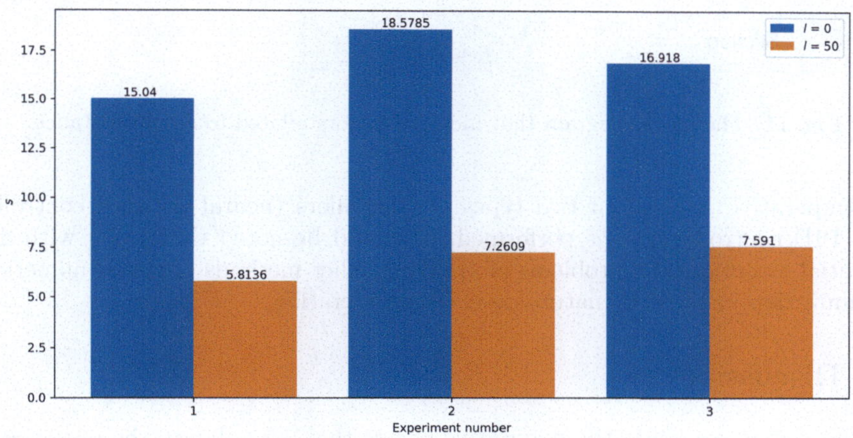

Fig. 10. Stabilization of the elevation angle in the system (1) with linear axial resistance (while using PID controller).

It should be noted that there is a similarity between Fig. 5 and Fig. 10. The insertion of axial resistance increases the stability of the elevation angle in the system (1) for a given parameters set.

The importance of axial resistance studying is due to the fact that this concept includes a variety of existing physical factors affecting dynamics of the system (1). The block diagram, which includes factors related to axial resistance, is shown in Fig. 11.

The analysis of the trajectory dynamics for other parameters of the system (1) is carried out. The results of this analysis are consistent with the results given in Sect. 3.

In the course of computational experiments, a study of trajectory dynamics is carried out for additional parameters sets of the system (1). The results of this study are consistent with the results given in Sect. 3. While verifying the results,

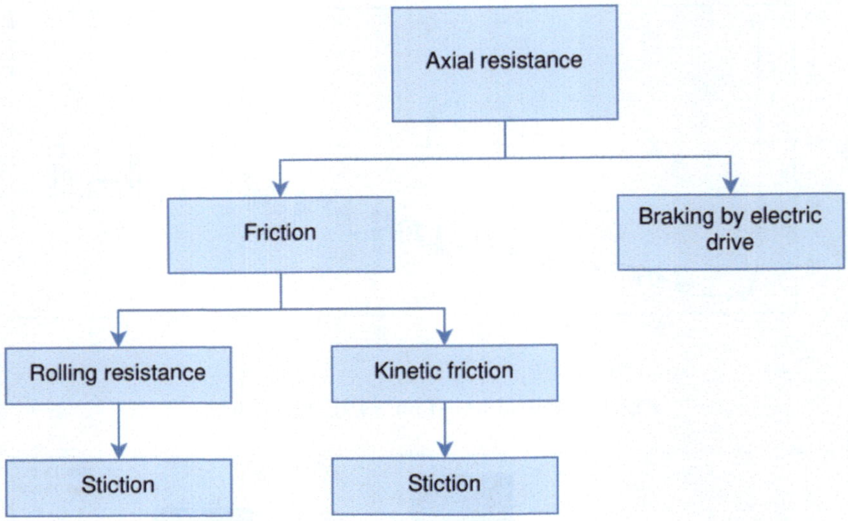

Fig. 11. The block diagram that includes factors related to axial resistance.

a comparative analysis of two types of controllers (neural network controller and PID controller) use is performed. It should be noted that along with differential evolution, the problems of applying other methods of global numerical optimization inspired by nature deserve consideration.

4 Discussion

In this paper, we study the qualitative effects that arise due to the presence of axial resistance. First of all, it should be noted that the presence of axial resistance in the system (1) is not taken into account when constructing controllers (training is carried out taking into account $l = 0$), however, the modified model has been investigated and verified taking into account $l \neq 0$. This fact indicates the adaptability of neural network control and PID control in relation to a linear increase in axial resistance.

Compared with the results obtained in [17,18], an increasing of the transient response settling time should be noted, regardless of the type of the control (Fig. 1, 2 and Fig. 6, 7), and the transient response settling time increases with increasing l. This circumstance is consistent with the physical sense, since the axial resistance tends to maintain a constant elevation angle of the conveyor. However, the axial resistance also has a less obvious effect on the stabilization of the elevation angle. Despite the fact that the switching of the loads mass is carried out instantly and with conservation of momentum (by analogy with [17]), the average deviation from a given position E_1 decreases with increasing l (Fig. 5 and Fig. 10). This result is consistent with the fact that an increase in linear axial resistance leads to an increase of the oscillations attenuation rate near the equilibrium state.

The results obtained in the process of comparative efficiency analysis correlate with the conclusions given in [17,18]. Despite the absence of an integral component, the neural network controller is faster and more accurate than the PID controller. Differences in the nature of the transient response are observed in Fig. 3 and Fig. 8. For neural network control (Fig. 3), the phase trajectory of the transient response has a parabolic form, while the phase trajectory for the PID controller (Fig. 8) at the final stage of the transition mode is close to a straight line. Structures of the synthesized signals for the neural network controller and the PID controller (Fig. 4 and Fig. 9 respectively) are similar.

Among the promising areas of research, it is necessary to highlight the modification of the model (1) taking into account the description of the following effects: transient processes of engagement of loads on the belt, moment of inertia, stretching of the belt. The case when the axial resistance linearly depends on the angular speed corresponds to a simplified version of the model description. In this regard, it is important to emphasize that the study of various types of axial resistance is also a promising area of research. The conveyor modeling method proposed in this paper (a method based on control theory and artificial intelligence) provides opportunities for studying complex models taking into account these effects.

5 Conclusion

Thus, the paper develops an approach to modeling conveyor transport systems with intelligent control. The construction of a new conveyor model with a dynamically variable angle between the horizontal plane and the conveyor belt plane is considered. The influence of axial resistance on the trajectory dynamics is studied in detail. The considered method of model analysis taking into account axial resistance can be used in the study of such generalized models of conveyor transport control that take into account various factors influencing the dynamics of the conveyor.

The developed software is focused on the use as part of a problem-oriented software package for modeling conveyor transport systems with intelligent control. This software is the basis of such specialized modules that complement and expand the specified software package.

An important aspect of further research of controlled conveyors models is the development of new methods of intelligent analysis (including methods of deep learning, reinforcement learning and combined types of machine learning). These methods can be implemented using open source software, high-level programming languages, as well as high-performance computing systems.

Taking into account the possibility of obtaining new qualitative effects, it seems appropriate to consider the control of the elevation angle in model (1) using fuzzy and neuro-fuzzy controllers. These types of controllers allow to formalize expert knowledge for the efficiency improvement of the control process.

References

1. Rachkov, E.V.: Continuous Transport Machines. Altair-MGAVT, Moscow (2014)
2. Subba Rao, D.V.: The Belt Conveyor: A Concise Basic Course. CRC Press, London, New York (2020)
3. Zhao, L., Lyn, Y.: Typical failure analysis and processing of belt conveyor. Procedia Eng. **26**, 942–946 (2011)
4. Andrejiova, M., Grincova, A., Marasova, D.: Measurement and simulation of impact wear damage to industrial conveyor belts. Wear **12**, 368–369 (2016)
5. Dmitrieva, V.V., Sizin, E.P.: Continuous belt conveyor speed control in case of reduced spectral density of load flow. Min. Inf. Anal. Bull. **2**, 130–138 (2020)
6. Zyuzicheva, Y.E.: Model of a belt conveyor located at an angle to the horizon determination of the optimal inclination angle for the transition process. Min. Inf. Anal. Bull. **7**, 212–216 (2006)
7. Kumar, R., Singh, V.P., Mathur, A. (eds.): Intelligent Algorithms for Analysis and Control of Dynamical Systems. AIS, Springer, Singapore (2021). https://doi.org/10.1007/978-981-15-8045-1
8. Yu, W.: Recent Advances in Intelligent Control Systems. Springer, Heidelberg (2009). https://doi.org/10.1007/978-1-84882-548-2
9. Jinyang, L., Xiaofeng, M.: Temperature decoupling control of double-level air flow field dynamic vacuum system based on neural network and prediction principle. Eng. Appl. Artif. Intell. **26**(4), 1237–1245 (2013)
10. Gu, W., Li, Y., Li, Z., Qian, Y.: An intelligent approach for dynamic AGV scheduling problem in the discrete manufacturing system. In: 2019 3rd Information Technology, Networking, Electronic and Automation Control Conference (ITNEC), pp. 1736–1739. IEEE, Chengdu, China (2019)
11. Mu, S.: Research on the control system of the multi-point driving belt conveyor tension device. In: International Conference on Big Data, Artificial Intelligence and Internet of Things Engineering (ICBAIE), pp. 321–326. IEEE, Fuzhou, China (2020)
12. Lee, D., Seo, H., Jung, M.W.: Neural basis of reinforcement learning and decision making. Annu. Rev. Neurosci. **35**(1), 287–308 (2012)
13. Kozhubaev, Y.N., Semenov, I.M.: Belt conveyor control systems. Sci. Tech. Bull. St. Petersburg State Polytechnical Univ. **2**(195), 181–186 (2014)
14. Ma, X.M., Gao, X.X.: Coal belt conveyor PID controller parameter regulation with neural network. Appl. Mech. Mater. **319**, 583–589 (2013)
15. Lv, Y., Liu, B., Liu, N., Zhao, M.: Design of automatic speed control system of belt conveyor based on image recognition. In: 2020 3rd International Conference on Artificial Intelligence and Big Data (ICAIBD), pp. 227–230. IEEE, Chengdu, China (2020)
16. Lutfy, O.F., Selamat, H., Mohd Noor, S.: Intelligent modeling and control of a conveyor belt grain dryer using a simplified type 2 neuro-fuzzy controller. Drying Technol. **33**(10), 1210–1222 (2015)
17. Masina, O.N., Druzhinina, O.V., Igonina, E.V., Petrov, A.A.: Synthesis and stabilization of belt conveyor models with intelligent control. In: Silhavy, R. (ed.) CSOC 2021. LNNS, vol. 228, pp. 645–658. Springer, Cham (2021). https://doi.org/10.1007/978-3-030-77448-6_63
18. Druzhinina, O.V., Masina, O.N., Petrov, A.A.: Modeling of the belt conveyor control system using artificial intelligence methods. J. Phys. Conf. Ser. **2001**, 012011 (2021)

19. Ang, K.H., Chong, G., Li, Y.: PID control system analysis, design, and technology. IEEE Trans. Contr. Syst. Technol. **13**(4), 559–576 (2005)
20. Rodriguez-Castellanos, J., Grisales-Palacio, V., Cote-Ballesteros, J.: A tuning proposal for direct fuzzy PID controllers oriented to industrial continuous processes. IFAC-PapersOnLine **51**(4), 657–662 (2018)
21. Yu, W.: PID Control with Intelligent Compensation for Exoskeleton Robots. Academic Press, Cambridge (2018)
22. Milovanović, M., et al.: Adaptive PID control based on orthogonal endocrine neural networks. Neural Netw. **84**, 80–90 (2016)
23. Gheisarnejad, M.: An effective hybrid harmony search and cuckoo optimization algorithm based fuzzy PID controller for load frequency control. Appl. Soft Comput. **65**, 121–138 (2018)
24. Chen, X.h., Rahman, M.L., Zhang, Y.: Adaptive fuzzy PID speed control of DC belt conveyor system. In: 2016 17th IEEE/ACIS International Conference on Software Engineering, Artificial Intelligence, Networking and Parallel/Distributed Computing (SNPD), pp. 1–4. IEEE, Shanghai, China, SNPD (2016)
25. Sundas, A., Bhatia, A., Saggi, M.: Reinforcement Learning, Machine Learning and Big Data: Concepts, Algorithms, Tools and Applications, pp. 281–303. Wiley, Hoboken (2020)
26. Ding, Z., Huang, Y., Yuan, H., Dong, H.: Introduction to reinforcement learning. In: Dong, H., Ding, Z., Zhang, S. (eds.) Deep Reinforcement Learning, pp. 47–123. Springer, Singapore (2020). https://doi.org/10.1007/978-981-15-4095-0_2
27. Mckinney, W.: Python for Data Analysis, 2e: Data Wrangling with Pandas, Numpy, and Ipython. O'Reilly, Boston (2017)
28. Rahmat, B., Nugroho, B.: Fuzzy and artificial neural networks-based intelligent control systems using python. Int. Semin. Res. Mon. Sci. Technol. People Empowerment **2018**, 152–170 (2019)
29. Druzhinina, O., Masina, O., Petrov, A.: The synthesis of the switching systems optimal parameters search algorithms. In: Evtushenko, Y., Jaćimović, M., Khachay, M., Kochetov, Y., Malkova, V., Posypkin, M. (eds.) OPTIMA 2018. CCIS, vol. 974, pp. 306–320. Springer, Cham (2019). https://doi.org/10.1007/978-3-030-10934-9_22
30. Druzhinina, O.V., Masina, O.N., Petrov, A.A., Lisovsky, E.V., Lyudagovskaya, M.A.: Neural network optimization algorithms for controlled switching systems. In: Silhavy, R. (ed.) CSOC 2020. AISC, vol. 1225, pp. 470–483. Springer, Cham (2020). https://doi.org/10.1007/978-3-030-51971-1_39
31. Tang, L., Zhao, Y., Liu, J.: An improved differential evolution algorithm for practical dynamic scheduling in steelmaking-continuous casting production. IEEE Trans. Evol. Comput. **18**(2), 209–225 (2014)
32. Storn, R., Price, K.: Differential evolution-a simple and efficient heuristic for global optimization over continuous spaces. J. Glob. Optim. **11**, 341–359 (1997). https://doi.org/10.1023/a:1008202821328

Program-Target Models for Optimizing Science-Intensive Projects to Ensure Sustainable Development

V. L. Rozaliev[1(✉)], V. V. Gilka[1], Y. V. Kulikova[1], and S. V. Pronichkin[2]

[1] Volgograd State Technical University, Volgograd, Russia
vlrozaliev@mail.ru

[2] Federal Research Center "Computer Science and Control" of Russian Academy of Sciences, Moscow, Russia

Abstract. One of the prerequisites for achieving sustainable development of complex socio-economic systems is scientific research, which is the most important decision-making tool. The elements of systems theory and systems approach to ensure the sustainability of complex socio-economic systems are considered. The main patterns of optimizing science-intensive projects to ensure sustainable development in environmental protection multi-level systems are revealed. A program-target model for ensuring sustainable development based on the optimization of science-intensive projects for environmental protection has been developed. Estimates of model parameters are analyzed.

Keywords: Sustainable development · Efficiency · Decision-making · Program-target model

1 Introduction

The basic principles of agriculture sustainable development, as a key component of the agro-food sector, focus on achieving harmony between increasing agricultural production, stability in the social sphere and environmental protection, i.e. emphasize the fact that economic growth should occur simultaneously with the environment preservation, be accompanied by social progress and contribute to an increase in the socio-cultural development level [1, 2]. One of the prerequisites for achieving sustainable development is scientific research, which at the same time are the most important tools for effective management and well-founded decision-making for environmental protection. At all stages of the implementation of the concept of sustainable development - from clarifying the need, formulating ideas to practical implementation, it requires a deep scientific and theoretical foundation [3, 4].

The sustainable development concept implementation requires a strategic approach based on an adaptive system of environmental protection. The implementation of science-intensive projects to ensure sustainable development will contribute to the formation of a new environmentally conscious society and value orientations that promote development

R. Silhavy (Ed.): CSOC 2022, LNNS 502, pp. 474–479, 2022.
https://doi.org/10.1007/978-3-031-09076-9_42

that is socially desirable, economically viable and environmentally balanced [5, 6]. It is necessary to reorient the existing program-target models of sustainable development to a comprehensive study of social, environmental, economic issues, taking into account local, national and regional conditions, as well as the global context.

The growth in the volume and cost of information leads to a corresponding complication of information systems to ensure sustainable development. Such systems become multilevel [7, 8]. The rise in the cost of these systems exacerbates the resources optimal use problem in science-intensive projects to ensure sustainable development. Optimization environmental and economic tasks are aimed at ensuring the best indicators of environmental safety and have two main directions: determining the optimal amount of investment in environmental protection; optimization of resource allocation between science-intensive projects to ensure sustainable development [9, 10].

In the process of finding a solution, one should take into account the changing conditions over time, associated with the "aging" of information and its updating, the emergence of new means of negative impact and the protection systems modernization. As a result, it is possible to set the task of dynamic resource management for science-intensive projects to ensure sustainable development in environment complex information structures. Such a scheme can represent both physical and electronic systems of the Internet of Things.

An example of a physical subsystem can be a program-target model, in which a common obstacle is a protected perimeter of the territory, objects are components of the natural environment, and individual obstacles are means protecting these objects. In an electronic system, objects are computers, servers protected by common and individual means. A complex program-target model can be reduced to simpler schemes with parallel or series-parallel arrangement of environmental protection elements.

A number of works are devoted to the consideration of these problems, in which the structure of the information system was not specified [11, 12], while in others only the simplest single-level systems were considered [13, 14]. In [15, 16], some general aspects of the multilevel structures analysis are considered and equations for calculating their indicators are presented. It should be noted that the transition to multilevel systems significantly expands the problem and raises a number of optimizing science-intensive projects issues to ensure sustainable development, due to the structure complication and the individual elements arrangement features. It is necessary to develop program-target models for optimizing science-intensive projects and to investigate the correlations between the resources amount optimal values in multi-level information systems for environmental protection.

2 Materials and Methods

Sustainable development is a scientific problem to be solved, which requires new technical means and technologies. The transition to such a development model has a scientific and technical content. With the aim of identifying the main patterns of optimizing science-intensive projects to ensure sustainable development in multi-level environmental protection systems, the following structures were considered:

- conducting additional research with a fixed budget;

- determination of the resources optimal distribution between general and individual research projects;
- establishing a correlation between the optimal distributions of resources aimed at individual research for given environmental protection schemes;
- development of recommendations for resource management in multilevel environmental information systems.

The target function is formalized, which determines the relative cost of science-intensive projects to ensure sustainable development:

$$e(i_1, i_2) = \frac{\sum_{j=1}^{N} k_j v_j(i_1, i_2)}{\sum_{j=1}^{N} s_j r^{(j)}(i_1, i_2)}, \tag{1}$$

where i_1 is the assessment of environmental pollution; i_2 are resources necessary for the restoration of the natural environment; j is the science-intensive project to ensure sustainable development; k_j is the relative importance of project j implementation; $v_j(i_1, i_2)$ is the density of the two-dimensional probability distribution of the effective implementation of the j project; s_j is the relative difficulty of implementing the j project; $r^{(j)}(i_1, i_2)$ is the proportion of lost information that determines the vulnerability of the j project.

Variables i_1 and i_2 are included in the functions $r^{(j)}(i_1, i_2)$ in the form of the ratio of environmental pollution and the resources necessary for its restoration i. The values of $r^{(j)}(i_1, i_2)$ are determined by the properties of the negative impact on the environment and must satisfy the following conditions: if $i \to 0$ then $r^{(j)}(i_1, i_2) \to 1$; if $i \to \infty$ then $r^{(j)}(i_1, i_2) \to 0$. A function that satisfies the specified conditions is a fractional power:

$$r^{(j)}(i_1, i_2) = (1 + q_j i)^{-l_j}, \tag{2}$$

For $l_j = 1$ Eq. (2) has a fractional-linear dependence, for $l_j > 1$ has a fractional-nonlinear one. The parameters l_j and q_j have the meaning of the productivity of the costs of protecting the natural environment and influence the form of dependencies between environmental pollution and the resources necessary for its restoration.

3 Results

The purpose of the calculations is to find the optimal distribution of the minimizing $e(i_1, i_2)$ for various variants of the parameters. It is necessary to determine the conditions under which the implementation of science-intensive projects to ensure sustainable development with a constant budget is expedient. The expediency criterion is to reduce the possible volume of lost information while ensuring the optimal distribution of protection resources between environmental objects. The simulation of the optimal distribution for various options of parameters has been carried out.

The results of the simulation carried out allow us to establish the optimal boundaries i_2 of the resources that should be allocated:

- $0 < i_2 < 1$ - there is an optimal value;

- $i_2 = 1$ - all resources should be directed to a science-intensive project;
- $i_2 = 0$ - it is inexpedient to implement a science-intensive project.

The desired distribution is found as the optimum of a function of three variables k_j, s_j, q_j. The optimum point is found by the method of evolutionary calculations of values possible combinations. Software has been developed to determine the parameters optimal values.

The optimal value of the parameters characterizes the situation when an increase in any of these values (at the expense of others) only leads to a worsening of the situation, i.e. an increase in the value of i. The each value of the quantities depends on the vulnerability of environmental protection objects and the productivity of the corresponding costs. The value of the derivatives $\frac{df}{di_2}$ affects the productivity of costs. The ratio of derivatives, taking into account weight coefficients for all possible values of i, determines the direction of the most efficient allocation of resources $\frac{dr}{di}$.

With the growth of q_j the vulnerability of the project $r^{(j)}(i_1, i_2)$ decreases, and with the growth of l_j it increases. Since the cost of environmental protection resources should be less than the cost of information. Therefore, the optimal value is reached in the system where q_j takes the minimum value. With average values of q_j and l_j the productivity of investments is low, and it is advisable to distribute them among individual science-intensive projects.

When changing both types of indicators q_j and l_j, it is necessary to take into account the influence of each of them. Comparing the indicators of different science-intensive projects, it should be taken into account that the indicators of the joint implementation of projects play an important role, since they have an impact on both objects of the natural environment. Under certain ratios of the values of q_j and l_j, the joint implementation of science-intensive projects is expedient and the value of i_2 is in the interval $0 < i_2 < 1$.

In semi-structured dynamic systems, the distribution of information can be uneven, which will also affect the optimal distribution of resources. The entire range of k_j values can be divided into three zones:

- with an increase in k_j, the values of s_j and q_j decrease monotonously and consistently, while the vulnerabilities of individual science-intensive projects increase, but at the same time the protective effect of joint projects grows - partial compensation occurs, as a result $e(i_1, i_2)$ changes slightly;
- s_j decreases to zero, q_j increases to the maximum possible (for a given k_j) value, respectively, $e(i_1, i_2)$ increases;
- q_j gradually decreases, k_j increases to the maximum possible value, $e(i_1, i_2)$ decreases.

Dependences s_j and q_j can take two fundamentally different forms. The first of them reproduces a consistent change in these quantities and is close to a linear dependence. This situation is observed when the vulnerabilities of environmental protection objects are described by linear-fractional functions. This conclusion is also valid for single-level systems with a large number of objects. The ratio between the necessary resources for the implementation of science-intensive projects to ensure sustainable development does not change significantly and does not require a significant restructuring of the environmental protection system. Starting from a value of 0.5, the value of s_j decreases to zero, and q_j increases to the maximum possible value.

4 Discussions

Optimization of resource allocation is aimed at achieving the minimum value of $e(i_1, i_2)$. Combinations of l_j and q_j are studied for which the value of $e(i_1, i_2)$ reaches the minimum value. We also studied the interval Δe of change in the value of $e(i_1, i_2)$ with an increase in k_j from 0 to 1. The best results for these indicators are achieved at $l_j = 2$ $e_{min}(i_1, i_2) = 0.46$, and in a significant part of the interval of changes in k_j, the values of $e(i_1, i_2)$ do not exceed 0.87. The smallest values of l_j express the least vulnerability of environmental protection objects. On passing to the values $l_j = 3$ $e_{min}(i_1, i_2)$ increases to 0.58, remaining close to this value throughout the range of k_j variation (at the boundaries of the interval $e_{min}(i_1, i_2) = 0.61$). For completely non-linear functions $l_j = 4$, the values of $e(i_1, i_2)$ increase, and the dependence itself becomes substantially non-uniform for given values of q_j. With high values of the parameters l_j and q_j, it is possible to achieve significant compensation for changes in the efficiency of individual and joint projects to ensure sustainable development.

The correlation between the values s_j and q_j is determined mainly by the value of l_j: at $l_j = 2$, these quantities correlate over the entire range of k_j, and at $l_j = 3$, the correlation coefficient at the value of $k_j = 0.3$ changes sign sharply from positive to negative.

5 Conclusion

The implementation of a joint project in addition to individual projects changes the balance between the reliability of the environmental protection system and requires an adjustment in the allocation of resources within the framework of a fixed budget. The feasibility of implementing joint projects and the optimal distribution of resources between them depends on the dynamic vulnerabilities of environmental objects and the distribution of information. These factors affect the extent to which resource allocation adjustments can be made, which can be negotiated for individual facilities, but can also translate into drastic resource reallocations. Determining the form of dependences $r^{(j)}(i_1, i_2)$ and parameters l_j, q_j - using the proposed methodology will make it possible to give recommendations on optimizing multi-level environmental protection systems.

Acknowledgments. The article is prepared with the financial support of the Russian Science Foundation, project №. 18-78-10047-P.

References

1. Laurett, R., Paco, A., Mainardes, W.: Antecedents and consequences of sustainable development in agriculture and the moderator role of the barriers: Proposal and test of a structural model. J. Rural Stud. **86**, 270–281 (2021)

2. Zhu, L., Li, F.: Agricultural data sharing and sustainable development of ecosystem based on block chain. J. Clean. Prod. **315**, 127869 (2021)
3. Pang, J., Liu, X., Huang, Q.: A new quality evaluation system of soil and water conservation for sustainable agricultural development. Agric. Water Manag **240**, 106235 (2020)
4. Yao, L., Li, Y., Chen, X.: A robust water-food-land nexus optimization model for sustainable agricultural development in the Yangtze River Basin. Agric Water Manag **256**, 107103 (2021)
5. Yue, Q., Wu, H., Guo, P.: Achieving sustainable development goals in agricultural energy-water-food nexus system: an integrated inexact multi-objective optimization approach. Resour. Conserv. Recycl. **174**, 105833 (2021)
6. Villarino, M., Garcia, L.: Adaptive design model for the integration of urban agriculture in the sustainable development of cities. A case study in northern Spain. Sustain. Cities Soc. **65**, 102595 (2020)
7. Holyoak, J., Pisaniello, J., Mayer, W.: The importance of accounting-integrated information systems for realising productivity and sustainability in the agricultural sector. Int. J. Account. Inf. Syst. **41**, 100512 (2021)
8. Maroli, A., Narwane, V., Gardas, B.: Applications of IoT for achieving sustainability in agricultural sector: a comprehensive review. J. Environ. Manage. **298**, 113488 (2021)
9. Claudet, J., Bopp, L., Gaill, F.: A roadmap for using the un decade of ocean science for sustainable development in support of science, policy, and action. One Earth. **2**, 34–42 (2019)
10. Gill, J., Mankelow, J., Mills, K.: The role of Earth and environmental science in addressing sustainable development priorities in Eastern Africa. Environ. Dev. **30**, 3–20 (2019)
11. Hong, S., Park, S., Chang, H.: An analysis of security systems for electronic information for establishing secure internet of things environments: focusing on research trends in the security field in South Korea. Future Gener. Comput. Syst. **82**, 769–782 (2017)
12. Waring, T., Skoumpopoulou, D.: Through the kaleidoscope: perspectives on cultural change within an integrated information systems environment. Int. J. Inf. Manage. **32**, 513–522 (2012)
13. Qian, J., Yang, X., Li, M.: Farm and environment information bidirectional acquisition system with individual tree identification using smartphones for orchard precision management. Comput. Electron. Agric. **116**, 101–108 (2015)
14. Yeo, I., Yoon, S., Yee, J.: Yee, J: Development of an environment and energy geographical information system (E-GIS) construction model to support environmentally friendly urban planning. Appl. Energy **104**, 723–739 (2013)
15. Munoz, E., Espuna, A., Puigjaner, L.: Integration of a multilevel control system in an ontological information environment. Comput. Aided Chem. Eng. **29**, 648–652 (2011)
16. Du, S., Zhang, X., Shi, X.: A review of multi-scale modelling, assessment, and improvement methods of the urban thermal and wind environment. Build Environ. Eng. **213**, 108860 (2022)

Multi-reservoir Echo State Networks with Encoders

Michal Čerňanský(✉) and Iveta Dirgová Luptáková

Department of Applied Informatics, University of SS. Cyril and Methodius, J. Herdu 2, 917 01 Trnava, Slovak Republic
michal.cernansky@ucm.sk
http://www.ucm.sk

Abstract. Echo state networks (ESNs) as a specific type of recurrent neural networks (RNNs) have gained o lot of attention within research community. Training of ESNs is much less computationally demanding since unlike more common fully trained RNNs only small part of ESN parameters is trained. Recently more and more research is aimed to study methods how to scale ESNs and multiple approaches to multi-reservoir ESNs have emerged such as Deep ESNs and Deep Projection-encoding ESNs. In this contribution we study several approaches to interconnect multiple reservoir layers to achieve rich timescale dynamics necessary for some task. We perform experiments on two widely used datasets on the prediction task and compare achieved results.

Keywords: Echo state network · Reservoir computing · Deep learning · Random networks

1 Introduction

Recurrent neural networks (RNNs) proved to be very successful in wide range of practical applications in various domains. Training process of RNNs is computationally expensive and suffers from various problems such as local minima and vanishing gradient problem. Successful application of state-of-the-art RNNs and in general of deep learning is related with the accessibility of high performance computing hardware mostly GPU-like processors. In common RNNs all or most weight parameters are trained usually with methods based on simple gradient descent.

Echo State Networks (ESNs) [2] were suggested to overcome some limitations of RNNs. They are probably the most prominent representative of so called Reservoir Computing (RC) methods, where "huge" randomly created and untrained reservoir serves as the source for trainable readout mechanism. The reservoir of ESN is the recurrent layer formed of a large number of sparsely interconnected units with non-trainable weights. ESN training procedure is a simple adjustment of output weights to fit training data. ESNs were successfully applied in sequence modelling tasks and performed exceptionally well [4,5].

R. Silhavy (Ed.): CSOC 2022, LNNS 502, pp. 480–489, 2022.
https://doi.org/10.1007/978-3-031-09076-9_43

The key feature of ESNs is so called "echo-state" property: under certain conditions ESN state is a function of finite history of inputs presented to the network - the state is the "echo" of the input history. It simply means that ESNs are based on contractive dynamics where recurrent units reflect the history of inputs presented to the network. The most recent input has the most important influence to the current network state and this influence gradually fades out.

But comparing to the current state-of-the art deep RNNs ESNs reservoirs are not that huge and the question how to successfully scale ESNs is still open. There are several benefits to address this issue: deep ESNs training and inference could use current computational resources commonly present in many devices, also understanding how to reasonably interconnect multiple reservoirs could also help to better understand the dynamics of ESNs and RNNs in general. A lot of attention in the RC community is devoted to the study of hierarchical ESNs and several multi-reservoir approaches have been both studied form theoretical point of view and also applied on real-world time series. Nevertheless, constructing a hierarchical ESN-based models is a challenging problem.

One of the earliest attempts to create hierarchical ESN is [6] where dynamical feature discoverer (DFD) was suggested. DFD consists of a hierarchy of reservoirs with gradient-based trained voting mechanism. Deep Echo State Network (DeepESN) were introduced and studied in [1] where authors studied properties of layered RNN architectures and stated the importance of the nature the layers are interconnected in order to observe multi-timescale reservoir dynamics beneficial for modelling complex time series. In [7] so called Deep Projection-encoding ESNs (DeePrESN) were introduced and studied in the context of the prediction tasks. Authors compared proposed model with several other multi-reservoir ESN approaches and DeePrESN performed very well. DeePrESN model consist of multiple common ESN reservoirs interleaved by projections layers with lower dimensionality that transform activities from the lower reservoir layer into input for higher reservoir layer. Authors studied two projection transformations, one based on PCA and the other based on extreme Learning Machine (ELM). Reservoirs of trained DeePrESN models exhibited distinct time scale properties comparing with other models which was by authors the key factor for higher model performance. The ability of multiple interconnected ESN reservoirs was thoroughly studied in [8]. Authors both theoretically and experimentally showed the effect of reservoir hyper-parameters to the ability of hierarchical models to exhibit multi-timescale dynamics.

The purpose of this contribution is to study the importance of interconnections between multiple reservoirs in deep ESN architecture. Several authors [1,8] stated that this is the key factor of model performance.

2 Multi-reservoir Echo State Networks

2.1 Echo State Networks

Echo State Network is a special type of RNN with a single hidden layer with recurrent interconnections called dynamical reservoir [2,5]. Reservoir of ESN is formed of large number of sparsely interconnected and randomly initialised units with nonlinear activation function. Input provided by input layer drives reservoir which should exhibit rich and potentially useful dynamics. As a readout mechanism output layer with linear units is used. To extract interesting features from dynamical reservoir and produce desired output these output connections are adapted usually by simple linear regression algorithm. Architecture is shown in Fig. 1.

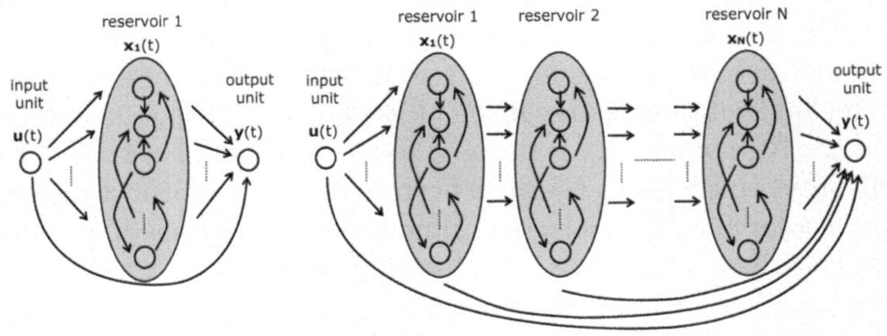

Fig. 1. The standard ESN architecture and deep ESN srchitecture

More precisely when $\mathbf{u}(t)$ is an input vector presented to the ESN at time step t, activations of hidden units $\mathbf{x}(t)$ and produced output $\mathbf{y}(t)$ are updated according to

$$\hat{\mathbf{x}}(t) = \tanh\left(\mathbf{W}^{\mathbf{in}} \cdot \mathbf{u}(t) + \mathbf{W}^{\mathbf{res}} \cdot \mathbf{x}(t-1)\right), \tag{1}$$

$$\mathbf{x}(t) = (1-\gamma)\mathbf{x}(t-1) + \gamma\hat{\mathbf{x}}(t), \tag{2}$$

$$\mathbf{y}(t) = \mathbf{W}^{\mathbf{out}} \cdot [\mathbf{u}(t), \mathbf{x}(t), 1], \tag{3}$$

where $\mathbf{W}^{\mathbf{res}}$, $\mathbf{W}^{\mathbf{in}}$ and $\mathbf{W}^{\mathbf{out}}$ are hidden-to-hidden, input-to-hidden, and hidden-to-output connection matrices, respectively. Parameter γ denotes so called leak rate where $\gamma = 1$ representing no leakage.

Echo state property means that for each internal unit x_i there exists an echo function e_i such that the current state can be written as $x_i(t) = e_i(u(t), u(t-1), \ldots)$, the network state is an "echo" of the input history [2]. The recent input presented to the network has more influence to the network state than an older input, the input influence gradually fades out. So the same input signal history $u(t), u(t-1), \ldots$ will drive the network to the same state $x_i(t)$ in time t regardless the network initial state. Echo states are crucial for successful operation of ESNs, their existence is usually ensured by rescaling recurrent weight matrix $\mathbf{W}^{\mathbf{res}}$ to specified spectral radius λ. This can be achieved by simply multiplying all elements of a randomly generated recurrent weight matrix with λ/λ_{max}, where λ_{max} is the spectral radius of the original matrix.

2.2 Deep Echo State Networks

Deep Echo State Network (DeepESN) [1] is a simple extension to the basic ESN created by stacking multiple reservoirs between input and output layer. Higher-order reservoir is not usually directly fed by the network input and possesses only connections from its direct precedent. On the other hand output layer has interconnections from all reservoirs. Deep ENS architecture with N reservoirs is shown in Fig. 1.

Output $\mathbf{y}(t)$ activations and activations of n-th reservoir $\mathbf{x}_{(\mathbf{n})}(t)$ where N is the count of reservoirs ($1 \leq n \leq N$) are updated as

$$\hat{\mathbf{x}}_1(t) = \tanh\left(\mathbf{W}_1^{\mathbf{in}} \cdot \mathbf{u}(t) + \mathbf{W}_1^{\mathbf{res}} \cdot \mathbf{x}_1(t-1)\right) \quad \text{for } n = 1, \tag{4}$$

$$\hat{\mathbf{x}}_n(t) = \tanh\left(\mathbf{W}_n^{\mathbf{in}} \cdot \mathbf{x}_{n-1}(t) + \mathbf{W}_n^{\mathbf{res}} \cdot \mathbf{x}_n(t-1)\right) \quad \text{for } n > 1, \tag{5}$$

$$\mathbf{x}_n(t) = (1 - \gamma_n)\mathbf{x}_n(t-1) + \gamma_n \hat{\mathbf{x}}_n(t), \tag{6}$$

$$\mathbf{y}(t) = \mathbf{W}^{\mathbf{out}} \cdot [\mathbf{u}(t), \mathbf{x}_1(t), ..., \mathbf{x}_N(t), 1], \tag{7}$$

where $\mathbf{W}_n^{\mathbf{res}}$ and $\mathbf{W}_n^{\mathbf{in}}$ are hidden-to-hidden and input-to-hidden, weights for reservoir n and $\mathbf{W}^{\mathbf{out}}$ is hidden-to-output connection matrix. The leak-rate parameter γ_n may be different for each reservoir.

2.3 Deep Projection-Encoding Echo State Networks

Deep Projection-encoding Echo State Network (DeePrESN) were proposed in [7]. This architecture is specific by stacking multiple reservoirs similarly to DeepESN but they are interleaved by encoder layers. The role of the encoder is to transform high dimensional activities of precedent reservoir to usually lower dimensional inputs that are projected with randomly initialised weights to the succeeding reservoir. The rational behind this architecture is that this encoder-projector may better separate reservoirs hence single reservoir specific multi-timescale dynamics can be better exploited. In [7] Principal Component Analysis (PCA) and ELM based autoencoder were used to find encoders weights. DeePrESN architecture with N reservoirs and $N-1$ encoders shown in Fig. 2.

Output activations $\mathbf{y}(t)$, n-th reservoir activations $\mathbf{x}_{(\mathbf{n})}(t)$ and encoder activations $\mathbf{z}_{(\mathbf{n})}(t)$ where N is the number of reservoirs and $N-1$ is the number of encoders are updated as

$$\hat{\mathbf{x}}_1(t) = \tanh\left(\mathbf{W}_1^{\mathbf{in}} \cdot \mathbf{u}(t) + \mathbf{W}_1^{\mathbf{res}} \cdot \mathbf{x}_1(t-1)\right) \quad \text{for } n = 1, \tag{8}$$

$$\mathbf{z}_n(t) = \mathbf{W}^{\mathbf{enc}} \cdot \mathbf{x}_n(t), \tag{9}$$

$$\hat{\mathbf{x}}_n(t) = \tanh\left(\mathbf{W}_n^{\mathbf{in}} \cdot \mathbf{z}_{n-1}(t) + \mathbf{W}_n^{\mathbf{res}} \cdot \mathbf{x}_n(t-1)\right) \quad \text{for } n > 1, \tag{10}$$

$$\mathbf{x}_n(t) = (1 - \gamma_n)\mathbf{x}_n(t-1) + \gamma_n \hat{\mathbf{x}}_n(t), \tag{11}$$

$$\mathbf{y}(t) = \mathbf{W}^{\mathbf{out}} \cdot [\mathbf{u}(t), \mathbf{x}_1(t), ..., \mathbf{x}_N(t), 1], \tag{12}$$

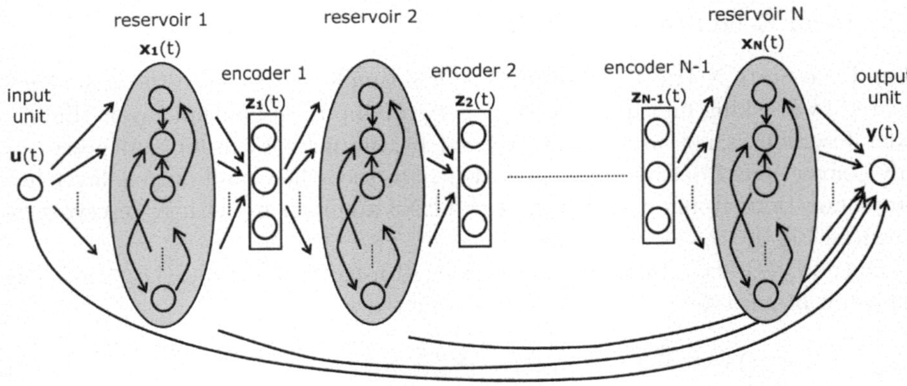

Fig. 2. Architecture of Deep Projection-encoding ESN architecture.

where $\mathbf{W}_n^{\mathbf{res}}$, $\mathbf{W}_n^{\mathbf{enc}}$ and $\mathbf{W}_n^{\mathbf{in}}$ are hidden-to-hidden and hidden-to-encoder and encoder-to-hidden, weights for layer n and $\mathbf{W}^{\mathbf{out}}$ is hidden-to-output connection matrix. Parameter γ_n is the leak-rate which may be different for each reservoir.

In [7] authors used carefully prepared encoders based on PCA and ELM which aims to capture the most prominent features that could be used to reconstruct original observations. Authors compared DeePrESN to DeepESN and also other many-reservoir approaches and evaluated the multi-timescale characterises within reservoir hierarchy of these methods. They showed that DeePrESN displayed superior ability to present the multi-timescale dynamics within reservoir hierarchy. According to [8] the key hyper-parameter related to timescale dynamics of the reservoir is the leak rate γ and in the trained DeePrESN models reservoirs leak rates differs significantly.

In this contribution several straightforward methods of interconnecting multiple dynamical reservoirs in deep ESNs will be studied. To reduce averaging effect of connecting reservoirs we allow input weights of regular DeepESN and studied deep ESN modifications to be zero, similarly to sparsity parameter for the reservoir initialisation. Sparser connections might result in better reservoir separation hence timescale characteristics specific to each of reservoirs might be better used by the readout. In studied deep ESN modifications the units that serve as inputs to the succeeding reservoir are preselected. For preliminary experiments we choose units selected randomly, units with activities exhibiting the highest variance and units with the smallest cross-covariance. Also activities from the precedent reservoirs are first normalised and only then serve as the inputs to the succeeding reservoirs. We compare these two methods with DeePrESN with PCA-based encoders and also regular ESN and DeepESN architectures.

3 Experiments

In [7] authors used encoder layers that were carefully prepared as PCA and ELM based autoencoders. In this contribution we suggest that the key factor might simply lie in better separation of reservoirs and we study several more straightforward methods of interconnecting reservoirs.

3.1 Models

The modified models we used in experiments is regular deep ESN where only selected units from preceding reservoir are interconnected to the following reservoir. The units selected as inputs were first chosen randomly, but also other methods of unit selection were tested such as units showing the highest variance or units exhibiting the lowest cross covariance. The motivation was that these selectively chosen units might produce diversifications of reservoirs so more interesting dynamics could be used by linear readout layer. We also perform normalisation of activities of units served as inputs to the following reservoir. Input weights in between reservoirs are chosen randomly form given interval and since reservoir activities might have magnitudes of different scales normalisation could help in producing more appropriate behaviour of the destination reservoirs.

Please note that all these modified deep ESN models with both specific input selection and activity normalisation can be described and modelled by deep ESN architecture with reservoirs interleaved with encoder layers - the very same architecture depicted in Fig. 2 of the DeePrESN model. Encoder layers consist of linear units with both input weights and biases. Reservoir unit selection with normalisation can be achieved by setting single nonzero weight and bias for every encoder unit. The weight magnitude is computed to achieve unit standard deviation and the bias is set to achieve zero mean value. Source of this encoder unit weight is the reservoir unit selected to be (indirect) input for the following reservoir. Hence we will use DeepEncESN abbreviation for this modified deep ESN architecture with encoders.

3.2 Datasets

We performed experiments with two datasets frequently used in the literature. Mackey-Glass (MG) system is a chaotic time series where ESNs showed excellent performance exceeding other approaches by several orders [2,5]. MG system is defined by the differential equationå

$$\partial x/\partial t = \big(0.2(x-\tau)\big)/\big(1+x(t-\tau)^{10} - 0.1x(t)\big) \tag{13}$$

and we used $\tau = 17$ as in [5]. The 9000 generated values were shifted by adding -1 and tanh function was used to squash values into appropriate interval as in [5]. The first 5000 values were used for the training set, the following 2000 values were used for validation and the remaining 2000 values for the test set. Initial 1000 from the training set were used only for forward propagation and corresponding hidden units activities were discarded from the adaptation process.

The second dataset is the 10th order nonlinear autoregressive moving average (NARMA) system from [3] defined as:

$$d(n+1) = 0.3d(n) + 0.05d(n)\left[\sum_{i=0}^{9} d(n-i)\right] + 1.5u(n-9)u(n) + 0.1 \qquad (14)$$

Values for input sequence $u(n)$ were generated uniformly from interval $[0, 0.5]$ and the task was to predict $d(n)$. Training set was composed of 5000 values and the first 1000 values were used as initial washout steps. The next 2000 values were used as validation set. Testing was performed on the remaining 2000 values. No noise was added to the sequence.

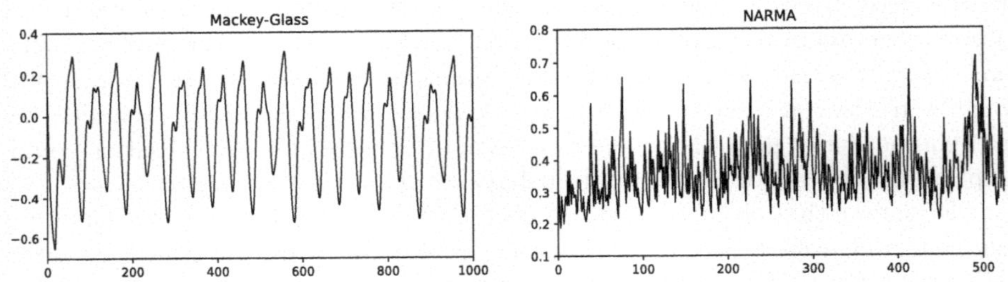

Fig. 3. Mackey-Glass and NARMA time series

3.3 Training

We performed experiments with regular ESN, DeepESN, DeePrESNs with PCA autoencoders and DeepEncESNs variations with specifically initialised encoders. Models with encoders that correspond to the random reservoir unit selection are denoted DeepEncESN-rnd, DeepEncESN-var is model where unit selection is based on the highest variance and DeepEncESN-corr is the model where units were selected to exhibit small cross-covariance.

We compare predictive performance using normalised root mean squared error (NRMSE). Inspired by results achieved in [7] we created models with reservoir(s) of 1200 units, deep architectures consisted of 4 reservoirs each with 300 units and architectures with encoders had 60 units in encoders. Similarly to [7] activities of encoders were not used as input for readout mechanisms hence all modes have the same number of trainable weights. Initial "washout" steps were used to stabilise reservoir dynamics and were not used for calculating readout weights. Also encoder weights were chosen using activities after the same number of initial steps. Ridge regression with regularisation constant set to 10^{-6} was used calculate output weights. Google™ AI platform hyper-parameter search with max. trials set to 200 was used to train all models with the following hyper-parameters for every reservoir: spectral radius, leak rate, input weight range and input weight probability, all bound into (0, 1] interval. For the first reservoir the input weight probability was fixed to 1. Sparsity parameter was fixed to 0.1.

3.4 Results

The NRMSE results are presented in Table 1. By small amount the best results on both datasets were achieved by DeePrESN-PCA. Although this observation is not decisive, it seems to be consistent among multiple experimental hype-parameter search runs - encoders based on PCA auto encoder tend to provide very good results. The lowest performance was achieved by regular single reservoir ESNs. These results strongly correspond to results achieved by [7].

On the other hand other DeepEncESN methods exhibited results only slightly inferior to the DeePrESN-PCA supporting the notion that the key element of the improved DeePrESN-PCA predictable performance is the reservoir separation using encoder layers. The role of specific encoder setup might still be important but also DeepEncESN-rand models with randomly selected reservoir units provide good results.

In our contribution we allowed stacked reservoirs to have zero input weights - one of the hyper-parameter was the probability of input weight being nonzero. We show that also DeepESN models achieved satisfactory performance by some margin inferior to the DeepEncESN and DeePrESN-PCA models and superior to the results provided in [7] on Mackey-Glass dataset.

Table 1. The mean and standard deviation values for NRMSE results on Mackey-Glass and NARMA datasets.

	Mackey-Glass dataset	NARMA dataset
ESN	5.44×10^{-2} (3.97×10^{-3})	1.61×10^{-1} (1.67×10^{-2})
DeepESN	6.70×10^{-3} (4.67×10^{-4})	1.24×10^{-1} (9.02×10^{-3})
DeePrESN-PCA	$\mathbf{6.13 \times 10^{-3}}$ $\mathbf{(1.37 \times 10^{-3})}$	$\mathbf{8.01 \times 10^{-2}}$ $\mathbf{(2.32 \times 10^{-3})}$
DeepEncESN-rnd	6.39×10^{-3} (1.40×10^{-3})	8.82×10^{-2} (4.21×10^{-3})
DeepEncESN-var	6.31×10^{-3} (1.17×10^{-3})	8.14×10^{-2} (3.35×10^{-3})
DeepEncESN-corr	6.49×10^{-3} (1.80×10^{-3})	8.11×10^{-2} (3.40×10^{-3})

To analyze reservoir separation we inspect multi-reservoir dynamics of all models. At time step 1000 the activity of the input signal is affected by a perturbation and the differences to the activities without perturbation averaged by reservoirs is shown in the charts in Fig. 4 for Mackey-Glass and Fig. 5 for NARMA time series. For regular ESN we show results for distinct blocks of units of size 300. Time step 0 at charts corresponds to the point perturbation was made.

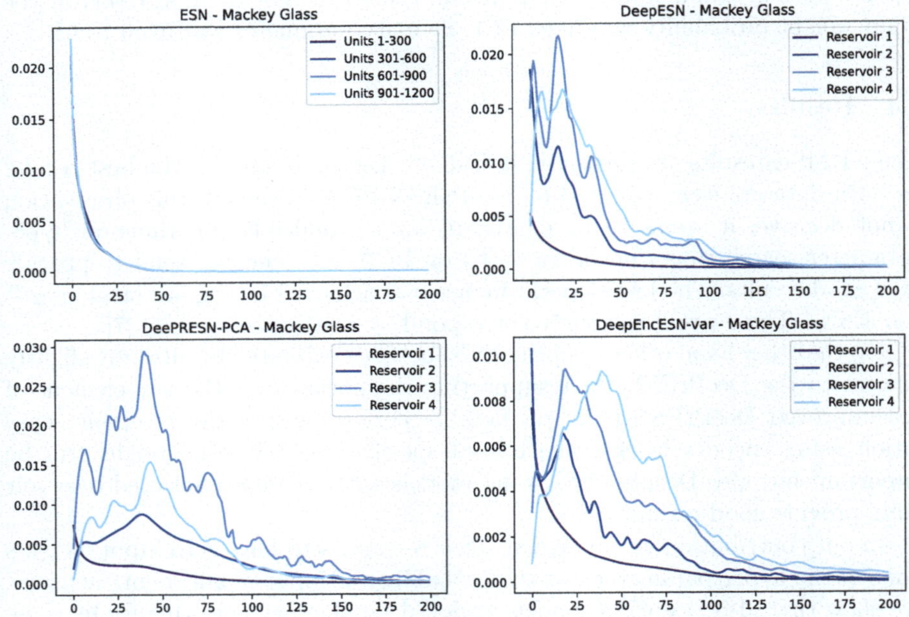

Fig. 4. Multi-timescale dynamics of models on Mackey-Glass time series.

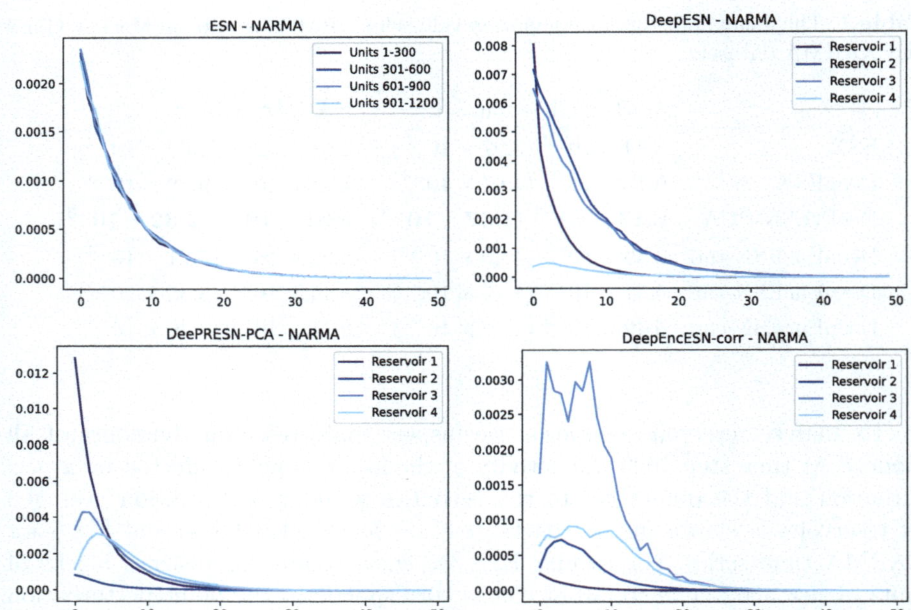

Fig. 5. Multi-timescale dynamics of models on NARMA time series.

Charts clearly show that the most complex multi-timescale dynamics is present in models with encoders: DeepEncESNs and DeePrESN-PCA. This is achieved by hyper parameter search favorizing different parameters for every reservoir in reservoir stack where the leak rate parameter γ plays a prominent role.

4 Conclusion

We've studied several deep ESN models consisting of multiple reservoirs stacked together and interconnected with encoder layers initialised using different techniques. All these DeepEncESN models achieved superior results to the regular ESNs or deepESNs and similar to DeePrESN-PCA. We showed that the key element of the successful predictive performance is the separation of the reservoirs with encoder layers. The specific encoder layer setup might be of less importance but this notion deserves further study. Complex multi-timescale dynamics is clearly present on models with encoders on both Mackey-Glass and NARMA datasets showing potential in creating larger multi reservoir models where reservoirs are interconnected with encoder layers.

Acknowledgments. The work was funded by the grant KEGA 012UCM-4/2021 Modern technologies and innovations in network security education.

References

1. Gallicchio, C., Micheli, A., Pedrelli, L.: Deep reservoir computing: a critical experimental analysis. Neurocomputing **268**, 87–99 (2017)
2. Jaeger, H.: The "echo state" approach to analysing and training recurrent neural networks. Technical report GMD 148, German National Research Center for Information Technology (2001)
3. Jaeger, H.: Adaptive nonlinear system identification with echo state networks. In: Proceedings of Neural Information Processing Systems, NIPS 2002, Vancouver, Canada (2002)
4. Jaeger, H.: Adaptive nonlinear system identification with echo state networks. In: Becker, S., Thrun, S., Obermayer, K. (eds.) Advances in Neural Information Processing Systems, vol. 15, pp. 593–600. MIT Press, Cambridge (2003)
5. Jaeger, H., Haas, H.: Harnessing nonlinearity: predicting chaotic systems and saving energy in wireless communication. Science **304**(5667), 78–80 (2004)
6. Jaeger, H.: Discovering multiscale dynamical features with hierarchical echo state networks. Technical report, Jacobs University Bremen (2007)
7. Ma, Q., Shen, L., Cottrell, G.W.: DeePr-ESN: a deep projection-encoding echo-state network. Inf. Sci. **511**, 152–171 (2020)
8. Manneschi, L., Ellis, M.O., Gigante, G., Lin, A.C., Del Giudice, P., Vasilaki, E.: Exploiting multiple timescales in hierarchical echo state networks. Front. Appl. Math. Stat. **6**, 76 (2021)

Towards Lithuanian Grammatical Error Correction

Lukas Stankevičius(✉) and Mantas Lukoševičius

Faculty of Informatics, Kaunas University of Technology, Kaunas, Lithuania
lukas.stankevicius@ktu.edu

Abstract. Everyone wants to write beautiful and correct text, yet the lack of language skills, experience, or hasty typing can result in errors. By employing the recent advances in transformer architectures, we construct a grammatical error correction model for Lithuanian, the language rich in archaic features. We compare subword and byte-level approaches and share our best trained model, achieving $F_{0.5} = 0.92$, and accompanying code, in an online open-source repository.

Keywords: Natural Language Processing · Grammatical error correction · Transformer models · ByT5 · Lithuanian

1 Introduction

Recent advances in neural Natural Language Processing (NLP) have pushed the frontiers. In particular, transformer-architecture-based models [25] surpassed human performance in various NLP benchmarks such as SQuAD2.0 [16], GLUE [27], and SuperGLUE [26]. This also opened new opportunities for the Grammatical Error Correction (GEC) task which we address in this work.

GEC is the task of correcting different kinds of errors in text such as spelling, punctuation, grammatical, and word choice errors. The abundance of such noise in the text can hinder not only the understanding by humans but also the performance of various downstream NLP systems. An error-free text is also more beautiful, clean, associated with a certain prestige. However, producing it may be problematic for non-native speakers, language learners, it requires additional time and effort.

NLP state of the art for GEC still has much room to improve. As of now, the best $F_{0.5}$ scores are only up to 0.72[1]. Moreover, the research is mostly focused on English and a few other popular languages. To reduce this gap and contribute to the GEC progress, in this work, we investigate it for the Lithuanian language.

The Lithuanian language is one of the oldest living languages in the world. It has retained most of the features of the Indo-European Protolanguage, i.e., it is characterized by a very ancient linguistic structure: declensions (of nouns, adjectives, and pronouns), short and long vowels, diphthongs, etc. Lithuanian also has many similarities with Sanskrit – the classical language of ancient India,

[1] http://nlpprogress.com/english/grammatical_error_correction.html.

R. Silhavy (Ed.): CSOC 2022, LNNS 502, pp. 490–503, 2022.
https://doi.org/10.1007/978-3-031-09076-9_44

still used today as a scholarly and liturgical language in Hinduism, Buddhism, and Jainism. Antoine Meillet (1886–1936), one of the most influential French linguists, once stated: "Anyone wishing to hear how Indo-Europeans spoke should come and listen to a Lithuanian peasant".

The Lithuanian language is synthetic and uses inflections to express syntactic relationships within a sentence. In other words, the relations in a sentence are expressed by word endings rather than with unbound morphemes and word order. This allows a lot of freedom in composing sentences. In contrast to agglutinative languages, which combine affixes by "gluing" them unchanged inside word ending, in Lithuanian inflectional categories are "fused". Meanwhile, prefixes, suffixes, and infixes are still used to derive words. Lithuanian verbs can be made from any onomatopoeia; phrasal verbs (e.g., go in, go out) are composed by adding the prefix to the verb. Lithuanian is unique for having 13 different participial forms of the verb [9] while modern English has only 2 (present and past participles). It is estimated that 47% of Lithuanian word forms are morphologically ambiguous [17], i.e., requiring context consideration to discern the meaning. All these mentioned features of the Lithuanian language make it interesting and important to analyze in the context of automatic GEC.

Our contributions are:

- We present the first GEC system for Lithuanian language based on deep neural networks.
- We compare sub-word and byte-level tokenization approaches for Lithuanian grammatical error correction.
- We share all the technical details, code, and model weights for open reuse and reproducibility.

2 Related Work

The simplest form of GEC is spellcheck. GNU Aspel[2] and Hunspell[3] are two widely-used open source spellcheckers. In particular, Hunspell [5] is the only system we found for Lithuanian GEC. Such systems work by keeping a large dictionary of possible words and detecting the non-words. During detection, the nearest alternatives from the dictionary are suggested. In Hunspell's case, the dictionary is made more compact by keeping only the main word forms with transformation rules, prefixes and suffixes. Spellchek systems are compact but limited to the correction of only non-words.

The first systems for a more complex GEC were based on Statistical Machine Translation (SMT) using a noisy channel model [2]. A significant contribution to GEC was the introduction of the CoNLL-2014 shared task [12]. Multiple systems were proposed, and among them, the phrase-based SMT setup was the most promising [8]. Yet neural approaches started to emerge, like [29]. As such systems advanced, hybrid statistical (SMT) and Neural Machine Translation

[2] http://aspell.net/.

[3] http://hunspell.github.io/.

(NMT) approaches [7] took the top. Only the introduction of the Transformer model [24] enabled neural approaches to supersede the statistical ones. As of now, the latter systems are claiming state-of-the-art results in GEC [13,18].

Novel less-supervised approaches are also emerging. A simple language model reaching a reasonable performance with minimal annotated training data was demonstrated in [3]. The proposed system used n-gram language model to score variations of a sentence until incremental inflections do not improve the score anymore. Such a system was again improved using transformer-based language models instead of the n-grams in [1]. It turns out that such a less-supervised approach can outperform fully-supervised systems that were claiming state-of-the-art results several years ago.

Currently, the main constraint for GEC is the lack of training data. Researchers make progress by including new data sources or using automatic grammatical error generation to synthesize them. A simple language-agnostic pre-training objective was proposed in [18]. The authors automatically corrupted sentences in character level: swapping, inserting, dropping spans; token level: swapping, dropping spans; word level: lower-casing, upper-casing the first letter. A bigger model and larger dataset allowed achieving state-of-the-art GEC results for 4 languages. Authors of [19] used a small corpus of spelling errors to derive statistics for typographical error generation and generate a large parallel synthetic corpus. Another way is to use the data that the model incorrectly predicted during the training. A fluency boost learning and inference mechanism was proposed in [6] that reuses less fluent model predictions as new inputs during subsequent epochs. Similar trends are emerging with other languages. For example, simply adding new data improved a Transformer GEC system for the Czech language [11]. To summarize, it is important for neural GEC systems to be trained on large and high-quality corpora.

3 Dataset

As mentioned, a large dataset is essential for training a neural GEC solution. The data also has to be of the highest quality so that we can take it as a gold standard of grammatically-correct text. The largest publicly available general-purpose dataset for the Lithuanian language is from OSCAR [14] at 5 GB of deduplicated text. However, it is obtained from a general Common Crawl[4] and makes trusting the grammatical correctness problematic.

To make sure that the text is of good quality, we crawled various Lithuanian websites ourselves. We manually checked how the text is structured in every webpage so that only the relevant parts: title, summary (optionally), and the main text paragraphs would be scrapped. We crawled the following types of web pages: news, literature, blogs, encyclopedias, others.

We added titles and summaries to the main text paragraphs as additional paragraphs. Finally, we split the data into paragraphs. As a result, a single paragraph became a single sample of our dataset.

[4] https://commoncrawl.org/.

3.1 Preprocessing

Although we performed a well-curated data scraping, there were still some artifacts in our data that had to be corrected or removed.

We had to remove some websites because of relatively high rates of spelling errors. This left us with a total of 34 final websites.

Some common error patterns can be easily corrected automatically. We looked at common mistakes in Lithuanian web texts [23] and performed the following corrections:

1. Incorrect quotation marks. In Lithuanian, the correct are “ABC”. Meanwhile, the English version “ABC” or others such as the universal “ABC” is often used instead.
2. The lack of space. It can happen before “m.” and “d.” abbreviations. For example, the text “1918m. vasario 16d.” must be corrected to “1918 m. vasario 16 d.”. Additionally, the space is often omitted after a full stop: “ir t.t.” and “A.Sabonis” must be corrected to “ir t. t.” and “A. Sabonis”.
3. An unnecessary space. The space must be omitted before most punctuation marks: “tik darbui, visiškai pamirštant poilsį,” is corrected to “tik darbui, visiškai pamirštant poilsį,”.

We also filtered the text samples based on some statistics:

1. The sample text length should be at least 20 characters.
2. The fraction of Lithuanian letters in a sample should be at least 0.98. This filters out text from other languages and with miscellaneous characters. In the end, we are solving a task for the Lithuanian language. We included the characters “€£$%wx” as Lithuanian since they are used quite often.
3. The fraction of spaces to non-spaces should be at most 0.02. This allowed us to filter out samples dominated by URL addresses.

Lastly, we deduplicated our text samples. We shuffled the resulting 29 312 785 samples and took a subset of 4 194 304 for this work. Some statistics for the subset are depicted in Table 1.

Table 1. Dataset sizes by various tokenizations. The total dataset size is 4 194 304 samples.

Tokenizer	Sample length, mean $\pm$ std	Tokens, $\times 10^6$	1Tokenization example
Characters	226 ± 194	947	Lietuva - graži šalis
ByT5 [30]	243 ± 194	1 017	Lietuva \xe2\x80\x93 gra\xc5\xbei \xc5\xa1alis
T5 [21]	48 ± 43	201	[_Lietuva] [_-] [_graži] [_šalis]
mT5 [31]	71 ± 61	298	[_] [Lietuva] [_-] [_] [graž] [i] [_šal] [is]
Words	30 ± 26	126	[Lietuva] [graži] [šalis]

The transformer model has a quadratic running time complexity $\mathcal{O}(n^2)$ with respect to the sequence length n (number of tokens). Usually, this is not a constraint as most text tasks are within the maximal sequence length of 512 (T5 [15]) sub-words or 1024 (ByT5 [30]) bytes. Yet in our training dataset, we had longer examples that we did not wish to truncate and, hence, lose. Instead, we split these too-long sequences to the length of 2100 characters for the T5 model and 700 characters for the ByT5. After that, we proceeded with the corresponding tokenization. As a result, the exact numbers of samples and tokens differ for both models, but the initial dataset and the amount of text (see Table 1) is the same.

For both runs, we set aside 0.05% of the data for validation and another 0.05% for testing.

4 Methods

4.1 Generating Grammatical Errors

We induce 3 groups of synthetic grammatical errors described below.

Typographical Errors. They are induced by modeling the way how humans mistype on the keyboard. We follow the exact same methodology as in [22]: take mistyping statistics between each pair of characters on a QWERTY keyboard from an English dataset and apply them probabilistically to our texts. Out of the all characters considered, this way we corrupted 2% of them; from which were: substitution, 36.1%; deletion, 31.7%; insertion, 17.8%; transposition, 14.4%.

Confusing Similar Sounding Letters. This is a very common source of spelling mistakes. We model them by defining sets of characters that sound alike, and randomly substituting a letter with one from the same set at a point of the generated error. For this, we use weighted sampling. The probabilities of the letters for substitution are proportional to how frequent they are in Lithuanian texts overall. For example, in a single set of letters "iįy" (where sounds differ only in their length), it is way more common to mistakenly write "i" instead of "į", rather than "į" instead of "i", as "i" is much more common. Only 2% of all such found occurrences were replaced. Groups of letters and detailed probabilities for the group members are derived from the raw (no preprocessing) subset of 2 909 403 samples, and are presented in Appendix A.

Other Errors. We also introduce errors in the text by the four specific rules described below. We, again, thus corrupt 2% of the matches of the rules.

1. Gemination are doubled consonant letters that sound like a single one and thus is prone to be typed only once. This also applies to any consecutive letters from "cčsšzž" group. For example, the words "pusseserė užsimerkė" may be mistakenly written as "puseserė usimerkė" as they sound similar due to the gemination.

2. Assimilation to an adjacent letter. This is specific to any letter of "ptksš" being before any of the "bdgzž" or vice versa. For example, the words "dirbti, lipdavo" may be mistakenly written as "dirpti, libdavo" as this is how they sound due to the assimilation.
3. Uppercasing or lowercasing the first letter in a word. For example, the word "ąžuolas" can start both with the lower or upper case depending on whether it is a tree (oak) or person's name. We exclude the first words in a sentence as these always have to start the upper case.
4. Delete and add space. We separately match all occurrences of spaces and all empty strings not at a word boundary.

Some samples of the corrupted sentences are presented in Appendix B.

4.2 Transformer Models

In this work, we compare T5 [21] and ByT5 [30] transformer models for grammatical error correction of Lithuanian. They are of sequence-to-sequence type. The encoder encodes the input sequence with attention operating on all input tokens while the decoder predicts output sequence tokens one by one, attending to tokens of both encoder (all) and decoder (only previous ones).

Below we further emphasize the properties of these models that make them appropriate for our task.

T5. The original T5 [15] was designed to be universal for multiple tasks. Authors showed that there is no difference whether a custom "head" is used (added on top of the pre-trained transformer) for fine-tuning purposes or a simple sequence-to-sequence formulation in text format is employed (no need to add additional weights to a pre-trained model). This way even tasks with outputs as float numbers can be formatted into a text-to-text format. Such generic task formulation made the T5 model very popular.

In previous work [21], we adapted the T5 model for the generation of summaries of Lithuanian news articles. We trained a SentencePiece [10] tokenizer on 10^6 and the main model on 2 027 418 news articles. As a result, this model should be familiar with the Lithuanian language (both tokenizer and model weights) and we use it as the basis for our fine-tuning purposes.

ByT5. ByT5 is a follow-up model from the multilingual mT5 [31] and T5 [15]. The authors showed that adapting byte-level tokenization can lead to a much more efficient use of model parameters. As an example, the multilingual mT5 had over 66% of its weights (for the base version) allocated to its multilingual word pieces (a total of 250 000) related weights (input embedding matrix and decoder softmax layer) which were only sparsely updating during the training. Meanwhile, ByT5 vocabulary has only 384 items and the model reuses the saved parameters in more massive layers rather than indexing tokens. These benefits allowed ByT5 to surpass the small and the base versions of mT5 [30].

The introduction of finer byte-level tokenization is especially important for grammatical error correction. Typos, variants in spelling and capitalization, and morphological changes can lead to completely different sub-word tokens, while byte tokens are affected the least. The authors of ByT5 showed that their model outperforms mT5 if various types of noise are introduced. Therefore, we use this model in our study of Lithuanian grammatical error correction.

4.3 Training Details

To train the models, we used a GeForce RTX 2080 Ti GPU. Following the best practices with the T5 family of models [15,18,30,31], we used the total batch size (number of samples to pass through the model before the gradient update) of 128 for fine-tuning. For ByT5 it was achieved by 128 gradient accumulation steps of batch size 1; while for T5, 64 gradient accumulation steps of batch size 2. We had to use multiple accumulation steps to process the total batch sequentially by smaller parts as the total batch did not fit into GPU memory at once. It took us approximately 100 h for ByT5 and 30 h for T5 fine-tuning. ByT5 took longer due to the longer sequences produced by finer byte-level tokenization.

We used the training script and Pytorch model implementation from the Hugging Face library [28]. For simplicity, we employed an Adafactor optimizer [20] with a constant learning rate of 0.001. If not stated otherwise, we used all the default parameters as in the Hugging Face library version 4.12.0.

4.4 Evaluation

One of the most popular grammatical error correction evaluation metrics is ERRANT [4]. It applies a set of rules operating over a set of linguistic annotations to construct the alignment and extract individual edits between corrupted, corrected, and gold-standard texts. This way precision, recall, and F-score can be calculated. We customized the original ERRANT by using Hunspell dictionaries [5], stemmer[5], spaCy version 3.2 pipeline `lt_core_news_lg`[6], and corresponding part-of-speech tags for the Lithuanian language.

During the inference, we used simple greedy decoding with a beam size of 1. That is, we simply selected for each next token the one that the model assigned the highest probability to.

5 Results

Training dynamics of T5 and ByT5 models are depicted in Fig. 1. Only after 6% of training, the ByT5 score $F_{0.5} = 0.85$ is already higher than the $F_{0.5} = 0.80$ T5 managed to reach after the full epoch. We can also see that the performance is steadily increasing during the fine-tuning and is expected to continue doing

[5] https://pypi.org/project/PyStemmer/.
[6] https://spacy.io/models/lt.

so. The same results are indicated by the training loss. It is much lower for the ByT5, hence the model is better.

We divided our synthesized errors into several groups and corrupted the test set with each group separately from the others. Evaluation results of such setup are presented in Table 2. We can see that the easiest task for both models was adding or deleting spaces, while the hardest task is correcting assimilation and gemination mistakes. This group may lag in performance due to the smaller abundance (2% of samples) in the training data.

We present some generated test samples in Appendix B.

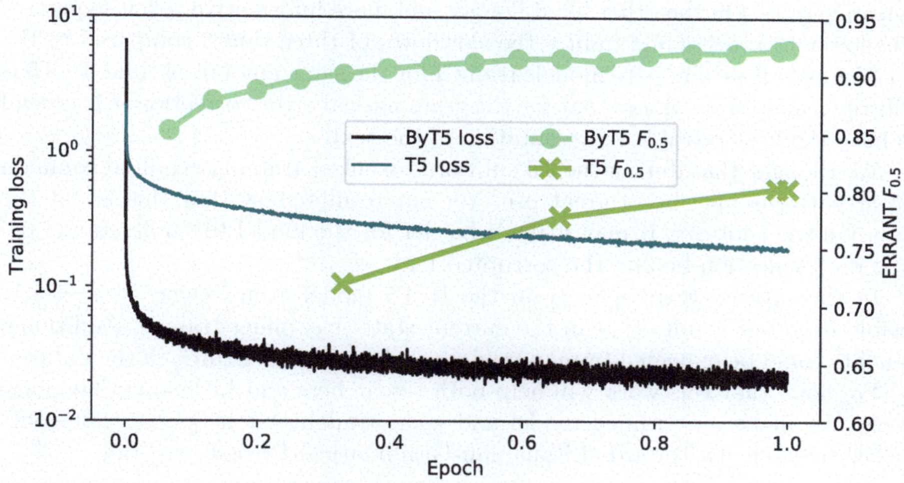

Fig. 1. Training loss and $F_{0.5}$ score for both T5 and ByT5 runs.

Table 2. Evaluation for the separate error categories with models trained for one epoch. We applied synthetic corruption for the test set of ByT5 (total of 2 155 samples) and T5 (total of 2 099 samples) with each error group separately. We show both ERRANT $F_{0.5}$ score and number of samples (#samples) affected and evaluated on.

Error group	ByT5		T5	
	$F_{0.5}$	#samples	$F_{0.5}$	#samples
Typographical	0.87	1 916	0.72	1 868
Punctuation	0.81	489	0.36	460
Similar sounding letters	0.88	1 115	0.55	1 143
Add/delete spaces	0.96	1 873	0.74	1 832
Assimilation/Gemination	0.79	56	0.30	43
Upper/Lower casing	0.86	785	0.47	781

6 Discussion

We trained the first reported deep-learning-based Lithuanian grammatical error correction system and compared two sequence-to-sequence transformer models for the task.

The ByT5 transformer model, based on byte-level tokenization, greatly outperformed the subword counterpart T5. We think that the main reason for this is that the fine-grained byte-level details allow the model to maximize acquired information about the sentence and thus calculate a more accurate representation. This way, the model sees a bigger picture and has to solve the task with less ambiguity. On the other hand, longer and more informative token sequences are slower to process and induce the slowdown of three times, compared to the T5. Yet even if we compare models trained for the same amount of time, ByT5 is still the leader. This shows that for the grammatical error correction it is crucial to have the best possible representation of the text.

We thought that during the T5 subword tokenizer training acquired common token patterns may be of great use. Yet our results show that this is not the case. On the contrary, it may make it harder for the model to "understand" the true representation behind the corrupted text.

In the future, we plan to train the ByT5 model even longer. It is clearly visible from our results that in the current state it is under-trained. Additional benefits could be expected from more data and more passes through the dataset.

We hope that this work will help both researchers and Lithuanian language users. We make our trained model and code available at https://github.com/LukasStankevicius/Towards-Lithuanian-Grammatical-Error-Correction.

Acknowledgement. We thank our project collaborators from Vytautas Magnus University, especially Jurgita Kapočiūtė-Dzikienė, for valuable discussions on related topics.

Fundings. The research is partially funded by the joint Kaunas University of Technology Research and Innovation Fund and Vytautas Magnus University project "Deep-Learning-Based Automatic Lithuanian Text Editor (Lituanistas)", Project no.: PP34/2108.

A Statistics for Corrupting Similar Letters and Punctuation

Table 3. Regex expressions to find specific patterns in texts and statistics of distinct finds used as probability weights for replacement.

<table>
<tr><th>Group (regular expression)</th><th colspan="6">Matches and counts</th></tr>
<tr><td>[,\.-]{0,1}␣</td><td>␣</td><td>79 695 056</td><td>.␣</td><td>5 125 941</td><td></td><td></td></tr>
<tr><td></td><td>,␣</td><td>9 876 726</td><td>-␣</td><td>1 347 515</td><td></td><td></td></tr>
<tr><td>[\.,;:\-\-?!\(\)\[\]\<\>/]</td><td>,</td><td>10 072 919</td><td>?</td><td>300 962</td><td>]</td><td>34 283</td></tr>
<tr><td></td><td>.</td><td>7 976 435</td><td>:</td><td>519 928</td><td>></td><td>5 759</td></tr>
<tr><td></td><td>-</td><td>1 453 095</td><td>!</td><td>106 333</td><td><</td><td>4 457</td></tr>
<tr><td></td><td>)</td><td>665 253</td><td>;</td><td>105 526</td><td></td><td></td></tr>
<tr><td></td><td>(</td><td>655 651</td><td>/</td><td>90 778</td><td></td><td></td></tr>
<tr><td></td><td>-</td><td>546 698</td><td>[</td><td>34 295</td><td></td><td></td></tr>
<tr><td>u{0,1}ou{0,1}</td><td>o</td><td>33 058 916</td><td>ou</td><td>41 509</td><td></td><td></td></tr>
<tr><td></td><td>uo</td><td>3 355 463</td><td>uou</td><td>34</td><td></td><td></td></tr>
<tr><td>ia|e</td><td>ia</td><td>6 733 731</td><td>e</td><td>35 509 427</td><td></td><td></td></tr>
<tr><td>[scz]</td><td>s</td><td>47 349 069</td><td>c</td><td>2 645 328</td><td>z</td><td>1 646 823</td></tr>
<tr><td>[ščž]</td><td>š</td><td>7 002 598</td><td>č</td><td>2 619 317</td><td>ž</td><td>5 044 500</td></tr>
<tr><td>[eęė]</td><td>e</td><td>35 509 427</td><td>ę</td><td>1 336 170</td><td>ė</td><td>9 781 460</td></tr>
<tr><td>[iįy]</td><td>į</td><td>3 490 952</td><td>y</td><td>8 347 510</td><td>i</td><td>82 431 807</td></tr>
<tr><td>[uųū]</td><td>ū</td><td>2 795 974</td><td>ų</td><td>7 826 828</td><td>u</td><td>28 978 236</td></tr>
<tr><td>[aą]</td><td>a</td><td>68 291 558</td><td>ą</td><td>4 471 872</td><td></td><td></td></tr>
<tr><td>[cč]</td><td>c</td><td>2 645 328</td><td>č</td><td>2 619 317</td><td></td><td></td></tr>
<tr><td>[zž]</td><td>z</td><td>1 646 823</td><td>ž</td><td>5 044 500</td><td></td><td></td></tr>
<tr><td>[td]</td><td>t</td><td>35 864 854</td><td>d</td><td>14 822 144</td><td></td><td></td></tr>
<tr><td>[kg]</td><td>k</td><td>26 461 947</td><td>g</td><td>10 626 341</td><td></td><td></td></tr>
<tr><td>[pb]</td><td>p</td><td>16 187 509</td><td>b</td><td>8 148 725</td><td></td><td></td></tr>
<tr><td>‘‘|,,|[,,‘‘"’’]|’’</td><td>"</td><td>436 378</td><td>,,</td><td>11 777</td><td>’’</td><td>87</td></tr>
<tr><td></td><td>’’</td><td>46 847</td><td>‘‘</td><td>817</td><td></td><td></td></tr>
</table>

B Corruption and Correction Examples

Table 4. Samples of the original, corrupted, and corrected text forms. Here fully-trained (1 epoch) ByT5 models were used.

Type	Text
Original	"Mes nenorime, kad jie keiktųsi, pyktųsi. Neleidžiame ne tik gerti, bet ir rūkyti. Taisyklės čia griežtos, rūkei, atleisime tau kartą, nepaklusai, eik iš kur atėjęs. Jei jau žmogus nusprendė keisti gyvenimą, tai turi būti daroma rimtai", - nuolaidų nežada M. Balčiūnas
Corrupted	"Mes nenorime, kad jie keiktųsi, byktųsi. Nleeidžiame ne tik gerti, bet ir rūkyti. Taisyklės čia griežtos, Rūkei, atle isime tau kartą, nepaklusai, eik iš kur atėjęs. Jei j au žmogus nuspr-endė keisti gyvenimą tai turi būti daromo rymtai", - nuolaidų nežada M. Balčiūnas]
ByT5	"Mes nenorime, kad jie keiktųsi, pyktųsi. Neleidžiame ne tik gerti, bet ir rūkyti. Taisyklės čia griežtos. Rūkei, atleisime tau kartą, nepaklusni, eik iš kur atėjęs. Jei jau žmogus nusprendė keisti gyvenimą, tai turi būti daroma rimtai", - nuolaidų nežada M. Balčiūnas
Original	Šeštadienio vakarą Klaipėdoje surengto "Eurovizijos" atrankos finalo dalyviai po renginio miegoti nėjo - dešimt savaičių trukusios kovos pabaigą atšventė uostamiesčio kokteilių bare "Oscar"
Corrupted	Šeštdienio vakarą Klaipėdoje surengto "Eurovizijos" atrankos finalo dalyvia i po rengicio miegoti nėjo - dešimt savaičių drukusio kovos pabaigą atšventė uostamiesčio kokteilių bare "oscar"
ByT5	Šeštadienio vakarą Klaipėdoje surengto "Eurovizijos" atrankos finalo dalyviai po renginio miegotinėjo - dešimt savaičių trukusio kovos pabaigą atšventė uostamiesčio kokteilių bare "Roscar"
Original	300 kg hašišo gabenimo į Lietuvą byla: vienas išteisintas, kitam sušvelninta bausmė
Corrupted	300 kg haši šo gabenimo į Lietuvą byla. vie nas i štsisintas, kitam sušverlninta buasmė
T5	300 kg hašišo gabenimo į Lietuvą byla: vienas išteisintas, kitam sušvelninta bausmė

References

1. Alikaniotis, D., Raheja, V.: The unreasonable effectiveness of transformer language models in grammatical error correction. In: Proceedings of the Fourteenth Workshop on Innovative Use of NLP for Building Educational Applications, pp. 127–133. Association for Computational Linguistics, Florence, August 2019. https://doi.org/10.18653/v1/W19-4412, https://aclanthology.org/W19-4412

2. Brockett, C., Dolan, W.B., Gamon, M.: Correcting ESL errors using phrasal SMT techniques. In: Proceedings of the 21st International Conference on Computational Linguistics and 44th Annual Meeting of the Association for Computational Linguistics, pp. 249–256. Association for Computational Linguistics, Sydney, July 2006. https://doi.org/10.3115/1220175.1220207, https://aclanthology.org/P06-1032
3. Bryant, C., Briscoe, T.: Language model based grammatical error correction without annotated training data. In: Proceedings of the Thirteenth Workshop on Innovative Use of NLP for Building Educational Applications, pp. 247–253. Association for Computational Linguistics, New Orleans, June 2018. https://doi.org/10.18653/v1/W18-0529, https://aclanthology.org/W18-0529
4. Bryant, C., Felice, M., Briscoe, T.: Automatic annotation and evaluation of error types for grammatical error correction. In: Proceedings of the 55th Annual Meeting of the Association for Computational Linguistics (Volume 1: Long Papers), pp. 793–805. Association for Computational Linguistics, Vancouver, July 2017. https://doi.org/10.18653/v1/P17-1074, https://aclanthology.org/P17-1074
5. Dadurkevičius, V.: Assessment data of the dictionary of modern lithuanian versus joint corpora (2020). http://hdl.handle.net/20.500.11821/36. CLARIN-LT digital library in the Republic of Lithuania
6. Ge, T., Wei, F., Zhou, M.: Reaching human-level performance in automatic grammatical error correction: an empirical study. arXiv preprint arXiv:1807.01270 (2018)
7. Grundkiewicz, R., Junczys-Dowmunt, M.: Near human-level performance in grammatical error correction with hybrid machine translation. In: Proceedings of the 2018 Conference of the North American Chapter of the Association for Computational Linguistics: Human Language Technologies, Volume 2 (Short Papers), pp. 284–290. Association for Computational Linguistics, New Orleans, June 2018. https://doi.org/10.18653/v1/N18-2046, https://aclanthology.org/N18-2046
8. Junczys-Dowmunt, M., Grundkiewicz, R.: Phrase-based machine translation is state-of-the-art for automatic grammatical error correction. In: Proceedings of the 2016 Conference on Empirical Methods in Natural Language Processing, pp. 1546–1556. Association for Computational Linguistics, Austin, November 2016. https://doi.org/10.18653/v1/D16-1161, https://aclanthology.org/D16-1161
9. Klimas, A.: Some unique features of Lithuanian. Lituanus **30**(3), 51–64 (1984)
10. Kudo, T., Richardson, J.: SentencePiece: a simple and language independent subword tokenizer and detokenizer for neural text processing. In: Proceedings of the 2018 Conference on Empirical Methods in Natural Language Processing: System Demonstrations, pp. 66–71 (2018)
11. Náplava, J., Straka, M., Straková, J., Rosen, A.: Czech grammar error correction with a large and diverse corpus. arXiv preprint arXiv:2201.05590 (2022)
12. Ng, H.T., Wu, S.M., Briscoe, T., Hadiwinoto, C., Susanto, R.H., Bryant, C.: The CoNLL-2014 shared task on grammatical error correction. In: Proceedings of the Eighteenth Conference on Computational Natural Language Learning: Shared Task, pp. 1–14. Association for Computational Linguistics, Baltimore, June 2014. https://doi.org/10.3115/v1/W14-1701, https://aclanthology.org/W14-1701
13. Omelianchuk, K., Atrasevych, V., Chernodub, A., Skurzhanskyi, O.: GECToR - grammatical error correction: tag, not rewrite. In: Proceedings of the Fifteenth Workshop on Innovative Use of NLP for Building Educational Applications, pp. 163–170. Association for Computational Linguistics, Seattle, July 2020. https://doi.org/10.18653/v1/2020.bea-1.16, https://aclanthology.org/2020.bea-1.16

14. Ortiz Suárez, P.J., Sagot, B., Romary, L.: Asynchronous pipeline for processing huge corpora on medium to low resource infrastructures. In: Bański, P., et al. (eds.) 7th Workshop on the Challenges in the Management of Large Corpora (CMLC-7). Cardiff, United Kingdom, July 2019. https://doi.org/10.14618/IDS-PUB-9021, https://hal.inria.fr/hal-02148693
15. Raffel, C., et al.: Exploring the limits of transfer learning with a unified text-to-text transformer. J. Mach. Learn. Res. **21**(140), 1–67 (2020). http://jmlr.org/papers/v21/20-074.html
16. Rajpurkar, P., Jia, R., Liang, P.: Know what you don't know: unanswerable questions for SQuAD. In: Proceedings of the 56th Annual Meeting of the Association for Computational Linguistics (Volume 2: Short Papers), pp. 784–789. Association for Computational Linguistics, Melbourne, July 2018. https://doi.org/10.18653/v1/P18-2124, https://aclanthology.org/P18-2124
17. Rimkutė, E.: Morfologinio daugiareikšmiškumo ribojimas kompiuteriniame tekstyne [Morphological Disambiguation of the Corpus of Lithuanian Language]. Ph.D. thesis, Vytautas Magnus University, Kaunas (2006). https://etalpykla.lituanistikadb.lt/object/LT-LDB-0001:E.02~2006~1367155963435/E.02~2006~1367155963435.pdf
18. Rothe, S., Mallinson, J., Malmi, E., Krause, S., Severyn, A.: A simple recipe for multilingual grammatical error correction. In: Proceedings of ACL-IJCNLP (2021)
19. Shah, K., de Melo, G.: Correcting the autocorrect: context-aware typographical error correction via training data augmentation. In: Proceedings of the 12th Language Resources and Evaluation Conference, pp. 6930–6936. European Language Resources Association, Marseille, May 2020. https://aclanthology.org/2020.lrec-1.856
20. Shazeer, N., Stern, M.: Adafactor: adaptive learning rates with sublinear memory cost. In: Dy, J., Krause, A. (eds.) Proceedings of the 35th International Conference on Machine Learning. Proceedings of Machine Learning Research, vol. 80, pp. 4596–4604. PMLR, 10–15 Jul 2018. https://proceedings.mlr.press/v80/shazeer18a.html
21. Stankevičius, L., Lukoševičius, M.: Generating abstractive summaries of lithuanian news articles using a transformer model. In: Lopata, A., Gudonienė, D., Butkienė, R. (eds.) ICIST 2021. CCIS, vol. 1486, pp. 341–352. Springer, Cham (2021). https://doi.org/10.1007/978-3-030-88304-1_27
22. Stankevičius, L., Lukoševičius, M., Kapočiūtė-Dzikienė, J., Briedienė, M., Krilavičius, T.: Correcting diacritics and typos with byt5 transformer model. arXiv preprint arXiv:2201.13242 (2022)
23. Tamulionienė, A., et al.: Būdingiausios rašybos klaidos mokiniu rašiniuose ir tinklaraščiuose. Bendrinė kalba (iki 2014 metu-Kalbos kultūra) (88), 1–24 (2015)
24. Vaswani, A., et al.: Attention is all you need. In: Guyon, I., et al. (eds.) Advances in Neural Information Processing Systems, vol. 30. Curran Associates, Inc. (2017). https://proceedings.neurips.cc/paper/2017/file/3f5ee243547dee91fbd053c1c4a845aa-Paper.pdf
25. Vaswani, A., et al.: Attention is all you need. In: Proceedings of the 31st International Conference on Neural Information Processing Systems, NIPS 2017, pp. 6000–6010. Curran Associates Inc., Red Hook (2017)
26. Wang, A., et al.: SuperGLUE: a stickier benchmark for general-purpose language understanding systems. In: Wallach, H., Larochelle, H., Beygelzimer, A., d'Alché-Buc, F., Fox, E., Garnett, R. (eds.) Advances in Neural Information Processing Systems, vol. 32. Curran Associates, Inc. (2019). https://proceedings.neurips.cc/paper/2019/file/4496bf24afe7fab6f046bf4923da8de6-Paper.pdf

27. Wang, A., Singh, A., Michael, J., Hill, F., Levy, O., Bowman, S.: GLUE: a multi-task benchmark and analysis platform for natural language understanding. In: Proceedings of the 2018 EMNLP Workshop BlackboxNLP: Analyzing and Interpreting Neural Networks for NLP, pp. 353–355. Association for Computational Linguistics, Brussels, November 2018. https://doi.org/10.18653/v1/W18-5446, https://aclanthology.org/W18-5446
28. Wolf, T., et al.: Transformers: state-of-the-art natural language processing. In: Proceedings of the 2020 Conference on Empirical Methods in Natural Language Processing: System Demonstrations, pp. 38–45. Association for Computational Linguistics, October 2020. https://doi.org/10.18653/v1/2020.emnlp-demos.6, https://aclanthology.org/2020.emnlp-demos.6
29. Xie, Z., Avati, A., Arivazhagan, N., Jurafsky, D., Ng, A.Y.: Neural language correction with character-based attention. arXiv preprint arXiv:1603.09727 (2016)
30. Xue, L., et al.: Byt5: towards a token-free future with pre-trained byte-to-byte models (2021)
31. Xue, L., et al.: mT5: a massively multilingual pre-trained text-to-text transformer. In: Proceedings of the 2021 Conference of the North American Chapter of the Association for Computational Linguistics: Human Language Technologies, pp. 483–498. Association for Computational Linguistics, June 2021. https://doi.org/10.18653/v1/2021.naacl-main.41, https://aclanthology.org/2021.naacl-main.41

A Simulation Modeling Method for Cooling Building Structures by Fire Robots

Irina Pozharkova[1](✉), Ayman Aljarbouh[2], Shahzool Hazimin Azizam[3], Aly Pasha Mohamed[4], Fazle Rabbi[5], and Roman Tsarev[6,7]

[1] FSBEE HE Siberian Fire and Rescue Academy EMERCOM of Russia, 1, Severnaya St., Zheleznogorsk, Russia
pozharkova@mail.ru
[2] University of Central Asia, Naryn, Kyrgyz Republic
[3] Universiti Sultan Zainal Abidin, Kuala Terengganu, Terengganu, Malaysia
[4] Ton Duc Thang University, Ho Chi Minh City 70000, Vietnam
[5] Australian Computer Society, 839, Collins Street, Docklands 3008, Australia
[6] International Academy of Science and Technologies, Moscow, Russia
rtsarev@acadst.org
[7] Siberian Federal University, Krasnoyarsk, Russia

Abstract. The research proposes a method for simulating the cooling of building structures during a fire by a jet of water supplied by a fire robot according to a given algorithm, using a Python script developed as part of the conducted research. A functional model for modeling fire dynamics in rooms with structures protected by fire robots is presented. An algorithm for generating Fire Dynamics Simulator code simulating the motion of a cooling agent jet in the Pyrosim software environment is developed. The results of the simulation of building truss cooling during a fire in a thermal power plant building are presented #CSOC1120.

Keywords: Fire robot · Simulation modeling · Computational Fluid Dynamics · Pyrosim · Python · SimInTech

1 Introduction

Fire-fighting robots are currently one of the most effective means of fire protection, which allows, without the direct presence of people in the zone of fire hazards, extinguishing with a jet of water or other extinguishing agents. At some facilities, the task of fire safety is often complicated by the danger of collapse of supporting building structures, such as metal trusses due to their heating to critical temperatures. In particular, for metal structures at 500 °C, the mechanical strength limit state may occur. In such cases, fire robots are also used for cooling building structures [1].

One of the ways to assess the effectiveness of fire protection means is simulation modeling [2]. In [3] the experience of using Fire Dynamics Simulator (FDS) to assess the effectiveness of protective equipment in fires at thermal power facilities is presented. The simulation of fire dynamics using fire robots is complicated by the need to simulate the

R. Silhavy (Ed.): CSOC 2022, LNNS 502, pp. 504–511, 2022.
https://doi.org/10.1007/978-3-031-09076-9_45

movement of their trunk according to a given algorithm. Most software solutions based on FDS, in particular Pyrosim [4], have a built-in implementation of stationary sources of extinguishing agents, which do not allow with high accuracy to simulate the jet of the robotic trunk moving in its coverage area along the critical heating protected building structures. In particular, "turning on" at a certain moment the supply to the protected metal structure of a cooling agent with a flow rate averaged over its entire area will lead to a uniform decrease in temperature in the corresponding coverage area. At the same time, for large-sized building structures, such a simulation result will differ significantly from the cooling of the jet moving along their surface, at which local overheating may occur, including to temperatures at which the loss of mechanical strength, formation of cracks, and destruction occur, at points significantly distant from the current position of the water supply location. Such a difference in results can be critical in certain situations. Therefore, it is advisable to use methods of simulation modeling of means of cooling of building structures, allowing to estimate their efficiency with high accuracy.

2 Method

To solve this problem, it is proposed to present a model of the movement of the water jet from the robotic nozzle in the Pyrosim software environment as a set of elements simulating the sequential delivery of the extinguishing agent to various points of the cooled building structures by switching on and off at points in time corresponding to the algorithm of the fire robot operation. The functional model for solving the problem of modeling fire dynamics in this approach is presented in Fig. 1.

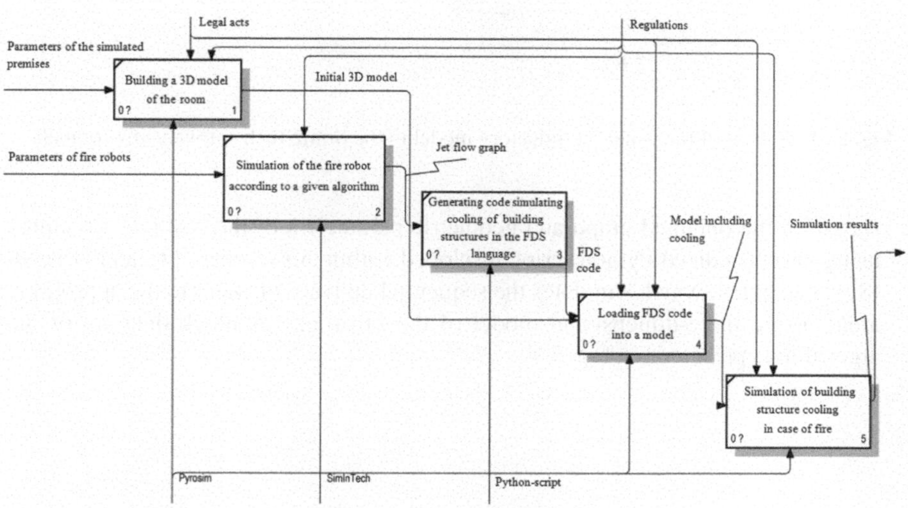

Fig. 1. Functional model of solving the problem of modeling the cooling of building structures using fire robots.

Modeling the movement of the water jet on the surface of building structures in the Pyrosim software environment, according to the presented scheme, can be divided into the following steps:

1. Modeling of water jet movement in SimInTech software environment [5]. Based on the mathematical model [6], describing the trajectory of the water jet movement from the fire monitors (Fig. 2), and the algorithm of the fire robot operation, a schedule of cooling agent supply to various sections of the surface of building structures is calculated - the geometric parameters of the corresponding sections at various points in time are recorded in a file.

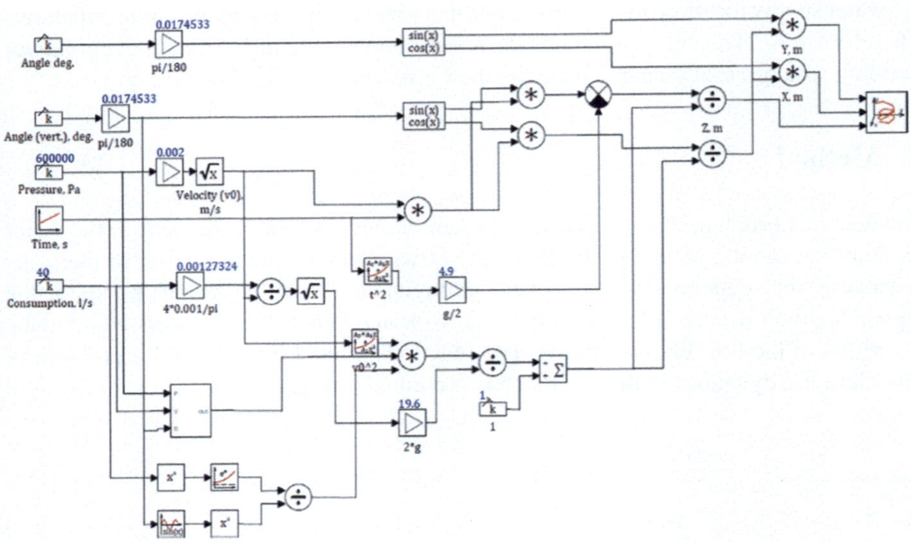

Fig. 2. Fragment of the water jet trajectory model in the SimInTech software environment

2. Based on the obtained graph and geometric parameters of the building structures, using the specialized Python script developed within this research [7], an FDS code [8] is generated, which simulates the sequential delivery of water to the appropriate areas of the three-dimensional model of the fire room. A block diagram of this algorithm is presented in Fig. 3.

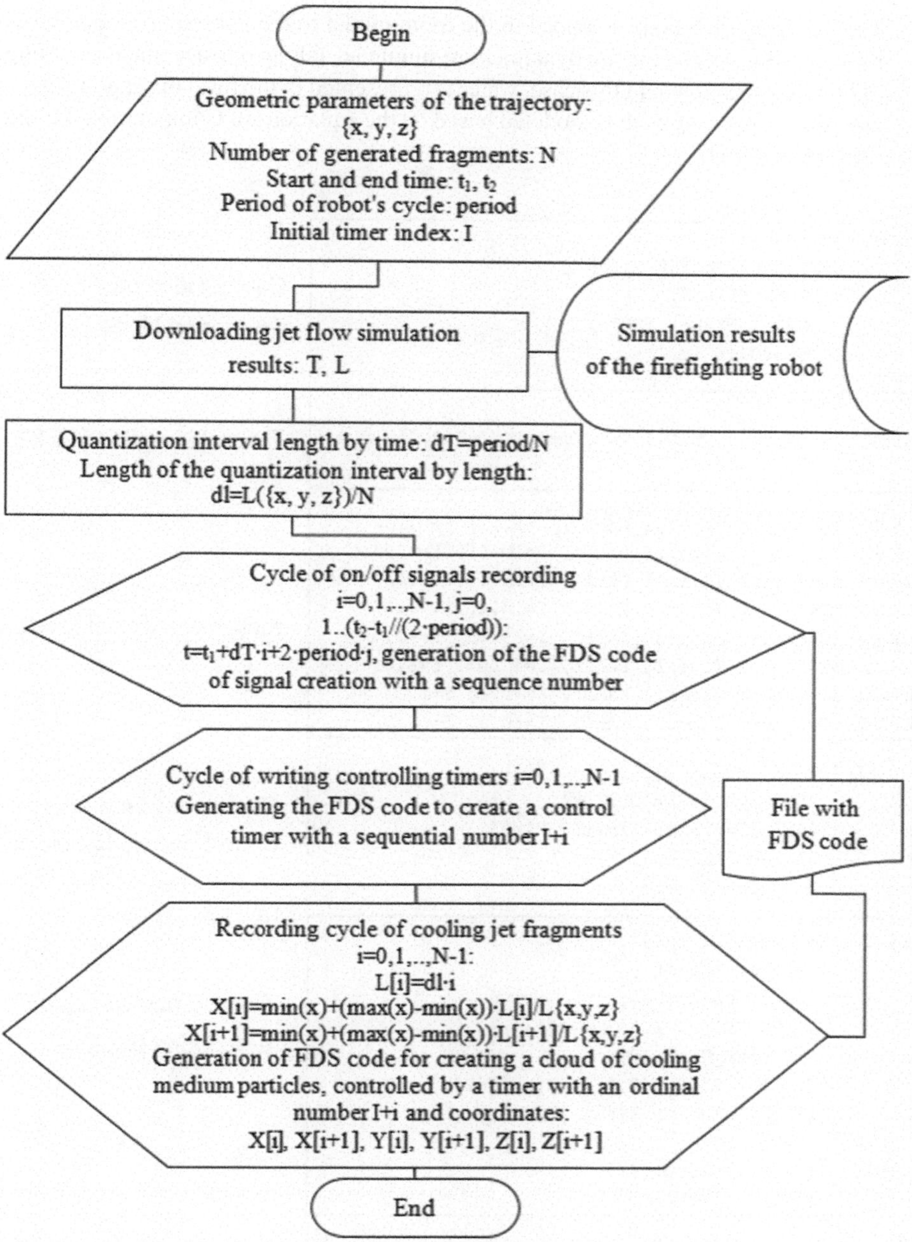

Fig. 3. Block diagram of algorithm of FDS code generation, simulating water jet motion

3. The resulting FDS code is loaded to the room model in the Pyrosim software environment (Fig. 4), and the fire dynamics are simulated, taking into account the cooling of building structures. In this case, water is represented in the form of separate particles, the motion of which is modeled based on the equations of Computational Fluid Dynamics (CFD).

```
&PART ID='Water',
       SPEC_ID='WATER VAPOR',
       DIAMETER=500.0,
       MONODISPERSE=.TRUE.,
       AGE=60.0/
```

Cooling medium description block

```
&DEVC ID='Time', QUANTITY='TIME', XYZ=0.0,0.0,0.0.
```

Timer description block

```
&RAMP ID='Water Timer12_RAMP', T=9.95, F=-1.0/
&RAMP ID='Water Timer12_RAMP', T=10.05, F=1.0/
&RAMP ID='Water Timer12_RAMP', T=10.1, F=1.0/
&RAMP ID='Water Timer12_RAMP', T=10.2, F=-1.0/
&RAMP ID='Water Timer12_RAMP', T=39.95, F=-1.0/
&RAMP ID='Water Timer12_RAMP', T=40.05, F=1.0/
&RAMP ID='Water Timer12_RAMP', T=40.1, F=1.0/
&RAMP ID='Water Timer12_RAMP', T=40.2, F=-1.0/
```

On/off coolant medium supply description block

```
&CTRL ID='Water Timer12',
       FUNCTION_TYPE='CUSTOM',
       RAMP_ID='Water Timer12_RAMP',
       LATCH=.FALSE., I
       NPUT_ID='Time'/
```

Coolant medium supply control description block

```
&INIT ID='Water Cloud12',
       PART_ID='Water',
       N_PARTICLES=1000,
       DT_INSERT=0.15,
       MASS_PER_TIME=40,
       XB=25.0,25.0,1.0,1.28,26.0,26.0,
       CTRL_ID='Water Timer12'/

&INIT ID='Water Cloud13',
       PART_ID='Water',
```

Cooling medium cloud description block

Fig. 4. Fragments of blocks of FDS code simulating the cooling of building structures in the Pyrosim environment

3 Results

To assess the effectiveness of cooling of building trusses by fire robots in the Pyrosim software environment, a model of the engine room of a combined heat and power plant was built (Fig. 5). To protect against overheating of building trusses in the building fire robots based on stationary fire monitors with water fire extinguishing, software and remote control, automatic pointing detector, and camera PR-LSD-S40U-IK-TV [6].

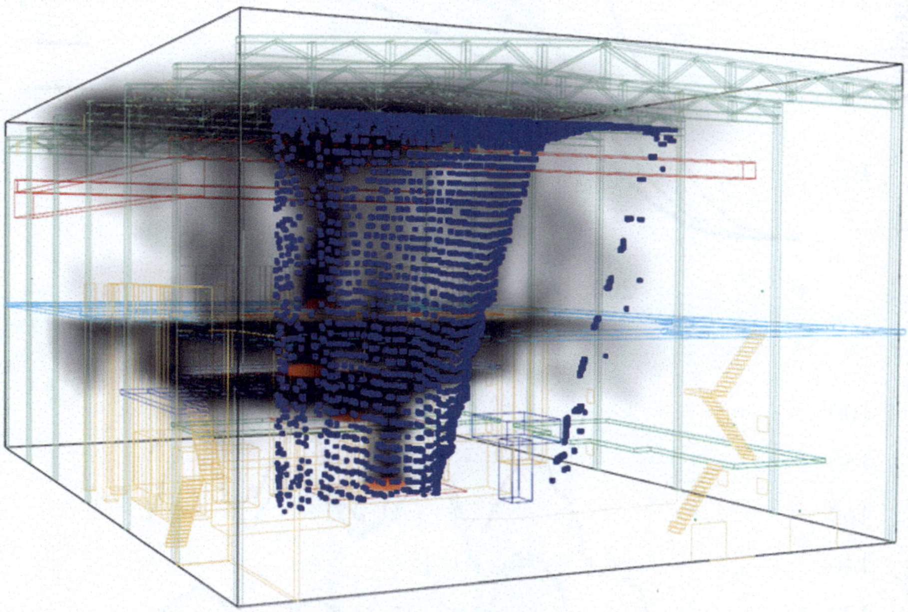

Fig. 5. Fragment of the 3D model of the engine room of the combined heat and power plant in the Pyrosim software environment

The investigated fire scenario is the damage to the oil lubrication system of the TP-100/110-8.8 turbine, oil spill, and its ignition. The total fire area, in this case, is 175 m^2.

Figure 5 shows a visualization of cooling by a fire robot of building trusses, located above the seat of fire in the area of the greatest, according to the results of modeling, heating, and in Fig. 6 shows graphs of temperature sensor readings located on this truss (a - without cooling, b - with cooling).

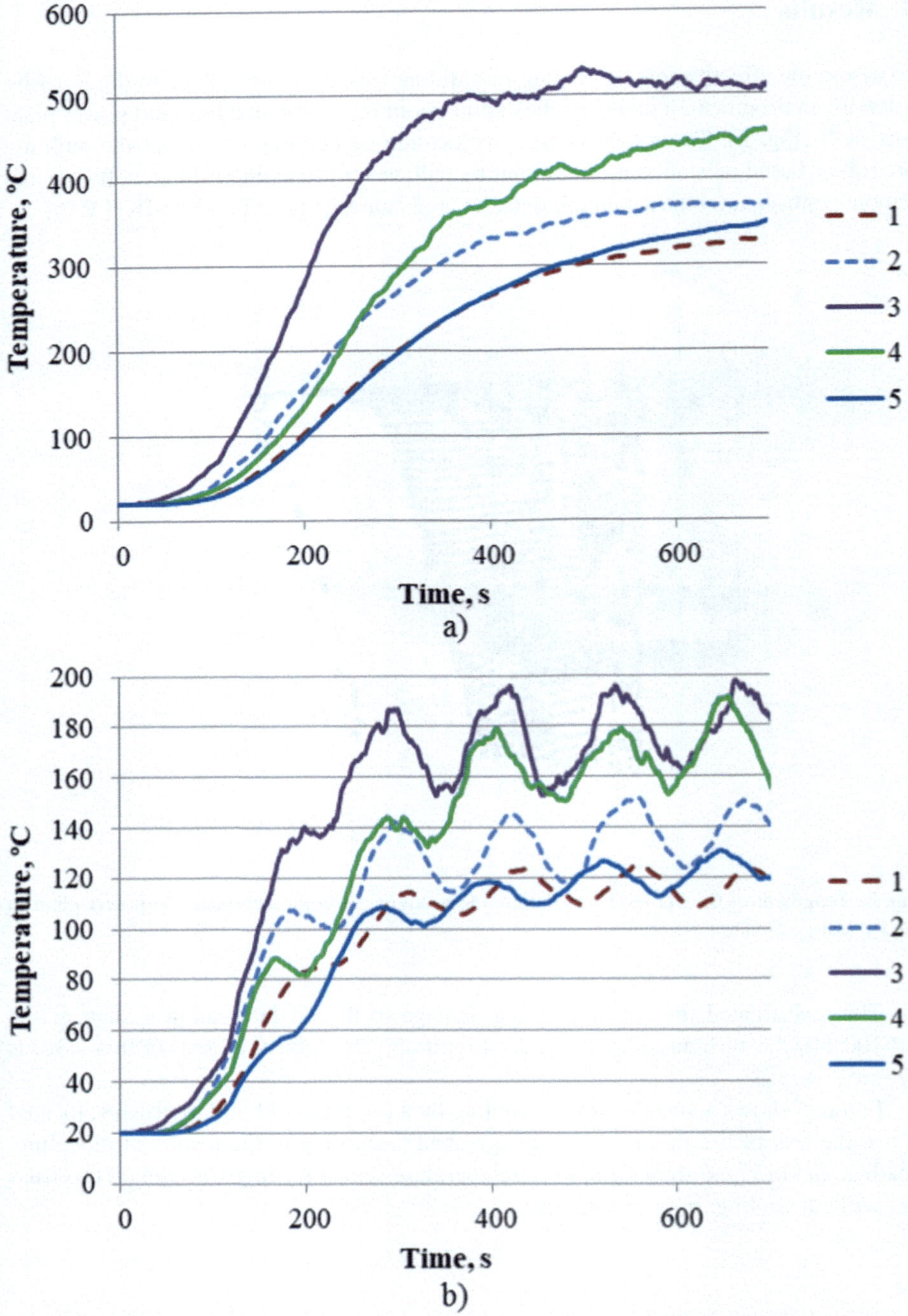

Fig. 6. Summary graphs of temperature sensor readings located on the farm: a) without cooling; b) with cooling.

As can be seen, without cooling, exceeding the critical temperature, the value of which is 500 °C, is observed on the seventh minute of the fire. When using a gun barrel heating above the 198 °C mark during the entire simulation period does not occur, which indicates the effectiveness of the use of robotic fire-extinguishing installation for cooling building trusses in a fire at the object under study.

4 Conclusions

On the graphs of temperature sensor readings, constructed from the results of modeling using simulated cooling according to the method presented in the article (Fig. 6b), there are periodic fluctuations of values corresponding to the fire robot's work cycle. The maximum amplitude of oscillations for sensor 3, located above the hearth of fire, is 47 °C - 24% of maximum temperature value, which is quite significant and in some cases may be critical in terms of loss as a result of heating the mechanical strength of building structures. Therefore, to assess the efficiency of cooling of building structures by a water jet of robotic shafts based on simulation modeling in FDS-based software tools, in particular Pyrosim, it is advisable to use methods that take into account the algorithm of robotic shafts.

The method presented in the research can also be used to simulate fire extinguishing with fire robots using various extinguishing agents, the motion, and interaction with the environment which is modeled based on the corresponding equations of computational fluid dynamics.

References

1. Pozharkova, I.: Construction metal structures simulation cooling by fire robots. In: AIP Conference Proceedings, vol. 2402, p. 050021 (2021). https://doi.org/10.1063/5.0071434
2. Zenyutkin, N.V., Kovalev, D.I., Tuev, E.V., Tueva, E.V.: On the ways of forming information structures for modeling objects, environments and processes. Mod. Innov. Syst. Technol. **1**(1), 10–22 (2021). https://doi.org/10.47813/2782-2818-2021-1-1-10-22
3. Mun, S.Y., Hwang, C.H., Park, J.S., Do, K.: Validation of FDS for predicting the fire characteristics in the multi-compartments of nuclear power plant (Part I: Over-ventilated fire condition). Fire Sci. Eng. **27**(2), 31–39 (2013)
4. Jones, R.T., Erwee, M.W.: Simulation of ferroalloy smelting in DC arc furnaces using Pyrosim and FactSage. Calphad **55**, 20–25 (2016)
5. Domakhin, E.A., Spopov, N., Vilberger, M.E., Anibroev, V.I., Singizin, I.I.: Comparative analysis and experimental verification of simulation modelling approach in MATLAB-Simulink and SimInTech. In: Journal of Physics: Conference Series, vol. 1661, no. 1, p. 012030 (2020)
6. Pozharkova, I.: Modelling the water jet trajectory of a robotic fire monitor in the SimInTech dynamic modelling environment. In: Silhavy, R., Silhavy, P., Prokopova, Z. (eds.) CoMeSySo 2020. AISC, vol. 1295, pp. 837–844. Springer, Cham (2020). https://doi.org/10.1007/978-3-030-63319-6_77
7. Vogelsang, J.: Python+ FDS: Vorstellung des FDSReader-packages zum Einlesen von FDS-Simulationsdaten in Python. Zivile Sicherheitsforschung (2021)
8. Ryder, N.L., Sutula, J.A., Schemel, C.F., Hamer, A.J., Van Brunt, V.: Consequence modeling using the fire dynamics simulator. J. Hazard. Mater. **115**(1–3), 149–154 (2004)

Natural Language Processing Technology for Spatial Information Recognition and Visualization

A. V. Vicentiy[1,2](✉)

[1] Institute for Informatics and Mathematical Modeling – Subdivision of the Federal Research Centre "Kola Science Centre of the Russian Academy of Science", 24A, Fersman st., 184209 Apatity, Murmansk Region, Russia
alx_2003@mail.ru

[2] Apatity Branch of Murmansk Arctic State University, Lesnaya st. 29, 184209 Apatity, Murmansk Region, Russia

Abstract. This paper is devoted to the problem of natural language texts processing for spatial information recognition. The paper describes the technology of recognizing named entities and extracting information about the spatial connectivity of geographical objects for natural language texts. A hybrid approach is used to process the texts. The hybrid approach allows to combine the capabilities of neural network approach, rule-based approach and lexico-semantic pattern-based approach for text processing within a single information technology. The peculiarity of the proposed technology is the original way to form and take into account the context for the recognition of named entities and spatial relationships of geographical objects. Also, this paper describes a way of creating and using lexico-semantic patterns, taking into account the peculiarities of language and subject area. This work is a development of technology for recognition and extraction of geo-attributed entities from natural language texts and the groundwork for creating cognitive geovisualization technology, taking into account the specifics of information perception by the end user. One of the prospects of the considered hybrid technology is its use as one of the main elements of cognitive geographical user interface.

Keywords: Natural language processing · Named entity recognition · Spatial relationship recognition · Lexical semantic patterns · Geovisualization · Text-to-map conversion · Automatic map generation · Based map generation

1 Introduction

The number of Internet and social network users has increased significantly in recent years. Many Internet audience studies note that in 2020–2021 the number of new users of various online services and applications grew extremely fast. Some researchers associate this fact with the Covid-19 pandemic. The increase in the number of Internet users is one important reason for the multifold increase in the number of natural-language texts

R. Silhavy (Ed.): CSOC 2022, LNNS 502, pp. 512–520, 2022.
https://doi.org/10.1007/978-3-031-09076-9_46

in the global information space. In turn, an increase in the number of available texts contributes to the attractiveness of these texts as a data source for automatic analysis and extraction of data and knowledge. Under these circumstances, developing new and improving existing natural language processing (NLP) methods is an urgent challenge [1, 2].

Currently, there are a huge number of texts of various subjects in different languages freely available. As a rule, these are unstructured or weakly structured texts in natural language, most of which are explicitly or implicitly associated with geodata, for instance, has a geographical reference, geotags, or contains a reference to geographical objects [3] There is a large number of studies aimed at the recognition of geographical information in natural-language texts [4–7]. However, the vast majority of them are focused on finding specific geographical objects, such as toponyms. At the same time, insufficient attention is paid to the problem of recognition and retrieval of information about geographical connectivity of objects [8, 9].

All existing methods of natural language texts processing for the named entities recognition can be classified into three groups: 1) methods based on the use of neural networks and machine learning; 2) methods based on the use of rules and gazetteers (special dictionaries); 3) hybrid methods based on the simultaneous use of machine learning and rules [10, 11, 12].

In this paper we propose a technology for processing natural language texts based on a hybrid approach. The technology includes several successive stages of text processing. In the first stage we use methods based on the use of neural networks and machine learning to recognize toponyms in texts. In the second stage, we use methods based on the use of rules and gazetteers to recognize complex toponyms that were not recognized in the first stage, and potential spatial relationships of geographical objects. The second stage of the analysis allows to increase the recognition accuracy for complex toponyms, as well as to identify lexical constructions that may be descriptions of spatial relationships of geographical objects, specific to the subject area. At the third stage we use specially developed lexical-semantic patterns to analyze the text. These patterns allow to take into account the peculiarities of the Russian language. The development of lexical-semantic patterns is a complex and time-consuming process. However, a good set of lexical-semantic patterns allows to recognize in texts not only toponyms and their relations, but also attributes of relations. In addition, taking into account the peculiarities of the subject area when developing lexical-semantic patterns has a positive effect on the final result of recognition [13–15].

To visualize the recognized spatial information, we use the capabilities of modern cloud geographic information systems and geoservices [16, 17]. The stages of geocoding and geo-visualization using GeoPy, ArcGis, Yandex.Maps and Google Maps web-services are described in more detail in our previous works [18, 19].

2 Materials and Methods

The goal of hybrid technology is to recognize geoattributed entities in text and visualize them in the form of schematic maps. We distinguish two types of geoattributed entities. The first kind is toponyms and other geographical named entities. The second type is spatial relations between several geographic named entities, or more specifically, descriptions of spatial relations in text. Existing natural language processing tools and frameworks are good at solving the problem of recognizing named entities of the first kind. We tested the capabilities of natural language processing tools such as SpaCy [20], Google Cloud NLP [21], Natasha [22], and AllenNLP [23]. All of these tools showed a high percentage (above 95%) of recognition of geographical named entities (toponyms). However, none of them was able to recognize spatial relationships between geographical named entities.

To recognize geographic named entities in our hybrid technology, we have chosen the library Natasha, which implements a neural network approach. This library is implemented based on the distillation of the BERT (Bidirectional Encoder Representations from Transformers) language model from Google [24], which is one of the most efficient models for natural language processing. But despite the fact that Natasha provides the recognition of about 98% of named entities for Russian-language texts, Natasha, like other frameworks cannot recognize the spatial relationships between geographical named entities. Therefore, the second step of the hybrid technology is lexico-syntactic analysis.

As the main tool for lexico-syntactic analysis, we chose Yargy parser [25], which is based on the Earley algorithm [26]. We chose this parser because it has a large set of ready-made rules to extract different kinds of named entities for the Russian language. This parser also allows supplementing the ready-made rules and creating our own rules with context-free grammars and dictionaries. The rule-based approach complements the neural network approach and provides more efficient recognition of named entities, increasing the recall and precision of the recognition. In addition, the use of the parser allows to take into account some peculiarities of the Russian language with the help of the rules. In particular, such features include free word order, rich inflection, and derivation [27, 28].

One of the main problems of the rule-based approach to text analysis is the development of complex, cumbersome, confusing grammars. These grammars are difficult to modify. Finding and correcting errors requires considerable effort and specialized knowledge from the user. In addition, even for a well-defined subject area, the gazetteers can be quite large, and keeping the gazetteers up to date requires additional costs. To some extent, these problems can be solved by applying lexical-semantic patterns to text analysis [29, 30]. We apply lexical-semantic patterns in the third stage of hybrid technology to recognize lexical constructions that describe spatial relations.

In the context of solving the problems of natural language processing, a lexical pattern is understood as a declarative structure, a certain structural pattern of a language construction that represents its lexical and syntactic properties. In other words, a lexical-semantic pattern is a description of some semantic phenomenon in the form of some "stable phrase" or "language formula" [31]. In our case, such a semantic phenomenon is a description of the spatial connection of geographical objects (toponyms).

A lexico-semantic pattern is defined as a structural pattern of the target linguistic construction with a certain composition and lexico-semantic properties. If the pattern is successfully matched with a fragment of the analyzed text, a lexical object is formed to which formal and semantic (e.g., class and properties) characteristics can be ascribed. Thus, a lexical-semantic pattern consists of a logical structure and a semantic description. Lexical-semantic patterns are characteristic expressions (phrases), constructions from certain language elements (collocations) and allow to build a semantic model corresponding to the text to which it applies [32].

We used a corpus of texts on Arctic logistics as material for our analysis. We have been collecting this corpus for the past three years, starting with the work [33] in which we described Arctic logistics projects and programs for the development of Russia's Arctic regions. The core of the corpus consists of texts describing the transport infrastructure of the Arctic zone of the Russian Federation. The corpus includes texts of different content and style, including texts of official documents, such as strategies for the development of the Murmansk region and the Arctic regions. The corpus also includes texts obtained from the official websites of the executive authorities. The corpus also includes a certain amount of texts obtained from social networks, as well as texts from the sites of local forums and sites for communication of local residents.

3 Results

Our experiments have shown that the use of the library Natasha allows to recognize about 96–99% of geographical named entities in Russian-language texts. For news texts the recognition rate is usually higher than for texts of official documents and texts written by users for social networks.

Thus, in the first step of text analysis we get a list of recognized geographical entities (toponyms) and information about their location in the text. However, this is not enough to recognize the spatial relationships between geographical entities. To solve this problem we propose to take into account the context of the recognized entity. This means that we store in a special data structure the sentence in which a geographical named entity was detected, as well as the neighborhood of this sentence. By neighborhood, we mean the sentences to the right and left of the sentence in which the named entity was detected. Within our technology, we call such data structures text blocks. The size of a neighborhood is specified as one of the parameters of the named entity recognition procedure as a pair of numbers like (l_n; r_n). The number of sentences that are included in the neighborhood and are to the left of the sentence containing the geographical named entity is given by l_n. The number of sentences that are included in the neighborhood and to the right of the sentence containing the geographic named entity is specified using r_n. A fragment of the Russian-language text (translated into English) with recognized geographic entities is shown in Fig. 1.

A new asphalt road will be built in 2021 as part of the federal program for the development of transport infrastructure. It will connect Murmansk with the village of Belokamenka. The federal program was approved by the Government of the Russian Federation in December 2017.

– geographical named entities (toponyms)
– spatial relationships
– spatial relationship attributes

Fig. 1. The first step of geographical named entities recognition.

It is important to note that one of the problems in recognizing spatial relations is related to the fact that references to a spatial relation and to geographical objects that are connected by this spatial relation may be located in different sentences. Most often, but not always, they are adjacent sentences. Empirically, we have found that the optimal values of neighborhood size for the "default setting" are (1; 1). In other words, most often the neighborhood of a sentence containing a geographic named entity consists of one sentence located in the text to its left and one sentence located in the text to its right. For example, for a fragment of the text shown in Fig. 1, for a given neighborhood (1; 1) the text block will include the whole fragment: "A new asphalt road will be built in 2021 as part of the federal program for the development of transport infrastructure. It will connect Murmansk with the village of Belokamenka. The federal program was approved by the Government of the Russian Federation in December 2017." It is easy to see that a sentence to the right of a sentence containing a geographical named entity is absolutely useless for recognizing the spatial relationship between geographical entities, i.e. it is superfluous or redundant. Unfortunately, we cannot predict in advance the optimal values of l_n and r_n for all possible text blocks. Therefore we assume that to increase the completeness of spatial relations recognition, it is better to include an extra sentence in the text block than to increase the risk of losing information about the spatial relationship of entities.

At the second stage of text processing the entire set of text blocks formed at the first stage is processed by the parser. Due to the fact that the second stage does not process the entire text, but only the set of generated text blocks, a certain number of redundant sentences does not affect the efficiency and speed of text processing too much.

Using the rule-based approach implemented in Yargy parser allows us to solve several important problems. One of them is to improve the precision and recall of the recognition of named entities through the use of domain-specific grammars and gazetteers. The second task is the recognition of key ("anchor") words, usually used in Russian texts to denote the spatial relationships of objects. In addition to direct recognition of spatial relationships in most cases we can determine it type. To determine the type of relation we additionally use the program module, which allows to refer the recognized relation to one of the classes of the developed ontology of spatial relations. A fragment of this ontology is shown in Fig. 2.

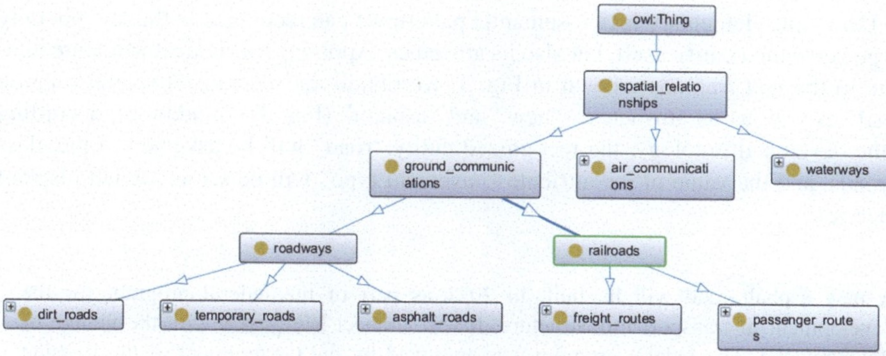

Fig. 2. The fragment of the ontology of spatial relations.

The parser analyzes each text block separately. At this step it is possible to recognize most of the spatial relations. But for some complex text blocks, where the description of the spatial relation in the text is implicit, it is impossible to solve the problem of spatial relation recognition by parsing. Such complex text blocks are stored in a separate subset and processed in the third step with lexico-semantic patterns. Also, a set of complex blocks is used for further analysis by experts to modify and develop new rules and lexico-semantic patterns. In this way, we can improve the efficiency of natural language processing technology for the recognition and visualization of spatial information in the future.

As mentioned above, we use handcrafted lexico-semantic patterns for complex cases. When creating lexico-semantic patterns, we use our knowledge of the features of the Russian language, the most characteristic "meta-patterns" of the language, as well as the framework of valences of different parts of speech and specific lexical units [34]. In addition, the handcrafted lexico-semantic patterns, as a rule, allow us to take into account the peculiarities of the subject area.

To create patterns, we use a modified version of the LSPL (Lexico-Syntactic Pattern Language) syntax, which has been greatly simplified by taking into account the peculiarities of the domain [35]. At the first step, the patterns are described as formal expressions in terms of set theory. After creation and verification, the patterns are converted into XML format for automatic computer processing. A simple example of a lexico-semantic pattern for extracting information about geoattributed entities is a pattern that describes a syntactic construction represented by several consecutive adjectives and a noun group (or a single noun). Such a pattern can be described as:

$LSP1 = \{A\_or\_P\}\ N(\{aw\})\ <A = N, P = N, A = P>$, where
LSP1 – the name of the pattern;
$\{A\_or\_P\}$ – the set of consecutive adjectives (the number of adjectives in this case can be any);
$N(\{aw\})$ – a noun (or noun group) from the set of "anchor words", which are a subset of keywords corresponding to the subject area;
$<A = N, P = N, A = P>$ - operator of grammatical matching of parts of speech, taking into account the peculiarities of word formation in Russian.

Using this elementary lexico-semantic pattern we can recognize in the text not only the geoattributed entity itself, but also its attributes. Applying this lexical-semantic template to the text fragment shown in Fig. 1, we obtain the recognized spatial relation "road" as well as its attributes - "new" and "asphalt" (Fig. 3). In addition, according to the developed ontology, the recognized entity "road" will be assigned to the class of roads, and the value of the attribute "pavement type" will be set as asphalt (asphalt concrete).

A new asphalt road will be built in 2021 as part of the federal program for the development of transport infrastructure. It will connect Murmansk with the village of Belokamenka. The federal program was approved by the Government of the Russian Federation in December 2017.

– geographical named entities (toponyms)
– spatial relationships
– spatial relationship attributes

Fig. 3. The result of applying the lexico-semantic pattern.

When the text analysis stage is complete and the geoattributed entities and their attributes are recognized, the process of mapping them is easy. To generate geoimages (schematic maps), we geocode the recognized entities using the software module developed in previous works that uses the capabilities of the GeoPy service. The result of this module is a list of geoattributed entities and coordinates. The list is generated in csv and geojson formats. Then, using this list of coordinates and ArcGIS API we generate a geoimage (schematic map) that we can work with interactively or export as a digital map. The developed software modules and the process of automatic geovisualization are described in more detail in our papers [36, 37].

4 Conclusions and Future Work

The hybrid technology of geoattributed entities recognition described in this paper allows to recognize not only point and area geographical objects, but also spatial relations between it. One of the features of the considered technology is the original way of forming the text blocks, which allow to take into account the context for the recognition of named entities and spatial relations between geographical objects. Consistent use of three different approaches (neural network, rule-based, lexico-semantic pattern-based) for recognition of geographical named entities allows increasing the accuracy and recall of recognition of both toponyms and geographical relationships.

We believe that this hybrid technology or its parts can be used as elements of multi-subject intelligent information systems or decision support systems. Some prospects of technology development we associate with taking into account the peculiarities of user's perception when working with geoinformation. In other words, we believe that the technology discussed in this paper can be used to create a cognitive user interface for working with interactive digital maps.

References

1. Agarwal, M.: An overview of natural language processing. Int. J. Res. Appl. Sci. Eng. Technol (IJRASET) **7**, 2811–2813 (2019)
2. Khurana, D., et al.: Natural Language Processing: State of the Art, Current Trends and Challenges, pp. 1–12 (2017). https://arxiv.org/ftp/arxiv/papers/1708/1708.05148.pdf
3. Hahmann, S., Burghardt, D.: How much information is geospatially referenced? Networks and cognition. Int. J. Geogr. Inf. Sci. **27**, 1171–1189 (2013)
4. Purves, R.S., Clough, P., Jones, C.B., Hall, M.H., Murdock, V.: Geographic information retrieval: progress and challenges in spatial search of text. Found. Trends Inf. Retrieval **12**, 164–318 (2018)
5. Stock, K.: Mining location from social media: a systematic review (2018). www.elsevier.com/locate/ceus. https://doi.org/10.1016/j.compenvurbsys.2018.05.007
6. Campelo, E.C., Baptista, C.S.: Mining influential terms for toponym recognition and resolution. Revista Brasileira de Cartografia **68**, 1123–1132 (2016)
7. Vicentiy, A.V., Dikovitsky, V.V., Shishaev, M.G.: Automated extraction and visualization of spatial data obtained by analyzing texts about projects of arctic transport logistics development. In: Silhavy, R., Silhavy, P., Prokopova, Z. (eds.) CoMeSySo 2019 2019. AISC, vol. 1046, pp. 419–433. Springer, Cham (2019). https://doi.org/10.1007/978-3-030-30329-7_37
8. Hall, M.M., Jones, C.B.: Generating geographical location descriptions with spatial templates: a salient toponym driven approach. Int. J. Geogr. Inf. Sci. (2021). https://doi.org/10.1080/13658816.2021.1913498
9. Vicentiy, A.V., Shishaev, M.G.: The technology of spatial relations visualization based on the analysis of natural language texts. In: Silhavy, R., Silhavy, P., Prokopova, Z. (eds.) CoMeSySo 2021. LNNS, vol. 232, pp. 971–980. Springer, Cham (2021). https://doi.org/10.1007/978-3-030-90318-3_78
10. Sun, P., Yang, X., Zhao, X., Wang, Z.: An overview of named entity recognition. In: Proceedings of the 2018 International Conference on Asian Language Processing, IALP 2018, pp. 273–278 (2019). https://doi.org/10.1109/IALP.2018.8629225
11. Goyal, A., Gupta, V., Kumar, M.: Recent named entity recognition and classification techniques: a systematic review. Comput. Sci. Rev. **29**, 21–43 (2018). https://doi.org/10.1016/J.COSREV.2018.06.001
12. Nadeau, D., Sekine, S.: A survey of named entity recognition and classification. Lingvisticae Investig. **30**, 3–26 (2007). https://doi.org/10.1075/li.30.1.03nad
13. Mozharova, V.A., Loukachevitch, N.V.: Combining knowledge and CRF-based approach to named entity recognition in Russian. In: Ignatov, D.I., et al. (eds.) AIST 2016. CCIS, vol. 661, pp. 185–195. Springer, Cham (2017). https://doi.org/10.1007/978-3-319-52920-2_18
14. Vicentiy, A.V.: The geoimage generation method for decision support systems based on natural language text analysis. In: Silhavy, R. (ed.) CSOC 2021. LNNS, vol. 230, pp. 609–619. Springer, Cham (2021). https://doi.org/10.1007/978-3-030-77442-4_51
15. Bogacheva, E., Puchkovskaia, A., Smetannikov, I.: Named entity recognition for Russian historical texts. In: ACM International Conference Proceeding Series, pp. 13–17. Association for Computing Machinery, New York (2019). https://doi.org/10.1145/3374587.3374637
16. ArcGis official. https://www.esri.com/en-us/arcgis/products/arcgis-pro/overview
17. Qgis official. https://qgis.org
18. Pilecki, B.M., Vicentiy, A.V.: Development of a method for extracting spatial data from texts for visualization and information decision-making support for territorial management. In: IOP Conference Series: Earth and Environmental Science. Institute of Physics Publishing (2020). https://doi.org/10.1088/1755-1315/539/1/012087

19. Vicentiy, A.V., Vicentiy, I.V.: The method of dynamic visualization of spatial data for cognitive interfaces of information systems supporting regional management. In: 19th International Multidisciplinary Scientific Geoconference, SGEM 2019, pp. 667–672 (2019). https://doi.org/10.5593/sgem2019/2.1/S07.087
20. Official SpaCy. https://spacy.io
21. Official Google Cloud NLP. https://cloud.google.com/natural-language
22. Official Natasha. https://natasha.github.io/
23. Official AllenNLP. https://allenai.org/allennlp
24. Devlin, J., Chang, M., Lee, K., Toutanova, K.: BERT: pre-training of deep bidirectional transformers for language understanding. In: Proceedings of the 2019 Conference of the North American Chapter of the Association for Computational Linguistics: Human Language Technologies (Long and Short Papers), vol. 1, pp. 4171–4186 (2019)
25. Official Yargy parser. https://github.com/natasha/yargy
26. Earley, J.: An efficient context-free parsing algorithm. Commun. ACM **13**, 94–102 (1970)
27. Cabrera-Diego, L.A., Moreno, J.G., Doucet, A.: Using a frustratingly easy domain and tagset adaptation for creating Slavic named entity recognition systems. In: Proceedings of the 8th BSNLP Workshop on Balto-Slavic Natural Language Processing, BSNLP 2021 - Co-located with 16th European Chapter of the Association for Computational Linguistics EACL 2021, pp. 98–104 (2021)
28. Piskorski, J., et al.: The second cross-lingual challenge on recognition, normalization, classification, and linking of named entities across Slavic languages. In: Proceedings of the 7th Workshop on Balto-Slavic Natural Language Processing, pp. 63–74. Association for Computational Linguistics, Stroudsburg (2019). https://doi.org/10.18653/v1/W19-3709
29. Naykhanova, L.V.: The Technology of Creating Methods for Automatically Constructing Ontologies Using Genetic and Automatic Programming. - UlanUde, p. 244 (2008). (in Russian)
30. Nenausnikov, K.V., Kuleshov, S.V.: Algorithm of automatic selection of collocations from the text. J. Instrum. Eng. **62**(11), 976–981 (2019). (in Russian)
31. Namestnikov, A.M., Arzamasceva, I.V.: Ontological approach to sentiment analysis of software systems. Autom. Control Process. **2**, 34–39 (2021). (in Russian)
32. Timofeev, P.S., Sidorova, E.A.: Lexico-semantic templates as a tool for declarative description of language constructions and linguistic analysis of text. Syst. Inform. **13**, 35–48 (2018). (in Russian)
33. Vicentiy, A.V., Dikovitsky, V.V., Shishaev, M.G.: The semantic models of arctic zone legal acts visualization for express content analysis. In: Silhavy, R. (ed.) CSOC2018 2018. AISC, vol. 763, pp. 216–228. Springer, Cham (2019). https://doi.org/10.1007/978-3-319-91186-1_23
34. Apresyan, Y.D.: Three-level control theory: lexicographical aspect. In: Typology of Language and Theory of Grammar, pp. 17–21 (2007). (in Russian)
35. Official LSPL. http://www.lspl.ru/index.php
36. Vicentiy, A.V.: Analysis and dynamic visualization of logistics flows in the Arctic based on geoservices. In: IOP Conference Series: Earth and Environmental Science. Institute of Physics Publishing (2020). https://doi.org/10.1088/1755-1315/539/1/012018
37. Pilecki, B.M., Vicentiy, A.V.: Automated geoimage generation of Arctic industrial and natural complexes for regional management information support systems. In: IOP Conference Series: Earth and Environmental Science, vol. 937, p. 042079 (2021). https://doi.org/10.1088/1755-1315/937/4/042079

A Polynomial Algorithm for Construction an Automaton for Checking the Equality of Infinite Iterations of Two Finite Languages

Boris Melnikov[1(✉)] and Aleksandra Melnikova[2]

[1] Department of Computational and Applied Mathematics, Shenzhen MSU–BIT University, No. 1, International University Park Road, Dayun New Town, Longgang District, Shenzhen 518172, Guangdong, China
bf-melnikov@yandex.ru

[2] Department of Mathematics, Dimitrovgrad Engineering and Technological Institute of the National Research Nuclear University "MEPhI", Kuybyshev str. 294, Dimitrovgrad 433511, Russia

Abstract. **Purpose:** In this paper, we continue the topic related to the special binary relation in a variety of formal languages, so called equivalence relation at infinity. We formulate a simpler binary relation for languages, the cover relation, whose double application is equivalent to application of the equivalence relation at infinity. After that, we considered the algorithm for checking the fulfillment of the coverage relation, defined auxiliary objects used to prove the correctness of this algorithm.

Methods: We have defined a specific such automaton for two given finite languages, i.e. primary automaton, PRI; it is deterministic, defined on sets of words, and each of these sets is a subset of the set of prefixes of the second of the given languages. Then we determine the corresponding to such an automaton several options for special non-deterministic automata, actually describing the construction of a given iterative morphism tree. We introduce a completely different object, the so-called simplified prime automaton, NSPRI, which also describes the construction of an iterative morphism tree, but is defined not on sets of words, but on words.

Results: One of investigated objects is the infinite iterative tree. In the considered infinite iterative trees, we combine equivalent states; in fact, in doing so, we obtain a deterministic finite automaton.

Conclusion: Thus, we are working with many sets of possible prefixes, which makes the algorithm polynomial impossible. So we are building a non-deterministic automaton, defined simply on the set of possible prefixes. This begs the question: when exactly this automaton gives a positive result. We solve this problem with what we demand so that the resulting non-deterministic automaton would be an analogue of an everywhere defined deterministic automaton.

Keywords: Formal languages · Language iterations · Morphisms · Binary relation · Infinite trees · Algorithms

R. Silhavy (Ed.): CSOC 2022, LNNS 502, pp. 521–530, 2022.
https://doi.org/10.1007/978-3-031-09076-9_47

1 Introduction, Motivation and Preliminaries

In this paper, we continue the topic of articles [1–4]. In them, we continued to work with a special binary relation in a variety of formal languages (considered primarily on the set of iterations of non-empty finite languages), so called equivalence relation at infinity. We have considered examples of the application of this relation (moreover, as examples of the need for its implementation, and examples of its use) in various areas of formal language theory, discrete mathematics and abstract algebra.

We formulated simpler (than mentioned equivalence at infinity) binary relation in many languages, so called coverage ratio, the double application of which is equivalent to application of the equivalence relation at infinity. After that, we considered the verification algorithm fulfillment of the coverage ratio, determined auxiliary objects used as evidence correctness of this algorithm, and for other problems of the theory of formal languages.

One of these objects is infinite coverage relation tree, i.e., a special case of an infinite iterative tree. With the help of such trees, we proved the correctness algorithm for checking the fulfillment of the coverage relation, and also estimated the complexity of this algorithm, in case of *formal* implementation of it.

In the considered infinite iterative trees we have combined the equivalent states, in fact, having received a deterministic finite automaton; we have also described some properties of such an automaton. We have described one of *concrete* such automata: for two given target languages, we have defined so called *primary* automaton, PRI. It is, as we have already noted, deterministic, defined on sets of words – and each of these sets is subset of the set of prefixes of the second of the given languages.

We then determine the appropriate him several options for special *non-deterministic* automata that actually describe the construction of this iterative morphism tree. We enter a completely different object, so called *simplified primary* automaton, NSPRI, which also describes the construction of an iterative morphism tree, but not defined on sets of words, but in words. Note that it "differs" from the automaton that can be obtained from PRI by applying algorithms for minimizing non-deterministic finite automata (see [5,6], as well as works by one of the authors [7–10] and etc.); however, we shall not consider the details of such minimization in this paper. The main idea of constructing a polynomial algorithm checking the equality of infinite iterations of two finite languages (this is actually a continuation of the algorithm, stated in [3,4]) is as follows. It is natural to construct for this problem *deterministic* automaton. However, we are working with many sets of possible prefixes, which makes it impossible for the algorithm to be polynomial: simply because such a set of sets has, in general, exponential size. Therefore, we build a *non-deterministic* automaton (in [4] denoted by us as NSPRI), defined simply on the set of possible prefixes (and not on a set of such sets). However, this raises the question of whether when exactly this automaton gives a positive result, and there was no such problem. in the case of a deterministic automaton PRI. We solve this problem by which we require that the resulting non-deterministic automaton would be an analogue of an everywhere defined deterministic automaton ("total automaton").

We note right away that *we do not claim that that the very work of the resulting automaton can be described using a polynomial algorithm.* However, we also do not claim the opposite:

- firstly, perhaps it will be possible to prove that resulting automaton NSPRI the number of states is limited; moreover, this limitation will make it possible to organize polynomial calculations;
- second, there may be fundamentally different algorithms linking a state automaton to check a relation $\preccurlyeq\!\!\!\!|$.

The algorithm proposed in the paper is important and irrespective of other problems, but there are other "points of motivation". First of all, we mean its application in two similar problems in the theory of formal languages:

- in the task of extracting the root of the given (or the maximum possible) degrees from a given language;
- and in the problem of constructing an optimal inverse morphism.

Certainly, in both of these problems, the possibility of polynomiality is important. We are going to consider these problems in detail in the next publication.

We present the contents of the paper. First, all the necessary designations have been discussed in detail earlier, and we repeat only the most important ones. Also we provide comments which explain the meaning of states of all finite automata considered in [3,4], both deterministic and non-deterministic. Then we provide a general description of the algorithm building the desired finite automaton: its auxiliary algorithms are indicated and shows exactly what needs to be demonstrated to prove the polynomiality of the entire algorithm for constructing an automaton.

Next, directly describe subalgorithms; *their polynomiality is based on these descriptions.* We consider a possible algorithm for constructing a set of states nondeterministic automaton NSPRI. Then we consider algorithms for constructing auxiliary objects, used by us in the description of the main algorithm. In fact, those are algorithms for constructing sets of edges of the transition graph of the automaton NSPRI.

In conclusion, directions for further work are briefly formulated.

All designations used by us have been considered in detail in previous articles [1–4], and we repeat only the most important of them, necessary for the present paper.

If for finite languages A and B, the condition

$$(\forall u \in A^*)\ (\exists v \in B^*)\ (u \in \text{opref}\,(v))$$

holds, then we write $A \preccurlyeq\!\!\!\!| B$ (or $B \succcurlyeq\!\!\!\!| A$). If conditions $A \preccurlyeq\!\!\!\!| B$ and $A \succcurlyeq\!\!\!\!| B$ are both satisfied, we write $A \preccurlyeq\!\!\!\!\succcurlyeq\!\!\!\!| B$.

Let us completely repeat [3, def. 7] – in it we define the function Φ. Namely, for a pair of non-empty finite languages $A, B \subseteq \Sigma^*$, such that $A \not\ni \varepsilon$ and $B \not\ni \varepsilon$, language

$$\underset{A-B}{\Phi}(u) \subseteq \Sigma^*$$

is constructed for some word $u \in \Sigma^*$ according to the following algorithm – it uses the auxiliary language D' and forms the response language D.

1. To start the algorithm, we set

$$D' = \{u\}A, \quad D = \varnothing.$$

2. If $D' = \varnothing$, then exit the algorithm with the answer D.
3. Choose some word $v \in D'$, excluding it from D':

$$D' \ -: v.$$

4. For the chosen v We consider all options for its presentation in the form $v = uv'$, where $u \in B^+$,

$$D \ +: v'$$

(i.e., we include each such v' in the formed language D).
5. If

$$(\exists w \in B)\,(v \in \text{opref}\,(w)),$$

then D +: v (i.e., we include v in the D language).
6. Go to item 2.

(We note again that according to the above algorithm about the final "output" language D one can assert that $D \subseteq \text{opref}\,(B)$. We also note that in this paper defining the value of the $\underset{A-B}{\Phi}(C)$ function $C \subseteq \Sigma^*$ is not used for the language.)

2 Results. The General Description of the Algorithm for Construction of the Automaton

As we noted in Introduction, the algorithm proposed in this paper is some modification of the algorithm, constructing finite automata in the article [3,4]. At the same time, it is clear that to check the fulfillment of the relation $A \mathrel{\substack{\triangleleft\\[-2pt]\triangleright}} B$ it is enough just to double-check the $\trianglelefteq$ relation: $A \trianglelefteq B$ and $B \trianglelefteq A$; thus, both algorithms are either simultaneously, or are not simultaneously polynomial. Therefore, it is important to note that based on the formulations of the algorithms, cited in our previous publications, about polynomial "it was too early to talk".

Based on one of these algorithms, in [3,4] we defined automaton PRI(A, B) for given languages A and B; however, it is obvious that any possible algorithm for its construction is exponential. So we move on to the non-deterministic finite automaton NSPRI, defined on words (i.e., prefixes of the language B), and not on sets of such words. In this case, a problem arises related to when exactly this automaton gives a positive result. We solve this problem with what we demand so that the resulting non-deterministic automaton would be an analogue to the total deterministic finite automaton. So that from any of its states for any letter of the considered alphabet would lead the edge (to some state that is not useless). And, certainly, the fulfillment of this *condition can be checked polynomially*. Therefore, we just need to polynomially construct the automaton NSPRI, which we do in the described algorithm.

And, certainly, to prove the polynomiality of the algorithm for constructing the set of automaton states it hardly makes sense to use a formal description some Turing machine that solves this problem, as well as the corresponding normal Markov algorithm, machine with unlimited registers [11] and other possible formalisms for this; apparently, it is enough to bring the program in some well-known programming language.

In this case, in principle, we can prove the polynomiality *based on a computer program given earlier*, in [1,2]: it actually solved the same problem, although it is titled differently. Object-oriented programming used in that program allowed, firstly, to make a high-quality program, secondly, use it as comments to the given algorithms ... however, this program is quite difficult to use to prove the polynomiality of the algorithm it implements.

Therefore, "we worse" this program, leaving practically nothing of the C++ elements, and get the following text; almost all of its designs are also translated in "non-plus" C, however, for convenience of presentation, the C++ language is formally chosen. The final construction required function Φ; we still describe not in the form of a program (it gets too cumbersome) and in the form of a detailed description of the algorithm, according to which this program is easy to build.

And, certainly, a proof of the existence of a polynomial algorithm for constructing the function Φ is enough to give a positive answer to the main question of the paper, about the existence of a polynomial algorithm building a non-deterministic finite automaton, designed to test equality infinite iterations of two finite languages. Indeed, we show that all states of the automaton are constructed by a polynomial algorithm; we also show that the number of such states in the automaton is proportional to N^2 for the input size N. Therefore, the number of pairs of such states is $\sim N^4$, and the presence of a polynomial algorithm for constructing a set of edges between each pair (we provide descriptions of the algorithm with complexity $\sim N^8$). Then there is the possibility of building an automaton in polynomial time $\sim N^{12}$.

3 Methods. The Polynomiality of the Algorithm for Constructing the Set of States of Nondeterministic Automaton

We repeat once again that non-determinism is very important to us. NSPRI automaton being built: for a deterministic automaton PRI construction algorithms are much simpler, however, from a formal point of view, they are non-polynomial. So, first consider the construction of a set of states automaton

NSPRI. We use strings only as a convenient notation for an array of characters without using string manipulation functions. However, "no procedural programming at all" (i.e., without the use of functions) is unlikely that it will be possible to do, although the "pure mathematicians", apparently, can criticize such a proof. Here is a description of the data:

```
const int N = 200;
const int NN = N*N;
const string A = "aaa|aabba|abba|bb|*";
const string B = "aaaa|abb|abba|bbb|*";
string Bpref;
int Begins[NN];
```

Some comments about data representation. Constant strings `A` and `B` describe given languages; in them, similar to printing the languages `lA` and `lB` in the program from [1,2], each word of the language ends with |, and the entire line is a symbol * (these symbols themselves are not included in the words under consideration). In the sample program, their values are same as similar lines in the examples given in [1,2]. We assume that the maximum possible length each of the lines `A` and `B` (including those special characters) limited to the value of a given constant `N`.

The result of the work (i.e., the whole set of custom string prefixes `A` and `B`) will be recorded to the output string `Bpref`, which is organized according to the same rules, as well as the lines `A` and `B`. It is important to note that the length of this output string certainly does not exceed N^2, it is easy to prove by the methods of elementary mathematics.

`Begins` is the generated array, where we will write byte numbers, each of which is the beginning of another word in a string (an array) `Bpref`; it is clear that there are at most N^2 words. After the beginning of the last word, we write the value -1.

Now, let us move on to the algorithms. To the generated the array (string `Bpref`), we use the following function `AddStrings()`; it adds all substrings, i.e. the string's own suffixes, which is implicitly passed to this function using parameters:

```
void AddStrings(int Beg, int End) {
  for (int i=Beg; i<=End; i++) {
    for (int k=Beg; k<=i; k++)
      Bpref += B[k];
    Bpref += '|';
  }
}
```

(Outer loop parameter `i` is the beginning of the substring in question).

We *do not look* for possible repetitions of included words: so the algorithm turns out to be simpler, and it does not cease to be polynomial. It is important for us that the execution of both nested loops consists of at most N iterations, so *complexity of the above algorithm* is $\sim N^2$. Further, in the main function `main()`, we are looking for words in the input string and apply the `AddStrings()` function described above:

```
Bpref = "";
for (int iBeg=0; iBeg<B.size(); iBeg++) {
  for (int iEnd=iBeg+1;
      iEnd<B.size(); iEnd++) {
    if (B[iEnd]!='|') continue;
    AddStrings(iBeg,iEnd-1);
    iBeg = iEnd;
    break;
  }
}
Bpref += '*';
cout << "Answer: " << Bpref << endl;
```

From the text of the function, it can be seen that each of the loops takes at most N iterations; thus, the algorithms we consider are polynomial.

4 Methods. The Polynomiality of the Algorithm for Constructing Auxiliary Objects

Certainly, the forming algorithm for the array`Begins` is also polynomial:

```
Begins[0] = 0;
int i = 1; // new index in Begins
for (int j=1; j<NN; j++) {
  if (Bpref[j]=='*') break;
  if (Bpref[j]!='|') continue;
  Begins[i++] = j+1;
}
Begins[--i] = -1;
```

And then we begin to describe the construction of the main function, previously denoted by us as either F or Φ, or, according to [3,4], marking the edge of the transition graph of the automaton NSPRI. First of all, we note that the argument of this function (or, using the notation from [3,4], the state of the automaton NSPRI) we can title not only by the word over the auxiliary alphabet Δ (sometimes referred to as Δ_A in cited papers), but also directly by the word of the language A (namely, $h_A(d)$ for some letter $d \in \Delta$).

In addition to the data structures described in the previous section, we add the following:

```
string FiTmp;
bool Nac[NN];
bool Otv[NN];
```

Some small comments:

- FiTmp is the intermediate representation for the formation of the function F; purely formally, the length of this string is limited by the value $N^2 + N$ (which, of course, is enough to prove the polynomiality of the algorithm); however, by means of elementary mathematics it can be shown that that it suffices to consider the value of N^2, which somewhat simplifies the writing of algorithms;
- Nac is the array of beginnings of the words of the decomposition of the considered word FiTmp according to language B; these beginnings of the words are indicated at the previous figures by green circles ([1, Fig. 2, 3]);
- Otv is the array of the beginnings of those words, which will go in response (to the generated value of the F function); the ends of such words are indicated at the same figures by red circles.

The following function forms the word FiTmp – by appending to each word included in Bpref, some (one) word of language A, given by its beginning and end in the string A.

```
void AddToBpref(int Beg, int End) {
  FiTmp = "";
  for (int i=0; ; i++) {
    char ch = Bpref[i];
    if (ch=='*') break;
    if (ch=='|')
      for (int j=Beg; j<=End; j++)
        FiTmp += A[j];
    FiTmp += ch;
  }
  FiTmp += '*';
}
```

In the above function, i is the number of the position in question in the line Bpref, j – number of the considered position in the string A.

In addition, we need to do the following:

```
Nac[0] = true;
for (int i=1; i<NN; i++)
  Nac[i] = false;
for (int i=0; i<NN; i++)
  Otv[i] = false;
```

According to [1,3,4] equivalent descriptions of the function Φ, it is obvious that that in the described algorithm the value of this function is constructed.

Now let us consider estimating the complexity of the algorithm. When describing its steps, we gave the maximum possible number of their repetitions, and this was done to estimate the complexity of the entire algorithm.

It is also obvious that this complexity is polynomial. Moreover, considering 4 nested loops, each of which consists of $\sim N^2$ repetitions (N is the input data size), it is easy to verify that the total complexity, as mentioned above, $\sim N^8$. Therefore, the total complexity of the entire resulting algorithm constructing an automaton NSPRI $\sim N^{12}$. However, we did not seek to minimize it (this degree): apparently, such a minimization hardly makes practical sense, because, again, for practice the object-oriented version is much better. The polynomiality of all described calculations (relative to the input length) is obvious.

5 Conclusion

Let us briefly formulate directions for further work, i.e. possible publications, related to the topic discussed in this paper.

The next possible post topic could be about semilattices of potential roots. Let us consider it in a little more detail. In previous publications we have repeatedly noted the similarity of the two tasks for some given finite language:

- the task of extracting the root of a given degree
- and task of constructing an optimal inverse morphism.

The main problem is the possible description of *polynomial* algorithms for these tasks.

In both cases, it is necessary to build the set of so called potential roots, usually for a given language A denoted in our previous papers by $\overline{\sqrt[n]{A}}$. The exponential algorithms for both of these problems are obvious: we just need to iterate over all subsets set of these potential roots.

For both these tasks, it is of interest the possible existence of two semilattices [12], built on a set of subsets of potential roots: semilattices by intersection and semilattices by union. At the same time, for the second of these tasks, generally speaking, there is none of these semilattices (examples were given in [3,4]). And for the second task, for any given language both semilattices are obtained.

Based on the second of these tasks we can formulate one more; apparently, it is much more interesting: to describe the *polynomial* algorithm constructing an optimal inverse morphism.

References

1. Melnikov, B., Melnikova, A.: Infinite trees in the algorithm for checking the equivalence condition of iterations of finite languages. Part I. Int. J. Open Inf. Technol. **9**(4), 1–11 (2021)
2. Melnikov, B., Melnikova, A.: Infinite trees in the algorithm for checking the equivalence condition of iterations of finite languages. Part II. Int. J. Open Inf. Technol. **9**(5), 1–11 (2021)
3. Melnikov, B.: Variants of finite automata corresponding to infinite iterative morphism trees. Part I. Int. J. Open Inf. Technol. **9**(7), 5–13 (2021)

4. Melnikov, B.: Variants of finite automata corresponding to infinite iterative morphism trees. Part II. Int. J. Open Inf. Technol. **9**(10), 1–8 (2021)
5. Kameda, T., Weiner, P.: On the state minimization of nondeterministic finite automata. IEEE Trans. Comput. **C-19**(7), 617–627 (1970)
6. Polák, L.: Minimizations of NFA using the universal automaton. Int. J. Found. Comput. Sci. **16**(5), 999–1010 (2005)
7. Melnikov, B., Sayfullina, M.: On some algorithms for equivalent transformation of nondeterministic finite automata. News of higher educational institutions. Mathematics **4**, 67–72 (2009)
8. Melnikov, B., Tsyganov, A.: The state minimization problem for nondeterministic finite automata: the parallel implementation of the truncated branch and bound method. In: Proceedings – 5th International Symposium on Parallel Architectures, Algorithms and Programming, PAAP 2012, pp. 194–201 (2012)
9. Dolgov, V., Melnikov, B.: Building a universal finite automaton. I. From the theory to the practical algorithms. Bull. Voronezh State Univ. Ser.: Phys. Math. (2), 173–181 (2013)
10. Dolgov, V., Melnikov, B.: Building a universal finite automaton. II. Examples of how algorithms work. Bull. Voronezh State Univ. Ser.: Phys. Math. (1), 78–85 (2014)
11. Igoshin, V.: Mathematical Logic and Theory of Algorithms. Akademia, Moscow (2004)
12. Skornyakov, L.: General Algebra, vol. 2. Nauka, Moscow (1991)

Grasping Point Detection Using Monocular Camera Image Processing and Knowledge of Center of Gravity

Dominik Stursa(✉), Petr Dolezel, and Daniel Honc

Faculty of Electrical Engineering and Informatics, University of Pardubice, 532 10 Pardubice, Czech Republic
{dominik.stursa,petr.dolezel,daniel.honc}@upce.cz
https://fei.upce.cz/en

Abstract. The ability to grasp objects is one of the basic functions of modern industrial robots. In this article, the focus is placed on a system for processing the image provided by a robot visual perception system leading to the detection of objects grasping points. The proposed processing system is based on a multi-step method using convolutional neural networks (CNN). The first step is to use the first CNN to transform the input image into a schematic image with labeled objects centers of gravity, which then serves as a supporting input to the second CNN. In this second CNN, original input and supporting input images are used to obtain a schematic image containing the grasping points of the objects. This solution is further compared with a network providing grasping points directly from the input image. As a result, the proposed method provided a 0.7% improvement in the average intersection over union for all of the models.

Keywords: Image processing · Grasping point detection · Convolutional neural network · Center of gravity

1 Introduction

In recent decades, many manufacturers have resorted to the use of automated robotic lines, which bring a number of advantages. The first and probably most obvious reason is the reduction in production costs. Although the initial cost may seem daunting, manufacturing companies will recoup their investment in the long run due to several key benefits such as the ability to work 24/7 without interruption or the high accuracy and repeat-ability of operations [1].

In addition, one of the biggest advantages is their predictable behaviour and precision of movements. This leads to a greater ability to consistently produce high quality products with little variation, all with minimal need for human control. The safety of workers, speeding up production and increasing profits are significantly influencing the number of robotic manipulators in industry [2].

R. Silhavy (Ed.): CSOC 2022, LNNS 502, pp. 531–541, 2022.
https://doi.org/10.1007/978-3-031-09076-9_48

Not only this pressure from the industry, but also advances in digital image processing are opening the door for all sorts of robotic applications. Machine vision is typically used in robotic industry to detect objects [3] or their grasping points [4], to locate obstacles [5] or to find product defects [6], and for quality control.

Based on this information about the surrounding world, the robotic manipulator can execute a controlled movement that, for example, ensures the removal of objects from the conveyor belt to defined locations in the new position (so-called pick and place), which is one of the typical robotic applications [7].

Over the past decades, approaches to the pick and place applications have been introduced. Thanks to the possibilities of individual parts of the pick and place system, it is possible to categorize them. For example, it is possible to categorize these approaches according to the robotic system used, the camera system used, the object detection method developed using an analytical procedure or a data-driven procedure.

Analytical approaches consider the geometric shape of the target object and try to find exactly possible positions to grasp the object. Thanks to deep learning methods, similar or even better results can be achieved by applying empirical approaches without the need for analytical processing [8]. Detailed descriptions of the benefits and differences of these approaches can be found in surveys [9,10].

Machine vision methods can be categorized according to many criteria that narrow down the problem. An example is the dimension of the solution needed, which delineates whether the problem needs to be solved in 2D or 3D space [11]. In our case, the application limitation is defined by a robotic manipulator with 3 degrees of freedom (DoF) and a suction cup end-effector. As the robotic manipulator is not able to grasp object randomly rotated in 3D space, but only object vertically accessible, this leads to limitation of computer vision solution in 2D space.

In this contribution, we consider the localization of the object's center of gravity as an accompanying information to the system to find a suitable grasping point, which is the initial part of the pick and place application.

1.1 Related Work

Finding object grasping points has recently become a widespread issue that helps find faster or more accurate solutions to make pick and place more efficient. Convolutional neural networks are increasingly being used to find grasping points [12]. One method that provides an improvement in the accuracy of grasping point estimation is to use neural networks to generate bounding boxes of important grasping points using point-wise convolution [13].

Other methods used in this area are deep learning methods that finds salient point coordinates by shape context information [14]. The use of object shape context information is also used in a more modern processing approach using image segmentation, where the grasping points are highlighted and used for further not so demanding processing [15].

It is also more common to combine different convolutional neural networks to add different information about object positions and grasping points locations with each other [16] or to obtain differently defined outputs that can play a role in the decision-making process to select the correct grasp point [17].

2 Problem Formulation

This section serves to correctly define the aim of the article. As mentioned, we are concerned with the first necessary step in the object pick and place task, which is machine vision. Specifically, determining the center of gravity of an irregular object with a homogeneous mass distribution using image data processing. The need to obtain the center of gravity of the object is related to the effort to find the most suitable grasping point, which will load the robotic manipulator in uncontrolled axes only minimally during the pick and place application.

Objects of interest are differently sized normalized triangular construction supports illustrated together with expected robotic manipulation implementation in Fig. 1.

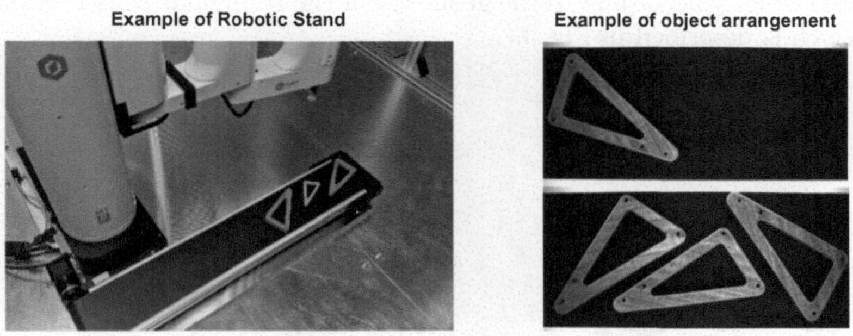

Fig. 1. Demonstration of the problem and arrangement of monitored objects on a conveyor belt.

According to the complexity of the shape and size of the objects along with the monotony of the surface on which the objects are placed, a monochromatic monocular camera was chosen to capture the image for processing.

2.1 Proposed Solution

Since we want to use image processing to find the center of gravity of the object, which we then use as auxiliary information for the evaluation of the grasping point, the following two procedures were chosen to determine the center of gravity. The first procedure is to construct a parameterized analytical solution. The second procedure is to obtain the center of gravity of the object using basic image processing methods. Both of these procedures will be used to label the data for convolutional neural network training.

Once the object center of gravity has been marked, this information is fed together with the input image into a second step in which a search for suitable grasping points is performed. Both steps are implemented by converting the input image into a schematic image using a convolutional neural network. In the output schematic image, the coordinates of possible grasping points are then calculated from the individual segmented entities. The whole process is illustrated in Fig. 2.

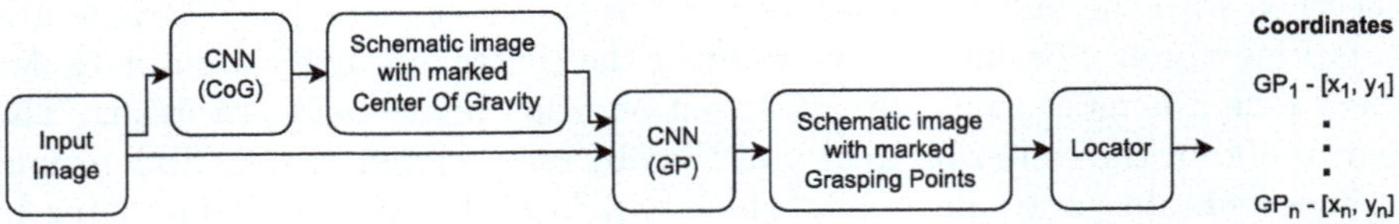

Fig. 2. Diagram of proposed processing approach.

Analytical Solution. Using the analytical solution, the center of gravity was calculated exactly according to the geometry of the triangular construction support, which is described in Fig. 3.

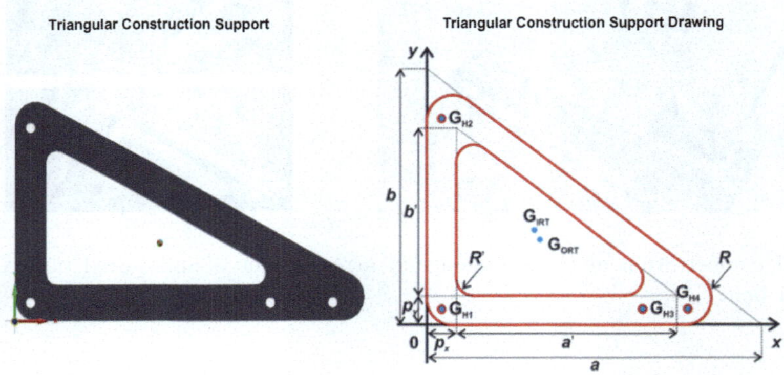

Fig. 3. Drawing of the triangular structure with all significant points, parameters and centers of gravity marked.

Used triangular beams contain rounded edges and cut-outs that cause the center of gravity to shift relative to a normal solid triangle. As a result, the analytical calculation of the center of gravity must be treated as a composition of the centers of gravity of the individual parts, leading to the following equations for the center of gravity coordinates of the object.

$$\mathrm{x}_G = \frac{x_{OTR} \cdot A_{OTR} - x_{ITR} \cdot A_{ITR} - \sum_{n=1}^{m} x_{H_n} \cdot A_{H_n}}{A_{OTR} - A_{ITR} - \sum_{n=1}^{m} A_{H_n}}, \tag{1}$$

$$y_G = \frac{y_{OTR} \cdot A_{OTR} - y_{ITR} \cdot A_{ITR} - \sum_{n=1}^{m} y_{H_n} \cdot A_{H_n}}{A_{OTR} - A_{ITR} - \sum_{n=1}^{m} A_{H_n}}, \tag{2}$$

where x_{OTR} and y_{OTR} are outer round shaped triangle coordinates of center of gravity; x_{ITR} and y_{ITR} are inner round shaped triangle coordinates of center of gravity; x_{H_n} and y_{H_n} are center coordinates of holes; and all A correspond to the contents of the area of the individual elements named identically as in the case of coordinates. Parameter m corresponds to the number of mounting holes in the support. Depending on these calculations, a center of gravity label is then created in the schematic image according to the size of the parameterized triangle.

Image Processing Solution. The application of basic image processing methods to calculate the center of gravity can be used when there is only the one object in the image, and only on objects that are fully filled [18].

Since the objects themselves contain cutouts, the problem in this case is also split to find multiple individual important centers of gravity, which will later be folded to obtain the main center of gravity. For the purpose of this method, the holes for the screws are neglected and only a large cut-out in the middle of the object is considered. The result is the determination of the center of gravity of the cut-out and the center of gravity of the filled object separately and the subsequent subtraction of the both individual coordinates in both axes.

The separate processing consists in converting the grayscale image into a binary image, in which the coordinates of the center of gravity are calculated. To obtain a binary image, the input image is processed using thresholding. The next step is to perform a pattern fill, which is used both to calculate center of gravity for filled object, but also after subtracting it from the original binary image to create the cut-out object and then calculating its center of gravity. The following is the composition of the centers of gravity of the components of the object and the marking of the center of gravity itself in the schematic figure. The size of the center of gravity marking is based on the size of the object area. The individual steps of image processing procedure for labeling the center of gravity of an object is shown in Fig. 4.

3 Experiments Procedure

In this section, the design of the system for estimating the center of gravity and for estimating the grasping points of triangular structural supports will be described. Both systems are based on convolutional neural networks performing image segmentation and producing a schematic image.

Two approaches have been chosen for comparison, the first is the aforementioned approach illustrated in Fig. 2, the second is the conventional approach that performs the processing of the whole image in one step providing directly the resulting grasping points. Moreover, due to the two methods for labeling the data, these approaches are also compared and evaluated with each other.

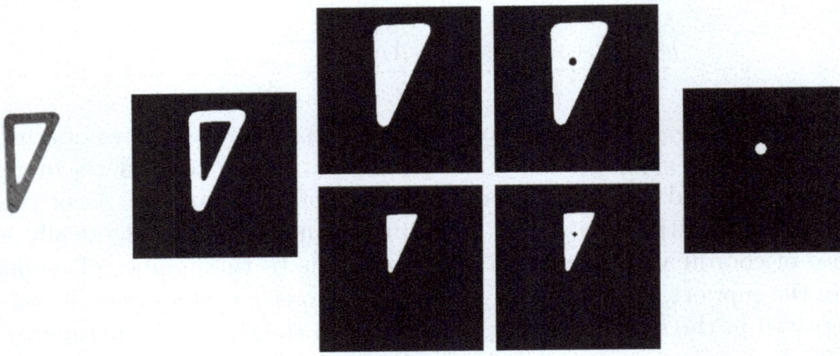

Fig. 4. Image processing procedure for marking the center of gravity of an object.

Both variants used the encoder-decoder principle, specifically a scalable version of the U-Net architecture that can be scaled according to provide the necessary segmentation. A scheme of this proposed scalable architecture based on the U-Net topology is shown in Fig. 5.

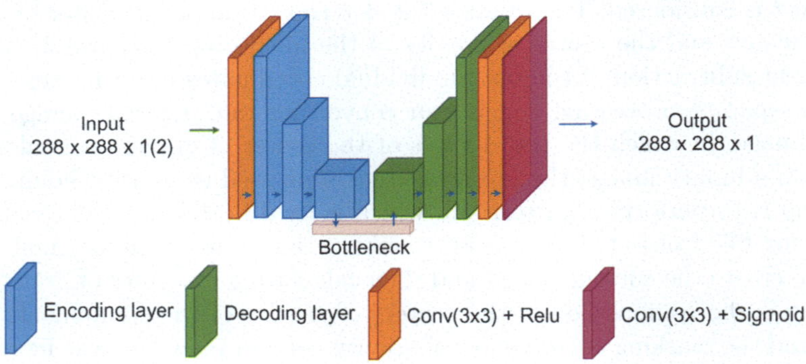

Fig. 5. The encoding layer halves the width and height of its layer while doubling its depth. The opposite function is performed by the decoding layer, which doubles the width and height of its inputs and reduces the depth to half of the previous decoding layer or bottleneck. Everything then depends on the parameters of the first convolution and the depth of the network (how many encoding or decoding modules are used).

3.1 Dataset Creation

In order to acquire training and testing data, the special stand was prepared. A Basler acA2500-60um industrial monochrome camera [19] was used as a proper sensor. This sensor was set to provide 2048 × 2048 px images. The camera was equipped with a Computar M3514-MP lens [20] in order to monitor the scan area of 300 × 300 mm from a distance of 500 mm, which was calculated using the basler configuration tool [21].

Data Acquisition. For the purpose of verifying the ability to adapt to the parameters (sizes) of triangular construction supports and due to the possibilities of the labeling techniques, the dataset was divided into 2 types of images. Specifically, the images where triangular construction support in different positions and rotation is present once in time, for 3 different supports sizes, where 200 images were taken for each size. The second group, used only for testing purposes, are images with multiple objects on a single image, of which 300 were created for these purposes. Thus, a total of 900 images were taken for training and testing neural network topologies. Examples of these images are shown in the following Fig. 6.

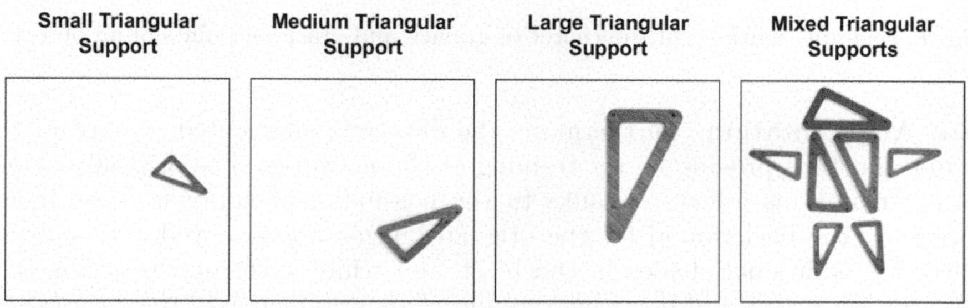

Fig. 6. Examples of different sizes of triangular construction support and of their mix.

Data Labeling. Each image from the training and test set was labeled for neural network use. Specifically, the positions of the center of gravity and the positions of the grasping points, for which corresponding schematic images were created according to the two techniques mentioned in Sect. 2.1. A custom labeling application was created, which for the first method was able to manually determine the important points (selected mounting holes of the triangular support), thanks to which the image is automatically labeled also with respect to the parameterization of the size of the marking determined by the distance of the selected points. The position of the center of gravity and grasping points is then calculated using calculations based on knowledge of the design drawing. The second part was the automatic marking purely using the described digital image processing to gain the center of gravity. An example of the given marking is shown in Fig. 7.

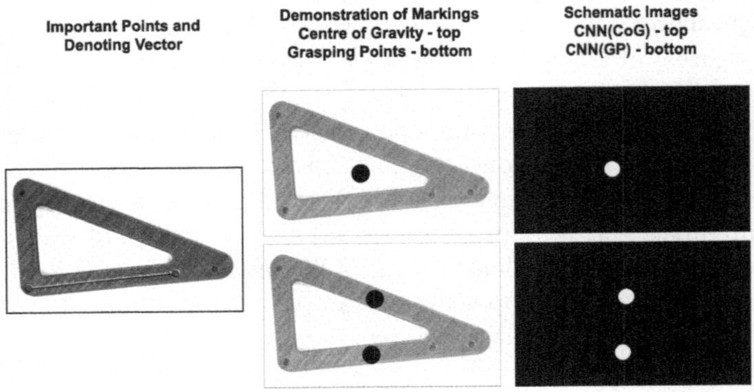

Fig. 7. Example marking of the center of gravity and grasping points of an object.

Data Augmentation. Furthermore, the data was augmented to expand the dataset and thus provide better training of the networks. The original dataset was augmented as follows. Thanks to the possibility of applying basic image processing, the background for the original images was removed and replaced with three additional shades in the black and white spectrum, resulting in a light grey, dark grey and black background. This step increased the dataset size to 2400 frames for training and 1200 frames for testing.

3.2 Neural Network Training

Overall, given the problem, the experiments with topologies can be divided into four individual parts. The first and second parts are training the neural network to gain the center of gravity, where for the first method the computed centers of gravity from manual labeling are used and for the second method the labels from image processing are used. The third part is a CNN training to determine grasping points, where the input information is not only the input image but also a supporting schematic image capturing the center of gravity of the objects. Fourth is to train a neural network to directly determine the grasp points from the input image only.

To solve the defined problem, separate training was performed within each problem. For each problem, 4 configurations of depth parameters and initial filter count of scalable topology were chosen. The ADAM optimizer was used as the optimization algorithm, and weights were set for each topology randomly with Gaussian distribution. Due to the stochastic nature of neural networks, each topology was trained five times to select the best performing model.

4 Results

For all the individual neural network solutions and topologies, the Intersection over Union (IoU) metric, which is commonly used to evaluate the performance of detection techniques, was selected.

A comparison of the methods for determining the center of gravity was carried out first. In this case, certain configurations of the scalable network were chosen for both labeling methods. Specifically, this was a combination of depths 2 and 4 along with the number of filters corresponding to values of 64 or 32. This resulted in 4 combinations of both that provided significantly different values for the testing set. The values of these results are shown in the following Table 1.

Table 1. Results of intersection over union for centers of gravity.

Topology	Model 1	Model 2	Model 3	Model 4	Model 5	Mean value
Computed - 2, 32	0.5941	0.5568	0.5685	0.5956	0.5355	0.5701
Computed - 2, 64	0.5935	0.6206	0.6267	0.6708	0.6014	0.6226
Computed - 4, 32	0.7460	0.8141	0.8391	0.7694	0.8347	0.8007
Computed - 4, 64	0.7632	0.7995	0.8361	**0.8554**	0.8326	0.8173
Processed - 2, 32	0.6041	0.5018	0.5510	0.5615	0.5532	0.5543
Processed - 2, 64	0.6015	0.5996	0.6671	0.6109	0.6244	0.6207
Processed - 4, 32	0.7545	0.8341	0.8415	0.7583	0.8422	0.8061
Processed - 4, 64	0.7752	0.7856	**0.8641**	0.8041	0.8221	0.8102

As second, the topology was tested to determine the grasping points from both types of inputs. These topologies used the best performing model of CoG determination providing input information. Results of IoU for tested topologies are summarized in Table 2.

Table 2. Results of grasping points estimation with supporting information.

Topology	Model 1	Model 2	Model 3	Model 4	Model 5	Mean val.
Comp. - 4, 64; 4,64	0.7613	0.7351	0.7815	0.7351	**0.8213**	0.7669
Comp. - 4, 64; 4,32	0.7515	0.7995	0.7319	0.7442	0.7726	0.7599
Proc. - 4, 64; 4,64	0.7503	0.7982	0.8111	0.8211	**0.8301**	0.8022
Proc. - 4, 64; 4,32	0.7611	0.7112	0.8078	0.8042	0.7896	0.7748

Lastly, the data were evaluated using the direct method to determine the grasping points, for which the 4 combinations of sizes and depths are shown in the Table 3.

Table 3. Results of grasping points estimation.

Topology	Model 1	Model 2	Model 3	Model 4	Model 5	Mean val.
T1 - 6, 32	0.6621	0.6011	0.6056	0.6202	0.6434	0.6265
T1 - 6, 64	0.6888	0.6055	0.5995	0.6332	0.6611	0.6376
T1 - 8, 32	0.7511	0.7684	0.7881	0.7687	0.7222	0.7597
T1 - 8, 64	0.8044	0.7800	0.7944	0.7992	**0.8155**	0.7987

5 Conclusion

The task of the first investigation was to determine whether generating data labels based on empirical computations provides higher accuracy than automatic data labeling. However, it can be determined from the data that the effect of the center of gravity method alone does not have a strong influence in the average values.

One of the other objectives of this work was to test whether preprocessing some key information about the object (the center of gravity in this case) can lead to more accurate estimation of the grasping points. For these cases, it is necessary to compare Table 2 and Table 3, which show the accuracy of the determination of the grasping points. These tables show that, on average, more accurate results are obtained with connected networks (with supporting information) containing a comparable number of parameters than with a separate network. Specifically, for parametrically comparable topologies, there is an average improvement of 0.7%.

In the future, the influence of preprocessing itself could be subjected to a more rigorous analysis that would precisely normalize the influence of the number of parameters against the resulting accuracy.

Acknowledgment. We would like to thank Dr. Lubos Rejfek for creating the images and their subsequent labeling for the dataset. We would also like to thank Associate Professor Jiri Tucek for providing the mathematical relations for accurate calculation of the center of gravity of triangular construction supports.

Fundings. The work was supported from ERDF/ESF "Cooperation in Applied Research between the University of Pardubice and companies, in the Field of Positioning, Detection and Simulation Technology for Transport Systems (PosiTrans)" (No. CZ.02.1.01/0.0/0.0/17_049/0008394).

References

1. Profozich, G.: California manufacturing technology consulting: ready or not, robotics in manufacturing is on the rise (2021). https://www.cmtc.com/blog/overview-of-robotics-in-manufacturing. Accessed 7 Feb 2022
2. International Federation of Robotics. Robot race: the world's top 10 automated countries (2021). https://ifr.org/ifr-press-releases/news/robot-race-the-worlds-top-10-automated-countries. Accessed 8 Feb 2022

3. Schwarz, M., Milan, A., Periyasamy, A.S., Behnke, S.: RGB-D object detection and semantic segmentation for autonomous manipulation in clutter. Int. J. Robot. Res. **37**(4–5), 437–451 (2018)
4. ten Pas, A., Gualtieri, M., Saenko, K., Platt, R.: Grasp pose detection in point clouds. Int. J. Robot. Res. **36**(13–14), 1455–1473 (2017)
5. Mancini, M., Costante, G., Valigi, P., Ciarfuglia, T.A.: J-MOD 2: joint monocular obstacle detection and depth estimation. IEEE Robot. Autom. Lett. **3**(3), 1490–1497 (2018)
6. Gong, X., Bai, Y., Liu, Y., Mu, H.: Application of deep learning in defect detection, vol. 1684 (2020)
7. Du, G., Wang, K., Lian, S., Zhao, K.: Vision-based robotic grasping from object localization, object pose estimation to grasp estimation for parallel grippers: a review. Artif. Intell. Rev. **54**(3), 1677–1734 (2020). https://doi.org/10.1007/s10462-020-09888-5
8. Sahbani, A., El-Khoury, S., Bidaud, P.: An overview of 3D object grasp synthesis algorithms. Robot. Auton. Syst. **60**(3), 326–336 (2012). Autonomous Grasping
9. Dharbaneshwer, S.J., Subramanian, S.J., Kohlhoff, K.: Robotic grasp analysis using deformable solid mechanics. Meccanica **54**(11), 1767–1784 (2019). https://doi.org/10.1007/s11012-019-01034-y
10. Kleeberger, K., Bormann, R., Kraus, W., Huber, M.F.: A survey on learning-based robotic grasping. Curr. Robot. Rep. **1**(4), 239–249 (2020). https://doi.org/10.1007/s43154-020-00021-6
11. Bjornsson, A., Jonsson, M., Johansen, K.: Automated material handling in composite manufacturing using pick-and-place systems - a review. Robot. Comput.-Integr. Manuf. **51**, 222–229 (2018)
12. Miao, W., Li, G., Jiang, G., Fang, Y., Ju, Z., Liu, H.: Optimal grasp planning of multi-fingered robotic hands: a review. Appl. Comput. Math. **14**(3), 228–247 (2015)
13. Teng, Y., Gao, P.: Generative robotic grasping using depthwise separable convolution. Comput. Electr. Eng. **94**, 107318 (2021)
14. Bohg, J., Kragic, D.: Learning grasping points with shape context. Robot. Auton. Syst. **58**(4), 362–377 (2010)
15. Luddecke, T., Kulvicius, T., Worgotter, F.: Context-based affordance segmentation from 2D images for robot actions. Robot. Auton. Syst. **119**, 92–107 (2019)
16. Kim, D.-Y., Sim, K.-H., Lee, G.-H.: Object detection by combining two different CNN algorithms and robotic grasping control. J. Inst. Control Robot. Syst. **25**(9), 811–817 (2019)
17. Corona, E., Alenyá, G., Gabas, A., Torras, C.: Active garment recognition and target grasping point detection using deep learning. Pattern Recogn. **74**, 629–641 (2018)
18. Van Assen, H.C., Egmont-Petersen, M., Reiber, J.H.C.: Accurate object localization in gray level images using the center of gravity measure: accuracy versus precision. IEEE Trans. Image Process. **11**, 1379–1384 (2002)
19. Basler. Basler ace (2022). https://www.baslerweb.com/en/products/cameras/area-scan-cameras/ace/aca2500-60um/. Accessed 18 Feb 2022
20. Computar. Computar lenses (2022). https://computar.com/product/705/M3514-MP.Accessed 18 Feb 2022
21. Lens Selector by Basler (2022). https://www.baslerweb.com/en/products/tools/lens-selector/. Accessed 15 Jan 2022

The Kazakhstan Republic Economy Three Sectoral Model Inter-sectoral Linkages Resource Assessment

Seyit Kerimkhulle(✉), Nauryz Baizakov, Assylzat Slanbekova, Zhanar Alimova, Gulmira Azieva, and Meruyert Koishybayeva

L.N. Gumilyov Eurasian National University, Nur-Sultan, Kazakhstan
kerimkhulle@gmail.com

Abstract. In this paper on the table "input-output" statistical data basis we estimate the national economy total resource units sum as gross output distribution a system by intermediate interaction and final attraction inter-branch links resources. The Leontief' model algebraic equations' system solution is reduced to difference equations, a system. To solve the latter, we used an input-output simulation model in Any Logic software environment. The input data parameters have been determined, and the gross output resources a balanced growth condition, intermediate and final demand for the national economy has been ensured.

Keywords: Input-output model · System dynamics · Resource · Economic system · Kazakhstan

1 Introduction

This study's purpose is to develop and implement a mathematical model, algorithm and intelligent information technology in mathematical decision support for improving an economic system inter-industry linkages efficiency using Leontief's input-output model theory and methodology [1], whose main provisions are detailed in [2], and a broad literature review in [3] which attempts to characterize the literature related to the input-output analysis in the period since.

While Input-Output models were initially used mainly to reveal the economic systems inter-industry relationships dependencies, today they are used in many science and technology branches to reveal the society's more complex systems and the intra-structural relationships hidden dependencies and business development, in particular, we provide them with a number in the following overview:

- Using the input-output model to study the energy-water-food nexus reveals a hidden dependency between resource supply chains and end products [4];
- It estimates the additional costs' impact associated with the Izmir mega-event due to i) additional hotels construction and ii) foreign visitors arriving on the regional economy using a closed regional input-output model based on a regional input-output table prepared by the Izmir Regional Development Agency IZKA [5];

R. Silhavy (Ed.): CSOC 2022, LNNS 502, pp. 542–550, 2022.
https://doi.org/10.1007/978-3-031-09076-9_49

– It is shown that a cost-benefit approach to measuring carbon dioxide emissions can be used to address the environmental resources sustainable consumption problem at different spatial levels: national, urban and households [6];
– proposed a new way to use the inter-sectoral balance Leontief model in solving the large dimension managing dynamic system problems in the economy [7].

2 Methods, Models, Data

To achieve this study goal - a mathematical model, algorithm and intelligent information technology development and implementation in the mathematical decision support to improve the economic system inter-industry links efficiency will be used the Leontief' Cost-Output model theory and methodology [1]:

$$X = AX + F, X \geq 0, \tag{1}$$

or as a matrix series for the productive matrix A [2]:

$$X = (E - A)^{-1}F = F + AF + A^2F + \ldots, \tag{2}$$

where $X = (X_1, X_2, \ldots, X_n)'$ – total resource units vector-column sums as a system for allocating gross output to sector output;

$F = (F_1, F_2, \ldots, F_n)'$ – the resource attracted units a vector-column sum as a system for the final demand distribution for the industry's products;

$AX = \left(\sum_{j=1}^{n} a_{1j}X_j, \sum_{j=1}^{n} a_{2j}X_j, \ldots, \sum_{j=1}^{n} a_{nj}X_j\right)'$ – resource units vector-column sums as a system for allocating intermediate demand for inter-industry products;

$A = \begin{pmatrix} a_{11} & a_{12} \ldots & a_{1n} \\ a_{21} & a_{22} \ldots & a_{2n} \\ \ldots & \ldots \ldots & \ldots \\ a_{n1} & a_{n2} \ldots & a_{nn} \end{pmatrix}$ – the additional resource units expenditure coefficients square matrix on total resource creation per unit; $a_{ij}, i, j = 1, n$ – the resource additional units expenditure a measure spent on interactions from the i-th industry to create total resource per unit from the j-th industry;

$(E - A)^{-1}$ – is the $E - A$ matrix inverse, E is the unit matrix.

From where we get that if we linearise the matrix series (2), then the Eq. (1) solution can be obtained iteratively using the first order differential-difference equation formula:

$$\frac{dX(t)}{dt} = F(t) - (E - A)X(t), X(0) = F(0), t = 0, 1, 2, \ldots, \tag{3}$$

in which the total resource units sum differential difference is equal to the final demand and final supply difference for branch products, which as the economic system inter-branch links gross output and efficiency a whole provide equilibrium growth.

The data information base on intersectoral linkages for the Kazakhstan Republic's three-sector national economy is the Input-Output table (see Table 1) in value terms in billion US dollars for 2019, which is borrowed from the OECD report [8].

Table 1. The input-output table Information database for the Kazakhstan Republic, USD billion, 2019.

	Produce resource			Final resource	Output resource
	Agriculture	Manufacturing	Service		
Agriculture	1.268	1.545	0.491	9.757	13.060
Manufacturing	0.610	21.935	10.238	67.596	100.388
Service	2.166	17.100	27.268	104.315	150.848

Then, according to the input-output model [1] and [2] Leontief's theory, we obtain the additional resource units expenditure coefficients numerical values to create total resource one unit:

$$A = \begin{pmatrix} 0.0971\ 0.1066\ 0.0348 \\ 0.0068\ 0.2185\ 0.1047 \\ 0.0155\ 0.1104\ 0.1808 \end{pmatrix}. \quad (4)$$

It will also make use of the system's dynamics theory methods [9], the models, methods, the open-source Any Logic software product information technologies and software codes [10], and several other papers' experience [11–20].

3 Results

3.1 Leontief' Input-Output Model Development

Let $t = 1, 2, \ldots, 52$, - forecasting weekly periods, then for the Leontiev's model "Input-output" (1)–(2) construction in economic system inter-branch relations efficiency decision making an increase mathematical support are used:

- first-order difference equations to ensure Input_t, reserves balanced growth, – attracted resources units inputs and net flows – NetInput_t t in the final demand distribution system for both sectoral and gross output in physical volume indices and (or) billion USD:

$$\frac{d(\text{Input}_t)}{dt} = \text{Net Input}_t,\ \text{Input}_\text{t}\big|_{\text{t}=0} = \text{InitialInput}, \quad (5)$$

- first-order difference equations for balanced stock growth $\text{ResourceOutputX1}_t$, – total resource gross output output units per first sector output in physical volume indices and/or USD billion:

$$\begin{aligned} \frac{d(\text{ResourceOutputX1}_t)}{dt} &= \text{ResourceProduceX1}_t - \text{ResourceAddedX1}_t, \\ \text{ResourceOutputX1}_t\big|_{t=0} &= \text{InitialResourceOutputX1}, \end{aligned} \tag{6}$$

- first-order difference equations for balanced stock growth $\text{ResourceOutputX2}_t$, – total resource gross output output units per second sector output in physical volume indices and (or) USD billion:

$$\begin{aligned} \frac{d(\text{ResourceOutputX1}_t)}{dt} &= \text{ResourceProduceX2}_t - \text{ResourceAddedX2}_t, \\ \text{ResourceOutputX2}_t\big|_{t=0} &= \text{InitialResourceOutputX1}, \end{aligned} \tag{7}$$

- first-order difference equations for balanced growth stocks $\text{ResourceOutputX3}_t$, – total resource gross output output units per third sector output in physical volume indices and/or USD billion:

$$\begin{aligned} \frac{d(\text{ResourceOutputX3}_t)}{dt} &= \text{ResourceProduceX3}_t - \text{ResourceAddedX3}_t, \\ \text{ResourceOutputX3}_t\big|_{t=0} &= \text{InitialResourceOutputX3}. \end{aligned} \tag{8}$$

3.2 The Leontief' Input-Output Model Implementation

Let the input parameters be given:

- Initial Input = 0.0683, – the units initial stock attracted in the final demand distribution system for both sectoral and gross output in physical volume indices and/or in USD 0.725, 5.027 and 7.758 billion, respectively, in the agricultural, industrial and services sectors;
- InitialResourceOutputX1 = 0.0683, – the gross output total resource units initial stock for the first sectoral output in physical volume indices and/or USD 0.725 billion in the agricultural sector;
- InitialResourceOutputX2 = 0.0683, – gross output total resource units' initial stock for the second sectoral output in volume indices and/or USD 5.027 billion in industry;
- InitialResourceOutputX3 = 0.0683, – gross output total resource units' initial stock for the third sector output in volume indices and/or USD 7.758 billion in the services sector
- NetInputRate = 0.0998, – the attracted resources units net flows rate in the final demand distribution system for both sectoral and gross output;
- ResourceProduceFactorA11 = 0.0971, ResourceProduceFactorA12 = 0.1066, ..., ResourceProduceFactorA33 = 0.1808, – the additional resource units consumption coefficient values spent on interaction from i-th industry for creation per unit of j-th industry total resource taken from the matrix corresponding elements (4).

Then, Leontief's mathematical input-output model (1)–(8) computer implementation in the AnyLogic software environment [15] allows the following output results – value estimation and prediction, the computer experiments results from simulation and visualisation:

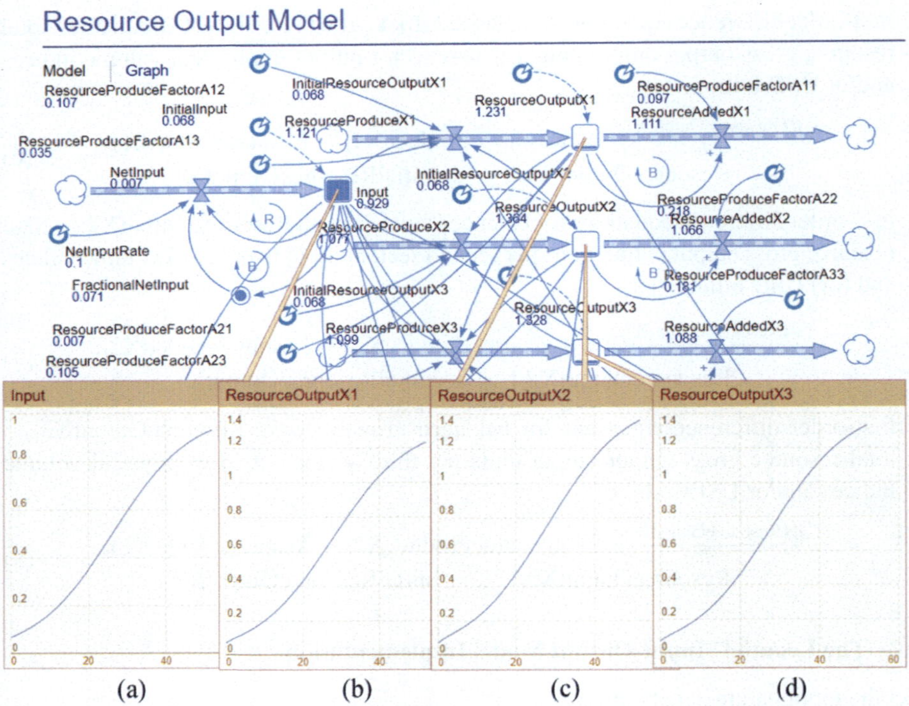

Fig. 1. Leontief's three-sector input-output model computer implementation results: input parameters and outputs data.

- Input_t, – balanced growth in the final demand attracted resources units number for the sector products:
- the agricultural sector obtained a balanced increase in resources from an initial 0.0683 physical volume units and/or in monetary terms, – from USD 0.725 billion to 0.9184 physical volume units and/or in monetary terms to USD 9.744 billion (see "S" of the figurative ascending logistic curve: Fig. 1, (a), Fig. 2, (b), Table 2),
- industry obtained resources a balanced growth from the initial 0.0683 billion USD in volume and/or in money terms, from 5.027 billion USD to 0.9184 billion USD in volume and/or in money terms, from 67.593 billion USD (see "S" of the ascending logistic curve: Fig. 1, (a), Fig. 2, (c), Table 2),
- at the services sector level, it was obtained resources a balanced growth with an initial volume of 0.0683 billion US dollars and/or in monetary terms from 7.758 billion US dollars to 0.9184 billion US dollars and/or in monetary terms to 104.323 billion US dollars (see "S" of the ascending logistic curve: Fig. 1, (a), Fig. 2, (d), Table 2)

Table 2. The three-sector input-output model Resource estimation results for the Kazakhstan Republic, physical resource volume index, 2019.

	Produce resource			Final resource (F_i)	Output resource (X_i)
	Agriculture (a_{i1})X_1	Manufacturing (a_{i2})X_2	Service (a_{i3})X_3		
Agriculture	0.130	0.158	0.051	1.000	1.339
Manufacturing	0.009	0.324	0.152	1.000	1.485
Service	0.021	0.164	0.261	1.000	1.446

- at the national economy level, resources a balanced growth was obtained from the initial 0.0683 of volume units and/or in monetary terms, from USD 13.510 billion to 0.9184 of volume units and/or in monetary terms to USD 181.660 billion (see "S" of the ascending logistic curve: Fig. 1, (a), Fig. 2, (a), Table 2);
- ResourceOutput$X1_t$, – the gross output resource units sum balanced growth in the agricultural sector, where we can observe resources a balanced growth from initial 0.0683 USD units of volume and/or in monetary terms from 0.725 to 1.231 USD units of volume and/or in monetary terms from 13.06 USD units (see "S" of the ascending logistic curve: Fig. 1, (b) and Fig. 2, (b), Table 2),
- ResourceOutput$X2_t$,- gross output total resource units a balanced growth per industry, in which there is resources a balanced growth from an initial 0. 0683 physical volume units and/or in monetary terms from USD 5.027 billion to USD 1.364 physical volume units and/or in monetary terms from USD 100.388 billion (see "S" of the ascending logistic curve: Fig. 1, (c), Fig. 2, (c), Table 2),
- ResourceOutput$X3_t$, - the gross output total resource units balanced growth in the services sector, where there is resources a balanced growth from an initial 0.0683 physical volume units and/or in money terms from USD 7.758 bn to USD 1.328 physical volume units and/or in money terms from USD 150.848 bn (see "S" of the ascending logistic curve: Fig. 1, (d), Fig. 2, (d), Table 2).

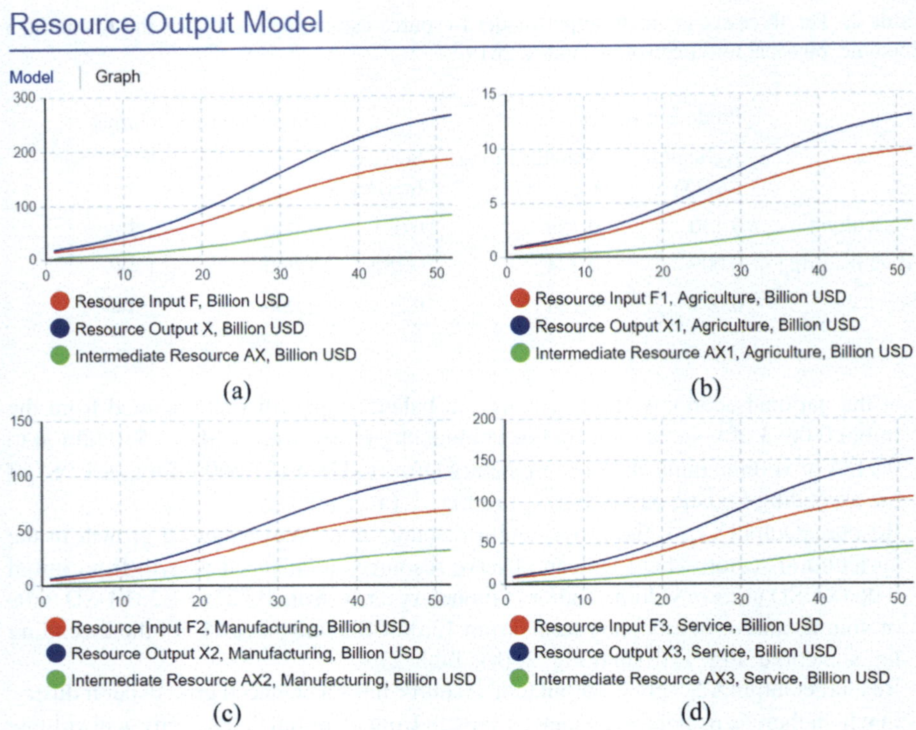

Fig. 2. The three sectors' Leontief's Input-Output model inputs and outputs resource estimates in current US dollars.

4 Discussions

Due to the intelligent computing technologies development, the need to develop mathematical models, algorithms and information technologies in decision-making mathematical support, not only in economic system inter-branch relations but also in human activity other areas a number, society and business development is increasing:

- on the input-output simulation model basis, the GHG emission intensity target sectors (GHG emission) are determined; the sectoral and national GHG emission peak volumes under different economic growth levels are estimated; the value-added and GHG emission intensity changes that affect their peak are estimated [21];
- established an optimal water allocation between industrial sectors based on water footprint model, which is both rational and has applications for sustainable water use in the PRC water management strategy future formulation [22];
- the interregional carbon emissions evolution characterization in the China northeastern industrial region [23] and carbon dioxide emissions distribution optimization based on the interprovincial trade cost-benefit analysis [24] and a hybrid cost-benefit method [25] which reduces annual carbon emissions using the energy consumption example in the five higher education institutions in the United States was proposed in the paper.

5 Conclusion

Thus, the paper, based on the input-output table statistical data, estimates the national economy total resource units sums as gross output distribution a system by the intermediate interaction and final attraction inter-branch links resources. The Leontief model algebraic equations system solution is reduced to differential-difference equations a system. To solve the latter, we used an input-output simulation model in the AnyLogic software environment. The input data parameters have been determined and the gross output, intermediate and final demand resources balanced growth condition for the country's economy has been ensured.

Acknowledgments. This research was funded by the Science Committee of the Ministry of Education and Science of the Republic of Kazakhstan (Grant No. AP09259435).

References

1. Leontief, W.: Quantitative input and output relations in the economic systems of the United States. Rev. Econ. Stat. **18**, 105–125 (1936)
2. ten Raa, T.: The Economics of Input-Output Analysis. Cambridge University Press, Cambridge (2006)
3. Xie, Y., Ji, L., Zhang, B., Huang, G.: Evolution of the scientific literature on input-output analysis: a bibliometric analysis of 1990–2017. Sustainability **10**(9), 3135 (2018)
4. Owen, A., Scott, K., Barrett, J.: Identifying critical supply chains and final products: an input-output approach to exploring the energy-water-food nexus. Appl. Energy. **210**, 632–642 (2018)
5. Aydogus, O., Deger, C., Çalişkan, E., Gunal, G.G.: Regional input-output analysis of a mega-event: possible impact of EXPO on Izmir economy. Anadolu Univ. J. Soc. Sci. **15**(2) (2015)
6. Munksgaard, J., Wier, M., Lenzen, M., Dey, C.: Using input-output analysis to measure the environmental pressure of consumption at different spatial levels. J. Ind. Ecol. **9**, 169–185 (2005)
7. Masaev, S.N.: Leontev input-output balance model as a dynamic system control problem. Herald of the Bauman Moscow State Tech. Univ. Ser. Instrum. Eng. **2**(135), 66–82 (2021). (in Russian)
8. OECD. https://stats.oecd.org. Accessed 07 Feb 2022
9. Sterman, J.D.: Business Dynamics: Systems Thinking and Modeling for a Complex World. McGraw-Hill Inc. (2000)
10. The AnyLogic Company. www.anylogic.com. Accessed 10 Jan 2022
11. Balabanova, A., Petrova, S., Fomenko, V., Keschyan, N.: Labor potential of youth for the development of ecology in the digital economy. In: E3S Web of Conferences, vol. 258, p. 06048 (2021)
12. Balabanova, A., Keschyan, N., Borisova, T., Hachemizova, E.: Using digital platforms for environmental management. In: E3S Web of Conferences, vol. 244, p. 07006 (2021)
13. Stone, M.: Developments in B to B and B to C marketing and sales automation systems. J. Bus. Bus. Mark. **28**(2), 203–222 (2021)
14. Low, S., Ullah, F., Shirowzhan, S., Sepasgozar, S.M.E., Lee, C.L.: Smart digital marketing capabilities for sustainable property development: a case of Malaysia. Sustain. (Switzerland) **12**(13), 5402 (2020)

15. Kadyraliev, A., et al.: The impact of cargo turnover on the ecology of Kyrgyzstan. In: IOP Conference Series: Earth and Environmental Science, vol. 937, no. 2, p. 022086 (2021)
16. Gnap, J., Konečný, V., Slávik, R., Beňová, D.: Possible impacts of regulating the weekly rest of road freight drivers on logistics in EU countries. Nase More. **65**(4 Special issue), 259–265 (2018)
17. Kerimkhulle, S., Aitkozha, Z.: A criterion for correct solvability of a first order difference equation. In: AIP Conference Proceedings, vol. 1880, p. 040016 (2017)
18. Kerimkhulle, S., Azieva, G., Saliyeva, A., Mukhanova, A.: Estimation of the volume of production of turbine vapor of a fuel boiler with stochastic exogenous factors. In: E3S Web of Conferences, vol. 339, p. 02006 (2022)
19. Jakubik, P., Kerimkhulle, S., Teleuova, S.: How to anticipate recession via transport indices? Ekonomicky Casopis **65**(10), 972–990 (2017)
20. Kerimkulov, S., Teleuova, S., Tazhbenova, G.: Measuring chaotic and cyclic fluctuations of CASS freight index: expenditures. Actual Prob. Econ. **171**(9), 434–445 (2015)
21. Song, J., Yang, W., Wang, S., Wang, X., Higano, Y., Fang, K.: Exploring potential pathways towards fossil energy-related GHG emission peak prior to 2030 for China: an integrated input-output simulation model. J. Clean. Prod. **178**, 688–702 (2018)
22. Xu, M., Li, C., Wang, X., Cai, Y., Yue, W.: Optimal water utilization and allocation in industrial sectors based on water footprint accounting in Dalian City. China. J. Clean. Prod. **176**, 1283–1291 (2018)
23. Cheng, H., Dong, S., Li, F., Yang, Y., Li, S., Li, Y.: Multiregional input-output analysis of spatial-temporal evolution driving force for carbon emissions embodied in interprovincial trade and optimization policies: case study of Northeast industrial district in China. Environ. Sci. Technol. **52**, 346–358 (2018)
24. Liu, S., Tian, X., Cai, W., Chen, W., Wang, Y.: How the transitions in iron and steel and construction material industries impact China's CO_2 emissions: comprehensive analysis from an inter-sector linked perspective. Appl. Energy. **211**, 64–75 (2018)
25. Dixit, M., Singh, S.: Embodied energy analysis of higher education buildings using an input-output – based hybrid method. Energy Build. **161**, 41–45 (2018)

Decision Support Models and Algorithms for Determining the Anthropogenic Load of Large Urban Agglomerations in the Context of Digitalization

I. B. Mamai[1](✉), Z. K. Vazirov[1], S. V. Pronichkin[2], and Y. S. Chernyshova[1]

[1] Federal Center of Theoretical and Applied Sociology of Russian Academy of Sciences, Moscow, Russia
ibmamai@mail.ru

[2] Federal Research Center "Computer Science and Control" of Russian Academy of Sciences, Moscow, Russia

Abstract. Modern decision support systems in ecology are aimed at resolving the contradiction between the inability of megacities ecosystems to meet the growing needs of society. Scientific and methodological approaches have been developed to determine the anthropogenic load of large urban agglomerations in the context of digitalization based on cellular automata and network models of quantitative and qualitative characteristics of ecosystems. A comparative analysis of environmental and socio-economic indicators has been carried out, which makes it possible to assess the level of variation of the analyzed target indicators to ensure the sustainable development of megacities ecosystems.

Keywords: Decision maker · Ecosystem · Digital transformation · Decision support system

1 Introduction

The development of world economic relations over the past few decades has been accompanied by a radical transformation of large urban agglomerations [1, 2]. Strengthening the integration processes of urbanized territories led to the emergence of both positive and negative effects. One of the most threatening challenges facing the world community at the present stage of development is leveling the consequences of anthropogenic impact on the ecosystem, since the lack of balanced and coordinated actions in this direction may have irreversible consequences in the short term. Thus, the solution of environmental problems both at the supranational and national levels is possible using decision support tools [3–5].

The state of megacities ecosystems in many countries of the world is characterized by the division of the anthropogenic load into regions with the dominance of a certain type of negative impact on the environment. In addition, the regions of large urban agglomerations are usually very stable and do not change over many years [6, 7].

R. Silhavy (Ed.): CSOC 2022, LNNS 502, pp. 551–556, 2022.
https://doi.org/10.1007/978-3-031-09076-9_50

Previous studies [8, 9] associated the anthropogenic load clear boundaries emergence with landscape features, geographical characteristics, and historical features of territories. Despite the existence of certain developments in the formalization of convergent relationships in terms of the overall level of anthropogenic load, the segment of anthropogenic load forecasting based on network structures has not been sufficiently studied. As a rule, studies of anthropogenic load are based on a comparative analysis of environmental indicators and do not imply a qualitative assessment of socio-economic effects.

2 Materials and methods

One of the approaches to modeling the anthropogenic load is cellular automata [10, 11]. They have become widespread due to the intensive development of information technologies, which made it possible to quickly and easily conduct numerical experiments with models of large urban agglomerations. In addition, the algorithms of such models are more transparent than differential equations, which require special implementation of numerical methods for their approximate solution.

The anthropogenic load formation process is considered as a set of elementary interactions between the actors of the megalopolis ecosystem [12, 13]. Actors are governed by the same set of rules. These rules set for each moment of time the state of the actor, which will be at the next moment of time, depending on the current state of other actors in the megalopolis ecosystem. The state of the actor will be considered the qualitative level of anthropogenic load (high, medium, low) at a certain point in time. Accordingly, the set of states is a set of alternatives to the decision support system for determining the anthropogenic load of large urban agglomerations. The set of actors is considered as not changing with time and fixed on a certain lattice. This means that for each actor, an ecosystem is specified, which is also fixed.

Anthropogenic load is determined by the negative impact on the environment by the actors of the immediate environment. Then the transition function of such an anthropogenic load cellular automaton is described by the equation

$$a_{kl}^{t+1} = \operatorname*{argmax}_{i} F_i^t(x_{kl}), \tag{1}$$

where a_{kl}^{t+1} is the anthropogenic load on the ecosystem in a cell with indices k and l at time $t+1$; x_{kl} are neighborhoods of the cell with indices k and l; $F_i^t(x_{kl})$ is the value of the cell optimality principle with the value of the anthropogenic load i in the neighborhoods of x_{kl}.

The model (1) with a deterministic function F_i^t is characterized by rapid convergence. The model forms stable areas of anthropogenic load with clear and fairly smooth boundaries of large urban agglomerations. In contrast, the stochastic model does not coincide with the deterministic model and forms rather complex areas, which are also characterized by the formation of groups of ecosystem actors, however, not as clear as in the deterministic model.

In addition to the considered model for determining the anthropogenic load, there are possibilities for its expansion, taking into account the uneven initial distribution, the influence of external factors and the dependence of various scenarios for reducing the anthropogenic load. However, the process of distribution of the anthropogenic load in such models occurs unevenly and is set only by the initial distribution.

The connections between the megalopolis ecosystem actors can be divided into weak and strong, taking as a definition that the more common neighbors two actors have, the stronger the connection between them. In this case, links of different types will play different roles in the anthropogenic load spread dynamics.

The conducted studies of the model showed that, in contrast to the basic deterministic model, the convergence rate turned out to be very high, while the final states indicate a significant quantitative superiority of the load from industrial areas of large urban agglomerations, which has a slight advantage at the beginning. However, such a model does not allow explaining the formation of regions resistant to anthropogenic load.

3 Results

It is proposed to jointly use cellular automata and network models for determining the anthropogenic load. Instead of a homogeneous lattice, the megalopolis ecosystem actors set is considered as a graph. Moreover, not a random graph is considered, but with parameters characteristic of the digitalization of urbanized territories.

The ecosystem actors' environment structure can influence and change in accordance with the position of the actor in the Internet of things (IoT) [14, 15]. This means that the network has many actors with a small number of connections and a small group of actors with a large number of connections. The connections between the ecosystem actors, in addition to the division into strong and weak, can be asymmetrical. The fact that some actors can influence the anthropogenic load of a particular actor does not mean that it also influences the choice of their strategy. Ecosystem actors with a large number of input connections usually have a small number of output connections.

In addition to direct communication, actors can be connected indirectly, that is, through an intermediary or a group of intermediaries. The shortest distance between the actors of the IoT network of a smart city and the average path are considered as the average of all pairs of actors of the shortest distance between them. Smart city IoT networks typically have a fairly short average path. This property is manifested precisely because of the presence of long-distance connections in the Internet of things.

Complex networks of the Internet of things are characterized by the presence of a community structure of ecosystem actors. In the digitalization context, groups of actors have a high density of their edges, despite the fact that the density of edges in the Internet of Things between individual groups is low. To identify the structure of subgroups of actors in the Internet of Things network, it is proposed to use a method based on modularity optimization [16, 17].

The value of modularity in the Internet of Things is equal to the difference between the share of edges within a group of actors in the megalopolis ecosystem and the expected share of connections if the edges were randomly placed. IoT networks of smart cities with high modularity have tight connections between ecosystem actors within a subgroup and weak connections between actors from different subgroups.

Let an Internet of Things network with actors divided into several groups be given. The edges between the actors are given by the adjacency matrix. Then modularity is defined by the following equation:

$$M = \frac{1}{4N} \sum\nolimits_{j=1}^{L} \left(y_j^T p\right)^2 s_j, \tag{2}$$

where y_j is the eigenvector of the adjacency matrix, p is a linear combination of normalized eigenvectors of the symmetric adjacency matrix, s_j are the eigenvalues, N is the total number of edges in the IoT network.

The task is to find such a partitioning vector of the Internet of Things actors network that would maximize the value of M.

Optimization consists of three phases:

- using cellular automata to identify areas of anthropogenic load on the ecosystem of large urban agglomerations;
- in the selected areas to identify subgroups of actors of the Internet of things network reaching local maxima of modularity;
- a new network of actors is built, the nodes of which are subgroups found at the first stage.

These phases are repeated until the maximum modularity is reached. The result of the algorithm is a hierarchy of subgroups of actors in the identified areas of anthropogenic load on the ecosystem of large urban agglomerations.

For each area, an analysis of environmental and socio-economic indicators is carried out, which makes it possible to assess the level of variation of the analyzed target indicators of sustainable development of ecosystems. An insignificant level of dispersion of the sample variables indicates the presence of convergent processes, while a significant dispersion of parameters, on the contrary, confirms the fact of the unstable state of the ecosystems of large urban agglomerations.

4 Discussion

Using the developed models and algorithms based on modularity, subgroups of ecosystem actors are found. The division into areas, although it passes along the boundaries of subgroups of actors, differs slightly from the division by modularity. The dependence between the degree of actors subgroups connectivity on the presence of a developed transport infrastructure is revealed. Where there is a greater intensity of traffic flows, the actors are more closely connected and overlap each other, and where there is not, there is more modularity. This proves the effectiveness of using modularity as a measure of the distribution of the anthropogenic load on the ecosystem, and the subgroups of actors obtained on its basis explain the formation of the boundaries of large urban agglomerations in the digitalization context. Despite the identified distances between subgroups of actors, the parameters defined within each subgroup of actors are characterized by insignificant differences. This confirms the thesis about a significant level of convergence of environmental management systems of industrial enterprises of large urban agglomerations.

There is a fairly high level of variation in socio-economic indicators of anthropogenic load between subgroups of actors in the megalopolis ecosystem, which decreases over time. However, in general, attention should be paid to the fact that this group of indicators is the lowest level of load, therefore, reducing their variability is important and indicates a further convergence of the parameters of environmental and socio-economic indicators of anthropogenic load.

As a module of the decision support system for the formation of strategies to reduce the anthropogenic load on the ecosystems of large urban agglomerations, various principles of optimality from taxation indicators are used:

- share of revenues from environmental taxes for emissions and discharges of pollutants;
- share of transport tax revenues in total tax revenues.

Transport taxes are characterized by a fairly high level of variability. At the same time, trends in the development of environmental taxes showed a fairly low level of variation.

5 Conclusion

The problem of analyzing the distribution of the anthropogenic load of large urban agglomerations in the context of digitalization is considered. Approaches to decision support based on cellular automata and network models are proposed. The results of the computational experiment confirmed the relationship between the subgroups of actors of the network model of the Internet of things and the limits of the distribution of the anthropogenic load. The proposed method based on modularity can be used to explain the existing boundaries of the anthropogenic load on the ecosystems of megacities and predict the emergence of new boundaries.

Acknowledgments. The article is prepared with the financial support of the Russian Science Foundation, project № 19–78-10035.

References

1. Chen, J., Wang, S., Zou, Y.: Construction of an ecological security pattern based on ecosystem sensitivity and the importance of ecological services: a case study of the Guanzhong plain urban agglomeration. China. Ecol. Indic. **136**, 108688 (2022)
2. Li, L., Ma, S., Xiao, X.: Integrated regional development: comparison of urban agglomeration policies in China. Land Use Policy **114**, 105939 (2021)
3. Delias, P., Manitsa, P., Karasavvoglou, A.: Robustness-oriented group decision support a case from ecology economics. Procedia Technol. **8**, 285–291 (2013)
4. Catford, J., Wilson, J., Duncan, R.: Addressing context dependence in ecology. Trends Ecol. Evol. **37**, 158–170 (2021)

5. Mamai, I., Savelyev, B., Pronichkin, S., Kholstov, A.: Using evolutionary algorithms to determine the environmental projects effectiveness. J. Phys. Conf. Ser. **2060**, 012009 (2021)
6. Li, X., Lu, Z.: Quantitative measurement on urbanization development level in urban Agglomerations: a case of JJJ urban agglomeration. Ecol. Indic. **133**, 108375 (2021)
7. Wang, Y., Yin, S., Chen, W.: Interaction of economic agglomeration, energy conservation and emission reduction: evidence from three major urban agglomerations in China. Energy **241**, 122519 (2022)
8. Ma, S., Cai, Y., Zhao, Y.: Towards balanced development stage: Regulating the spatial pattern of agglomeration with collaborative optimal allocation of urban land. Cities **126**, 103645 (2022)
9. Wang, Q., Liu, S., Yu, L.: Effects of urban agglomeration and expansion on landscape connectivity in the river valley region. Qinghai-Tibet Plateau. Glob. Ecol. Conserv. **34**, e02004 (2022)
10. Dong, Y., Zhao, G.: A spatiotemporal chaotic system based on pseudo-random coupled map lattices and elementary cellular automata. Chaos Solitons Fractals. **151**, 111217 (2021)
11. Romero, N., Rovella, A., Vilamajo, F.: Remark on cellular automata and shift preserving maps. Appl. Math. Lett. **19**, 576–580 (2006)
12. Silva, E., Ahern, J., Wileden, J.: Strategies for landscape ecology: an application using cellular automata models. Prog. Plann. **70**, 133–177 (2008)
13. Wu, X., Lin, A., Song, D.: Simulating spatiotemporal land use change in middle and high latitude regions using multiscale fusion and cellular automata: the case of Northeast China. Ecol. Indic. **133**, 108449 (2021)
14. Zhao, W., Yi, L.: Research on the evolution of the innovation ecosystem of the internet of things: a case study of Xiaomi (China). Procedia Comput. Sci. **199**, 56–62 (2022)
15. Kang, L.: Street architecture landscape design based on wireless internet of things and GIS system. Microprocess. Microsyst. **80**, 103362 (2020)
16. Bruckmann, C., Stadler, P., Hellmuth, M.: From modular decomposition trees to rooted median graphs. Discret. Appl. Math. **310**, 1–9 (2022)
17. Jacoby, D., Freeman, R.: Emerging network-based tools in movement ecology. Trends Ecol. Evol. **31**, 301–314 (2016)

The Effect of Additive Regularization for Topic Modeling of Social Media Communities

Andrey M. Fedorov and Igor O. Datyev(✉)

Institute for Informatics and Mathematical Modeling, Kola Science Centre of the Russian Academy of Sciences, 14, Fersman St., Apatity 184209, Russia
{fedorov,datyev}@iimm.ru

Abstract. This article presents the experience of improving the results of social networks communities topic modeling using the Additive Regularization for Topic Modeling (ARTM) mathematical approach. The improvement of the results is achieved through the use of basic regularizers available in the open-source software BigARTM. Topic models obtained using regularized ARTM are compared with topic models obtained by Latent Dirichlet Allocation and Probabilistic Latent Semantic Analysis. The experiments were carried out on a dataset, containing preprocessed texts of posts from communities of an online social network. In particular, the values of the UMass metric, the purity of topics, and the sparsity of the distribution matrices are compared. Disadvantages of UMass-type coherence metrics for assessing the quality of topic models obtained using the ARTM method are discussed. Metrics are proposed that can be used for assessing the quality of topic models. Conclusions are drawn about the suitability of the ARTM approach for modeling communities of online social networks in the framework of our future research.

Keywords: Social network communities · Probabilistic topic modeling · ARTM · LDA · PLSA · Sparseness · Smoothing · Decorrelation

1 Introduction

Heterogeneous information placed in social media is the object of analysis of modern intelligent information systems. In addition to commercial use by marketers, the topic analysis of online social networking communities can be the basis for reporting on the retrospective and current state of society, as well as help predict its development. For automated extraction of the meaning of textual and related information, topic modeling methods are used. The task of modeling communities (groups) of users in social media is an urgent task in the framework of information support for decision-making at different levels of state and municipal (regional) government, as well as for the development of modern digital sciences.

Our research is devoted to the use of topic modeling to build models of social media communities. In a previous work [1] we investigated several probabilistic topic modeling methods on our dataset, consisting of posts from the communities of the online social network VKontakte.

R. Silhavy (Ed.): CSOC 2022, LNNS 502, pp. 557–567, 2022.
https://doi.org/10.1007/978-3-031-09076-9_51

Based on the experimental results, we have shown that the ARTM approach is one of the fastest model training times. Another advantage of BigARTM is multimodality, which suits us for modeling social media communities, since they contain related information, and not always plain text. In addition, we were interested in the automatic evaluation of the quality of topics obtained using various topic modeling methods. We chose the UMass metric as such a quality assessment, since researchers noted the highest correlation of this metric with human judgments [2, 3]. As an additional quality criterion, we used the sum of the probabilities of the top ten topic tokens (the so-called topic purity) [4]. However, in our past experiments, the values of the UMass coherence metric of ARTM topic models turned out to be no better than those of other methods. The LDA method significantly outperformed ARTM and other methods in terms of the UMass coherence criterion. We explained this by using ARTM without regularizers, in this mode the ARTM approach is the usual PLSA [5].

For further work on social network communities topic modeling and the development of the corresponding software framework [6] we need to confirm the benefits of the ARTM approach. We are experimenting with tuning the basic regularizers available in the BigARTM library and the hyperparameters of the LDA method. As a baseline for comparison, we use the LDA method, since it showed the best results in our previous experiments [1], as well as the ARTM method without regularization, i.e., actually PLSA. The main research question of this work: Can the ARTM method using basic regularizers show results comparable to the LDA method (by the UMass metric, the sum of the top ten topic tokens) on our dataset?

2 Regularization Features of the ARTM Approach

The main question of this study concerns the BigARTM library and the ARTM mathematical approach to the construction and combination of topic models that is programmatically implemented in it [7]. Therefore, the authorship of most of the works considered in our brief review belongs to the creators of this approach and the library. The main author of the ARTM approach is K.V. Vorontsov is one of the leading Russian specialists in the field of probabilistic topic modeling and information retrieval.

Konstantin Vorontsov and colleagues offer an alternative to the Bayesian approach - Additive Regularization of Topic Models, ARTM [8]. ARTM is an application of the classical regularization theory of ill-posed problems [9] to topic modeling. Training a topic model from a document collection is reduced to the problem of stochastic matrix factorization. In the general case, it has an infinite set of solutions, that is ill-posed. To regularize the problem, terms are added to the logarithm of the likelihood that formalize additional requirements for the model.

ARTM has two fundamental differences [4] from the Bayesian approach. First, the task of building a fully generative probabilistic model of text is not posed. Many topic model requirements are formalized using optimization criteria rather than prior distributions. In this case, regularizers are not required to have a probabilistic interpretation. Secondly, instead of Bayesian inference, a simpler approach is used - a regularized EM algorithm. ARTM does not require integration over the model parameter space and is reduced to differentiating regularizers with respect to parameters. Building of multi-purpose multi-objective topic models [10] is greatly simplified due to the additivity of the regularizers. Adding a regularizer to the model requires a slight modification of the M-step in the EM-like algorithm. ARTM also differs from previously proposed regularization methods [11–13]. Each of these methods used a specific regularizer: KL-divergence, Dirichlet distribution, L1- or L2-norm. ARTM is a generalized approach to topic modeling as a multiobjective optimization problem [14].

It should be noted the possibility of building multimodal models using ARTM, which allows taking into account related, often non-textual information contained in document collections [15, 16]. Examples of such information in social media are: time, source, categories, links, tags, phrases, named entities (names of user communities, captions for multimedia elements), objects in images, records of user actions, emoticons, hashtags, etc.

ARTM was further developed in the works [17–19], where a lot of attention is paid to tuning regularizers or creating new regularizers, as well as [20] studies the ARTM algorithm convergence problem.

In [4, 14, 21] it was shown that the combination of smoothing, sparse, decorrelation regularizers increase the semantic interpretability of topics and form the minimum required set of regularizers for most topic modeling problems.

In this work, we also conduct experiments with three main ARTM regularizers: smoothing, sparseness, decorrelation. Separately for the LDA method, we are trying to tune the two parameters alpha and beta, which control the degree of approximation of the matrices Phi, Theta, by the Dirichlet distribution.

3 Regularizers Tuning and Quality Metrics

We evaluate our models on a real-life dataset mined from the regional communities open data of the most popular Russian online social network VKontakte. The dataset contains 15754 tokens (we used only unigrams), 9084 lemmatized posts, and described in more detail in [1]. In our experiments, we have trained a series of topic models. In total we have evaluated nine models with a range of topics from 1 to 400 for each model. For all models, except for LDA_auto and ARTM without regularization (PLSA), we have chosen regularization coefficients manually based on the results of several test experiments.

We tuned not only the basic ARTM regularizers, but also tried to find the values of the LDA method parameters that give the best topic models on our dataset. For this we used several software libraries: BigARTM, Gensim, Gensim(Mallet). In these libraries, it is possible to configure two main (with the exception of Gensim(Mallet), where we found only one parameter - alpha) LDA hyperparameters - alpha and beta.

At the next stage of the experiments, we tried to find the best values of the regularization coefficients for the ARTM approach. The total number of iterations (passes through the collection of documents) for each model is 30. We placed the results of the two stages of our experiments on general graphs (see Figs. 1, 2, 3 and 4). Designations for all figures:

- LDA_auto - LDA model from Gensim package with automatic initialization of alpha and beta parameters.

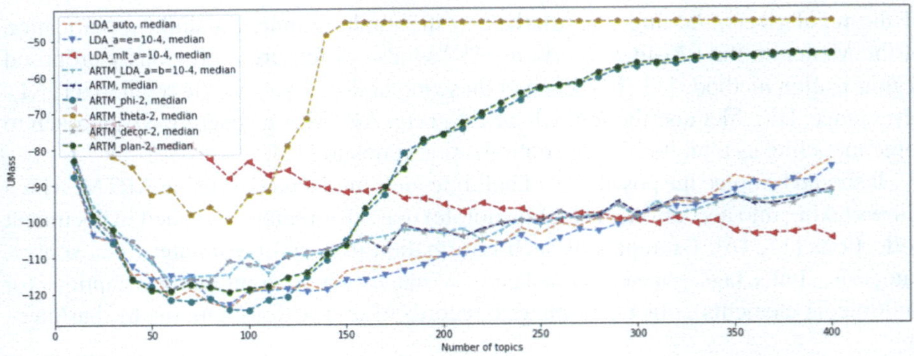

Fig. 1. Values of the UMass coherence metric for topic models obtained by LDA and ARTM methods depending on the number of topics.

- LDA_a = e = 10^{-4} - LDA model from Gensim package with alpha and beta parameters initialized with values 10^{-4}.
- LDA_mlt_a = 10^{-4} - LDA model from Gensim package with Mallet[1] modification with alpha parameter initialized to 10^{-4}.
- ARTM_LDA_a = b = 10^{-4} - LDA model from the BigARTM package with alpha and beta parameters initialized to 10^{-4}.
- ARTM - ARTM model without regularization (PLSA).
- ARTM_phi-2 - ARTM model with Phi matrix sparse, tau = –2.
- ARTM_theta-2 - ARTM model with Theta matrix sparse, tau = –2.
- ARTM_decor-2 - ARTM model with topic decorrelating, tau = –2.
- ARTM_plan-2 - ARTM model using a sequence of sparse regularizers with tau = –2 (10 iterations each) for Phi, Theta matrices, and topic decorrelating regularizer (tau = –2).

Figures 1, 2, 3 and 4 show the values of various metrics (UMass coherence, the probabilities sum of the top ten tokens of the Phi matrix topic, the proportion of near-zero (located in the interval [0, 10^{-4}] values of the Phi and Theta matrices elements) depending on the number of topics for topic models obtained using the LDA method and the ARTM approach. Metrics are calculated for each topic separately. The graphs show the median values for each topic model.

The proportion of near-zero elements in the Phi matrix for different models demonstrates similar behavior (see Fig. 3). There is a rather sharp increase in the overall sparsity and a plateau near the value of 1.0, which corresponds to the complete zeroing of the Phi matrix.

Only on the basis of the metrics presented in Figs. 1, 2, 3 and 4, it is not possible to unambiguously judge the advantage of one of the topic models for the following reason. With an increase in the number of topics in the model, the sparseness of the Phi and Theta matrices increases. In the matrices, the number of zero values increases in the

[1] Python wrapper for Latent Dirichlet Allocation (LDA) from MALLET, the Java topic modelling toolkit - https://radimrehurek.com/gensim_3.8.3/models/wrappers/ldamallet.html.

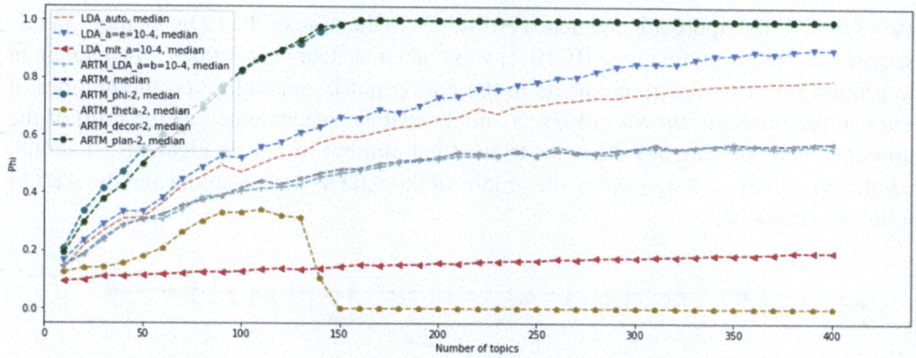

Fig. 2. Median of the probabilities sum of the top ten tokens (Phi) vs the number of topics.

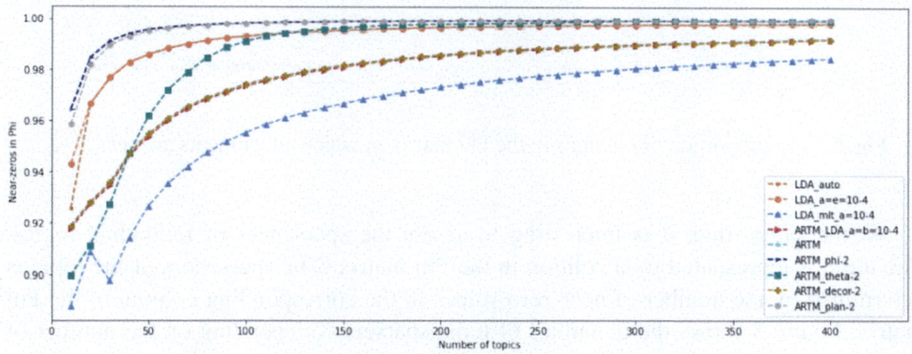

Fig. 3. The proportion of near-zero elements in the matrix Phi.

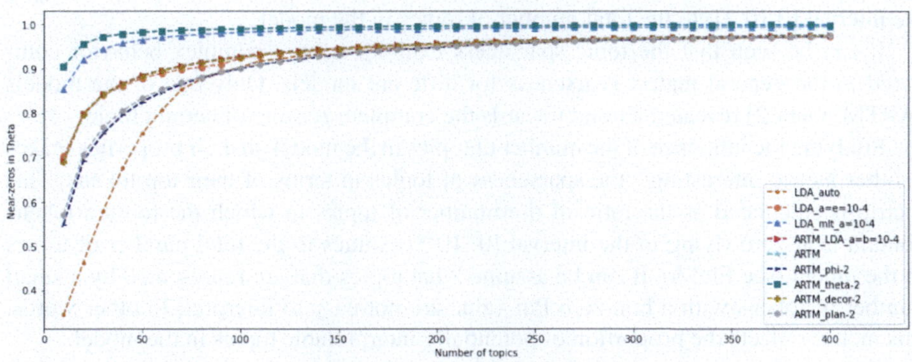

Fig. 4. The proportion of near-zero elements in the matrix Theta.

case of the ARTM approach and near-zero values in the case of the LDA method. In this experiment, values in the range [0, 10^{-4}] were taken as near-zero values. The change in the proportion of near-zero elements in the entire matrix depending on the number of topics in the model is shown in Figs. 3 and 4. Sparsity is calculated as the ratio of the number of near-zero matrix elements to the total number of matrix elements. It should be noted that in Figs. 3, 4, 5 and 6, the graphs of the ARTM_decor model and the ARTM model are the same.

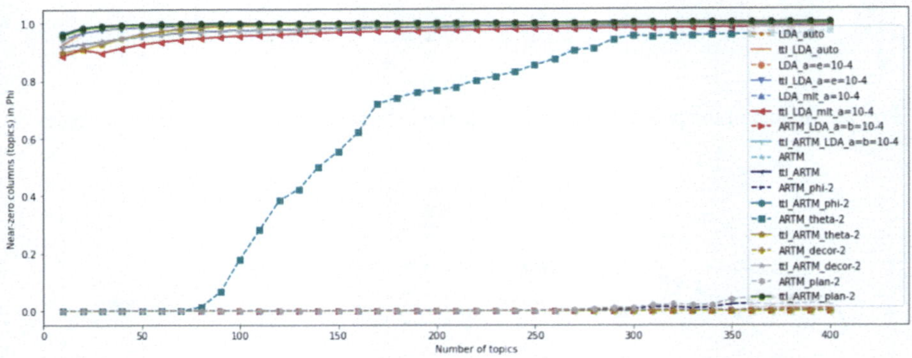

Fig. 5. The proportion of columns in the Phi matrix in which all elements are near-zero.

At the same time, it is interesting to assess the sparseness of individual topics. The topic is represented by a column in the Phi matrix. The sparseness of the topic is determined by the number of near-zero values in the corresponding column of the Phi matrix. Figure 5 shows the dynamics of topic sparseness depending on the number of topics in the model.

For clarity, this graph (see Fig. 5) duplicates the proportion of near-zero elements in the Phi matrix of topic models (shown in Fig. 3), which in the legend of Fig. 5 are marked with the prefix "ttl_". The rest of the legend corresponds to those in Fig. 1. Topic sparseness is calculated as the ratio of topics in which all values are near-zero (lying in the interval [0.10^{-4}]) to the total number of topics in the model.

It can be seen that the topic sparseness exhibits a more complex behavior compared to the general matrix sparseness for different models. Only one of the models (ARTM_theta-2) revealed a trend towards the complete zeroing of certain topics.

Studying the influence of the number of topics in the model on their properties makes another metric interesting - the sparseness of topics in terms of their top tokens. This metric is calculated as the ratio of the number of topics in which the top ten tokens contain near-zero (lying in the interval [0, 10^{-4}]) values to the total number of topics in the model (see Fig. 6). It can be assumed that topics that are represented by a small number of tokens with a non-zero Phi value are not easy to interpret. In other words, this metric reflects the proportion of potentially interpretable topics in the model.

Another interesting metric of the topic model is the engagement in topics of all vocabulary tokens. In the Phi matrix, this fact is represented by the absence of zero rows. The presence of zero rows in the Phi matrix indicates that the Phi matrix is sparse

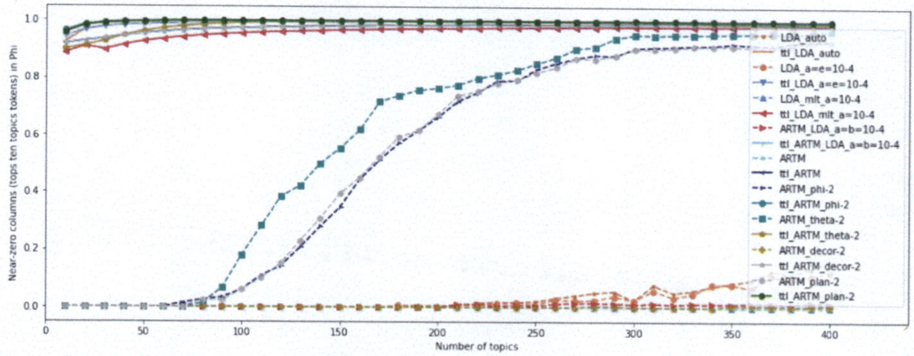

Fig. 6. The proportion of columns in the Phi matrix whose top ten tokens contain near-zero elements.

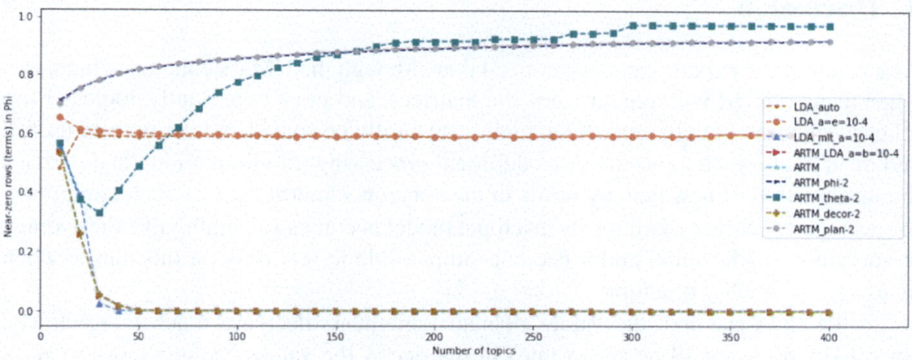

Fig. 7. The proportion of near-zero rows in the Phi matrix (i.e., the proportion of dictionary tokens that are not represented in any of the model topics).

by tokens. The sparseness of the Phi matrix by tokens is calculated as the ratio of the number of rows filled with near-zero elements to the total number of rows. Figure 7 shows the dynamics of the sparsity of the Phi matrix by tokens depending on the number of topics in the model. As in previous experiments, not absolutely zero values are used to estimate sparsity, but the interval $[0, 10^{-4}]$. With this in mind, the sparseness of the Phi matrix by row characterizes the proportion of dictionary tokens that are almost never used in topics, i.e., have an extremely low probability of occurring in any topic of the topic model.

In our study, we again tested the fit time of the model. Unlike the previous experiment [1], we have more models and the Gensim library (Mallet, denoted by LDA_mallet), as well as the maximum number of topics on the chart is 400 (see Fig. 8). Now ARTM is in second place, but the fit time is still comparable to the leader and is not critically high. It should be noted that the time depends on the number of passes through the collection (in our case it is 30), i.e., ARTM results can be improved, and even if ARTM is not considered the fastest, it is undoubtedly one of the fastest.

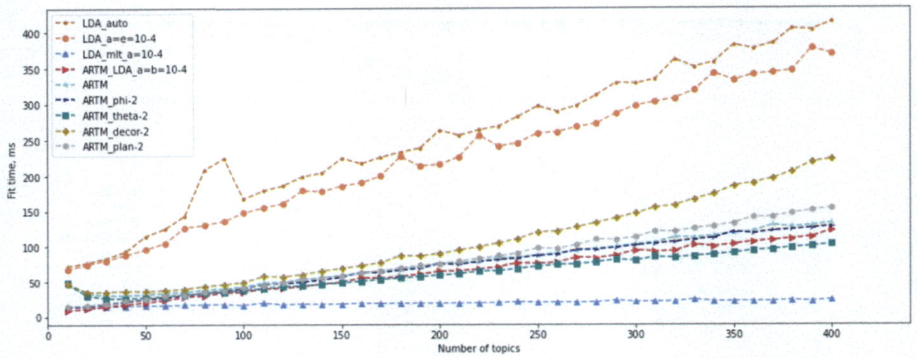

Fig. 8. Model fit time (milliseconds) for different methods, varying the number of topics.

4 Discussion

As a result of the experiments, we noticed that although the UMass coherence increases when using ARTM with regularizers, the matrices, and most importantly, topics or top tokens, can turn out to be zero. Such topics can hardly be considered correct and useful, and most likely, such topics require additional processing (or elimination), and perhaps the introduction of new penalty terms in the coherence metric. Under such conditions, the average coherence of all topics in a topic model becomes something like the average temperature in a hospital, and it becomes impossible to rely only on this metric when assessing the quality of a topic.

In the LDA method, the values of matrix elements decrease much more slowly, since LDA does not allow zero values in matrices - the values become close to zero, which we tried to display on the graphs (empirically setting the proximity threshold to 10^{-4}). ARTM is a relatively new approach and not all coherence metrics are suitable for evaluating topic models generated by ARTM. This unsuitability can be explained by the fact that coherence metrics such as UMass and UCI are not initially designed for the presence of zero probabilities in matrices.

As a result, we identified possible reasons for the coherence metrics incorrectness:

- According to the calculation formula, the value of the Umass metric of the zero topic may be greater than that of the non-zero topic, which makes an incorrect contribution to the metric.
- Features of social media documents, in particular, too short documents - posts, which, for example, may reduce the co-occurrence of words.
- The topic coherence metric is quite correct for calculating the coherence of a single topic. When we have multiple topics, how do we apply this metric to assess the quality of the topic model as a whole? Previously, we used the average. This works for LDA, for PLSA without regularization. However, during the regularization process, some topics are nullified. In other words, the metric is suitable for individual topics, but when we want to evaluate the model as a whole, i.e., set of topics, for social media communities we encounter good (higher) coherence values for bad null topics.

- Authors [22] note that coherence metrics like UMass and UCI usually take only the top ten tokens for a topic, and these top ten tokens in the text collection corpus are 1–2 percent, and the proportion of text covered by top words does not controlled. Perhaps in short texts of social networks, this proportion is even less.

We tried to take into account the above reasons when developing quality metrics for topic models proposed in the experimental part of this article.

5 Conclusion

This work is the final stage of the rationale for choosing the ARTM approach for modeling social media communities. Previously, using ARTM without tuning the regularizers (out of the box), we failed to obtain topic models that exceed the topic models obtained using the LDA method in terms of automated coherence metrics. However, since ARTM has a key advantage for modeling social network communities - multimodality, we decided to conduct a study using basic ARTM regularizers. The experiments were carried out on our dataset, containing posts from the communities of the online social network. As a baseline method for comparison, we used LDA, which showed the best results in our previous study, and the ARTM approach without regularization (i.e., PLSA).

We have demonstrated that the ARTM approach with regularization is comparable to other topic modeling methods in terms of the UMass coherence metric and the probabilities sum of the top ten topic tokens. This finally confirmed the suitability of the BigARTM library as part of our future research for topic modeling of social network communities. However, as a result of experiments, we found that the UMass coherence metric is not always suitable for automated quality evaluation of topic models obtained using the ARTM method with regularizers. We have proposed several metrics that may be useful in evaluating the quality of a topic model.

Acknowledgments. The work is supported by the Ministry of Science and Higher Education of the Russian Federation. Topic title: Methodology for creating information and analytical systems to support the management of regional development based on formative artificial intelligence and big data (reg.n. FMEZ-2022-0007).

References

1. Fedorov, A.M., Datyev, I.O., Shchur, A.L.: Social media communities topic modeling. In: Silhavy, R., Silhavy, P., Prokopova, Z. (eds.) CoMeSySo 2021. LNNS, vol. 231, pp. 605–614. Springer, Cham (2021). https://doi.org/10.1007/978-3-030-90321-3_50
2. Mimno, D., Wallach, H., Talley, Ed., Leenders, M., McCallum, A: Optimizing semantic coherence in topic models. In: Proceedings of the 2011 Conference on Empirical Methods in Natural Language Processing, Edinburgh, Scotland, pp. 262–272. Association of Computational Linguistics (2011)
3. Newman, D., Lau, J.H., Grieser, K., Baldwin, T.: Automatic evaluation of topic coherence. In: Human Language Technologies: The 2010 Annual Conference of the North American Chapter of the Association for Computational Linguistics (HLT 2010), pp. 100–108. Association for Computational Linguistics, Stroudsburg (2010)

4. Vorontsov, K., Potapenko, A.: Additive regularization of topic models. Mach. Learn. **101**(1–3), 303–323 (2014). https://doi.org/10.1007/s10994-014-5476-6
5. Hofmann, T.: Probabilistic latent semantic indexing. In: Proceedings of the 22nd Annual International ACM SIGIR Conference on Research and Development in Information Retrieval (SIGIR 1999), pp. 50–57. Association for Computing Machinery, New York, NY (1999). https://doi.org/10.1145/312624.312649
6. Datyev, I.O., Fedorov, A.M., Shchur, A.L.: Framework for civic engagement analysis based on open social media data. In: Silhavy, R. (ed.) CSOC 2020. AISC, vol. 1225, pp. 586–597. Springer, Cham (2020). https://doi.org/10.1007/978-3-030-51971-1_48
7. Kochedykov, D., Apishev, M., Golitsyn, L., Vorontsov, K.: Fast and modular regularized topic modelling. In: 21st Conference of Open Innovations Association (FRUCT), pp. 182–193. FRUCT Oy, Helsinki, Uusimaa (2017). https://doi.org/10.23919/FRUCT.2017.8250181
8. Vorontsov, K.V.: Additive regularization for topic models of text collections. Dokl. Math. **89**(3), 301–304 (2014). https://doi.org/10.1134/S1064562414020185
9. Tikhonov, A.N., Arsenin, V.Y.: Solution of Ill-Posed Problems. Winston, Washington DC (1977)
10. Khalifa, O., Corne, D.W., Chantler, M., Halley, F.: Multi-objective topic modeling. In: Purshouse, R.C., Fleming, P.J., Fonseca, C.M., Greco, S., Shaw, J. (eds.) EMO 2013. LNCS, vol. 7811, pp. 51–65. Springer, Heidelberg (2013). https://doi.org/10.1007/978-3-642-37140-0_8
11. Si, L., Jin, R.: Adjusting mixture weights of Gaussian mixture model via regularized probabilistic latent semantic analysis. In: Ho, T.B., Cheung, D., Liu, H. (eds.) PAKDD 2005. LNCS (LNAI), vol. 3518, pp. 622–631. Springer, Heidelberg (2005). https://doi.org/10.1007/11430919_72
12. Chien, J.-T., Wu, M.-S.: Adaptive Bayesian latent semantic analysis. IEEE Trans. Audio Speech Lang. Process. **1**(16), 198–207 (2008)
13. Larsson, M.O., Ugander, J.: A concave regularization technique for sparse mixture models. In: ShaweTaylor, J., Zemel, R., Bartlett, P., Pereira, F., Weinberger, K. (eds.) Advances in Neural Information Processing Systems 24 (NIPS 2011), pp. 1890–1898 (2011)
14. Vorontsov, K., Potapenko, A., Plavin, A.: Additive regularization of topic models for topic selection and sparse factorization. In: Gammerman, A., Vovk, V., Papadopoulos, H. (eds.) SLDS 2015. LNCS (LNAI), vol. 9047, pp. 193–202. Springer, Cham (2015). https://doi.org/10.1007/978-3-319-17091-6_14
15. Chirkova, N.A.: Additive regularization for hierarchical multimodal topic modeling. Mach. Learn. Data Anal. **2**(2), 187–200 (2016). https://doi.org/10.21469/22233792.2.2.05
16. Ianina, A.: Multimodal topic modeling for exploratory search in collective blog. Mach. Learn. Data Anal. **2**(2), 173–186 (2016). https://doi.org/10.21469/22233792.2.2.04
17. Apishev, M., Koltcov, S., Koltsova, O., Nikolenko, S., Vorontsov, K.: Additive regularization for topic modeling in sociological studies of user-generated texts. In: Sidorov, G., Herrera-Alcántara, O. (eds.) MICAI 2016. LNCS (LNAI), vol. 10061, pp. 169–184. Springer, Cham (2017). https://doi.org/10.1007/978-3-319-62434-1_14
18. Bulatov, V., et al.: TopicNet: making additive regularisation for topic modelling accessible. In: Proceedings of the 12th Language Resources and Evaluation Conference, pp. 6745–6752, European Language Resources Association, Marseille (2020). https://aclanthology.org/2020.lrec-1.833.pdf. Accessed 21 Jan 2022
19. Veselova, E., Vorontsov, K.: Topic balancing with additive regularization of topic models. In: Proceedings of the 58th Annual Meeting of the Association for Computational Linguistics: Student Research Workshop, pp. 59–65, Association for Computational Linguistics (2020). https://doi.org/10.18653/v1/2020.acl-srw.9
20. Irkhin, I.A., Vorontsov, K.V.: Convergence of the algorithm of additive regularization of topic models. Trudy Instituta Matematiki i Mekhaniki URO RAN **3**(26), 56–68 (2020). https://doi.org/10.21538/0134-4889-2020-26-3-56-68

21. Sukhareva, A.V., Vorontsov, K.V.: Building a complete set of topics of probabilistic topic models. Intell. Syst. Theory Appl. **4**(23), 7–23 (2019)
22. Alekseev, V.A., Bulatov, V.G., Vorontsov, K.V.: Intra-text coherence as a measure of topic models' interpretability. In: Komp'juternaja Lingvistika i Intellektual'nye Tehnologii, pp. 1–13, Moscow (2018). https://www.dialog-21.ru/media/4281/alekseevva.pdf. Accessed 21 Jan 2022

Optimal Number of Ants Determination for Ant Colony Optimization Image Segmentation Method for Complexly Structured Images

Samer El-Khatib[1], Yuri Skobtsov[2], Sergey Rodzin[1], and Valerii Zakharov[3](✉)

[1] Southern Federal University, Rostov-on-Don, Russia
[2] St. Petersburg State University of Aerospace Instrumentation, Saint Petersburg, Russia
[3] St. Petersburg Federal Research Center of the RAS, Saint Petersburg, Russia
valeriov@yandex.ru

Abstract. Segmentation is the process of dividing an initial image of size M × N into K clusters so that they differ as much as possible from each other and represent objects of the original image. Given the initial images in the form of a set of pixels with visual properties such as brightness, color, texture, as well as a certain size, noise level, contrast and quality, it is necessary, within the available resources to find such a marking of digital images for a certain number of K clusters, which provides high accuracy and subsequent recognition quality. Complexly structured images are one of the most complex and significant objects. Complexly structured images are understood as images with a non-deterministic and non-linear structure, which consist of a large number of composite elements with significant characteristics. A large number of variations of input complex-structured images does not allow finding an unambiguous universal method for solving the segmentation problem. A suitable way to effetively solve the problem of image segmentation is to use mathematical transformations that describe the collective behavior of a decentralized self-organized system consisting of many agents that locally interact with each other and with the environment to achieve a given goal. In image processing and computer vision, segmentation is still a challenging problem in many real time applications and hence more research work is required. The number of ants affects the computational complexity of the method. It is a mistake to believe that the more ants, the better. With a large number of ants in a colony, productivity drops significantly, since the operating time of the method greatly increases, an overabundance of pheromones occurs, and the method gets stuck at the local optimum. The Hybrid Ant Colony Optimization (ACO) – k-means and optimal number of ants detection for MRI images are considered in this paper.

Keywords: Complexly structured images segmentation · Ant colony optimization · K-means algorithm · Swarm intelligence

1 Introduction

The demanded areas of digital image recognition are medicine, space, industry, art and many others [1]. Complexly structured images are one of the most complex and

R. Silhavy (Ed.): CSOC 2022, LNNS 502, pp. 568–574, 2022.
https://doi.org/10.1007/978-3-031-09076-9_52

significant objects. Complexly structured images are understood as images with a non-deterministic and non-linear structure, which consist of a large number of composite elements with significant characteristics. Complexly structured images are semantically rich images. Complexly structured images have the following properties:

- The presence of a large number of dissimilar objects in the image;
- Objects in the image are of different types;
- Each type of object has its own significant characteristics that must be taken into account;
- Objects are morphologically complex structures;
- The color distribution of the object and the background most often does not intersect.

Examples of such images can be geographical and topographic maps, images of the Earth's surface from space, etc. The most obvious example of complex structured images are medical images, in particular, images obtained using magnetic resonance imaging (MRI images).

Segmentation is the process of dividing the initial complexly structured image of size M $\times$ N into K clusters so that they differ as much as possible from each other and represent the objects of the original image. Given the initial images in the form of a set of pixels with visual properties such as brightness, color, texture, as well as a certain size, noise level, contrast and quality, it is necessary, within the available resources to find such a marking of digital images for a certain number of K clusters, which provides high accuracy and subsequent recognition quality.

A cluster means a certain isolated area, consisting of separate elements (pixels), such that the distance between its elements is minimal, and the distance between two adjacent areas is maximal.

Ant methods have shown their competitiveness in solving many NP-complete transcomputational problems [2] and are of interest for use in the problem of image segmentation [3, 4].

The input data for the method are the following parameters:

- K - the number of clusters into which the original image must be split;
- m is the number of ants that directly perform segmentation;
- n_{t0}- the maximum number of iterations of the method for finding a solution;
- α - parameter that controls the degree of influence of the pheromone level;
- β is a heuristic parameter that controls the degree of belonging of a pixel to a certain cluster.

In this article we will investigate parameter m – number of ants of proposed algorithm.

2 Hybrid Ant Colony Optimization K-Means Method for Image Segmentation

A distinctive feature of the proposed bio-inspired ant method of image segmentation is the use of fast cluster analysis (k-means) [6] to recalculate the center of each segment and the use of a superposition of several criteria for the optimality of the solutions obtained, taking into account both color and geometric characteristics of the image [7–11]. The image can be either color or grayscale, with different initial conditions (noise, high contrast, good quality, blur).

At the first stage of the method, it is necessary to set the value of the number of clusters and initialize their centers. Further, the belonging of each pixel of the image to a certain cluster is determined, according to the probability, which is inversely proportional to the distance between the pixel, the center of the cluster and the variable τ, which represents the pheromone level:

$$P_i(X_n) = \frac{[\tau_i(X_n)]^\alpha [\eta_i(X_n)]^\beta}{\sum_{j=0}^{K} [\tau_j(X_n)]^\alpha [\eta_j(X_n)]^\beta} \tag{1}$$

where $\tau_i(X_n)$ and $\eta_i(X_n)$ - information about pheromone and heuristic variable values of belonging of a pixel to the cluster i respectively, α and β - heuristic coefficients of ACO algorithm, k – number of clusters. Heuristic information is obtained as follows:

$$\eta_i(X_n) = \frac{k}{CDist(X_n, CC_i) * PDist(X_n, PC_i)} \tag{2}$$

where X_n nth pixel, CC_i - ith spectral cluster center, PC_i - ith spatial cluster center, $CDist(X_n, CC_i)$ - distance between (X_n, CC_i) according to color characteristics – (3), $PDist(X_n, PC_i)$ - the geometrical distance between (X_n, PC_i) - (4), k – constant value.

$$CDist(X_n, CC_i) = |Int(X_n) - Int(CC_i)| \tag{3}$$

where $Int(X_n)$ is the intensity of the pixel X_n.

$$PDist(X_n, PC_i) = \sqrt{(X_n.x - PC_i.x)^2 + (X_n.y - PC_i.y)^2} \tag{4}$$

where $X_n.x$ and $X_n.y$ are x and y coordinates respectively for the pixel X_n.

The target function (fitness function) is important for the bio-inspired ant method. In the bio-inspired ant method proposed in the article for solving the problem of segmentation of complex-structured images, the following set of rules is used as a criterion for finding an optimal solution:

- $\max\limits_{i=1..m}\left(\sum_{k=1..K-1} \sum_{j=k+1..K} CDist\left(C_k, C_j\right)\right)$ - the maximum sum value of the color distance between the cluster centers for all ants (the distance between the clusters, in terms of color characteristics, should be at a maximum, then clusters will be different from each other), where CDist – the color distance between two pixels, C_k - center of the cluster k.

- $\min_{i=1..m}\left(\sum_{k=1..K}\sum_{p=1..S_k} PDist(C_k, X_p)\right)$ - the minimum value of the sums of geometric distances between the centers of the clusters and the pixels belonging to the cluster (the sum of the Euclidean distances between the center of the cluster and each of its pixels should be minimal, according to the spatial characteristics, respectively, the cluster will be more homogeneous), where S_k - the number of pixels in the cluster k, PDist – Euclidian distance between two pixels, C_k - center of the cluster k.
- $\min_{i=1..m}\left(\sum_{k=1..K}\sum_{p=1..S_k} CDist(C_k, X_p)\right)$ - the minimum sum value of the color distances between the centers of the clusters and the pixels belonging to the cluster (sum of the Euclidean distances between the center of the cluster and its each pixel, according to color characteristics, should be minimal, then, the cluster will be more compact), where S_k- the number of pixels in the cluster k, CDist – the color distance between two pixels, C_k - center of the cluster k.

The fitness function of the ant colony is 3-criterial and can be obtained as follows:

$$f(m_i) = \begin{cases} f_1 = \sum_{k=1..K-1}\sum_{j=k+1..K} CDist(C_{m_ik}, C_{m_ij}) \\ f_2 = \sum_{k=1..K}\sum_{p=1..S_{m_ik}} PDist(C_{m_ik}, X_{m_ip}) \\ f_3 = \sum_{k=1..K}\sum_{p=1..S_{m_ik}} CDist(C_{m_ik}, X_{m_ip}) \end{cases} \tag{5}$$

Choice of the best solution can be presented as:

$$f(best) = \sum_{i=1..m} \begin{cases} max(f_1) \\ min(f_2) \\ min(f_3) \end{cases} \tag{6}$$

Pheromone updating is calculated as in standard ACO algorithm according to:

$$\tau_i(X_n) \leftarrow (1-\rho)\tau_i(X_n) + \sum_i \Delta\tau_i(X_n) \tag{7}$$

and pheromone evaporation is calculated as:

$$\Delta\tau_i(X_n) = \begin{cases} \frac{Q*Min(k')}{AvgCDist(k',i)*AvgPDist(k',i)}, \; if\ X_n \subset clusteri \\ 0, \; otherwise \end{cases} \tag{8}$$

where Q is a constant value, $Min(k\prime)$ - the minimum color outer cluster distance found by the most successful ant, $AvgCDist(k\prime, i)$ and $AvgPDist(k', i)$ - the average value of the spatial Euclidean and color distances between each pixel and cluster centers for the most successful ant.

Hybrid ACO algorithm for image segmentation consists of the following steps:

Algorithm 1. Hybrid ACO-k-means segmentation algorithm

```
Begin
 Initialize([number clusters], [number ants]);
 Repeat
 For each ant do
  M: For each pixel do
   Calc(probability belonging pixel to cluster) (1);
  End
  Update (cluster center);
  If (NewCenter <> OldCenter) then
    goto M;
  Else
    Save(current solution);
 End
 Select Best Solution From All Ants (5);
 Update(for each pixel) (6, 7);
 Correct(common solution);
 Until criteria not reached
End
```

3 Determination of the Optimal Settings for the Hybrid Ant Method

The effectiveness of the ant method depends on a number of initial parameters discussed below. In this article we will investigate m - the number of ants. The number of ants affects the computational complexity of the method. It is a mistake to believe that the more ants, the better. With a large number of ants in a colony, productivity drops significantly, since the operating time of the method greatly increases, an overabundance of pheromones occurs, and the method gets stuck at the local optimum.

To analyze and calculate changes in the main heuristic parameters of the proposed method, complexly structured images from the Ossirix set [5] were used. The original images were divided into several groups. Let us describe here the general research methodology for determining the optimal parameters of the hybrid ant method. As an example, a complex snapshot from the BRAIN collection of the Ossirix set called IM-0001–0002.dcm is used. This complexly structured image has the following parameters: width - 109; height - 106; good quality; format - 8 bits per pixel; size - 38 Kb; the group is the brain. The clustering parameters are the number of clusters $K = 5$; the maximum number of method iterations. To identify the optimal values of the parameters of the hybrid ant method of segmentation, the images were studied under various initial conditions (noise, contrast, blur). The influence of the values of the initial cluster centers on the behavior of the method and the final result was also investigated.

So, with automatic segmentation, we investigated the effect of changing such a parameter of the method as the number of ants on the value of the objective function of the best ants (2.3). The values are presented in the format of formula (5), where the first value matches and the last one. As this table large enough we will not publish it in this article.

The same experiments as for described image we provided for all the rest complexly structured images from Ossirix dataset. The results were processed and presented as diagram Fig. 1.

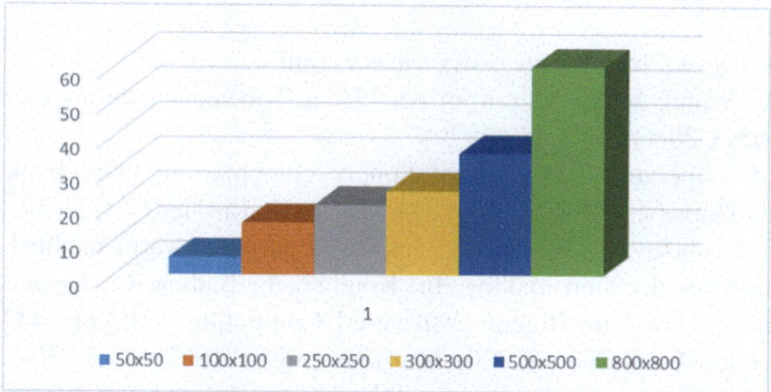

Fig. 1. The range of values for the optimal number of ants depending on the size of the image.

The data in Fig. 1 are summarized results for automatic segmentation by the hybrid ant method.

So, presented diagram shows optimal range of ants in dependency of image sizes. Figure 1 shows that with an image size of up to 50 × 50 pixels, it is rational to use no more than 5 ant agents. At large values of the image size (up to 800–800 pixels), it is advisable to limit the number of ants to no more than 60. These values were obtained experimentally, however, due to the stochastic nature of the method and various variants of combinations of initial arrangements of ants, optimal or close to optimal solutions were achieved with both a smaller and a larger number of ants, but much less frequently.

4 Conclusion

In the article was presented Hybrid ACO-k-means method for complexly structured images segmentation. Optimal number of ants investigation was performed that showed that it is rationally to use not more than 60 ants for images greater than 800 × 800 pixels.

Acknowledgements. The reported study was funded by Russian Foundation for Basic Research according to the research project № 19-07-00570 "Bio-inspired models of problem-oriented systems and methods of their application for clustering, classification, filtering and optimization problems, including big data".

The research described in this paper is partially supported by the Russian Foundation for Basic Research grant **№ 20–08-01046**, state research **FFZF-2022–0004.**

References

1. Gonzalez, R.C., Woods, R.E.: Digital Image Processing, 3rd edn. Prentice-Hall, Hoboken (2008)

2. Kennedy, J., Eberhart, R.C.: Particle swarm intelligence. In: Proceedings of the IEEE International Joint Conference on Neural Networks, p. 1942–1948 (1995)
3. Bozhenyuk, A., El-Khatib, S., Kacprzyk, J., Knyazeva, M., Rodzin, S.: Hybrid ant fuzzy algorithm for MRI images segmentation. In: Rutkowski, L., Scherer, R., Korytkowski, M., Pedrycz, W., Tadeusiewicz, R., Zurada, J.M. (eds.) ICAISC 2019. LNCS (LNAI), vol. 11509, pp. 127–137. Springer, Cham (2019). https://doi.org/10.1007/978-3-030-20915-5_12
4. Saatchi, S., Hung, C.C.: Swarm intelligence and image segmentation swarm intelligence. ARS J (2007)
5. Ossirix image dataset. http://www.osirix-viewer.com/
6. Hartigan, J.A., Wond, M.A.: Algorithm AS 136: a K-means clustering algorithm. J. Royal Stat. Soc. Series **C28**(1), 100–108 (1979)
7. Skobtsov, Y.A., Speransky, D.V.: Evolutionary computation: hand book.-Moscow: The National Open University "INTUIT", p.331 (2015). (in Russian)
8. El-Khatib, S., Skobtsov, Y., Rodzin, S.: Improved particle swarm medical image segmentation algorithm for decision making. In: Kotenko, I., Badica, C., Desnitsky, V., El Baz, D., Ivanovic, M. (eds.) Intelligent Distributed Computing XIII, pp. 437–442. Springer International Publishing, Cham (2020). https://doi.org/10.1007/978-3-030-32258-8_51
9. El-Khatib, S.A., Skobtsov, Y.A., Rodzin, S.I.: Theoretical and experimental evaluation of hybrid ACO-k-means image segmentation algorithm for MRI images using drift-analysis. Procedia Comput. Sci. **150**, 324 (2019)
10. El-Khatib, S., Skobtsov, Y., Rodzin, S., Zakharov, V.: Comparison of modified object detected graph cut and hybrid ant colony optimization - k-means for MRI images segmentation. In: Silhavy, R. (ed.) CSOC 2021. LNNS, vol. 230, pp. 701–708. Springer, Cham (2021). https://doi.org/10.1007/978-3-030-77442-4_58
11. El-Khatib, S., Skobtsov, Y., Rodzin, S., Potryasaev, S.: Comparison of hybrid ACO-k-means algorithm and graph cut for MRI images segmentation. In: Silhavy, R. (ed.) CSOC 2020. AISC, vol. 1225, pp. 73–80. Springer, Cham (2020). https://doi.org/10.1007/978-3-030-51971-1_6

Ontological and Fuzzy-Possibility Approach to the Synthesis of the DM Functional Equivalent for Management of Hierarchical Systems

Igor T. Kimyaev[1(✉)] and Alexander V. Spesivtsev[2]

[1] LLC IC "Sibintek", 1, Building 1, Zagorodnoe sh., 117152 Moscow, Russia
igor95a@mail.ru

[2] St. Petersburg Institute for Informatics and Automation of the Russian Academy of Sciences (SPIIRAS), 39, 14th Line, St. Petersburg 199178, Russia

Abstract. Modern integrated Control and management systems for Business Entity are extremely complex, but require active participation in management decisions by a qualified operator. The choice of a role from the way of describing his knowledge about the choice of managerial decisions is an urgent task.

Keywords: Integrated Control Systems · Decision Maker · Fuzzy possibility approach · Viable System Model · Operation Technology · Information Technology

1 Introduction

A modern, successful manufacturing company as a business entity (BE) is a complex of functional subsystems that are linked by material, energy, financial, and other flows with many feedbacks in vertical and horizontal dimensions. With purpose to maintain such complexes in a working condition the Integrated Control Systems (ICS) of appropriate complexity are required.

But creating an ICS, i.e. synthesizing all the multilevel functional and cybernetic complexity of the BE, under any conceptual approach, using available modeling tools and within a reasonable cost is almost impossible. This leads to widespread use at the key points of management process of creative and cognitive abilities of a person with appropriate qualifications and skills decision-making, the so-called Decision Maker (DM).

Practical experience proves that functional DM within ICS is applied mostly in the areas and subsystems that are characterized by significant uncertainty of their current state. This approach is conditioned by poor formalization of the subject area, lack or incompleteness of data, multi-criteria of objective functions, as well as an overabundance of possible decision-making options ("curse of dimension").

However, nowadays there are no systematic and standardized approaches to justifying and localizing DM within ICS as a consolidated function. There are also

R. Silhavy (Ed.): CSOC 2022, LNNS 502, pp. 575–584, 2022.
https://doi.org/10.1007/978-3-031-09076-9_53

no approaches and formalized rules for identifying the completeness of structured knowledge for operating a specific subject area (subsystem) within BE.

The task of identifying and formalizing the heuristic rules used by the most experienced specialists-experts for choosing correct and effective operational decisions is very urgent. This urgency is aggravating significantly under conditions of leaving of qualified DM the working structure of ICS due to the natural rotation of production personnel.

This publication claims to substantiate the advisability of including DM as functional units in the standardized ICS structures. The method of knowledge representation "semantic networks on a basic ontology" is proposed in order to solve the above problems, as well as a method of formalizing explicit and implicit expert knowledge based on the fuzzy possibility approach (FPA).

2 Management of BE as a Viable Hierarchical System

Many researchers [1–3] suggest that any management structure of a complex BE can be represented as a hierarchy, at each level of which functional production, financial and other plans are developed, resources, incl. equipment are allocated, tasks and operations for execution are described. All activities within each level of this hierarchy, including the information and control flows between levels, implement appropriate management of BE.

However, the following factors and limitations are to be considered:

a) The formation of plans and tasks, as well as their operational execution should be strictly synchronized between the levels of the hierarchy;
b) Internal life activity of BE should be balanced in terms of resources, i.e. tasks are not assigned without allocated resources;
c) Ensuring a mutually beneficial exchange of BE with external environment, i.e. producing production so much as to get in return a sufficient amount of resources necessary for its further functioning (financial, material, labor, etc.).

In turn, S. Beer [4] proposed to consider hierarchical BE as an analogue of a living organism, and actions and procedures for organizing the life of a company (in fact, managing it) – as analogy with individual functions of a living organism with several designated areas of responsibility.

Therefore, viability of BE is defined as ability of BE to adapt to changes in the external environment while maintaining its functionality to solve a set of target tasks. The key factor for implementation of this concept is the so-called The Viable System Model (VSM) shown in Fig. 1.

Managing BE can be approached from the standpoint of ensuring its viability as a purposeful approach of the complex BE $\leftrightarrow$ ICS to the reference VSM.

Modern BE that manufactures commercial products are very complex in terms of their internal structure; they include many separate processing units (PU), united into conditionally self-sufficient technological stages (areas), specialized organizational structures and local technical systems to ensure work of PU and production areas [5, 6].

The transition of management functions from one layer (subsystem) to another (in most cases "top-down") is provided by DMs having special skills and qualifications.

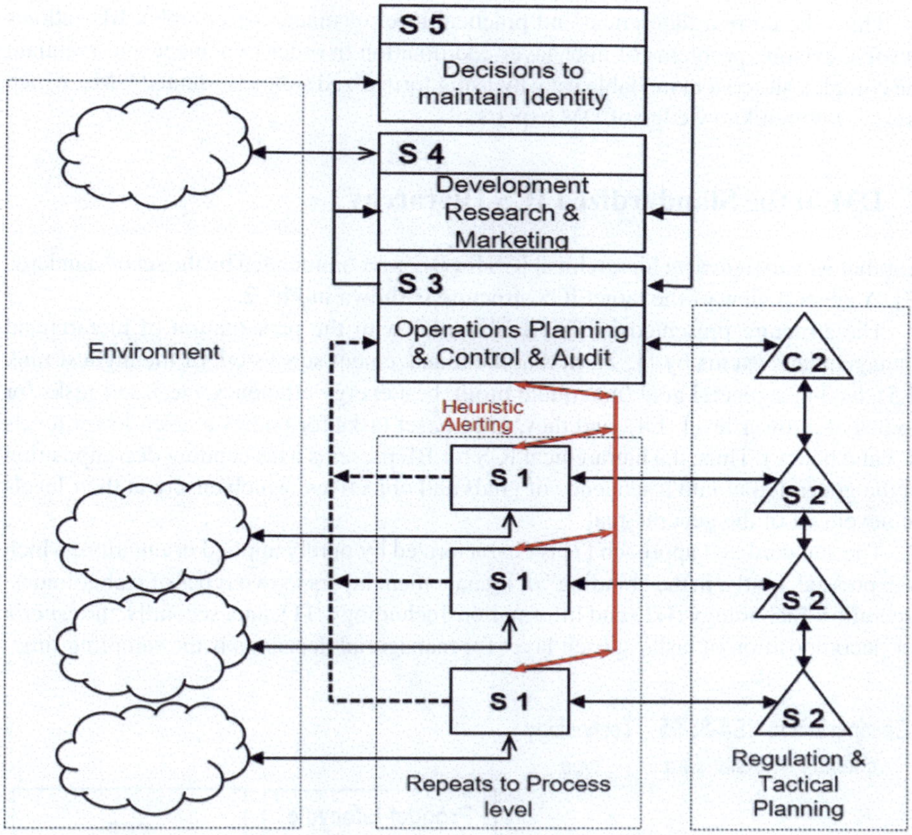

Fig. 1. A cybernetic description of the main VSM functions.

One of the key features of such human-machine complex subsystems is the ability of DMs to perform purposeful manipulations, which lead to the planned results.

Scant information about the mechanisms of how the "supervised" subsystem is operated (like a "Black Box") does not mean that the comprehensive knowledge (in fact, ontologies) are absent. Thus, the operability of each separate BE subsystem is ensured by the rules and instructions formulated according to the principle – "input – output", which constitute the subject matter dependent metalanguage (ML). ML is actually a subject-oriented ontological model (OM), which is more or less presented in the format of process operating procedures (POP) and other manuals. In addition, ML contains rules for setting goals, for implementing production and technological operations under certain restrictions.

The application of such rules and instructions in the specific production environment is performed by the DM at his own discretion, relying on experience and intuition, using decisive "heuristic alerting". The points of application in the VSM structure are shown in Fig. 1.

Thus, the current theoretical and practical level of managing complex BEs allows to solve existing problems of inter-layer coordination in order to manage and maintain the complex subsystem in viable state by using formalized subject-oriented OMs, which present implicit knowledge of DM experts.

3 DM in the Standardized ICS Hierarchy

Another way to represent hierarchical ICS for BE is recommended by the set of standards [7]. A general view of the target ICS structure is shown in Fig. 2.

The structure presented in Fig. 2 is similar with the presentation of hierarchical management systems by [1, 2], in that the management subsystem of the highest rank (L5), having a general goal (maximum profit, best energy efficiency, etc.), sets tasks for bodies of a lower level (L4), and they, in turn, set tasks for bodies at even lower levels (L3 and below). Thus, the hierarchical ICS for BE presents a descending decomposition of the general goal into a sequence of goals and objectives, which ensure at their levels achievement of the general goal.

The standardized approach [7] is characterized by purely applied orientation, which is expressed, firstly, in the "binding" of management layers to two types of technologies: Operation Technology (OT) and Information Technology (IT), and, secondly, the criteria for decomposition of tasks at each layer for managerial decisions is the sampling time.

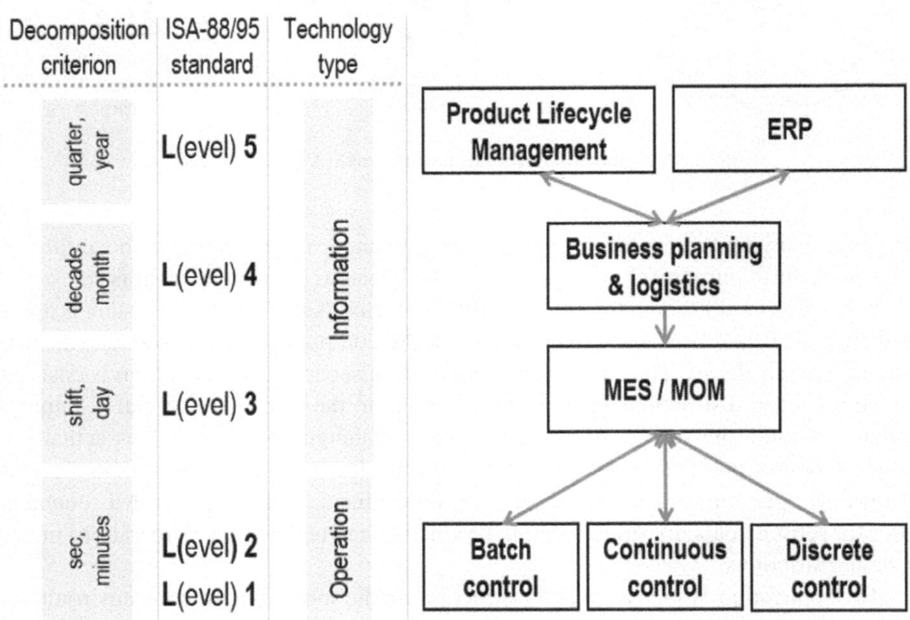

Fig. 2. The standardized ICS hierarchy of IT/OT.

However, as noted above, the achievement of the general goal of a BE can be performed not only by tools of OT/IT for solving hierarchically linked tasks, but also by a heuristic choice of management decisions based on subject-oriented OM.

For assessment of the requirements to VSM with included heuristic OMs for synthesizing the ICS structure according to [7] standards, suggested by S. Beer [4], we have to revise the functional and ideological equivalence of both approaches.

Since the authors could not find direct references and borrowings between the two approaches, the Table 1 presents logical and functional comparable levels of VSM (S1… 5) and ICS hierarchy for BE (Level 1… 5).

Table 1. Comparison of the functional characteristics of the VSM and standardized ICS levels

VSM Hierarchy by S. Beer		Hierarchy by ISA-88/95	
Subsystem	VSM function	ICS function	Hierarchy levels
S1	Implementation of several key functions for BE's core activities. Each local S1 is viable, recursive	Local control of several separate technical devices and units	Level 1, 2
S2	Channels and bodies for exchange of information for communication between several S1 and the superior S3 to control and coordinate the activities of the subordinate S1	Industrial and local data exchange networks, provision of a database for technological parameters common for L2, L4, L5	Level 3
S3	Formation of rules for resource provision, restrictions for S1, and interaction with S4, S5	Basic plan of production (rules), material use, set points (settings) for subsystems on L2	Level 4
S4	Monitoring changes in the environment with aim to form the ways for BE to adapt and remain viable	Commerce in the environment, collecting information about current/forecast prices, market needs to adapt	Level 5
S5	Generating political decisions to manage the BE as a whole in order to balance the needs and demands of its various parts	Generating business plans for investments, long-medium-term production volumes, pricing policy, etc.	–

Comparing the functions implemented on the levels of the VSM and standardized ICS, it is obvious that, apart from S5, each subsystem has its equivalent counterpart. The revealed significant equivalence to description of ICS between these two approaches proves that they have a deep ontological similarity and suggests that when designing a viable complex BE $\leftrightarrow$ ICS according to [7] standards, subject matter OM filled with heuristic rules should be included in the OT/IT control loops.

In other words, despite the widespread use of mathematical and software-hardware tools in OT/IT, the DM should play a significant role to effectively monitor the state of BE and make many key managerial decisions in order to ensure its end-to-end performance and viability. Moreover, the multicriteria context, resource constraints, incomplete information about the situation and its prospects, DM can be the only functional unit possessing capabilities for making correct and rational decisions [8].

Nevertheless, the active participation of the DMs in the management of a BE does not mean that their role in the ICS is once and for all. With evolvement of the mathematical methods for modeling subsystems and "enlightenment" of "Black Boxes", further development of techniques for formalizing knowledge of DMs (experts), elaboration of formal rules for choosing right decisions, the role and functions of a man are subject to evolution from mere execution of typical routine operations to the intelligent behavior.

4 Functions of a DM as a Decision Maker in the ICS Hierarchy

As noted in [3, 5, 6, 9], modern cybernetic approaches to synthesis and modeling of both the physical processes and human economic activities involve the use of a variety of mathematical apparatus, which, thanks to the progress in computing and telecommunication technologies, makes it possible to successfully design ICS components of the high complexity degree.

But as shown above, even the most complex ICS components, which are implemented according to the modern standards, require the coordination of their functioning through the DM.

The DM while performing coordination and management functions for a specific BE subsystem, has to use a subject matter OM suited for the DM's role and a formalized process operating procedure (POP).

An efficient OM [9–11] built on the provisions of the current POP can be represented in the form:

$$O = <C, R, F> \tag{1}$$

where C is a finite set of concepts (notions) of ICS as described in POP; R is a finite set of relations between concepts C; F is a finite set of functions for interpretation (axiomatization) defined on the concepts and relations of the ontology space O.

One of the workable approaches to design a subject matter restricted ontology for solving local tasks is shown e.g. in [12–15]. This approach suggests to divide the ontology space O into sets of entities, mappings, tasks, solutions, and identify the axiomatization functions for the set F.

Defining the set of F for the OM of the existing human-machine subsystem within the ICS is the most responsible and difficult task, since this set e.g. even in the verbal form, may be incomplete and contradictory.

In addition to the method of designing OM by Diamond, one can use a fuzzy-possibility approach [3, 13] to define the set of F, which suggests the synthesis of a decision-making function (DMF) in fuzzy-plural form based on explicit and implicit expert data.

At the conceptual level, the general structure and content of the tasks related to design of models and methods for extracting, formalizing, and structuring explicit and implicit expert knowledge for assessing the state of ICS or its subsystems can be presented in the following form [3, 13, 14]:

$$< X, \Phi_Z, E^Z_{rep}, F_Z, M_Z, MZ^{DM}(q, t, \mu), Z^{DM}_{pr}(q, t, \mu); Z^{DM}_{adj}(q, t, \mu); K; Y/\Xi >, \quad (2)$$

where X is the set of ICS states, which after being identified must be assigned to one of the defined classes Y/Ξ;

Φ_z – a set of methods for knowledge extraction;

$E^z{}_{rep}$ – a set of methods for knowledge representation;

F_z – a set of methods and algorithms for formalizing knowledge;

M_z – a set of conditions necessary for performing all operations when building models;

$MZ^{DM}(q, t, \mu)$ – a lot of meta-knowledge, general erudition, explicit and implicit expert knowledge, the ability to make decisions under conditions of uncertainty;

$Z^{DM}{}_{pr}(q, t, \mu)$ – a lot of professional knowledge, comprising the ability to understand and formulate the problem, implicit knowledge, skills, experience, intuition.

$Z^{DM}{}_{adj}(q, t, \mu)$ – a lot of knowledge in adjacent subject areas related to the profession;

Y – is the computed generalized assessment of the ICS state, which is used to assign it to one of the defined classes of states;

K – is a set of criteria for evaluating the quality of models;

Y/Ξ – is the set of ICS state classes, one of which should be used to characterize the ICS state based on the result of its estimation.

A commutative scheme is proposed to form a DMF representing a sequence of interrelated processes of extraction, representation and formalization of explicit and implicit expert knowledge (see Fig. 3).

Disambiguation of the ICS state by means of explicit and implicit expert knowledge enables to form the following mapping:

$$\mu : T \times U \times Y \rightarrow Y/\Xi \quad (3)$$

where T is the set of time values t at which the object is observed;

U, Y – the sets of input U and output Y parameters, respectively;

Y/Ξ – factor-set of ICS states.

This mapping itself can be represented by a combination of the mappings

$$\mu = g_1 \circ g_2 \circ g_3, \quad (4)$$

where g_1 – is the set of parameters characterizing the state of a specific ICS $Z^{э}_{\kappa}(q, t, \mu)$ extracted from the expert's meta knowledge $MZ^{DM}(q, t, \mu)$, including professional explicit knowledge $Z^{DM}_{pr}(q, t, \mu)$, which comprises the expert's ability to understand and carry out a problem statement, implicit knowledge $Z^{DM}_{sk}(q, t, \mu)$, skills, experience, intuition, as well as knowledge in the adjacent subject areas $Z^{э}_{conp}(q, t, \mu)$.

g_2 – presentation of the characteristics $Z^{DM}_{\kappa}(q, t, \mu)$ in the form of linguistic variables and formation of the factor space $Z^{DM}_{fs}(q, t, \mu)$, in which the expert makes a decision on the state of the ICS for a specific task;

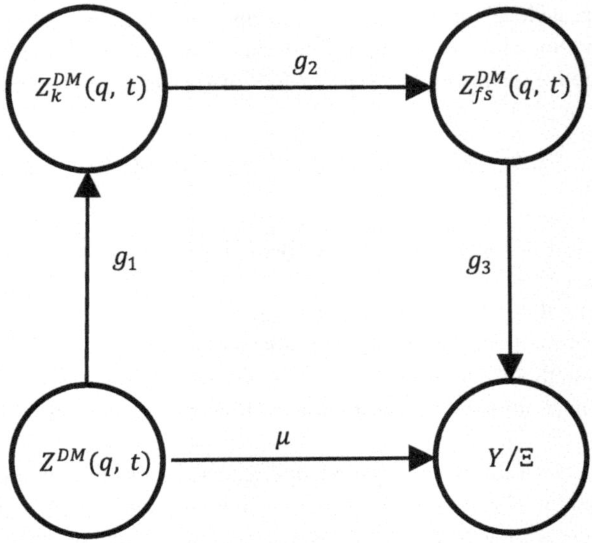

Fig. 3. A commutative scheme of interrelated processes of extraction, representation and formalization of explicit and implicit expert knowledge.

g_3 – formalized explicit and implicit expert knowledge. At this g_3 comprises the models of formalized representation of explicit and implicit expert knowledge about the ICS state, including, e.g. a polynomial expression using methods of Design of experiments theory on the set of fuzzy production rules for the factor space $Z_{fs}^{DM}(q, t, \mu)$ and assessment on the factor space Y/Ξ of ICS state, which must be assigned to one of the state classes after having been identified.

Therefore, we can argue that while solving a management problem, the DM can skillfully and reliably estimate the current situation in real time and make an effective decision. However, to provide the ICS with efficiency that is comparable to DM in decision making, the verbal expert knowledge for OM must be represented by an analytical expression. For this purpose, a fuzzy-possibilities approach has been developed, which allows to consistently perform the procedures for extracting, presenting and formalizing DM knowledge, as shown in Fig. 3 [3].

5 Conclusion

Two methodological approaches to the design of managing structures for a complex BE have been considered: on the one hand, the existing practice to present BE as a hierarchy, at each level of which functional, production, financial and other plans are formed, and on the other hand, a view to consider a hierarchical BE as an analogue of a living organism with responsible areas which aim is to ensure BE viability when environmental conditions change.

In order to fully link both approaches to the design of managing structures for a complex BE, a more thorough assessment of their ontological conformability and similarity is required, which will be the subject of further research.

The current research has proved the functional and ideological equivalence of both approaches. Moreover, each of the approaches uses its own ontological features, what causes their "imaginary divergence". However, both approaches declare the importance of the DM in making key managerial decisions in order to ensure the viability and vitality of a BE. Thereat, it is shown that under conditions of significant uncertainty, the DM is the only actor who has the opportunity and right to make a correct and rational decision in real time.

Such a high importance of DM is conditioned by specific properties of human intelligence. It is known that human intellectual property to solve difficult-to-formalize problems with fuzzy or inaccurate information surpasses any computational means in terms of efficiency. Such a property of a human expert as DM actor contributes to overcoming the complexity of managing structures for a complex BE by using DM's experience and knowledge of the professional field. In essence, expert knowledge formalized in the form of a polynomial model has a unique property: independence of retrospective statistical data and predicted experimental data. This advantage makes it possible to bypass the methodological difficulties of traditional mathematical methods arising from their application.

The way to formalize verbal expert knowledge by means of analytical models based on the fuzzy-possibilities approach is presented. One of the main advantages of the proposed approach is clarity, using only the necessary knowledge and experience of an expert in his professional language, without use of the specific mathematician language.

Acknowledgements. This research work was carried out with partial financial support from the Russian Foundation for Basic Research (Grant No. 1807-01272) within the framework of the state budget theme No. FFZF-2022-0004.

References

1. Mesarovich, M.D., Macko, D., Takahara, Y.: Theory of Hierarchical, Multilevel, Systems. Academic Press, NY and London (1970)
2. Pospelov, G.S.: Introduction to the Theory of System Planning and Management. Lectures/Ministry of Higher and Special Education RSFSR. MFTI, Moscow (1974)
3. Spesivtsev, A.V., Kimyaev, I.T.: Advanced processes control (APC) for complex technological objects on the fuzzy logic rules based modeling and analysis of safety and risk in complex systems. In: International Scientific Conference Proceedings. 23–25 June, Saint Petersburg (2020)
4. Beer, S.: Brain of the Firm, 2nd edn. Wiley, Manchester (1981)
5. King, M.: Process Control. A Practical Approach. Wiley, Chichester (2011)
6. VanDoren, V.: Techniques for Adaptive Control, 275 p. Elsevier Science, Amsterdam (2002)
7. ANSI/ISA-95.00.01-2010 (IEC 62264-1 Mod) Enterprise-Control System Integration – Part 1: Models and Terminology (2010)
8. Simon, H.A.: Rationality as process and as product of thought. Am. Econ. Rev. **68**(2), 1–16 (1978)
9. Mikoni, V.S.: Comprehensibility of the ontological model as a characteristic of its quality. Ontol. Des. **11**(1), 20–34 (2021)
10. Dean, A., Hendler, J., Gandon, F.: SemanticWeb for the Working Ontologist. Effective Modeling for Linked Data, RDFS, and OWL, 495 p. ACM Books (2020)

11. Stylios, C.D., Groumpos, P.P.: Modeling complex systems using fuzzy cognitive maps. IEEE Trans. Syst. Man Cybern.: Syst. **34**(1), 155–162 (2004)
12. Dimand, I.N., Spesivtsev, A.V.: Semantic networks based on ontology. In: XIV International Conference on Soft Computing and Measurements, vol. 1, pp. 76–81. 23–25 May. Saint-Petersburg (2013)
13. Spesivtsev, A.V.: Fuzzy-possibility approach to formalization and usage of expert knowledge for assessment of complex objects. Instrum. Eng. **63**(11), 985–994 (2020)
14. Ignatyev, M.B., Marley, V.E., Michaylov, V.V., Spesivtsev, A.V.: Models for Weakly Formalized Systems Based on Explicit and Implicit Knowledge, 430 p. Politech-Express, St. Petersburg (2018)
15. Prokopchina, S.V., et al.: Management Under Uncertainty Study, p. 304. Printing House SPbGTU «LETI», St. Petersburg (2014)
16. Tara, A., Butean, A., Zamfirescu, C., Learney, R.: An ontology model for interoperability and multi-organization data exchange. In: Silhavy, R. (ed.) CSOC 2020. AISC, vol. 1225, pp. 284–296. Springer, Cham (2020). https://doi.org/10.1007/978-3-030-51971-1_23
17. Al-Fedaghi, S., Makdessi, M.: Modeling business process and events. In: Silhavy, R. (ed.) CSOC 2020. AISC, vol. 1224, pp. 366–379. Springer, Cham (2020). https://doi.org/10.1007/978-3-030-51965-0_32

Symbolic Modeling (S-Modeling): an Introduction to Theory

Vladimir D. Ilyin(✉)

The Federal Research Center "Computer Science and Control" of the Russian Academy of Sciences, 44/2, Vavilova str., 119333 Moscow, Russia
vdilyin@yandex.ru

Abstract. The theory of *symbolic modeling of arbitrary objects in a human-machine environment* (*S-modeling*) includes languages for a formalized description of an extensible system of S-modeling notions, a description of the core of this system and classes of basic tasks for constructing and manipulating S-models. The theory of S-modeling is considered as a methodological platform for the scientifically based development of information technologies and *the human-machine environment of S-modeling and digitalization of various types of activities* (*S-environment*). S-modeling uses all kinds of symbols (audio, visual, etc.) implementable in the S-environment. S-models are studied as entities having three interrelated representations in the S-environment: symbolic, code and signal. The construction of S-models is carried out according to the rules corresponding to the classes of basic S-modeling tasks. Classes of basic tasks of S-(representation, transformation, recognition, construction, interpretation, interaction, saving, accumulation, search, information security) are defined. #CSOC1120.

Keywords: Symbolic Modeling (S-modeling) · S-modeling Theory · S-symbol · S-code · S-signal · S-environment · Basic Tasks of S-modeling

1 Introduction

Without effective symbolic modeling of the studied entities, the development of science, technology and other types of intellectual activity is impossible [1, 2]. Achieving a goal in the course of some activity implies a clear idea of the problems to be solved. When achieving non-trivial goals (to design a machine, develop information technology, etc.), the symbolic representation of the idea in the human-machine environment for problem solving (S-environment) is the most productive [3–5].

This approach has a number of proven advantages. Firstly, by analyzing the symbolic model of the idea (device scheme, task specification, etc.), we can check whether the model corresponds to the idea, and if it does not, make corrections to the model. Secondly, on a model recognized as corresponding to the idea, it is possible to verify the validity of the idea itself. And if verification is successful, it is possible to make a decision on the feasibility of implementing the idea. Otherwise – to engage in a change of idea. The

R. Silhavy (Ed.): CSOC 2022, LNNS 502, pp. 585–591, 2022.
https://doi.org/10.1007/978-3-031-09076-9_54

idea of achieving a goal using symbolic models is embodied in many technologies for various types of activities. The most successful technologies prohibit behavior that does not comply the rules (attempts of unacceptable behavior have as much chance of success as attempts to play poker against a chess program).

A distinctive features of *S-objects* that exist in the *S-environment* (files of books, articles, videos, electronic maps, computer programs, etc.) are easy copying without distortion, easy distribution and storage of copies (in comparison with physical models, layouts of scientific and technical facilities, etc.) [3–5].

1.1 Markuping Text Fragments and Writing Formulas

For markuping text fragments, the S-modeling language *TSM* (*Textual Symbolic Modeling*) is used:

□<text fragment>□ means a definition;
:: is a separator which is placed after the defined notion (before the definition);
≈ is a substitute for the phrase "the same as";
○<text fragment>○ means an example;
◇<text fragment>◇ means a note.

Italics are used to highlight the concepts and sentences, to which the author wants to attract the attention, and also for variables names [5].

2 S-Modeling: Basic Notions

2.1 S-(Symbol, Code, Signal)

□ *S-symbol* :: a substitute for a natural or invented object, denoting this object and being an element of the system for constructing symbolic messages (texts, computer programs, etc.), designed for human or robot perception. □

○ In S-modeling, the Russian alphabet together with punctuation marks is considered as *a system of text symbols* for constructing messages according to the rules of the grammar of the Russian language (each element of the alphabet is a substitute for the sound used in *speech messages*); Braille for the blind – as *a system of textured symbols* for constructing text messages designed to be perceived with the touch of the fingers; musical notation, *a system of musical symbols* – as a means of constructing graphically presented *musical messages*; *a system of chess graphic symbols* – as a means of visual representation of chess positions. ○

In computers, smartphones, and other *S-machines* [3–5], the S-symbol is represented in the form of *S-code*, intended for constructing, storing, transmitting and interpreting symbolic messages.

□ *S-code* :: a substitute for an S-symbol or symbolic message, designed to construct, store, transmit and interpret symbolic messages using S-machines. □

○ Morse codes, codes of The Unicode Standard. ○

□ *S-signal* :: a physically realized representation of S-symbol, designed for perception by human senses (or robot sensors), or an S-code representation, designed to be received by hardware of S-machines. □

2.2 Systems of Notions and Knowledge Systems

In S-modeling, the definition of a system of notions is considered as a description of its S-model, accompanied by an indication of the application area.

□ *S-model of a system sc of notions* :: < collection *set* ^{sc} of notions >, < family *rel* (*set* ^{sc}) of dependencies given on *set* ^{sc} >.

For each notion of system *sc*, a set of values is defined.

The application area of the definition is specified by the description of the types:

- *the correspondent* (for whom the definition is intended to be interpreted);
- *the purpose*, in the process of achieving the one the definition is advisable to be applied (○ tasks, in the study of which the definition can be useful ○);
- *the stage* where it makes sense to use the definition (○ problem statement, development of a solution method ○). □

○ System *tr* of notions named *a triangle* :: < *set* ^{tr} ≈ collection of notions >, < *rel* (*set* ^{tr}) ≈ family of dependencies defined on *set* ^{tr} >.

In *tr*, the elements of *set*^{tr} are the sides (a, b, c), angles (α, β, γ), perimeter p, etc. The family *rel* (*set* ^{tr}) includes $p = a + b + c$; $\alpha + \beta + \gamma = \pi$, etc. ○

○ The system $tr^{\pi/2}$ of notions of a right triangle can be defined as a specialization of *tr*: $tr^{\pi/2} \approx tr$ [$\#\alpha = \pi/2$] (by adding the restriction $\alpha = \pi/2$, which distinguishes a subset of triangles where the value of one of the angles is equal to $\pi/2$). ○

The representation of dependencies between notions in the form of solvable tasks is a necessary condition for constructing quantitative S-models of notions systems.

□ *S-task* :: { *Formul, Rulsys, Alg, Prog*}, where *Formul* is the statement of the task; *Rulsys* is the set of systems of *mandatory and orienting requirements for solving the task*, aligned with *Formul*; *Alg* is the union of sets of algorithms, each of which corresponds to one element from *Rulsys*; *Prog* is the union of sets of programs, each of which is assigned to one of the elements of *Alg* [4].

The task statement *Formul* is a pair { *Mem, Rel*}, where *Mem* is the set of notions of the task, where the partition *Mem* = *Inp* U *Out* (*Inp* ^ *Out* = 0) and the set *Rel* of dependencies between notions are specified, where *Rel* defines the binary relation *Rel* < *Inp* * *Out*. The set *Mem* is named *the task memory*; *Inp* and *Out* are its input and output, the values of which are supposed to be set and found, respectively. □

□ *S-algorithm* :: a system of rules for solving a task (corresponding to one element of *Rulsys*), which allows, in a finite number of steps, to put the resulting set belonging to *Out* to one-to-one correspondence with a given data set belonging to *Inp*. □

□ *S-program* :: S-algorithm implemented (in a high-level programming language, machine-oriented language and/or in a system of machine instructions), presented in the form of a message that determines the behavior of S-machine task solver with specified properties. The S-algorithm exists in symbolic, code and signal forms connected by the translation relations]. □

◇ In general case, the sets *Rulsys*, *Alg* and *Prog* can be empty: the number of their elements depends on the degree of knowledge of the task. ◇

Descriptions of the use of set elements:

- *Rulsys* includes a specification of the task solver type (autonomous S-machine, networked S-machines cooperation, human-S-machine cooperation, etc.); requirement for information security, etc.;
- *Alg* includes data on the admissible modes of the task solver (automatic local, automatic distributed, interactive local, etc.), requirements for the result obtained, etc.;
- *Prog* includes data on programming languages, operating systems, etc.

◇ Each program is accompanied by links to sets of test cases. ◇

□ The S-model of the knowledge system *sk* :: < *ca* ≈ S-model of the notions system *sc* >, < set^{lng} ≈ S-model of the set of message languages interpreted on *ca* >, < set^{intr} ≈ S-model of the set of interpreters of *ca*-messages composed in languages from set^{lng} >. □

Interpreting the message on *ca*:

1. constructing an output message based on a given input message (messages are presented in languages from the set set^{lng});
2. analysis of the output message (whether changes are required in *ca*);
3. if required, initiate changing the *ca*; if not, end.

2.3 S-(Message, Data, Information)

□ *S-message* :: a finite sequence of S-symbols designed for recognition and interpretation by the recipient, or its S-code that meets the requirements for basic S-tasks solvers (see Sect. 3). □

○ S-models of systems of notions and knowledge systems, which present the results of the study of certain entities (objects of research); programs that determine the behavior of S-machines; web pages and document files – all these are S-messages. ○

□ *S-file* :: a named unit of storage of the S-machine message code (data or program) on a drive (SSD, hard disk, etc.) of computer device (desktop, laptop, smartphone, digital camera, etc.). □

□ *S-data* :: S-message required to solve a certain task or a family of tasks, presented in a form designed for recognition, transformation and interpretation by a task solver (a program or person). □

□ *S-information* :: the result of interpreting a message on the S-model of a system of notions. To extract information from a message, it is necessary to have:

- an accepted message presented in a form designed for recognition and interpretation by the message recipient;
- models of systems of notions stored in memory, among which – the necessary one for interpretation of the received message;
- mechanisms for finding the necessary model, interpreting the message, presenting the result of interpretation in the form of a message and writing it to memory. □

○ The result of interpretation of the m^a message presented in language *a*, received by the translator (human or robot) – translated to the m^b message in language *b*, is the information extracted from the m^a message. ○

3 The General Method and Classes of Basic *S-Modeling Tasks* (*S-tasks*)

□ The general method of S-modeling :: a constructive proof of the existence of S-model of an arbitrary object, representable in S-environment. □

Studying the properties and regularities of S-modeling at each stage of the S-modeling theory development allows to determine the classes of basic S-modeling tasks. Nowadays, the S-tasks presented in Table 1 are the most relevant.

Table 1. Classes of basic S-tasks.

Name of the class of basic S-tasks	Basic S-tasks
S-representation	Creation of interconnected systems of S-(symbols, codes, signals), specification languages, programming languages, queries languages; representation of S-(messages, data, information), S-models of systems of notions and knowledge systems [4–23]
S-transformation	Converting S-messages (○ speech ↔ text; analog ↔ digital; uncompressed ↔ compressed; *.doc ↔ *.pdf, etc. ○) [4, 5]
S-recognition	A necessary but insufficient condition for recognition - is the presentation of S-message in a format known to recipient. When this condition is met, the tasks of matching the sample models or matching the properties of the recognized model with the properties of sample models are solved [4, 5]
S-construction	Construction of new S-objects from previously created S-objects (○ tasks specifications, programs, systems of notions, knowledge systems, etc. ○), presented as constructive S-objects [4–6]
S-interpretation	Interpretation supposes the existence of an accepted S-message, a S-model of the system of notions on which it should be interpreted, and an interpretation mechanism. ○ To interpret a web page presented on a monitor screen, a person uses systems of notions stored in his memory. For S-machine microprocessor, the S-messages to be interpreted are the codes of the S-machine commands and data; for the compiler, the source code of the program is the S-message to be translated [4, 5]. ○

(*continued*)

Table 1. (*continued*)

Name of the class of basic S-tasks	Basic S-tasks
S-interaction	In this class, the tasks of interaction in S-environment (man – S-machine; S-machine – S-machine) are studied. Senders and recipients of S-messages, means of sending, transmitting and receiving messages are classified. Systems of messaging rules (S-network protocols), S-network architectures, service-oriented architectures, document management systems are being developed [4–23]
S-(saving, accumulating and searching)	This class includes the related S-tasks of saving, accumulating, and searching for S-messages. Memory devices, their management mechanisms, forms of storage and accumulation, methods of accumulation and search, databases and program libraries are studied here and typed [4, 5]
S-security	S-tasks of this class are designed to prevent and detect vulnerabilities, perform access control, protect against unauthorized use, malware, and message interception [4, 5, 24, 25]

References

1. Newell, A., Simon, H.: Computer science as empirical inquiry: symbols and search. Commun. ACM **19**(3), 113–126 (1976). https://doi.org/10.1145/360018.360022
2. Ilyin, V.: The program generating system. Nauka, Moscow (1989). https://www.elibrary.ru/item.asp?id=24889586, Accessed 15 Jan 2022
3. Ilyin, V.: Symbolic modeling (S-modeling). The Great Russian Encyclopedia – electronic version (2017). https://bigenc.ru/technology_and_technique/text/4010980, Accessed 15 Jan 2022
4. Ilyin, A., Ilyin, V.: Fundamentals of the Theory of S-Modeling. Institute of Informatics Problems of the Russian Academy of Sciences, Moscow (2009). https://www.elibrary.ru/item.asp?id=25784971. Accessed 15 Jan 2022
5. Ilyin, A., Ilyin, V.: S-modeling of tasks and construction of programs. Institute of Informatics Problems of the Russian Academy of Sciences, Moscow (2012). https://www.elibrary.ru/item.asp?id=25816802, Accessed 15 Jan 2022
6. Ilyin, A.V., Ilyin, V.D.: Updated methodology for task knowledge based development of parallel programs. In: Silhavy, R., Silhavy, P., Prokopova, Z. (eds.) CoMeSySo 2021. LNNS, vol. 231, pp. 319–328. Springer, Cham (2021). https://doi.org/10.1007/978-3-030-90321-3_25
7. Licklider, J., Clark, W.: On-line man-computer communication. In: AIEE-IRE 1962 (Spring) Proceedings of the Spring Joint Computer Conference, 1–3 May 1962, pp. 113–128 (1962)
8. Cerf, V., Kahn, R.: A protocol for packet network intercommunication. IEEE Trans. Commun. **22**(5), 637–648 (1974)
9. Jamsa, K.: Cloud computing. Jones & Bartlett, Learning Burlington (2013)

10. Kay, A.: Personal computing. Learning research group. Xerox Palo Alto Research Center, Palo Alto (1975)
11. Berners-Lee, T.: Information management: a proposal. In: CERN, March 1989 (1990)
12. Berners-Lee, T.: Long live the web. Sci. Am. **303**(6), 80–85 (2010)
13. Kim, R.: Efficient wireless communications schemes for machine to machine communications. Comm. Com. Inf. Sc. **181**(3), 313–323 (2011)
14. Lien, S., Liau, T., Kao, C., Chen, K.: Cooperative access class barring for machine-to-machine communications. IEEE Trans. Wirel. Commun. **11**(1), 27–32 (2012)
15. Pereyra, C., Liu, C., Jayawardena, S.: The emerging Internet of Things marketplace from an industrial perspective: a survey. IEEE Trans. Emerg. Topics Comput. **3**(4), 585–598 (2015). https://doi.org/10.1109/TETC.2015.2390034
16. Kravchenko, V., Shirapov, D.: Logic-functional modeling of nonlinear radio engineering systems. In: 2018 International Multi-Conference on Industrial Engineering and Modern Technologies (FarEastCon), pp. 1–6. IEEE (2018). https://doi.org/10.1109/FarEastCon.2018.8602769
17. Rojek, I., Mikołajewski, D., et al.: Digital twins in product lifecycle for sustainability in manufacturing and maintenance. Appl. Sci. **11**(1), 31 (2020). https://doi.org/10.3390/app11010031
18. Semeraro, C., Lezoche, M., et al.: Digital twin paradigm: a systematic literature review. Comput. Ind. **130**, 103469 (2021). https://doi.org/10.1016/j.compind.2021.103469
19. Liu, K., Song, L., Han, W., Cui, Y., Wang, Y.: Time-varying error prediction and compensation for movement axis of CNC machine tool based on digital twin. IEEE Trans. Ind. Inf. **18**(1), 109–118 (2022). https://doi.org/10.1109/TII.2021.3073649
20. Nguyen, H., Trestian, R., et al.: Digital twin for 5G and beyond. IEEE Commun. Mag. **59**(2), 10–15 (2021). https://doi.org/10.1109/MCOM.001.2000343
21. Jia, P., Wang, X., et al.: Digital-twin-enabled intelligent distributed clock synchronization in industrial IoT systems. IEEE Internet Things J. **8**(6), 4548–4559 (2021). https://doi.org/10.1109/JIOT.2020.3029131
22. Rathore, M., Shah, S., et al.: The role of ai, machine learning, and big data in digital twinning: a systematic literature review, challenges, and opportunities. IEEE Access **9**, 32030–32052 (2021). https://doi.org/10.1109/ACCESS.2021.3060863
23. Zhang, S., Kang, C., et al.: A product quality monitor model with the digital twin model and the stacked auto encoder. IEEE Access **8**, 113826–113836 (2020). https://doi.org/10.1109/ACCESS.2020.3003723
24. Yang, B., Huang, R., et al.: Efficient lattice-based cryptosystems with key dependent message security. Appl. Sci. **11**(24), 12161 (2021). https://doi.org/10.3390/app112412161
25. Lu, X., Wang, F., et al.: A universal malicious documents static detection framework based on feature generalization. Appl. Sci. **11**(24), 12134 (2021). https://doi.org/10.3390/app112412134

Diffusion Innovation, Adoption, and Acceptance of Recommender Systems in Higher Education

Ntsako Morgen Hlungwane(✉), Odilia M. M. Sehume, and Tranos Zuva

Faculty of Applied and Computer Science, Department of ICT, Vaal University of Technology, Vanderbijlpark 1911, South Africa
Morgen.hlungwane@gmail.com

Abstract. Recommender systems are widely used to help in reducing information overload through suggesting relevance items to users proactively or reactively. The usefulness of RS has made them to be used in different sectors such as e-commerce, retail environment and even in education sector. RS is penetrating different sectors with different speed and its adoption and acceptance vary in different sectors. Therefore, this study investigated the diffusion, adoption, and acceptance of RS in higher education. A quantitative method was used in this study. The data was collected using questionnaires and a model was proposed. Two hundred and sixty-one usable questionnaires received from participants. Reliability and validity of the study were satisfied. The results have shown that diffusion, adoption, and acceptance of RSs in higher education is not by chance. It also indicated that diffusion has a positive influence on the adoption with $\beta = 0.625$, Sig. = 0.000 and adoption has positive influence on the acceptance with $\beta = 0.606$, sig. = 0.000 of RSs in higher education sector. The benefits being accrued by those using RSs are having a positive effect on the diffusion ($\beta = 0.695$, Sig. 0.000) and adoption ($\beta = 0.579$, Sig. = 0.000) of RSs, this is enabling growth of RSs in higher education sector. RSs are still in their early diffusion stage in higher education as an education tool. There is need to intentionally promote RSs benefits in higher education to enable a speed diffusion of the technology.

Keywords: Rodgers's theory of innovation · Technology acceptance model · Recommender systems · Higher education

1 Introduction and Background Information

Diffusion, adoption, and acceptance of technology have been studied for many years. Some researchers have explained difference between adoption and acceptance, others have used the terms interchangeably [1]. Adoption is defined as acquiring of new innovation whereas acceptance as the ability of a person to adopt to new innovation [2]. Additionally [3] viewed adoption as approaching and using new technology fully and acceptance as an adaptation process to new technology. The definitions given by [2] and [3] proves that there is no universal accepted definition or usage, despite their continuous use in technology research.

R. Silhavy (Ed.): CSOC 2022, LNNS 502, pp. 592–603, 2022.
https://doi.org/10.1007/978-3-031-09076-9_55

An internet user's biggest challenge is not only getting enough information to make a decision, but also how to make better decision when faced with so much information. People these days use internet to search for products and services they need. To cope with information overload, they rely on recommender systems, whether consciously or unconsciously. Information overload can be addressed by providing users with more proactive, personalised information services.

As per [4], the study, usability has become increasingly important because of the web's access structure, which is especially true when we factor in the problem of massive amount of information. Recommender systems have emerged to help users cope with the sheer amount of information [5]. The recommender systems purpose is to sort through vast amounts of information efficiently and bring the most relevant bits of information to the user's intension. As a result, RS facilitates the navigation of complex information spaces, allowing users to be more efficient and coping with overload of data. The recommender system was originally proposed by [6] and has supported users for the last 28 years by providing interesting items based on user's profile and item attributes. The purpose of this research is to analyse how users perceive different qualities of recommender systems, ultimately lead them to accept the new technology and ultimately adopt it.

2 Problem Statement

Complex technologies such as recommender systems are being used to help users with information overload. Users get to know about these technologies through technology diffusion. Users must make decisions whether to adopt the technologies or not. In the event that a user decides to experiment with the system, they either accept or reject the technology [7]. This research, whilst focusing on the technology adoption and acceptance model (TAM) among university students, the study adds to the current understanding of how technology is diffused. To gain a better understanding of the whole diffusion cycle, it is necessary to study the recommended changes of technology acceptance in the context of other populations, technologies, and modes. This study endeavors to investigate how the knowledge of systems manages to influence users to decide to use new technologies, and how the use of new technology leads to adoption and acceptance of technology.

3 Research Question

How does the knowledge of a systems influence new users to want to use new technologies, and how does innovation lead to the adoption and acceptance of new technologies?

3.1 Research Objective

To answer the research questions as mentioned above, the study sought to address four broad research objectives namely, to:

- To investigate factors that influencing the diffusion, adoption, and acceptance of RS in higher education.
- To propose a model that can be used for the diffusion, adoption, and acceptance of RS in higher education.
- To evaluate the influence of different factors in the diffusion, adoption, and acceptance of RS in higher education.

4 Literature Review

4.1 Diffusion, Adoption and Acceptance Recommendation System in Higher Education

Recommender systems is a complex technology, a relatively new one, and because of the portion of indecision in people's minds about the best way to employ one, they are trying to absorb the new technology and put it into practice prior to beginning efforts focused at using it [5]. Attitudes and intension toward practice or use may be ill-informed or absent in persuasion. Some attitudes about usage may even emerge only after initial attempts to absorb the system have been made. In other words, the real practice might not be straight or immediate description of the system. One of the challenges explored in the study is how users behave and feel when confronted with new technologies. When users seldom pass through the entrance barrier, the system's value is seriously diminished. In addition, as explained in the preamble, web services are rapidly changing, with popular systems appearing for short periods, but disappearing again shortly thereafter. This presents a major challenge. The key to a successful service is to get users to come to it, accept it and stick with it.

4.2 Theoretical Framework

The concept of adoption appears in diffuse innovation literature as one of the oldest [8]. "Adoption can be referred to a process, an event, or stage that occurs at the same time. Those who adopt have the positive value of having a finality attached to them, unlike rejectors who decide not to adopt or non- adopters who have not yet started the adoption process" [9]. The have been many studies that used adoption as the main variable, and this has provided the basic platform for its generalization [10].

The process of scaling up the adoption of new interventions is commonly referred to in the research literature as the diffusion of innovations [8]. This term has been popularised through Everett Rogers' seminal model of diffusion of innovations [8]. The factors discussed in diffusion of innovation focus exclusively on technology related factors [5]. In addition [8] mapped out this process, noting that most people take to the new idea and adapt it after the initial few are open to it. The goal of early innovators is to spread the word so that more and more people are exposed to it, leading to a critical mass. Over time, innovative ideas or products become diffused throughout the population to the point of saturation.

Scholar [7] investigated how user's perspectives influenced their acceptance of recommended recommendations and ultimately led to their adoption of recommender systems. Throughout the investigation, one essential feature kept coming up: the diversity of suggestions. When an e-commerce presents a set of recommendations to the users, the latter is readily able to discern the degree of diversity. A more detailed study of this effect was conducted in order to learn how diversity contributes to user acceptance of recommendations. In addition [7] six experiments were conducted to assess the participants reactions to a variety of systems. There were 306 real-life scenarios in which 306 users were surveyed, collecting large amounts of data pertaining to the perceptions of recommendation. Thus, a baseline model of technology acceptance was used throughout the transversally piloted investigations. The study demonstrated how accuracy and diversity recommendations are interconnected, with time being a new element added to the user satisfaction assessment. The investigations provided in-depth understanding of how users perceive and act upon recommendations when faced with them first time while using recommender system.

DOI and TAM theories are similar in some constructs and complement each another to examine the adoption of new innovation. Researchers indicate that the ideas employed in technology acceptance model are essentially a subsection of perceived innovation features; thus, by integrating both theories, we can obtain a model that is stronger the either alone [11]. Past studies integrated two theories, providing good results [7]. The theoretical framework for this study is drawn from the DIO theory and TAM model.

5 Methodology

5.1 Research Design

This research used a quantitative approach together with questionnaires to determine if the hypothesis was supported. A university of technology (UoT) was selected because it offers similar programmes. The set of measurement items in the questionnaire was developed specifically for this study on the diffusion, adoption, and acceptance of recommender systems within academic institutions. The theoretical framework for this study is presented as follows (Fig. 1):

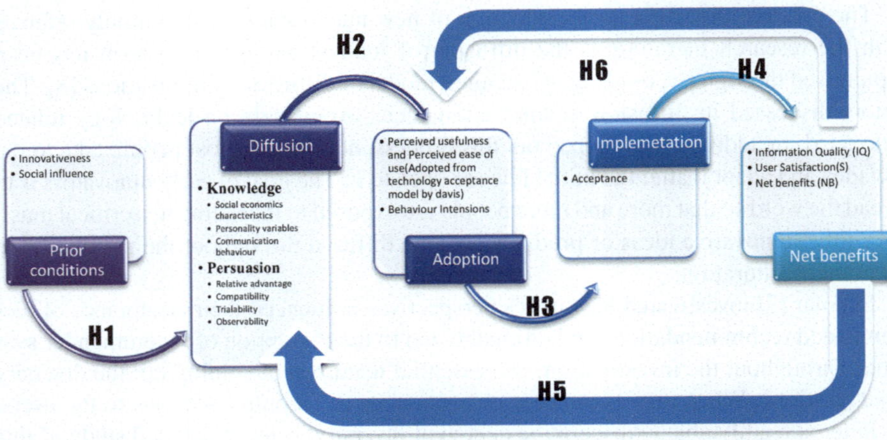

Fig. 1. Proposed research model derived from innovation of diffusion theory [8]

5.2 Sampling and Data Collection

This study aimed at providing insights on the diffusion, adoption, and acceptance of recommender systems in universities. Therefore, targeted population of the study were undergraduate and post graduate students at a university of technology (UoT). The validity of questionnaires was examined in a preliminary study. In total, there were 261 participants who filled in the survey. Approximately 261 of students filled out the survey. Of the 261 responses, 58 were incomplete and were deleted. Another 4 responses were dropped due to lack of interest, thus leading to number of participants of 199 completed dataset, with which the data analysis is conducted. All data was inputted using SPSS version 27.

5.3 Data Analysis

Data collected in this study were analysed using the latest SPSS version 27. Descriptive analysis was done for demographic data. Cronbach's alpha and factor analysis were used to assess reliability and validity, respectively. Regression analysis was conducted to check the correctness of the hypothesis.

6 Results

6.1 Demographic Data

The response rate was 76% as 199 usable responses were obtained. Among the usable returns from the survey, 51.8% were male and 48.2% were female. The age group ranged from 18–25 years old with 39.7%, 26–40 years old with 46.2% and above 40 years with 14.1%. Majority of the participants were undergraduates with 60.3% and Post-graduate with 39,7%. See Table 1.

Table 1. Demographics

Total participants = 199	Labels	Percentages
Gender	Male	51.8
	Female	48.2
Age	18–25	39.7
	26–40	46.2
	Above 40	14.1
Education level	Postgraduates	60.3
	Undergraduates	39.7
Access to online material	Daily	27.1
	Once a week	22.1
	Every fortnight	6.5
	Once a month	44.2
Self-Assessment using recommender systems	Low experience	15.6
	Moderate experience	45.7
	High experience	38.7

6.2 Reliability and Validity

The Cronbach's alpha for this study is 0.805, indicating acceptable reliability. Table 2 displays factor analysis results based on the KMO and Bartlett test, which is 0.870 with the P-value of 0.00, this indicates that factor analysis is appropriate. Five items with eigen value greater or equal to one were considered for the study.

Table 2. Factor analysis results

KMO and Bartlett's Test

Kaiser-Meyer-Olkin Measure of Sampling Adequacy.		.870
Bartlett's Test of Sphericity	Approx. Chi-Square	1971.080
	df	171
	Sig.	.000

Variables	Rotated component Matrix				
	Components				
	1	2	3	4	5
Prior condition 1 (Innovativeness)	.627				
Prior condition 2 (Social Influence)	.596				
Diffusion 1 (Socio economic status)		.589			
Diffusion 2 (Personality variable)		.820			
Diffusion 3 (Relative advantage)		.624			
Diffusion 4 (Trialability)		.721			
Adoption 1 (Perceived usefulness)			.552		
Adoption 2 (Behaviour intension)			.669		
Implementation 1 (Acceptance)				.491	
Implementation 2 (Acceptance)				.538	
Implementation 3 (Acceptance)				.725	
Net benefit 1 (Information quality)					.646
Net benefit 2 (User satisfaction)					.814
Net benefit 3 (Others)					.677

The averages were used as level of diffusion, adoption, and acceptance and T-test was used to check whether averages are by chance or not. The averages can be seen in Table 3, Table 4 and Table 5 for diffusion, adoption, and acceptance respectively.

Table 3. One sample statistics shows the results from the actual t-test.

T-Test

One-Sample Statistics

	N	Mean	Std. Deviation	Std. Error Mean
Diffusion	196	3.8288	.51190	.03656

One-Sample Test

	Test Value = 3					
					95% Confidence Interval of the Difference	
	t	df	Sig. (2-tailed)	Mean Difference	Lower	Upper
Diffusion	22.668	195	.000	.82884	.7567	.9010

Diffusion mean value is 3.8 and was compared with the neutral value 3. The result showed that the average is significantly different to 3.

Table 4. One sample statistics shows the results from the actual t-test.

T-Test

One-Sample Statistics

	N	Mean	Std. Deviation	Std. Error Mean
Adoption	199	4.0166	.52805	.03743

One-Sample Test

	Test Value = 3					
					95% Confidence Interval of the Difference	
	t	df	Sig. (2-tailed)	Mean Difference	Lower	Upper
Adoption	27.159	198	.000	1.01661	.9428	1.0904

Adoption level is approximately 4.0 and was compared with the neutral value 3. It was noted that it is significantly different.

Acceptance level is approximately 4.2 and is significantly different from 3 which is neutral.

Table 5. One sample statistics shows the results from the actual t-test.

T-Test

One-Sample Statistics

	N	Mean	Std. Deviation	Std. Error Mean
Acceptance	199	4.1508	.51862	.03676

One-Sample Test

	Test Value = 3					
					95% Confidence Interval of the Difference	
	t	df	Sig. (2-tailed)	Mean Difference	Lower	Upper
Acceptance	31.301	198	.000	1.15075	1.0783	1.2233

The analysis of hypothesizes was done using multiple regression analysis. The results are shown in Table 6. The resulting path coefficients of the suggested research model are shown in Table 6.

Table 6. Multiple regression analysis

Model			Unstandardized coefficient		Standardized coefficients	Sig
Independent variable	**Dependent variables**		**B**	**Std. error**		
	(Constant)	**1.333**		**.187**		**.000**
Prior conditions	**Diffusion**		**.696**	**.051**	**.696**	**.000**
	(Constant)	**1.623**		**.228**		**.000**
Diffusion	**Adoption**		**.625**	**0.59**	**.606**	**.000**
	(Constant)	**1.729**		**.224**		**.000**
Adoption	**Implementation(acceptance)**		**.603**	**.055**	**.614**	**.000**
	(Constant)	**1.813**		**.211**		**.000**
Net benefits	**Acceptance**		**.579**	**.052**	**.623**	**.000**
	(Constant)	**1.212**		**.186**		**.000**
Net benefits	**Adoption**		**.695**	**.046**	**.735**	**.000**
	(Constant)	**2.229**		**.242**		**.000**
Net benefits	**Diffusion**		**.396**	**.059**	**.432**	**.000**

The significance levels in Table 6 are less than .05 meaning all the hypothesis are supported as summarized in the Table 7.

Table 7. Hypothesis testing results.

	Results
H1: The prior condition positively influence diffusion	Accepted
H2: Diffusion positively influence adoption	Accepted
H3: Adoption positively influence acceptance	Accepted
H4: Acceptance have an impact on net benefits	Accepted
H5: Net benefit positively affected adoption	Accepted
H6: Net benefit positively affected by on diffusion	Accepted

7 Ethical Considerations

An important ethical issue is an informed consent. In informed consent, the individual voluntarily participates in the survey fully understanding the risks and benefits of doing so [12]. The participant consent form stated that participation is voluntary. The informed consent allowed the participants to have full knowledge about the study and withdraw at any time. The purpose of the study and highlight that all collected data was used for research purposes. It was explained that all responses were anonymous as no personal identifying information about the respondent, or the organisation was requested.

8 Discussion

This research focused on developing a broad adoption model combining key ideas from diffusion innovation and technology acceptance theories for the following reasons main reasons. 1st, in spite of some facts about explaining personality traits[13], there is no structured combination of ides related to this theory, but researchers are frequently simulating other theories to expand our understanding of it [14]. 2nd, particularly the complexity of complex technology [15] similar to the recommender systems with its various technologies for different situations [16] to better understand the adoption process, different concepts grouped together. H1 is supported to knowledge where researchers can imply by experience to an innovation and increases some understanding of how it functions. An interesting observation is that among the two items that made up innovativeness and social influence, Innovativeness has the highest score. This means that the students in the campus are basically motivated to use the new technologies. That they think are persuasive to them.

In addition to their relatively readiness to adopt new technologies, innovative people tend to imagine high performance from innovation. This supports the combined measurement, as seen in [14]. The students believe that new system will benefit them, and they anticipate they will give an outstanding performance through the use of it.

The hypotheses H2 is supported. The principal is required to continue the innovation in order to ensure the recommender system activities carried out by students. The hypotheses H3 is supported. Among all the DOI factors, relative advantage, compatibility, and trialability has a strong influence on adoption. Furthermore, diffusion has a significant impact on the adoption of use perception [13]. In other studies, diffusion has been identified as a significant predictor of the technology adoption, and understanding its diffusion is therefore essential [13], also, the present study found a significant effect students desire products that simplify and improve their lives without causing significant changes in their state of affairs or routine, as evidenced by the solid influence of diffusion on adoption benefits. When recommender systems are well integrated with current experience and principles, changes are perceived as lest dramatic, and innovation are more easily adopted.

The hypotheses H4, H5 and H6 is also supported. According to the results, adoption has a positive impact on implementation, suggesting that potential users may continue to use RS if they find it useful. To conclude, we found that the current studies showed the importance of liking and combining different concepts to create new perceptions, which may not have been apparent when perusing each model individually [14] for a context on users innovativeness. According to the results, diffusion of innovation has a smaller impact on usage intension compared to technology acceptance model. Researchers would have misjudged the significance of these predictors without this effect being seen in a single theory study. It is therefore recommended that existing theories such as the diffusion of innovation theory be broadened.

9 Conclusion

Several scholars have derived the models and theories from TAM and DOI theory. However, the focus is mainly on the constructs and characteristics that influence technology acceptance. It is important for researchers to focus on key constructs in different stages of technology. Despite the term's diffusion, adoption, and acceptance being used globally for information systems research, their definitions have yet to be determined. Research should get a clearer understanding of how their definitions differ in order to guide their studies. This paper demonstrates the importance of including the process of technology adoption rather than just focusing on factors affecting it. In light of the fact many theories lack process aspect, future research on information system should attempt to include the environment and process in which technologies are employed.

10 Recommendations

The study has found that adoption and acceptance of recommender system is still in early stages. Possibly future scholars will focus on educational factors, but more importantly they must gather facts from a greater number of universities in the county so as to illustrate if there are differences in states of adopting recommender systems. Researchers are encouraged in the future to examine factual usage of RS, and to compare different RS since recommender systems can have a variety of uses. Dou to the fact that RS application is new, its adoption and diffusion process will need to empirically explore at different points in time to enable more cooperative and beneficial understanding.

References

1. Kiwanuka, A.: Acceptance process: the missing link between UTAUT and diffusion of innovation theory. Am. J. Inf. Syst. **3**(2), 40–44 (2015)
2. Kollmann, T.: Attitude, adoption or acceptance? measuring the market success of telecommunication and multimedia technology. Int. J. Bus. Perform. Manag. **6**(2), 133–152 (2004)
3. Renaud, K.: Predicting technology acceptance and adoption by the elderly: a qualitative study. In: Proceedings of the 2008 Annual Research Conference of the South African Institute of Computer Scientists and Information Technologists on IT Research in Developing Countries: Riding the Wave of Technology, pp. 210–219 (2008)
4. Nielsen, J.: Usability 101: introduction to usability. **9**, 35 (2012)
5. Bagozzi, R., Davis, F., Warshaw, P.: Development and test of a theory of technological learning and usage. Hum. Relat. **45**(7), 659–686 (2016). https://doi.org/10.1177/001872679204500702
6. Goldberg, D., Nichols, D., Oki, B.M., Terry, D.: Using collaborative filtering to weave an information tapestry. Commun. ACM **35**(12), 61–70 (1992)
7. Jones, N.: User perceived qualities and acceptance of recommender systems. In: EPFL (2010)
8. Rogers, E.M.: Diffusion of innovations. Simon and Schuster (2010)
9. Zenobia, B.A.: A grounded agent model of the consumer technology adoption process. Portland State University (2008)
10. Vagnani, G., Volpe, L.: Innovation attributes and managers' decisions about the adoption of innovations in organizations: a meta-analytical review. Int. J. Innov. Stud. **1**(2), 107–133 (2017)
11. Campbell, A., Ryley, T., Thring, R.: The influence of knowledge and persuasion on the decision to adopt or reject alternative fuel vehicles (2018)
12. Fouka, G., Mantzorou, M.: What are the major ethical issues in conducting research? is there a conflict between the research ethics and the nature of nursing? Health Sci. J. **5**(1), 3 (2011)
13. Lee, Y.-H., Hsieh, Y.-C., Hsu, C.-N.: Adding innovation diffusion theory to the technology acceptance model: supporting employees' intentions to use e-learning systems. J. Educ. Technol. Soc. **14**(4), 124–137 (2011)
14. Venkatesh, M., Davis, D.: User acceptance of information technology: toward a unified view. MIS Q. **27**(3), 425 (2003). https://doi.org/10.2307/30036540
15. Tung, F.-C., Chang, S.-C., Chou, C.-M.: An extension of trust and TAM model with IDT in the adoption of the electronic logistics information system in HIS in the medical industry. Int. J. Med. Inf. **77**(5), 324–335 (2008)
16. Yang, H., Lee, H., Zo, H.: User acceptance of smart home services: an extension of the theory of planned behavior. Ind. Manag. Data Syst. **117**, 68–89 (2017)

Approximation of the Jacobian in Solving Harmonic Balance Equations

Vladimir Lantsov(✉)

Vladimir State University, Vladimir 600000, Russia
lantsov@ieee.org

Abstract. A new algorithm is presented to reduce computational costs in solving harmonic balance equations obtained by separating state variables. The harmonic balance method is widely used in RF electronics CAD systems. In the author's previous works, an approach was proposed where the vector (matrix) of unknowns is replaced by two matrices of small dimension, which leads to two systems of balance equations that are solved iteratively. The first equation reduces the number of harmonics in the balance equations, the second equation reduces the number of circuit nodes. Equations with reduced dimension are solved sequentially by Newton's method. This algorithm made it possible to reduce computer memory for storing model equations and reduce computational costs when solving high-dimensional problems. In this paper, it is proposed to further reduce computational costs by approximating part of the elements of the balance equations using the decomposition procedure based on singular values. It is proposed to construct a matrix of sets of responses of nonlinear dependencies of circuit models before solving the problem by an iterative method. This matrix reflects all the main changes in nonlinear dependencies with changes in the amplitudes of the input effect and over time. The resulting matrix is then approximated by applying decomposition based on singular values. The matrix of averaged values reduced in this way is substituted into the balance equations. Comparison of the proposed algorithm with the standard harmonic balance method and algorithms developed by the author earlier showed its high efficiency. #CSOC1120.

Keywords: Harmonic balance method · Electronic CAD · Approximation · Singular values · Algorithm of solution

1 Introduction

In [1], a new method for solving harmonic balance equations (GB) was proposed, based on dividing the equations into two components of very small dimension. The reduction of memory and computational costs in the new method was determined by the significantly smaller dimension of the equations and the fact that they were used in solving sequentially. In [1], a simple iteration method was used to solve the equations. Further development of the method was proposed in [2], where relations were obtained and an algorithm for implementation in software tools based on the use of Newton's iterative

R. Silhavy (Ed.): CSOC 2022, LNNS 502, pp. 604–609, 2022.
https://doi.org/10.1007/978-3-031-09076-9_56

method was developed. Further development of the method and implementation algorithms was proposed in [3], where detailed algorithms for implementing of method in CAD software tools for electronics are presented.

The method and algorithms of its implementation in the software for modeling complex electronic circuits have shown high efficiency, especially when solving problems of very high dimension [1–3]. The experience of using previously developed software tools has shown that the main costs of computer memory and simulation time are associated with the storage and processing of the Jacobian of full dimension. In this paper, an algorithm is proposed to significantly reduce computational costs using an effective method of approximation of the Jacobian of the harmonic balance equations when solving by the Newton method.

2 Basic Equations of the Harmonic Balance Equations

The basic relations and algorithms for solving the GB equations for electronic circuits by the new method are given in detail in [2, 3], so here we will give only some basic relations necessary for the presentation of the proposed approach.

Let the system of nonlinear GB equations in matrix-vector form in the frequency domain have the form

$$F(V) = I(V) + YV - I_E = 0, \tag{1}$$

where V is a vector of unknowns (nodal voltages in the circuit); the first term $I(V)$ describes nonlinear elements; the second term YV characterizes the linear part of the circuit; the element I_{E} is a vector of input sources. The dimension of the equations is equal to $[N \times (2K + 1)]$, where N is the number of nodes in the circuit, K is the number of harmonics taken into account [1].

The idea of the method described in [1, 2] was to replace the vector of variables V of the GB equations with two matrices of reduced dimension

$$V = V_H \times V_N, \tag{2}$$

where the matrix V_{H} reduces the number of harmonics and has dimension $[N \times R]$, V_{N} reduces the number of nodes of the circuit and has dimension $[R \times (2K + 1)]$, R is the reduced dimension of the equations, $R << N$, $R << (2K + 1)$.

Replacement of (2) leads to two systems of balance equations of reduced dimension as it was obtained in [1]

$$F(V_H) = I(V_H V_N) \cdot V_N^T + YV_H - I_E \cdot V_N^T = 0 \tag{3}$$

and

$$F(V_N) = V_H^T \cdot I(V_H V_N) + YV_N - V_H^T I_E. \tag{4}$$

The solution of the system of Eqs. (3) by Newton's method gives us the following iterative formula [2]

$$J\left(V_H^i\right) \cdot V_H^{i+1} = I_H^i. \tag{5}$$

Here the Jacobi matrix is defined as

$$J\left(V_H^i\right) = {}^{\partial F}/_{\partial V_H} = {}^{\partial I}/_{\partial V} * V_N^T + Y. \tag{6}$$

The vector of the right parts of the system of Eqs. (5) will have the form [2]

$$I_H^i = I_E V_N^T - {}^{\partial I}/_{\partial V} V_H^i - I(V) V_N^T. \tag{7}$$

Similarly, to solve the system of Eqs. (4), we have the following basic iterative formulas [2]

$$J\left(V_N^i\right) \cdot V_N^{i+1} = I_N^i, \tag{8}$$

$$J\left(V_N^i\right) = {}^{\partial F}/_{\partial V_N} = V_H^T * {}^{\partial I}/_{\partial V} + Y, \tag{9}$$

$$I_N^i = V_H^T I_E - {}^{\partial I}/_{\partial V} V_N^i - V_H^T I(V). \tag{10}$$

3 Introduction to the New Algorithm

Note that in Eqs. (7) and (10) there is an element $I(V)$ that determines the dependence of current on voltages on nonlinear elements in the frequency domain. In standard of circuit simulation tools, all models of nonlinear devices (elements) are described by dependencies in the time domain $i_{NE}(t) = f[v_{NE}(t)]$. The basic GB equations are solved in the frequency domain. Therefore, at each iteration of Newton method, when solving equations, it is necessary to make transformations from the frequency domain to the time domain and back using the Fast Fourier Transform [3–5]

$$I(V) = \Gamma\{i(t) = f[v(t)]\}\Gamma^{-1}. \tag{11}$$

Here Γ is the forward and Γ^{-1} is the inverse Fourier transform. The relationship between the representation of the signal in the time and frequency domains will be defined as $V = \Gamma v$ and $v = \Gamma^{-1} V$.

Similarly, in the iterative formulas there is an element of calculating the derivatives of currents of nonlinear elements according to the corresponding voltages

$$ {}^{\partial I}/_{\partial V} = \Gamma\left\{{}^{\partial i(t)}/_{\partial v(t)}\right\}\Gamma^{-1}. \tag{12}$$

Earlier, in previous works [1–3], the standard part of traditional GB-based software tools and the full dimensions of these equation elements were used to calculate (11) and (12).

Unfortunately, the experience of using software tools implementing previously developed methods and algorithms [1–3] has shown that the main computational costs (memory and time) are associated precisely with the calculation of element $I(V)$, and especially with the calculation of the derivative for the Jacobian ${}^{\partial I}/_{\partial V}$.

4 Approximation Algorithm for Calculating Derivatives of Nonlinear Dependencies

Note that the element ${}^{\partial I}/_{\partial V}$. will be the same for Eqs. (6 and 7), (9 and 10). It is proposed to approximate this element and reduce its dimension using a method widely used in recent years based on Singular Value Decomposition (SVD) [6–9].

This approximation will correspond to the methods of averages used in the algorithms for solving the GB equations to approximate the Jacobian in [10, 11]. It is important to note that the approximation is performed once and before the start of the main iterations.

Let's form a matrix of responses of nonlinear elements in the form of sets of snapshots [12].

$$G(t)\cdot = \left[g_1(t), \ldots, g_n(t)\right]^T \in R^n, \tag{13}$$

where $g_k(t) = {}^{di(t)}/_{du(t)}|_{u_{k(t)}}$ are derived currents of nonlinear elements (usually in circuit simulators, these derivatives are in analytical form); time $t \in [0, T]$, T is the period of the lowest harmonic spectrum at the output of the circuit; $n = N_{\text{NE}} * N_{\text{IN}}$; N_{NE} is a number of nonlinear dependencies; N_{IN} is a number of analysis when we change the amplitude of the input.

The total dimension of the matrix of sets G will be defined as $n * N\text{t}$, where $N\text{t}$ is the number of time (discrete) samples on the period T.

The correct and sufficiently large for selection of a set of snapshots is a decisive factor in building of the SVD algorithm. This choice can greatly affect the approximation of the initial description [12]. It is assumed that the sample set will correspond to the dominant states of the model and the sample set is large enough. The SVD method creates a reduced basis, which is optimal in the sense that the approximation error with respect to data sets is minimized [12].

The SVD algorithm with respect to the matrix of sets G gives the following decomposition

$$G = U\,\Sigma\,V^{\text{T}}, \tag{14}$$

where the matrices $U\cdot = (u_1, \ldots, u_r) \in R^{r\times r}$ and $V\cdot = (v_1, \ldots, v_r) \in R^{Nt\times r}$ are orthogonal, i.e. $U\,V = I$, where I is the unit matrix, $\Sigma = diag(\sigma_1 \geq \sigma_2 \geq \ldots \geq \sigma_r \geq 0)$ is a diagonal matrix of singular values.

A reduced matrix of averaged values of G_{av} is obtained using the following relation [12].

$$G_{\text{av}} = U\,\Sigma\,V^{\text{T}} = U * diag(\sigma_1, \sigma_2, \ldots, \sigma_r, 0, 0, 0) * V^{\text{T}}. \tag{15}$$

To coordinate the dimensions in the balance equations, the value of N can be taken as the value of r, which is easily done in most cases.

The final expressions for calculation by iterative formulas will look like this.

For Eqs. (6–7).

$$J\left(V_H^i\right) = Gav * V_N^T + Y, \tag{16}$$

$$I_H^i = I_E V_N^T - Gav * V_H^i - I(V)V_N^T. \tag{17}$$

For Eqs. (9–10).

$$J\left(V_N^i\right) = V_H^T * Gav + Y, \tag{18}$$

$$I_N^i = V_H^T I_E - Gav * V_N^i - V_H^T I(V). \tag{19}$$

Verification of the proposed algorithms based on the circuit simulation software tools [1–3] was performed on a well-known example from the author's previously published works and examples known from publications in the world literature (Fig. 3 [13], Fig. 1 [14]). This example is characterized by the fact that by changing the number of nonlinear cascades of the circuit, it is possible to increase the dimension of the circuit to very high dimensions determined by the memory of a particular computer. The comparison was performed both with the standard *GB* method and with previously published results. Since the convergence in all the examples coincided with the calculations by the standard *GB* method, we do not provide graphs of the results.

The table below shows the results of the comparison (Table 1).

Table 1. Comparison of methods by calculation time.

	Calculation options: the number of repeating cascades of the circuit, the number of nodes, the number of harmonics taken into account	New algorithm	Standard GB method	The algorithm is in paper [1]
1	3, 11, 3	5 s	4 s	6 s
2	3, 11, 99	9 s	9 s	11 s
3	14,42, 3	20 s	22 s	23 s
4	14, 42, 99	112 s	135 s	128 s
5	45, 135, 3	175 s	206 s	196 s
6	45, 135, 99	271 s	342 s	302 s

5 Conclusion

1. The results of modeling of circuits that differ in the number of electronic circuit nodes and the number of harmonics taken into account showed that the gain was obtained for almost all variants.
2. The new algorithm is suitable for solving large and super-large harmonic balance equations.

References

1. Lantsov, V.: A new algorithm for solving of harmonic balance equations by using the model order reduction method. In: Proceedings of 2020 Ural Symposium on Biomedical Engineering, Radioelectronics and Information Technology (USBEREIT), Ekaterinburg, pp. 295–297 (2020)
2. Lantsov, V.N.: A new algorithm of solving harmonic balance equations. J. IOP Conf. Ser. Mater. Sci. Eng. **862**(5), 05203 (2020)
3. Lantsov, V., Papulina, A.: Algorithms for software tools with harmonic balance method. In: Proceedings of 2021 IEEE Ural Symposium on Biomedical Engineering, Radioelectronics and Information Technology (USBERREIT), Ekaterinburg, 13–14 May 2021, pp. 228–230 (2021)
4. Nastov, O., Telichevsky, R., Kundert, K., White, J.: Fundamentals of fast simulation algorithms for simulation algorithms for RF circuits. Proc. IEEE. **95**(3), 600–621 (2007)
5. Carvalho, N.B., Pedro, J.C., Jang, W., Steer, M.B.: Nonlinear RF circuits and systems simulation when driven by several modulated signals. IEEE Trans. MTT **54**(2), 572–579 (2006)
6. Lantsov, V.N.: Order-reduction methods for models of complex electronic circuits (review). Radio Telecommun. Syst. No. **3**, 59–65 (2012)
7. Charumathi, V., Ramakrishna, M., Vasudevan, V.: Fast proper orthogonal decomposition using improved sampling and iterative techniques for singular value decomposition. Cornell University Library. https://arxiv.org/abs/1905.05107 (2019)
8. Antoulas, A.C., Sorensen, D.C.: Approximation of large-scale dynamical systems: an overview. Int. J. Appl. Math. Comput. Sci. **11**(5), 1093–1121 (2001)
9. Chaturantabut, S., Sorensen, D.C.: Nonlinear model reduction via discrete empirical interpolation. SIAM J. Sci. Comput. **32**(5), 2737–2764 (2010)
10. Kundert, K., White, J., Sangiovanni-Vincentelli, A.: Steady-State Methods for Simulating Analog and Microwave Circuits, p. 455. Klawer Academic Publishers, New York (1990)
11. Ushida, A., Adachi, T., Chua, I.O.: Steady-state analysis of nonlinear circuits based on hybrid methods. IEEE Trans. **CAS-39**(8), 649–661 (1992)
12. Ramalingam, A., Singh, A.K., Nassif, S.R.: Accurate waveform modeling using Singular Value Decomposition with applications to timing analysis. In: Proceedings of IEEE Design Automation Conference, pp. 148–153 (2007)
13. Bond, B.N., Daniel, L.: A piecewise-linear moment-matching approach to parameterized model-order reduction for highly nonlinear systems. IEEE Trans. Comput.-Aided Des. **26**(12), 2116–2129 (2007)
14. Lantsov, V.N., Dolinina, A.A.: Method of lowering the order of models based on Volterra series. Dyn. Complex Syst. **10**(3), 50–54 (2016). (in Russian)

Author Index

R. Silhavy (Ed.): CSOC 2022, LNNS 502, pp. 611–613, 2022.
https://doi.org/10.1007/978-3-031-09076-9